THE OXFO[RD ...]

General Edito[r]

FRANCIS BACON was born in London in 1561, the sixth and youngest son of Sir Nicholas Bacon, the second most important counsellor to Queen Elizabeth. After attending Trinity College Cambridge, Bacon took up a career in law, soon gaining respect as a legal adviser and becoming an MP in his twenties. He spent his whole life in public service, reaching the highest legal offices: Solicitor-General, Attorney-General, Lord Keeper, Lord Chancellor, and being created successively Baron Verulam and Viscount St Alban. In his spare time Bacon studied natural philosophy and a wide variety of other subjects, pursuing an ambitious scheme to reform the whole of human learning. His *Essays, Advancement of Learning*, and *New Atlantis* have assured him a distinguished place in English literature. He died in 1626, having caught a chill experimenting with the refrigeration of food.

BRIAN VICKERS, educated at St Marylebone Grammar School and Trinity College Cambridge, has been Professor of English Literature and Director of the Centre for Renaissance Studies at the ETH Zurich since 1975. He has written extensively on Francis Bacon, on Renaissance science and philosophy, and on classical rhetoric. His other books include *Towards Greek Tragedy* (1973), *In Defence of Rhetoric* (1988), *Returning to Shakespeare* (1989), and *Appropriating Shakespeare: Contemporary Critical Quarrels* (1993).

THE OXFORD AUTHORS

FRANCIS BACON

EDITED BY

BRIAN VICKERS

Oxford New York

OXFORD UNIVERSITY PRESS

1996

Oxford University Press, Walton Street, Oxford OX2 6DP

Oxford New York
Athens Auckland Bangkok Bombay
Calcutta Cape Town Dar es Salaam Delhi
Florence Hong Kong Istanbul Karachi
Kuala Lumpur Madras Madrid Melbourne
Mexico City Nairobi Paris Singapore
Taipei Tokyo Toronto

and associated companies in
Berlin Ibadan

Oxford is a trade mark of Oxford University Press

Published in the United States
by Oxford University Press Inc., New York

British Library Cataloguing in Publication Data
Data available

Library of Congress Cataloguing-in-Publication Data

Bacon, Francis, 1561–1626.
[Selections. 1996]
Francis Bacon / edited by Brian Vickers.
p. cm. — (The Oxford authors)
Includes bibliographical references and index.
I. Vickers, Brian. II. Title. III. Series.
824'.3—dc20 PR2205.V53 1996 95–16588

ISBN 0–19–254198–6 (hb).
ISBN 0–19–282025–7 (pb)

1 3 5 7 9 10 8 6 4 2

Typeset by Pure Tech India Ltd, Pondicherry, India.
Printed in Great Britain
on acid-free paper by
Biddles Ltd
Guildford and King's Lynn

PREFACE

Originally this selection from Bacon was intended to include many of his scientific works, which formed the basis of his fame in the seventeenth century, and which have been positively revalued in recent times (see Further Reading, § 7). However, these were mostly written in Latin, then the international scholarly language, and the English translations by Francis Headlam and James Spedding in the standard edition on which this selection is based (that produced by James Spedding, R. L. Ellis, and D. D. Heath between 1857 and 1874) are sometimes inconsistent and anachronistic in rendering key concepts. That deficiency could be corrected by selective re-translation, but as I worked on annotating the texts I realized that I was having to evolve two different types of annotation: one for the translated texts in Victorian English, directed to explaining the scientific concepts and argument, and a quite different one for Bacon's original Elizabethan and Jacobean English writings, which turned out to require a large amount of basic semantic and philological explication. The contrast between these two modes of annotation became so extreme that I found myself in effect editing two different books; so I decided to leave the scientific works for some other occasion and devote this volume solely to Bacon's writings in English.

This change of plan, while delaying its publication for several years, meant that I could now give a much wider coverage of Bacon's English writings than any single volume has yet attempted. In addition to the obvious major works, the *Advancement of Learning* (1605), the *Essays* of 1625, and those posthumously published *New Atlantis*, I could include the earliest version of the *Essays* (1597), selections from the 1612 *Essays*, and those important and neglected early proto-dramatic court entertainments that he produced in the 1590s. But I could also include representative examples of his writings in politics, law, and theology, not reprinted since Spedding's time. For the first category, politics, I have chosen to illustrate Bacon's independence as a counsellor, his ability to make a rational analysis of a given situation, rather than simply telling his superiors what they wanted to hear. My selection includes his plea for civil tolerance, *An Advertisement touching the Controversies of the Church of England* of *c.*1589–91 (pp. 1 ff.), which criticizes the policies of both the reformers and the Established Church, and his justification of Queen

Elizabeth's foreign and domestic policies in the concluding speech *Of Tribute*, 1592 (pp. 22 ff.). These two pieces were produced in a more-or-less private capacity as an independent adviser, and I balance them with two legal works deriving from his official position as Attorney-General: first, his *Charge touching duels* of 1614 (pp. 304 ff.), which spells out the attitude of the government—or indeed any responsible member of society—to that source of conflict and waste; secondly, his *Charge against Somerset* of 1616 in the notorious Overbury poisoning case (pp. 314 ff.), an example of Bacon's legal rhetoric which, for reasons explained in the notes, was deliberately subdued on this occasion. Most of Bacon's professional writings on jurisprudence are too technical for the general reader, but these two speeches can be appreciated without such professional knowledge.

For the third subject area, theology, where Bacon wrote not as an expert but as a well-informed layman, I have chosen excerpts from the *Religious Meditations* (1597), which link up with his secular writings on ethics, and the important *Confession of Faith* (*c.*1602), a unique document which shows the extent to which Bacon's Protestant (Calvinist) inheritance infused the whole of his work, including natural philosophy. Finally, I have even found room for his poems and verse translations of the Psalms.

Reprinting these little-known works will, I hope, help readers appreciate the range and diversity of Bacon's work. However, it brought with it three further requirements. First, the need to provide some background information if these texts were to be properly understood. Few readers, other than specialists, have a working knowledge of ecclesiastical controversies in the 1590s, or Calvinist theology, or the legal status of duelling; so, in the head-notes below I have tried to give the necessary minimum context for us to understand Bacon's choice of position within the range of options then available. From Bacon's hundreds of surviving letters (still neither fully identified nor edited), I have included three which record crucial stages in his life: that to Burghley in 1592 (p. 20), announcing his intention to devote himself to the advancement of intellectual enquiry; that to King James in 1621 (p. 326), in the middle of the political crisis which resulted in him being deprived of all public offices; and another letter of the following year to his lifelong friend Lancelot Andrewes (p. 328), showing Bacon resigned to a new existence outside public life, and devoting all his energies to intellectual activities. Annotating these letters also allowed me to sketch in parts of Bacon's biography that have been badly misinterpreted since the Victorian age, particularly the circumstances surrounding his removal from office. Since these misrepresentations, many of them malicious, continue to blacken Bacon's reputation and distract attention from his real achievements as a public servant and philosopher, it is a matter of some urgency to re-create the complex political situation in which he found himself. Meeting these demands has resulted in head-notes that are longer and more detailed than those in other volumes in this series, but Bacon's life and works interpenetrated to a much greater extent than other writers, most of whom were not leading public figures.

The second requirement needed before these writings could be properly understood was to provide more help at the basic level of vocabulary. Bacon's English is in many ways as rich, and as remote from our own, as Shakespeare's. But whereas Shakespeare's plays have been edited over and over again for nearly three hundred years, building up a massive reservoir of linguistic annotation, responsive to every shade of meaning and restoring many lost allusions to contemporary social life or intellectual fashions, Bacon has so far received only one serviceable edition of his writings, that by Spedding, which did not attempt detailed annotation. In effect, but two of his works, the *Essays* and (to a lesser extent) the *Advancement of Learning*, have ever been edited with the attention to detail lavished on Shakespeare's plays. Other editors in this series have been able to refer readers to excellent modern editions (the Twickenham Pope, William Ringler's edition of Sidney's *Poems*) for further information and annotation, but I have had to provide this service myself, to the best of my abilities, drawing on the marvellous *Oxford English Dictionary*. As I explain in the head-note to the *Notes* (p. 493), many of Bacon's words have totally changed their meaning since he wrote, and not to be aware of their intended sense means that readers would receive at best a vague impression, if not a completely misleading one. Inevitably, then, there are more notes than in other volumes in this series, but, as I discovered when reading closely, modern readers need more linguistic information for Bacon's rich and complex texts than they do for writers nearer in time, and who did not work in so many different fields.

Finally, this volume differs from others in this series in the amount of secondary literature cited. Here again Bacon is a special case, for the fact that his output was so many-sided means that specialists in one area may be ignoramuses in another, and important similarities (and differences) within his work go unnoticed. This may be one reason why no solid tradition of commentary and criticism has yet established itself. I have cited relevant studies as often as I could, particularly those dealing with Bacon's debt to classical and Renaissance writers. Paradoxical though it may seem for such a famous author, we actually know very little about Bacon's sources, and are thus in no position to evaluate his individuality or originality. I hope that this modest edition may serve as a rallying-point for old and new readers, stimulating fresh enquiry.

ACKNOWLEDGEMENTS

The miniature by Nicholas Hilliard, now at Belvoir Castle, Grantham, is reproduced by kind permission of His Grace the Duke of Rutland. (For the history of this painting see Roy Strong in the *Burlington Magazine*, 106 (1964), 337.)

I am most grateful to the Kodama Memorial Library of Meisei University, Tokyo, for supplying me with a microfilm of their manuscript of *Of Tribute*, and for permission to use it as the basis of my edition. This manuscript volume previously belonged to Peter Beal, whose *Index of English Literary Manuscripts* has proved a milestone in research for so many authors, including Bacon. At one point Dr Beal made for me a transcript of the second speech, 'In Praise of Knowledge', and had hoped to transcribe the whole work, but was in the event unable to do so. Luckily, Dr Henry Woudhuysen, of University College London, was able to step in, using his extensive experience with Elizabethan manuscripts to provide me with a transcript of the Kodama MS and a collation of the other two manuscripts, on the basis of which I have made my own edition. I accept full responsibility for editorial decisions.

I thank *Huntington Library Quarterly* for allowing me to base my edition of Bacon's *Advice to Fulke Greville on his Studies* on the transcript made by Vernon F. Snow in his essay, 'Francis Bacon's Advice to Fulke Greville on Research Techniques', *HLQ* 23 (1960), 369–78.

My understanding of Bacon's *Confession of Faith* has been immeasurably increased thanks to the collaboration of a former colleague from my Cambridge days, Basil Hall, who went on to become Professor of Ecclesiastical History at the University of Manchester, and subsequently Fellow and Dean of St John's College Cambridge. Basil put his great knowledge of sixteenth-century Protestant theology freely at my disposal, not only reading and commenting on Bacon's text but critically examining several drafts of my annotation, making many constructive suggestions. His sudden death in October 1994, shortly after I had completed the manuscript, was a grievous loss.

My understanding of Bacon's *Advertisement touching the Controversies of the Church of England* owes a great deal to Patrick Collinson, Regius Professor of Modern History, University of Cambridge, and Fellow of Trinity College: not just to his extensive publications on

Elizabethan religion but also to the happy coincidence that we both took part in the Folger Institute conference on Richard Hooker in August 1993. In his paper on 'Hooker and the Establishment' Professor Collinson made a convincing identification of Richard Bancroft as one of the unnamed targets attacked by Bacon; he subsequently allowed me to quote from that paper, and kindly criticized a draft of my introduction to this treatise.

Michael Kiernan, who is editing the *Advancement of Learning* for the Clarendon Press, looked at my text and annotations, and helped improve them. The general editor, Professor Sir Frank Kermode, has read much of this book and offered characteristically shrewd and helpful suggestions. None of these colleagues is, of course, responsible for errors and failures of understanding that remain.

For help in tracking down some of Bacon's more obscure Latin quotations I am particularly indebted to Mr Frank Gerber of the University of Zurich. Other colleagues who helped here include Professors Hermann Tränkle (Zurich), Michael Winterbottom (Oxford), and Tony O'Rourke and Jamie Robinson (Chadwyck-Healey Ltd., Cambridge), and Luc Jocqué of the editorial board *Corpus Christianorum* (Brepols, Bruges). Despite all our efforts, a few quotations have eluded us, and I would be glad to receive any identifications and corrections.

At Oxford University Press I thank Judith Luna for her patient forbearance over what must have seemed an inordinately long period, as I drastically changed the volume's contents, and kept breaking off to fulfil other commitments. For copy-editing I am once again grateful to Alice Park for her eagle-eyed scrutiny of the typescript, which has eliminated many errors and inconsistencies, and to Elizabeth Stratford and her staff for most scrupulous proofreading.

My greatest debt is to my assistant, Dr Margrit Soland, who has worked on this edition for several years, presiding over many changes in its make-up, and showing remarkable patience and cheerfulness at all times. Her meticulous care for detail has been of enormous help, and to her this edition is warmly and gratefully dedicated.

B.V.

Zurich
October 1994

CONTENTS

INTRODUCTION

I

Francis Bacon, the glory of his age and nation, the adorner and ornament of learning, was born in York House, or York Place, in the Strand, on the two and twentieth day of January, in the year of our Lord 1561. His father was that famous counsellor to Queen Elizabeth, the second prop of the kingdom in his time, Sir Nicholas Bacon, knight, lord-keeper of the great seal of England; a lord of known prudence, sufficiency, moderation, and integrity. His mother was Anne, one of the daughters of Sir Anthony Cooke (unto whom the erudition of King Edward the Sixth had been committed); a choice lady, and eminent for piety, virtue, and learning; being exquisitely skilled, for a woman, in the Greek and Latin tongues. These being the parents, you may easily imagine what the issue was like to be; having had whatsoever nature or breeding could put into him. (*Works*, i. 3)[1]

So begins the first biography of Bacon, written in 1657 by William Rawley, his chaplain, and editor of various posthumously published works. It is a good beginning, to the life as to the biography, yet the privileges of birth and place did not work out as gloriously as they promised. Educated at first at home, Bacon went up to Trinity College Cambridge in April 1573, where he was under the special care of the Master, John Whitgift, subsequently Archbishop of Canterbury. Due to the outbreak of plague Bacon had only been in residence for thirty-two months when he decided to leave Cambridge at Christmas 1575, just before his fifteenth birthday, but he had seen enough of the academic curriculum, he told Rawley later, to conceive a strong dislike for Aristotelian philosophy as then taught. After university the next step in the training for a public career would have been to go to one of the Inns of Court, and Bacon was duly admitted to Gray's Inn (where his father had been trained) in June 1576. But a better opportunity of gaining political experience arose, and in September he was sent by his father, the second most important statesman in England, to serve with Sir Amias Paulet, the Queen's ambassador to France. Bacon stayed with him from 1576 to 1579, presumably studying statecraft and performing routine diplomatic duties.

[1] All references to Bacon's *Works* are to the edition produced by James Spedding and others in 14 vols. (London, 1857–74), comprising the 'Philosophical Works' (vols. i–v), the 'Literary and Professional Works' (vi–vii), and the *Letters and Life* (viii–xiv).

The first set-back to Bacon's career occurred on 22 February 1579, when his father suddenly died. Bacon was the second son of a second marriage, thus the youngest of six sons. Sir Nicholas had settled estates on the first five, and was in the course of doing so for his youngest when he died. With no position, no land, and no income, Bacon returned to England and took up the law as a profession, advancing with remarkable speed, graduating from Gray's Inn in 1582, becoming Bencher in 1586, Reader (a lecturer entrusted with expounding the law) in 1588, and Double Reader in 1600, an extraordinary honour for one so young. At the same time he began his parliamentary career in 1581, as member for Bossiney in Cornwall, and he was to sit in every parliament from then until he was expelled from office in 1621. While Bacon is celebrated today as a writer and a natural philosopher (or scientist, as we should say), for the main part of his life he was constantly occupied with the law (he lived for long periods in Gray's Inn, and played a full part in its activities), and with public affairs at a high level. Legal and political works came from his pen without interruption throughout his life; indeed, amazing though it may seem, Bacon's scientific, philosophical, and literary works were the product of his spare time, evenings, weekends, above all the vacations from the law terms, or when parliament was not sitting.

As a lawyer, parliamentarian, and politician Bacon was an expert in several fields, not surprisingly, perhaps, given his enormous industry and his family background. Bacon's father served both Henry VIII and Queen Elizabeth I, who made him a knight, a Privy Counsellor, and Lord Keeper of the Great Seal; his uncle, Sir William Cecil, later Lord Burghley, was for many years the most powerful politician in England. Through both sides of the family Bacon was connected with influential figures in Tudor humanism (Sir John Cheke, Roger Ascham) and with many others who put their great learning to the service of the state as teachers and counsellors. The ideology of the *vita activa*, the dedicating of one's energies *pro bono publico*, runs all the way through Bacon's career. Yet his life illustrates peculiarly poignantly the fundamental uncertainties of such a career. The theorists urged all able men to go to the court and serve the monarch: but neither Queen Elizabeth nor James I took much notice of unappointed counsellors. The political courtier was doomed to years of writing notes of advice which the monarch might not even read, and to the practice of the baser arts of ingratiation and flattery. Many of Bacon's letters record the unavoidable, but ignoble and pathetic attempts to catch the eye of people in power, and in those addressed to James I's favourite George Villiers, Duke of Buckingham, it is impossible not to feel that Bacon wasted his abilities on a man in every way his inferior. 'The rising unto place is laborious', he wrote in one of his *Essays*, 'and by pains men come to greater pains; and it is sometimes base; and by indignities men come to dignities' (*Works*, vi. 399 f.)— or not, as the case may be.

It was in his parliamentary career that Bacon suffered his second set-back. The Queen may have ignored advice, but she certainly did not ignore criticism. In the Parliament of 1593 Bacon opposed what seemed to him, and to many other members, an excessive tax raised over too short a period. The vote went against Bacon, but the Queen was furious, denying him access and reducing him to an abject suitor (*Works*, viii. 214–41). This piece of plain-speaking cost him his hopes of advancement to the leading legal offices, Solicitor-General and Attorney-General, where his lifelong rival Sir Edward Coke moved up the ladder in front of him. Yet the Queen continued to employ him in various legal capacities, and accepted some of his writings which expounded or defended the government position. In the 1580s and 1590s Bacon had become attached to the Earl of Essex, advising his patron on several political issues, and writing letters and speeches on his behalf. Bacon's advice was always sensible, yet it had little effect on Essex, who pursued a violent course and finally attempted to raise a rebellion against the Crown. Easily defeated, he was formally prosecuted, and ironically enough, after years of neglect the Queen made use of Bacon for the prosecution and in writing the official account (*Works*, ix. 247–321), an affair which inevitably damaged his reputation and caused him to publish a self-defence (*Works*, x. 139–60). While some critics will always regret that Bacon put his loyalty to England first, they should also consider what the rule of Essex would have represented.

Under King James, Bacon's merits were recognized more quickly. Knighted in 1603, he became King's Counsel in 1604, Solicitor-General in 1607, Attorney-General in 1613, a Privy Counsellor in 1616, Lord Keeper in 1617, Lord Chancellor in 1618. Having more than emulated his father in public office, he excelled him in rank, being elevated to the House of Lords as Baron Verulam in 1618, and created Viscount St Albans in 1621. Although the promotion was remarkable, it was well earned, for Bacon was celebrated by his contemporaries for his forensic skills, his memory of cases and procedure, and his capacity to grasp all the complexities of the issue at stake. He was particularly in demand as a reporter of complicated parliamentary proceedings, where his clear thinking showed at its best. As the leading legal officer he took part in many state prosecutions for treason, including, alas, that of Sir Walter Ralegh (*Works*, xiii. 379 ff.), and played a major part in the trials following the murder of Sir Thomas Overbury. As one reads through the five volumes of Spedding's monumental *Letters and Life* devoted to this period, it is clear that Bacon was always in the thick of the most important public business. Yet he was still only a tool of James, and the image that comes across is, rather like Eliot's 'Prufrock', of one 'Deferential, glad to be of use': conscientious, yet always implementing others' policy.

In only two areas could Bacon lead an independent existence. One was as a scientist planning to reform natural philosophy, and even though the pressure of public work between 1612 and 1620 almost extinguished his scientific activities, the period between 1603 and 1610 had been tremendously productive, and he had assembled enough material then, and developed his ideas sufficiently far to be able to publish in 1620 the beginnings of what he called a *Great Instauration*, designed to put the whole of natural philosophy on a new footing. But even here the first part, he announced, was not yet written, and the second part, the *Novum Organum*, although it is one of the works by which he became famous, is itself only a fragment of what he had planned. As the 'uomo universale' of the English Renaissance, Bacon was paying a high price for maintaining the *vita activa* on so many fronts.

The other area where Bacon was largely independent of crown policy was as a judge. As state prosecutor he necessarily expressed official attitudes, but in his own court—that of Chancery, set up as a court of appeal against injustices perpetrated by other courts—he could administer the law according to his knowledge and beliefs. He did so with some authority, as we can see both from the tone of his official speeches and from the comments of his contemporaries. He was also remarkably efficient as a judge in clearing off huge backlogs of business in a short time, and he gave fair judgments. Yet here occurred the third and biggest set-back of his career: in 1621 he was accused of corruption, found guilty by the House of Lords, dismissed from parliament and from all offices, temporarily imprisoned in the Tower, fined £40,000, and banned from coming within the verge (the radius of twelve miles around the court). Some two years earlier Bacon had taken presents from two men whose cases had been tried in his court: his decision was not affected by the presents, in fact it went against both suitors, so that he cannot be accused of perverting justice.[2] Also, his offence has to be seen against the norms of his society, where men in power (who received no official salaries) habitually sold offices to suitors and accepted lavish presents, indeed expected and depended on them. Bacon was unfortunate in that a scandal earlier that year had aroused a wish among opponents of the government to purge corruption in public places; he was unfortunate in thus giving a weapon to enemies of the government, who located the disgruntled suitors and encouraged them to denounce him; and he was unfortunate in that both James and Buckingham coolly sacrificed him in order to preserve their own positions. In effect Bacon was the scapegoat of a parliamentary opposition group bent on punishing several widespread abuses, from which many public figures profited, particularly the widely detested favourite Buckingham (see below, pp. 692 ff.). Bacon was careless, naïve, and foolish in accepting the gifts, and he was morally wrong to do so. No one would wish to deny that he was at fault, even though a comparable inquiry into the rest of the English ruling class would have removed many men from office.

Bacon's fall was an enormous physical and mental shock for him, and it darkened his reputation for centuries. Yet in removing him at one go from the ever-increasing demands of public office it also liberated him for his own work. The five years of life remaining saw an enormous output in the most varied fields: history, poetry, the essay, a collection of apophthegms (witty sayings), and above all natural philosophy. It is to this period that we owe the *Essays* in their final revision, the *New Atlantis*, and the *History of Henry the Seventh*, three English works which no other writer of that century was capable of producing and which would alone guarantee his place in English literature. To this period also belong the major scientific works, written in Latin, which were intended to comprise the *Historia Naturalis et Experimentalis*; also the vast *Sylva Sylvarum*, written in English, a curious compound of borrowings from other men's books and of his own experiments and observation. It was in the pursuit of an experiment that he died, the amateur scientist being finally caught in a professional role. As John Aubrey recorded it:

> Mr. Hobbes told me that the cause of his Lordship's death was trying an Experiment; viz. as he was taking the aire in a Coach with Dr. Wither-

[2] See Daniel R. Coquillette, *Francis Bacon* (Edinburgh, 1992, in the series 'Jurists: Profiles in Legal Theory').

borne . . . towards High-gate, snow lay on the ground, and it came to my Lord's thoughts, why flesh might not be preserved in snow, as in Salt. They were resolved they would try the Experiment presently. They alighted out of the Coach and went into a poore woman's house . . . and bought a Hen, and made the woman exenterate [disembowel] it, and then stuffed the body with Snow, and my Lord did help to doe it himselfe. The Snow so chilled him that he immediately fell so extremely ill that he could not return to his Lodging . . . but went to the Earl of Arundel's house at High-gate where they putt him into . . . a damp bed that had not been layn-in about a yeare before, which gave him such a colde that in 2 or 3 dayes . . . he dyed of Suffocation.[3]

He died on 9 April 1626, and was buried, as he had requested, in St Michael's church at St Albans, where his mother lay.

II

For indeed to write at leisure that which is to be read at leisure matters little; but to bring about the better ordering of man's life and business, with all its troubles and difficulties, by the help of sound and true contemplations—this is the thing I aim at. (Francis Bacon to Isaac Casaubon, 1609: *Works*, xi. 147)

Bacon's writings have long enjoyed a firm place in English literature. The qualities that continue to attract readers—a powerful intellectual grasp, analytical penetration, a mastery of the expressive resources of language, the ability to adapt style to subject-matter and purpose—have been celebrated by many distinguished writers, from Ben Jonson and Abraham Cowley to Dr Johnson and Pope, Coleridge and Hazlitt, De Quincey, Shelley, and Ruskin.[4] Bacon's high standing in the English literary canon seems assured, yet it rests on a strange paradox, namely that he never wrote a single piece of 'pure' literature. Everything that he ever produced was dedicated to his lifelong goal of improving the amount and quality of knowledge available to mankind, so as to alleviate the miseries of human existence. As he wrote in his programmatic letter to Burghley in 1592 (echoing Cicero's prescription for the perfect orator), 'I have taken all knowledge to be my province; and if I could purge it of two sorts of rovers [distracting influences] . . . I hope I should bring in industrious observations, grounded conclusions, and profitable inventions and

[3] *Aubrey's Brief Lives*, ed. O. L. Dick (London, 1958; Penguin Books, 1962), 124.

[4] See Brian Vickers, *Francis Bacon and Renaissance Prose* (Cambridge, 1968), ch. 8, 'Judgments of Bacon's Style' (pp. 232–61). For other literature on Bacon's reception and reputation from the 17th to the 19th centuries see Further Reading, §6.

discoveries' (below, p. 20). They would be 'profitable' not in the sense of financial gain but in terms of his distinctive conception of the true goal of intellectual pursuits, to be applied for 'the benefit and use of men', knowledge itself being 'a rich storehouse, for the glory of the Creator and the relief of man's estate' (below, p. 147). All Bacon's writings are dedicated to this functional, charitable goal, whatever categories editors subsequently assign them to—and the categories are many: natural philosophy, or what we would call the physical sciences (embracing the logic of scientific discovery, the forms of matter, physics, astronomy), the life sciences (studies of life and its prolongation, human and animal biology), psychology, sociology, the communication sciences (logic, rhetoric, ciphers, emblems), and much else.

The *Essays* have been traditionally valued as a literary work, but, as R. S. Crane first showed,[5] they should also be seen as a contribution to the study of human life from what we would describe as a psychological and sociological viewpoint. In the *Advancement of Learning* (1605) Bacon called for more research into 'the several characters and tempers of men's natures and dispositions', in particular the differences between people caused by 'those impressions of nature which are imposed on the mind by the sex, by the age, by the region, by health and sickness, by beauty and deformity, and the like, which are inherent and not extern; and again those which are caused by extern fortune, as sovereignty, nobility . . . riches . . . prosperity, adversity . . . and the like' (below, p. 258). Also, he suggested, we need studies of 'those points which are within our own command, and have force and operation upon the mind to affect the will and appetite and to alter manners', such as 'custom, habit . . . friends . . . fame . . . laws . . . studies' (below, p. 260). For each of these topics Bacon wrote an essay, to collect his own observations from life and reading into just such a 'storehouse', not arranged systematically but in the form of aphorisms, disconnected utterances which illuminate different, sometimes opposed aspects of the topic studied.[6] If the 1597 *Essays* present this aim in embryo, glimpsed but not fully articulated, those of 1625 are its fulfilment, and can be linked at every point to Bacon's wider project for a science of man.

[5] R. S. Crane, 'The Relation of Bacon's *Essays* to His Program for the Advancement of Learning', in Brian Vickers (ed.), *Essential Articles for the Study of Francis Bacon* (Hamden, Conn., 1968; London, 1972), 272–92.

[6] For further comment see the head-notes below to the *Essays* of 1597 (pp. 542 ff.), 1612 (pp. 672), and 1625 (pp. 709 ff.).

But even while we make this point, that none of his works is purely literary, we have to add its corollary, that throughout his life Bacon constantly used literary genres to embody and disseminate his ideas. If we review his career, for once in reverse order, we may be surprised at the range of genres he drew on. The posthumously published *Sylva Sylvarum, or a Natural History* (1627), was divided into ten books, each of ten centuries, a mode of organization common in meditational literature, such as Joseph Hall's *Meditations and Vows, Divine and Morall, III Centuries* (1605, 1606). At the end of that volume appeared *New Atlantis*, a hybrid work combining the genres of voyage literature, utopian fable, and secular apocalypse (a revelation handed down orally, from teacher to pupil or initiate). In 1624 Bacon published his *Translation of Certain Psalms into English Verse* (see below, pp. 333 ff.), and made at least one other verse translation (from the classics: see below, p. 332); also in that year he released his collection of *Apophthegms New and Old*, pithy sayings, usually by famous historical personages. The psalms were celebrated throughout the Christian Church as the supreme example of divine poetry, also as an example of sacred epideictic rhetoric, praising God, while the apophthegm was one of several rhetorical storehouses that Bacon recommended all writers to have in readiness (see below, pp. 118, 223 ff., 240 ff.). In 1624 he drafted *Considerations Touching a War with Spain*, the last of his many treatises of advice to the Sovereign, a central genre in Renaissance political literature. As for historiography, his *History of the Reign of King Henry the Seventh* (1622) was a major work, the only one of the several histories that he drafted (from 1602 onwards) to be actually completed. In 1622 he began his only work in the dialogue genre, the *Advertisement Touching a Holy War*, a fragment first published in 1629, in which, true to one of the conventions of Renaissance dialogue, each of the speakers represents a different philosophy or religious creed, so that their discussion comes to resemble a debate in which personalities clash, as well as ideas.

These experiments with different genres were all products of the frantically busy period after he was driven out of office (see below, pp. 688 ff.). In the preceding twenty-five years Bacon's written output derived mainly from his two main careers, politics and law. Many of his speeches in both Houses of Parliament survive, and after the enormous developments of parliamentary history over the last thirty years the time may soon be ripe for a full study of his contribution to both the substance and the form of parliamentary debate. The literary genre that Bacon used most frequently of all was the oration, as could be expected from his choice of the *vita activa*. His political and legal speeches largely fall into the two dominant forms of the classical oration, the deliberative, giving advice to a constituted authority on matters of public policy, and the judicial, arguing for the guilt or innocence of a person on trial. But Bacon also left distinguished examples of the epideictic (or demonstrative) oration, which praised virtuous behaviour and censured evil. Such is his tribute to Queen Elizabeth I, *In felicem memoriam Elizabethae, Angliae Reginae* (*Works*, vi. 291–318), a composition which he valued so highly that he gave express instructions in his will for it to be published (*Works*, xiv. 228). Another separate piece of epideictic rhetoric was his *In Henricum Principem Walliae Elogium* (1612; *Works*, vi. 323–9), one of many such eulogies produced after the untimely death of England's favourite prince.

Given Bacon's professional skills in rhetoric,[7] it is quite fitting that he should have supplied examples of another popular Renaissance genre (dominated by Erasmus), aids and models for practising speakers, what he called the 'provision or preparatory store for the furniture of speech and readiness of invention' (below, p. 240). These include the 'Antitheses of Things', or 'Theses argued *pro et contra*', sketched in the *Advancement of Learning* (below, pp. 240 ff.) and expanded in its Latin translation, the *De Augmentis Scientiarum*, *Libri IX* (1623: *Works*, iv. 472–93). These were intended to provide material useful 'not only to the judicial kind of oratory'—already done by Cicero—'but also to the deliberative and demonstrative' (iv. 472). Indeed, Bacon subsequently recognized that his collection provided most material for the 'moral and demonstrative kind' (493), which was, after all, the only one of the three oratorical genres that could be applied to literary and philosophical ends. The parallel treatise in the *De Augmentis* was the *Colores boni et mali*, 'a collection of the popular signs or colours of apparent good and evil . . . which are really the sophisms of rhetoric', and which amounted to a sizeable composition (*Works*, iv. 458–72). This, too, was sketched in the *Advancement*, but its origins go much further back, for Bacon's first published work, the *Essays* of 1597, included 'Of the Colours of Good and Evil, a Fragment' (*Works*, vii. 75–92; see below, pp. 97 ff. for a selection).

This continuity of interest in intellectual matters across a period of more than thirty years of public life testifies to Bacon's tenacity, but also to the limited amount of time that he could expend on the

[7] See Brian Vickers, 'Bacon and Rhetoric', in M. Peltonen (ed.), *The Cambridge Companion to Francis Bacon* (Cambridge, 1996).

projects so majestically announced in his 1592 letter to Burghley. The *De Augmentis* is in effect a vast repository of fragmentary treatises in many genres. One of them is an 'Example of a Treatise on Universal Justice or the Fountains of Equity, by Aphorisms; one Title of it' (*Works*, v. 88–110), highlighting another genre important throughout Bacon's career, the aphorism. Bacon first used the aphorism in two works dating back to 1597, the *Essays*, which he published, and the *Maxims of the Law* (*Works*, vii. 313–87) which he did not. There he first outlined a coherent and original theory of the aphorism, further developed in the *Advancement of Learning* (below, pp. 234 ff.), and used as the basic form in most of his mature scientific works. The main value of the aphorism for Bacon was not so much its brevity as the fact that it could contain original observations of nature or human life in separate, uncoordinated units. This meant that they could not be prematurely forced into a system, which would then close down further development. 'Knowledge, while it is in aphorisms, is in growth', as he put it (below, p. 145).

The *De Augmentis* also includes another genre first used by Bacon many years earlier, the fable. It gives three 'Examples of Philosophy according to the Fables of the Ancients' (*Works*, iv. 316–35), one in natural philosophy ('Of the Universe, according to the fable of Pan'), one in politics ('Of War, according to the story of Perseus'), and one in moral philosophy ('Of Desire, according to the fable of Dionysus'). Here Bacon drew on his *De Sapientia Veterum* (1609; *Works*, vi. 617–764), translated into English by Sir Arthur Gorges as *The Wisdom of the Ancients* (1619), which is in one sense a tribute to the great Renaissance mythographers, such as Natalis Comes and Vincenzo Cartari. Yet this collection of thirty-one fables is more than a docile reproduction of known material, being in effect paradigmatic of Bacon's whole attitude to literary genres. For although he appropriates these fables, too, for supra-literary purposes (natural, moral, and political philosophy), he invests in them much more energy and imagination than any narrowly functional or utilitarian purpose could justify. The voyage and utopic frame of the *New Atlantis*, similarly, is grossly disproportionate if we take its main purpose as being to introduce the description of 'Salomon's House, or the College of the Six Days' Work'. Bacon was at least as much involved with the fable itself as with its ostensible philosophical function.

The most striking example of Bacon's involvement with a literary genre over and above its rhetorical usefulness as a vehicle for his ideas is his lifelong flirtation with the masque. The masque, a popular court entertainment between 1590 and 1642, was a proto-dramatic form which united theatrical spectacle, dancing, singing, and instrumental music around a stylized literary theme, often an encounter or temptation fable (where virtue regularly defeats pleasure), or a debate between exponents of varying forms of life, good and bad. The performers were partly professional (the musicians, painters, and choreographers), and partly amateur, but all saw their activity as directed to the Sovereign or high-ranking courtier being honoured by this single performance. Taking them in reverse order again, Bacon's essay 'Of Masques and Triumphs' in the 1625 volume simultaneously deprecates the genre and justifies it, as a legitimate form of entertainment for princes and rulers to indulge in during their leisure intervals from the serious activities, 'Civil and Moral', to which the rest of the volume is dedicated: 'These things are but toys, to come amongst such serious observations. But yet, since princes will have such things, it is better they should be graced with elegancy than daubed with cost' (below, p. 416). The page that follows displays a detailed, first-hand knowledge of the genre, giving a quite individual evaluation of its various ingredients. (In this essay, as in those 'Of Buildings' and 'Of Gardens', Bacon uses the first-person pronoun more than in any others, as if recognizing these as topics on which it was proper to record individual tastes.)

As for the 'elegancy' and 'cost' involved, a dozen years earlier Bacon had taken a leading role in the organization and financing of *The Masque of Flowers*, performed on Twelfth Night, 1614, in the Banqueting House, Whitehall, to celebrate 'the marriage of the right honourable Earl of Somerset and the Lady Frances [Howard], daughter of the Earl of Suffolk'. This masque had a most elaborate stage-set, dozens of dancers and musicians, and ran up costs of over £2,000. (For today's equivalent, multiply by 100.) Bacon, who had recently been appointed Solicitor-General, felt obliged to show his gratitude to the King and his new favourite Somerset by accepting 'the whole charge, with the honour', as a contemporary letter-writer put it, even refusing a contribution of £500 from Sir Henry Yelverton, his successor as Attorney-General: 'Marry, his obligations are such, as well to his majesty as to the great lord and to the whole house of Howards, as he can admit no partner.'[8] (All the more

[8] See E. K. Chambers, *The Elizabethan Stage*, 4 vols. (Oxford, 1923), iv. 59–60, i. 211, iii. 213–14, for the original documents describing the spectacle, and Bacon's *Works*, xi. 391–5.

poignant, then, that two years later Bacon had to lead the prosecution case against the newly-wed couple on the charge of having poisoned Sir Thomas Overbury: see below, pp. 314 ff.) Even more magnificent, fortunately not at his expense, was the entertainment of which he was 'the chief contriver' a year earlier, *The Masque of the Inner Temple and Grayes Inn . . . or The Marriage of the Thames and the Rhine*.[9] To celebrate the wedding of Princess Elizabeth and the Elector Frederick V of the Palatinate, this masque was acted in the Banqueting House, Whitehall, on 20 February 1613, before King James, Queen Anne, Prince Charles, and the rest of the court, together with foreign ambassadors. At this time Bacon, as Attorney-General and an eminent figure within Gray's Inn (where his first theatrical activities had taken place as early as 1587[10]) was sharing in a joint act of loyalty and devotion to the Crown on behalf of the two most powerful Inns of Court, some of whose graduates would go on to hold high office in law and government.

That brief reminder of the masque's socio-political context helps to bring out its function within a closely-knit system of gift-exchanges, favours, and rewards. Whereas the public theatre was open to all who could buy a ticket, and only rarely had any impact on the court and its politics, masques were devices which, while discharging obligations between clients and patrons, provided an arena in which individual participants came very close to the centres of power and

[9] See Chambers, ibid. iii. 233–5 for contemporary accounts (including the spectacular arrival of the masquers by water, from Southwark to Whitehall, which alone cost over £300), and *Works*, xi. 343–5. The text of both masques can be found, with useful commentary, in T. J. B. Spencer and S. W. Wells (edd.), *A Book of Masques in Honour of Allardyce Nicoll* (Cambridge, 1967), 127–77.

[10] Bacon's first appearance in print was as one of a group of Gray's Inn students who devised dumb-shows performed before each act of Thomas Hughes's *The Misfortunes of Arthur* (1588), a play which made a peculiar blend of Arthurian legend (Geoffrey of Monmouth) and Senecan tragedy (*Thyestes*). From the title-page of its only edition it would seem that, apart from its role as an offering to the monarch, spectacle had been the main attraction: *Certaine Devises and shewes presented to her Majestie by the Gentlemen of Grayes-Inne at her Highnesse Courte in Greenewich, the twenty-eighth day of Februarie in the thirtieth yeare of her Majestie's most happy Reigne*. See Chambers, *Elizabethan Stage*, iii. 348–9, and the text as reprinted in John W. Cunliffe (ed.), *Early English Classical Tragedies* (Oxford, 1912), 217–96. A note at the end of the text records: 'The dumbe showes were partly devised by Maister Christopher Yelverton, Maister Frauncis Bacon, Maister [Frauncis] Flower' (p. 296). Whatever Bacon's contribution, scrutiny of these five extended stage-directions (pp. 225, 240, 255–6, 271, 283–4) discloses no individual hand, and I have not thought them worth reprinting.

influence. While Bacon evidently had a considerable expertise in all the arts needed to stage a masque, when he came to write scripts for three devices in the 1590s[11] he chose to emphasize political and intellectual goals. In 1595 he wrote a device for Essex to present to the Queen in celebration of her accession day, 17 November, 'Of Love and Self-Love' (below, pp. 61 ff.). The ostensible plot, in which Philautia tries to persuade Erophilus not to love the Queen, obviously represents Essex's own wooing of Elizabeth. Bacon's real interest lay in the three speeches delivered within this framework, which constitute a serious intellectual debate on the areas of public life to which the sovereign might be attracted: 'Contemplation or Studies', 'Fame' derived from military excellence, and 'Experience' or government. In the device that he wrote for the Gray's Inn Christmas Revels of 1594–5 Bacon had used a larger cast, bringing on five counsellors, each advising the Prince to make a 'Choice of Life' in their favour, representing in turn 'the Exercise of War', 'the Study of Philosophy', 'Eternizement and Fame by Buildings and Foundations', 'Absoluteness of State and Treasure', 'Virtue and a gracious Government' (below, pp. 52 ff.). His earliest device, apparently written in 1592 (no evidence survives of it ever being performed), had a title which sums up the whole social nexus of hopes and favours on which the masque depended, 'Of Tribute; or, giving that which is due' (below, pp. 22 ff.). This has four speakers, who perform a competition in epideictic rhetoric, praising successively 'the Worthiest Virtue' (Fortitude), 'the Worthiest Affection' (Love), 'the Worthiest Power' (Knowledge), and 'the Worthiest Person' (Queen Elizabeth).

These early devices are little known outside the pages of Spedding's seven-volume edition of *The Letters and Life of Francis Bacon*, which also had to depend on some imperfect texts (here partly remedied). I have chosen to give them special prominence because they represent the 'choice of life' that Bacon himself faced in the 1590s, a fundamental decision, confronting (in theory, at least) every educated man. All had to choose between the *vita activa* of serving the state, according to the Ciceronian ethos so actively promoted by Renaissance humanism, and the *vita contemplativa*—in the secular

[11] See Chambers, *Elizabethan Stage*, iii. 212–13; A. Wigfall Green, *The Inns of Court and Early English Drama* (New Haven, 1931); Brian Vickers, 'Bacon's Use of Theatrical Imagery', in W. A. Sessions (ed.), *Francis Bacon's Legacy of Texts* (New York, 1990), 171–213; Kenneth Alan Hovey, 'Bacon's Parabolic Drama: Iconoclastic Philosophy and Elizabethan Politics', ibid. 215–36.

sense that Bacon invariably gave it—of intellectual activity. The opposition had been undermined long ago, in effect, for already in ancient Rome the retired life of *otium* in the sense of leisure was only felt to be justifiable if it produced written works of philosophy, literature, or other forms of knowledge that would benefit mankind at large. Otherwise 'fruitless' leisure could be dismissed as *otium* in the sense of idleness. Bacon was following a long-established tradition when he announced in his letter to Burghley that his then 'Choice of Life' was one with 'vast contemplative ends' and 'moderate civil ends: for I have taken all knowledge to be my province' (below, p. 20). Where scholars previously gave Bacon's goals the anachronistic label of 'utilitarian', we can now see them as motivated by a typically Renaissance wish to improve human life by applying the fruits of his studies *pro bono publico*. In the process Bacon was to achieve a distinctly original and influential redirecting of the goal of natural philosophy towards an 'operative science', which would differ from all previous forms of knowledge by producing works for the benefit of mankind.[12]

In these devices Bacon's personae act out the goals and values between which gentlemen and courtiers—those who had been fortunate enough to have a proper education, and to be placed near the sources of power—had to choose. Their arguments in favour of the civic life, its burdens and its splendours, show in embryo Bacon's wide knowledge of the constraints on political activity, the workings of power, and the importance of what we would today call 'the presentation of the self in everyday life' displayed so impressively in the *Advancement of Learning*, the *Essays*, and the flood of counsel and advice that he poured out, to so little effect, to both the monarchs he served. These arguments for the intellectual life are particularly fascinating, since they show Bacon having already formed some of the ideas for the renewal of knowledge that were not to appear again in his public work for many years. The programme outlined at the Gray's Inn Revels for the reform of natural philosophy, for instance, including the setting up of a comprehensive research library, a botanical garden, a museum of inventions, and an experimental laboratory, was not touched on again in public until the posthumously published *New Atlantis*, a quarter of a century later.

[12] For an outstanding re-evaluation of Bacon's programme for science see Antonio Pérez-Ramos, *Francis Bacon's Idea of Science and the Maker's Knowledge Tradition* (Oxford, 1988).

Despite announcing his 'moderate civil ends' in 1592, as we have seen, Bacon spent the rest of his life pursuing a career in law and politics. He had no other choice, lacking both a personal fortune and the institutional context within which he could have pursued his projects for the reform of knowledge. (No such context existed then.) One of the recurring motifs in his life was the need for scientific research to be supported by some adequate source of money and power, which in practice meant the monarch, since no one else could set such a large change in progress. (Moderns who dislike Bacon's praise of King James should reflect on the client–patron relationship as we see it in Galileo, or Beethoven.) Yet in a sense those early devices made a tactical error by presenting distinct personae, each pleading for their own careers. Bacon should have brought on, for instance, a counsellor arguing that all forms of intellectual enquiry are essential to the health of any society, or a scientist arguing that civil government needs to be exercised with prudence for the good of its citizens. Separation of roles, although an admirable basis for debate, meant that the sovereign was able to choose one and reject others.

While the ephemeral splendours of the court masques and shows evaporated during their single performance, the words remain, and Bacon's texts show his already considerable skills in argument. He first establishes each speaker with all his topics fluently organized, then allows every subsequent speaker to damage or even disqualify the preceding arguments, giving the concluding decision to the Prince or moderator. These devices remain both intellectually stimulating and eloquent, with many memorable passages. The speaker praising Fortitude in 1592 gives a penetrating account of one completely lacking that virtue, 'the timorous man', who when confronted with 'his pleasures and desires' is afraid to hope, and even more afraid to enjoy, for

> he ever imagineth some ill is hid in every good: so as his pleasure is as solid as the sands, being interrupted with continual fears and doubts. And when the pleasure is passed, then he thinketh it a dream, a surfeit of desire, a false joy. He is ungrateful to nature, for still the sense of grief printeth so deep, and the sense of delight so lightly, as the one seemeth unto him a truth, the other a deceit. (below, p. 25)

The psychological penetration there is but one of several impressive elements in that speech. But its speaker, having stated his case so vigorously, is instantly routed by his successor, who argues that while fortitude may free us 'from the tyrannies of fortune, yet doth it not in such perfection as doth love. For Fortitude indeed strengtheneth the mind, but it giveth no filling. It leaveth it empty' (below, p. 30):

only love can give it a worthwhile content. To intellectual and psychological penetration these devices add more emotional appeals, as in the speech by the 'Hermit' (contemplative philosopher) in *Of Love and Self-Love*, urging Erophilus to abandon earthly love and become, rather, a Philosophus, for

> The gardens of love wherein he now playeth himself are fresh to-day and fading to-morrow, as the sun comforts them or is turned from them. But the gardens of the Muses keep the privilege of the golden age; they ever flourish and are in league with time. The monuments of wit survive the monuments of power. . . . (below, p. 62)

That eloquent argument is given still more resonance when it reappears in the conclusion to Book One of the *Advancement of Learning*.

III

Bacon's place in English imaginative literature, we can argue, has been earned by his unceasing experimentation with so many different genres, from aphorisms and apophthegms to dialogues and speeches, essays and treatises, fables and masques. But it also derives from his excellence as a writer within these often hybrid works. From his own times to the present day he has been praised for his power with metaphor, that ability to see resemblances between things which Aristotle defined as a mark of genius (*Poetics*, ch. 22). For Bacon, as for all writers within the Renaissance rhetorical tradition, the use of metaphor was not some optional literary grace but an organic part of intentional utterance. According to theorists in rhetoric and poetics, from the Greeks to the seventeenth century, metaphors properly used could both illuminate an argument and give it more force, appealing simultaneously to the intellect and to the affections (passions, feelings). Aristotle, followed by the author of *Rhetorica ad Herennium*, among many others, said that metaphor could be used to raise or lower the value of our subject-matter; Quintilian wrote that 'metaphor is designed to move the feelings, give special distinction to things, and place them vividly before the eye'; many Renaissance rhetoricians agreed that metaphors and similes can produce greater 'vehemency' than other verbal resources, that they 'confirm our understanding and fastest cleave unto the memory', and do 'not onely bewtifie our tale, but also very much inforce & inlarge it'.[13]

[13] For an account of Bacon's use of metaphor, together with other elements in his prose style, see Vickers, *Francis Bacon and Renaissance Prose*.

Metaphor is particularly important in epideictic rhetoric, juxtaposing praise and blame. This can be seen most vividly from Book One of the *Advancement of Learning*, which is in effect an extended epideictic oration, defending the pursuit of knowledge from its enemies (below, pp. 122 ff.), attacking those movements and fashions (medieval scholasticism, Renaissance Ciceronianism) which pervert true knowledge (below, pp. 138 ff.), and ending with a still inspiring celebration of the power of knowledge to survive the vicissitudes of time and chance (below, pp. 167 ff.). Bacon uses metaphor freely for persuasive ends, often anchoring his expressed evaluation in a comparison with some familiar natural phenomenon: 'For as water will not ascend higher than the level of the first spring-head from whence it descendeth, so knowledge derived from Aristotle, and exempted from liberty of examination, will not rise again higher than the knowledge of Aristotle' (below, p. 144). The first part of that comparison being incontrovertible, the second part comes to seem even more likely. Intellectual energy carries the reader's mind with Bacon, across the implied division, making us see what the two things have in common. This persuasive effect is achieved most brilliantly in his account of the medieval schoolmen, living in monasteries and studying exclusively the works of Aristotle, an unhealthy diet, Bacon suggests, which inevitably began to 'putrefy into worms', reducing the substance of their knowledge to 'vermiculate questions'. If not worms, then their dusty isolation transformed the scholastics into spiders, spinning out 'those laborious webs of learning that are extant in their books . . . admirable for the fineness of thread and work, but of no substance or profit' (below, p. 140).

Bacon's metaphors in the opposite mode of praise, what a Renaissance writer might call the *pars construens* of his work, often fall into coherent groups or clusters. Metaphors of vegetation and growth comprise one of the main sequences of imagery that run through Bacon's works calling for a renewal of science. Other major motifs include the building of new structures of knowledge, which involves clearing the ground of previous edifices, designing the plans, collecting the materials, carrying the building through, and finally 'removing the scaffolding and ladders out of sight' (e.g. *Works*, iv. 103–4, 105–6, 107, 110, 111). Images of a journey on land and sea serve as a metaphor for the search for knowledge, or pilgrimage for truth, which can all too easily lose its way. These positive and encouraging sequences of imagery also include negative ones, obstacles to be overcome, labyrinths or dead ends which represent the difficulties that mankind will inevitably experience in trying to understand and control nature. It is this constant dramatization of the pursuit of knowledge, the recognition of difficulties together with the certainty that they could be overcome, that made Bacon such an inspiring advocate for intellectual endeavour in his own age and afterwards. Two particularly effective image-patterns concern water (especially fountains) as a metaphor for the transmission of knowledge, with all the attendant difficulties graphically depicted in the *Advancement of Learning* (e.g. below, pp. 170, 189, 205), and the related classical and Christian conception of knowledge in terms of a God-given light, as put so memorably in Essay 1, 'Of Truth' (below, p. 342). In the *Advancement* Bacon deploys a whole range of related light metaphors deriving from natural phenomena, juxtaposing accurate reflection with innate distortion (e.g. below, pp. 123, 142, 191, 205) as in the famous metaphor of the mind of man as being an 'enchanted glass', a mirror subject to all kinds of distortions, internal and external (below, p. 227).

Metaphors and similes belong to the tropes of rhetoric, devices which transfer and extend the meaning of words. The other main expressive resource classified by rhetoric was that of the figures or schemes, devices which emphasized the physical form of words by various types of repetition. This technique could involve the repetition of whole words at the beginning or ending of phrases or sentences; the repetition of parts of words (echoing their beginnings or endings, as in alliteration or rhyme); or repeating the whole or the parts in inverse order, and so forth. Here again theories of rhetoric and poetics held that such devices were always to be used intentionally, purposefully, arousing the reader's emotions by techniques of contrast and emphasis. The effect of using several such figures simultaneously is to set up a whole range of parallelisms in the structure and syntax of prose (or poetry), which in turn can be used for an unpredictable range of persuasive effects. The first Book of the *Advancement* is one of the most carefully composed pieces of writing in Bacon's whole output, as if he sensed that the accession of King James offered a decisive chance to realize the reforms of knowledge to which he hoped to devote his life. Bacon uses these rhetorical structures creating syntactical symmetry completely effectively in the celebratory mode, as in discriminating the proper ends for which learning should be pursued (below, pp. 147–8), but in the *Advancement of Learning* they appear most brilliantly for two extended sequences of *vituperatio*. One attacks the scholastics (below, pp.

140–2), the other that peculiar group of writers who perverted knowledge in the early Renaissance. These were the over-zealous imitators of Cicero, who devoted so much energy to the form of language that they neglected its content.[14] Their 'affectionate ['over-zealous', but also 'affected'] study of eloquence and copie [copiousness] of speech' led to a form of decadence, for men began to hunt

more after words *than* matter; and
more after the choiceness of the phrase, and
the round and clean composition of the sentence, and
the sweet falling of the clauses, and
the varying and illustration of their works with tropes and figures
than after the weight of matter,
worth of subject,
soundness of argument,
life of invention, or
depth of judgment. (p. 16; my italics)

There the two clauses in parallel, structured on a 'more *x* than *y*' pattern, consist of 5 and 54 words, respectively, as the self-indulgence of the Ciceronians' concern for verbal effect is allowed to spread and extend itself, before being brought to account by the five concluding parallel phrases which remind us of what a writer's real values should be.

Bacon used the resources of rhetorical figures as a way of organizing argument throughout his writing career, disproving claims that he practised a 'baroque', 'asymmetrical', or 'Senecan' prose style. They are just as effective in the early device 'Of Tribute' (e.g. below, pp. 23 f., 32 f.) as in the *Essays*, indeed a comparison of the double revision of that collection, in 1612 and 1625, will show that Bacon made previous symmetries more pointed, and introduced new ones. Such are the parallels and contrasts in 'Of Regiment of Health' (below, p. 404), 'Of Studies' (below, pp. 439–40) or 'Of Goodness and Goodness of Nature', where Bacon recurs, as so often, to the fall of man:

The desire of power in excess
caused the angels to fall;

[14] It is worth underlining the fact that Bacon's satire was directed against Ciceronianism, an excessive phase of *imitatio* that violated some of the main principles of Renaissance humanism, not against humanism itself. This misidentification has been made by Douglas Bush, *English Literature in the Earlier Seventeenth Century* (Oxford, 1962), 276; Charles Webster, *The Great Instauration: Science, Medicine and Reform 1626–1660* (London, 1975), 105; Anthony Quinton, *Francis Bacon* (Oxford, 1980), 13, 23; and Neil Rhodes, *The Power of Eloquence and English Renaissance Literature* (London, 1992), 61–2, 182, among others.

the desire of knowledge in excess
    caused man to fall;
but in charity there is no excess;
    neither can angel or man come in danger by it. (p. 363)

Set out visually such symmetries seem more obvious; in the experience of reading they are unpredictable, regularities which make any deviations from them more striking, as in this little sequence from 'Of Friendship':

For a crowd is not company;
and faces are but a gallery of pictures;
    and talk but a tinkling cymbal,
where there is no love. (below, p. 391)

These and other literary resources, adapted so often and so effectively to context and intention—whether readers absorb them unconsciously or become aware of their operation—help account for Bacon's power as a writer over the last three and a half centuries.

IV

This volume includes not only Bacon's major English writings but also a dozen lesser-known works, which show his range and diversity of output, and will, I hope, stimulate a long-overdue reassessment of his contribution to the English Renaissance. In selecting these lesser-known works I have deliberately included many from the 1590s, that crucial decade when Bacon began to emerge in the public eye, and to define his own 'Choice of Life'. Some of these belong to the masque tradition, as we have seen; others—such as the *Advice to the Earl of Rutland on his travels* of 1596 (below, pp. 69 ff.), the *Advice to Fulke Greville on his studies* of *c.*1599 (below, pp. 102 ff.), and the *Letter and Discourse to Sir Henry Savile, touching helps for the intellectual powers* of *c.*1602 (below, pp. 114 ff.)—belong to the genre of counsel, as indeed do many of the *Essays*. Actually, all the masque 'devices' could also be subsumed under that head, since they act out conflicting advice on how the prince should allocate his energies and resources. (It is perhaps no accident that these advice treatises date from after 1593, when his opposition to Elizabeth's tax demands cost him royal favour. Instead of the sovereign, he henceforth addressed whoever might benefit from his wisdom.) The crucial point about these works is that Bacon, in advising others, reveals and defines his own attitudes and values. True, there is a difference in status between himself and some of the recipients of his advice, such as Essex or

Villiers, and only in fantasy could Bacon's own resources have stretched to the grandiose conceptions laid down in the essays 'Of Building' and 'Of Gardens' (although accounts of his occasionally lavish expenditure suggest that he could indeed lose touch with the realities of credit and debit). But all the same, much of the advice given touches on issues important to all men and women, in particular that perennial problem in classical and Christian ethics, 'how to live'.

Several of these early writings illuminate the development of Bacon's ethics. In the 1592 device 'Of Tribute' the speaker praising Fortitude (below, pp. 22 ff.) presents a detailed and coherent account of that major virtue in classical and Renaissance moral philosophy, one worthy to be set beside Cicero's *Tusculan Disputations*. Some of his arguments recur in the first letter of advice to Rutland (below, pp. 69 ff.), and reappear, as one would expect, when Bacon subsequently deals with the fear of death, one of the major issues addressed by classical philosophy. But when we follow this kind of thinking through his writings we see it being joined and ultimately overshadowed by Christian ethics, especially the concept of Charity. Briefly discussed in the 1597 *Religious Meditations* (below, pp. 91–2), Charity becomes a major concern in the *Advancement of Learning*, first as a safeguard against the vanities to which knowledge can lead (below, p. 124). Charity forms the centre of the concluding, and so decisive paragraphs discussing the 'Culture of the Mind' in its 'most noble and effectual' form, that is, 'the reducing [ordering] of the mind unto virtue and good estate [condition]'. Dismissing the classical tradition of cultivating individual virtues, Bacon argues that if a man sets himself 'honest and good ends', and follows them constantly, 'he shall mould himself into all virtue at once'. He cites Aristotle in support of this theory, and another classic instance as exemplifying it in practice, only to dismiss both as 'heathen and profane passages, having but a shadow of that divine state of mind which religion and the holy faith doth conduct men unto, by imprinting upon their souls Charity, which is excellently called the bond of Perfection, because it comprehendeth and fasteneth all virtues together. . . . so certainly if a man's mind be truly inflamed with Charity, it doth work him suddenly into greater perfection than all the doctrine of morality [moral philosophy] can do' (below, pp. 262–4). As the supreme Christian concept, Bacon most often invokes charity to denounce all forms of egocentric, antisocial behaviour, notably in the *Essays*. In Essay 13, 'Of Goodness and Goodness of Nature', he leaves us in no doubt about what this concept represents: 'This of all virtues and dignities of the mind is the greatest; being the character of the Deity: and without it man is a busy, mischievous, wretched thing; no better than a kind of vermin' (below, p. 363). The same evaluation, indeed the same association of images, recur in Essay 23, 'Of Wisdom for a Man's Self':

> An ant is a wise creature for itself, but it is a shrewd [destructive] thing in an orchard or garden. And certainly men that are great lovers of themselves waste the public. Divide with reason between self-love and society; and be so true to thyself, as thou be not false to others; specially to thy king and country. It is a poor centre of a man's actions, *himself*. It is right earth. For that only stands fast upon his own centre; whereas all things that have affinity with the heavens, move upon the centre of another, which they benefit. (p. 386)

To use a recent term from sociology, Bacon's ethical beliefs are allocentric, other-centred, and resolutely attack all forms of selfishness, from everyday egoism to the occult tradition's cultivation of mystery and obscurity as a way of restricting knowledge to a closed group of initiates.

A related issue which this volume is intended to clarify is Bacon's religion. His religious beliefs infuse one of his first publications, those *Meditationes Sacrae* or *Religious Meditations*, published with the 1597 *Essays* (*Works*, vii. 233–54). These consist of twelve essays on such topics as 'Of the works of God and man', 'Of the miracles of our Saviour', 'Of the innocence of the dove, and wisdom of the serpent' (see below, pp. 89 ff.). Bacon set out his complete system of belief in highly compressed form in the *Confession of Faith* (*c.*1602), a remarkable document that has received very little discussion so far (below, pp. 107 ff.). As my annotation shows, this is at one level a digest of Calvinist thought, at times drawing closely on Calvin's *Christianae Religionis Institutio*. Yet it is also a highly original work, conforming to none of the standard patterns within the confessional genre, and expressing a distinctly individual attitude on several key issues (see below, pp. 559 ff.). Many years ago James Spedding, Bacon's greatest editor, briefly described Bacon's religious beliefs as embracing Christianity's moral and social emphases, accepting the authority and benevolence of God, and recognizing the life of Christ as the highest proof of divine care for man. But over and above these beliefs, Spedding felt, for Bacon 'the entire scheme of Christian theology,—creation, temptation, fall, mediation, election, reprobation, redemption,—is constantly in his thoughts; underlies everything; defines for him the limits of the province of human speculation' (*Works*, vii. 21). The particular individuality of Bacon's *Confession* is its concern with the laws of nature, and the extent to which they, too, were affected by the Fall. Here his interest in natural philosophy informs the theology, as Bacon begins to entertain the new idea, which was to be developed by Kepler and Galileo, of the laws of nature as physical regularities which can be established by observation and experiment. Such a collocation of theology and science, quite common in the Renaissance, raises issues which lead far beyond the scope of this book; but at least one of the key texts in Bacon's writing is now generally available.

To reach a better understanding of Bacon's place in this, as in the other intellectual traditions with which he aligned himself, and the extent of his originality within them, we need to know much more about his reading, both in the wider sense of his general education and as providing sources for particular works, ideas, and attitudes. Since Emil Wolff published his pioneering, but uncompleted survey of *Francis Bacon und seine Quellen*,[15] knowledge of his reading has accumulated intermittently, but as yet we have no reliable general picture. One of the first things that a modern reader notices about the texture of Bacon's writing is the frequency with which he quotes in Latin, not only Cicero and Tacitus, as you might expect, but also Greek writers (Plato, Aristotle, Demosthenes), and the Bible. Few of us have enough training in the classics to recognize these quotations, recall their context, or even to understand them, which is why editors necessarily spend so much time identifying their sources and offering translations. Their frequency (over a thousand in the *Advancement of Learning* alone), and their importance—not just as illustrations but as integral parts of the argument—means that we have to be aware of the very different role that quotation had in Renaissance literature than it has today.

If we try to reconstruct Bacon's intellectual formation, we must start by observing that he had the standard humanist training, acquiring completely fluent Latin and rather less Greek, based on a detailed study of classical texts, which was regarded as the grounding

[15] 'Literarische Forschungen', vols. xl and lii (Berlin, 1910, 1912; facs. repr. Nendeln, Liechtenstein, 1977). This invaluable tool has hardly been used by Bacon scholars. Certainly its incomplete state (vol. iii never appeared), its awkward arrangement according to classical authors rather than to Bacon's works, and its lack of any indexes, have not made it easy to use.

for all education. His mother, Lady Ann Bacon, was one of the five remarkable daughters of Sir Anthony Cooke, a dedicated classicist, responsible with Sir John Cheke for educating the young King Edward VI. Like her sisters, Ann was fortunate to receive the classical education from which women were otherwise excluded, and undoubtedly encouraged her sons' early education. As modern research has shown, she was instrumental in appointing John Walsall, newly graduated from Christ Church, as chaplain to her husband, Sir Nicholas Bacon, the Lord Keeper, and tutor to his sons, between 1566 (when Anthony was 8 years old, Francis 5) and at least 1569.[16] The two boys went up to Trinity College Cambridge, in April 1573, at the rather early ages of 15 and 12 respectively (the median age of admission then being 17.2).[17] From the accounts kept by their tutor Whitgift we know that he bought for them the major classical texts and commentaries, including '2 aristotells' (i.e. 2 copies), for 36 shillings (the price suggests a substantial edition); two copies of Plato, for 24*s*.; 'tullies workes' (another large edition, probably Cicero's philosophical writings, and perhaps his letters) for 30*s*., and a separate edition of 'ciceronis rheto' (the rhetorical works, probably including the *Rhetorica ad Herennium*), 4*s*. Other books on rhetoric that they acquired included 'one commentarie of tullis orations', 13*s*., the *Orations* of Demosthenes, 1*s*., and 'hermogenes in greke and laten' (various treatises forming his *Ars rhetorica*), 7*s*. 6*d*. Historical works included the *Commentarii* of Julius Caesar (2*s*.), 'one salust', 1*s*. 4*d*., and 'one zenophon gre. and latin', 18*s*. Whitgift also expended, 'for a latin bible for antho[ny]', 7*s*.[18] This triple emphasis, on philosophy, rhetoric, and history, was a typical feature of Renaissance humanism, especially for anyone entering public life, where ability in speaking was vital.

From Cambridge Bacon went not to Gray's Inn, as we have seen, but to France, serving between 1576 and 1579 as a secretary to the English ambassador, mostly in Paris and Poitiers. While having

[16] See Virgil B. Heltzel, 'Young Francis Bacon's Tutor', *Modern Language Notes*, 63 (1948), 483–5, and Robert Tittler, *Nicholas Bacon: The Making of a Tudor Statesman* (Athens, Oh., 1976), 49–51, 61–2.

[17] Cf. David Cressey, 'Francis Bacon and the Advancement of Schooling', *History of European Ideas*, 2 (1981), 68–74, at 73 n. 5.

[18] Whitgift's accounts were first printed by S. R. Maitland, 'Archbishop Whitgift's College Pupils', *The British Magazine*, 32–33 (1847–8); see now Philip Gaskell, 'Books bought by Whitgift's Pupils in the 1570s', *Transactions of the Cambridge Bibliographical Society*, 7 (1979), 284–93.

certain public duties Bacon undoubtedly pursued his studies, and it is significant that many of the humanist editions that Wolff identified him as using were issued by the great scholar and publisher Stephanus (Henri Estienne) in Paris or Basle. It may seem strange to modern readers, but Bacon read Homer in the Latin translation by Eobanus Hesse (Basle, 1540), one of the translations used by George Chapman for his celebrated English versions of Homer published between 1598 and 1616—indeed, as F. L. Schoell's pioneering studies showed, Chapman frequently translated the commentator's footnotes as part of the supposedly Homeric text.[19] Like many Renaissance humanists, Bacon got to know Greek authors in a Latin version (for form's sake they occasionally cite the Greek): so he read Hippocrates in the translation by Xylander (e.g. Leunclavius, Frankfurt, 1592), Theocritus in the Latin of Eobanus Hesse (Paris, 1546), Demosthenes in Wolf's version (Basle, 1572), and Aeschines in the same edition. Bacon read Plutarch, in part at least, in the Latin translation by Sambucus (e.g. Antwerp, 1566), also in the other Latin version by Cruserius, published by Stephanus. Herodian and Epictetus he probably read in Politian's translation. In most cases these identifications are clear-cut, in others they are complicated by the existence of Latin compilations, such as the many genre-collections produced by Erasmus (proverbs, similitudes, apophthegms, rhetorical devices). Just as Keats derived a lot of his knowledge from *Lemprière's Classical Dictionary*, so Bacon, like many Renaissance writers, could gain an enormous amount of information from the great reference works by Robert Stephanus, the *Thesaurus Linguae Latinae* (1531; 1543), or the *Dictionarium Historicum Geographicum, Poeticum* (1553; many revisions) by his brother Charles Stephanus, or the *Thesaurus Linguae Graecae* by his son Henry.[20] Bacon's debt to such sources has yet to be investigated.

Bacon read Greek authors not only in Latin but also in French translations. He certainly read Montaigne in French, who is himself a fund of quotations from the classics, and he evidently knew Charron's *De la Sagesse* (see below, p. 655). Thanks to his using the unusual word 'impoyning' in the 1592 device 'Of Tribute', I have been able to show that on this occasion he used Plutarch's 'Life' of

[19] F. L. Schoell, *Études sur l'humanisme continental en Angleterre* (Paris, 1926), esp. 'L'Hellénisme français en Angleterre à la fin de la Renaissance', pp. 133–77.

[20] See D. T. Starnes and E. W. Talbert, *Classical Myth and Legend in Renaissance Dictionaries* (Chapel Hill, NC, 1955).

Caesar in the French version of Jacques Amyot (see below, p. 518). Bacon's debt to French intellectual culture in the late sixteenth century, humanist and Huguenot, needs further exploration. We know that Sir Amias Paulet had been 'governor of the openly presbyterian community in Jersey' before becoming the Queen's ambassador, and that Francis made contact with many Huguenot leaders during his years in France, and even studied with the eminent Reformed theologian Lambert Daneau. His brother Anthony had extended dealings with the French Protestant cause, being a personal friend of Theodore Beza, with whom he lodged for a while, and also studying with Daneau. No one knows how much Bacon absorbed from the French Reformation, but Calvin was certainly a major influence, as the *Confession of Faith* proves by its detailed absorption of Calvinist doctrine. A further revealing detail of Calvin's influence, as the late Basil Hall pointed out to me (see below, p. 564), is Bacon's interpretation of Jacob's ladder as a symbol of Christ (by which angels minister to man, and we ascend to heaven), rather than (according to Jewish tradition) as a symbol of divine providence. Bacon certainly used the main English Calvinist edition of the Scriptures, the so-called 'Geneva Bible' (1560),[21] as we see from his preferred spelling, Salomon. When quoting in Latin he mostly used Jerome's Vulgate, but he also used the Protestant Latin translation by Tremellius and Junius. As Aldis Wright first showed, one of Bacon's favourite quotations was a passage in 1 Kings 4: 33 describing how King Salomon 'spake of trees, from the cedar tree that is in Lebanon unto the hyssop that springeth out of the wall'; only instead of 'hyssop' (which is found in the Septuagint, the Vulgate, and previous English versions) he preferred 'moss', as rendered by Tremellius and Junius.[22] (In the *Sylva Sylvarum*, vi. 536, Bacon again uses the word 'moss', adding 'for so the best translations have it': *Works*, ii. 510.)

[21] During Queen Elizabeth's reign, of the 85 editions of the Holy Scriptures printed, 60 were of the Geneva Bible, which included long marginal notes and summaries embodying (moderate) Calvinist doctrine. See C. D. Cremeans, *The Reception of Calvinist Thought in England* (Urbana, Ill., 1949).

[22] W. Aldis Wright, *Bacon's Advancement of Learning* (Oxford, 1889), 279. See *Testamenti veteris Biblia sacra sive libri canonici, Latini recens ex Hebreo fact. ab I. Tremellio & F. Junio* (London, 1580/1579; STC 2056). This seems to have been the main Latin translation of the Bible in England, registering some 18 editions or reissues by 1640, as against only one, incomplete edition of the Vulgate in 1535 (STC 2055). See Basil Hall, 'Biblical Scholarship: Editions and Commentaries', in S. L. Greenslade (ed.), *The Cambridge History of the Bible*, vol. iii, *The West from the Reformation to the Present Day* (Cambridge, 1963), 38–93, at 72–3.

We tend to think of writers having their roots in their own country, but in a global culture like the Renaissance we need to look further afield.

The crucial point in all such enquiries is to realize that the communication of knowledge is greatly affected by the form in which it is communicated. There is an obvious difference between reading an anthology of wise sayings from the classics and using a detailed edition of an author, together with a commentary. Such commentaries regularly included summaries, expositions, and interpretations by all the major commentators, who might include (for an author like Aristotle) Hellenistic Neoplatonists, Alexandrian philologists and allegorists, medieval Arabic philosophers translated into Latin, and a host of Renaissance scholars busy synthesizing all these traditions. (Bacon's ideas about the imagination, for instance, were probably shaped by the huge commentary tradition on Aristotle's *De anima*, supplemented by his own reading in the increasingly eclectic sixteenth century sources.) This cumulative accretion process meant that all knowledge was simultaneously present, and that any quotation enjoyed the same status, in illustration or argument, as any other. This explains why Bacon can, in one paragraph, quote side by side Plato, Tacitus, the Bible, Machiavelli, and Montaigne. However incongruous these collocations might appear to us, for Bacon and his contemporaries they were all equally relevant, with the Bible obviously more authoritative in certain fields. But even the Bible was not treated solely as a source of religious belief and practice: Bacon cites it for apt material in politics, history, ethics, natural history, astronomy, and many other forms of knowledge (see e.g. below, pp. 151, 267 ff.). For some modern readers the eclecticism of Renaissance writers, their pragmatic attitude to quotations (ignoring huge differences in the original languages, the author's intentions, and the very different genres involved), is problematic. It is indeed a strange paradox that Renaissance scholars, who by their great skills in historical philology were the first people able to distinguish original Latin texts from later forgeries, and could give a reliable chronology of ancient history, in their actual writings simultaneously jumbled up all authors into one vast sea of quotations, to be used on any occasion, for any purpose.

The explanation of this strange contradiction, and one particularly relevant in Bacon's case, is the great importance Renaissance humanism attached to the notebook. From the first influential schoolmasters in fifteenth-century Italy, those pioneers who did so much to establish humanism as a discipline that could be taught in school (Vergerius, Guarino da Verona, Vittorino da Feltre), up to such polymaths as Erasmus, Vives, and Melanchthon in the sixteenth century, and onwards throughout Europe in the next two centuries, the notebook played a crucial role in the transmission of knowledge. All educationalists taught that reading was to be carried out pen in hand, ready to note in the margin metaphors, similes, *exempla*, *sententiae*, apophthegms, proverbs, or any other transportable units of literary composition. These were then to be copied out into one or more notebooks, divided either alphabetically or by topics, and to be reused in one's own writing. Although most attention was given to the establishing of notebooks during school years, many authorities emphasized that they were to be used throughout one's life. The Renaissance was fundamentally a notebook culture, its greatest literary productions displaying what has been called a *stile a mosaico*.[23] Many passages in Montaigne or Rabelais, Bacon or Burton, Chapman or Webster, are tissues of quotations held together by a thin thread of argument. Modern readers must learn to see quotations as simultaneously foreign, the result of an individual author's reading, and yet an integral part of the text, having been appropriated for and indeed by it. A whole theory of *imitatio* was developed, teaching how such material should be digested, integrated into the body or metabolism of the new work. Renaissance readers could certainly tell the difference between learning integrated and learning flaunted, or not properly understood.

Bacon's debt to this notebook culture was enormous. He must have kept files of his own from an early age, for the British Library possesses a strange compilation (Harleian MS 7017) called 'Promus of Formularies and Elegancies'[24]—a 'Promus' is a butler or storekeeper (cf. below, pp. 248, 648). Bacon began to compile this notebook on 5 December 1594, the start of the Christmas vacation that was to culminate in the Gray's Inn Revels. In it he seems to have set himself an exercise, writing out from memory all the quotations he could remember, a total of over 1,600 items, including 255 from

[23] Cf., e.g., P. Porteau, *Montaigne et la vie pédagogique de son temps* (Paris, 1935), 178–84; W. T. Costello, *The Scholastic Curriculum at early seventeenth-century Cambridge* (Cambridge, Mass., 1958), 181; Robert Bolgar, *The Classical Heritage and its Beneficiaries* (Cambridge, 1954), esp. 87–8, 268–301, 321–6.

[24] Spedding printed excerpts from it in *Works*, vii. 197–211; the complete text was edited (unfortunately with deluded attempts to prove Bacon's authorship of Shakespeare) by Mrs Henry Pott (London, 1883).

Erasmus' *Adagia*, 110 from Virgil, 107 from the Bible, 46 from Ovid, a huge collection (443) of proverbs in Latin, French, Spanish, and Italian, together with a large number of metaphors, similes, and 'mottoes', or brief phrases. To begin with they are set down without any order, but as the compilation grew Bacon began to add section- or topic-headings, such as 'Upon Impatience of Audience', 'Upon question to reward evil with evil', 'Benedictions and Maledictions', 'Play', 'Formularies and Elegancies', and 'Semblances or popularities of good and evill with their regulations for deliberacions'—which soon became *The Colours of Good and Evil*. Many of these quotations reappear in his writings over the following thirty years, sometimes in illuminating configurations.

Proof that Bacon continued to keep notebooks, now on a far larger scale, comes from another manuscript that has survived, the so-called 'Comentarius Solutus', a miscellany of notes and ideas also set down during a vacation, this time from 25 to 31 July 1608 (*Works*, xi. 18–95). From the notes made on the first day we can get some idea of the scale on which Bacon (by then Solicitor-General, an MP of many years' standing, with a busy legal practice, and still hoping to reform knowledge) was working. He describes this volume as being 'like a merchant's waste book, where to enter all manner of remembrance of matter, form, business, study, touching my self, services, others, either *sparsim*, or in schedules'. This volume was to be divided in future into two further books, one a digest of those notes worth preserving, the other to contain things which should be registered 'briefly . . . at the first leisure, and while memory can supply it', such material to be subsequently sorted into separate notebooks, 'wherein things of [the same] nature may be (by the labour of a servant, in part) entered in order, and under fit titles' (ibid. 59–62). We know that Bacon had many servants working for him in the role of secretaries and copyists, all educated men, who must have been kept very busy, for on the same day Bacon listed the other notebooks that he proposed to keep up. These included six volumes containing his own writings (one of theology, one of politics and ethics, one of natural philosophy, one of logic, rhetoric, and philology, one of orations, and one of letters); four books of notes, two containing his own ideas, and two taken from his reading ('to receive such parts and passages of Authors as I shall note and underline in the books themselves, to be written forth by a servant', and subsequently arranged into topics); no less than nine books of notes on law and jurisprudence; four notebooks concerned with parliamentary and lawcourt affairs; and five notebooks of personal affairs (including finances, people to contact, problems with his health, etc.). Some of these may still be in existence, their true nature unrecognized.

The concrete existence of such notebooks helps us to appreciate the realities of Bacon's commitment to ancient and modern culture. As he wrote to the great French humanist Isaac Casaubon in 1609, 'I seem to have my conversation among the ancients more than amongst those with whom I live' (*Works*, xi. 147). His vast reading was ever-present, either in his memory or in his notes, and could be used for any purpose. This explains the density of quotation that we find (and, therefore, the amount of annotation it now calls for), and also the curious timelessness of these references, as if all knowledge was simultaneously present. Victorian editors sometimes complain that Bacon was using this or that quotation 'out of context', and it is true that in some cases he does bend the sense to his own purposes, which were not those of Lucan, or whoever it might be. The point is, though, that once all this reading has been digested into the notebook, that now constitutes the relevant context: learning digested for use.

Bacon's cultivation of all this knowledge, and his passing it on to posterity together with his own observations and discoveries, was on a truly heroic scale. Fragmentary though many of his writings are, they show an energy and conviction of the value of intellectual endeavour to which one might apply his own celebration of the immortality of learning:

But the images of men's wits and knowledge remain in books, exempted from the wrong of time and capable of perpetual renovation. Neither are they fitly to be called images, because they generate still, and cast their seeds in the minds of others, provoking and causing infinite actions and opinions in succeeding ages. (below, p. 168)

CHRONOLOGY

1561 22 Jan., born at York House, the Strand, the youngest of Sir Nicholas Bacon's two sons by his marriage to Lady Ann Bacon (née Cooke), his second wife (there being four sons from the first marriage).

1573 5 Apr., goes up to Trinity College Cambridge with his elder brother Anthony; matriculated 10 June.

1575 Mar., leaves Cambridge.

1576 27 June, is entered at Gray's Inn; admitted 21 Nov., but in Sept. had accompanied Sir Amias Paulet, English ambassador to France.

1579 22 Feb., death of his father. 20 Mar., returns from France. Trinity Term, admitted to Gray's Inn.

1581 Elected to Parliament as member for Bossiney, Cornwall.

1582 27 June, admitted as Utter Barrister of Gray's Inn.

1584 23 Nov., MP for Weymouth and Melcombe Regis.

1586 29 Oct., MP for Taunton. Becomes a Bencher of Gray's Inn.

1587 Lent Term, elected Reader at Gray's Inn. Privy Council consults him on legal matters.

1588 Aug., appointed to government committee examining recusants. Dec., appointed to select committee of 16 lawyers (4 from each Inn of Court) to review parliamentary statutes.

1589 4 Feb., MP for Liverpool. Asked to prepare official document justifying the Queen's religious policies. 29 Oct., granted reversion of Clerkship of Star Chamber.

1591 Easter Term, first appearance as pleader in court.

1593 19 Feb., MP for Middlesex. 2 and 8 Mar., speaks against the Queen's demand for additional taxes, and loses royal favour.

1594 25 Jan., argument in Chudleigh's case.

1596 Appointed Queen's Counsel Extraordinary (honorific title).

1597 Feb., first edition of *Essays* (10 essays, with *Colours of Good and Evil*, and *Religious Meditations*). Reprinted 1597, 1598, 1606, 1612. 18 Oct., MP for Ipswich; speaks against enclosures.

1599 Mar., acts as prosecuting counsel in trial of Essex for Irish débâcle.

1600 24 Oct., Double Reader at Gray's Inn.

1601 19 Feb., acts as state prosecutor in trial of Essex for rebellion. Commanded by the Queen to write *A Declaration of the Practises &*

Treasons attempted and committed by Robert late Earle of Essex. . . . May, death of his brother Anthony. 27 Oct., MP for Ipswich. Nov., introduces bill for repealing superfluous laws.

1603 24 Mar., death of Queen Elizabeth; accession of King James I. 23 July, Bacon knighted by James, along with 300 others. Publishes (anonymously) *A Brief Discourse, touching the Happie Union of the Kingdomes of England, and Scotland.* Member of the ecclesiastical commission.

1604 Mar., MP for Ipswich. June, publishes *Sir Francis Bacon His Apologie, in certaine imputations concerning the late Earl of Essex*, and (anonymously) *Certain Considerations touching the better Pacification and Edification of the Church of England*, which is suppressed by the Bishop of London. 18 Aug., appointed King's Counsel.

1605 Oct., publishes *The Twoo Bookes of Francis Bacon. Of the proficience and aduancement of Learning, diuine and human.* Reprinted 1629.

1606 10 May, marries Alice Barnham, daughter of wealthy London alderman.

1607 17 Feb., makes important speeches in Parliament supporting union of the kingdoms and naturalization of Scottish citizens. 25 June, appointed Solicitor-General.

1608 Becomes Clerk of the Star Chamber. Appointed Treasurer of Gray's Inn. Argument in the Postnati case.

1609 Publishes *De Sapientia Veterum.* English trans. by Sir Arthur Gorges, 1619.

1610 Feb., MP for Ipswich. June, speaks in defence of the royal prerogative. Aug., death of his mother, Lady Ann Bacon.

1612 Nov., publishes enlarged edition of the *Essays* (38 essays). Reprinted 1612, 1613 (three editions so dated), 1614, 1624.

1613 26 Oct., appointed Attorney-General.

1614 Jan., publishes *The Charge of Sir Francis Bacon, Knight, His Majesties Attourney-generall, touching Duells* . . . Apr., MP for Cambridge University.

1616 25 May, acts as state prosecutor against the Earl and Countess of Somerset for the poisoning of Sir Thomas Overbury. 9 June, appointed Privy Counsellor.

1617 7 Mar., appointed Lord Keeper of the Seal; reforms workings of Chancery.

1618 Jan., appointed Lord Chancellor. 12 July, created Baron Verulam.

1619 Oct., involved in prosecution of the Earl of Suffolk for illegal exaction.

1620 12 Oct., publishes in part *Instauratio Magna*: Preface, 'Plan of the Work', and part II, *Novum Organum* (two books only). This volume includes *Parasceve ad Historiam Naturalem et Experimentalem*. Nov., involved in prosecution of Sir Henry Yelverton, Attorney-General, for unlawfully amending the charter of the City of London.

1621 27 Jan., created Viscount St Alban. 3 May, sentenced by the House of Lords on charge of taking bribes. Dismissed from office as Lord Chancellor, fined £40,000, and temporarily imprisoned; but retains other titles and is given a limited pardon. Retires to Gorhambury.

1622 Mar., publishes *The Historie of the Raigne of King Henry the Seuenth*. Nov., publishes *Historia Naturalis et Experimentalis ad condendam Philosophiam*, part I: *Historia Ventorum*.

1623 Jan., publishes part II: *Historia Vitae et Mortis*. Oct., publishes *De Dignitate & Augmentis Scientiarum Libri IX*.

1624 Mar., composes treatise concerning a war with Spain. Dec., publishes *Apophthegms New and Old*, and *The Translation of Certaine Psalmes into English verse*.

1625 27 Mar., death of King James; accession of King Charles I. Apr., publishes third edition of the *Essayes or Counsels, Civill and Morall . . . Newly enlarged* (58 essays), reprinted 1625, 1629, 1632, etc. 19 Dec., makes last will and testament.

1626 9 Apr., dies at Highgate, having caught a chill experimenting with the effect of refrigeration on preserving food.

NOTE ON THE TEXT

This edition is largely based on that monument to Victorian scholarship, *The Works of Francis Bacon*, edited by James Spedding, with the assistance of Robert L. Ellis for the scientific works, and Douglas D. Heath for the legal works, in seven volumes (London, 1857–9). Subsequently Spedding alone edited *The Letters and Life of Francis Bacon, including all his Occasional Works, namely Letters, Speeches, Tracts, State papers, Memorial Devices, and all Authentic Writings not already printed among his Philosophical, Literary and Professional Works, newly collected and set forth in chronological order with a Commentary Biographical and Historical*, in seven volumes (London, 1861–74), which reprints over 200 legal and political writings, major and minor, and some 550 letters. Spedding's editorial work derived from an extensive knowledge of the early printed texts (often in their partly corrected states) supplemented by reference to all the surviving manuscripts then known, many of which he was the first to identify and collate. Since his day more manuscripts have turned up, and the principles of editing have suffered several (not always fruitful) revolutions. It is good news that an old-spelling edition of Bacon is being prepared for Clarendon Press by an international team, but it will take many years to appear, and even longer before anyone can attempt a one-volume edition such as this. In the mean time Spedding's modernized text will meet the needs of most readers.

Two of my texts have been edited from newly discovered manuscripts. When Spedding included Bacon's 1592 device *Of Tribute; or, giving that which is due* in Volume One of the *Letters and Life* he only had access to two defective and incomplete manuscripts, and even the fuller version which subsequently turned up in Northumberland House, and which Spedding published in 1870 as *A Conference of Pleasures*, was badly damaged by fire. Fortunately, in the course of researching his monumental *Index of English Literary Manuscripts* (1980–) Dr Peter Beal identified, and for a while owned, a superior manuscript of the complete text, which now belongs to the Kodama Memorial Library of Meisei University, Tokyo. Dr Henry Woudhuysen kindly transcribed this manuscript for me from a microfilm copy, collating it against the two other manuscripts. From these texts I have prepared my edition, modernizing the spelling and punctuation (a process which involves many editorial decisions), and making several conjectural emendations. I am alone responsible for the resulting text.

For Bacon's *Advice to Fulke Greville on his studies* I have been able to use the transcript made by V. F. Snow of the manuscript he discovered in the Public Record Office, and published in *Huntington Library Quarterly* for 1960: I have made some corrections.

For the most chaotically printed of Bacon's works, the *Advancement of Learning* (especially Book Two) I have listed in the notes some variant readings from the 1605 edition and its later reprints, and on a few occasions I have emended Spedding's text.

An Advertisement° Touching the Controversies of the Church of England°

It is but ignorance if any man find it strange that the state of religion (especially in the days of peace) should be exercised° and troubled with controversies. For as it is the condition of the church militant° to be ever under trials, so it cometh to pass that when the fiery trial of persecution ceaseth there succeedeth another trial, which as it were by contrary blasts of doctrine doth sift and winnow° men's faith, and proveth° them whether they know God aright, even as that other of afflictions discovereth° whether they love him better than the world. Accordingly was it foretold by Christ, saying, that 'in the latter times it should be said, "Lo here, lo there is Christ" ':° which is to be understood, not as if the very person° of Christ should be assumed° and counterfeited, but his authority and pre-eminence (which is to be Truth itself) that should be challenged and pretended. Thus have we read and seen to be fulfilled that which followeth, 'Ecce in deserto, ecce in penetralibus':° while some have sought the truth in the conventicles[u] and conciliables[u] of heretics[u] and sectaries,[u] and others in the extern° face and representation of the church; and both sorts been seduced.° Were it then that the controversies of the Church of England were such as did divide the unity of the spirit, and not such as only do unswathe her of her bonds[u] (the bonds of peace), yet could it be no occasion for any pretended Catholic to judge us, or for any irreligious person to despise us. Or if it be, it shall but happen to us all as it hath used° to do; to them to be hardened, and to us to endure the good pleasure of God. But now that our contentions are such, as we need not so much that general canon° and sentence of Christ pronounced against heretics, 'Erratis, nescientes Scripturas, nec potestatem Dei',° as we need the admonition of St. James, 'Let every man be swift to hear, slow to speak, slow to wrath';° and that the wound is no way dangerous, except we poison it with our remedies; as the former sort of men have less reason to make themselves music in our discord,° so I have good hope that nothing shall displease ourselves which shall be sincerely and modestly propounded for the appeasing of these dissensions. For if any shall be offended at this voice, 'Vos estis fratres; ye are brethren, why strive ye?'° he shall give a great presumption against himself, that he is 'the party that doth his brother wrong'.°

The controversies themselves I will not enter into,° as judging that the disease requireth rather rest than any other cure. Thus much we all know and confess, that they be not of the highest nature; for they are not touching° the high mysteries of faith, such as detained° the churches after their first peace° for many years; what time° the heretics moved° curious° questions, and made strange anatomies° of the natures and person° of Christ; and the Catholic fathers° were compelled to follow them with all subtility° of decisions and determinations,° to exclude them from their evasions° and to take° them in their labyrinths; so as it is rightly said, 'illis temporibus ingeniosa res fuit esse Christianum';° 'in those days it was an ingenious and subtle matter to be a Christian'. Neither are they concerning the great° parts of the worship of God, of which it is true that 'non servatur unitas in credendo, nisi eadem adsit in colendo';° 'there will be kept no unity in believing, except it be entertained in worshipping'; such as were the controversies of the east and west churches° touching images;° and such as are many of those between the Church of Rome and us; as about the adoration of the Sacrament,° and the like. But we contend° about ceremonies° and things indifferent;° about the extern policy° and government of the church. In which kind, if we would but remember that the ancient and true bonds of unity are 'one faith, one baptism',° and not one ceremony, one policy; if we would observe the league° amongst Christians that is penned° by our Saviour, 'he that is not against us is with us':° if we could but comprehend that saying, 'differentia rituum commendat unitatem doctrinae';° 'the diversity of ceremonies doth set forth the unity of doctrinae'; and that 'habet religio quae sunt aeternitatis, habet quae sunt temporis';° 'religion hath parts which belong to eternity, and parts which pertain to time': and if we did but know the virtue of silence and slowness to speak, commended by St. James;° our controversies of themselves would close up and grow together.° But most especially, if we would leave the over-weening° and turbulent° humours° of these times, and revive the blessed proceeding of the Apostles and Fathers of the primitive church,° which was, in the like and greater cases, not to enter into assertions and positions,° but to deliver counsels and advices, we should need no other remedy at all. 'Si eadem consulis, frater, quae affirmas, debetur consulenti reverentia, cum non debeatur fides affirmanti';° 'brother, if that which you set down as an assertion, you would deliver by way of advice, there were reverence due to your counsel, whereas faith is not due to your affirmation.' St. Paul was content to speak thus, 'Ego, non Dominus';° 'I, and not the Lord': 'Et, secundum consilium meum';° 'according to my counsel.' But now men do too lightly say, 'Non ego, sed Dominus'; 'not I, but the Lord': yea, and bind° it with heavy denunciations of his judgments, to terrify the simple,° which have not sufficiently understood out of Salomon, that 'the causeless curse shall not come.'°

Therefore seeing the accidents° are they which breed the peril, and not the things themselves in their own nature, it is meet the remedies be applied unto them, by opening° what it is on either part, that keepeth the wound green,° and formalizeth° both sides to a further opposition, and worketh an indisposition° in men's minds to be reunited. Wherein no accusation is pretended;° but I find in reason,° that peace is best built upon a repetition° of wrongs: and in example, that the speeches which have been made by the wisest men *de concordia ordinum*° have not abstained from reducing to memory° the extremities° used on both parts. So as it is true which is said, 'Qui pacem tractat non repetitis conditionibus dissidii, is magis animos hominum dulcedine pacis fallit, quam aequitate componit.'°

And first of all, it is more than time that there were an end and surseance° made of this immodest° and deformed manner of writing lately entertained, whereby matters of religion are handled in the style of the stage.° Indeed, bitter and earnest writing may not hastily be condemned; for men cannot contend coldly and without affection° about things which they hold dear and precious. A politic° man may write from his brain, without touch and sense° of his heart, as in a speculation° that pertaineth° not unto him; but a feeling Christian will express in his words a character either of zeal or love. The latter of which as I could wish rather embraced, being more fit for these times, yet is the former warranted° also by great examples. But to leave° all reverent and religious compassion towards evils, or indignation towards faults, and to turn religion into a comedy or satire; to search° and rip up° wounds with a laughing countenance; to intermix Scripture and scurrility sometime in one sentence; is a thing far from the devout reverence of a Christian, and scant beseeming° the honest regard of a sober man. 'Non est major confusio, quam serii et joci';° 'there is no greater confusion, than the confounding of jest and earnest.' The majesty of religion, and the contempt and deformity of things ridiculous,° are things as distant as things may be. Two principal causes have I ever known of Atheism;° curious controversies, and profane scoffing.° Now that these two are joined in one, no doubt that sect° will make no small progression.

And here I do much esteem the wisdom and religion of that bishop° which replied to the first pamphlet of this kind, who remembered that 'a fool was to be answered, but not by becoming like unto him';° and considered the matter° that he handled, and not the person with whom he dealt. Job, speaking of the majesty and gravity of a judge in himself, saith, 'If I did smile, they believed it not':° as if he should have said, 'If I diverted,° or glanced° unto conceit of mirth, yet men's minds were so possessed with a reverence of the action in hand, as they could not receive° it.' Much more ought this to be amongst bishops and divines disputing about holy things. And therefore as much do I dislike the invention° of him° who (as it seemeth) pleased himself in it as in no mean policy, that these men are to be dealt withal at their own weapons, and pledged in their own cup.° This seemed to him as profound a device, as when the Cardinal Sansovino counselled Julius II. to encounter° the Council of Pisa with the Council Lateran;° or as lawful a challenge as Mr. Jewel° made to confute the pretended° Catholics by the Fathers.° But these things will not excuse the imitation of evil in another. It should be contrariwise with us, as Caesar said, 'Nil malo, quam eos similes esse sui, et me mei.'° But now, 'Dum de bonis contendimus, in malis consentimus';° 'while we differ about good things, we resemble in evil.' Surely, if I were asked of° these men who were the more to be blamed, I should percase° remember the proverb, that 'the second blow maketh the fray',° and the saying of an obscure° fellow, 'Qui replicat, multiplicat';° 'he that replieth, multiplieth.' But I would determine the question with this sentence; 'Alter principium malo dedit, alter modum abstulit';° 'by the one's means we have a beginning, and by the other's we shall have none end.' And truly, as I do marvel that some of those preachers which call for reformation (whom I am far from wronging so far as to join° them with these scoffers)° do not publish some declaration whereby they may satisfy the world that they dislike their° cause should be thus solicited;° so I hope assuredly that my lords of the clergy have none intelligence° with this other libeller,° but do altogether disallow that their credit° should be thus defended. For though I observe in him many glosses,° whereby the man would insinuate himself into their favours, yet I find it to be ordinary,° that many pressing° and fawning° persons do misconjecture of the humours° of men in authority, and many times 'Veneri immolant suem',° 'they seek to gratify them with that which they most dislike'. For I have great reason to satisfy° myself touching the judgments of my lords the bishops in this matter, by that which was written by one° of them, which I mentioned before with honour.° Nevertheless I note, there is not an indifferent° hand carried° towards these pamphlets as they deserve. For the one sort flieth in the dark,° and the other is uttered° openly; wherein I might advise that side° out of a wise writer, who hath set it down that 'punitis ingeniis gliscit auctoritas'.° And indeed we see it ever° falleth out° that the forbidden writing is thought to be certain sparks of a truth that fly up in the faces of those that seek to choke it and tread it out; whereas a book authorized is thought to be but 'temporis voces', 'the language of the time'. But in plain truth I do find (to my understanding) these pamphlets as meet° to be suppressed as the other. First, because as the former sort doth deface° the government of the church in the persons of the bishops and prelates, so the other° doth lead into contempt the exercises° of religion in the persons of sundry preachers; so as° it disgraceth an higher matter, though in the meaner° person. Next, I find certain indiscreet and dangerous amplifications,° as if the civil government itself of this estate had near° lost the force of her sinews,° and were ready to enter into some convulsion,° all things being full of faction and disorder; which is as unwisely acknowledged° as untruly affirmed. I know his° meaning is to enforce this unreverent and violent impugning° of the government of bishops to be a suspected forerunner of a more general contempt. And I grant there is sympathy° between the states; but no such matter in the civil policy,° as deserveth so dishonourable a taxation.° To conclude this point: as it were to be wished that these writings had been abortive, and never seen the sun; so the next is, since they be comen abroad,° that they be censured° (by all that have understanding and conscience) as the intemperate extravagancies° of some light° persons. Yea further, that men beware (except they mean to adventure to deprive themselves of all sense of religion, and to pave° their own hearts, and make them as the highway)° how they be conversant in them, and much more how they delight in that vein;° but rather to turn their laughing into blushing, and to be ashamed, as of a short madness,° that they have in matters of religion taken their disport and solace.° But this perchance is of° those faults which will be soonest acknowledged; though I perceive nevertheless that there want not some who seek to blanch° and excuse it.

But to descend to a sincere° view and consideration of the accidents° and circumstances of these controversies, wherein either part deserveth blame or imputation;° I find generally, in causes of church controversies, that men do offend in some or all of these five points.

1. The first is, the giving of occasion unto the controversies: and also the inconsiderate and ungrounded taking of occasion.

2. The next is, the extending and multiplying the controversies to a more general opposition or contradiction than appeareth at the first propounding of them, when men's judgments are less partial.°

3. The third is, the passionate and unbrotherly practices and proceedings of both parts towards the persons each of others, for their discredit and suppression.

4. The fourth is, the courses° holden° and entertained° on either side, for the drawing of their partizants° to a more strait° union within themselves, which ever importeth° a further distraction° of the entire body.

5. The last is, the undue and inconvenient propounding,° publishing, and debating of the controversies. In which point the most palpable° error hath been already spoken of; as that which through the strangeness and freshness of the abuse first offereth itself to the conceits° of all men.

1. Now concerning the occasion of controversies, it cannot be denied but that the imperfections in the conversation° and government of those which have chief place in the church have ever been principal causes and motives of schisms and divisions. For whilst the bishops and governors of the church continue full of knowledge and good works; whilst they feed the flock indeed; whilst they deal with the secular states° in all liberty and resolution, according to the majesty of their calling, and the precious care of souls imposed upon them; so long the church 'is situate as it were upon an hill';° no man maketh question° of it, or seeketh to depart from it. But when these virtues in the fathers and leaders of the church have lost their light, and that they wax° worldly, 'lovers of themselves,° and pleasers of men',° then men begin to grope° for the church as in the dark; they are in doubt whether they be the successors of the Apostles, or of the Pharisees;° yea, howsoever° they sit in Moses' chair,° yet they can never speak 'tanquam auctoritatem habentes',° 'as having authority', because they have lost their reputation in the consciences of men, by declining° their steps from the way which they trace out° to others. So as men had need continually have sounding in their ears this saying: 'Nolite exire',° 'go not out'; so ready are they to depart from the church upon every voice.° And therefore it is truly noted by one

that writeth as a natural° man, 'that the hypocrisy of freres° did for a great time maintain and bear out° the irreligion of bishops and prelates'. For this is the double policy of the spiritual enemy,° either by counterfeit holiness of life to establish and authorize° errors; or by corruption of manners to discredit and draw in question truth and things lawful. This concerneth my lords the bishops, unto whom I am witness to myself that I stand affected° as I ought.° No contradiction° hath supplanted in me the reverence I owe to their calling; neither hath any detractation° or calumny embased° mine opinion of their persons. I know some of them, whose names are most pierced° with these accusations, to be men of great virtues; although the indisposition° of the time, and the want° of correspondence° many ways, is enough to frustrate the best endeavours in the edifying° of the church. And for the rest generally, I can condemn none. I am no judge of them that belong to so high a master; neither have I 'two witnesses'.° And I know it is truly said of fame, 'Pariter facta, atque infecta canebat'.° Their taxations° arise not all from one coast;° they have many and different enemies, ready to invent slander, more ready to amplify it, and most ready to believe it. And 'Magnes mendacii credulitas';° 'credulity is the adamant of lies.' But if any be,° against whom the supreme bishop 'hath not a few things but many things';° if any have 'lost his first love';° if any 'be neither hot nor cold';° if any have stumbled too foully at the threshold, in sort that he cannot sit well° which entered ill; it is time 'they return whence they are fallen, and confirm the things that remain'.° Great is the weight of this fault; 'et eorum causa abhorrebant homines à sacrificio Domini':° 'and for their cause did men abhor the adoration of God.' But howsoever it be, those which have sought to deface them, and cast contempt upon them, are not to be excused.

It is the precept of Salomon, 'that the rulers be not reproached; no, not in thought':° but that we draw our very conceit° into a modest interpretation of their doings. The holy angel would give no sentence of blasphemy against the common slanderer, but said, 'Increpet te Dominus';° 'the Lord rebuke thee'. The Apostle St. Paul, though against him that did pollute sacred justice with tyrannous violence he did justly denounce the judgment of God, in saying 'Percutiet te Dominus';° 'the Lord will strike thee'; yet in saying 'paries dealbate',° he thought he had gone too far, and retracted° it: whereupon a learned father said, 'ipsum quamvis inane nomen et umbram sacerdotis cogitans expavit'.° The ancient councils and synods° (as is noted by the ecclesiastical story),° when they deprived° any bishop, never recorded the offence, but buried it in perpetual silence. Only Ham° purchased his curse with revealing his father's disgrace. And yet a much greater fault is it to ascend from their person to their calling,° and draw that in question. Many good fathers spake rigorously and severely of the unworthiness of bishops, as if presently° it did forfeit and cease° their office. One saith, 'Sacerdotes nominamur et non sumus';° 'we are called priests, but priests we are not.' Another saith, 'Nisi bonum opus amplectaris, episcopus esse non potes';° 'except thou undertake the good work, thou canst not be a bishop'. Yet they meant nothing less than to make doubt of their calling or ordination.

The second occasion of controversies, is the nature and humour° of some men. The church never wanteth° a kind of persons which love 'the salutation of Rabbi, master';° not in ceremony or compliment,° but in an inward authority which they seek over men's minds, in drawing them to depend upon their opinion, and 'to seek knowledge at their lips'.° These men are the true successors of 'Diotrephes, the lover of pre-eminence',° and not lords bishops. Such spirits do light upon° another sort of natures, which do adhere° to them; men 'quorum gloria in obsequio';° 'stiff followers, and such as zeal marvellously for those whom they have chosen for their masters.' This latter sort, for the most part, are men of young years and superficial understanding, carried away with partial respect of persons, or with the enticing appearance of goodly names and pretences. 'Pauci res ipsas sequuntur, plures nomina rerum, plurimi nomina magistrorum';° 'few follow the things themselves, more the names of the things, and most the names of their masters.' About these general affections are wreathed° accidental° and private emulations and discontentments, all which together break forth into contentions; such as either violate truth, sobriety,° or peace. These° generalities° apply themselves. The universities are the seat and continent° of this disease, whence it hath been and is derived° into the rest of the realm. There some will no longer be *è numero*, of the number. There some others side° themselves before 'they know their right hand from their left'.° So it is true which is said, 'transeunt ab ignorantia ad praejudicium',° 'they leap from ignorance to a prejudicate opinion', and never take a sound judgment° in their way.° But as it is well noted, 'inter juvenile judicium et senile praejudicium, omnis veritas corrumpitur':° 'when men are indifferent, and not partial, then their judgment is weak and unripe through want of years; and when it groweth to strength and ripeness, by that time it is forestalled° with such a number of prejudicate opinions, as it is made unprofitable: so as between these two all truth is corrupted.' In the meanwhile, the honourable names of sincerity, reformation,° and discipline° are put in the foreward:° so as contentions and evil zeals cannot be touched,° except° these holy things be thought first to be violated. But howsoever they shall infer° the solicitation for the peace of the church to proceed from carnal sense,° yet I will conclude ever with the Apostle Paul, 'Cum sit inter vos zelus et contentio, nonne carnales estis?'° 'Whilst there is amongst you zeal and contention, are ye not carnal?' And howsoever they esteem the compounding° of controversies to savour° of man's wisdom and human policy, and think themselves led by the wisdom which is from above, yet I say with St. James, 'Non est ista sapientia de sursum descendens, sed terrena, animalis, diabolica: ubi enim zelus et contentio, ibi inconstantia et omne opus pravum.'° Of this inconstancy, it is said by a learned father, 'Procedere volunt non ad perfectionem, sed ad permutationem';° 'they seek to go forward still, not to perfection, but to change.'

The third occasion of controversies I observe to be, an extreme and unlimited detestation of some former heresy or corruption of the church already acknowledged and convicted. This was the cause that produced the heresy of Arius, grounded chiefly upon detestation of Gentilism,° lest the Christians should seem, by the assertion of the co-equal divinity of our Saviour Christ, to approach unto the acknowledgment° of more gods than one. The detestation of the heresy of Arius produced that of Sabellius;° who, holding for execrable the dissimilitude° which Arius pretended in the Trinity, fled so far from him, as he fell upon that other extremity, to deny the distinction of persons;° and to say they were but only names of several° offices° and dispensations.° Yea, most of the heresies and schisms of the church have sprung up of this root; while men have made it as it were their scale,° by which to measure the bounds° of the most perfect religion; taking it by the furthest distance from the error last condemned. These be 'posthumi haeresium filii';° 'heresies that arise out of the ashes of other heresies that are extinct and amortized.'° This manner of apprehension° doth in some degree possess many in our times. They think it the true touchstone° to try° what is good and holy, by measuring what is more or less opposite° to the institutions of the Church of Rome; be it ceremony, be it policy or government, yea be it other institution of greater weight, that is ever most perfect which is removed° most degrees from that church; and that is ever polluted and blemished which participateth in any appearance° with it. This is a subtle° and dangerous conceit° for men to entertain,° apt to delude themselves, more apt to seduce the people, and most apt of all to calumniate their adversaries. This surely (but° that a notorious condemnation° of that position was before our eyes) had long since brought us to the rebaptising of children° baptised according to the pretended catholic religion. For I see that which is a matter of much like reason,° which is the re-ordaining of priests,° is a matter already resolutely maintained.° It is very meet° that men beware how they be abused° by this opinion; and that they know that it is a consideration of much greater wisdom and sobriety° to be well advised, whether in the general demolition of the institutions of the Church of Rome there were not (as men's actions are imperfect) some good purged° with the bad, rather than to purge the church, as they pretend, every day anew; which is the way to make a wound in her bowels,° as is already begun.

The fourth and last occasion of these controversies (a matter which did also trouble the Church in former times), is the partial° affectation° and imitation of foreign churches. For many of our men, during the time of persecution and since, having been conversant in° churches abroad, and received a great impression of the form of government there ordained, have violently sought to intrude the same upon our Church. But I answer, 'Consentiamus in eo quod convenit, non in eo quod receptum est';° 'let us agree in this, that every church do that which is convenient for the estate of itself, and not in particular customs.' Although their churches had received the better form, yet many times it is to be sought, 'non quid optimum, sed è bonis quid proximum';° 'not what is best, but of good things what is next and readiest to be had.' Our church is not now to plant;° it is settled and established. It may be, in civil° states, a republic is a better policy° than a kingdom: yet God forbid that lawful kingdoms should be tied° to innovate and make alteration. 'Qui mala introducit, voluntatem Dei oppugnat revelatam in verbo; qui nova introducit, voluntatem Dei oppugnat revelatam in rebus';° 'he that bringeth in evil customs, resisteth the will of God revealed in his word; he that bringeth in new things, resisteth the will of God revealed in the things themselves.' 'Consule providentiam Dei, cum verbo Dei';° 'take counsel of the providence of God, as well as of his word.' Neither yet do I admit° that their form° (though° it were possible and convenient) is better than ours, if some abuses were taken away. The parity and equality of ministers° is a thing of wonderful great confusion;° and so is an ordinary° government by synods,° which doth necessarily ensue upon the other. It is hard in all causes, but especially in matters of religion, when voices shall be 'numbered and not weighed'.° 'Equidem' (saith a wise father) 'ut vere quod res est scribam, prorsus decrevi fugere omnem conventum episcoporum; nullius enim concilii bonum exitum unquam vidi; concilia enim non minuunt mala, sed augent potius':° 'To say the truth, I am utterly determined never to come to any council of bishops: for I never yet saw good end of any council; for councils abate° not ill things, but rather increase them': which is to be understood not so much of general councils, as of synods gathered for the ordinary government of the church; as for deprivation of bishops, and such-like causes; which mischief° hath taught the use° of archbishops, patriarchs,° and primates; as the abuse of them since hath taught men to mislike them. But it will be said, 'Look to the fruits° of the churches abroad and ours.' To which I say, that I beseech the Lord to multiply his blessings and graces upon those churches an hundredfold. But yet it is not good, that we fall on numbering of them. It may be our peace hath made us more wanton:° It may be also (though I would be loath to derogate° from the honour of those churches, were it not to remove scandals,) that their fruits are as torches in the dark, which appear greatest° afar off. I know they may have some more strict orders for the repressing of sundry excesses. But when I consider of the censures of some persons, as well upon particular men as upon churches,° I think of the saying of a Platonist,° who saith 'Certe vitia irascibilis partis animae sunt gradu praviora quam concupiscibilis, tametsi occultiora';° a matter that appeared well° by the ancient contentions° of bishops. God grant that we may contend with other churches, as the vine with the olive, which of us beareth best fruit; and not as the brier with the thistle, which of us is most unprofitable. And thus much touching the occasion of these controversies.

2. Now, briefly to set down the growth and progression of these controversies; whereby will be verified the wise counsel of Salomon, that the course of contentions is to be stopped at the first; being else 'as the waters',° which if they gain a breach,° it will hardly be ever recovered. It may be remembered, that on their part° which call for reformation, was first propounded some dislike of certain ceremonies supposed to be superstitious; some complaint of dumb ministers° who possessed rich benefices;° and some invectives against the idle and monastical° continuance within the universities,° by those who had livings to be resident upon;° and suchlike abuses. Thence they went

on to condemn the government of bishops as an hierarchy remaining to us of the corruptions of the Roman church, and to except° to sundry institutions as not sufficiently delivered° from the pollutions of the former times. And lastly, they are advanced° to define of° an only° and perpetual form of policy° in the church; which (without consideration of possibility, or foresight of peril and perturbation of the church and state) must be erected° and planted by the magistrate.° Here they stay.° Others (not able to keep footing° in so steep a ground) descend further;° That the same must be entered into and accepted of the people, at their peril,° without the attending of° the establishment of authority: and so in the meantime they refuse to communicate° with us, reputing us to have no church. This hath been the progression° of that side: I mean of the generality. For I know, some persons (being of the nature, not only to love extremities, but also to fall to° them without degrees) were at the highest strain° at the first.

The other part,° which maintaineth the present government of the church, hath not kept one tenor° neither. First, those ceremonies which were pretended to be corrupt they maintained to be things indifferent, and opposed° the examples of the good times° of the church to that challenge which was made unto them, because they were used in the later superstitious times.° Then were they also content mildly to acknowledge many imperfections in the church: as tares° come up amongst the corn; which yet (according to the wisdom taught° by our Saviour) were not with strife to be pulled up, lest it might spoil and supplant° the good corn, but to grow on together until the harvest. After,° they grew to a more absolute° defence and maintenance of all the orders of the church, and stiffly° to hold that nothing was to be innovated; partly because it needed not,° partly because it would make a breach upon the rest. Thence (exasperate° through contentions) they are fallen to a direct condemnation of the contrary part, as of° a sect.° Yea, and some indiscreet persons have been bold in open preaching to use dishonourable and derogative° speech and censure of the churches abroad; and that so far, as some of our men (as I have heard) ordained in foreign parts have been pronounced to be no lawful ministers.° Thus we see the beginnings were modest, but the extremes are violent; so as° there is almost as great a distance now of either side from itself, as was at the first of one from the other. And surely, though my meaning and scope be not (as I said before)° to enter into the controversies themselves, yet I do admonish° the maintainers of the alone° discipline to weigh and consider seriously and attentively, how near° they are unto those with whom I know they will not join. It is very hard° to affirm that the discipline° which they say we want° is one of the essential parts of the worship of God, and not to affirm withal° that the people themselves upon peril of salvation, without staying for° the magistrate, are to gather themselves into it. I demand,° if a civil state should receive° the preaching of the word and baptism, and interdict and exclude the sacrament of the supper,° were not men bound upon danger of their souls to draw° themselves to congregations, wherein they might celebrate that mystery, and not to content themselves with that part of the worship of God which the magistrate hath authorized? This I speak, not to draw° them into the mislike° of others, but into a more deep consideration of themselves: 'Fortasse non redeunt, quia suum progressum non intelligunt.'° Again, to my lords the bishops I say, that it is hard for them to avoid blame (in the opinion of an indifferent° person) in standing so precisely° upon altering nothing. 'Leges, novis legibus non recreatae, acescunt';° 'laws, not refreshed with new laws, wax sour.' 'Qui mala non permutat, in bonis non perseverat°': 'without change of the ill, a man cannot continue the good.' To take away abuses supplanteth° not good orders,° but establisheth them. 'Morosa moris retentio res turbulenta est, aeque ac novitas'; 'a contentious retaining of custom is a turbulent thing, as well as innovation.' A good husbandman° is ever proyning° and stirring° in his vineyard or field; not unseasonably (indeed) nor unskilfully. But lightly° he findeth ever somewhat to do. We have heard of no offers of the bishops of bills in parliament;° which (no doubt) proceeding from them to whom it properly pertaineth,° would have everywhere received acceptation.° Their own constitutions° and orders have reformed little. Is nothing amiss? Can any man defend the use of excommunication° as a base process° to lackey° up and down for duties and fees;° it being the greatest judgment next the general judgment of the latter day?° Is there no means to train up and nurse° ministers (for the yield of the universities° will not serve,° though they were never so well governed),—to train them, I say, not to preach (for that every man confidently adventureth° to do), but to preach soundly, and handle the Scriptures with wisdom and judgment? I know prophesying° was subject to great abuse, and would be more abused now; because heat of contentions° is increased. But I say the only reason of the abuse was, because there was admitted to it a popular auditory,° and it was not contained° within a private conference of ministers. Other things might be spoken of. I pray God to inspire the bishops with a fervent love and care of the people; and that they may not so much urge° things in controversy,° as things out of controversy, which all men confess to be gracious and good. And thus much for the second point.

3. Now, as to the third point, of unbrotherly proceeding on either part, it is directly contrary to my purpose to amplify wrongs: it is enough to note° and number them; which I do also to move° compassion and remorse on the offending side, and not to animate challenges and complaints on the other. And this point (as reason is) doth chiefly touch° that side which can do most. 'Injuriae potentiorum sunt':° 'injuries come from them that have the upper hand.'

The wrongs of them which are possessed of° the government of the church towards the other,° may hardly be dissembled° or excused. They have charged° them as though 'they denied tribute to Caesar',° and withdrew from the civil magistrate their obedience which they have ever performed and taught. They have ever sorted° and coupled° them with the Family of love,° whose heresies they have laboured to descry° and confute. They have been swift of credit to receive accusations against them from those that have quarrelled with them but° for speaking against sin and vice. Their examinations and inquisitions have been strait.° Swearing men° to blanks° and generalities° (not included within a compass° of matter certain,° which the party that is to take the oath may comprehend) is a thing captious° and strainable.° Their urging of subscription° to their own articles° is but 'lacessere et irritare morbos ecclesiae',° which otherwise would spend° and exercise° themselves. 'Non consensum quaerit sed dissidium, qui quod factis praestatur in verbis exigit':° 'he seeketh not unity, but division, which° exacteth° in words that which men are content to yield in action.' And it is true, there are some which (as I am persuaded) will not easily offend by inconformity,° who notwithstanding make some conscience° to subscribe.° For they know this note° of inconstancy and defection° from that which they have long held shall disable them to do that good which otherwise they would do: for such is the weakness of many that their ministry should be thereby discredited. As for their° easy° silencing° of them, in° such great scarcity of preachers, it is to punish the people, and not them. Ought they not (I mean the bishops) to keep one eye open to look upon the good that these men do, but to fix them both upon the hurt that they suppose° cometh by them? Indeed, such as are intemperate and incorrigible, God forbid they should be permitted to teach. But shall every inconsiderate word, sometimes captiously° watched,° and for the most part hardly° enforced,° be a forfeiture of their voice and gift° of teaching? As for sundry particular molestations,° I take no pleasure to recite them. If a minister shall be troubled for saying in baptism, 'do you believe?' for, 'dost thou believe?': If another shall be called in question for praying for her Majesty without the addition of her style;° whereas the very form of prayer in the book of common prayer° hath 'Thy servant Elizabeth', and no more: If a third shall be accused, upon these words uttered touching the controversies, 'tollatur lex et fiat certamen'° (whereby was meant that the prejudice of the law removed, either's reasons° should be equally compared), of calling the people to sedition and mutiny, as if he had said, 'Away with the law, and try it out by force': If these and sundry other like° particulars be true, which I have but by rumour,° and cannot affirm; it is to be lamented that they should labour amongst us with so little comfort.° I know 'restrained governments are better than remiss';° and I am of his mind that said, 'Better is it to live where nothing is lawful, than where all things are lawful.'° I dislike that laws be contemned,° or disturbers be unpunished. But laws are likened to the grape, that being too much pressed° yield an hard and unwholesome wine. Of these things I must say: 'Ira viri non operatur justitiam Dei';° 'the wrath of man worketh not the righteousness of God.'

As for the injuries of the other part,° they be 'ictus inermium'; as it were, 'headless arrows'; they are fiery and eager° invectives, and in some fond° men uncivil° and unreverent behaviour towards their persons. This last invention also, which exposeth them to derision and obloquy by libels, chargeth° not (as I am persuaded)° the whole side: neither doth that other, which is yet more odious, practised by the worst sort of them, which is, to call in as it were to their aids certain mercenary bands,° which impugn bishops and other ecclesiastical dignities, to have° the spoil° of their endowments and livings. Of this I cannot speak too hardly.° It is an intelligence° between incendiaries° and robbers, the one to fire° the house, the other to rifle° it. And thus much touching the third point.

4. The fourth point wholly pertaineth to them which impugn the present ecclesiastical government; who, although they have not cut themselves off from the body and communion of the church, yet do they affect° certain cognizances° and differences,° wherein they seek to correspond° amongst themselves, and to be separated from others. And it is truly said, 'tam sunt mores quidam schismatici, quam

dogmata schismatica';° 'there be as well schismatical fashions as opinions.' First, they have impropered° to themselves the names of zealous,° sincere, and reformed; as if all others were cold,° minglers of holy things and profane,° and friends of abuses. Yea, be a man endued with great virtues and fruitful in good works, yet if he concur not with them, they term him (in derogation) a civil and moral° man, and compare him to Socrates or some heathen philosopher: whereas the wisdom of the Scriptures teacheth us contrariwise to judge and denominate° men religious according to their works of the second table;° because they of the first are often counterfeited° and practised in hypocrisy. So St. John saith, that 'a man doth vainly boast of loving God whom he hath not seen, if he love not his brother whom he hath seen.'° And St. James saith, 'This is true religion, to visit the fatherless and the widow',° etc. So as that which is with them but philosophical° and moral, is, in the phrase of the Apostle, 'true religion' and Christianity. As in affection° they challenge° the said virtues of 'zeal' and the rest, so in knowledge° they attribute unto themselves 'light' and 'perfection'. They say, the Church of England in King Edward's° time, and in the beginning of her Majesty's reign,° was but in the cradle; and the bishops in those times did somewhat for daybreak,° but that maturity and fullness of light proceeded from themselves. So Sabinus, bishop of Heraclea, a Macedonian, said that the fathers in the Council of Nicaea° 'were but infants and ignorant men'; and that 'the church was not so to persist in their decrees as to refuse that further ripeness of knowledge which the time had revealed.' And as they censure° virtuous men by the names of civil and moral, so do they censure men truly and godly wise (who see into the vanity of their assertions) by the name of politiques;° saying that their wisdom is but carnal and savouring of man's brain. So likewise if a preacher preach with care and meditation (I speak not of the vain scholastical° manner of preaching, but soundly indeed, ordering° the matter he handleth distinctly° for memory, deducing and drawing it down° for direction,° and authorizing° it with strong proofs and warrants),° they censure it as a form of speaking not becoming° the simplicity of the Gospel, and refer it to the reprehension of St. Paul, speaking of 'the enticing speech of man's wisdom'.°

Now for their own manner of teaching, what is it? Surely they exhort° well, and work compunction° of mind, and bring men well to the question, 'Viri, fratres, quid agemus?'° But that is not enough, except they resolve° that question. They handle matters of controversy weakly and 'obiter',° and as before° a people that will accept of anything. In doctrine of manners° there is little but generality and repetition. The word (the 'bread of life'°) they toss up and down, they break° it not. They draw not their directions° down 'ad casus conscientiae';° that a man may be warranted° in his particular actions whether they be lawful or not. Neither indeed are many of them able to do it, what through° want of grounded° knowledge, what through want of study and time. It is an easy and compendious° thing to call for the observation of the Sabbath-day, or to speak against unlawful gain; but what actions and works may be done upon the Sabbath, and in what cases; and what courses of gain are lawful, and what not; to set this down, and to clear° the whole matter with good distinctions and decisions, is a matter of great knowledge and labour, and asketh° much meditation and conversation in° the Scriptures, and other helps which God hath provided and preserved for instruction. Again, they carry not an equal hand° in teaching the people their lawful liberty, as well as their restraints and prohibitions: but they think a man cannot go too far in that that hath a show of° a commandment. They forget that there are 'sins on the right hand, as well as on the left';° and that 'the word is double-edged',° and cutteth on both sides, as well the superstitious observances as the profane transgressions. Who doubteth but it is as unlawful to shut where God hath opened, as to open where God hath shut? to bind where God hath loosed, as to loose where God hath bound?° Amongst men it is commonly as ill taken to turn back° favours as to disobey commandments. In this kind of zeal (for example) they have pronounced generally, and without difference,° all untruths unlawful; notwithstanding that the midwives are directly reported to have been blessed° for their excuse; and Rahab is said 'by faith' to have concealed the spies;° and Salomon's selected judgment° proceeded upon a simulation; and our Saviour, the more to touch the hearts of the two disciples with a holy dalliance, made° as if he would have passed Emmaus.° Further, I have heard some sermons of mortification, which I think (with very good meaning) they have preached out of their own experience and exercise,° and things in private counsels not unmeet; but surely no sound conceits;° much like to Parson's 'Resolution',° or not so good; apt to breed in men rather weak opinions and perplexed despairs, than filial° and true repentance which is sought. Another point of great inconvenience and peril, is to entitle° the people to hear controversies and all points of doctrine. They say no part of the counsel of God must be suppressed, nor the people defrauded: so as the difference which the Apostle maketh between 'milk and strong meat'° is confounded:° and his precept 'that the weak be not admitted unto questions and controversies'° taketh no place. But most of all is to be suspected, as a seed of further inconvenience°, their manner of handling the Scriptures; for whilst they seek express Scripture° for everything; and that they have (in manner) deprived themselves and the church of a special help and support by embasing° the authority of the fathers; they resort to naked examples, conceited° inferences, and forced allusions, such as do mine into° all certainty of religion. Another extremity° is the excessive magnifying of that which, though it be a principal and most holy institution, yet hath it limits as all things else have. We see wheresoever (in manner) they find in the Scriptures the 'word' spoken of, they expound it of preaching.° They have made it almost of the essence of the sacrament of the supper, to have a sermon precedent.° They have (in sort) annihilated° the use of liturgies,° and forms of divine service, although the house of God be denominated of the principal,° 'domus orationis', a house of prayer, and not a house of preaching. As for the life of the good monks and the hermits in the primitive church, I know they will condemn a man as half a Papist, if he should maintain them as other than profane, because they heard no sermons. In the meantime, what preaching is, and who may be said to preach, they make no question.° But as far as I see, every man that presumeth to speak in chair° is accounted a preacher. But I am assured that not a few that call hotly for a preaching ministry deserve to be of the first themselves that should be expelled. These and some other errors and misproceedings they do fortify and entrench° by being so greatly addicted to their opinions, and impatient to hear contradiction or argument. Yea, I know some of them that would think it 'a tempting of God'° to hear or read what might be said against them; as if there could be a 'quod bonum est tenete', without an 'omnia probate'° going before.

This may suffice to offer unto themselves a view and consideration, whether in these things they do well or no, and to correct and assuage° the partiality of their followers and dependants. For as for any man that shall hereby enter into a contempt of their ministry, it is but his own hardness of heart.° I know the work of exhortation doth chiefly rest° upon these men, and they have zeal and hate of sin. But again, let them take heed that it be not true which one of their adversaries said, 'that they have but two small wants, knowledge and love.'° And so I conclude this fourth point.

5. The last point, touching the due publishing and debating of these controversies, needeth no long speech. This strange abuse of antics° and pasquils° hath been touched before. So likewise I repeat that which I said before, that a character° of love is more proper for debates of this nature than that of zeal. As for all indirect or direct glances or levels at° men's persons, they were ever in these cases disallowed. Lastly, whatsoever be pretended, the people is no meet° judge nor arbitrator, but rather the quiet, moderate, and private assemblies and conferences of the learned. 'Qui apud incapacem loquitur, non disceptat, sed calumniatur'.° The press° and pulpit would° be freed and discharged of these contentions. Neither promotion° on the one side, nor glory and heat on the other, ought to continue those challenges and cartels° at the Cross° and other places. But rather all preachers, especially all such as be of good temper,° and have wisdom with conscience, ought to inculcate and beat upon° a peace, silence, and surseance.° Neither let them fear Solon's law,° which compelled in factions 'every particular° person to range himself on the one side'; nor yet the fond° calumny° of 'neutrality'; but let them know that is true which is said by a wise man, that 'neuters in contentions are either better or worse than either side.'°

These things have I in all sincerity and simplicity° set down, touching the controversies which now trouble the Church of England; and that without all art and insinuation,° and therefore not like to be grateful° to either part. Notwithstanding, I trust what hath been said shall find a correspondence in their minds which are not embarked in° partiality,° and which love the whole better than a part. Whereby I am not out of hope that it may do good. At the least I shall not repent myself of the meditation.

Letter to Lord Burghley°

My Lord,

With as much confidence as mine own honest and faithful devotion unto your service and your honourable correspondence° unto me and my poor state can breed in a man, do I commend myself unto your Lordship. I wax° now somewhat ancient; one and thirty years is a great deal of sand in the hour-glass. My health, I thank God, I find confirmed;° and I do not fear that action shall impair it, because I account my ordinary course of study and meditation to be more painful° than most parts of action are. I ever bare a mind (in some middle place° that I could discharge)° to serve her Majesty; not as a man born under Sol,° that loveth honour; nor under Jupiter, that loveth business° (for the contemplative planet° carrieth me away wholly); but as a man born under an excellent Sovereign, that deserveth the dedication of all men's abilities. Besides, I do not find in myself so much self-love,° but that the greater parts of my thoughts are to deserve well (if I were able) of my friends, and namely of your Lordship; who being thc Atlas° of this commonwealth, the honour of my house, and the second founder of my poor estate,° I am tied by all duties, both of a good patriot, and of an unworthy kinsman, and of an obliged servant, to employ whatsoever I am to do you service. Again, the meanness of my estate doth somewhat move° me: for though I cannot accuse myself that I am either prodigal or slothful, yet my health is not to spend, nor my course to get.°

Lastly, I confess that I have as vast contemplative ends, as I have moderate civil ends: for I have taken all knowledge to be my province;° and if I could purge° it of two sorts of rovers,° whereof the one with frivolous° disputations,° confutations, and verbosities, the other with blind° experiments and auricular° traditions and impostures,° hath committed so many spoils,° I hope I should bring in industrious° observations, grounded conclusions, and profitable° inventions and discoveries; the best state of that province. This, whether it be curiosity,° or vain-glory, or nature, or (if one take it favourably) *philanthropia*,° is so fixed in my mind as it cannot be removed. And I do easily see, that place° of any reasonable countenance° doth bring commandment of more wits° than of a man's own;

which is the thing I greatly affect.° And for your Lordship, perhaps you shall not find more strength and less encounter° in any other. And if your Lordship shall find now, or at any time, that I do seek or affect any place whereunto any° that is nearer unto your Lordship shall be concurrent,° say then that I am a most dishonest man. And if your Lordship will not carry me on, I will not do as Anaxagoras° did, who reduced himself with contemplation unto voluntary poverty:° but this I will do; I will sell the inheritance that I have, and purchase some lease° of quick revenue,° or some office of gain that shall be executed by deputy, and so give over° all care of service,° and become some sorry book-maker,° or a true pioner° in that mine of truth, which (he° said) lay so deep.

This which I have writ unto your Lordship is rather thoughts than words,° being set down without all art, disguising, or reservation.° Wherein I have done honour both to your Lordship's wisdom, in judging that that will be best believed of your Lordship which is truest, and to your Lordship's good nature, in retaining° nothing from you. And even so I wish your Lordship all happiness, and to myself means and occasion to be added to my faithful desire to do you service.

From my lodging at Gray's Inn.°

Of Tribute; or, giving that which is due

The Praise of the Worthiest Virtue
The Praise of the Worthiest Affection
The Praise of the Worthiest Power
The Praise of the Worthiest Person

[Four speakers:] *A*, *B*, *C*, *D*.

A. 'Since we are met let me govern our leisure.'

B, *C*, *D*. 'Command.'

A. 'Let every man honour that which he esteemeth most and celebrate it with praise.'

B. 'O vain motion° and ignorance of the times. Are not satires° more in price than hymns?'

A. 'Obey!'

The Praise of Fortitude

B. 'My praise shall be dedicated to the noblest of the virtues.° Prudence to discern between good and evil; Justice to stand indifferent between self-love and society; Temperance to decide aright between passion and reason: these be good innocent things,° but the virtue of merit, the virtue of resolution, the virtue of effect,° is Fortitude. Present unto a man largely endued with Prudence the tempest of a sudden and great danger, and let Fortitude absent herself, what use hath he of his wisdom? Hath he power either to behold the danger or intend° the remedy? Or rather doth not the first impression° disable him to take a true view of the peril, and the apprehension of the peril so attach° and seize his senses that he cannot employ them for his deliverance? Where be the goodly grounds° of reason, observations of experience, rules, precepts, and cautions whereby he was wont at leisure to consider, compare, and conclude? All these digested° thoughts are confounded,° their prints° are defaced. A tumult, an alarm of peril, hath, as Berecynthia's horn,° drowned Orpheus's music,° or as a blast of wind disordered Sibylla's leaves.° His very wisdom is the first thing that flies. His spirits° that should sit in council in his brain are gone to succour his heart, and so he is left abandoned to his perils by the treason of his fears. And yet his wisdom could have told him that he that fronteth° a danger and looketh it in the face, while he considereth how to receive and bear it, doth often descry how to avoid it. His wisdom could have told him again what a madness it is in men to treble every adversity by preventing° the sense thereof by fear, and repeating it by remembrance. Many other points it might suggest unto him. But all these well-collected° principles are to seek,° or if they occur they appear to be vain speculations, idle discourse, good entertainments to persuade men well of the strength of their nature, but deceitful in the execution and trial. What price therefore or regard can wisdom carry, which tireth a man's thoughts with forecasting and providing for perils which never come (as if one could embrace all accidents), but when a danger cometh unexpected, it leaveth a man in prey to his adventures? But now let Prudence, this weak lady, ravished° by every invasion and assault of sudden danger, obtain for her champion and knight Fortitude, and then see how she entertaineth° the challenges of fortune! Doth a man fly before he knoweth, or suffer before he feeleth? No, but straightways the discovery of the peril maketh him more than himself. It awaketh his senses; it quickeneth his motions;° it redoubleth his forces. He looketh through and through the peril; he taketh hold of every light of remedy;° he discerneth what must be concluded,° what may be deferred; he ceaseth not to devise for the rest while he executeth that which is instant, nor to execute the present while he deviseth for that to come. But he is always in his power, rejoicing in the proof of himself and welcoming necessity.

Thus is Fortitude the marshal of thoughts, the armour of the will, and the fort of reason. Let us turn our consideration and behold Justice, the sacred virtue, the virtue of refuge, the virtue of society. Doth not she also shroud herself under the protection of Fortitude? Let a man be abstinent from wrong, exact in duty, grateful in obligation, and yet dismantled° and open to fear or dolour, what will ensue? Will not the menace of a tyrant make him to condemn the innocent? Will not the sense of torture make him appeal° his dearest friend, and that untruly? But pain hath taught him a new philosophy. He beginneth to be persuaded that it is Justice to pay tribute to nature, to yield to the rigour of pain, to be merciful to himself. He would give others leave to do the like by him: he would forgive them if they did so. So as now his balances° wherewith he was wont to weigh out every man his due°

are stricken out of his hands. He is at the devotion° of the mightiest. His Justice remaineth with him, but as a fury to upbraid his fault and to increase his torment.

As for Temperance, divide° it from Fortitude and Magnanimity, and what shall I say of it? It is the prison of nature, the abuse of a man's self, and nothing but pure envy. But because you cannot obtain, you will condemn. Is this a virtue? You esteem it impossible; you have not patience to expect° it; you cannot submit yourself to the conditions of obtaining. Will you therefore fall to despise? Will you affect° to be admirable because you neither follow others nor spare yourself? Will you make yourself nothing but an accusation and a censure of others? "O, but I mean no such matter, no vain-glory,° no malignity, no diffidence, no impatience. I desire but a relief from perturbations. I seek but an equal tenor of mind.° I will not use because I will not desire. I will not desire because I will not fear to want."° Lo, we see all these circumstances.° All his preparation is but to keep afore° of fear and grief, which Fortitude rejoiceth to challenge and to chase. But when a true fear and grief cometh, such as all men are subject to, I mean a fear and grief which ariseth not of the destitution° of a pleasure but of the access of a disfortune,° then what use hath he of his Temperance? Doth he not then esteem it a great folly that he hath provided against the hurt of the sunshine, but not of fire?° Doth he not take it for a mad conclusion that if a man could make himself impassible° of pleasure, he should make himself in one labour impassible of pain? Whereas contrariwise, it is an introduction to bear stronger griefs, to desire often with want.° But let Fortitude and strength of mind assist Temperance and what followeth then? A man is able to use pleasures and to spare them;° to contain° himself in the entry,° or greatest downfall; yea and to entertain himself ever in pleasure, having in prosperity sense of solace, and in adversity sense of strength. Therefore it is Fortitude that doth either guard or consummate or ennoble all virtues.

Of pleasure now let us inquire, which being limited° and governed, no severity of conceit, no height of language shall mar,° but it is the blessing of nature, the marriage of the senses, the feast and holiday of this our laborious and unquiet life. Only let men discern the present signal° and token of nature from the bait° of affection,° let them discern that which is pleasant in the sum and total from that which is pleasant for a time. Now what true and solid pleasure can there be without Fortitude? See, I pray you, what sport fear maketh with the timorous man in the course of his pleasures and desires.° Hope he will, and that lightly° and vainly, else he were undone, else he would die of melancholy. And when will he hope, when the effect is far off? But when° it approacheth, and he would address himself to obtain, then all his hopes sink, and if he be very near, then doth he think time such a traitor that, according to the proverb° of the Grecians, "He waxeth old in a day". When he cometh to the sense of fruition of his pleasure, then he is in a maze:° he is as a deer that is come unto an unwonted good pasture and stands at a gaze and dare scantly feed. So he ever imagineth some ill is hid in every good: so as his pleasure is as solid as the sands, being interrupted with continual fears and doubts. And when the pleasure is passed, then he thinketh it a dream, a surfeit° of desire, a false joy. He is ungrateful to nature, for still the sense of grief printeth° so deep, and the sense of delight so lightly, as the one seemeth unto him a truth, the other a deceit. Judge then how native° and perfect pleasures are to him to whom expectation is a rack,° enjoying is an amazement,° remembrance is a distaste and bitterness. Again, what doth so much increase and enrich all pleasures as novelty? But it is a rule that to a fearful man, whatsoever is new is suspect. So as that which should season° and enrich pleasures doth taint° and embase them.

But now let us take breath a while, and look about whether we can see anything else good in nature. Virtue, the perfection of nature; pleasure, the fruit of nature; is there anything else?—O beauty, the ornament of nature! I cannot say that Fortitude will make a crooked man straight,° nor a foul person fair. But this I may say, that fear is the mother of deformity, and that I never saw a man comely° in fear. So it is Fortitude that giveth a grace, a majesty, a beauty to all actions. But why do we stay so long upon the merits of Fortitude in showing how it is a protector and benefactor to all that is good, and do not hasten to the conquests and victories thereof? Have not we done well, because it is more meritorious to succour than to subdue, and more excellent to compound° civil dissensions° than to defeat foreign enemies? Therefore now we have showed how Fortitude enableth° the mind to the works and actions of virtue, to the taste and fruition of pleasure, it is time to set forth what it can do against these extern° things. What be these evils? Let them be mustered. Are they pain of body, grief of mind, slander of name, scarcity of means, solitude of friends, loss of life? Why, none of these are ill with° Fortitude, which can bear them without violating the repose of our minds in themselves, or our duty to others. It conditeth° them, it taketh away their venomous quality, it reconcileth them to nature. Let no man quarrel with nature and the divine providence, which hath included or ordained in every ill the remedy. It hath given stupefaction° of the parts,° and the weakness of the spirits° against pain of the body; apprehension° and custom° against the grief of the mind; conscience° against slander; appetite° against baseness of means;° the condition of humanity against the loss of friends; and no feeling of a full remedy against not° being.° Religion ministereth higher comforts, but those are in nature.° But it is fear and impatience that are the sergeants° of fortune, that arrest and subdue us to those things, being otherwise free men. So as when any of these draweth from men lamentations, outcries, excess of grief, it is not the outward enemy but the inward traitor. Nothing is to be feared, but fear itself.° Nothing grievous, but to yield to grief.

For let us remember how men endued with this virtue Fortitude have entertained° death, the mightiest of these enemies. Let us consider whether it hath wrought any alteration in them, whether it hath disordered and put out of frame their ordinary fashions and behaviour. And therein I muse at the Stoics,° that accounted themselves to hold° the masculine philosophy, esteeming other sects delicate, tender, and effeminate; what they meant, to publish and advise men so much to the meditation of death, which they professed to assuage. Must it not be a terrible enemy against whom there is no end of preparation?° Ought they not to have taught men to die as if they lived, and not to live as though they continually died?° Much more manfully thought the voluptuous sect,° that counted it as one of the ordinary actions of nature.

But to return, let us lead about our consideration to take view of those which have been men of known value and Fortitude, and see whether death presented hath so much as untuned their ordinary manner of conceit and custom.° Julius Caesar,° the worthiest man that ever lived, the bravest soldier, a man of the greatest honour, and one that had the most real and effectual° eloquence that ever man had; not a sounding° and delightful eloquence, for a continuate speech,° but an eloquence of action, an eloquence of affairs, an eloquence that had suppressed a great mutiny by the only word "Quirites",° an eloquence to imprint°

and work upon any to whom he spake. See now whether he varied from himself at his death.° The first blow that was given him on the neck by Casca, that stood behind his chair, he turned about and caught hold of his arm. "Traitor, Casca, what dost thou?"° The words were but plain, but yet what could have been devised upon study more apt to daunt the conspirator and to invite succours? Should he have cried "Help"? He would rather have lost a thousand lives. Should he have cried "Treason"? That had been also an imploring of aid. Should he have said "What, in the Senate, in the temple of the gods?" It was not decent° for Caesar to claim sanctuary, his person was more venerable than the place. Therefore he said "Traitor", which was as effectual to invite succours, but yet retained the majesty of his courage. He added "Casca". He was nothing astonished:° he singled him out by his name. Who knoweth not that is anything skilful in the weight and force of words, but that the compellation° by name giveth, as it were, a point to a man's speech, to make it enter and penetrate? And what saith he to him? Doth he ask "For what desert of mine?" That had been to ask mercy. Doth he ask "Is this the reward of my clemency?"—as if it had been a time to expostulate,° or as if virtue did seek the reward of life? And therefore he saith "What dost thou? Thy offences want a name. It is beyond all cause. It is mere fury. Descend into thyself,° thou and thy complices." All came about him being unarmed, and as a stag at bay, yet he never ceased to put himself in defence and to avoid the blows by the declining of his body, impoyning° of their weapons, and all the means of an unarmed man. A form° excellently becoming a military man, though he knew it would not help. At last when Marcus Brutus gave him a wound ("And thou my son!"), noble Caesar, he had no weapon to wound Brutus again. But this word wounded him, this word pierced him, this word enchanted him, this word made him ever despair of a final good success of the war, though his cause was just and his proceeding at the first prosperous. This word inspired him once at his birthday, when his affairs were in most prosperous terms, to break out causeless° into this verse:

At me sors misera et Latonae perdidit infans.°

This word afterwards turned itself into the likeness of an ill spirit that appeared unto him in his tent.° In the end when his strength failed him, yet he took an honourable regard to fall in comely manner, and covered° after the manner of the apparel of that time.

So as that compliment,° that point of honour which it had been much for a lady to have remembered, unto whom modesty and honour of person° were *summum bonum*, so great a monarch, so great a captain, in so strange and violent an assault, forgot not at the point of death. Augustus° Caesar, his nephew, a man nothing of that strength of courage, but of great assurance and serenity of mind, he that by the calmness and repose of his countenance had° appalled a barbarous conspirator, he that would ever wish himself *euthanasia*,° in sum a dainty° and fine man, was he not the same man at his end? "Livia, bear in mind our marriage, live and farewell":° a farewell at length for a long absence. Vespasian, a man exceedingly given to the humour of dicacity° and jesting, his last words were "An' I be not much deceived, I am upon the point to be made a god°"; scoffing° at death, at himself, and at the times. Severus (Septimius, I mean), a man of infinite pursuit of action and dispatch°—"An' there be anything for me to do"°—and further he could not go. The like° words he would have used if he had been but going to sleep. Socrates, that would never affirm anything, in his last words to the judges said "It is time that I conclude, that I may be dismissed to die and you to live. But whether° is better knoweth Jupiter.°" He left° not his irony, for himself had told his opinion to his friends before. So the Roman senator° that had delighted so much in the inquisition of the truth, when he went to his death by commandment of Caius Caligula, took certain philosophers that he was wont to reason and dispute with all gently by the hand, and said "Sirs, you shall dispute of death and the immortality of the soul, but I shall know by and by how it standeth.°" So this desire of knowledge left him not. He studied upon his death. He took care to understand his death, as if he had been but to try a conclusion.°

So that by all these examples it appeareth how Fortitude is able to steel° men's minds in such sort that every strict° habit or fashion is stronger than the fear of death or sense of his approaches. Neither will I so much extenuate° the praise of this noble virtue Fortitude, that I make it the masterpiece° and the principal coat of honour° thereof to overcome the fear of death. For we see barbarous customs, false superstitions, violent passions are able to effect as much. But all these do it as madness sometimes doth it: they overcome one tyrant by another, but they leave not the mind in entire liberty, for that is the only work of fortitude. Other virtues deliver us from the servitude of vices, but Fortitude alone delivereth us from the servitude of fortune.°'

A. 'Your speech were able to warm the heart of a coward, for either it would put honour into him, or else if nothing could prevail with him but fear, it would make him more afraid of fear itself than of any peril.'

C. 'He deserveth to be knighted for his speech.'

A. 'Let us hear what you will deserve.'

The Praise of Love

C. 'My praise shall be dedicated to the happiest state of mind, to the highest elevation of mind, to the noblest affection. These virtues are mediocrities:° they are laws of the mind, they restrain it, they limit it—they raise it not, they amplify it not. They are as the mill° when it is set° upon a rich diamond. Here it grindeth out a race° and there a grain,° to make it more neat and paragon.° But in the meanwhile the stone loseth carats,° loseth substance. For these virtues, they polish the mind, they make it without blemish, they give it an excellent form, but commonly they take off much of the natural greatness. They be the affections° which make the mind heroical,° that give it a vigour to exceed itself and to fascinate and bind° others. Do we not see it in agility of body? No sleight°—no, no practice° can bring a man to do that which sometimes fear or fury maketh him to do. In the melting of an hard metal, can a mighty dead fire do as much as a small fire blown?° In beating out metals, can a huge weight do as much as the blow of a hammer? It is motion, therefore,° that animateth all things. It is in vain to think that any strength of matter can countervail a violent motion. The affections are the motions of the mind, the virtues pray in aid of° the affections. Admiration and wondering is the life of Prudence; modesty is the life of Temperance; indignation is the life of Fortitude. All virtues take measure from themselves, but power and strength from the affections. Therefore amongst affections will I seek happiness and height of mind.°

But why do I dissemble? Why do I alter and invert° the true steps of my thoughts, as if the worthiness of the affections [were]° as a goodly pinnacle or seamark, seen long before a man discovereth° the shore? As for the other affections, they be but sufferings° of nature,° they seek ransoms and rescues from that which is evil. Not enjoying an union with that which is good, they seek to expel that which is contrary, not to attract that which is agreeable. Fear and grief, the traitors of nature; bashfulness, a thraldom° to every man's conceit and countenance; pity, a confederacy with the miserable; desire of revenge, the supplying° of a wound: all these they endeavour to keep the main stock of nature, to preserve her from loss and diminution. But love is pure gain and advancement in nature. It is not a good by comparison, but a true good;° it is not an ease of pain, but a true purchase of pleasure. And therefore when our minds are soundest, when they are as it were not in sickness and therefore out of taste,° but when we be in prosperity, when we want° nothing—then is the season, the opportunity and the spring of love.° And as it springeth not out of ill, so it is not intermixed with ill. It is not like the virtues, which by a steep and craggy way conduct us to a plain,° and are hard taskmasters at the first, and after give an honourable hire.° But the first aspect of love and all that followeth is gracious and pleasant.

And now to you sir, your so much commended virtue, Fortitude—and therein chiefly commended because it doth enfranchise us from the tyrannies of fortune,° yet doth it not in such perfection as doth love. For Fortitude indeed strengtheneth the mind, but it giveth no filling. It leaveth it empty. It ministereth° unto it no apt contemplation° to fix itself upon, that it may the more easily be diverted from the sense of dolours. And that is the reason why those things which you would in no wise admit to be competitors with Fortitude in this honour—as barbarous customs and false superstitions—do this notwithstanding more easily and effectually than that virtue. But love doth so fill and possess all the powers of the mind as it sweeteneth the harshness of all disfortunes. Let no man fear the yoke of fortune that is in the yoke of love. What fortune can be such an Hercules as shall be able to overcome two?° When two souls are joined in one, when one hath another to divide his fortune withal, no force can depress him.

Therefore, since love hath not her seat in ill, as have other affections; since it hath no part ill, as virtue hath the beginning;° since it admitteth no extern ill, and therein excelleth Fortitude, now let us see if it be not as rich in good as exempt from ill. Now therefore will I teach lovers to read. They have all this while loved by rote! I will give them the alphabet of love, and show them how it is spelled. For this is a principle: "The nature of man is very compound", and there is no great deep pleasure to be found but is also very compound and full of multiplicity. So as it is not so much the exquisite° sense of any simple° pleasure that affecteth as the concurrence of divers satisfactions. I affirm then, and that truly (as it is much to be said for love), that it is very confluence, Pandora° and an assembly° of the most affecting delights incident° to the nature of man. Curiosity and desire to know is natural, and carrieth a great delight.° These we see chiefly in contemplative° men, in whom all other affections are made tributary° to this. We find it very generally in peregrinations° to strange countries, concourses° to unwonted actions,° listening after news, games of chance and sundry other particulars. This delight doth also wind itself like a woodbine° about other affections, in sort that few delights have grace° longer than novelty commendeth them to the apprehension.

Now therefore love is a fountain° of curiosity, a most sweet ground° set with infinite changes,° and of the strongest° and most various adventures. I demand in love what are these excitations° by absence, these redintegrations° by unkindnesses, these refreshings by alteration of attires° and change of presences but, as it were, quaverings° upon this stop?° But of all others, new merit and demonstration of devotion is the gratefullest novelty. And it is not only the variety of knowledge that pleaseth, but the certainty. For assuredly no person knoweth at any time the mind of another, but in love. Love is the only window of the heart. So as we see what a rich tribute curiosity and desire to know pay unto love, being indeed if not the highest, yet the sweetest contemplation of all other.

Now turn we our view upon ambition, an affection mighty and general. Dionysius, when he was chased from his tyranny, would° be a *pedantius.*° A child will lord it over his dog and bird. Is not this humour so mighty as it infecteth the sense? Have not we heard of *ambitiosa fames*,° when men desire not the meat of the best taste, but that which is dearest bought or hardliest° procured, not unlike to the receipt *aurum potabile*?° Consider then, is not love a goal of ambition, a perfection of commandment,° including not only the commandment of the person but of the will?° Do we not see that in popular° states ambition is more sweet, because honour is more voluntary?° Do we not observe how the *heresiarchae*° and beginners of sects make it the *summum bonum*° to reign in men's minds, and therefore are justly called *violatores mentium*,° the deflowerers of understandings? So that as it is the nature of such extravagant and strange spirits to seek a commandment of men's reason and beliefs, so it is natural in a man to aspire to commandment of persons, and especially of affections and wills. Another delight ministered unto the nature of man by this condition is to have such as may be compassible° with him. Many are the griefs and disasters whereto men's states are exposed; and the very representation of them by foresight doth disrelish° our greatest prosperities. But then when one foreseeth withal that is unto his misfortune,° there cannot be added solitude, but that he shall have a partner, that settleth and quieteth the mind.

A further inward and deep affection bred in the mind of man is the continuing and, if it might be, perpetuating of himself.° Hence issueth the common and natural desire of children and posterity,° the desire to carry° and advance favourites and creatures, the affecting same° by inventions of our own or celebration of others, the planting [of]° memory, the foundations and monuments. But assuredly the perfectest continuance, and next to life itself, is the living in the thoughts of them upon whose hearts our image is engraven. So also in men, the delight of pride and of taking an high° and comfortable° impression of a man's self from the admiration and endeared estimation of others. Was not flattery ever in grace?° But there is no flattery like to that of a lover. One said well that a man's self was the "archflatterer":° but he should have excepted his lover. For the proudest man that ever was, never thought so well of himself as the lover thinketh of the person loved. Consider again the delight of concurrence° in desire, without emulation. If two be but set at one game they love, or labour together in some one work or invention, mark how well pleased, how well disposed, how contented they be. So then if minds are sharpened against minds, as iron against iron, in every action, what shall we think of that union and conjunction of minds which love worketh? What vigour, what alacrity must it give?

Behold further the nature of the mind of man. It is every man's observation that remission° and relaxation of mind is a most necessary part of life. It is noted also that absolute idleness and leisure, when the mind is altogether without object, is but languishing and weariness.° How precious then is love, which is the sweetest repose from travails and affairs, and the sweetest employment in leisure and idleness. So as in one respect it is like the earth to Antaeus,° that giveth fresh forces; in the other it is like Penelope's web,° which entertaineth° time, and putteth off expectation. For it is no ill commendation to say that love is an idle man's occupation. Not but it but catcheth° the busiest. Can a tyrant be idle the first year of his usurpation? See Appius and Virginia.° Could the state and enleagued enemy of Octavius Caesar want what to think?° See Antony and Cleopatra.° So that it is not the fruit of idleness° but the remedy.

Lastly, to leave where Love beginneth, who discerneth not that the eye is the most affecting sense?° They be tales, the propositions to the contrary. The humour of melancholy importuneth those that are overcome with it, with the memory of the most affecting delights. Confer° with one that is entering to be melancholy. Shall you hear him complain of harsh sounds or odious savours represented to his imagination? No, but always meditating° of fearful and disliking forms. Now who denieth that the eye is first contented in love, being fed and feasted with fresh proportionable° shapes, and° decent° motions? Therefore if all delights of sense, affection and understanding be tributary to love; if love effecteth the sweetest contemplation to him that desireth to know, the exactest commandment to him that desireth to rule, the comfortablest promise to him that looketh into his fortune, a second life to him that seeketh to survive himself, the most flattering glass° to him that delighteth to view himself with advantage, the greatest union of minds to him that is sociable, the most refreshing repose from action, the most acceptable entertainment of leisure, the most pleasing object to the most imprinting sense: let us not marvel if [the]° burning glass° that gathereth the beams of so many pleasures into one point do so mightily inflame the soul. And let us conclude that there be many excellent delights, but not so complete.

Non deus ut perhibent amor est, sed amorosus et error.'°

A. 'I was thinking what you did deserve. And hearing you speak so distinctly of a passion that is of that nature as a man cannot well tell who should know it best, he that hath tried it, or he that hath not tried it, I thought you deserved a patent° of that which hath been granted but seldom, and that is *amare et sapere*.'°

B. 'He had rather have his lady's favour! But here sitteth one as if he neither gave much ear, nor meant to speak.'

D. 'I was never niggard of my ear, but I would be pardoned of° my speech.'

A. 'The wrong were not to us but to that you honour most, if you shall deceive it of your praise and celebration.'

The Praise of Knowledge

D. 'Silence were the best celebration of that which I mean to commend; for who would not use silence where silence is not made, and what crier can make silence in such a noise and tumult of vain and popular opinions? My praise shall be dedicated to the mind itself. The mind is the man, and the knowledge is the mind. A man is but what he knoweth. The mind itself is but an accident° to knowledge, for knowledge is a double of that which is.° The truth of being and the truth of knowing is all one.°

Are [not]° the pleasures of the affections greater than the pleasures of the senses? And are not the pleasures of the intellect greater than the pleasures of the affections? Is not that only a true and natural pleasure, whereof there is no satiety?° Is it not knowledge that doth alone clear the mind of all perturbations?° How many things be there which we imagine that are not! How many things do we esteem and value otherwise than they are! These vain imaginations,° these ill-proportioned estimations,° these be the clouds of error that turn into the storms of perturbation. Is there then any such happiness as for a man's mind to be raised above the confusion of things, where he may have the prospect° of the order of nature and the errors of men?

Is this but a view only of delight, and not of discovery? of contentment, and not of benefit? Shall he not as well discern the riches of nature's warehouse,° as the beauty of her shop? Is truth ever barren? Shall he not be able thereby to produce worthy effects, and to endow the life of man with infinite new commodities?°

But shall I make this garland to be put upon a wrong head? Would anybody believe me, if I should verify° this upon the knowledge which is now in use? Are we the richer by one poor invention, by means of all the learning that hath been these many hundred years? The industry of artificers° maketh some small improvements of things invented; and chance sometimes in experimenting maketh us stumble upon somewhat which is new; but all the disputations of the learned° never brought to light one effect of nature before unknown. When things are known and found out, then they can descant upon° them, they can knit° them into certain courses,° they can reduce° them to their principles. If any instance of experience stand up against° them, they can rank it in order by some distinctions. But all this is but a web° of the wit, it can work° nothing. I do not doubt but that common notions, which we call reason, and the knitting of them together, which we call logic or the art of reason, may have use in popular studies. But they rather cast obscurity than give light to the contemplation° of nature.

All the philosophy of nature which is now received, is either the philosophy of the Grecians,° or that other of the alchemists.° That of the Grecians hath the foundation in words,° in ostentation, in confutation, in sects, in auditories,° in schools, in disputations. The Grecians were (as one of themselves saith), "you Grecians, ever children".° They knew little antiquity; they knew (except fables) not much above five hundred years before themselves; they knew but a small portion of the world. That of the alchemists hath the foundation in imposture,° in auricular° traditions, and obscurity.° It was catching hold° of religion: but the best principle of it is, "populus vult decipi".° So as I know no great difference between these great philosophies, but that the one is a loud crying folly, and the other is a whispering folly; the one gathered out of a few vulgar° observations, and the other out of a few experiments of the furnace.° The one never faileth to multiply words, and the other oft faileth to multiply gold.

Who would not smile at Aristotle,° when he admireth the eternity and invariableness of the heavens,° as if there were not the like in the bowels of the earth? There be° the confines and borders of these two great kingdoms, where the continual alterations and incursions° are. The superficies and upper parts of the earth are full of variety. The superficies and lower parts of the heavens (which we call the middle region of the air) is full of variety. There is much spirit° in the one place that cannot be brought into mass;° there is much massy body in the other place that cannot be refined to spirit. The common air is as the waste ground between the borders.

Who would not smile at the astronomers,° I mean not these new car-men° which drive the earth about, but the ancient astronomers, that feign the moon to be the swiftest of the planets in motion, and the rest in order, the higher the slower; and so are compelled to imagine a double motion?° Whereas how evident it is, that that which they call a contrary motion is but an abatement° of motion. The fixed stars overgo° Saturn, and Saturn leaveth behind him Jupiter, and so in them and the rest all is but one motion, and the nearer the earth the slower; a motion also whereof the air and the water doth participate, though much interrupted.°

But why do I in a conference of pleasure° enter into these great matters, in sort that pretending to know much, I should not know what is seasonable?° Pardon me, it was because almost all things may be endued and adorned with speech, but knowledge itself is more beautiful than any apparel of words that can be put upon it.

And let me not seem arrogant, or without respect to this° great reputed° author. Let me so give every man his due, as I give Time his due, which is to discover truth.° Many of these men had great wits,° far above mine own, and so are many in the universities of Europe at this day. But alas, they learn nothing there but to believe:° first to believe that others know that which they know not; and after, that themselves know that which they know not. But indeed facility° to believe, impatience to doubt, temerity to assever,° glory to know,° doubt to contradict, end to gain,° sloth to search, seeking things in words,° resting in a part of nature°—these and the like have been the things which have forbidden the happy match° between the mind of man and the nature of things, and in place thereof have married it to vain° notions and blind experiments. And what the posterity and issue° of so honourable a match may be, it is not hard to consider. Printing, a gross° invention; artillery, a thing that lay not far out of the way; the needle,° a thing partly known before: what a change have these three° made in the world and in these times; the one in the state of learning, the other in the state of the wars, the third in the state of treasure, commodities, and navigation. And these were (as I say) but stumbled upon and lighted on by chance. Therefore, no doubt the sovereignty of man lieth hid in knowledge, wherein many things are reserved° which kings with their treasures cannot buy, nor with their forces command. Their spies and intelligencers can give no news of them, their seamen and discoverers cannot sail where they grow. Now we govern nature in opinions, but we are thrall° to her in necessities. But if we would be led by her in invention, we should command her in action.'

A. 'This speech deserveth to be understood.'

B. 'Now sir, you that have made this motion, I wish you no greater revenge, but that some of us three should have intercepted° your choice.'

A. 'It were small revenge, for then I would be silent.'

B. 'Nay, that were against your own law. But I should smile to see you put° to go over the same matter again.'

The Praise of his Sovereign

A. 'No praise of magnanimity,° nor of love, nor of knowledge, can intercept her praise that planteth and nourisheth magnanimity° by her example, love by her person, and knowledge by the peace and serenity of her times; and if these rich pieces be so fair unset,° what are they set,° and set in all perfection?

Magnanimity no doubt consisteth in contempt of peril, in contempt of profit, and in meriting° of the times wherein one liveth.

For contempt of peril, see a lady that cometh to a crown after the expiration of some adverse fortune,° which for the most part extenuateth° the mind, and maketh it apprehensive of fears. No sooner she taketh the sceptre in her sacred hands, but she putteth on a resolution to make the greatest, the most important, the most dangerous alteration that can be in a state, the alteration of religion.° This she doth, not after a sovereignty established and continued for many years, when custom might have bred in her people a more absolute obedience, when trial of her servants might have made her more assured whom to employ, when the reputation of her policy and virtue might have made her policy more redoubted;° but at the very entrance of her reign, when she was green in authority, her servants scant known unto her, the adverse° part not weakened, her own part not confirmed.° Neither doth she reduce° or reunite her realm to the religion of the states about her, that the evil inclination of the subject might be countervailed by the good correspondence° in foreign parts; but contrariwise she introduceth a religion exterminated° and persecuted both at home and abroad. Her proceedings herein is not by degrees and by stealth, but absolute° and at once. Was she encouraged thereto by the strength she found in leagues and alliances with great and potent confederates? No, but she found her realm in wars with her nearest and mightiest neighbours;° she stood single and alone, in league only with one,° that after the people of her nation had made his wars, left her to make her own peace; one that could never be by any solicitation moved to renew the treaties; and one that hath since proceeded from doubtful terms of amity to the highest acts of hostility.° Yet notwithstanding all this, the opposition so great, the support so weak, the season so improper; yet, I say, because it was a religion that freed her subjects from pretence° of foreign powers, a religion wherein she was nourished and brought up, and indeed the true religion, she brought to pass this great work with success worthy so noble a resolution. See again a Queen that, when a deep and secret conspiracy° was plotted against her sacred person, practised by subtle instruments, embraced by violent and desperate humours, strengthened and bound by vows and sacraments, and the same was revealed unto her (and yet the nature of the affairs required further ripening before the apprehension of any of the parties), was content to put herself in guard of the divine providence and her own prudence, to have some of the conspirators in her eye, to suffer them to approach to her person, to take a petition of the hand that was conjured° for her death; and that with such majesty of countenance, such mildness and serenity of gesture, such art and impression of words,° as had been sufficient to have repressed and bound the hand, and turned and altered the heart of a conspirator, that had not been discovered. Lastly, see a Queen, that when her realm was to be invaded by an army,° the preparation whereof was like the travail of an elephant, the provisions whereof were infinite, the setting forth whereof was the terror and wonder of Europe; it was not seen that her cheer, her fashion, her ordinary manner was anything altered; not a cloud of that storm did appear in that countenance wherein peace doth ever shine. But with excellent assurance and advised security she inspired her council,° animated her nobility,° redoubled the courage of her people; still having this noble apprehension, not only that she would communicate° her fortune with them, but that it was she that should protect them, and not they her; which she testified with no less demonstration than her presence in camp.° Therefore that Magnanimity that neither feareth greatness of alteration, nor the vows of conspirators, nor the power of enemies, is more than heroical.

For contempt of profit, consider her offers,° consider her purchases.° She hath reigned in a most populous and wealthy peace, her people greatly multiplied, wealthily appointed,° and singularly devoted. She wanted not the example of the power of her armies, in the memorable voyages and invasions prosperously made and achieved by sundry her noble progenitors.° She hath not wanted pretences, as well of claim and right as of quarrel and revenge. She hath reigned during the minority of some of her neighbour princes,° and during the factions and divisions of their people upon deep and irreconcilable quarrels, and during the embracing greatness of some one° that hath made himself as weak through too much burthen as others are through decay of strength; and yet see her sitting within the compass° of her sands. Scotland, that doth in a manner eclipse her land; the United Provinces of the Low Countries,° which for site, wealth, commodity of traffic,° affection to our nation, were most meet to be annexed to the crown; she left the possession of the one, and refused the sovereignty of the other. So that notwithstanding the greatness of her means, the justice of her pretences,° and the rareness of her opportunities, she hath continued her first mind; she hath made the possessions which she received the limits of her dominions, and the world the limits of her name, by a peace that hath stained all victories.

For her merit, who doth not acknowledge that she hath been as a star of most fortunate influence upon the age wherein she hath shined? Shall we speak of merit of clemency, or merit of beneficence? Where shall a man take the most proper and natural trial of her royal clemency? Will it not best appear in the injuries which were done unto her before she attained the crown? For after she is seated in the throne, and that the commonwealth is incorporate in her person, then clemency is drawn in question,° as a dangerous encounter° between Justice and policy.° And therefore who did ever note that she did resent (after she was established in her kingdom) of the wrongs done in her former state? Who doth not remember how she revenged the rudeness and rigour of her jailor° by a word, and that not bitter but salt,° and such as showed rather the excellency of her wit than any impression of her wrong? Yea and further, is it not too manifest that since her reign, notwithstanding that principle, "that princes should not neglect the Commonwealth's wrongs included in themselves", yet when it is a question of drawing the sword, there is ever a conflict between the justice of her place joined with the necessity of her state, and her own royal clemency, which as a sovereign and precious balm° continually distilleth from her fair hands, and falleth into the wounds of many that have incurred the offence of her laws.

Now for her beneficence, what kind of persons have breathed° during her most happy reign, but have had the benefit of her virtues conveyed into them? Take a view and consider whether they have not extended to subjects, to neighbours, to remote strangers, yea to her greatest enemies.

For her subjects, where shall we begin in such a maze of benefits as presenteth itself to our remembrance? Shall we speak of the purging away the dross of religion, the heavenly treasure; or that of monies, the earthly treasure? The greater was touched before, and the latter deserveth not to be forgotten. For who knoweth not (that knoweth anything in matter of state) of the great absurdities and frauds that arise of the divorcing of legal estimation° of monies from the general and (as I may term it) natural estimation of the metals;° and again, the uncertain and wavering values of coins,° a very labyrinth of cozenages° and abuses, and yet such as great princes have made their profit of towards their own people? Pass on from the mint to the revenues and receipts. There shall you find no raising of rents, notwithstanding the alteration of prices and the usage of the times; but the overvalue,° besides a reasonable fine,° left for the relief of tenants and the reward of servants; no raising of customs, notwithstanding her continual charges of setting to the sea; no extremities taken of forfeitures° and penal laws, a means used by some kings for the gathering of great treasures: a few forfeitures indeed, not taken to her own purse but set over° to some others, for trial only, whether gain° could bring these laws to be well executed which the ministers of justice did neglect. But after it was found that only compositions° were used, and the law never the nearer the execution, the course was straight suppressed and discontinued. Yea further, there have been made laws more than one in her time for the restraint of the vexation of informers and promoters;° nay a course taken by her own direction for the repeal of all heavy and snaring° laws, if it had not been crossed° by those to whom the benefit should have redounded. There shall you find no new taxes, impositions, nor devices; but the benevolence of the subject freely offered by assent of parliament, according to the ancient rates,° and with great moderation in assessment; and not so only, but some new forms of contribution offered likewise by the subjects in parliament, and the demonstration of their devotion only accepted,° but the thing never put in ure.° There shall you find loans, but honourably answered and paid, as if it were the contract of a private man. To conclude, there shall you find monies levied° upon sale of lands, alienations° (though not of the ancient patrimony) yet of the rich and commodious purchases° and perquisites° of the crown, only because she would not be grievous and burdensome to the people.

This treasure, so innocently levied,° so honourably gathered and raised, with such tenderness° to the subject, without any baseness or dryness° at all, how hath it been expended and employed? Where be the wasteful buildings, the exorbitant and prodigal donatives,° the sumptuous dissipations in pleasures and vain ostentations, which we find have exhausted the coffers of so many kings? It is the honour of her house, the royal remunerating of her servants, the preservation of her people and state, the protection of her suppliants and allies, the encounter, breaking, and defeating the enemies of her realm, that have been the only pores° and pipes° whereby her treasure hath issued. Hath it been the sinews of glorious and ambitious war? No, but it hath been the sinews of a blessed and prosperous peace. But how of peace? Hath she lent the King of Spain money upon some cavillation° not to be repeated, and so bought his favour? Or hath she given large pensions to corrupt his council?° No, but she hath used the most honourable diversion of troubles that can be in the world. She hath kept the fires from her own walls by seeking to quench it in her neighbours'. That poor brand° of the states of Burgundy,° and that other of the crown of France that remaineth, had been in ashes but for that ready fountain of her continual benignity.° For the honour of her house, it is well known that although the universal manner of the times doth incline to a certain parsimony and dryness in that kind of expense, yet she retaineth the ancient magnificence,° the allowance° as full, the charge° greater than they were in the time of her father or any king before. The books appear;° the computations will not flatter. And for the remunerating and rewarding of her servants and the attendance of her court, let a man cast° and sum up all the books° of gifts, fee-farms,° leases, and custodies° that have passed her bountiful hand; let him consider again what a number of commodious and gainful offices heretofore bestowed upon men of other° education and profession have been withdrawn and conferred upon her court; let him remember what a number of other gifts, disguised by divers names but in effect as good as money given out of her coffers, have been granted by her; and he will conclude that her royal mind is far above her means. The other benefits of her politic, clement, and gracious government towards the subjects are without number. The state of justice good, notwithstanding the great subtlety° and humorous° affections of these times. The security of peace greater than can be described by that verse,

Tutus bos etenim rura perambulat:
Nutrit rura Ceres, almaque Faustitas:°

or that other,

Condit quisque diem collibus in suis.°

The opulency of the peace such, that if you have respect (to take one sign for many) to the number of fair houses that have been built since her reign, as Augustus said that he had received the city of brick and left it of marble,° so she may say she received a realm of cottages and hath made it a realm of palaces. The state of traffic° rich and great. The customs,° notwithstanding these wars and interruptions,° not fallen. Many profitable trades; many honourable discoveries. And lastly, to make an end where no end is, the shipping of this realm so advanced and made so mighty and potent, as this island is become (as the natural site thereof deserved) the lady of the sea;° a point of so high consequence, as it may be truly said that the commandment of the sea is an abridgement° or quintessence of an universal monarchy.° Thus and much more hath she merited of her subjects.

Now to set forth her merit of her neighbours° and the states about her. It seemeth the times have been purveyors of continual new and noble occasions for her to show her benignity, and that the fires of troubles abroad have been ordained as lights and tapers° to make her virtue and magnanimity more apparent. For when that one, stranger born, the family of the Guise,° being as a hasty weed sprung up in a night, had spread itself to a greatness, a greatness, I say, not civil but seditious; a greatness, not of encounter° of the ancient nobility, not of pre-eminence in the favour of kings, not remiss° of affairs from kings; but a greatness of innovation of state, of usurpation of authority, of affectation of crowns; and that accordingly, under the colour° of consanguinity and religion, they had brought French forces into Scotland,° in the absence of their king and queen being in their usurped tutele;° and that the ancient nobility of that realm, seeing the imminent danger of reducing° that kingdom under the tyranny of foreigners and their faction,° had, according to the good intelligence° between the two crowns, prayed her neighbourly succours; she undertook the action, expelled the strangers, restored the nobility to their degrees and the state to peace.° And lest any man should think her intent was to unnestle° ill° neighbours, and not to aid good neighbours, or that she was readier to rescue what was invaded by others than to render what was in her own hands, see if the times provided not a new occasion afterwards,° when through their own divisions (without the entermise° of strangers) her forces were again° sought and required. She forsook them not, prevailed so far as to be possessed of the castle of Edinburgh, the principal strength of that kingdom;° which place continually without cunctation° or cavillation° (the preambles of a wavering faith) she rendered° with all honour and sincerity, and restored the young king's government to authority and his person to safe and faithful hands;° and so ever after during his minority continued his principal guardian and protector.

In the times between these two occasions° of Scotland, when the same faction of Guise, covered° still with the pretence of religion, and strengthened by the desire of retaining government in the queen mother° of France, hath raised and moved civil wars° in that kingdom, only to extirpate the ancient nobility by shocking° them one against another, and to waste° that realm as the candle which is lighted at both ends; and that those of the religion,° being near of the blood royal and otherwise of the greatest houses in France, and great officers of the crown, opposed themselves only against their insolency, and to their support called in their aids,° giving unto them Newhaven for a place of security:° see with what alacrity, in tender regard towards the fortune of that young king,° whose name was used towards the supplanting° of his own strength, she embraced the enterprise; by the support and reputation whereof the same party made suddenly great proceedings, and in conclusion made their peace° as they would° themselves. And although they joined themselves against her, and performed the parts rather of good patriots than good confederates, and that after great demonstration of valour in her subjects (as the French to this day will report), especially through the great mortality by the hand of God,° and the rather because as it is known she did never much affect° the holding of that town to her own use, it was left, and her forces withdrawn; yet that did nothing diminish her merit of that crown, and namely of that party,° who recovered by it such strength as by that and no other thing they subsisted long after. And lest any man should sinisterly and maliciously interpret that she did nourish those divisions,° who knoweth not what faithful advice, continual and earnest solicitation she used by her ambassadors and ministers to the French kings successively, and to their mother, to move them to keep their edicts° of pacification, and to retain their own authority and greatness by the union of their subjects? Which counsel, if it had been as happily followed as it was prudently and sincerely given, France had been at this day a most flourishing kingdom, which is now a theatre of misery.

And now at last, when the same house of Guise, being one of the whips of God, whereof themselves are but the cords° and Spain the stock,° had by their infinite aspiring practices° wrought that miracle of state, to make a king in possession long established to play° again for his crown, without any title° of a competitor, without any invasion of a foreign enemy, yea without any combination° in substance° of any blood-royal or nobility; but only by furring in° audacious persons into sundry governments, and by making the populace of towns drunk with seditious preachers; and that King Henry the Third,° awaked by these pressing° dangers, was compelled to execute the Duke of Guise without ceremony, and yet nevertheless found the despair° of so many persons embarked and engaged in that conspiracy so violent that the flame thereby was little assuaged, so as he was inforced to implore her° aid and succours; consider how benign care and good correspondence° she gave to the distressed requests of that king; and he soon after being by that sacrilegious hand of a wretched Jacobin,° lifted up against the sacred person of his natural sovereign, taken away (wherein not the criminous° blood of Guise, but the innocent blood which he had often spilled by the instigation of him and his house, was revenged), and that this worthy gentleman who now reigneth came to the crown;° it will not be forgotten by so grateful a king and so observing° an age, how ready, how opportune and seasonable, how royal and sufficient her succours were, whereby she enlarged° him at that time and preserved him to his better fortune; and ever since in these tedious wars, wherein he hath to do with a Hydra, or monster with many heads, she hath supported him with treasure, with forces, and with employment of one which she favoureth most. What shall I speak of the offering of Don Antonio° to his fortune; a devoted Catholic, only commended to her by his oppressed state? What should I speak of that great storm of a mighty invasion,° not in preparation but in act, by the Turk upon the King of Poland, lately dissipated only by the beams of her reputation, which with the Grand Signior° is greater than that of all the states of Europe put together?

But let me rest° upon the honourable and continual aid and relief which she hath given to the distressed and desolate people of the Low Countries; a people recommended° unto her by ancient confederacy° and daily intercourse,° by their cause so innocent, and their fortune so lamentable. And yet notwithstanding, to keep the conformity of her own proceedings, never stained with the least note of ambition or malice, she refuseth the sovereignty of divers of those goodly provinces, offered° unto her with great instance,° to have been accepted with greater contentment both of her own people and theirs, and justly to be received either in respect of the hostility of Spain, or in respect of the conditions, liberties, and privileges of those subjects, and without charge, danger, and offence to the King of Spain and his partisans.° She hath taken upon her only their defence and protection, without any further avail or profit unto herself than the honour and merit of her benignity to a people that have been pursued by their natural king only upon passion and wrath, in sort that he hath, according to the proverb° "Aratro jaculari", consumed his means upon revenge.° And herein, to verify that which I said, that her merits have extended to her greatest enemies, let it be remembered what hath passed in that matter between the King of Spain and her; first, how in the beginning of the troubles° there she gave and imparted to him faithful and friendly advice touching the course that was to be taken for quieting and appeasing them; then she interposed herself to move and treat most just and reasonable capitulations, wherein always should have been reserved unto him as ample interest, jurisdiction and superiority in those countries as he in right could claim, or a prince moderately minded would seek to have; and (which is the greatest point) she did by her advice, credit and policy, and all good means, interrupt and impeach° that the same people by despair should not utterly alien and distract° themselves from the obedience of the King of Spain and cast themselves into the arms of a stranger; insomuch as it is most true that she did ever dissuade the Duke of Anjou° from that action, notwithstanding the affection she bare unto the said duke, and the obstinacy which she saw daily grow in the King of Spain.

Lastly, to touch the mighty and general merit of this Queen, bear in mind that her benignity and beneficence have been as large as the oppression and ambition of Spain. For to begin with the Church of Rome, the pretended° apostolic° see° is become a donative cell° of the King of Spain;° the Vicar of Christ is become the King of Spain's chaplain; he parteth° the coming of a new Pope for the treasure of the old; he was wont to exclude but some two or three cardinals, and to leave the election free of the rest;° but now he doth include and present directly some small number, all incapable° and incompatible° with the conclave,° put in only for colour,° except one or two.° The states of Italy, they be like quillets° of freehold° lying intermixt in the midst of a great honour° or lordship.° France is turned upside down, the subject above the king, cut and mangled infinitely, a country of Rodamonts and Roytelets,° farmers of the wars. Portugal usurped by no other title° than strength and vicinity. The Low Countries warred upon, because he seeketh, not to possess them, for they were possessed by him before, but to plant there an absolute° and martial° government, and to suppress their liberties. The like at this day attempted upon Aragon.° The poor Indies,° whereas the Christian religion generally brought enfranchisement of slaves in places where it came, in a contrary course by it men are brought from freemen to be slaves, and slaves of the most miserable condition. The sundry practices and tyrannies of the king's ambition in Germany, Denmark, Scotland, the east towns, are not unknown. Then it is her government, and her government alone, which hath been the sconce° and fort of all Europe, which hath let° this proud nation from overrunning all. If any state be yet free from his factions erected in the bowels thereof; if there be any state where this faction is entered that is not fired° with civil troubles; if there be any state under his protection upon whom he usurpeth not;° if there be any state subject to him, that enjoyeth not moderate liberty, upon whom he tyrannizeth not; let them all know it to be the merit of this renowned Queen, that standeth between them and their disfortunes. These be some of the beams of her noble and radiant magnanimity, in contempt° of peril which so many fly,° in contempt of profit which so many admire, and in merit of the world which so many include in themselves; set forth in my simplicity of speech with much loss of lustre, but with near approach of truth, as the sun is seen in water.

Now to pass to the excellency of her person. The view of them whole and not taken severally° doth make so sweet a wonder, as I fear to divide them. A nobility extracted out of the royal and victorious line of the kings of England; yea both roses, white and red,° do as well flourish in her nobility as in her beauty. A health, such as it is like she should have that was brought forth between° two of the most goodly princes of the world, in the strength of their years, in the heat of their love; that hath not been injured neither with any over-liberal neither with any over-curious° diet; that hath not been softened by an umbratile° life still under the roof, but strengthened by the use of the pure and open air; that still retaineth flower and vigour of youth. For the beauty and many graces of her presence, what colours are fine enough for such a portraiture? Let no light° poet be used for such a description, but the chastest and the royalest.°

Of her gait,	*Et vera incessu patuit Dea;°*
of her voice,	*Nec vox hominem sonat;°*
of her eye,	*Et laetos oculis afflarat honores;°*
of her colour,	*Indum sanguineo veluti violaverit ostro* *Si quis ebur;°*
of her neck,	*Et rosea cervice refulsit;°*
of her breasts,	*Veste sinus collecta fluentes;°*
of her hair,	*Ambrosiaeque comae divinum vertice odorem* *Spiravere.°*

If this be presumption, let him bear the blame which oweth° the verses.

What shall I speak of her rare qualities of compliment?° which as they be excellent in the things themselves, so they have always besides somewhat of a queen; and as queens use shadows and veils with their rich apparel, so methinks in all her qualities there is somewhat that flieth from ostentation, and yet inviteth the mind to contemplate the more. What should I speak of her excellent gift of speech, bearing a character° of the greatness of her conceit,° the height of her degree, and the sweetness of her nature? What life, what edge, what grace is there in those words and glances wherewith at pleasure she can give a man long to think,° be it that she mean to daunt him, to encourage him, or to amaze him. How admirable is her discourse, whether it be in learning or love! What variety of knowledge; what rareness of conceit; what choice of words; what grace of utterance! Doth it not appear that though her wit be as the adamant° of excellencies, which draweth out of any book ancient or new, out of any writing or speech the best, yet she refineth it, she enricheth it far above the value wherein it was received?° And is her speech only that language which the child learneth with pleasure, and not those which the studious learn with industry? Hath she not besides her rare eloquence in her own language, infinitely polished since her happy times, changes of other languages° both learned and modern? so that she is able to negotiate with divers ambassadors in their own languages, and that with no disadvantage unto her but rather with disadvantage to them: who I think cannot but have a great part of their wits distracted from their matters in hand to the contemplation and admiration of such perfection.

What should I wander on to speak of the excellencies of her nature, which cannot endure to be looked on with a discontented eye? of the constancy of her favours, which maketh her service as a journey by land, whereas the service of other princes is like an embarking by sea? As for her royal wisdom and policy of government, he that shall note and observe the prudent temper she useth in admitting access, on the one side maintaining the majesty of her degree, and on the other side not prejudicing herself° by looking into her state through too few windows; her exquisite judgment in choosing and finding good servants (a point wherein her father did excel), and her politic skill in making and training good servants (a point beyond the former); her profound discretion in assigning and appropriating every of them to their aptest employment; her penetrating sight in discovering every man's ends and drifts; her wonderful art in keeping her servants in satisfaction, and yet in appetite;° her intentive° wit in contriving plots and covertures;° her exact caution in censuring° the propositions of others; her secrecy; her foreseeing events; her usage of occasions°—he that shall consider of these, and other things that may not well be touched, as he shall never cease to wonder at such a queen, so he shall wonder the less that in so dangerous times, when wits are so cunning, humours so extravagant, passions so violent, corruptions so great, dissimulations so deep, factions so many, she hath notwithstanding done so° great things and reigned in such felicity.

For to speak of her fortune, let no man object to me as a defect in her fortune (that which I did reserve° for a garland of her honour), and that is that she liveth a virgin and hath no children,° for it is that which maketh all her other virtues and acts more sacred, more august, more divine. Let them leave children that leave no other memory in their times: "Brutorum aeternitas soboles."° Revolve° in histories the memories of happy men, and you shall not find any of rare felicity but either he died childless, or his line° spent° soon after his death, or else was infortunate in his children. Should a man have children to be slain by his vassals, as the *posthumus* of Alexander the Great was?° or to call them his impostumes,° as Augustus Caesar called his? Peruse the catalogue:

Cornelius Sulla, Julius Caesar, Flavius Vespasianus, Septimius Severus, Constantinus the Great, and many more. And the rule holdeth: "Generare et liberi, humana: creare et opera, divina."° And therefore, this objection removed, let us proceed to take a view of her felicity.

A mate° of fortune° she never took; only some adversity she passed° at the first, to give her quicker° sense of the prosperity that should follow, and to make her more reposed in the divine providence. Well, she cometh to the crown. It was no small fortune to find at her entrance some such servants and counsellors as she then found.° The French king, who at that time by reason of the peace concluded with Spain and the interest he had in Scotland might have proved a dangerous neighbour, by how strange an accident° was he taken away! The King of Spain—who, if he would have inclined to have reduced° the Low Countries by lenity, considering the goodly revenues which he drew from those countries, the great commodity° to annoy her state from them, might have made a mighty and perilous machination° against her repose—putteth on a resolution not only to use the means of those countries, but to spend and consume all his other means, the treasure of his Indies, and the forces of his ill-compacted° dominions, there and upon them. The Earls that rebelled in the North,° before the Duke of Norfolk's plot° (which indeed was the strength and the steel of that commotion) was fully ripe, brake forth and prevented° their time. The King Sebastian of Portugal,° whom the King of Spain would fain have persuaded that it was a devouter enterprise to purge Christendom than to enlarge it (though I know some° think that he did artificially° nourish° him in that voyage), is cut in pieces with his army in Africa.° Then hath the King of Spain work cut out to make all things in a readiness during the old Cardinal's° time for the conquest of Portugal; whereby his desire of invading England was slackened and put off some years, and by that means was put in execution at a time for° some respect much more for his disadvantage. And the same invasion,° like as if it had been attempted before, it had had the time much more proper and favourable, so likewise had it in true discourse had a better season afterwards; for if it be° deferred till time that the League° had been better confirmed° in France—which no doubt it would have been, if the Duke of Guise, who was the only man of worth on that side, had lived (and the French king durst never have laid hand upon him, had he not been animated° by the English victory° against the Spaniard precedent)°—and then some maritime towns had been gotten into the hands of the League, it had been a great surety and strength to that enterprise. The Popes, to consider of them whose course and policy it had been (knowing her Majesty's natural clemency) to have temporized and to have dispensed° with the Papists coming to church, that through° the mask of their hypocrisy they might have been brought into places of government in state and in the country: they contrariwise, by the instigation of some fugitive scholars that advised them, not what was best for the see of Rome but what agreed best with their own eager° humours and desperate states, discover and declare° themselves so far by sending in of seminaries° and taking reconcilements,° [so]° as there is new security of laws introduced for the repressing of that sort, and men of that religion are become the more suspect.

What should I say of so many great conspiracies° miraculously detected? The records show the treasons: but it is yet hidden in many of them how they came to light. What shall I speak of the opportune deaths of her enemies, and the evil instruments° towards her state? Don Juan° died not amiss. D'Aubigny, Duke of Lennox,° who was used as an instrument to divorce Scotland from the amity of England, died in no ill season; a man withdrawn indeed at that time into France, but not without great hopes. I may not mention the death of some° that occur to my mind; but still methinks they live that should live, and they die that should die. I would not have the King of Spain die yet; he is *seges gloriae*:° but when he groweth dangerous, or any other besides him, I am persuaded° they will die.

What shall I speak of the fortune and honour of her armies, which, notwithstanding the inward peace of this nation, were never more renowned? What should I recount Leith,° Newhaven,° I say Newhaven, for the honourable skirmishes and services there are no blemish at all to the *militiae* of England. In the Low Countries, the Lammas day,° the retreat at Ghent,° the day at Zutphen,° and the prosperous progress of this summer;° then the bravado in Portugal,° and the honourable exploits in the aid of the French king, besides the memorable voyages into the Indies;° and lastly, the good entertainment of the Invincible Navy,° which, chased until the chasers were weary, after infinite loss, without taking a cock-boat, without firing° a sheep-cot,° sailed on, at the mercy of the wind, at the discretion of their adventurers, making a perambulation° or pilgrimage about the northern seas, and ennobling° many shores and points of land by shipwrecks: and so returned home with scorn and dishonour much greater than the terror and expectation of their setting forth.

These virtues and perfections, joined with so great felicity, have made her the honour of her times, the admiration of the world, the suit and aspiring of the greatest kings and princes, who yet durst never have aspired to her as worthy of her, but as their minds were raised by love.

But why do I forget that words do extenuate° and embase matters of this height?° Time is her best commender, which never brought forth such a prince; whose imperial virtues contend with the excellencies of her person, both virtues contend with her fortune, and both virtue and fortune contend with her fame.

Orbis amor, famae carmen, coelique pupilla;
Tu decus omne tuis, tu decus ipsa tibi!°

A Device for the Gray's Inn Revels

When the King at Arms had read all these Articles of the Order of the Knighthood, and finished the ceremonies belonging to the same, and that every one had taken their places as before, there was variety of consort-music;° and in the mean while, the Knights of the Order, which were not strangers,° brought into the hall a running banquet° in very good order,° and gave it to the Prince, and Lords, and other strangers, in imitation of the feast that belongeth to all such honourable institutions.

This being done, there was a table set in the midst of the stage, before the Prince's seat; and there sat six of the Lords of his Privy Council, which at that time were appointed to attend, in Council, the Prince's leisure. Then the Prince spake to them in this manner.

The Prince of Purpoole

'My Lords,

We have made choice of you, our most faithful and favoured counsellors, to advise° with you, not any particular action of our state but in general, of the scope and end° whereunto you think it most for our honour and the happiness of our state that our government [should] be rightly bent and directed. For we mean not to do as many princes use, which conclude of their ends out of their own humours° and take counsel only of the means, abusing for the most part the wisdom of their counsellors [to] set them in the right way° to the wrong place. But we, desirous to leave as little to chance or humour as may be, do now give you liberty and warrant to set before us° to what port, as it were, the ship of our government should be bounden. And this we require you to do without either respect to our affections° or your own; neither guessing what is most agreeable with our disposition,° wherein we may easily deceive you, for Princes' hearts are inscrutable;° nor on the other side putting the case by yourselves,° as if you would present us with a robe whereof measure were taken by yourselves. Thus you perceive our mind and we expect your answer.'

The First Counsellor, advising the Exercise° of War

'Most excellent Prince,

Except there be such amongst us, as I am fully persuaded there is none, that regardeth more his own greatness under you than your greatness over others, I think there will be little difference° in choosing for you a goal worthy your virtue and power. For he that shall set before him your magnanimity and valour, supported by the youth and disposition of your body; your flourishing Court, like the horse of Troy, full of brave commanders and leaders; your populous and man-rife° provinces, overflowing with warlike people; your coffers, like the Indian° mines when that they were first opened; your storehouses and arsenals, like to Vulcan's° cave; your navy like to an huge floating city; the devotion of your subjects to your crown and person, their good agreement amongst themselves, their wealth and provision; and then your strait° and unrevocable° confederation with these noble and honourable personages, and the fame and reputation without° of so rare a concurrence, whereof all the former regards° do grow; how can he think any exercise worthy of your means but that of conquest? For in few words, what is your strength, if you find it not? your fortune, if you try° it not? your virtue, if you show it not? Think, excellent Prince, what sense of content° you found in yourself, when you were first invested° in our state; for though I know your Excellency is far from vanity and lightness,° yet it is the nature of all things to find rest when they come to due and proper places.° But be assured of this, that this delight will languish and vanish; for power will quench appetite and satiety will induce tediousness. But if you embrace the wars, your trophies and triumphs shall be as continual coronations, that will not suffer your glory and contentment to fade and wither. Then when you have enlarged your territories, ennobled your country, distributed fortunes, good or bad, at your pleasure, not only to particulars° but to cities and nations; marked the computations of times with your expeditions and voyages, and the memory of places by your exploits and victories; in your later years you shall find a sweet respect° into the adventures of your youth; you shall enjoy your reputation; you shall record your travels; and after your own time you shall eternise your name, and leave deep footsteps of your power in the world. To conclude, excellent Prince, and most worthy to have the titles of victories added to your other high and deserved titles, Remember: the divines find nothing more glorious to resemble our state unto than a warfare. All things in earnest and jest do affect° a kind of victory; and all other victories are but shadows to the victories of the wars. Therefore embrace the wars, for they disparage you not; and believe that if any Prince do otherwise it is either in the weakness of his mind or means.'

The Second Counsellor, advising the Study of Philosophy

'It may seem, most excellent Prince, that my Lord which now hath spoken did never read the just censures of the wisest men, who compared great conquerors to great rovers° and witches, whose power is in destruction and not in preservation;° else would he never have advised your Excellency to become as some comet or blazing star, which should threaten and portend nothing but death and dearth, combustions and troubles of the world. And whereas the governing faculties° of men are two, force and reason, whereof the one is brute and the other divine,° he wisheth you for your principal ornament and regality the talons of the eagle to catch the prey, and not the piercing sight which seeth into the bottom of the sea. But I contrariwise will wish unto your Highness the exercise of the best and purest part of the mind, and the most innocent and meriting conquest, being the conquest of the works of nature; making this proposition, that you bend the excellency of your spirits to the searching out, inventing, and discovering of all whatsoever is hid and secret° in the world; that your Excellency be not as a lamp that shineth to others and yet seeth not itself, but as the Eye° of the World, that both carrieth and useth light. Antiquity, that presenteth unto us in dark visions the wisdom of former times, informeth us that the governments of° kingdoms have always had an affinity with the secrets and mysteries of learning. Amongst the Persians, the kings were attended on by the Magi.° The Gymnosophists° had all the government under the princes of Asia; and generally those kingdoms were accounted most happy, that had rulers most addicted to philosophy. The Ptolemies° in Egypt may be for instance; and Salomon° was a man so seen° in the universality of nature that he wrote an herbal° of all that was green upon the earth. No conquest of Julius Caesar made him so remembered as the Calendar.° Alexander the Great wrote to Aristotle,° upon the publishing of the *Physics*, that he esteemed more of excellent men in knowledge than in empire. And to this purpose I will commend to your Highness four principal works and monuments° of yourself: First, the collecting of a most perfect° and general library, wherein whatsoever the wit of man hath heretofore committed to books of worth, be they ancient or modern, printed or manuscript, European or of the other parts, of one or other language, may be made contributory to your wisdom. Next, a spacious, wonderful garden, wherein whatsoever plant the sun of divers climates, out of the earth of divers moulds,° either wild or by the culture of man brought forth, may be with that care that appertaineth to the good prospering thereof set° and cherished:° This garden to be built about with rooms to stable in all rare beasts and to cage in all rare birds; with two lakes adjoining, the one of fresh water the other of salt, for like variety of fishes. And so you may have in small compass a model of universal nature made private. The third, a goodly huge cabinet,° wherein whatsoever the hand of man by exquisite art° or engine° hath made rare in stuff, form, or motion; whatsoever singularity chance and the shuffle of things° hath produced; whatsoever Nature hath wrought in things that want life° and may be kept; shall be sorted and included. The fourth such a still-house,° so furnished with mills, instruments, furnaces, and vessels, as may be a palace fit for a philosopher's stone.° Thus, when your Excellency shall have added depth of knowledge to the fineness of [your] spirits and greatness of your power, then indeed shall you be a Trismegistus;° and then when all other miracles and wonders shall cease, by reason that you shall have discovered their natural causes, yourself shall be left the only miracle and wonder of the world.'

The Third Counsellor, advising Eternizement and Fame by Buildings and Foundations°

'My Lords that have already spoken, most excellent Prince, have both used one° fallacy,° in taking that for certain and granted which was most uncertain and doubtful; for the one hath neither drawn in question° the success and fortune of the wars, nor the other the difficulties and errors in the conclusions of nature.° But these immoderate hopes and promises do many times issue forth, those of the wars into tragedies of calamities and distresses; and those of mystical° philosophy into comedies of ridiculous frustrations and disappointments of such conceipts° and curiosities.° But on the other side, in one point my Lords have well agreed; that they both according to their several intentions counselled your Excellency to win fame and to eternize your name; though the one adviseth it in a course of great peril, and the other of little dignity and magnificence. But the plain and approved way, that is safe and yet proportionable to the greatness of a monarch, to present himself to posterity, is not rumour and hearsay, but the visible memory of himself in the magnificence of goodly and royal buildings and foundations, and the new institutions° of orders,° ordinances,° and societies; that is, that [as] your coin be stamped with your own image, so in every part of your state there may be somewhat new, which by continuance may make the founder and author remembered. It was perceived at the first, when men sought to cure° mortality by fame, that buildings was the only way; and thereof proceeded the known holy antiquity of building the Tower of Babel;° which, as it was a sin in the immoderate appetite of fame, so it was punished in the kind; for the diversities of languages have imprisoned fame ever since. As for the pyramids, the colosses, the number of temples, colleges, bridges, aqueducts, castles, theatres, palaces, and the like, they may show us that men ever mistrusted any other way to fame than this only, of works and monuments. Yea, even they which had the best choice of other means. Alexander did not think his fame so engraven in his conquests, but that he thought it further shined in the buildings of Alexandria.° Augustus Caesar thought no man had done greater things in military actions than himself, yet that which at his death ran most in his mind was his building, when he said, not, as some mistake it, metaphorically, but literally, "I found the city of brick but I leave it of marble".° Constantine the Great was wont to call with envy the Emperor Trajan, *wallflower*,° because his name was upon so many buildings; which notwithstanding he himself did embrace in the new founding of Constantinople,° and sundry other buildings; and yet none greater conquerors than these two. And surely they had reason; for the fame of great actions is like to a landflood° which hath no certain head° or spring; but the memory and fame of buildings and foundations hath, as it were, a fountain in an hill, which continually feedeth and refresheth the other waters. Neither do I, excellent Prince, restrain° my speeches to dead buildings only, but intend it also to other foundations, institutions, and creations; wherein I presume the more to speak confidently, because I am warranted herein by your own wisdom, who have made the first-fruits of your actions of state to institute the honourable Order of the Helmet;° the less shall I need to say, leaving your Excellency not so much to follow my advice as your own example.'

The Fourth Counsellor, advising Absoluteness° of State and Treasure

'Let it not seem pusillanimity for your Excellency, mighty Prince, to descend a little from your high thoughts to a necessary consideration of your own estate. Neither do you deny, honourable Lords, to acknowledge safety, profit, and power to be of the substance of policy,° and fame and honour rather to be as flowers of well-ordered actions than as good ends. Now if you examine the courses propounded according to these respects, it must be confessed that the course of wars may seem to increase power, and the course of contemplations° and foundations not prejudice safety. But if you look beyond the exterior you shall find that the first breeds weakness and the latter nurses peril. For certain it is, during wars your Excellency will be enforced° to your soldiers and generally to your people, and become less absolute and monarchical than if you reigned in peace; and then if your success be good, that° you make new conquests, you shall be constrained to spend° the strength of your ancient and settled provinces to assure your new and doubtful, and become like a strong man that by taking a great burden upon his shoulders maketh himself weaker than he was before. Again, if you think you may intend° contemplations with security,° your Excellency will be deceived; for such studies will make you retired° and disused° with your business, whence will follow a diminution of your authority. As for the other point, of erecting in every part of your state something new derived from yourself, it will acquaint your Excellency° with an humour° of innovation° and alteration, which will make your reign very turbulent and unsettled; and many times your change will be for [the] worse, as in the example last touched of Constantine, who by his new translation of his estate ruinated° the Roman Empire.° As for profit, there appeareth a direct contrariety between that and all the three courses; for nothing causeth such a dissipation of treasure as wars, curiosities,° and buildings; and for all this to be recompensed in a supposed honour, a matter apt to be much extolled in words, but not greatly to be prized in conceipt,° I do think it a loser's bargain. Besides that many politic princes have received as much commendation for their wise and well-ordered government as others have done for their conquests and glorious affections; and more worthy, because the praise of wisdom and judgment is less communicated° with fortune.

Therefore, excellent Prince, be not transported° with shows. Follow the order of nature, first to make the most of that you possess, before you seek to purchase more. To put the case by° a private man (for I cannot speak high), if a man were born to an hundred pounds by the year,° and one show° him how with charge° to purchase an hundred pounds more, and another should show him how without charge to raise° that hundred pounds unto five hundred pounds, I should think the latter advice should be followed. The proverb is a country proverb, but significative, "Milk the cow that standeth still; why follow you her that flieth away?"° Do not think, excellent Prince, that all the conquests you are to make be foreign. You are to conquer here at home the overgrowing° of your grandees° in factions, and too great liberties of your people; the great reverence and formalities° given to your laws and customs, in derogation° of your absolute prerogatives:° these and such-like be conquests of state, though not of war. You want a Joseph,° that should by advice make you the only proprietor of all the lands and wealth of your subjects. The means how to strain up° your sovereignty, and how to accumulate treasure and revenue, they are the secrets of your state; I will not enter into them at this place: I wish your Excellency as ready to [desire] them, as I have the means ready to perform them.'

The Fifth Counsellor, advising him Virtue and a gracious Government

'Most excellent Prince,

I have heard sundry plats° and propositions offered unto you severally;° one to make you a great Prince, another to make you a strong Prince, and another to make you a memorable Prince, and a fourth to make you an absolute Prince. But I hear of no invention° to make you a good and a virtuous Prince; which surely my Lords have left out in discretion,° as to arise of your own motion and choice; and so I should have thought, had they not handled their own propositions so artificially° and persuadingly, as doth assure me their speech was not formal.° But, most worthy Prince, fame is too light, and profit and surety are too low, and power is either such as you have or ought not so to seek to have. It is the meriting° of your subjects, the making of golden times, the becoming of a natural parent to your state; these are the only [fit] and worthy ends of your Grace's virtuous reign. My Lords have taught you to refer all things to yourself, your greatness, memory,° and advantage; but whereunto shall yourself be referred? If you will be heavenly you must have influence.° Will you be as a standing° pool, that spendeth and choketh his spring within itself, and hath no streams nor current to bless and make fruitful whole tracts of countries whereby it runneth? Wherefore, first of all, most virtuous Prince, assure yourself of an inward peace, that the storms without do not disturb any of your repairers° of state within. Therein use and practise all honourable diversions.° That done, visit all the parts of your state, and let the balm° distil everywhere from your sovereign hands, to the medicining of any part that complaineth. Beginning with your seat of state, take order that the faults of your great ones do not rebound upon yourself; have care that your intelligence,° which is the light of your state, do not go out or burn dim or obscure; advance men of virtue and not of mercenary minds; repress all faction be it either malign or violent. Then look into the state of your laws and justice of your land; purge out multiplicity of laws,° clear the incertainty of them, repeal those that are snaring,° and press° the execution of those that are wholesome and necessary; define the jurisdiction of your courts, repress all suits° and vexations, all causeless delays and fraudulent shifts and devices, and reform all such abuses of right and justice; assist the ministers thereof, punish severely all extortions and exactions° of officers, all corruptions in trials and sentences of judgment.

Yet when you have done all this, think not that the bridle and spur will make the horse to go alone° without time and custom. Trust not to your laws for correcting the times, but give all strength to good education; see to the government of your universities and all seminaries of youth, and to the private order of families, maintaining due obedience of children towards their parents, and reverence of the younger sort towards the ancient. Then when you have confirmed° the noble and vital parts of your realm of state, proceed to take care of the blood and flesh and good habit of the body. Remedy all decays° of population, make provision for the poor, remove all stops in traffic,° and all cankers° and causes of consumption° in trades and mysteries;° redress all—But whither do I run, exceeding the bounds of that perhaps I am now demanded? But pardon me, most excellent Prince, for as if I should commend unto your Excellency the beauty of some excellent Lady, I could not so well express it with relation° as if I showed you her picture; so I esteem the best way to commend a virtuous government, to describe and make appear what it is; but my pencil perhaps disgraceth it; therefore I leave it to your Excellency to take the picture out of your wise observation, and then to double it and express it in your government.'

The Sixth Counsellor, persuading° Pastimes and Sports

'When I heard, most excellent Prince, the three first of my Lords so careful to continue your fame and memory, methought it was as if a man should come to some young prince as yourself is, and immediately after his coronation be in hand° with him to make himself a sumptuous and stately tomb. And, to speak out of my soul, I muse how any of your servants can once endure to think of you as of a prince past. And for my other Lords, who would engage you so deeply in matters of state, the one persuading you to a more absolute, the other to a more gracious government, I assure your Excellency their lessons were° so cumbersome, as if they would make you a king in a play, who, when one would think he standeth in great majesty and felicity, he is troubled° to say his part. What! nothing but tasks, nothing but working-days? No feasting, no music, no dancing, no triumphs,° no comedies, no love, no ladies? Let other men's lives be as pilgrimages,° because they are tied to divers necessities and duties; but princes' lives are as progresses,° dedicated only to variety and solace. And [as] if your Excellency should take your barge° in a summer evening, or your horse or chariot, to take the air; and if you should do any the favour to visit him; yet your pleasure is the principal, and that is but as it falleth out;° so if any of these matters which have been spoken of fall out in the way of your pleasure, it may be taken, but no otherwise. And therefore leave your wars to your lieutenants, and your works and buildings to your surveyors, and your books to your universities, and your state-matters to your counsellors, and attend you that in person which you cannot execute by deputy: use the advantage of your youth: be not sullen° to your fortune; make your pleasure the distinction° of your honours, the study of your favourites, the talk of your people, and the allurement of all foreign gallants to your Court. And in a word, sweet Sovereign, dismiss your five counsellors, and only take counsel of your five senses.'

The Prince's Answer and Conclusion° to the Speeches of the Counsellors

'My Lords,

We thank you for your good opinions; which have been so well set forth, as we should think ourselves not capable of good counsel if in so great variety of persuading° reasons we should suddenly resolve. Meantime it shall not be amiss to make choice of the last, and upon more deliberation to determine° of the rest; and what time° we spend in long consulting, in the end we will gain by prompt and speedy executing.'

The Prince having ended his speech, arose from his seat, and took that occasion of revelling. So he made choice of a Lady to dance withal; so likewise did the Lord Ambassador, the pensioners° and courtiers attending the Prince. The rest of that night was passed in those pastimes.

Of Love and Self-Love°

[*The ambassadors of Philautia to Erophilus present their arguments to Erophilus' squire, before the Queen.*]

The Squire's Speech in the Tiltyard

'Most excellent and most glorious Queen, give me leave, I beseech your Majesty, to offer my master his° complaint and petition; complaint that coming hither to your Majesty's most happy day, he is tormented with the importunity of a melancholy° dreaming Hermit, a mutinous brain-sick° Soldier, and a busy° tedious° Secretary. His petition is that he may be as free as the rest, and at least whilst he is here, troubled with nothing but with care how to please and honour you.'

The Hermit's Speech in the Presence,° in wish of Contemplation or Studies

'Though our ends be diverse, and therefore may be one more just than another, yet the complaint of this Squire is general, and therefore alike unjust against us all. Albeit he is angry that we offer ourselves to his master uncalled, and forgets we come not of ourselves but as the messengers of Self-love, from whom all that comes should be well taken. He saith when we come we are importunate. If he mean that we err in form, we have that of his master, who being a lover useth no other form of soliciting. If he will charge us to err in matter, I for my part will presently prove that I persuade him to nothing but for his own good. For I wish him to leave turning over the book of fortune, which is but a play for children, where there be so many books of truth and knowledge better worthy° the revolving,° and not fix his view only upon a picture in a little table,° where there be so many tables of histories, yea to life, excellent to behold and admire. Whether he believe me or no, there is no prison to° the prison of the thoughts, which are free under the greatest tyrants. Shall any man make his conceit° as an anchor, mured° up with the compass of one beauty or person, that may have the liberty of all contemplation? Shall he exchange the sweet travelling° through the universal variety° for one wearisome and endless round or labyrinth?

Let thy master, Squire, offer his service to the Muses. It is long since they received any into their court. They give alms° continually at their gate, that many come to live upon; but few have they ever admitted into their palace. There shall he find secrets not dangerous to know, sides and parties not factious to hold, precepts and commandments not penal to disobey. The gardens of love° wherein he now playeth himself are fresh to-day and fading to-morrow,° as the sun comforts them or is turned from them. But the gardens of the Muses keep the privilege of the golden age;° they ever flourish and are in league with time. The monuments of wit survive the monuments of power: the verses of a poet endure without a syllable lost, while states and empires pass many periods.° Let him not think he shall descend, for° he is now upon a hill as a ship is mounted upon the ridge of a wave; but that hill of the Muses is above tempests, always clear and calm; a hill of the goodliest discovery that man can have, being a prospect upon all the errors and wanderings of the present and former times.° Yea, [as from a] cliff it leadeth the eye beyond the horizon of time, and giveth no obscure divinations of times to come. So that if he will indeed lead "vitam vitalem",° a life that unites safety and dignity, pleasure and merit; if he will win admiration without envy; if he will be in the feast and not in the throng,° in the light and not in the heat;° let him embrace the life of study and contemplation. And if he will accept of no other reason, yet because the gift of the Muses will enworthy him in his love,° and where he now looks on his mistress's outside with the eyes of sense,° which are dazzled and amazed, he shall then behold her high perfections and heavenly mind with the eyes of judgment, which grow stronger by more nearly and more directly viewing such an object.'

The Soldier's Speech in wish of Fame

'Squire, the good old man hath said well to you, but I dare say thou wouldst be sorry to leave to° carry thy master's shield, and to carry his books, and I am sure thy master had rather be a falcon, a bird of prey, than a singing-bird in a cage. The Muses are to serve martial men, to sing their famous actions, and not to be served by them. Then hearken to me.

It is the wars that giveth all spirits of valour not only honour but contentment. For mark whether ever you did see a man grown to any honourable commandment in the wars, but whensoever he gave it over he was ready to die with melancholy? Such a sweet felicity is in that noble exercise,° that he that hath tasted it throughly is distasted° for all other. And no marvel; for if the hunter take such solace in his chase, if the matches° and wagers of sport pass away with such satisfaction and delight, if the looker-on be affected with pleasure in the representation of a feigned tragedy,° think what contentment a man receiveth when they that are equal to him in nature, from the height of insolency and fury are brought to the condition of a chased prey, when a victory is obtained whereof the victories of games are but counterfeits and shadows, and when in a lively° tragedy a man's enemies are sacrificed before his eyes to his fortune.

Then for the dignity of military profession, is it not the truest and perfectest practice of all virtues? of wisdom, in disposing those things which are most subject to confusion and accident; of justice, in continual distributing rewards; of temperance, in exercising of the straitest° discipline; of fortitude, in toleration of all labours and abstinence from effeminate delights; of constancy, in bearing and digesting the greatest variety of fortune. So that when all other places° and professions require but their several° virtues, a brave leader in the wars must be accomplished with all. It is the wars that are the tribunal seat, where the highest rights and possessions are decided; the occupation of kings, the root of nobility, the protection of all estates. And lastly, lovers never thought their profession sufficiently graced, till they have compared it to a warfare.° All that in any other profession can be wished for is but to live happily: but to be a brave commander in the field, death itself doth crown the head with glory. Therefore, Squire, let thy master go with me, and though he be resolved in the pursuit of his love, let him aspire to it by the noblest means. For ladies count it no honour to subdue them with their fairest eyes, which will be daunted° with the fierce encounter of an enemy; and they will quickly discern a champion fit to wear their glove, from a page not worthy to carry their pantofle.° Therefore I say again, let him seek his fortune in the field, where he may either lose his love, or find new arguments to advance it.'

The Statesman's Speech in wish of Experience

'Squire, my advice to thy master shall be as a token° wrapped up in words; but° then will it show itself fair, when it is unfolded° in his actions. To wish him to change from one humour to another, were but as if for the cure of a man in pain one should advise him to lie upon the other side, but not enable him to stand on his feet. If from a sanguine° delightful° humour of love he turn to a melancholy retired° humour of contemplation, or a turbulent boiling humour of the wars, what doth he but change tyrants? Contemplation is a dream, love a trance, and the humour of war is raving. These be shifts of humour,° but no reclaiming to reason. I debar him not studies nor books, to give him store and variety of conceit, to refresh his mind, to cover° sloth and indisposition, and to draw° to him from those that are studious respect and commendation. But let him beware lest they possess not too much of his time, that they abstract not his judgment from present experience, nor make him presume upon knowing much to apply the less. For the wars, I deny him no enterprise that shall be worthy in greatness, likely in success, or necessary in duty; not mixed with any circumstance of jealousy, but duly laid upon him. But I would° not have him take the alarm° from his own humour, but from the occasion; and I would again he should know an employment from a discourting.° And for his love, let it not so disarm his heart within, as it make him too credulous° to favours, nor too tender° to unkindnesses, nor too apt to depend upon the heart he knows not. Nay in his demonstration° of love let him not go too far; for these silly lovers, when they profess such infinite affection and obligation, they tax themselves at so high a rate that they are ever under arrest.° It makes their service° seem nothing, and every cavil° or imputation° very great.

But what, Squire, is thy master's end? If to make the prince happy he serves, let the instructions to employed men,° the relations° of ambassadors, the treaties between princes, and actions of the present time, be the books he reads: let the orations of wise princes or experimented° counsellors in council or parliament, and the final sentences° of grave and learned judges in weighty and doubtful causes,° be the lectures he frequents. Let the holding of affection° with confederates° without charge,° the frustrating of the attempts of enemies without battles, the entitling° of the Crown to new possessions without show of wrong, the filling of the prince's coffers without grudging,° the appeasing tumults and seditions without violence, the keeping of men in appetite° without impatience, be the inventions he seeks out. Let policy and matter of state be the chief, and almost the only thing he intends. But if he will believe Philautia, and seek most° his own happiness, he must not of them embrace all kinds, but make choice, and avoid all matter of peril, displeasure, and charge,° and turn them over to some novices that know not manacles° from bracelets, nor burdens from robes. For himself, let him set for° matters of commodity° and strength,° though they be joined with envy. Let him not trouble himself too laboriously to sound into any matter deeply, or to execute anything exactly; but let him make himself cunning rather in the humours and drifts° of persons than in the nature of business and affairs.° Of that it sufficeth to know only so much as may make him able to make use of other men's wits, and to make again a smooth and pleasing report. Let him entertain° the proposition° of others, and ever rather let him have an eye to the circumstances° than to the matter itself; for then shall he ever seem to add somewhat of his own:° and besides, when a man doth not forget so much as a circumstance, men do think his wit° doth superabound° for the substance.° In his counsels let him not be confident, for that will rather make him obnoxious° to the success; but let him follow the wisdom of oracles, which uttered that which might ever be applied to the event.° And ever rather let him take the side which is likeliest to be followed, than that which is soundest and best,° that everything may seem to be carried by his direction. To conclude, let him be true to himself,° and avoid all tedious reaches of state° that are not merely° pertinent to his particular.° And if he will needs pursue his affection, and go on his course, what can so much advance him in his own way?

The merit of war is too outwardly glorious to be inwardly grateful,° and it is the exile of his eye, which looking with such affection upon the picture, cannot but with infinite contentment behold the life. But when his mistress shall perceive that his endeavours are [to] become a true supporter of her, a discharge of her care, a watchman of her person, a scholar of her wisdom, an instrument of her operation, and a conduit° of her virtue, this with his diligences,° accesses,° humility, and patience, may move her to give him further degrees and approaches to her favour. So that I conclude I have traced him the way to that which hath been granted to some few, "amare et sapere",° to love and be wise.'

The Reply of the Squire

'Wandering° Hermit, storming° Soldier, and hollow° Statesman, the enchanting° orators of Philautia, which have attempted by your high charms to turn resolved Erophilus into a statua° deprived of action, or into a vulture attending about dead bodies,° or into a monster with a double heart;° with infinite assurance,° but with just indignation and forced° patience, I have suffered you to bring in play° your whole forces. For I would not vouchsafe° to combat you one by one, as if I trusted to the goodness of my breath and not the goodness of my strength, which little needeth the advantage of your severing,° and much less of your disagreeing.

Therefore, first, I would° know of you all what assurance you have of the fruit whereto you aspire. You (Father) that pretend to truth and knowledge, how are you assured that you adore not vain chimeras° and imaginations? that in your high prospect,° when you think men wander up and down, that they stand not indeed still in their place, and it is some smoke or cloud between you and them which moveth, or else the dazzling° of your own eyes? Have not many which take themselves to be inward° counsellors with Nature, proved but idle believers, that told us tales which were no such matter?

And, Soldier, what security° have you for these victories and garlands which you promise to yourself? Know you not of many which have made provision of laurel for the victory, and have been fain to exchange it with cypress° for the funeral? of many which have bespoken° fame to sound their triumphs, and have been glad to pray her to say nothing of them, and not to discover° them in their flights?°

Corrupt Statesman, you that think by your engines° and motions° to govern the wheel of fortune; do you not mark that clocks cannot be long in temper,° that jugglers are no longer in request when their tricks and sleights are once perceived? Nay do you not see that never any man made his own cunning and practice° (without religion, honour, and moral honesty) his foundation, but he over-built° himself, and in the end made his house a windfall?°

But give ear now to the comparison of my master's condition, and acknowledge such a difference as is betwixt the melting hail-stone and the solid pearl. Indeed it seemeth to depend° as the globe of the earth seemeth to hang in the air; but yet it is firm and stable in itself. It is like a cube or die-form,° which toss it or throw it any way, it ever lighteth upon° a square.° Is he denied the hopes of favours to come? He can resort to the remembrance of contentments past: destiny cannot repeal that which is past. Doth he find the acknowledgment° of his affection° small? He may find the merit° of his affection the greater: fortune cannot have power over that which is within. Nay his falls are like the falls of Antaeus;° they renew his strength. His clouds are like the clouds of harvest,° which make the sun break forth with greater force; his wanes° and changes are like the moon, whose globe is all light towards the sun when it is all dark towards the world; such is the excellency of her nature and of his estate.

Attend,° you beadsman° of the Muses, you take your pleasure in a wilderness of variety; but it is but of shadows.° You are as a man rich in pictures, medals, and crystals.° Your mind is of the° water, which taketh all forms and impressions,° but is weak of substance. Will you compare shadows with bodies, picture with life, variety of many beauties with the peerless excellency of one? the element of water with the element of fire?° And such is the comparison between knowledge and love.

Come out (man of war), you must be ever in noise.° You will° give laws, and advance° force, and trouble nations, and remove landmarks of kingdoms, and hunt men, and pen° tragedies in blood: and that which is worst of all, make all the virtues accessary to° bloodshed. Hath the practice of force so deprived you of the use of reason, as that you will compare the interruption° of society with the perfection of society, the conquest of bodies with the conquest of spirits,° the terrestrial fire which destroyeth and dissolveth with the celestial fire° which quickeneth and giveth life? And such is the comparison between the soldier and the lover.

And as for you, untrue° Politique,° but truest bondman° to Philautia, you that presume to bind occasion° and to overwork° fortune, I would ask you but one question. Did ever any lady, hard to please, or disposed to exercise her lover, enjoin him so hard tasks and commandments, as Philautia exacteth of you? While your life is nothing but a continual acting upon a stage; and that your mind must serve your humour, and yet your outward person must serve your end; so as you carry in one person two several° servitudes to contrary masters.° But I will leave you to the scorn of that mistress whom you undertake to govern; that is, to fortune, to whom Philautia hath bound you.

And yet, you commissioners° of Philautia, I will proceed one degree further. If I allowed both of° your assurance and of your values as you have set° them, may not my master enjoy his own felicity, and have all yours for advantage? I do not mean that he should divide himself in both pursuits, as in your fainting° tales towards the conclusion you did yield° him; but because all these are in the hands of his mistress more fully to bestow than they can be attained by your addresses,° knowledge, fame, and fortune.° For the Muses, they are tributary to her Majesty° for the great liberties° they have enjoyed in her kingdom during her most flourishing reign; in thankfulness whereof they have adorned and accomplished her Majesty with the gifts of all the sisters. What library can present such a story° of great actions as her Majesty carrieth in her royal breast by the often return of this happy day?° What worthy author or favourite of the Muses is not familiar with her? Or what language wherein the Muses have used to speak is unknown to her?° Therefore, the hearing of her, the observing of her, the receiving instructions from her, may be to Erophilus a lecture exceeding all dead monuments of the Muses. For Fame, can all the exploits of the war win him such a title, as to have the name of favoured and selected servant of such a Queen? For Fortune, can any insolent politique promise to himself such a fortune by making his own way, as the excellency of her nature cannot deny to a careful, obsequious, and dutiful servant? And if he could, were it equal honour to obtain it by a shop° of cunning as by the gift of such a hand?

Therefore Erophilus' resolution is fixed: he renounceth Philautia, and all her enchantments. For her° recreation, he will confer with his muse: for her defence and honour, he will sacrifice his life in the wars, hoping to be embalmed in the sweet odours of her remembrance; to her service will he consecrate all his watchful endeavours; and will ever bear in his heart the picture of her beauty, in his actions of her will, and in his fortune of her grace and favour.'

Advice to the Earl of Rutland on his travels

Letter I

My Lord,

I hold it for a principle in the course of intelligence of state,° not to discourage men of mean sufficiency° from writing unto me, though I had at the same time very able advertisers;° for either they sent me some matter which the other had omitted, or made it clearer by delivering the circumstances; or if they added nothing, yet they confirmed that which coming single° I might have doubted. This rule therefore I have prescribed to others, and now give it to myself. Your Lordship hath many friends who have more leisure to think and more sufficiency to counsel than myself, yet doth my love to you dedicate these few free hours to study of you and your intended course; in which study if I find out nothing but that which you have from others, yet I shall perhaps confirm the opinion of wiser than myself.

Your Lordship's purpose is to travel, and your study must be what use to make of your travel. The question is ordinary,° and there is to it an ordinary answer; that is, your Lordship shall see the beauty of many cities, know the manners of the people of many countries, and learn the language of many nations. Some of these may serve for ornaments, and all of them for delights; but your Lordship must look further than these; for the greatest ornament is the inward beauty of the mind, and when you have known as great variety of delight as the world will afford, you will confess that the greatest delight is 'sentire te in dies fieri meliorem';° to feel that you do every day become more worthy; therefore your Lordship's end and scope should be that which in moral philosophy we call 'cultum animi',° the tilling° and manuring° of your own mind.

The gifts or excellencies of the mind are the same as those are of the body;° Beauty, Health, and Strength. Beauty of the mind is showed in graceful and acceptable forms,° and sweetness of behaviour; and they that have that gift cause those to whom they deny anything to go better contented away, than men of contrary° disposition do them to whom they grant. Health consisteth in an unmovable constancy and a freedom from passions, which are indeed the sicknesses of the mind.° Strength of mind is that active power which maketh us perform good things and great things, as well as° health and even temper of mind keeps from those that are evil and base. All these three are to be sought for, though the greatest part of° men have none of them; some have one and lack the other two; a few attain to have two of them and lack the third; and almost none have all.

The first way to attain experience of forms or behaviour, is to make the mind itself expert. For behaviour is but a garment,° and it is easy to make a comely garment for a body that is itself well-proportioned, whereas a deformed body can never be so helped by tailor's art but the counterfeit° will appear; and in the form of our mind it is a true rule, that a man may mend° his faults with as little labour as cover° them.

The second way is by imitation, and to that end good choice is to be made of those with whom you converse; therefore your Lordship should affect° their company whom you find to be worthiest, and not partially° think them most worthy whom you affect.

To attain to health of mind, we must use the same means that we do for the health of our bodies; that is, to take observation what diseases we are aptest to fall into, and to provide against them, for physic° hath not more medicines against the diseases of the body, than reason hath preservatives° against the passions of the mind. The Stoics were of opinion that there was no way to attain to this even temper of the mind but to be senseless,° and so they sold their goods to ransom themselves from their evils; but not only Divinity, our schoolmistress, doth teach us the effect of grace,° but even Philosophy, her handmaid, doth condemn our want of care and industry if we do not win very much upon° ourselves. To prove which I will only use one instance: there is nothing in nature more general or more strong than the fear of death,° and to a natural° man there is nothing seems more impossible than to resolve° against death. But both martyrs for religion, heathen for glory, some for love of their country, others for affection to one special person, have encountered death without fear, and suffered it without show of alteration; and therefore, if many have conquered passion's chiefest and strongest fortress, it is lack of undertaking in him that getteth not an absolute victory.

To set down the ways how a man may attain to the active power mentioned in this place (I mean strength of mind), is much harder than to give rules in the other two; for behaviour or good form may be gotten by education, and health or even temper of mind by observation. But if there be not in nature some partner to this active strength, it can never be obtained by any industry; for the virtues° which are proper unto it are liberality or magnificence, and fortitude or magnanimity; and some are by nature so covetous or cowardly, as it is as much in vain to seek to enlarge or inflame their minds, as to go about to plough the rocks. But where these active virtues are but budding, they must be ripened by clearness of judgment and custom of well-doing. Clearness of judgment makes men liberal,° for it teacheth men to esteem of the goods of fortune° not for themselves, for so they are but jailors to them, but for their use, for so they are lords over them; and it makes us to know that it is 'beatius dare quam accipere',° the one being a badge° of sovereignty, the other of subjection. Also it leadeth us to fortitude, for it teacheth us that we should not too much prize life which we cannot keep, nor fear death which we cannot shun; that he which dies nobly doth live for ever, and he that lives in fear doth die continually; that pain and danger be great only by opinion, and that in truth nothing is fearful but fear itself;° that custom makes the thing used natural as it were to the user. I shall not need to prove these two things, since we see by experience it holds true in all things, but yet those that give with judgment° are not only encouraged to be liberal by the return of thankfulness from those to whom they give, but find in the very exercise of that virtue a delight to do good. And if custom be strong to confirm° any one virtue more than another, it is the virtue of fortitude, for it makes us triumph over the fear which we have conquered, and anew to challenge danger which happily° we have encountered, and hold more dear the reputation of honour which we have increased.

I have hitherto set down what desire or wish I would have your Lordship to take into your mind, that is to make yourself an expert man, and what are the general helps which all men may use which have the said desire; I will now move° your Lordship to consider what helps your travel may give you.

First, when you see infinite variety of behaviour and manners of men, you may choose and imitate the best;° when you see new delights° which you never knew, and have passions stirred in you which before you never felt, you shall both know what disease° your mind is aptest to fall into, and what the things are that breed the disease; when you come into armies, or places where you shall see anything of the wars (as I would wish you to see them before your return), you shall both confirm your natural courage, and be made more fit for true fortitude, which is not given to man by nature, but must grow out of discourse of reason;° and lastly, in your travel you shall have great help to attain to knowledge, which is not only the excellentest thing in man, but the very excellency of man.

In manners or behaviour, your Lordship must not be caught with° novelty, which is pleasing to young men; nor infected with custom, which makes us keep our own ill graces,° and participate of° those we see every day; nor given to affection° (a general fault of most of our English travellers), which is both displeasing and ridiculous.

In discovering your passions and meeting with them, give not way to yourself nor dispense with yourself in little,° though resolving to conquer yourself in great; for the same stream that may be stopped with one man's hand at the spring head, may drown whole armies of men when it hath run long. In your being in the wars, think it better at the first to do a great deal too much than anything too little; for a young man's, especially a stranger's, first actions are looked upon, and reputation once gotten is easily kept, but an evil impression conceived at the first is not quickly removed.

The last thing that I am to speak of, but the first that you are to seek, is conceived° knowledge. To praise knowledge, or to persuade your Lordship to the love of it, I shall not need to use many words; I will only say, that where that wants° the man is void of all good; without it there can be no fortitude, for all other darings° come of fury, and fury is a passion, and passions ever turn into their contraries; and therefore the most furious men, when their first blaze is spent, be commonly the most fearful; without it there can be no liberality, for giving is but want of audacity to deny, or of discretion to prize;° without it there can be no justice, for giving to a man that which is his own is but chance, or want° of a corrupter or seducer; without it there can be no constancy or patience, for suffering is but dullness or senselessness; without it there can be no temperance, for we shall restrain ourselves from good as well as from evil, for that° they that cannot discern, cannot elect or choose; nay without it there can be no true religion, all other devotion being but blind zeal, which is as strong in heresy as in truth.

To reckon up all parts of knowledge, and to show the way to attain to every part, is a work too great for me at any time, and too long to discourse at this; therefore I will only speak of such knowledge as your Lordship should have desire to seek, and shall have means to compass. I forbear also to speak of divine knowledge, which must direct your faith, both because I find my own constancy insufficiency,° and also because I hope your Lordship doth still nourish the seeds of religion which during your education at Cambridge were sown in you. I will only say this; as the irresolute man can never perform any action well, so he that is not resolved in soul and conscience, can never be resolute in anything else. But that civil knowledge, which will make you do well by yourself, and do good unto others, must be sought by study, by conference,° and by observation.

Before I persuade your Lordship to study, I must look to answer an argument drawn from the nobility° of all places of° the world, which now is utterly unlearned, if it be not° some very few; and an authority of an English proverb, made in despite of° learning, that the greatest clerks° are not the wisest men. The first I answer, that this want of learning hath been in good countries ruined by civil wars, or in states corrupted through wealth or too great length of peace.° In the one sort men's wits were employed in their necessary defence, in the other drowned in the study of 'artes luxuriae'.° But in all flourishing states learning hath ever flourished. If it seem strange that I account no state flourishing but that which hath neither civil wars nor too long peace, I answer, that politic bodies are like our natural bodies, and must as well have some exercise to spend their humours,° as to be kept from too violent or continual outrages° which spend their best spirits.° The proverb I take to be made in that age when the nobility of England brought up their sons but as they entered their whelps,° and thought them wise enough if they could chase their deer; and I answer it with another proverb made by a wise man, 'Scientia non habet inimicum praeter ignorantem'.°

All men that live are drawn° either by book or example, and in books your Lordship shall find (in what course° soever you propound to yourself) rules prescribed by the wisest men, and examples left by the wisest men that have lived before us. Therefore knowledge is to be sought by your private study; and opportunity you shall have to study, if you do not often remove from place to place, but stay some time and reside in the best. In the course of your study and choice of your books, you must first seek to have the grounds of learning, which are the liberal arts;° for without them you shall neither gather other knowledge easily, nor make use of that you have; and then use studies of delight but° sometimes for recreation, and neither drown yourself in them, nor omit those studies whereof you are to have continual use. Above all other books be conversant in the Histories, for they will best instruct you in matter moral, military, and politic,° by which and in which you must ripen and settle your judgment.

In your study you are to seek two things: the first to conceive or understand; the second to lay up° or remember; for as the philosopher saith, 'discere est tanquam recordari'.° To help you to conceive, you may do well in those things which you are to read to draw° yourself to read with somebody that may give you help, and to that end you must either carry over with you some good general scholar, or make some abode in the universities abroad, where you may hear the professors in every art.° To help you to remember, you must use writing, or meditation,° or both; by writing I mean making of notes and abridgments of that which you would remember. I make conference° the second help to knowledge in order,° though I have found it the first and greatest in profiting, and I have so placed them because he that hath not studied knows not what to doubt° nor what to ask: but when the little I had learned had taught me to find out mine own emptiness, I profited more by some expert man in half a day's conference, than by myself in a month's study.° To profit much by conference, you must first choose to confer with expert men, I mean expert in that which you desire to know; next with many, for expert men will be of diverse and contrary opinions, and every one will make his own probable, so as if you hear but one you shall know in all questions but one opinion; whereas by hearing many, you shall, by seeing the reasons of one, confute the reasons of the other, and be able to judge of the truth. Besides, there is no one man that is expert in all things, but every great scholar is expert in some one, so as your wit shall be whetted° with conversing with many great wits, and you shall have the cream and quintessence of every one of theirs. In conference be neither superstitious, nor believing all you hear (what opinion soever you have of the man that delivereth it), nor too desirous to contradict. For of the first grows a facility to be led into all kind of error; since you shall ever think that he that knows all that you know, and somewhat more, hath infinite knowledge, because you cannot sound° or measure it. Of the second grows such a carping° humour, as you shall without reason censure all men, and want reason to censure yourself. I do conclude this point of conference with this advice, that your Lordship shall rather go a hundred miles out of the way to speak with a wise man, than five to see a fair town.

The third way to attain knowledge is observation, and not long life or seeing much; because, as he that rides a way often, and takes no care of marks or notes to direct him if he come the same again, or to make him know where he is if he come unto it, shall never prove a good guide; so he that lives long and sees much, but observes nothing, shall never prove a wise man. The use of observation° is in noting the coherence° of causes and effects, counsels° and successes,° and the proportion and likeness between nature and nature, force and force, action and action, state and state, time past and time present. The philosopher did think that all knowledge doth much depend on the knowledge of causes; as he said, 'id demum scimus cuius causam scimus';° and therefore a private man cannot prove so great a soldier as he that commands an army, nor so great a politique° as he that rules a state, because the one sees only the events and knows not the causes, the other makes the causes that govern the events. The observation of proportion or likeness between one person or one thing and another, makes° nothing without example, nor nothing new: and although 'exempla illustrant non probant',° examples may make things plain that are proved, but prove not themselves; yet when circumstances agree, and proportion is kept, that which is probable in one case is probable in a thousand, and that which is reason once is reason ever.

Your Lordship now seeing that the end of study, conference, and observation, is knowledge; you must know also that the true end of knowledge is clearness and strength of judgment, and not ostentation or ability to discourse;° which I do the rather put your Lordship in mind of, because the most part of our noblemen and gentlemen of our time have no other use of their learning but their table-talk; and the reason is because they before setting down their journey's end ere they attain to it they rest, and travel not so far as they should; but God knows they have gotten little that have only this discoursing gift; for though, like empty casks, they sound loud when a man knocks upon their outside, yet if you pierce into them you shall find them full of nothing but wind. This rule holds not only in knowledge, or in the virtue of knowledge, or in the virtue of prudence, but in all other virtues; that is, that we should both seek and love virtue for itself, and not for praise; for, as one said, 'turpe est proco ancillam sollicitare, est autem virtutis ancilla laus':° it is a shame for him that woos the mistress to court the maid, for praise 'is the handmaid of virtue'.°

I will here break off, for I have both exceeded the convenient length of a letter, and come short of such a discourse as this subject doth deserve. Your Lordship may perhaps find in this paper many things superfluous, most things imperfect and lame; I will, as well as I can, supply that defect upon a second advertisement,° if you call me to account. What confusion soever you find in my order or method, is not only my fault, whose wits are confounded with too much business, but the fault of this season, this being written in Christmas, in which confusion and disorder° hath by tradition not only been winked at but warranted.° If there be but any one thing that your Lordship may make use of, I think my pain well bestowed in all; and how weak soever my counsels be, my wishes shall be as strong as any man's for your Lordship's happiness. And so I rest, your Lordship's very affectionate cousin and loving friend,

E.

Greenwich, Jan. 4.

POSTSCRIPT

My Lord,

If any curious° scholar happening to see this discourse shall quarrel with my divisions of the gifts of the mind, because he findeth it not perhaps in his book,° and says that health and even temper of the mind is a kind of strength, and so I have erred against the rule 'membra dividenda non debent confundi',° I answer him: the quality of health and strength, as I have set them down, are not only unlike but mere° contrary, for the one binds in the mind and confines it, the other raises and enlarges it.

LETTER II

My good Lord,

The last I sent to your Lordship was so long, as it is no more than needful to give you a breathing-time before I send another: yet for the love I bear your Lordship I cannot be silent, being desirous both to satisfy myself and others how you prosper in your travels, and how you find yourself bettered thereby either in knowledge of God or of the world: the rather because the days you have already spent abroad are now sufficient both to give you light how to fix yourself an end with counsel, and accordingly shape your course constantly unto it. Besides, it is a vulgar scandal° of travellers, that few return more religious than they went out; wherein both my hope and request to your Lordship is that your principal care be to hold your foundation,° and to make no other use of informing yourself in the corruptions and superstitions of other nations than only thereby to engage your own heart more firmly to the truth. You live in a country° of two several professions,° and you shall return a novice from thence if you be not able to give an account of their ordinances, progress, and strength of each in reputation and party, and how both are supported, balanced, and managed by the state, as being the contrary humours° in the strength and predominancy whereof the health or disease of the body doth consist.

These things, my Lord, you will observe, not only as an Englishman whom it may concern to know what interest° his country may expect in the consciences of his neighbours; but also as a Christian to consider both the benefits and blemishes, the hopes and dangers, of the Church in all places.

Now for the world, I know it too well to persuade you to dive into the practices° thereof: rather stand upon your guard against them all that tempt you thereunto, or may practise upon° you in your conscience, your reputation, or your purse: resolve that no man is wise or safe but he that is honest; and let this persuasion turn your studies and observations from the compliment and impostures of this deboshed° age to more real° grounds of wisdom gathered out of the stories of times past, and out of the government of the present state. Your best guide to these is the knowledge of the country and people amongst whom you live. For the country, though you cannot see all places, yet if as you pass along you shall inquire carefully, and further help yourself with books that are written of° the cosmography of those parts, you shall thereby sufficiently gather the strength, riches, traffic,° havens,° shipping, commodities, vent,° and the wants and disadvantages of all places; wherein also for your own use hereafter, and for your friends, it will be fit to note their buildings, their furnitures, their entertainments, all their husbandry° and ingenious inventions in whatsoever concerneth either pleasure or profit.

For the people, your traffic° among them while your learn their language will sufficiently instruct you in their abilities, dispositions, and humours, if you a little enlarge the privacy of your own nature to seek acquaintance of the best sort of strangers, and restrain your affection and participation from your own countrymen of whatsoever condition.

In the history of France, you have a large and pleasant field in the lives of their kings to observe their alliances and successions, their conquests and their wars, especially with us; their counsels, their treaties, and all rules and examples of experience and wisdom; which may be lights and remembrances to you hereafter to judge all occurrences at home and abroad.

Lastly, for government, your end must not be, like an Intelligencer,° to spend all your time in fishing after the present news, humours, graces, or disgraces of the Court, which happily° may change before you come home: but your Lordship's better and more constant ground will be to know the consanguinities, alliances, and estates of their princes, the proportion° betwixt the nobility and the magistracy, the constitution of the courts of justice, the state of their laws, as well for the making as for the execution thereof; how the sovereignty of the King infuseth itself into all acts and ordinances; how many° ways they lay impositions° and taxations, and gather revenues to the Crown; what be the liberties and servitudes of all degrees;° what discipline and preparation for wars; what inventions° for increase of traffic at home, for multiplying their commodities,° encouraging arts or manufactures of worth of any kind; also what good establishments to prevent the necessities and discontentments of the people, to cut off suits at law and quarrels, to suppress thieves and all disorders.

To be short, because my purpose is not to bring all your observations to heads,° but only by these few to let you know what manner of return your friends will expect from you; let me for these and all the rest give you this one note, which I desire you to observe as the counsel of a friend: not to spend your spirits° and the precious time of your travels in a captious° prejudice and censuring of all things, nor in an infectious collection° of base vices of men and women, and general corruption of these times, which will be of use only amongst humourists° for jests and table-talk; but rather strain your wits and industry soundly to instruct yourself in all things between heaven and earth which may tend to virtue, wisdom, and honour; and which may make yourself more profitable to your country, more comfortable° to your friends, and acceptable to God; and to conclude, let all these riches be treasured up, not only in your memory (where time may lessen your stock), but rather in good writings and books of account, which will keep them safe for your use hereafter. And if in this time of your liberal traffic° you will give me any advertisement° of your commodities° in this kind, I will make you as liberal return from myself and your friends here as I shall be able. And so desiring your Lordship's pardon for this boldness, I commend your good endeavours to him that must either wither or prosper them; and so do most humbly take my leave, ever resting,

Your Lordship's in all duty to serve you.

Letter III

My Lord,

Since you have required of me some advice now at the very instant of your going, I must not refuse you, though my want of leisure and health will make that which you receive from me little worth.

My first letter to your Lordship did contain generalities. My second was particular to direct you in course of study, and this shall only tell you what are the notes I would wish you to gather in your travel; which being but a posting° night's work after everybody is gone to bed, I desire may be private to yourself, and may serve to awake you in some things though it cannot instruct you in all.

When your Lordship comes into any country, I would wish you to observe the nature of the climate and temperature of the air: for so you shall both judge of the healthfulness of the place and may have some inducement to guess at the disposition of the people. Also, to mark the condition of the soil, whether it be fertile or barren, mountainous or even, full of woods or champain.° And to note the principal rivers, their beginnings and course, the straits° and passages° that do sever one province or piece of province from another, and what their length or breadth is; the circuit and the diameter or length of the country; how it is peopled and inhabited; what are the commodities with which it abounds, and which it vents;° and of the other side, what it wants and draws to it from foreign parts; what ports it hath, what shipping, and how their traffic lies; how the people are armed and trained; what fortified towns or castles, what revenue, what arsenal, what alliances and what known enemies the state hath; for these things will lead you to know whether any country be rich or poor, strong or weak. But above all things I would have you understand the manner of government of the place where you are; where the sovereignty is,—in one, as in a monarchy, in a few, or in the people;° or, if it be mixed,° to which of these forms it most inclines. Next, what ministers of state and subalternate° governors, as council and magistrates. Thirdly, by what laws or customs it is governed. And lastly, what is the execution of justice in peace and their discipline in war.

If your Lordship will tell me that these things will be too many to remember, I answer that I had rather you trusted your note-book than your memory. If you object that some of these things being martial, and others points of state,° you shall not be able to collect them nor judge of them, I must ask you whether you would not get° a pilot on a strange coast, and guide in an unknown way. And so if where you come you seek after these things, you shall as soon find directors to guide you to them as to any matters of sport or vanity.

The first thing your Lordship must seek in all this course is Industry: for as great difference is betwixt it and idleness, or betwixt an active sprightful° man and a slothful, as betwixt a living man and a dead. The second is to direct that industry to good things; for else the more you do the more ill you do, and the faster you go the further you go out of the way.° The last is that you be rather endeavouring to do well than believing you do well: for besides that all self-conceited young men do grow infinitely vain, when once out of opinion that they are wise or good enough they hold themselves pleased with themselves, they fall more backward in a month than they grow forward in a year.

St. Albans, 16 October

Essays° (1597)

Of Studies

Studies serve for pastimes, for ornaments, and for abilities.° Their chief use for pastime is in privateness and retiring;° for ornament is in discourse,° and for ability is in judgment. For expert° men can execute,° but learned men are fittest to judge or censure.°

¶ To spend too much time in them is sloth;° to use them too much for ornament is affectation; to make judgment wholly by their rules is the humour° of a scholar.°

¶ They perfect Nature, and are perfected by experience.

¶ Crafty° men contemn° them, simple° men admire° them, wise° men use them. For they teach not their own use;° but that is a wisdom without them, and above them,° won by observation.

¶ Read not to contradict, nor to believe, but to weigh and consider.°

¶ Some books are to be tasted, others to be swallowed, and some few to be chewed and digested. That is, some books are to be read only in parts; others to be read, but cursorily; and some few to be read wholly and with diligence and attention.

¶ Reading maketh a full man, conference° a ready man, and writing an exact man. And therefore if a man write little, he had need have a great memory; if he confer° little, he had need have a present wit;° and if he read little, he had need have much cunning, to seem to know that he doth not.

¶ History makes men wise, poets witty;° the mathematics subtle, natural philosophy deep; moral grave,° logic and rhetoric able to contend.

Of Discourse

Some in their discourse desire rather commendation of wit° in being able to hold° all arguments, than of judgment in discerning° what is true; as if it were a praise to know what might be said, and not what should be thought.° Some have certain commonplaces and themes° wherein they are good, and want variety;° which kind of poverty is for the most part tedious, and now and then ridiculous.

¶ The honourablest part° of talk is to give the occasion,° and again to moderate and pass to somewhat else.

¶ It is good to vary and mix speech of the present occasion° with argument,° tales with reasons,° asking of questions with telling of opinions, and jest° with earnest.

¶ But some things are privileged° from jest, namely religion, matters of state,° great persons,° any man's present° business of importance, and any case that deserveth pity.

¶ He that questioneth much shall learn much, and content° much, specially if he apply° his questions to the skill of the person of whom he asketh, for he shall give them occasion to please themselves in speaking, and himself shall continually gather knowledge.

¶ If you dissemble sometimes your knowledge of that° you are thought to know, you shall be thought another time to know that you know not.

¶ Speech of a man's self is not good often; and there is but one case wherein a man may commend himself with good grace,° and that is in commending virtue in another, especially if it be such a virtue as whereunto himself pretendeth.°

¶ Discretion of speech is more than eloquence, and to speak agreeably° to him with whom we deal is more than to speak in good words or in good order.

¶ A good continued° speech, without a good speech of interlocution, sheweth slowness; and a good reply or second speech, without a good set speech, sheweth shallowness and weakness: as we see in beasts, that those that are weakest in the course are yet nimblest in the turn.°

¶ To use too many circumstances° ere one come to the matter is wearisome; to use none at all is blunt.

OF CEREMONIES AND RESPECTS°

HE that is only real° had need° have exceeding great parts of virtue,° as the stone had need be rich that is set without foil.°

¶ But commonly it is in° praise as it is in gain. For as the proverb is true, that 'light gains° make heavy purses', because they come thick,° whereas great come but now and then, so it is as true that small matters° win great commendation, because they are continually in use and in note,° whereas the occasion of any great virtue cometh but on holidays.°

¶ To attain° good forms° it sufficeth not to despise them; for so° shall a man observe them in others, and° let him trust himself with the rest.° For if he care° to express them, he shall leese° their grace, which is to be natural and unaffected. Some men's behaviour is like a verse, wherein every syllable is measured.° How can a man comprehend° great matters, that breaketh° his mind too much to small observations?

¶ Not to use ceremonies at all is to teach others not to use them again,° and so diminish his respect; especially they be° not to be omitted to strangers and strange° natures.

¶ Amongst a man's peers a man shall be sure of familiarity, and therefore it is a good title° to keep state.° Amongst a man's inferiors one shall be sure of reverence, and therefore it is good a little to be familiar.

¶ He that is too much° in any thing, so that he give another occasion of satiety, maketh himself cheap.

¶ To apply° one's self to others is good; so it be with demonstration that a man does it upon regard,° and not upon facility.°

¶ It is a good precept generally in seconding° another yet to add somewhat of one's own; as if you will grant his opinion, let it be with some distinction; if you will follow his motion,° let it be with condition; if you allow his counsel,° let it be with alleging° further reason.

OF FOLLOWERS AND FRIENDS

COSTLY° followers are not to be liked, lest while a man maketh his train° longer he make his wings shorter. I reckon to be costly not them alone which° charge° the purse, but which are wearisome and importune° in suits.° Ordinary following ought to challenge° no higher conditions° than countenance,° recommendation, and protection from wrong.

¶ Factious° followers are worse to be liked, which follow not upon affection to him with whom they range° themselves, but upon discontentment conceived against some other; whereupon commonly ensueth that ill intelligence° that we many times see between great personages.

¶ The following by certain states,° answerable° to that° which a great person himself professeth, as of soldiers to him° that has been employed in the wars, and the like, has ever been a thing civil,° and well taken even in monarchies, so° it be without too much pomp or popularity.°

¶ But the most honourable kind of following is to be followed as one that apprehendeth° to advance° virtue and desert° in all sorts of persons; and yet where there is no eminent odds in sufficiency,° it is better to take with° the more passable° than with the more able. In government° it is good to use men of one rank° equally,° for to countenance some extraordinarily is to make them insolent and the rest discontent, because they may claim a due.° But in favours° to use men with much difference and election° is good, for it maketh the persons preferred more thankful and the rest more officious,° because all is of favour.°

¶ It is good not to make too much° of any man at first, because one cannot hold out° that proportion.°

¶ To be governed by one° is not good, and to be distracted° with many is worse; but to take advice of friends is ever honourable. 'For lookers-on many times see more than gamesters;° and the vale best discovereth the hill'.°

¶ There is little friendship in the world, and least of all between equals; which was wont to be magnified.° That that is, is between superior and inferior, whose fortunes may comprehend° the one the other.

Of Suitors°

Many ill matters° are undertaken, and many good matters with ill minds. Some embrace suits which° never mean to deal° effectually in them. But if they see there may be life in the matter by some other means,° they will be content to win a thank or take a second° reward. Some take hold of suits only for an occasion to cross° some other, or to make° an information whereof they could not otherwise have an apt pretext, without care what become of the suit when that turn is served. Nay some undertake suits with a full purpose to let them fall, to the end to gratify the adverse party or competitor.

¶ Surely there is in sort° a right in every suit, either a right of equity,° if it be a suit of controversy;° or a right of desert,° if it be a suit of petition.° If affection° lead a man to favour the wrong side in justice,° let him rather use his countenance° to compound° the matter than to carry° it. If affection lead a man to favour the less worthy in desert, let him do it without depraving° or disabling° the better deserver.

¶ In suits a man doth not well understand, it is good to refer them to some friend of trust and judgment, that may report whether he may deal° in them with honour.°

¶ Suitors are so distasted° with delays and abuses° that plain dealing in denying° to deal in suits at first,° and reporting the success barely,° and in challenging° no more thanks than one has deserved, is grown not only honourable but also gracious.°

¶ In suits of favour° the first coming° ought to take little place.° So far forth consideration may be had of his° trust, that if intelligence° of the matter could not otherwise have been had but by him, advantage be not taken of the note.°

¶ To be ignorant of the value° of a suit is simplicity,° as well as to be ignorant of the right° thereof is want of conscience.

¶ Secrecy in suits is a great means of obtaining, for voicing° them to be in forwardness° may discourage some kind of suitors, but does quicken and awake others.

¶ But timing° of the suits is the principal, timing I say not only in respect of the person that should grant it, but in respect of those which° are like to cross it.

¶ Nothing is thought so easy° a request to a great person as his letter, and yet if it be not in° a good cause it is so much out of his reputation.

Of Expense

RICHES are for spending, and spending for honour° and good actions. Therefore extraordinary expense must be limited° by the worth of the occasion; for voluntary undoing° may be as well for a man's country as for the kingdom of heaven.° But ordinary expense ought to be limited by a man's estate, and governed with such regard as it be within his compass,° and not subject to deceit and abuse° of servants; and ordered to the best shew,° that° the bills may be less than the estimation abroad.°

¶ It is no baseness° for the greatest to descend and look into their own estate. Some forebear it not upon negligence alone, but doubting° to bring themselves into melancholy in respect° they shall find it broken.° But 'wounds cannot be cured without searching'.°

¶ He that cannot look into his own estate had need° both choose well those whom he employeth, yea, and change them often. For new° are more timorous and less subtle.°

¶ In clearing° of a man's° estate, he may as well° hurt° himself in being too sudden° as in letting it run on too long, for hasty selling is commonly as disadvantageable as interest.°

¶ He that has a state° to repair may not despise small things; and commonly it is less dishonourable to abridge° petty° charges than to stoop to petty gettings.

¶ A man ought warily to begin charges which once begun must continue. But in matters that return not° he may be more magnificent.

Of Regiment° of Health

There is a wisdom in this° beyond the rules of physic.° A man's own observation what he finds good of, and what he finds hurt of,° is the best physic to preserve health. But it is a safer conclusion to say this agrees not well with me, therefore I will not continue it, than this I find no offence, of this therefore I may use it. For strength of nature in youth passeth over many excesses, which are owing° a man till his age.

¶ Discern of the coming on of years, and think not to do the same things still.°

¶ Beware of any sudden change in any great point of diet, and if necessity enforce it, fit the rest to it.

¶ To be free-minded° and cheerfully disposed at hours of meat, and of sleep, and of exercise, is the best precept of long lasting.

¶ If you fly physic in health altogether,° it will be too strange° to your body when you shall need it; if you make it too familiar, it will work no extraordinary effect when sickness cometh.

¶ Despise no new accident° in the body, but ask opinion of it.

¶ In sickness respect° health principally, and in health action.° For those that put their bodies to endure° in health, may in most sicknesses which are not very sharp° be cured only with diet and tendering.°

¶ Physicians are some of them so pleasing° and conformable to the humours° of the patient, as they press° not the true cure of the disease; and some other are so regular° in proceeding according to art° for the disease, as they respect not sufficiently the condition of the patient. Take one of a middle temper; or if it may not be found in one man, compound two of both sorts, and forget not to call as well the best acquainted with your body as the best reputed of for his faculty.°

Of Honour and Reputation

The winning° of honour is but the revealing° of a man's virtue and worth without disadvantage.° For some in their actions do affect° honour and reputation, which sort of men are commonly much talked of, but inwardly° little admired; and some darken° their virtue in the show of it, so as they be undervalued in opinion.

¶ If a man perform that which hath not been attempted before, or attempted and given over,° or has been achieved but not with so good circumstance, he shall purchase° more honour than by effecting a matter of greater difficulty or virtue,° wherein he is but a follower.°

¶ If a man so temper° his actions as in some one of them he do content every faction or combination° of people, the music will be the fuller.

¶ A man is an ill husband° of his honour that entereth into° any action, the failing wherein may disgrace him more than the carrying of it through can honour him.

¶ Discreet followers help much to reputation.

¶ Envy, which is the canker° of honour, is best extinguished by declaring° a man's° self in his ends, rather to seek merit than fame, and by attributing a man's successes rather to divine providence and felicity° than to his virtue or policy.°

¶ The true marshalling° of the degrees of sovereign honour are these. In the first place are *Conditores*,° founders of states. In the second place are *Legislatores*, lawgivers, which are also called second founders, or *Perpetui principes*,° because they govern by their ordinances° after they are gone. In the third place are *Liberatores*,° such as compound° the long miseries of civil wars, or deliver their countries from servitude of strangers or tyrants. In the fourth place are *Propagatores* or *Propugnatores imperii*,° such as in honourable wars enlarge their territories, or make noble defence against invaders. And in the last place are *Patres patriae*,° which° reign justly and make the times good° wherein they live.

¶ Degrees of honour in subjects are first *Participes curarum*,° those upon whom princes do discharge the greatest weight of their affairs, their *Right hands* (as we call them). The next are *Duces belli*,° great leaders, such as are princes' lieutenants,° and do them notable services in the wars. The third are *Gratiosi*, favourites, such as exceed not this scantling,° to be solace to the sovereign and harmless to the people. And the fourth *Negotiis pares*,° such as have great place under princes, and execute their places° with sufficiency.°

OF FACTION°

MANY have a new wisdom,° indeed a fond° opinion, that for a prince to govern his estate,° or for a great person to govern his proceedings,°

according to the respects of° factions, is the principal part of policy.° Whereas contrariwise, the chiefest wisdom is either in ordering° those things which are general,° and wherein men of several° factions do nevertheless agree, or in dealing with correspondence to° particular persons one by one. But I say not that the consideration of factions is to be neglected.

¶ Mean° men must adhere,° but great° men, that have strength in themselves,° were better to maintain themselves indifferent° and neutral; yet even in beginners, to adhere so moderately, as he be a man of the one faction which is passablest° with the other, commonly giveth best way.°

¶ The lower and weaker faction is the firmer in conjunction.°

¶ When one of the factions is extinguished° the remaining subdivideth, which is good for a second faction. It is commonly seen that men once placed° take in° with the contrary faction to that by which they enter.°

¶ The traitor in factions lightly goeth away with it,° for when matters have stuck long in balancing,° the winning of some one man casteth° them, and he getteth all the thanks.

Of Negotiating°

It is generally better to deal by speech than by letter, and by the mediation of a third° than by a man's self.° Letters are good when a man would draw° an answer by letter back again, or when it may serve for a man's justification afterwards to produce his own letter. To deal in person is good when a man's face breeds regard,° as commonly with inferiors.

¶ In choice of instruments° it is better to choose men of a plainer° sort, that are like° to do that that is committed to them, and to report back again faithfully the success,° than those that are cunning to contrive out of other men's business somewhat° to grace° themselves, and will help the matter in report for satisfaction's sake.°

¶ It is better to sound° a person with whom one deals afar off,° than to fall upon the point at first; except you mean to surprise him by some short° question.

¶ It is better dealing with men in appetite,° than with those which are where they would be.

¶ If a man deal with another upon conditions,° the start or first performance° is all;° which a man cannot reasonably demand,° except either the nature of the thing be such, which° must go before,° or else a man can persuade the other party° that he shall still need him in some other thing, or else that he° be counted the honester° man.

¶ All practice° is to discover° or to work.° Men discover themselves in trust,° in passion,° at unawares,° and of necessity, when they would° have somewhat done and cannot find an apt pretext. If you would work any man, you must either know his nature and fashions,° and so lead him; or his ends, and so win him; or his weaknesses or disadvantages,° and so awe° him; or those that have interest in° him, and so govern him.

¶ In dealing with cunning persons we must ever consider their ends to interpret their speeches; and it is good to say little to them, and that which they least look for.°

Religious Meditations [*Selections*]

OF THE WORKS OF GOD AND MAN

'GOD beheld all things which his hands had made, and lo they were all passing good.'° But when man turned him about, and took a view of the works which his hands had made, he found 'all to be vanity and vexation of spirit.'° Wherefore if thou shalt work in the works of God, thy sweat shall be as an ointment of odours,° and thy rest as the Sabbath of God.° Thou shalt travail in the sweat of a good conscience, and shalt keep holiday in the quietness and liberty of the sweetest contemplations. But if thou shalt aspire after the glorious acts of men, thy working shall be accompanied with compunction° and strife, and thy remembrance followed with distaste and upbraidings.° And justly doth it come to pass towards thee (O man), that since thou which art God's work dost him no reason in yielding him well-pleasing service, even thine own works also should reward thee with the like fruit of bitterness.

OF THE MIRACLES OF OUR SAVIOUR

'He hath done all things well.'°

A true confession and applause: God when he created all things, saw that everything in particular, and all things in general were exceeding good. God the word in the miracles which he wrought (now every miracle is a new creation and not according to the first creation)° would do nothing which breathed not towards men favour and bounty. Moscs wrought miracles and scourged the Egyptians with many plagues.° Elias° wrought miracles and shut up heaven, that no rain should fall upon the earth;° and again, brought down from heaven the fire of God upon the captains° and their bands.° Elizeus° wrought also, and called bears out of the desert to devour young children.° Peter struck Ananias the sacrilegious hypocrite with present° death,° and Paul, Elymas the sorcerer with blindness.° But no such thing did Jesus. The spirit of God descended down upon him in the form of a dove,° of whom he said, 'You know not of what spirit you are'.° The spirit of Jesus is the spirit of a dove. Those servants of God were as the oxen of God, treading out the corn and trampling the straw° down under their feet, but Jesus is the lamb of God without wrath or judgements.

All his miracles were consummate° about man's body, as his doctrine° respected the soul of man. The body of man needeth these things: sustenance, defence from outward wrongs, and medicine. It was he that drew a multitude of fishes into the nets,° that he might give unto men more liberal provision. He turned water, a less worthy nourishment of man's body, into wine,° a more worthy, that gladdens the heart of man.° He sentenced the fig tree to wither° for not doing that duty whereunto it was ordained, which is to bear fruit for men's food. He multiplied the scarcity of a few loaves and fishes° to a sufficiency, to victual a host of people. He rebuked the winds° that threatened destruction to the seafaring men. He restored° motion to the lame,° light to the blind,° speech to the dumb,° health to the sick,° cleanness to the leprous,° a right mind to those that were possessed,° and life to the dead.° No miracle of his is to be found to have been of judgement or revenge, but all of goodness and mercy, and respecting man's body. For as touching riches, he did not vouchsafe to do any miracle save one only, that tribute might be given to Caesar.°

Of the Innocence of the Dove, and Wisdom of the Serpent

'The fool receiveth° not the words of wisdom, except thou discover to him what he hath in his heart.'°

To a man of a perverse and corrupt judgement all instruction or persuasion is fruitless and contemptible, which begins not with discovery, and laying open of the distemper° and ill complexion° of the mind which is to be recured,° as a plaster is unseasonably applied before the wound be searched.° For men of corrupt understanding, that have lost all sound discerning of good and evil, come possessed with this prejudicate° opinion, that they think all honesty and goodness proceedeth out of a simplicity° of manners, and a kind of want of experience and unacquaintance with the affairs of the world. Therefore, except they may perceive that those things which are in their hearts, that is to say, their own corrupt principles, and the deepest reaches of their cunning and rottenness, to be throughly° sounded and known to him that goes about to persuade with them, they make but a play of the words of wisdom.

Therefore it behoveth him which aspireth to a goodness not retired or particular° to himself, but a fructifying and begetting goodness,° which should draw on° others to know those points which he called in the Revelation 'the deeps of Satan',° that he may speak with authority and true insinuation.° Hence is the precept: 'Try all things and hold that which is good',° which induceth a discerning election° out of an examination whence nothing at all is excluded. Out of the same fountain° ariseth that direction: 'Be you wise as serpents, and innocent as doves'.° There are neither teeth, nor stings, nor venom, nor wreaths and folds° of serpents which ought not to be all known, and as far as examination does lead, tried. Neither let any man here fear infection or pollution, for the sun entereth into sinks° and is not defiled. Neither let any man think that herein he tempteth God, for this diligence and generality° of examination is commanded, and 'God is sufficient to preserve you immaculate and pure.'°

Of the Exaltation° of Charity

'If I have rejoiced at the overthrow of him that hated me, or took pleasure when adversity did befall him.'°

The detestation or renouncing of Job.° For a man to love again where he is loved, it is the charity of publicans, contracted by mutual profit and good offices; but to love a man's enemies° is one of the cunningest points of the law of Christ, and an imitation of the divine nature. But yet again, of this charity there be divers degrees, whereof the first is to pardon our enemies when they repent. Of which charity there is a shadow and image even in noble beasts, for of lions it is a received opinion that their fury and fierceness ceaseth towards anything that yieldeth and prostrateth itself. The second degree is to pardon our enemies, though they persist, and without satisfactions° and submissions. The third degree is not only to pardon and forgive and forbear our enemies, but to deserve well of them, and to do them good. But all these three degrees either have or may have in them a certain bravery° and greatness of the mind,° rather than pure charity. For when a man perceives virtue to proceed and flow from himself, it is possible that he is puffed up, and takes contentment rather in the fruit of his own virtue than in the good of his neighbour's. But if any evil overtake the enemy from any other coast than from thyself, and thou in the inwardest motions of thy heart beest grieved and compassionate, and dost no ways insult° as if thy days of right and revenge were at the last come: this I interpret to be the height and exaltation of charity.

Of Earthly Hope

'Better is the sight of the eye than the apprehension of the mind.'°

Pure sense, receiving everything according to the natural impression,° makes a better state and government° of the mind than these same imaginations and apprehensions° of the mind. For the mind of man hath this nature and property, even in the gravest and most settled wits, that from the sense of every particular° it doth as it were bound and spring forward, and take hold of other matters, foretelling to itself that all shall prove like unto that which beateth upon° the present sense.° If the sense be of good, it easily runs into an unlimited hope, and into a like fear when the sense is of evil, according as is said: 'The oracles of hopes doth oft abuse.'° And that contrary: 'A froward° soothsayer is fear in doubts.'° But yet of fear there may be made some use, for it prepareth patience, and awaketh industry.

No shape of ill comes new or strange to me.
All sorts set down, yea and prepared be.°

But hope seemeth a thing altogether unprofitable, for to what end serveth this conceit of good? Consider and note a little: if the good fall out less than thou hopest (good though it be, yet less because it is), it seemeth rather loss than benefit, through thy excess of hope. If the good prove equal and proportionable in event to thy hope, yet the flower thereof by thy hope is gathered; so as when it comes, the grace of it is gone, and it seems used and therefore sooner draweth on satiety. Admit thy success prove better than thy hope, it is true gain seems to be made. But had it not been better to have gained the principal° by hoping for nothing, than the increase° by hoping for less? And this is the operation of hope in good fortunes; but in misfortunes it weakeneth all force and vigour of the mind. For neither is there always matter of hope; and if there be, yet if it fail but in part, it doth wholly overthrow the constancy and resolution of the mind. And besides, though it doth carry us through, yet is it a greater dignity of mind to bear evils by fortitude° and judgement, than by a kind of absenting and alienation of the mind from things present to things future—for that it is to hope. And therefore it was much lightness° in the poets, to feign hope° to be as a counterpoison° of human diseases, as to mitigate and assuage the fury and anger of them; whereas indeed it doth kindle and enrage them, and causeth both doubling of them and relapses. Notwithstanding, we see that the greatest number of men give themselves over to their imaginations of hope, and apprehensions of the mind, in such sort that, ungrateful towards things past° and in a manner unmindful of things present, as if they were ever children and beginners, they are still in longing for things to come. 'I saw all men walking under the sun resort and gather to the second person, which was afterwards to succeed; this is an evil disease and a great idleness of the mind.'°

But perhaps you will ask the question, whether it be not better, when things stand in doubtful terms, to presume the best, and rather hope well than distrust, specially seeing that hope does cause a greater tranquillity of mind?

Surely I do judge a state of mind which in all doubtful expectations is settled and floateth° not, and doth this out of a good government and composition° of the affections, to be one of the principal supports of man's life. But that assurance and repose of the mind which only rides at anchor° upon hope, I do reject as wavering and weak. Not that it is not convenient° to foresee and presuppose out of a sound and sober conjecture as well the good as the evil, that thereby we may fit our actions to the probabilities and likelihoods of their event, so that° this be a work of the understanding and judgement, with a due° bent and inclination of the affection. But which of you hath so kept his hopes within limits, as when it is so that you have, out of a watchful and strong discourse of the mind° set down the better success to be in apparency the more likely, you have not dwelt upon the very muse and forethought of the good to come; and, giving scope and favour to your mind, to fall into such cogitations as into a pleasant dream. And this it is which makes the mind light, frothy, unequal° and wandering.° Wherefore all our hope is to be bestowed upon the heavenly life to come. But here on earth, the purer our sense is from the infection and tincture° of imagination, the better and the wiser soul.

The sum of life to little doth amount,
And therefore doth forbid a longer count.°

Of Impostors

'Whether we be transported in mind it is to Godward. Or whether we be sober it is to youwards.'°

This is the true image and true temper of a man, and of him that is God's faithful workman. His carriage° and conversation° towards God is full of passion, of zeal, and of trances.° Thence proceed groans unspeakable and exultings, likewise in comfort, ravishment° of spirit and agonies. But contrariwise, his carriage and conversation towards men is full of mildness, sobriety, and appliable° demeanour. Hence is that saying, 'I am become all things to all men',° and such like. Contrary it is with hypocrites and impostors. For they in the church and before the people set themselves on fire, and are carried as it were out of themselves, and becoming as men inspired with holy furies, they set heaven and earth together. But if a man did see their solitary and separate meditations, and conversation, whereunto God is only privy,° he might towards God find them not only cold and without virtue, but also full of ill nature, and leaven:° 'Sober enough to God, and transported only towards men.'°

Of the Several Kinds of Imposture

'Avoid profane strangeness of words and oppositions of knowledge, falsely so called.'° 'Avoid fond and idle fables.'° 'Let no man deceive you by high speech.'°

There are three forms of speaking which are as it were the style and phrase° of imposture. The first kind is of them who, as soon as they have gotten any subject or matter, do straight cast it into an art, inventing new terms of art,° reducing all into divisions and distinctions, thence drawing assertions or positions, and so framing oppositions by questions and answers: hence issueth the cobwebs and clatterings° of the Schoolmen.°

The second kind is of them who, out of the vanity of their wit (as church poets),° do make and devise all variety of tales, stories, and examples, whereby they may lead men's minds to a belief, from whence did grow the legends° and infinite fabulous inventions and dreams of the ancient heretics.

The third kind is of them who fill men's ears with mysteries, high parables, allegories, and illusions: which mystical and profound° form many of the heretics have also made choice of. By the first kind of these, the capacity and wit° of man is fettered and entangled; by the second it is trained on° and inveigled; by the third it is astonished and enchanted; but by every of them the while it is seduced and abused.°

Of Atheism°

'The fool hath said in his heart there is no God.'°

First it is to be noted that the Scripture says, the fool hath said in his heart, and not he hath thought in his heart. That is to say, he doth not so fully think it in judgement, as he hath a good will° to be of that belief. For seeing it makes not° for him that there should be a God, he doth seek by all means accordingly to persuade and resolve° himself, and studies to affirm, prove and verify it to himself as some theme or position.° All which labour, notwithstanding that sparkle of our creation light,° whereby men acknowledge a Deity burneth still within, and in vain doth he strive utterly to alienate it or put it out, so that it is out of the corruption of his heart and will, and not out of the natural apprehension of his brain and conceit that he doth set down his opinion, as the comical poet saith: 'Then came my mind to be of mine opinion',° as if himself and his mind had been two diverse things. Therefore the atheist hath rather said and held it in his heart, than thought or believed in his heart, that there is no God.

Secondly it is to be observed that he hath said in his heart, and not spoken it with his mouth. But again you shall note, that this smothering of this persuasion within the heart cometh to pass for fear of government° and of speech amongst men: for as he saith, 'To deny God in a public argument were much, but in a familiar conference were current enough'.° For if this bridle were removed there is no heresy which would contend more to spread, and multiply, and disseminate itself abroad than atheism. Neither shall you see those men which are drenched° in this frenzy of mind to breathe almost anything else, or to inculcate° even without occasion anything more than speech tending to atheism, as may appear in Lucretius the Epicure, who makes of his invectives against religion° as it were a burden or verse of return to all his other discourses. The reason seems to be, for that the atheist not relying sufficiently upon himself, floating in mind, and unsatisfied and enduring° within many faintings, and as it were false of° his opinion, desires by other men's opinions agreeing with his to be recovered and brought again. For it is a true saying: 'Who so laboureth earnestly to prove an opinion to another, himself distrusts it.'°

Thirdly, it is a fool that hath so said in his heart, which is most true, not only in respect that he hath no taste in those things which are supernatural and divine: but in respect of human and civil wisdom. For first of all, if you mark the wits and dispositions which are inclined to atheism you shall find them light, scoffing, impudent, and vain; briefly, of such a constitution as is most contrary to wisdom and moral gravity. Secondly, amongst statesmen and politiques,° those which have been of greatest depths and compass,° and of largest and most universal understanding, have not only in cunning° made their profit in seeming religious to the people but in truth have been touched with an inward sense of the knowledge of Deity, as they which you shall evermore note to have attributed much to fortune and providence.°

Contrariwise, those who ascribed all things to their own cunnings and practices, and to the immediate and apparent causes, and as the prophet saith, 'have sacrificed to their own nets',° have been always but petty counterfeit statesmen, and not capable of the greatest actions.

Lastly, this I dare affirm in knowledge of nature,° that a little natural philosophy, and the first entrance into it, doth dispose the opinion to atheism. But on the other side, much natural philosophy, and wading deep into it, will bring about men's minds to religion.° Wherefore atheism every way seems to be joined and combined with folly and ignorance, seeing that nothing can be more justly allotted to be the saying of fools than this, 'there is no God.'

Of the Colours of Good and Evil [*Selections*]

In deliberatives° the point is, what is good and what is evil, and of good what is greater, and of evil what is the less.

So that the persuader's labour is to make things appear good or evil, and that in higher or lower degree; which as it may be performed by true and solid reasons, so it may be represented also by colours, popularities° and circumstances,° which are of such force, as they sway the ordinary judgment either of a weak° man, or of a wise man not fully and considerately attending and pondering the matter. Besides their power to alter the nature of the subject in appearance, and so to lead to error, they are of no less use to quicken° and strengthen the opinions and persuasions which are true: for reasons plainly delivered, and always after one manner, especially with fine and fastidious° minds, enter but heavily and dully: whereas if they be varied and have more life and vigour put into them by these forms and insinuations,° they cause a stronger apprehension,° and many times suddenly win the mind to a resolution.° Lastly, to make a true and safe judgment, nothing can be of greater use and defence to the mind, than the discovering and reprehension° of these colours, shewing in what cases they hold,° and in what they deceive: which as it cannot be done, but out of a very universal knowledge of the nature of things, so being performed, it so cleareth man's judgment and election, as it is the less apt to slide into any error.

A table of Colours or appearances of Good and Evil, and their degrees, as places of Persuasion and Dissuasion, and their several Fallaxes,° and the Elenches° of them.

III

Quod ad veritatem refertur majus est quam quod ad opinionem. Modus autem et probatio ejus quod ad opinionem pertinet haec est, quod quis si clam putaret fore, facturus non esset.°

So the Epicures° say of the Stoics' felicity° placed in virtue; that it is like the felicity of a player,° who if he were left° of his auditory and their applause, he would straight be out of heart and countenance;° and therefore they call virtue 'bonum theatrale'.° But of riches the poet saith:

Populus me sibilat, at mihi plaudo.°

And of pleasure,

Grata sub imo
Gaudia corde premens, vultu simulante pudorem.°

The fallax of this colour is somewhat subtile,° though the answer to the example be ready; for virtue is not chosen 'propter auram popularem';° but contrariwise, 'maxime omnium teipsum reverere,'° so as a virtuous man will be virtuous *in solitudine*,° and not only *in theatro*,° though percase° it will be more strong by glory and fame, as an heat which is doubled by reflexion. But that denieth the supposition,° it doth not reprehend° the fallax, whereof the reprehension is: Allow that virtue (such as is joined with labour and conflict) would not be chosen but for fame and opinion,° yet it followeth not that the chief motive of the election° should not be real and for it self; for fame may be only *causa impulsiva*,° and not *causa constituens* or *efficiens*.° As if there were two horses, and the one would do better without the spur than the other: but again, the other with the spur would far exceed the doing° of the former, giving him the spur also; yet the latter will be judged to be the better horse. And the form° as to say, 'Tush, the life of this horse is but in the spur,' will not serve as to a wise judgment: for since the ordinary instrument of horsemanship is the spur, and that it is no manner of impediment nor burden, the horse is not to be accounted the less of which° will not do well without the spur, but rather the other is to be reckoned a delicacy° than a virtue: so glory and honour are as spurs to virtue: and although virtue would languish without them, yet since they be always at hand to attend° virtue, virtue is not to be said the less chosen for itself because it needeth the spur of fame and reputation: and therefore that position, 'nota ejus rei quod propter opinionem et non propter veritatem eligitur, haec est, quod quis si clam putaret fore facturus non esset',° is reprehended.

VII

'*Quod bono vicinum, bonum; quod a bono remotum, malum.*'°

Such is the nature of things, that things contrary and distant in nature and quality are also severed° and disjoined in place, and things like and consenting° in quality are placed and as it were quartered°

together: for partly in regard of the nature to spread, multiply, and infect in similitude,° and partly in regard of the nature to break, expel, and alter that which is disagreeable and contrary, most things do either associate and draw near to themselves the like, or at least assimilate to themselves that which approacheth near them, and do also drive away, chase, and exterminate their contraries. And that is the reason commonly yielded, why the middle region° of the air should be coldest, because the sun and stars are either hot by direct beams or by reflexion. The direct beams heat the upper region, the reflected beams from the earth and seas heat the lower region. That which is in the midst, being furthest distant in place from these two regions of heat, are most distant in nature, that is, coldest; which is that they term cold or hot *per antiperistasin*,° that is, environing° by contraries: which was pleasantly taken hold of° by him that said that an honest man in these days must needs be more honest than in ages heretofore, *propter antiperistasin*, because the shutting of him in the midst of contraries must needs make the honesty stronger and more compact in itself.

The reprehension of this colour is, first, many things of amplitude° in their kind° do as it were engross° to themselves all, and leave that which is next° them most destitute: as the shoots or underwood that grow near a great and spread° tree is the most pined° and shrubby wood of the field, because the great tree doth deprive and deceive them of sap and nourishment. So he saith well, 'divitis servi maxime servi',° and the comparison was pleasant of him that compared courtiers attendant in the courts of princes, without great place or office, to fasting-days, which were next° the holy-days, but otherwise were the leanest days in all the week.

Another reprehension is, that things of greatness and predominancy, though they do not extenuate° the things adjoining in substance, yet they drown them and obscure them in show and appearance. And therefore the astronomers° say, that whereas in all other planets conjunction° is the perfectest amity; the sun contrariwise is good by aspect,° but evil by conjunction.

A third reprehension is, because evil approacheth to good sometimes for concealment, sometimes for protection; and good to evil for conversion and reformation. So hypocrisy draweth near to religion for covert and hiding itself; 'saepe latet vitium proximitate boni,'° and sanctuary-men,° which were commonly inordinate° men and malefactors, were wont to be nearest to priests and prelates, and holy men; for the majesty of good things is such, as the confines° of them are revered. On the other side, our Saviour, charged with nearness of° publicans and rioters,° said, 'The physician approacheth the sick rather than the whole.'°

VIII

Quod quis culpa sua contraxit, majus malum, quod ab externis imponitur, minus malum.'°

The reason is, because the sting and remorse of the mind accusing itself doubleth all adversity: contrariwise, the considering and recording inwardly that a man is clear and free from fault and just imputation° doth attemper outward calamities. For if the evil be in the sense and in the conscience both, there is a gemination° of it; but if evil be in the one and comfort in the other, it is a kind of compensation. So the poets in tragedies do make the most passionate lamentations, and those that fore-run final despair, to be accusing, questioning, and torturing of a man's self.

Seque unum clamat causamque caputque malorum.°

And contrariwise, the extremities° of worthy persons have been annihilated in the consideration of their own good deserving. Besides, when the evil cometh from without,° there is left a kind of evaporation° of grief, if it come by human injury, either by indignation and meditating of revenge from ourselves, or by expecting or fore-conceiving that Nemesis° and retribution will take hold of the authors of our hurt; or if it be by fortune or accident, yet there is left a kind of expostulation° against the divine powers;

Atque Deos atque astra vocat crudelia mater.°

But where the evil is derived from a man's own fault, there all strikes deadly inwards and suffocateth.°

The reprehension of this colour is first in respect of hope; for reformation of our faults is *in nostra potestate*,° but amendment of our fortune simply is not. Therefore Demosthenes in many of his orations saith thus to the people of Athens: 'That which having regard to the time past is the worst point and circumstance of all the rest, that as to the time to come is the best. What is that? Even this, that by your sloth, irresolution, and misgovernment, your affairs are grown to this declination° and decay. For had you used and ordered your means and forces to the best, and done your parts every way to the full, and notwithstanding your matters should have gone backward in this manner as they do, there had been no hope left of recovery or reparation; but since it hath been only by your own errors,"° &c. So Epictetus in his degrees° saith, 'The worst state of man is to accuse extern° things; better than that to accuse a man's self; and best of all to accuse neither."°

Another reprehension of this colour is in respect of the well bearing of evils wherewith a man can charge nobody but himself, which maketh them the less.

Leve fit quod bene fertur onus.°

And therefore many natures that are either extremely proud, and will take no fault to themselves, or else very true and cleaving to themselves,° (when they see the blame of anything that falls out ill° must light upon themselves,) have no other shift° but to bear it out well, and to make the least of it; for as we see when sometimes a fault is committed, and before it be known who is to blame, much ado° is made of it, but after, if it appear to be done by a son or by a wife or by a near friend, then it is light made of; so much more when a man must take it upon himself. And therefore it is commonly seen, that women that marry husbands of their own choosing against their friends' consents,° if they be never so ill used,° yet you shall seldom see them complain, but to set a good face on it.

Advice to Fulke Greville on his studies

Cousin Fulke, you tell me you are going to Cambridge, and that the ends of your going are, to get° a scholar to your liking to live with you, and some two or three others to remain in the University, and gather° for you; and you require my opinion, what instruction you shall give these gatherers. To which I yield, more out of affection for your satisfaction to do what I can, than out of confidence that I can do anything; and though you get nothing else by this idle discourse, yet you shall learn this, that if you will have your friend perform what you require, you must require nothing above his strength.

He that shall out of his own reading gather for the use of another, must (as I think) do it by epitome, or abridgment, or under heads and common places.° Epitomes° also may be of two sorts: of any one art, or part of knowledge out of many books; or of one book by itself. Of the first kind we have many patterns: as for civil law, Justinian;° Littleton° for our own; Ramus, logic;° Valerius, physics;° Lipsius, politics,° and Machiavel's Art of War.° Some in every kind, and divers in some one. In matter of history° I will not cite Carion,° Functius, Melanchthon, nor the new French Bibliotheque Historien;° because they are rather calendars° to direct a man to histories than abridgments of history. But the reading of the best of these (and these to be the best we have) will no more make a man a good civilian,° common lawyer, logician, naturalist,° politician, nor soldier, than the seeing of the names of London, Bristol, York, Lincoln, and some few other places of note in a Mercator's general Map° will make a stranger understand the cosmography of England.

And if the works of so excellent men be so fruitless, what shall their abridgments be? who can give us no great proportion of knowledge if they gave us all they understood themselves. I do not deny, but he that hath such abridgments of all arts shall have a general notion of all kinds of knowledge. But he shall be like a man of many trades, that thrives less than he that seriously follows one. For it is Seneca's rule 'Multum non multa'.°

It may be objected that knowledge is so infinite, and the writers of every sort of it so tedious,° as it is reason° to allow a man all helps to go the shortest and nearest way. But they that only study abridgments, like men that would° visit all places, pass through every place in such post° as they have no time to observe as they go, or make profit of their travel.

The epitome of any special book is but a short narration of that which the book itself doth discourse at large: where commonly in matter of art° the positions° are set down without their proofs; and in matter of history the things done without the councils and circumstances, which indeed are of a thousand times more use than the examples themselves. Such abridgments may make us know the places where great battles have been fought, and the names of the conquerors and conquered, and will minister arguments° of discourse, but cannot breed soundness of judgement, which is the true use of all learning.

As for example: let him that never read Livy tell me what he is the wiser for Florus's epitome;° or he that never studied the Mathematics, what he would learn of a title of Euclid's definitions and divisions, or (if you will) axioms also, without seeing the demonstrations which must lead himself to conceive° them. So as I think epitomes, of the one or other kind, of themselves of little profit.

It may be thought the slowness of mine own conceit° in this point corrupts my judgement. But surely I do not measure all men by myself: for the wants I have shall make me more honour great gifts in them.

I confess, excellent wits will make use of every little thing. But yet against all slothful students I learn this rule out of Livy, 'Nusquam nec opera sine emolumento, nec emolumentum ferme sine impensa opera est. Labor voluptasque, dissimillima natura, societate quadam inter se naturali sunt conjuncta.'°

I hold collections under heads and common places of far more profit and use; because they have in them a kind of observation, without the which neither long life breeds experience, nor great reading great knowledge: for 'id demum scimus, cujus causam scimus'.°

As for example, he that out of Curtius° or Plutarch° will make an epitome of the life of Alexander, considers but the numbers of years he lived, the names of places he conquered, the humours° and affections he had, and the variety of accidents° he met withal in the course of his life. But he that will draw notes out of his life under heads, will shew under the title° of conqueror, that to begin in the strength and flower of his age, to have a way made to greatness by his father, to find an army disciplined and a council of great captains, and to procure himself to be made head of a league against a common enemy (whereby both his quarrel may be popular° and his assistance great) are necessary helps to great conquests.

Under the title of war, that the invader hath ever the advantage of the invaded: for he comes resolved, strikes terror, hath all if he win, and if he lose loseth but his hopes; that [it] is not the number of soldiers so much as the goodness of them, and the conduct of the leaders, that is cause of victory; and that before any man make foreign wars, he must, according to Alexander's example, be sure to settle all near home.

Under the title of periods or revolutions of state: that the uncertainty of succession, the equal greatness of divers grandees,° and the overmuch (if I may so speak) unwieldiness of a state, are sufficient causes to ruin the greatest monarchy.

The last may seem somewhat strange: but as I believe it to be a true cause of the dissipation or loss of the Grecian monarchy, so sure it was of the Roman, which made Livy in his time, foreseeing the Fall of it, say of that commonwealth; 'Quae ab exiguis profecta initiis eo creverit ut iam magnitudine laboret sua';° and in another place, 'Tempora, quibus praevalentis populi vires se ipse conficiunt'.°

I name these few heads only to shew the difference between these kinds of collections, which I think is soon discerned.

But though I prefer this second kind of collections before the first, yet neither is that anything if these three circumstances be not observed: choice to be made of the heads under which all things are reduced;° of the books out of which they are to be taken; and above all things, of the notes themselves that shall be set down. For the first, you will say your abridgers shall follow patterns; for the second, that they may read the books that are in reputation in the university; for the third, you will trust their judgement.

But I answer: the patterns they shall find either are made by young students like themselves, or by common book-makers° that follow an alphabet and fill the Index with many idle heads,° enough to make him that shall follow their pattern to fill his paper-book° as full of idle marks. Therefore special advice I would wish him to take that should make such heads, and to make far fewer than you shall find in any pattern.

The judgement of the university I do infinitely reverence: but the general reputation of books (I think) will little direct them. My reasons are, that all or most of grounded judgement do only follow one of three professions, divinity, law or physic,° and are strangers to the books your abridgers should read, because they despise them; and other young students are more satisfied with a flowering easy style than with excellent matter in harsh words.°

Of the choice (because you mean° the study of humanity)° I think history of most, and I had almost said of only use. For poets, I can commend none, being resolved to be ever a stranger to them. Of orators, if I must choose you any, it shall be Demosthenes, both for the argument he handles, and for that° his eloquence is more proper to a statesman than Cicero's.° Of all histories I think Tacitus° simply the best; Livy is very good; Thucydides above any of the writers of Greek matters;° and the worst of these, and divers others of the ancients, to be preferred before the best of our moderns.

The third and the hardest point is the choice of the notes themselves, which must be natural, moral,° politic or military. Of the two first your gatherers may have good judgement, but you shall have little use; of the two latter your use is greatest and their judgement least. I doubt not but in the university you shall find choice of many excellent wits, and in things wherein they have waded,° many of good understanding. But they that have the best eyes° are not always the best lapidaries;° and according to the proverb,° the greatest clerks are not always the wisest men. A mere scholar in state or military matters will no more satisfy you than Phormio did Hannibal.° Therefore, to speak plainly of the gathering of heads or common places I think, first, that in general one man's notes will little profit another, because one man's conceit° doth so much differ from another's; and also because the bare note itself is nothing so much worth as the suggestion° it gives the reader. Next, I think no profit is gotten by his notes that is not judicious° in that whereof he makes his notes.

But, you will say, I exceed my commission: for instead of advice I do dehort.° I do confess, I would have you gather the chiefest things, and out of the chiefest books yourself; and to use your other collectors in gathering arguments and examples to prove or illustrate any particular position or question. For they should, like labourers, bring stone, timber, mortar and other necessaries to your building. But you should put them together, and be the master-workman yourself; and instruction is easier given, and will be better followed, in one point than in many.

As I began so must I end, assuring you I have no end° but your satisfaction; no, not of thanks from you! for you cannot be so much pleased with seeing proof of the credit you have with me as you will be distasted° with the insufficiency of that which you have sought to be satisfied in.

Make° you this private, to conceal my weakness, and I will by many better trials than this, publish to the world that affection with which it is undertaken.

A Confession of Faith°

I BELIEVE that nothing is without beginning but God; no nature, no matter, no spirit,° but one only and the same God. That God as he is eternally almighty, only wise, only good, in his nature, so he is eternally Father, Son, and Spirit, in persons.°

I believe that God is so holy, pure, and jealous,° as it is impossible for him to be pleased in any creature, though the work of his own hands; So that neither Angel, Man, nor World, could stand, or can stand, one moment in his eyes, without beholding the same in the face of a Mediator;° And therefore that before him with whom all things are present, the Lamb of God was slain before all worlds;° without which eternal counsel° of his, it was impossible for him to have descended° to any work of creation; but he should have enjoyed the blessed and individual society of three persons in Godhead only° for ever.

But that out of his eternal and infinite goodness and love purposing to become a Creator, and to communicate° with his creatures, he ordained in his eternal counsel, that one person of the Godhead° should in time be united to one nature and to one particular° of his creatures: that so in the person of the Mediator the true ladder might be fixed, whereby God might descend to his creatures, and his creatures might ascend to God:° so that God, by the reconcilement of the Mediator, turning his countenance towards his creatures (though not in the same light° and degree), made way unto the dispensation of his most holy and secret will; whereby some of his creatures might stand and keep their state, others might possibly fall and be restored, and others might fall, and not be restored in their state,° but yet remain in being, though under wrath and corruption:° all in the virtue of° the Mediator; which is the great mystery and perfite° centre of all God's ways° with his creatures, and unto which all his other works and wonders do but serve and refer.

That he chose (according to his good pleasure) Man to be that creature, to whose nature the person of the eternal Son of God should be united;° and amongst the generations of men, elected a small flock,° in whom (by the participation of himself) he purposed to express° the riches of his glory;° all the ministration of angels,° damnation of devils and reprobate,° and universal administration° of all creatures, and dispensation of all times,° having no other end, but as the ways and ambages° of God to be further glorified in his Saints, who are one° with the Mediator, who is one with God.

That by the virtue of this his eternal counsel touching a Mediator, he descended at his own good pleasure, and according to the times and seasons to himself known,° to become a Creator; and by his eternal Word° created all things, and by his eternal Spirit doth comfort and preserve them.

That he made all things in their first estate° good, and removed from himself the beginning of all evil and vanity into the liberty of the creature;° but reserved in himself the beginning of all restitution° to the liberty of his grace; using nevertheless and turning the falling and defection of the creature, (which to his prescience° was eternally known) to make way to his eternal counsel touching a Mediator, and the work he purposed to accomplish in him.

That God created Spirits,° whereof some kept their standing, and others fell. He created heaven and earth, and all their armies° and generations,° and gave unto them constant and everlasting laws, which we call Nature,° which is nothing but the laws of the creation; which laws nevertheless have had three changes or times,° and are to have a fourth and last. The first, when the matter of heaven and earth was created without forms:° the second, the *interim*° of every day's work: the third, by the curse,° which notwithstanding was no new creation, but a privation° of part of the virtue° of the first creation: and the last, at the end of the world,° the manner whereof is not yet revealed. So as the laws of Nature, which now remain and govern inviolably till the end of the world, began to be in force when God first rested from his works and ceased to create; but received a revocation° in part by the curse, since which time they change not.

That notwithstanding God hath rested and ceased from creating since the first Sabbath,° yet nevertheless he doth accomplish and fulfil his divine will in all things great and small, singular and general, as fully and exactly by providence,° as he could by miracle and new creation, though his working be not immediate and direct, but by compass;° not violating Nature, which is his own law upon° the creature.

That at the first the soul of Man was not produced° by heaven or earth, but was breathed° immediately from God; so that the ways and proceedings of God with spirits° are not included in Nature, that is, in the laws of heaven and earth; but are reserved to the law of his secret will and grace: wherein God worketh still, and resteth not from the work of redemption, as he resteth from the work of creation: but continueth working till the end of the world; what time that work also shall be accomplished, and an eternal sabbath shall ensue. Likewise that whensoever God doth break the law of Nature by miracles (which are ever new creations),° he never cometh to that point or pass,° but° in regard of the work of redemption, which is the greater, and whereto all God's signs and miracles do refer.

That God created Man in his own image,° in a reasonable soul, in innocency, in free-will, and in sovereignty:° That he gave him a law and commandment, which was in his power to keep, but he kept it not: That man made a total defection° from God, presuming to imagine that the commandments and prohibitions of God were not the rules of Good and Evil, but that Good and Evil had their own principles and beginnings;° and lusted after the knowledge of those imagined° beginnings, to the end to depend no more upon God's will revealed, but upon himself and his own light,° as a god; than the which there could not be a sin more opposite to the whole law of God: That yet nevertheless this great sin was not originally moved° by the malice of man, but was insinuated by the suggestion° and instigation of the devil, who was the first defected° creature, and fell of° malice and not by temptation.

That upon the fall of Man, death and vanity° entered by the justice of God, and the image of God in man was defaced,° and heaven and earth which were made for man's use were subdued to corruption° by his fall; but then that instantly and without intermission of time, after the word of God's law became through the fall of man frustrate° as to obedience, there succeeded the greater word of the promise,° that the righteousness of God might be wrought° by faith.

That as well the law of God as the word of his promise endure the same for ever: but that they have been revealed in several° manners, according to the dispensation of times. For the law was first imprinted in that remnant of light of nature,° which was left after the fall, being sufficient to accuse:° then it was more manifestly expressed in the written law;° and was yet more opened° by the prophets; and lastly expounded in the true perfection by the Son of God, the great prophet and perfect interpreter of the law.° That likewise the word of the promise was manifested and revealed, first by immediate revelation and inspiration; after by figures,° which were of two natures: the one, the rites and ceremonies of the law;° the other, the continual° history of the old world, and Church of the Jews,° which though it be literally true, yet is it pregnant° of a perpetual allegory and shadow° of the work of the Redemption to follow. The same promise or evangile° was more clearly revealed and declared by the prophets, and then by the Son himself, and lastly by the Holy Ghost, which illuminateth the Church to the end of the world.

That in the fullness of time, according to the promise and oath of God, of a chosen lineage descended the blessed seed of the woman, Jesus Christ, the only begotten Son of God and Saviour of the world; who was conceived by the power and overshadowing° of the Holy Ghost, and took flesh of the Virgin Mary: that the Word did not only take flesh, or was joined to flesh, but was made flesh,° though without confusion of substance or nature:° so as the eternal Son of God and the ever blessed Son of Mary was one person; so° one, as the blessed Virgin may be truly and catholicly° called *Deipara*,° the Mother of God; so one, as there is no unity in universal nature, not that of the soul and body of man, so perfect; for the three heavenly unities (whereof that is the second) exceed all natural unities: that is to say, the unity of the three persons in Godhead; the unity of God and Man in Christ; and the unity of Christ and the Church: the Holy Ghost being the worker° of both these latter unities; for by the Holy Ghost was Christ incarnate and quickened in flesh, and by the Holy Ghost is man regenerate and quickened° in spirit.

That Jesus the Lord became in the flesh a sacrificer and sacrifice° for sin; a satisfaction° and price° to the justice of God; a meriter of glory and the kingdom; a pattern of all righteousness; a preacher of the word which himself was; a finisher of the ceremony;° a corner-stone° to remove the separation between Jew and Gentile; an intercessor for the Church; a Lord of Nature° in his miracles; a conqueror of death and the power of darkness in his resurrection; and that he fulfilled the whole counsel of God, performed his whole sacred offices and anointing on earth, accomplished the whole work of the redemption and restitution of man to a state superior to the Angels,° whereas the state of his creation was inferior; and reconciled or established all things according to the eternal will of the Father.

That in time, Jesus the Lord was born in the days of Herod, and suffered under the government of Pontius Pilate, being deputy of the Romans,° and under the high priesthood of Caiaphas, and was betrayed by Judas, one of the twelve apostles, and was crucified at Jerusalem, and after a true and natural death, and his body laid in the sepulchre, the third day he raised himself° from the bonds of death, and arose and shewed himself to many chosen witnesses, by the space° of divers° days; and at the end of those days, in the sight of many, ascended into heaven; where he continueth his intercession; and shall from thence at the day appointed come in greatest glory to judge the world.

That the sufferings and merits° of Christ, as they are sufficient to do away° the sins of the whole world, so they are only effectual° to those that are regenerate° by the Holy Ghost; who breatheth where he will of free grace;° which grace, as a seed incorruptible, quickeneth° the spirit of man, and conceiveth° him anew the son of God and the member° of Christ: so that Christ having man's flesh, and man having Christ's spirit, there is an open passage and mutual imputation;° whereby sin and wrath is conveyed to Christ from man, and merit and life is conveyed to man from Christ: which seed of the Holy Ghost first figureth° in us the image of Christ slain or crucified, in a lively° faith; and then reneweth in us the image of God in holiness and charity; though both imperfectly, and in degrees far differing even in God's elect,° as well in regard of° the fire of the Spirit, as of the illumination, which is more or less° in a large proportion: as namely, in the Church before Christ; which yet nevertheless was partaker of one and the same salvation and one and the same means of salvation with us.

That the work of the Spirit, though it be not tied to any means in heaven or earth, yet it is ordinarily dispensed by the preaching of the word, the administration of the sacraments,° the covenant of the fathers upon the children,° prayer, reading,° the censures° of the Church, the society of the godly,° the cross and afflictions, God's benefits, his judgments upon others, miracles, the contemplation of his creatures, all which (though some be more principal) God useth as the means of vocation° and conversion of his elect; not derogating° from his power to call immediately by his grace, and at all hours and moments of the day (that is, of man's life), according to his good pleasure.

That the word of God, whereby his will is revealed, continued in revelation and tradition° until Moses; and that the Scriptures were° from Moses' times to the times of the Apostles and Evangelists; in whose age, after the coming of the Holy Ghost, the teacher of all truth, the book of the Scriptures is shut and closed,° as to° receive any new addition; and that the Church hath no power over the Scriptures to teach or command anything contrary to the written word, but is as the Ark, wherein the tables of the first testament were kept and preserved:° that is to say, the Church hath only the custody and delivery over of the Scriptures committed unto the same; together with the interpretation of them.

That there is an universal or catholic° Church of God, dispersed over the face of the earth; which is Christ's spouse,° and Christ's body; being gathered° of the fathers of the old world, of the Church of the Jews, of the spirits of the faithful dissolved,° of the spirits of the faithful militant,° and of the names yet to be born, which are already written in the book of life. That there is also a visible Church, distinguished by the outward works of God's covenant, and the receiving of the holy doctrine, with the use of the mysteries° of God, and the invocation and sanctification of his holy name. That there is also a holy succession° in the prophets of the new testament and fathers of the Church, from the time of the apostles and disciples which saw our Saviour in the flesh, unto the consummation of the work of the ministry; which persons are called from° God by gift,° or inward anointing,° and the vocation of God followed by an outward calling and ordination of the Church.

I believe that the souls of those that die in the Lord are blessed, and rest from their labours,° and enjoy the sight of God, yet so as they are in expectation of a further revelation of their glory in the last day; at which time all flesh of man shall arise and be changed,° and shall appear and receive from Jesus Christ his eternal judgment; and the glory of the saints shall then be full, and the kingdom shall be given up to God the Father, from which time all things shall continue for ever in that being and state which they shall receive; so as there are three times (if times they may be called) or parts of eternity: The first, the time before beginnings, when the Godhead was only,° without the being of any creature: The second, the time of the mystery,° which continueth from the time of creation to the dissolution of the world: And the third, the time of the revelation of the sons of God;° which time is the last, and is everlasting without change.

Two Prayers

The first Prayer, called by his Lordship

THE STUDENT'S PRAYER

TO God the Father, God the Word, God the Spirit, we pour forth most humble and hearty supplications; that He, remembering the calamities of mankind and the pilgrimage° of this our life, in which we wear out days few and evil,° would please to open to us new refreshments out of the fountains of his goodness, for the alleviating of our miseries. This also we humbly and earnestly beg, that Human things may not prejudice° such as are Divine; neither that from the unlocking of the gates of sense,° and the kindling of a greater natural light, anything of incredulity or intellectual night° may arise in our minds towards the Divine Mysteries. But rather that by our mind throughly cleansed and purged from fancy and vanities, and yet subject and perfectly given up to the Divine Oracles, there may be given unto Faith the things that are Faith's.° Amen.

The second Prayer, called by his Lordship

THE WRITER'S PRAYER

THOU, O Father! who gavest the Visible Light as the first-born of thy Creatures,° and didst pour into man the Intellectual Light° as the top and consummation of thy workmanship, be pleased to protect and govern this work, which coming from thy Goodness returneth to thy Glory. Thou, after thou hadst reviewed the works which thy hands had made, beheldest that 'everything was very good';° and thou didst rest with complacency° in them. But Man, reflecting on the works which he had made, saw that 'all was vanity and vexation of Spirit',° and could by no means acquiesce° in them. Wherefore if we labour in thy works with the sweat of our brows,° thou wilt make us partakers of thy Vision and thy Sabbath.° We humbly beg that this mind may be steadfastly in us, and that thou, by our hands and also by the hands of others on whom thou shalt bestow the same spirit, wilt please to convey a largeness° of new alms to thy family of Mankind.° These things we commend to thy everlasting love, by our Jesus, thy Christ, God with us. Amen.

A Letter and Discourse to Sir Henry Savile,° touching helps for the intellectual powers

Mr. Savile.

Coming back from your invitation at Eton, where I had refreshed myself with company which I loved, I fell into a consideration of that part of policy,° whereof philosophy speaketh too much and laws° too little; and that is of Education of youth. Whereupon fixing my mind a while, I found straightways and noted, even in the discourses of philosophers, which are so large° in this argument, a strange silence concerning one principal part of that subject. For as touching the framing and seasoning of youth to moral virtues, tolerance of labours, continency from pleasures, obedience, honour, and the like, they handle it; but touching the improvement and helping of the intellectual powers, as of conceit,° memory, and judgment, they say nothing. Whether it were that they thought it to be a matter wherein nature only prevailed; or that they intended° it as referred to the several and proper arts which teach the use of reason and speech. But for the former of these two reasons, howsoever it pleaseth them to distinguish of habits and powers, the experience is manifest enough that the motions and faculties of the wit and memory may be not only governed and guided, but also confirmed° and enlarged, by custom and exercise duly applied: as if a man exercise shooting, he shall not only shoot nearer the mark° but also draw a stronger bow.° And as for the latter, of comprehending these precepts within the arts of logic and rhetoric, if it be rightly considered, their office° is distinct altogether from this point. For it is no part of the doctrine of the use or handling of an instrument to teach how to whet or grind the instrument to give it a sharp edge, or how to quench it or otherwise, whereby to give it a stronger temper.° Wherefore finding this part of knowledge not broken,° I have but 'tanquam aliud agens'° entered into it, and salute you with it, dedicating it after the ancient manner, first as to a dear friend, and then as to an apt° person, for as much as you have both place to practise it, and judgment and leisure to look deeper into it than I have done. Herein you must call to mind ’Αριστον μὲν ὕδωρ.° Though the argument be not of great height and dignity, nevertheless it is of great and universal use. And yet I do not see why (to consider it rightly) that should not be a learning of height, which teacheth to raise the highest and worthiest part of the mind. But howsoever that be, if the world take any light and use by this writing, I will that the gratulation° be, to the good friendship and acquaintance between us two. And so I commend you to God's divine protection.

A Discourse Touching Helps for the Intellectual Powers

I did ever hold it for an insolent and unlucky° saying, 'Faber quisque suae fortunae',° except it be uttered only as a hortative° or spur to correct sloth. For otherwise, if it be believed as it soundeth, and that a man entereth into a high° imagination that he can compass and fathom° all accidents,° and ascribeth all successes to his drifts and reaches° and the contrary to his errors and sleepings, it is commonly seen that the evening fortune of that man is not so prosperous, as of° him that without slackening of his industry attributeth much to felicity° and providence° above him. But if the sentence were turned to this, 'Faber quisque ingenii sui',° it were somewhat more true and much more profitable; because it would teach men to bend° themselves to reform those imperfections in themselves, which now they seek but to cover;° and to attain those virtues and good parts, which now they seek but to have only in shew and demonstration.° Yet notwithstanding° every man attempteth to be of the first trade of carpenters, and few bind° themselves to the second: whereas nevertheless the rising in fortune seldom amendeth° the mind; but on the other side the removing of the stonds° and impediments of the mind doth often clear° the passage and current of a man's fortune. But certain it is, whether it be believed or no, that as the most excellent of metals, gold, is of all other the most pliant and most enduring to be wrought; so of all living and breathing substances, the perfectest° (Man) is the most susceptible of help, improvement, impression,° and alteration. And not only in his body, but in his mind and spirit. And there again not only in his appetite and affection, but in his power of wit and reason.

For as to the body of man, we find many and strange experiences how nature is overwrought by custom, even in actions that seem of most difficulty and least possible. As first in Voluntary Motion; which though it be termed voluntary, yet the highest degrees of it are not voluntary; for it is in my power and will to run; but to run faster than according to my lightness or disposition of body, is not in my power nor will. We see the industry and practice of tumblers° and funambulos,° what effects of great wonder it bringeth the body of man unto. So for suffering of pain and dolour, which is thought so contrary to the nature of man, there is much example of penances° in strict orders of superstition,° what they do endure; such as may well verify the report° of the Spartan boys, which were wont to be scourged° upon the altar so bitterly as sometimes they died of it, and yet were never heard complain. And to pass to those faculties which are reckoned to be more involuntary, as long fasting and abstinence, and the contrary extreme (voracity); the leaving and forbearing the use of drink for altogether; the enduring vehement cold; and the like; there have not wanted, neither do want, divers examples of strange victories over the body in every of these. Nay in respiration, the proof hath been of some, who by continual use of diving and working under the water have brought themselves to be able to hold their breath an incredible time. And others that have been able without suffocation to endure the stifling breath of an oven or furnace so heated, as, though it did not scald nor burn, yet it was many degrees too hot for any man, not made° to it, to breathe or take in. And some impostors and counterfeits likewise have been able to wreath and cast° their bodies into strange forms and motions: yea and others to bring themselves into trances and astonishments. All which examples do demonstrate how variously, and to how high points and degrees, the body of man may be (as it were) moulded and wrought. And if any man conceive that it is some secret propriety° of nature that hath been in those persons which have attained to these points, and that it is not open for every man to do the like, though he had been put to° it; for which cause such things come but very rarely to pass; it is true, no doubt but some persons are apter than other; but so as the more aptness causes perfection, but the less aptness doth not disable; so that for example, the more apt child that is taken to be made a 'funambulo', will prove more excellent in his feats; but the less apt will be 'gregarius funambulo'° also. And there is small question but that these abilities would have been more common, and others of like sort not attempted would likewise have been brought upon the stage, but for two reasons. The one because of men's diffidence° in prejudging° them as impossibilities; for it holdeth° in those things, which the poet saith, 'Possunt quia posse videntur';° for no man shall know how much may be done, except he believe much may be done. The other reason is, because they be but practices base and inglorious, and of no great use; and therefore sequestered° from reward of value; and on the other side, painful;° so as the recompence balanceth not with the travel° and suffering.

And as to the will of man, it is that which is most maniable° and obedient; as that which admitteth most medicines to cure and alter it. The most sovereign of all is Religion, which is able to change and transform it in the deepest and most inward inclinations and motions. And next to that is opinion° and apprehension;° whether it be infused° by tradition° and institution,° or wrought° in by disputation° and persuasion. And the third is example, which transformeth the will of man into the similitude of that which is much obversant° and familiar towards it. And the fourth is, when one affection° is healed and corrected by another; as when cowardice is remedied by shame and dishonour, or sluggishness and backwardness by indignation and emulation; and so of the like. And lastly, when all these means, or any of them, have new framed or formed human will, then doth custom and habit corroborate and confirm° all the rest.

Therefore it is no marvel though° this faculty of the mind of will and election,° which inclineth° affection and appetite, being but the inceptions° and rudiments° of will, may be so well governed and managed, because it admitteth access to so divers remedies to be applied to it and to work upon it. The effects whereof are so many and so known° as require no enumeration; but generally they do issue, as medicines do, into two kinds of cures; whereof the one is a just or true cure, and the other is called palliation.° For either the labour and intention is to reform the affections really and truly, restraining them if they be too violent, and raising them if they be too soft and weak, or else it is to cover them; or if occasion be, to pretend° and represent them: of the former sort whereof the examples are plentiful in the schools of philosophers, and in all other institutions° of moral virtue; and of the other sort the examples are more plentiful in the courts of princes, and in all politic traffic,° where it is ordinary to find not only profound dissimulations and suffocating° the affections that° no note or mark appear of them outwardly, but also lively simulations and affectations, carrying the tokens of passions which are not, as 'risus jussus'° and 'lachrymae coactae',° and the like.

Of Help of the Intellectual Powers

The intellectual powers have fewer means to work upon them than the will or body of man; but the one that prevaileth, that is exercise, worketh more forcibly in them than in the rest.°

The ancient habit of the philosophers; 'Si quis quaerat in utramque partem de omni scibili'.°

The exercise of scholars making verses ex tempore;° 'Stans pede in uno.'°

The exercise of lawyers in memory narrative.°

The exercise of sophists,° and 'Jo. ad oppositum',° with manifest effect.

Artificial memory° greatly holpen° by exercise.

The exercise of buffons,° to draw° all things to conceits ridiculous.

The means that help the understanding and faculties thereof are:—

Not example, as in the will, by conversation;° and here the conceit of imitation, already disgested, with the confutation 'obiter, si videbitur',° of Tully's opinion, advising a man to take some one to imitate.° Similitude of faces analysed.

Arts, Logic, Rhetoric. The Ancients, Aristotle; Plato:° *Thaetetus*, *Gorgias*, *Litigiosus vel Sophista*, query.° *Protagoras*, Aristotle, Schola sua:° *Topics*, *Elenchs*,° *Rhetorics*, *Organon.*° Cicero, Hermogenes.° The Neoterics,° Ramus,° Agricola,° 'Nil sacri',° Lullius,° Typocosmia;° studying Cooper's Dictionary;° Mattheus° Collection of proper words for Metaphors; Agrippa° *de Vanitate*, &c.

Query if not here of imitation?°

Collections preparative. Aristotle's similitude° of a shoemaker's shop, full of shoes of all sorts; Demosthenes, *Exordia Concionum.*° Tully's precept° of Theses of all sorts preparative.

The relying upon exercise, with the difference° of using and tempering° the instrument; and the similitude of prescribing against the laws of nature and of estate.

Five Points

1. That exercises are to be framed to the life;° that is to say, to work° ability in that kind, whereof a man in the course of actions shall have most use.

2. The indirect and oblique exercises which do *per partes*° and *per consequentiam*° inable those faculties, which perhaps direct exercise at first would but distort. And those have chiefly place where the faculty is weak not *per se*° but *per accidens.*° As if want of memory grow through lightness° of wit and want of stayed° attention, then the mathematics° or the law helpeth; because they are things wherein if the mind once roam° it cannot recover.

3. Of the advantages of exercise; as to dance with heavy shoes, to march with heavy armour and carriage; and the contrary advantage (in natures very dull and unapt) of working alacrity by framing° an exercise with some delight and affection;

veluti pueris dant crustula blandi
Doctores, elementa velint ut discere prima.°

4. Of the cautions of exercise; as to beware lest by evil doing, as all beginners do weakly, a man grow and be inveterate° in an ill habit; and so take not the advantage of custom in perfection, but in confirming ill.

Slubbering° on the lute.

5. The marshalling and sequel of sciences and practices: Logic and Rhetoric should be used to be read after Poesy, History, and Philosophy. First exercise to do things well and clean;° after, promptly and readily.

I

The exercises in the universities and schools are of memory and invention; either to speak by heart that which is set down *verbatim*, or to speak *ex tempore*; whereas there is little use in action of either of both: but most things which we utter are neither verbally premeditate, nor merely extemporal. Therefore exercise would° be framed to take a little breathing;° and to consider of heads;° and then to form and fit the speech *ex tempore*. This would be done in two manners, both with writing and tables,° and without: for in most actions it is permitted and passable to use the note; whereunto if a man be not accustomed, it will put him out.

There is no use of a Narrative Memory in academies, viz. with circumstances of times, persons, and places, and with names; and it is one art to discourse,° and another to relate and describe; and herein use and action is most conversant.

Also to sum up and contract is a thing in action of very general use.

The Advancement of Learning

Book One

There were under the Law (excellent King) both daily sacrifices and freewill offerings;° the one proceeding upon° ordinary observance, the other upon a devout cheerfulness. In like manner there belongeth to kings from their servants both tribute of duty and presents of affection. In the former of these I hope I shall not live to be wanting,° according to my most humble duty, and the good pleasure of your Majesty's employments: for the latter, I thought it more respective° to make choice of some oblation° which might rather refer to the propriety° and excellency of your individual person, than to the business of your crown and state.

Wherefore representing your Majesty many times unto my mind, and beholding you not with the inquisitive° eye of presumption° to discover that which the Scripture telleth me is inscrutable,° but with the observant eye of duty and admiration; leaving aside the other parts of your virtue and fortune, I have been touched, yea and possessed with an extreme wonder at those your virtues and faculties which the philosophers call intellectual; the largeness of your capacity,° the faithfulness of your memory, the swiftness of your apprehension, the penetration of your judgment, and the facility and order of your elocution: and I have often thought that of all the persons living that I have known, your Majesty were the best instance to make a man of Plato's opinion, that all knowledge is but remembrance,° and that the mind of man by nature knoweth all things, and hath but her own native° and original notions (which by the strangeness and darkness of this tabernacle° of the body are sequestered)° again revived and restored: such a light of nature I have observed in your Majesty, and such a readiness to take flame and blaze from the least occasion presented, or the least spark of another's knowledge delivered. And as the Scripture saith of the wisest king, that 'his heart was as the sands of the sea';° which though it be one of the largest bodies yet it consisteth of the smallest and finest portions;° so hath God given your Majesty a composition of understanding admirable,° being able to compass and comprehend the greatest matters, and nevertheless to touch and apprehend the least;

whereas it should° seem an impossibility in nature for the same instrument to make itself fit for great and small works. And for your gift of speech, I call to mind what Cornelius Tacitus saith of Augustus Caesar, 'Augusto profluens, et quae principem deceret, eloquentia fuit';° for if we note it well, speech that is uttered with labour and difficulty, or speech that savoureth° of the affectation° of art and precepts, or speech that is framed° after the imitation of some pattern° of eloquence, though never so excellent,—all this has somewhat servile, and holding of the subject.° But your Majesty's manner of speech is indeed prince-like, flowing as from a fountain, and yet streaming and branching itself into nature's order, full of facility and felicity,° imitating none, and inimitable by any. And as in your civil estate° there appeareth to be an emulation and contention of your Majesty's virtue with your fortune;° a virtuous disposition with a fortunate regiment;° a virtuous expectation (when time was)° of your greater fortune, with a prosperous possession thereof in the due time; a virtuous observation of the laws of marriage, with most blessed and happy fruit° of marriage; a virtuous and most Christian desire of peace, with a fortunate inclination in your neighbour princes thereunto: so likewise in these intellectual matters, there seemeth to be no less contention between the excellency of your Majesty's gifts of nature and the universality° and perfection of your learning.

For I am well assured that this which I shall say is no amplification° at all, but a positive and measured truth; which is, that there hath not been since Christ's time any king or temporal monarch which hath been so learned in all literature and erudition, divine and human.° For let a man seriously and diligently revolve° and peruse° the succession of the emperors of Rome, of which Caesar the dictator,° who lived some years before Christ, and Marcus Antoninus° were the best learned; and so descend to the emperors of Graecia,° or of the West, and then to the lines° of France, Spain, England, Scotland, and the rest; and he shall find this judgment is truly made. For it seemeth much° in a king, if by the compendious extractions° of other men's wits and labours he can take hold of any superficial ornaments and shews° of learning, or if he countenance° and prefer° learning and learned men: but to drink indeed of the true fountains of learning, nay to have such a fountain of learning in himself, in a king, and in a king born, is almost a miracle. And the more, because there is met in your Majesty a rare conjunction as well of divine and sacred literature as of profane and human; so as° your Majesty standeth invested of that triplicity° which in great veneration was ascribed to the ancient Hermes;° the power and fortune of a King, the knowledge and illumination of a Priest, and the learning and universality° of a Philosopher. This propriety inherent° and individual attribute in your Majesty deserveth to be expressed not only in the fame and admiration of the present time, nor in the history or tradition of the ages succeeding; but also in some solid work, fixed memorial, and immortal monument, bearing a character or signature both of the power of a king and the difference° and perfection of such a king.

Therefore I did conclude with myself, that I could not make unto your Majesty a better oblation than of some treatise tending to that end; whereof the sum will consist of these two parts: the former concerning the excellency of learning and knowledge, and the excellency of the merit and true glory in the augmentation and propagation thereof; the latter, what the particular acts and works are which have been embraced and undertaken for the advancement of learning, and again what defects and undervalues° I find in such particular acts; to the end that though I cannot positively° or affirmatively° advise your Majesty, or propound unto you framed particulars,° yet I may excite° your princely cogitations° to visit the excellent treasure of your own mind, and thence to extract particulars for this purpose agreeable to your magnanimity and wisdom.

In the entrance to the former of these,—to clear the way, and as it were to make silence to have the true testimonies concerning the dignity of learning to be better heard without the interruption of tacit objections,—I think good to deliver° it from the discredits and disgraces which it hath received; all from ignorance; but ignorance severally° disguised; appearing sometimes in the zeal° and jealousy of divines, sometimes in the severity and arrogancy° of politiques,° and sometimes in the errors and imperfections of learned men themselves.

I hear the former sort say, that knowledge is of° those things which are to be accepted of° with great limitation and caution; that the aspiring to over-much knowledge was the original temptation and sin, whereupon ensued the fall of man; that knowledge hath in it somewhat of the serpent,° and therefore where it entereth into a man it makes him swell,—'Scientia inflat',° that Salomon gives a censure,° that 'there is no end of making books', and that 'much reading is weariness of the flesh';° and again in another place, that 'in spacious knowledge there is much contristation',° and that 'he that increaseth knowledge increaseth anxiety';° that St. Paul gives a caveat, that 'we be not spoiled° through vain philosophy';° that experience demonstrates how learned men have been arch-heretics,° how learned times have been inclined to atheism,° and how the contemplation of second causes° doth derogate from° our dependence upon God, who is the first cause.

To discover° then the ignorance and error of this opinion and the misunderstanding in the grounds° thereof, it may well appear these men do not observe or consider that it was not the pure knowledge of nature and universality,° a knowledge by the light whereof man did give names° unto other creatures in Paradise, as they were brought before him, according unto their proprieties,° which gave the occasion to the fall; but it was the proud knowledge of good and evil,° with an intent° in man to give law unto himself and to depend no more upon God's commandments, which was the form° of the temptation. Neither is it any quantity of knowledge how great soever that can make the mind of man to swell; for nothing can fill, much less extend, the soul of man, but God and the contemplation of God; and therefore Salomon° speaking of the two principal senses of inquisition,° the eye and the ear, affirmeth that the eye is never satisfied with seeing, nor the ear with hearing; and if there be no fullness, then is the continent° greater than the content:° so of knowledge itself and the mind of man, whereto the senses are but reporters, he defineth likewise in these words, placed after that calendar or ephemerides° which he maketh of the diversities of times and seasons for all actions and purposes; and concludeth thus: 'God hath made all things beautiful, or decent,° in the true return of their seasons: Also he hath placed the world in man's heart, yet cannot man find out the work which God worketh from the beginning to the end':° declaring not obscurely that God hath framed the mind of man as a mirror or glass capable of° the image of the universal world, and joyful to receive the impression thereof, as the eye joyeth° to receive light; and not only delighted in beholding the variety of things and vicissitude° of times, but raised° also to find out and discern the ordinances and decrees° which throughout all those changes are infallibly observed. And although he doth insinuate that the supreme or summary° law of nature, which he calleth 'the work which God worketh from the beginning to the end', is not possible to be found out by man; yet that doth not derogate from the capacity of the mind, but may be referred° to the impediments, as of° shortness of life, ill conjunction° of labours, ill tradition° of knowledge over from hand to hand, and many other inconveniences whereunto the condition of man is subject. For that nothing parcel° of the world is denied to man's inquiry and invention he doth in another place° rule over,° when he saith, 'The spirit of man is as the lamp of God, wherewith he searcheth the inwardness of all secrets.'°

If then such be the capacity and receit° of the mind of man, it is manifest that there is no danger at all in the proportion or quantity of knowledge, how large soever, lest it should make it swell or out-compass° itself; no, but it is merely the quality of knowledge, which be it in quantity more or less, if it be taken° without the true corrective° thereof, hath in it some nature° of venom or malignity, and some effects of that venom, which is ventosity° or swelling. This corrective spice, the mixture whereof maketh knowledge so sovereign,° is Charity, which the apostle immediately addeth to the former clause; for so he saith, 'knowledge bloweth up,° but charity buildeth up';° not unlike unto that which he delivereth° in another place: 'If I spake' (saith he) 'with the tongues of men and angels, and had not charity, it were but as a tinkling cymbal';° not but that it is an excellent thing to speak with the tongues of men and angels, but because if it be severed from charity, and not referred° to the good of men and mankind, it hath rather a sounding° and unworthy glory° than a meriting° and substantial virtue.

And as for that censure° of Salomon° concerning the excess of writing and reading books and the anxiety of spirit which redoundeth° from knowledge, and that admonition of St. Paul, that 'we be not seduced by vain philosophy';° let those places° be rightly understood, and they do indeed excellently set forth the true bounds and limitations whereby human knowledge is confined and circumscribed; and yet without any such contracting or coarctation,° but that it may comprehend all the universal nature of things. For these limitations are three. The first, that 'we do not so place our felicity in knowledge, as we forget our mortality.' The second, that 'we make application of our knowledge to give ourselves repose and contentment, and not distaste° or repining.'° The third, that 'we do not presume by the contemplation of nature to attain to the mysteries of God.' For as touching° the first of these, Salomon doth excellently expound° himself in another place° of the same book, where he saith, 'I saw well that knowledge recedeth° as far from ignorance as light doth from darkness, and that the wise man's eyes keep watch in his head, whereas the fool roundeth° about in darkness: but withal I learned that the same mortality involveth them both'.° And for the second, certain it is, there is no vexation or anxiety of mind which resulteth from knowledge otherwise than merely by accident;° for all knowledge and wonder (which is the seed of knowledge)° is an impression of pleasure in itself: but when men fall to framing° conclusions out of their knowledge, applying it to their particular,° and ministering° to themselves thereby weak fears or vast desires, there groweth that carefulness° and trouble of mind which is spoken of: for then knowledge is no more 'Lumen siccum',° whereof Heraclitus the profound said, 'Lumen siccum optima anima',° but it becometh 'Lumen madidum' or 'maceratum',° being steeped and infused in the humours of the affections.°

And as for the third point, it deserveth to be a little stood upon° and not to be lightly passed over: for if any man shall think by view and inquiry into these sensible° and material things to attain that light whereby he may reveal unto himself the nature or will of God, then indeed is he spoiled by vain philosophy: for the contemplation of God's creatures and works produceth (having regard to the works and creatures themselves) knowledge; but having regard to God, no perfect knowledge, but wonder, which is broken knowledge.° And therefore it was most aptly said by one of Plato's school,° that 'the sense of man carrieth a resemblance with the sun, which (as we see) openeth and revealeth all the terrestrial globe;° but then again it obscureth and concealeth the stars and celestial globe:° so doth the sense discover natural things, but it darkeneth and shutteth up divine.' And hence it is true that it hath proceeded° that divers° great° learned men have been heretical, whilst they have sought to fly up to the secrets of the Deity by the waxen wings° of the senses.

And as for the conceit° that too much knowledge should incline a man to atheism, and that the ignorance of second causes should make a more devout dependence upon God which° is the first cause; first, it is good to ask the question which Job asked of his friends, 'Will you lie for God, as one man will do for another, to gratify him?'° For certain it is that God worketh nothing in nature but by° second causes; and if they would have it otherwise believed, it is mere imposture, as it were in favour towards God;° and nothing else but to offer to the author of truth the unclean sacrifice° of a lie. But farther, it is an assured truth and a conclusion of experience, that a little or superficial knowledge of philosophy may incline the mind of man to atheism, but a farther proceeding therein doth bring the mind back again to religion;° for in the entrance° of philosophy, when the second causes, which are next° unto the senses, do offer themselves to the mind of man, if it dwell and stay there, it may induce° some oblivion of the highest cause; but when a man passeth on farther, and seeth the dependence° of causes° and the works of Providence;° then, according to the allegory of the poets, he will easily believe that the highest link of nature's chain° must needs be tied to the foot of Jupiter's chair. To conclude therefore, let no man, upon a weak conceit° of sobriety° or an ill-applied moderation, think or maintain that a man can search too far or be too well studied in the book of God's word or in the book of God's works;° divinity or philosophy; but rather let men endeavour° an endless progress or proficience° in both; only let men beware that they apply both to charity, and not to swelling;° to use, and not to ostentation; and again, that they do not unwisely mingle or confound these learnings together.

And as for the disgraces which learning receiveth from politiques,° they be° of this nature; that learning doth soften° men's minds, and makes them more unapt for the honour and exercise of arms; that it doth mar° and pervert° men's dispositions for matter of government and policy,° in making them too curious° and irresolute° by variety of reading, or too peremptory or positive° by strictness° of rules and axioms, or too immoderate and overweening by reason of the greatness of examples, or too incompatible° and differing from° the times by reason of the dissimilitude of examples;° or at least that it doth divert men's travails° from action and business, and bringeth them to a love of leisure and privateness;° and that it doth bring into states a relaxation of discipline, whilst every man is more ready to argue than to obey and execute. Out of this conceit Cato° surnamed the Censor, one of the wisest men indeed that ever lived, when Carneades the philosopher came in embassage° to Rome, and that the young men of Rome began to flock about him, being allured with the sweetness and majesty of his eloquence and learning, gave counsel in open senate that they should give him his dispatch° with all speed, lest he should infect and inchant the minds and affections of the youth, and at unawares° bring in an alteration of the manners and customs of the state. Out of the same conceit or humour did Virgil, turning his pen to the advantage of his country and the disadvantage of his own profession, make a kind of separation between policy and government and between arts and sciences, in the verses so much renowned, attributing and challenging° the one to the Romans, and leaving and yielding the other to the Grecians;° 'Tu regere imperio populos, Romane, memento, Hae tibi erunt artes', &c.° So likewise we see that Anytus,° the accuser of Socrates, laid it as an article of charge and accusation against him that he did with the variety and power of his discourses and disputations withdraw young men from due reverence to the laws and customs of their country; and that he did profess a dangerous and pernicious science,° which was to make the worse matter seem the better, and to suppress truth by force of eloquence and speech.

But these and the like imputations have rather a countenance of gravity° than any ground of justice: for experience doth warrant° that both in persons and in times there hath been a meeting and concurrence in learning and arms, flourishing and excelling in the same men and the same ages. For as for men, there cannot be a better nor the like instance, as of that pair,° Alexander the Great and Julius Caesar the dictator; whereof the one was Aristotle's scholar° in philosophy, and the other was Cicero's rival in eloquence; or if any man had rather call for scholars that were great generals than generals that were great scholars, let him take Epaminondas° the Theban, or Xenophon° the Athenian; whereof the one was the first that abated° the power of Sparta, and the other was the first that made way to the overthrow of the monarchy of Persia. And this concurrence is yet more visible in times than in persons, by how much° an age is a° greater object than a man. For both in Egypt, Assyria, Persia, Graecia, and Rome, the same times that are most renowned for arms are likewise most admired for learning; so that the greatest authors and philosophers and the greatest captains and governors have lived in the same ages. Neither can it otherwise be: for as in man the ripeness of strength of the body and mind cometh much about an age,° save that the strength of the body cometh somewhat the more early; so in states, arms and learning, whereof the one correspondeth to the body, the other to the soul of man, have a concurrence or near sequence in times.

And for matter of policy° and government, that learning should rather hurt than enable° thereunto, is a thing very improbable. We see it is accounted an error to commit a natural body to empiric physicians,° which commonly have a few pleasing° receits° whereupon they are confident and adventurous,° but know neither the causes of diseases, nor the complexions° of patients, nor peril of accidents,° nor the true method of cures. We see it is a like error to rely upon advocates or lawyers which are only men of practice° and not grounded° in their books, who are many times easily surprised when matter falleth out besides° their experience, to the prejudice of the causes they handle. So by like reason it cannot be but a matter of doubtful consequence,° if states be managed by empiric statesmen, not well mingled with men grounded in learning. But contrariwise,°

it is almost without instance contradictory,° that ever any government was disastrous that was in the hands of learned governors.

For howsoever it hath been ordinary° with politic men° to extenuate and disable° learned men by the names of *Pedantes*;° yet in the records of time it appeareth in many particulars, that the governments of princes in minority° (notwithstanding the infinite disadvantage of that kind of state) have nevertheless excelled the government of princes of mature age, even for that reason which they seek to traduce,° which is, that by that occasion the state hath been in the hands of *Pedantes*. For so was the state of Rome for the first five years, which are so much magnified,° during the minority of Nero, in the hands of Seneca,° a *Pedanti*: so it was again for ten years space or more, during the minority of Gordianus the younger, with great applause and contentation° in the hands of Misitheus,° a *Pedanti*: so was it before that, in the minority of Alexander Severus,° in like happiness, in hands not much unlike, by reason of the rule of the women, who were aided by the teachers and preceptors. Nay let a man look into the government of the bishops of Rome, as by name into the government of Pius Quintus° and Sextus Quintus° in our times, who were both at their entrance esteemed but as pedantical friars, and he shall find that such popes do greater things, and proceed upon truer principles of estate,° than those which have ascended to the papacy from an education and breeding in affairs of estate and courts of princes. For although men bred in learning are perhaps to seek° in points of convenience° and accommodating° for the present, which the Italians call 'ragioni di stato',° whereof the same Pius Quintus could not hear spoken with patience, terming them° inventions against religion and the moral virtues; yet on the other side, to recompense° that, they are perfect in those same plain grounds of religion, justice, honour, and moral virtue; which if they be well and watchfully pursued, there will be seldom use of those other, no more than of physic° in a sound or well-dieted body. Neither can the experience of one man's life furnish examples and precedents for the events of one man's° life: for as it happeneth sometimes that the grandchild or other descendant resembleth the ancestor more than the son; so many times occurrences of present times may sort° better with ancient examples than with those of the later or immediate° times: and lastly, the wit of one man can no more countervail° learning than one man's means can hold way° with a common purse.°

And as for those particular seducements° or indispositions of the mind for policy and government, which learning is pretended° to insinuate;° if it be granted that any such thing be, it must be remembered withal, that learning ministereth° in every° of them greater strength of medicine or remedy, than it offereth cause of indisposition or infirmity. For if by a secret operation it make men perplexed and irresolute, on the other side by plain precept it teacheth them when and upon what ground to resolve;° yea, and how to carry things in suspense without prejudice till they resolve. If it make men positive° and regular,° it teacheth them what things are in their nature demonstrative,° and what are conjectural; and as well the use of distinctions and exceptions, as the latitude of principles and rules. If it mislead by disproportion or dissimilitude of examples, it teacheth men the force of circumstances, the errors of comparisons, and all the cautions of application;° so that in all these it doth rectify° more effectually than it can pervert. And these medicines it conveyeth into men's minds much more forcibly by the quickness and penetration of examples. For let a man look into the errors of Clement° the seventh, so lively° described by Guicciardine, who served under him, or into the errors of Cicero° painted out° by his own pencil in his epistles to Atticus, and he will fly apace° from being irresolute. Let him look into the errors of Phocion,° and he will beware how he be obstinate or inflexible. Let him but read the fable of Ixion,° and it will hold° him from being vaporous or imaginative.° Let him look into the errors of Cato the second,° and he will never be one of the Antipodes,° to tread opposite° to the present world.

And for the conceit that learning should dispose men to leisure and privateness, and make men slothful; it were a strange thing if that which accustometh the mind to a perpetual motion and agitation should induce slothfulness. Whereas contrariwise, it may be truly affirmed that no kind of men love business for itself but those that are learned. For other persons love it for profit, as an hireling° that loves the work for the wages; or for honour, as because it beareth them up° in the eyes of men, and refresheth their reputation which otherwise would wear;° or because it putteth them in mind of their fortune, and giveth them occasion° to pleasure and displeasure; or because it exerciseth some faculty wherein they take pride, and so entertaineth them in good humour and pleasing conceits toward themselves; or because it advanceth any other their ends. So that as it is said of untrue valours° that some men's valours are in the eyes of them that look on, so such men's industries are in the eyes of others, or at least in regard of their own designments;° only learned men love business as an action according to° nature, as agreeable to health of mind as exercise is to health of body, taking pleasure in the action itself, and not in the purchase:° so that of all men they are the most indefatigable, if it be towards any business which can hold or detain their mind.

And if any man be laborious in reading and study and yet idle in business and action, it groweth from some weakness of body or softness° of spirit, such as Seneca speaketh of; 'Quidam tam sunt umbratiles, ut putent in turbido esse quicquid in luce est',° and not of learning. Well may it be that such a point of a man's nature may make him give himself to learning, but it is not learning that breedeth any such point in his nature.

And that learning should take up too much time or leisure;° I answer, the most active or busy man that hath been or can be hath (no question) many vacant times of leisure, while he expecteth° the tides and returns of business, (except he be either tedious° and of no dispatch,° or lightly and unworthily ambitious to meddle in things that may be better done by others;) and then the question is but how those spaces and times of leisure shall be filled and spent; whether in pleasures or in studies; as was well answered by Demosthenes° to his adversary Aeschines, that was a man given to pleasure, and told him that 'his orations did smell of the lamp':° 'Indeed' (said Demosthenes) 'there is a great difference between the things that you and I do by lamp-light.' So as° no man need doubt° that learning will expulse° business; but rather it will keep and defend the possession of the mind against idleness and pleasure, which otherwise at unawares may enter to the prejudice° of both.°

Again, for that other conceit that learning should undermine the reverence of° laws and government, it is assuredly a mere depravation° and calumny without all° shadow of truth. For to say that a blind custom of obedience should be a surer obligation than duty taught and understood, it is to affirm that a blind man may tread° surer by° a guide than a seeing man can by a light. And it is without all controversy that learning doth make the minds of men gentle,° generous, maniable,° and pliant to government; whereas ignorance makes them churlish,° thwart,° and mutinous: and the evidence of time doth clear° this assertion, considering that the most barbarous, rude, and unlearned times have been most subject to tumults, seditions, and changes.

And as to the judgment of Cato° the Censor, he was well punished for his blasphemy° against learning, in the same kind wherein he offended; for when he was past threescore years old, he was taken with an extreme desire to go to school again and to learn the Greek tongue,° to the end° to peruse the Greek authors; which doth well demonstrate, that his former censure of the Grecian learning was rather an affected gravity,° than according to the inward sense of his own opinion. And as for Virgil's verses,° though it pleased him to brave° the world in taking° to the Romans the art of empire, and leaving to others the arts of subjects; yet so much is manifest, that the Romans never ascended to that height of empire till the time they had ascended to the height of other arts. For in the time of the two first Caesars,° which had the art of government in greatest perfection, there lived the best poet,° Virgilius Maro; the best historiographer,° Titus Livius; the best antiquary,° Marcus Varro; and the best, or second° orator, Marcus Cicero, that to the memory of man are known. As for the accusation of Socrates,° the time must be remembered when it was prosecuted; which was under the thirty tyrants, the most base, bloody, and envious° persons that have governed; which revolution° of state° was no sooner over, but Socrates, whom they had made a person criminal, was made a person heroical,° and his memory accumulate° with honours divine and human; and those discourses of his, which were then termed corrupting of manners, were after acknowledged for° sovereign medicines of the mind and manners, and so have been received ever since till this day. Let this therefore serve for answer to politiques, which in their humorous° severity or in their feigned gravity have presumed to throw imputations upon learning; which redargution° nevertheless (save that we know not whether our labours may extend to other ages) were not needful for the present, in regard of the love and reverence towards learning which the example and countenance of two so learned princes, queen Elizabeth and your Majesty, being as Castor and Pollux,° *lucida sidera*,° stars of excellent light and most benign influence,° hath wrought in all men of place and authority in our nation.

Now therefore we come to that third sort of discredit or diminution of credit, that groweth unto° learning from learned men themselves, which commonly cleaveth° fastest.° It is either from their fortune, or from their manners, or from the nature of their studies. For the first, it is not in their power; and the second is accidental;° the third only is proper to be handled.° But because we are not in hand with° true measure, but with popular estimation and conceit, it is not amiss to speak somewhat of the two former. The derogations° therefore which grow to learning from the fortune or condition of learned men, are either in respect of scarcity of means, or in respect of privateness of life and meanness of employments.

Concerning want, and that it is the case of learned men usually to begin with little and not to grow rich so fast as other men, by reason° they convert not their labours chiefly to lucre° and increase; it were good to leave the common place° in commendation of poverty to some friar to handle, to whom much was attributed by Machiavel° in this point, when he said, that 'the kingdom of the clergy had been long before at an end, if the reputation and reverence towards the poverty of friars had not borne out° the scandal of the superfluities and excesses of bishops and prelates.' So a man might say that the felicity and delicacy° of princes and great persons had° long since turned to rudeness° and barbarism, if the poverty of learning had not kept up civility° and honour of life. But without any such advantages, it is worthy the observation what a reverend and honoured thing poverty of fortune was for some ages in the Roman state, which nevertheless was a state without paradoxes.° For we see what Titus Livius saith in his introduction: 'Caeterum aut me amor negotii suscepti fallit, aut nulla unquam respublica nec major, nec sanctior, nec bonis exemplis ditior fuit; nec in quam tam serae avaritia luxuriaque immigraverint; nec ubi tantus ac tam diu paupertati ac parsimoniae honos fuerit.'°

We see likewise, after that the state of Rome was not itself but did degenerate, how that person that took upon him to be counsellor to Julius Caesar° after his victory, where to begin his restoration of the state, maketh it of all points the most summary° to take away the estimation° of wealth: 'Verum haec et omnia mala pariter cum honore pecuniae desinent; si neque magistratus, neque alia vulgo cupienda, venalia erunt'.° To conclude this point, as it was truly said that 'rubor est virtutis color',° though sometime it come from vice; so it may be fitly said that 'paupertas est virtutis fortuna',° though sometime it may proceed from misgovernment and accident. Surely Salomon hath pronounced it, both in censure, 'Qui festinat ad divitias non erit insons',° and in precept, 'Buy the truth, and sell it not; and so of wisdom and knowledge';° judging that means° were to be spent upon learning, and not learning to be applied to means. And as for the privateness or obscureness° (as it may be in vulgar estimation accounted) of life of contemplative° men; it is a theme so common to extol a private life, not taxed with° sensuality and sloth, in comparison and to the disadvantage of a civil life,° for safety, liberty, pleasure, and dignity, or at least freedom from indignity, as no man handleth it but handleth it well; such a consonancy° it hath to men's conceits° in the expressing and to men's consents in the allowing.° This only I will add, that learned men forgotten in states, and not living in the eyes of men, are like the images of Cassius and Brutus° in the funeral of Junia; of which not being represented, as many others were, Tacitus saith, 'Eo ipso praefulgebant, quod non visebantur'.°

And for meanness° of employment, that which is most traduced to contempt° is that the government of youth is commonly allotted to them; which age, because it is the age of least authority, it is transferred° to the disesteeming° of those employments wherein youth is conversant, and which are conversant about youth. But how unjust this traducement is (if you will reduce° things from popularity of opinion° to measure of reason) may appear in that we see men are more curious° what they put into a new vessel than into a vessel seasoned,° and what mould they lay about a young plant than about a plant corroborate;° so as the weakest terms° and times of all things use° to have the best applications° and helps. And will you hearken to the Hebrew Rabbins?° 'Your young men shall see visions, and your old men shall dream dreams';° say they youth is the worthier age, for that visions are nearer apparitions° of God than dreams. And let it be noted, that howsoever the conditions of life of *Pedantes* have been scorned upon theatres, as the ape° of tyranny; and that the modern looseness or negligence hath taken no due regard to the choice of school-masters and tutors;° yet the ancient wisdom of the best times did always make a just complaint that states were too busy with their laws and too negligent in point of education: which excellent part of ancient discipline hath been in some sort revived of late times by the colleges of the Jesuits; of whom, although in regard of their superstition I may say, 'quo meliores, eo deteriores',° yet in regard of this, and some other points concerning human learning and moral matters, I may say, as Agesilaus said to his enemy Pharnabazus, 'talis quum sis, utinam noster esses'.° And thus much touching the discredits drawn from the fortunes of learned men.

As touching the manners of learned men, it is a thing personal and individual: and no doubt there be amongst them, as in other professions, of all temperatures:° but yet so as it is not without truth which is said, that 'abeunt studia in mores',° studies have an influence and operation upon the manners of those that are conversant in them.

But upon an attentive and indifferent° review, I for my part cannot find any disgrace to learning can proceed from the manners of learned men; not° inherent to° them as they are learned; except it be a fault (which was the supposed fault of Demosthenes, Cicero, Cato the second, Seneca, and many more) that because the times they read of are commonly better than the times they live in, and the duties taught better than the duties practised, they contend° sometimes too far to bring things to perfection, and to reduce° the corruption of manners to honesty° of precepts or examples of too great height.° And yet hereof they have caveats° enough in their own walks.° For Solon,° when he was asked whether he had given his citizens the best laws, answered wisely, 'Yea of such as they would receive': and Plato,° finding that his own heart could not agree with the corrupt manners of his country, refused to bear place or office; saying that 'a man's country was to be used as his parents were, that is, with humble persuasions, and not with contestations':° and Caesar's counsellor° put in the same caveat, 'Non ad vetera instituta revocans quae jampridem corruptis moribus ludibrio sunt',° and Cicero noteth this error directly in Cato the second, when he writes to his friend Atticus; 'Cato optime sentit, sed nocet interdum reipublicae; loquitur enim tanquam in republica Platonis, non tanquam in faece Romuli,'° and the same Cicero° doth excuse and expound° the philosophers for going too far and being too exact in their prescripts,° when he saith, 'Isti ipsi praeceptores virtutis et magistri videntur fines officiorum paulo longius quam natura vellet protulisse, ut cum ad ultimum animo contendissemus, ibi tamen, ubi oportet, consisteremus':° and yet himself might have said, 'Monitis sum minor ipse meis';° for it was his own fault, though not in so extreme a degree.

Another fault° likewise much of this kind hath been incident to learned men; which is, that they have esteemed the preservation, good, and honour of their countries or masters before their own fortunes or safeties. For so saith Demosthenes° unto the Athenians: 'If it please you to note it, my counsels unto you are not such whereby I should grow great amongst you, and you become little amongst the Grecians: but they be of that nature, as° they are sometimes not good for me to give, but are always good for you to follow.' And so Seneca, after he had consecrated that *Quinquennium Neronis*° to the eternal glory of learned governors, held on° his honest and loyal course of good and free counsel, after his master grew extremely corrupt in his government. Neither can this point otherwise be; for learning endueth men's minds with a true sense of the frailty of their persons, the casualty° of their fortunes, and the dignity of their soul and vocation;° so that it is impossible for them to esteem that any greatness of their own fortune can be a true or worthy end° of their being and ordainment;° and therefore are desirous to give their account to God, and so likewise to their masters under God (as kings and the states that they serve), in these words; 'Ecce tibi lucrefeci', and not 'Ecce mihi lucrefeci',° whereas the corrupter sort of mere politiques, that have not their thoughts established by learning in the love and apprehension of duty, nor never look abroad into universality,° do refer all things to themselves, and thrust themselves into the centre of the world, as if all lines° should meet in them and their fortunes; never caring in all tempests what becomes of the ship of estates,° so they may save themselves in the cockboat° of their own fortune; whereas men that feel the weight of duty, and know the limits of self-love, use to make good° their places and duties, though with peril. And if they stand in° seditious and violent alterations, it is rather the reverence which many times both adverse parts° do give to honesty, than any versatile° advantage of° their own carriage.° But for this point of tender sense° and fast° obligation of duty, which learning doth endue the mind withal, howsoever fortune may tax° it and many in the depth of their corrupt principles may despise it, yet it will receive an open allowance,° and therefore needs the less disproof or excusation.°

Another fault incident commonly to learned men, which may be more probably° defended than truly denied, is that they fail sometimes in applying° themselves to particular persons; which want of exact application° ariseth from two causes; the one, because the largeness of their mind can hardly confine itself to dwell in the exquisite° observation or examination of the nature and customs of one person: for it is a speech for a lover and not for a wise man, 'Satis magnum alter alteri theatrum sumus'.° Nevertheless I shall yield,° that he that cannot contract the sight of his mind as well as disperse° and dilate it, wanteth° a great faculty. But there is a second cause, which is no inability but a rejection upon choice and judgment. For the honest and just bounds of observation by one person upon another extend no farther but to understand him sufficiently, whereby not to give him offence, or whereby to be able to give him faithful counsel, or whereby to stand upon reasonable guard and caution in respect of a man's self. But to be speculative° into another man, to the end to know how to work him or wind° him or govern him, proceedeth from a heart that is double and cloven,° and not entire and ingenuous;° which as in friendship it is want of integrity, so towards princes or superiors is want of duty. For the custom° of the Levant,° which is, that subjects do forbear to gaze or fix their eyes upon princes, is in the outward ceremony barbarous, but the moral is good: for men ought not by cunning and bent° observations to pierce and penetrate into the hearts of kings, which the Scripture° hath declared to be inscrutable.

There is yet another fault (with which I will conclude this part) which is often noted in learned men, that they do many times fail to observe decency° and discretion° in their behaviour and carriage, and commit errors in small and ordinary points of action; so as the vulgar sort of capacities do make a judgment of them in greater matters by that which they find wanting in them in smaller. But this consequence° doth oft deceive men; for which I do refer them over to that which was said by Themistocles,° arrogantly and uncivilly° being applied to himself out of his own mouth, but being applied to the general state of this question pertinently and justly; when being invited to touch a lute, he said 'he could not fiddle, but he could make a small town a great state.' So no doubt many may be well seen° in the passages° of government and policy, which are to seek° in little and punctual° occasions. I refer them also to that which Plato° said of his master Socrates, whom he compared to the gallypots° of apothecaries, which on the outside had apes and owls and antiques,° but contained within sovereign and precious liquors and confections;° acknowledging that to an external report he was not without superficial levities° and deformities, but was inwardly replenished with excellent virtues and powers. And so much touching the point of manners of learned men.

But in the mean time I have no purpose to give allowance to some conditions and courses° base and unworthy, wherein divers° professors of° learning have wronged themselves and gone too far; such as were those trencher philosophers,° which in the later age of the Roman state were usually in the houses of great persons, being little better than solemn parasites; of which kind, Lucian° maketh a merry description of the philosopher that the great lady took to ride with her in her coach, and would needs have him carry her little dog, which he doing officiously and yet uncomely,° the page scoffed, and said that 'he doubted° the philosopher of a Stoic would turn to be a Cynic.'° But above all the rest, the gross and palpable flattery whereunto many (not unlearned) have abased and abused their wits and pens, turning (as Du Bartas° saith) Hecuba into Helena and Faustina° into Lucretia, hath most diminished the price° and estimation of learning. Neither is the modern dedications of books and writings, as° to patrons, to be commended: for that° books (such as are worthy the name of books) ought to have no patrons but truth and reason; and the ancient custom was to dedicate them only to private and equal friends, or to intitle° the books with their names; or if to kings and great persons, it was to some such as the argument of the book was fit and proper for. But these and the like courses may deserve rather reprehension than defence.

Not that I can tax or condemn the morigeration° or application° of learned men to men in fortune.° For the answer was good that Diogenes° made to one that asked him in mockery, 'How it came to pass that philosophers were the followers of rich men, and not rich men of philosophers?' He answered soberly, and yet sharply, 'Because the one sort knew what they had need of, and the other did not'. And of the like nature was the answer which Aristippus° made, when having a petition to Dionysius and no ear given to him, he fell down at his feet, whereupon Dionysius staid and gave him the hearing and granted it; and afterward some person tender° on the behalf of philosophy, reproved Aristippus that he would offer the profession of philosophy such an indignity, as for a private suit to fall at a tyrant's feet: but he answered, 'It was not his fault, but it was the fault of Dionysius, that had his ears in his feet.' Neither was it accounted weakness, but discretion, in him° that would not dispute his best° with Adrianus Caesar; excusing himself, that 'it was reason° to yield to him that commanded thirty legions.' These and the like applications and stooping to points of necessity and convenience cannot be disallowed;° for though they may have some outward baseness, yet in a judgment truly made they are to be accounted submissions to the occasion and not to the person.

Now I proceed to those errors° and vanities° which have intervened° amongst the studies themselves of the learned; which is that which is principal and proper to the present argument; wherein my purpose is not to make a justification of the errors, but, by a censure° and separation° of the errors, to make a justification of that which is good and sound, and to deliver that from the aspersion° of the other. For we see that it is the manner of men to scandalize° and deprave° that which retaineth the state° and virtue,° by taking advantage upon° that which is corrupt and degenerate: as the Heathens in the primitive church used to blemish° and taint° the Christians with the faults and corruptions of heretics. But nevertheless I have no meaning at this time to make any exact animadversion° of the errors and impediments in matters of learning which are more secret and remote from vulgar opinion; but only to speak unto° such as do fall under, or near unto, a popular° observation.

There be therefore chiefly three vanities in studies, whereby learning hath been most traduced. For those things we do esteem vain,° which are either false or frivolous,° those which either have no truth or no use: and those persons we esteem vain, which are either credulous or curious;° and curiosity is either in matter or words. So that in reason° as well as in experience,° there fall out° to be these three distempers° (as I may term them) of learning; the first, fantastical° learning; the second, contentious° learning; and the last, delicate° learning; vain imaginations, vain altercations,° and vain affectations; and with the last I will begin.

Martin Luther,° conducted (no doubt) by an higher Providence, but in discourse of reason° finding what a province° he had undertaken against the Bishop of Rome and the degenerate traditions of the church, and finding his own solitude,° being no ways° aided by the opinions of his own time, was enforced to awake all antiquity, and to call former times to his succors to make a party° against the present time; so that the ancient authors, both in divinity and in humanity,° which had long time slept in libraries, began generally to be read and revolved.° This by consequence did draw on a necessity of a more exquisite travail° in the languages original wherein those authors did write, for the better understanding of those authors and the better advantage of pressing° and applying their words. And thereof grew again a delight in their manner of style and phrase, and an admiration of that kind of writing;° which was much furthered and precipitated° by the enmity and opposition that the propounders of those (primitive° but seeming new) opinions had against the schoolmen;° who were generally of the contrary part, and whose writings were altogether in a differing style and form; taking liberty to coin and frame new terms of art° to express their own sense and to avoid circuit of speech,° without regard to the pureness,° pleasantness, and (as I may call it) lawfulness of the phrase or word. And again, because the great labour then was with the people (of whom the Pharisees° were wont to say, 'Execrabilis ista turba, quae non novit legem),° for the winning and persuading of them, there grew of necessity in chief price° and request eloquence and variety of discourse, as the fittest and forciblest access into the capacity of the vulgar sort.°

So that these four causes concurring, the admiration of ancient authors, the hate of the schoolmen, the exact study of languages, and the efficacy of preaching, did bring in an affectionate° study of eloquence and copie° of speech, which then began to flourish. This grew speedily to an excess; for men began to hunt more after words than matter;° and more after the choiceness° of the phrase, and the round and clean° composition of the sentence, and the sweet falling° of the clauses, and the varying° and illustration of their works with tropes and figures, than after the weight of matter, worth of subject, soundness of argument, life of invention, or depth of judgment. Then grew the flowing and watery° vein of Osorius,° the Portugal bishop, to be in price. Then did Sturmius° spend such infinite and curious pains upon Cicero the orator and Hermogenes the rhetorician, besides his own books of periods° and imitation° and the like. Then did Carr° of Cambridge, and Ascham,° with their lectures and writings, almost deify Cicero and Demosthenes, and allure all young men that were studious unto that delicate and polished kind of learning. Then did Erasmus° take occasion to make the scoffing echo; 'Decem annos consumpsi in legendo Cicerone',° and the echo answered in Greek, ὄνε, *Asine*. Then grew the learning of the schoolmen to be utterly despised as barbarous. In sum, the whole inclination and bent of those times was rather towards copie than weight.

Here therefore is° the first distemper of learning, when men study words and not matter: whereof though I have represented an example of late times, yet it hath been and will be 'secundum majus et minus'° in all time. And how is it possible but this should have° an operation° to discredit learning, even with vulgar capacities, when they see learned men's works like the first letter of a patent or limned book;° which though it hath large flourishes,° yet it is but a letter? It seems to me that Pygmalion's frenzy° is a good emblem or portraiture of this vanity: for words are but the images of matter; and except they have life of reason and invention, to fall in love with them is all one° as to fall in love with a picture.

But yet notwithstanding it is a thing not hastily to be condemned, to clothe and adorn the obscurity even of philosophy itself with sensible° and plausible° elocution. For hereof we have great examples in Xenophon, Cicero, Seneca, Plutarch, and of Plato also in some degree; and hereof likewise there is great use; for° surely to the severe inquisition of truth, and the deep progress into philosophy, it is some hinderance; because it is too early satisfactory to the mind of man, and quencheth the desire of further search, before we come to a just period;° but then if a man be to° have any use of such knowledge in civil occasions,° of conference,° counsel,° persuasion,° discourse,° or the like; then shall he find it prepared to his hands in those authors which write in that manner. But the excess of this is so justly contemptible, that as Hercules, when he saw the image of Adonis, Venus' minion,° in a temple, said in disdain, 'Nil sacri es',° so there is none of Hercules' followers in learning, that is, the more severe and laborious sort of inquirers into truth, but will despise those delicacies and affectations, as indeed capable of no divineness. And thus much of the first disease or distemper of learning.

The second, which followeth, is in nature worse than the former; for as substance of matter is better than beauty of words, so contrariwise vain matter is worse than vain words: wherein it seemeth the reprehension of St. Paul was not only proper for those times, but prophetical for the times following; and not only respective° to divinity, but extensive° to all knowledge: 'Devita profanas vocum novitates, et oppositiones falsi nominis scientiae'.° For he assigneth two marks and badges° of suspected° and falsified° science; the one, the novelty and strangeness of terms; the other, the strictness of positions,° which of necessity doth induce oppositions, and so questions and altercations. Surely, like as many substances in nature which are solid° do putrefy and corrupt into worms, so it is the property of good and sound knowledge to putrefy and dissolve into a number of subtile,° idle, unwholesome, and (as I may term them) vermiculate° questions, which have indeed a kind of quickness° and life of spirit,° but no soundness of matter or goodness of quality. This kind of degenerate° learning did chiefly reign amongst the schoolmen; who having sharp and strong wits, and abundance of leisure, and small variety of reading; but their wits being shut up in the cells of a few authors (chiefly Aristotle their dictator) as their persons were shut up in the cells of monasteries and colleges; and knowing little history, either of nature or time; did out of no great quantity of matter, and infinite agitation° of wit, spin out unto us those laborious webs of learning which are extant in their books. For the wit and mind of man, if it work upon matter, which is the contemplation of the creatures° of God, worketh according to the stuff, and is limited thereby; but if it work upon itself, as the spider worketh his web,° then it is endless, and brings forth indeed cobwebs° of learning, admirable for the fineness of thread and work, but of no substance or profit.

This same unprofitable subtility or curiosity° is of two sorts; either in the subject itself that they handle, when it is a fruitless speculation or controversy, (whereof there are no small number both in divinity and philosophy), or in the manner or method of handling of a knowledge;° which amongst them was this: upon every particular position or assertion to frame objections, and to those objections, solutions;° which solutions were for the most part not confutations, but distinctions: whereas indeed the strength of all sciences is, as the strength of the old man's faggot,° in the bond. For the harmony of a science, supporting each part the other, is and ought to be the true and brief confutation and suppression of all the smaller sort of objections; but on the other side, if you take out every axiom, as the sticks of the faggot, one by one, you may quarrel with them and bend them and break them at your pleasure: so that as was said of Seneca, 'Verborum minutiis rerum frangit pondera',° so a man may truly say of the schoolmen, 'Quaestionum minutiis scientiarum frangunt soliditatem'.° For were it not better for a man in a fair° room to set up one great° light, or branching candlestick of lights, than to go about with a small watch candle° into every corner?

And such is their method, that rests not so much upon evidence of truth proved° by arguments, authorities, similitudes, examples, as upon particular confutations and solutions of every scruple, cavillation,° and objection; breeding for the most part one question as fast it solveth another; even as in the former resemblance, when you carry the light into one corner, you darken the rest. So that the fable and fiction of Scylla° seemeth to be a lively image of this kind of philosophy or knowledge, which was transformed into a comely virgin for the upper parts, but then 'Candida succinctam latrantibus inguina monstris'.° So the generalities of the schoolmen are for a while good and proportionable;° but then when you descend into their distinctions and decisions,° instead of a fruitful womb for the use and benefit of man's life, they end in monstrous altercations and barking questions. So as it is not possible but this quality of knowledge must fall under popular contempt, the people being apt to contemn truth upon occasion of controversies and altercations, and to think they are all out of their way° which never meet:° and when they see such digladiation° about subtilities and matter of no use nor moment, they easily fall upon that judgment of Dionysius of Syracusa, 'Verba ista sunt senum otiosorum'.°

Notwithstanding certain it is, that if those schoolmen to their great thirst of truth and unwearied travail of wit° had joined variety and universality of reading and contemplation, they had proved excellent lights, to the great advancement of all learning and knowledge. But as they are, they are great undertakers° indeed, and fierce with dark keeping;° but as in the inquiry of the divine truth their pride inclined to leave the oracle of God's word and to vanish in the mixture of their own inventions, so in the inquisition of nature they ever left the oracle of God's works and adored the deceiving and deformed images which the unequal° mirror of their own minds or a few received authors or principles did represent unto them. And thus much for the second disease of learning.

For the third vice or disease of learning, which concerneth deceit or untruth, it is of all the rest the foulest;° as that which doth destroy the essential form of knowledge, which is nothing but a representation° of truth. For the truth of being and the truth of knowing are one,° differing no more than the direct beam° and the beam reflected. This vice therefore brancheth itself into two sorts; delight in deceiving, and aptness to be deceived; imposture and credulity; which, although they appear to be of a diverse nature, the one seeming to proceed of cunning, and the other of simplicity,° yet certainly they do for the most part concur.° For as the verse noteth,

Percontatorem fugito, nam garrulus idem est,°

an inquisitive man is a prattler, so upon the like reason a credulous man is a deceiver: as we see it in fame,° that he that will easily believe rumours will as easily augment rumours and add somewhat to them of his own; which Tacitus wisely noteth, when he saith, 'Fingunt simul creduntque',° so great an affinity hath fiction and belief.

This facility° of credit,° and accepting or admitting things weakly authorized or warranted, is of two kinds, according to the subject: for it is either a belief of history (or as the lawyers speak, matter of fact),° or else of matter of art° and opinion. As to the former, we see the experience and inconvenience of this error in ecclesiastical history; which hath too easily received° and registered° reports and narrations of miracles wrought by martyrs, hermits, or monks of the desert, and other holy men, and their relics, shrines, chapels, and images: which though they had a passage° for a time, by the ignorance of the people, the superstitious simplicity of some, and the politic° toleration of others, holding them but as divine poesies;° yet after a period of time, when the mist began to clear up, they grew to be esteemed but as old wives' fables, impostures of the clergy, illusions of spirits, and badges° of antichrist,° to the great scandal° and detriment of religion.

So in natural history, we see there hath not been that choice and judgment used as ought to have been; as may appear in the writings of Plinius,° Cardanus,° Albertus,° and divers of the Arabians; being fraught with° much fabulous matter, a great part not only untried but notoriously untrue, to the great derogation of the credit of natural philosophy with the grave and sober kind of wits. Wherein the wisdom and integrity of Aristotle° is worthy to be observed; that having made so diligent and exquisite a history° of living creatures, hath mingled it sparingly with any vain or feigned matter; and yet on the other side hath cast all prodigious narrations which he thought worthy the recording into one book; excellently discerning that matter of manifest truth, such whereupon observation and rule was to be built, was not to be mingled or weakened with matter of doubtful credit; and yet again that rarities and reports that seem uncredible are not to be suppressed or denied to the memory of men.

And as for the facility of credit which is yielded to arts and opinions, it is likewise of two kinds; either when too much belief is attributed to the arts themselves, or to certain authors in any art. The sciences themselves which have had better intelligence° and confederacy with the imagination° of man than with his reason, are three in number; Astrology, Natural Magic,° and Alchemy; of which sciences nevertheless the ends or pretences° are noble. For astrology pretendeth° to discover that correspondence or concatenation° which is between the superior globe and the inferior:° natural magic pretendeth to call and reduce° natural philosophy from variety of speculations to the magnitude of works: and alchemy pretendeth to make separation of all the unlike parts of bodies which in mixtures° of nature are incorporate. But the derivations and prosecutions to these ends,° both in the theories and in the practices, are full of error and vanity; which the great professors themselves have sought to veil over and conceal by enigmatical writings,° and referring themselves to auricular° traditions, and such other devices to save the credit of impostures. And yet surely to alchemy this right is due, that it may be compared to the husbandman whereof Aesop° makes the fable, that when he died told his sons that he had left unto them gold buried under ground in his vineyard; and they digged over all the ground, and gold they found none, but by reason of their stirring and digging the mould° about the roots of their vines, they had a great vintage the year following: so assuredly the search and stir to make gold hath brought to light a great number of good and fruitful inventions and experiments, as well for the disclosing of nature as for the use of man's life.

And as for the overmuch credit that hath been given unto authors in sciences, in making them dictators,° that their words should stand,°

and not consuls° to give advice; the damage is infinite that sciences have received thereby, as the principal cause that hath kept them low, at a stay° without growth or advancement. For hence it hath comen that in arts mechanical° the first deviser comes shortest,° and time addeth and perfecteth; but in sciences the first author goeth furthest, and time leeseth° and corrupteth. So we see, artillery,° sailing, printing, and the like, were grossly° managed at the first, and by time accommodated and refined; but contrariwise the philosophies and sciences of Aristotle, Plato, Democritus, Hippocrates, Euclides, Archimedes, of most vigour at the first, and by time degenerate and imbased;° whereof the reason is no other, but that in the former many wits and industries have contributed in one;° and in the latter many wits and industries have been spent about the wit of some one,° whom many times they have rather depraved° than illustrated.° For as water will not ascend higher than the level of the first spring-head from whence it descendeth, so knowledge derived from Aristotle, and exempted from liberty of examination, will not rise again higher than the knowledge of Aristotle. And therefore, although the position be good, 'Oportet discentem credere',° yet it must be coupled with this, 'Oportet edoctum judicare',° for disciples do owe unto masters only a temporary belief and a suspension of their own judgment until they be fully instructed, and not an absolute resignation or perpetual captivity: and therefore to conclude this point, I will say no more but, so let great authors have their due, as time° which is the author of authors be not deprived of his due, which is further and further to discover truth. Thus have I gone over these three diseases of learning; besides the which, there are some other rather peccant humours° than formed diseases, which nevertheless are not so secret and intrinsic° but that they fall under a popular observation and traducement,° and therefore are not to be passed over.

The first of these is the extreme affecting of° two extremities; the one Antiquity, the other Novelty: wherein it seemeth the children of time° do take after the nature and malice° of the father. For as he devoureth his children, so one of them seeketh to devour and suppress the other; while antiquity envieth° there should be new additions, and novelty cannot be content to add but it must deface.° Surely the advice of the prophet is the true direction in this matter, 'State super vias antiquas, et videte quaenam sit via recta et bona, et ambulate in ea.'° Antiquity deserveth that reverence,° that men should make a stand° thereupon, and discover what is the best way;

but when the discovery is well taken,° then to make progression. And to speak truly, 'Antiquitas saeculi juventus mundi'.° These times are the ancient times, when the world is ancient, and not those which we account ancient 'ordine retrogrado', by a computation backward from ourselves.

Another error, induced by the former, is a distrust that any thing should be now to be found out, which the world should have missed and passed over so long time; as if the same objection were to be made to time that Lucian° maketh to Jupiter and other° the heathen gods, of which he wondereth that they begot so many children in old time° and begot none in his time, and asketh whether they were become septuagenary,° or whether the law *Pappia*,° made against old men's marriages, had restrained them. So it seemeth men doubt lest time is become past children and generation;° wherein contrariwise we see commonly the levity and unconstancy of men's judgments, which, till a matter be done, wonder that it can be done; and as soon as it is done, wonder again that it was no sooner done; as we see in the expedition of Alexander into Asia, which at first was prejudged° as a vast and impossible enterprise; and yet afterwards it pleaseth Livy to make no more of it than this, 'Nil aliud quam bene ausus vana contemnere'.° And the same happened to Columbus° in the western navigation. But in intellectual matters it is much more common; as may be seen in most of the propositions of Euclid, which till they be demonstrate, they seem strange to our assent; but being demonstrate, our mind accepteth of them by a kind of relation° (as the lawyers speak) as if we had known them before.

Another error, that hath also some affinity with the former, is a conceit that of former opinions or sects, after variety and examination,° the best hath still° prevailed and suppressed the rest; so as if a man should begin the labour of a new search, he were but like° to light upon somewhat formerly rejected, and by rejection brought into oblivion: as if the multitude, or the wisest for the multitude's sake, were not ready to give passage° rather to that which is popular° and superficial than to that which is substantial and profound. For the truth is, that time seemeth to be of the nature of a river or stream, which carrieth down to us that which is light and blown up, and sinketh and drowneth that which is weighty and solid.

Another error, of a diverse nature from all the former, is the over-early and peremptory° reduction° of knowledge into arts and methods;° from which time commonly sciences receive small or no augmentation. But as young men, when they knit and shape°

perfectly, do seldom grow to a further stature; so knowledge, while it is in aphorisms° and observations, it is in growth; but when it once is comprehended° in exact methods, it may perchance be further polished and illustrate, and accommodated for use and practice; but it increaseth no more in bulk and substance.

Another error, which doth succeed° that which we last mentioned, is that after the distribution of particular arts and sciences, men have abandoned universality,° or 'philosophia prima';° which cannot but cease° and stop all progression. For no perfect discovery° can be made upon a flat or a level: neither is it possible to discover the more remote and deeper parts of any science, if you stand but upon the level of the same science, and ascend not to a higher science.

Another error hath proceeded from too great a reverence, and a kind of adoration of the mind and understanding of man; by means whereof men have withdrawn themselves too much from the contemplation of nature and the observations of experience, and have tumbled up and down° in their own reason and conceits. Upon these intellectualists,° which are notwithstanding commonly taken for the most sublime and divine philosophers, Heraclitus° gave a just censure, saying, 'Men sought truth in their own little worlds, and not in the great and common world'; for they disdain to spell and so by degrees to read in the volume of God's works; and contrariwise by continual meditation and agitation of wit do urge and as it were invocate° their own spirits to divine° and give oracles unto them, whereby they are deservedly deluded.

Another error that hath some connexion with this latter is, that men have used° to infect their meditations, opinions, and doctrines, with some conceits which they have most admired, or some sciences which they have most applied;° and given all things else a tincture° according to them, utterly untrue and unproper. So hath Plato intermingled his philosophy with theology, and Aristotle with logic, and the second school of Plato, Proclus° and the rest, with the mathematics. For these were the arts which had a kind of primogeniture° with them severally.° So have the alchemists made a philosophy out of a few experiments of the furnace; and Gilbertus,° our countryman, hath made a philosophy out of the observations of a loadstone. So Cicero, when, reciting the several opinions of the nature of the soul, he found a musician that held the soul was but a harmony, saith pleasantly, 'Hic ab arte sua non recessit, &c'.° But of these conceits° Aristotle speaketh seriously and wisely, when he saith, 'Qui respiciunt ad pauca de facili pronunciant'.°

Another error is an impatience of doubt, and haste to assertion without due and mature suspension of judgment. For the two ways of contemplation are not unlike the two ways of action° commonly spoken of by the ancients; the one plain and smooth in the beginning, and in the end impassable; the other rough and troublesome in the entrance, but after a while fair and even. So it is in contemplation; if a man will begin with certainties, he shall end in doubts; but if he will be content to begin with doubts, he shall end in certainties.

Another error is in the manner of the tradition° and delivery of knowledge, which is for the most part magistral° and peremptory, and not ingenuous and faithful;° in a sort as may be soonest believed, and not easiliest examined. It is true that in compendious treatises for practice° that form is not to be disallowed.° But in the true handling of knowledge, men ought not to fall either on the one side into the vein° of Velleius the Epicurean, 'Nil tam metuens, quam ne dubitare aliqua de re videretur,'° nor on the other side into Socrates his ironical doubting of all things; but to propound things sincerely, with more or less asseveration,° as they stand in a man's own judgment proved more or less.

Other errors there are in the scope° that men propound to themselves, whereunto they bend their endeavours; for whereas the more constant and devote kind of professors° of any science ought to propound to themselves to make some additions to their science, they convert their labours to aspire to certain° second prizes; as to be a profound interpreter or commenter,° to be a sharp champion or defender, to be a methodical compounder or abridger; and so the patrimony° of knowledge cometh to be sometimes improved, but seldom augmented.

But the greatest error of all the rest is the mistaking or misplacing of the last or furthest end of knowledge. For men have entered into a desire of learning and knowledge, sometimes upon° a natural curiosity and inquisitive appetite; sometimes to entertain their minds with variety and delight; sometimes for ornament and reputation; and sometimes to enable° them to victory of wit and contradiction; and most times for lucre and profession;° and seldom sincerely to give a true account of their gift of reason, to the benefit and use of men:° as if there were sought in knowledge a couch, whereupon to rest a searching and restless spirit; or a terrace, for a wandering° and variable mind to walk up and down with a fair prospect; or a tower of state, for a proud mind to raise itself upon; or a fort or commanding ground, for strife and contention; or a shop, for profit or sale; and not a rich storehouse, for the glory of the Creator and the relief of man's estate.° But this is that which will indeed dignify and exalt knowledge, if contemplation and action may be more nearly and straitly° conjoined and united together than they have been; a conjunction like unto that of the two highest planets, Saturn the planet of rest and contemplation, and Jupiter° the planet of civil society and action. Howbeit, I do not mean, when I speak of use and action, that end before-mentioned of the applying of knowledge to lucre and profession: for I am not ignorant how much that diverteth and interrupteth the prosecution and advancement of knowledge; like unto the golden ball thrown before Atalanta,° which while she goeth aside and stoopeth to take up, the race is hindered,

Declinat cursus, aurumque volubile tollit.°

Neither is my meaning, as was spoken of Socrates,° to call philosophy down from heaven to converse° upon the earth; that is, to leave natural philosophy aside, and to apply knowledge only to manners and policy.° But as both heaven and earth do conspire and contribute to the use and benefit of man, so the end ought to be, from both philosophies to separate and reject vain speculations and whatsoever is empty and void, and to preserve and augment whatsoever is solid and fruitful; that knowledge may not be as a curtesan, for pleasure and vanity only, or as a bond-woman,° to acquire and gain to her master's use; but as a spouse, for generation, fruit, and comfort.

Thus have I described and opened,° as by a kind of dissection, those peccant humours° (the principal of them) which have not only given impediment to the proficience° of learning, but have given also occasion to the traducement thereof: wherein if I have been too plain, it must be remembered 'Fidelia vulnera amantis, sed dolosa oscula malignantis.'° This I think I have gained, that I ought to be the better believed in that which I shall say pertaining to commendation, because I have proceeded so freely° in that which concerneth censure. And yet I have no purpose to enter into a laudative° of learning, or to make a hymn to the muses, (though I am of opinion that it is long since their rites were duly celebrated): but my intent is, without varnish or amplification,° justly to weigh the dignity of knowledge in the balance with other things, and to take the true value thereof by testimonies and arguments divine and human.

First therefore, let us seek the dignity of knowledge in the arch-type° or first platform,° which is in the attributes and acts of God, as far as they are revealed to man and may be observed with sobriety;° wherein we may not seek it by the name of learning; for all learning is knowledge acquired, and all knowledge in God is original:° and therefore we must look for it by another name, that of wisdom° or sapience, as the Scriptures call it.°

It is so then, that in the work of the creation we see a double emanation of virtue° from God; the one referring more properly to power, the other to wisdom; the one expressed in making the subsistence° of the matter, and the other in disposing° the beauty of the form. This being supposed, it is to be observed, that for any thing which appeareth° in the history of the creation, the confused mass and matter of heaven and earth was made in a moment, and the order and disposition° of that chaos or mass was the work of six days; such a note of difference° it pleased God to put upon the works of power and the works of wisdom; wherewith concurreth, that in the former it is not set down that God said, 'Let there be heaven and earth', as it is set down of the works following; but actually, that God made° heaven and earth: the one carrying the style of a manufacture, and the other of a law, decree, or counsel.

To proceed to that which is next in order, from God to spirits; we find, as far as credit is to be given to the celestial hierarchy of that supposed° Dionysius° the senator of Athens, the first place or degree is given to the angels of love, which are termed Seraphim; the second to the angels of light, which are termed Cherubim; and the third and so following places to thrones, principalities, and the rest, which are all angels of power and ministry;° so as the angels of knowledge and illumination are placed before the angels of office and domination.

To descend from spirits and intellectual forms° to sensible and material forms; we read the first form that was created was light,° which hath a relation and correspondence in nature and corporal things, to knowledge in spirits and incorporal° things.

So in the distribution of days,° we see the day wherein God did rest and contemplate his own works, was blessed above all the days wherein he did effect and accomplish them.

After the creation was finished, it is set down° unto us that man was placed in the garden to work therein; which work so appointed to him could be no other than work of contemplation;° that is, when the end of work is but for exercise° and experiment,° not for necessity; for there being then no reluctation° of the creature, nor sweat of the brow,° man's employment must of consequence have been matter of delight in the experiment, and not matter of labour for the use. Again, the first acts° which man performed in Paradise consisted of the two summary° parts of knowledge; the view of creatures, and the imposition° of names. As for the knowledge which induced the fall, it was, as was touched° before, not the natural knowledge of creatures, but the moral knowledge of good and evil; wherein the supposition° was, that God's commandments or prohibitions were not the originals of good and evil, but that they had other beginnings, which man aspired to know, to the end° to make a total defection from God, and to depend wholly upon himself.°

To pass on: in the first event or occurrence after the fall of man, we see (as the Scriptures have infinite mysteries, not violating at all the truth of the story or letter),° an image of the two estates, the contemplative state and the active state, figured° in the two persons° of Abel and Cain, and in the two simplest and most primitive° trades of life; that of the shepherd (who, by reason of his leisure, rest in a place, and living in view of heaven, is a lively image of a contemplative life),° and that of the husbandman:° where we see again the favour and election° of God went to the shepherd, and not to the tiller of the ground.

So in the age before the flood, the holy records within° those few memorials which are there entered and registered have vouchsafed to mention and honour the name of the inventors and authors of music and works in metal.° In the age after the flood, the first great judgment of God upon the ambition of man was the confusion of tongues;° whereby the open trade and intercourse of learning and knowledge was chiefly imbarred.°

To descend to Moses the lawgiver, and God's first pen:° he is adorned by the Scriptures° with this addition and commendation, that he was 'seen in° all the learning of the Egyptians'; which nation we know was one of the most ancient schools of the world: for so Plato° brings in the Egyptian priest saying unto Solon: 'You Grecians are ever children; you have no knowledge of antiquity, nor antiquity of knowledge'. Take a view of the ceremonial law of Moses; you shall find, besides the prefiguration° of Christ, the badge or difference° of the people of God, the exercise and impression° of obedience, and other divine uses and fruits thereof, that some of the most learned Rabbins have travailed profitably and profoundly to observe,° some of them a natural, some of them a moral sense or reduction° of many of the ceremonies and ordinances. As in the law of the leprosy, where it is said, 'If the whiteness have overspread the flesh, the patient may pass abroad for clean; but if there be any whole flesh remaining, he is to be shut up for unclean';° one of them noteth° a principle of nature,° that putrefaction is more contagious before maturity than after: and another noteth a position of moral philosophy, that men abandoned to vice do not so much corrupt manners, as those that are half good and half evil. So in this and very many other places in that law, there is to be found, besides the theological sense, much aspersion° of philosophy.

So likewise in that excellent book of Job, if it be revolved with diligence, it will be found pregnant and swelling with natural philosophy; as for example, cosmography and the roundness of the world; 'Qui extendit aquilonem super vacuum, et appendit terram super nihilum';° wherein the pensileness° of the earth, the pole of the north, and the finiteness° or convexity of heaven are manifestly touched. So again matter of astronomy; 'Spiritus ejus ornavit coelos, et obstetricante manu ejus eductus est Coluber tortuosus'.° And in another place, 'Nunquid conjungere valebis micantes stellas Pleiadas, aut gyrum Arcturi poteris dissipare?'° where the fixing of the stars, ever standing at equal distance, is with great elegancy noted. And in another place, 'Qui facit Arcturum, et Oriona, et Hyadas, et interiora Austri';° where again he takes knowledge of° the depression° of the southern pole, calling it the secrets of the south, because the southern stars were in that climate° unseen. Matter of generation;° 'Annon sicut lac mulsisti me, et sicut caseum coagulasti me?' &c. Matter of minerals; 'Habet argentum venarum suarum principia: et auro locus est in quo conflatur, ferrum de terra tollitur, et lapis solutus calore in aes vertitur:'° and so forwards° in that chapter.

So likewise in the person of Salomon the king, we see the gift or endowment of wisdom and learning, both in Salomon's petition° and in God's assent thereunto, preferred before° all other terrene° and temporal felicity. By virtue of which grant or donative° of God, Salomon became enabled not only to write those excellent parables or aphorisms concerning divine and moral philosophy, but also to compile a natural history of all verdure,° from the cedar upon the mountain to the moss° upon the wall (which is but a rudiment° between putrefaction and an herb), and also of all things that breathe or move. Nay, the same Salomon the king, although he excelled in the glory of treasure and magnificent buildings, of shipping and navigation, of service and attendance, of fame and renown, and the like, yet he maketh no claim to any of those glories, but only to the glory of inquisition° of truth; for so he saith expressly, 'The glory of God is to conceal a thing, but the glory of the king is to find it out';° as if, according to the innocent play of children, the Divine Majesty took delight to hide his works, to the end to have them found out; and as if kings could not obtain a greater honour than to be God's playfellows in that game, considering the great commandment° of wits and means, whereby nothing needeth to be hidden from them.

Neither did the dispensation of God vary in the times after our Saviour came into the world; for our Saviour himself did first shew his power to subdue ignorance, by his conference° with the priests and doctors of the law, before he shewed his power to subdue nature by his miracles. And the coming of the Holy Spirit was chiefly figured and expressed in the similitude and gift of tongues,° which are but 'vehicula scientiae.'°

So in the election° of those instruments which it pleased God to use for the plantation of the faith,° notwithstanding that at the first he did employ persons altogether unlearned otherwise than by inspiration, more evidently to declare his immediate° working, and to abase all human wisdom or knowledge;° yet nevertheless that counsel of his was no sooner performed, but in the next vicissitude° and succession he did send his divine truth into the world waited on° with° other learnings° as with servants or handmaids: for so we see St. Paul, who was only learned° amongst the apostles, had his pen most used in° the scriptures of the New Testament.

So again we find that many of the ancient bishops and fathers of the Church were excellently read and studied in all the learning of the heathen; insomuch that the edict° of the emperor Julianus (whereby it was interdicted unto Christians to be admitted into schools, lectures, or exercises of learning) was esteemed and accounted a more pernicious engine° and machination° against the Christian faith, than were all the sanguinary° prosecutions of his predecessors; neither could the emulation° and jealousy° of Gregory the first° of that name, bishop of Rome, ever obtain the opinion° of piety or devotion; but contrariwise received the censure° of humour,° malignity, and pusillanimity,° even amongst holy men; in that he designed to obliterate and extinguish the memory of heathen antiquity and authors. But contrariwise° it was the Christian Church, which amidst the inundations of the Scythians° on the one side from the north-west, and the Saracens° from the east, did preserve in the sacred lap and bosom thereof the precious relics even of heathen learning, which otherwise had been extinguished as if no such thing had ever been.

And we see before our eyes, that in the age of ourselves and our fathers, when it pleased God to call the church of Rome to account for their degenerate manners and ceremonies, and sundry doctrines obnoxious° and framed to uphold the same abuses; at one and the same time it was ordained by the Divine Providence that there should attend withal a renovation° and new spring of all other knowledges:° and on the other side we see the Jesuits,° who partly in themselves and partly by the emulation and provocation° of their example, have much quickened and strengthened the state of learning,—we see (I say) what notable service and reparation° they have done to the Roman see.°

Wherefore to conclude this part, let it be observed that there be two principal duties and services, besides ornament and illustration,° which philosophy and human learning do perform to faith and religion. The one, because they are an effectual inducement to the exaltation of the glory of God. For as the Psalms and other Scriptures do often invite us to consider and magnify° the great and wonderful works of God, so if we should rest only in the contemplation of the exterior of them as they first offer themselves to our senses, we should do a like injury unto the majesty of God as if we should judge or construe° of the store of some excellent jeweller by that only which is set out toward the street in his shop. The other, because they minister a singular help and preservative against unbelief and error: For our Saviour saith, 'You err, not knowing the Scriptures, nor the power of God',° laying before us two books° or volumes to study, if we will be secured° from error; first the Scriptures, revealing the will of God, and then the creatures expressing his power; whereof the latter is a key unto the former; not only opening our understanding to conceive the true sense of the Scriptures, by the general notions of reason and rules of speech; but chiefly opening our belief, in drawing us into a due meditation of the omnipotency of God, which is chiefly signed° and engraven upon his works. Thus much therefore for divine testimony and evidence concerning the true dignity and value of learning.

As for human proofs, it is so large a field, as in a discourse of this nature and brevity it is fit rather to use choice of those things which we shall produce, than to embrace the variety of them. First therefore, in the degrees of human honour amongst the heathen it was the highest, to obtain to° a veneration and adoration as a god. This unto the Christians is as the forbidden fruit. But we speak now separately° of human testimony: according to which that which the Grecians call *apotheosis*,° and the Latins 'relatio inter divos',° was the supreme honour which man could attribute unto man; specially when it was given, not by a formal decree or act of state, as it was used among the Roman emperors, but by an inward assent° and belief; which honour being so high, had also a degree or middle term;° for there were reckoned above human honours, honours heroical and divine; in the attribution and distribution of which honours we see antiquity made this difference: that whereas founders and uniters of states and cities, lawgivers, extirpers° of tyrants, fathers of the people, and other eminent persons in civil merit, were honoured but with the titles of worthies° or demi-gods; such as were Hercules, Theseus, Minos, Romulus, and the like; on the other side, such as were inventors and authors of new arts, endowments, and commodities towards man's life, were ever consecrated amongst the gods themselves; as was Ceres, Bacchus, Mercurius, Apollo, and others; and justly; for the merit of the former is confined within the circle° of an age or a nation; and is like fruitful showers, which though they be profitable and good, yet serve but for that season, and for a latitude of° ground where they fall; but the other is indeed like the benefits of heaven, which are permanent and universal. The former again is mixed with strife and perturbation; but the latter hath the true character of divine presence, coming in 'aura leni',° without noise or agitation.°

Neither is certainly that other merit of learning, in repressing the inconveniencies which grow from man to man,° much inferior to the former, of relieving the necessities which arise from nature; which merit was lively° set forth by the ancients in that feigned relation° of Orpheus' theatre;° where all beasts and birds assembled, and forgetting their several appetites, some of prey, some of game,° some of quarrel, stood all sociably together listening unto the airs and accords° of the harp; the sound whereof no sooner ceased, or was drowned by some louder noise, but every beast returned to his own nature: wherein is aptly described the nature and condition of men; who are full of savage and unreclaimed desires, of profit, of lust, of revenge, which as long as they give ear to precepts, to laws, to religion, sweetly touched with eloquence and persuasion of books, of sermons, of harangues, so long is society and peace maintained. But if these instruments be silent, or that sedition and tumult make them not audible, all things dissolve into anarchy and confusion.

But this appeareth more manifestly, when kings themselves, or persons of authority under them, or other governors in commonwealths and popular estates,° are endued with learning. For although he might be thought partial to his own profession, that said 'then should people and estates be happy, when either kings were philosophers, or philosophers kings';° yet so much is verified by experience, that under learned princes and governors there have been ever the best times. For howsoever kings may have their imperfections in their passions and customs,° yet if they be illuminate° by learning, they have those notions of religion, policy, and morality, which do preserve them and refrain° them from all ruinous and peremptory° errors and excesses; whispering evermore in their ears, when counsellors and servants stand mute and silent. And senators or counsellors likewise which be learned, do proceed upon more safe and substantial principles than counsellors which are only men of experience; the one sort keeping dangers afar off, whereas the other discover them not till they come near hand,° and then trust to the agility of their wit to ward° or avoid them.

Which felicity of times under learned princes (to keep still the law of brevity, by using the most eminent and selected examples) doth best appear in the age which passed from the death of Domitianus the emperor until the reign of Commodus;° comprehending a succession of six princes, all learned or singular favourers and advancers of learning; which age, for° temporal respects,° was the most happy and flourishing that ever the Roman empire (which then was a model° of the world) enjoyed: a matter revealed and prefigured unto Domitian° in a dream the night before he was slain; for he thought there was grown behind upon his shoulders a neck and a head of gold, which came accordingly to pass in those golden times which succeeded: of which princes we will make some commemoration; wherein although the matter° will be vulgar,° and may be thought fitter for a declamation° than agreeable to a treatise infolded° as this is, yet because it is pertinent to the point in hand, 'neque semper arcum tendit Apollo',° and to name them only° were too naked and cursory, I will not omit it altogether.

The first was Nerva;° the excellent temper of whose government is by a glance° in Cornelius Tacitus touched to the life: 'Postquam divus Nerva res olim insociabiles miscuisset, imperium et libertatem'.° And in token of his learning, the last act of his short reign left to memory was a missive° to his adopted son Trajan, proceeding upon° some inward discontent at the ingratitude of the times, comprehended in a verse of Homer's;

> Telis, Phoebe, tuis lacrymas ulciscere nostras.°

Trajan,° who succeeded, was for his person not learned:° but if we will hearken to the speech of our Saviour, that saith, 'He that receiveth a prophet in the name of a prophet, shall have a prophet's reward',° he deserveth to be placed amongst the most learned princes: for there was not a greater admirer of learning or benefactor of learning; a founder of famous libraries, a perpetual advancer of learned men to office, and a familiar converser° with learned professors and preceptors, who were noted° to have then most credit in court. On the other side, how much Trajan's virtue and government was admired and renowned, surely no testimony of grave and faithful history doth more lively set forth, than that legend° tale of Gregorius Magnus,° bishop of Rome, who was noted for the extreme envy° he bare towards all heathen excellency:° and yet he is reported, out of the love and estimation of Trajan's moral virtues, to have made unto God passionate and fervent prayers for the delivery of his soul out of hell; and to have obtained it, with a caveat° that he should make no more such petitions. In this prince's time also the persecutions against the Christians received intermission, upon the certificate° of Plinius Secundus,° a man of excellent learning and by Trajan advanced.

Hadrian,° his successor, was the most curious° man that lived, and the most universal inquirer; insomuch as it was noted for an error in his mind, that he desired to comprehend all things, and not to reserve himself for the worthiest things; falling into the like humour that was long before noted in Philip of Macedon, who when he would needs over-rule and put down an excellent musician in an argument touching music, was well answered by him again, 'God forbid, Sir,' (saith he), 'that your fortune should be so bad, as to know these things better than I'.° It pleased God likewise to use the curiosity° of this emperor as an inducement° to the peace of his church in those days. For having Christ in veneration, not as a God or Saviour, but as a wonder or novelty, and having his picture in his gallery matched° with Appollonius° (with whom in his vain imagination he thought he had some conformity), yet it served the turn to allay the bitter hatred of those times against the Christian name; so as the church had peace during his time. And for his government civil, although he did not attain to that of Trajan's in glory of arms or perfection of justice, yet in deserving of the weal° of the subject° he did exceed him. For Trajan erected many famous monuments and buildings; insomuch as° Constantine the Great in emulation was wont to call him *Parietaria*, wall-flower,° because his name was upon so many walls: but his buildings and works were more of° glory and triumph than use and necessity. But Hadrian spent his whole reign, which was peaceable, in a perambulation or survey° of the Roman empire; giving order and making assignation° where he went for re-edifying° of cities, towns, and forts decayed, and for cutting of rivers and streams, and for making bridges and passages,° and for policing° of cities and commonalties° with new ordinances and constitutions, and granting new franchises and incorporations;° so that his whole time was a very restoration of all the lapses° and decays of former times.

Antoninus Pius,° who succeeded him, was a prince excellently learned; and had the patient and subtile° wit of a schoolman;° insomuch as in common speech (which leaves no virtue untaxed)° he was called 'cymini sector',° a carver or divider of cumin seed, which is one of the least° seeds; such a patience he had and settled° spirit to enter into the least and most exact differences of causes; a fruit no doubt of the exceeding tranquillity and serenity of his mind; which being no ways charged or incumbered either with fears, remorses, or scruples,° but having been noted for a man of the purest goodness, without all° fiction° or affectation, that hath reigned or lived, made his mind continually present and entire. He likewise approached a degree nearer unto Christianity, and became, as Agrippa said unto St. Paul, 'half a Christian';° holding their religion and law in good opinion, and not only ceasing persecution, but giving way to° the advancement of Christians.

There succeeded him the first 'Divi fratres',° the two adoptive° brethren, Lucius Commodus Verus, son to Aelius Verus, who delighted much in the softer° kind of learning, and was wont to call the poet Martial° his Virgil; and Marcus Aurelius Antoninus; whereof the latter, who obscured his colleague and survived him long,° was named the Philosopher: who as he excelled all the rest in learning, so he excelled them likewise in perfection of all royal virtues; insomuch as Julianus° the emperor, in his book intitled *Caesares*, being as a pasquil° or satire to deride all his predecessors, feigned that they were all invited to a banquet of the gods, and Silenus the jester sat at the nether° end of the table and bestowed a scoff on every one as they came in; but when Marcus Philosophus° came in, Silenus was gravelled° and out of countenance, not knowing where to carp at° him; save at the last he gave a glance° at his patience towards his wife. And the virtue of this prince, continued with that of his predecessor, made the name of Antoninus so sacred in the world, that though it were extremely dishonoured in Commodus, Caracalla, and Heliogabalus,° who all bare the name, yet when Alexander Severus refused the name because he was a stranger to the family, the Senate with one acclamation said, 'Quomodo Augustus, sic et Antoninus':° in such renown and veneration was the name of these two princes in those days, that they would have it as a perpetual addition in all the emperors' style.° In this emperor's time also the church for the most part was in peace; so as in this sequence of six princes we do see the blessed effects of learning in sovereignty, painted forth° in the greatest table° of the world.

But for a tablet or picture of smaller volume (not presuming to speak of your Majesty that liveth), in my judgment the most excellent is that of queen Elizabeth,° your immediate predecessor in this part of Britain; a prince that, if Plutarch° were now alive to write lives by parallels, would trouble him, I think, to find for her a parallel amongst women. This lady was endued with learning in her sex singular, and rare even amongst masculine princes; whether we speak of learning of language or of science; modern or ancient; divinity or humanity. And unto the very last year of her life she accustomed° to appoint set hours for reading, scarcely any young student in an university more daily or more duly.° As for her government, I assure myself I shall not exceed° if I do affirm that this part of the island never had forty-five years of better times; and yet not through the calmness of the season, but through the wisdom of her regiment.° For if there be considered of the one side, the truth of religion established;° the constant peace and security; the good administration of justice; the temperate use of the prerogative,° not slackened, nor much strained; the flourishing state of learning, sortable° to so excellent a patroness; the convenient° estate of wealth and means, both of crown and subject; the habit of obedience, and the moderation of discontents;° and there be considered on the other side, the differences of religion, the troubles of neighbour countries, the ambition of Spain, and opposition of Rome; and then that she was solitary and of herself:° these things I say considered, as I could not have chosen an instance so recent and so proper, so I suppose I could not have chosen one more remarkable or eminent, to° the purpose now in hand; which is concerning the conjunction of learning in the prince with felicity in the people.

Neither hath learning an influence and operation only upon civil merit and moral virtue, and the arts or temperature° of peace and peaceable government; but likewise it hath no less power and efficacy in enablement towards° martial and military virtue and prowess; as may be notably represented in the examples of Alexander the Great and Caesar the Dictator, mentioned before, but now in fit place to be resumed; of whose virtues and acts in war there needs no note or recital, having been the wonders of time in that kind; but of their affections° towards learning, and perfections in learning, it is pertinent to say somewhat.

Alexander was bred and taught under Aristotle the great philosopher, who dedicated divers of his books of philosophy unto him. He was attended with° Callisthenes° and divers other learned persons, that followed him in camp, throughout his journeys and conquests. What price and estimation he had learning in doth notably° appear in these three particulars: first, in the envy he used to express that he bare° towards Achilles, in this that he had so good a trumpet of his praises as Homer's verses; secondly, in the judgment or solution° he gave touching° that precious cabinet° of Darius,° which was found among his jewels, whereof question was made what thing was worthy to be put into it, and he gave his opinion for Homer's works; thirdly, in his letter to Aristotle, after he had set forth° his books of nature,° wherein he expostulateth with him for publishing the secrets or mysteries of philosophy, and gave him to understand that himself esteemed it more to excel other men in learning and knowledge than in power and empire. And what use he had of learning doth appear, or rather shine, in all his speeches and answers, being full of science° and use° of science, and that in all variety.

And herein again it may seem a thing scholastical,° and somewhat idle, to recite things that every man knoweth; but yet since the argument I handle° leadeth me thereunto, I am glad that men shall perceive I am as willing to flatter (if they will so call it) an Alexander or a Caesar or an Antoninus, that are dead many hundred years since, as any that now liveth: for it is the displaying of the glory of learning in sovereignty that I propound to myself, and not an humour° of declaiming in any man's praises. Observe then the speech he used of Diogenes,° and see if it tend not to the true state° of one of the greatest questions of moral philosophy; whether the enjoying of outward things or the contemning of them be the greatest happiness; for when he saw Diogenes so perfectly contented with so little, he said° to those that mocked at his condition, 'Were I not Alexander, I would wish to be Diogenes'. But Seneca inverteth° it, and saith, 'Plus erat quod hic nollet accipere, quam quod ille posset dare'.° 'There were more things which Diogenes would have refused, than those were which Alexander could have given or enjoyed.'

Observe again that speech which was usual with him, that 'he felt his mortality chiefly in two things, sleep and lust';° and see if it were not a speech extracted out of the depth of natural philosophy, and liker° to have comen out of the mouth of Aristotle or Democritus° than from Alexander.

See again that speech° of humanity and poesy; when upon the bleeding of his wounds, he called unto him one of his flatterers that was wont to ascribe to him divine honour, and said, 'Look, this is very blood; this is not such a liquor° as Homer° speaketh of, which ran from Venus' hand when it was pierced by Diomedes'.

See likewise his readiness in reprehension of logic,° in the speech he used to Cassander upon a complaint that was made against his father Antipater: for when Alexander happed° to say,° 'Do you think these men would have come from so far to complain, except they had just cause of grief?' and Cassander answered, 'Yea, that was the matter,° because they thought they should not be disproved'; said Alexander laughing, 'See the subtelties of Aristotle, to take a matter both ways,' *pro et contra*, &c.

But note again how well he could use the same art which he reprehended, to serve his own humour, when bearing a secret grudge to Callisthenes because he was against the new ceremony of his adoration,° feasting one night where the same Callisthenes was at the table, it was moved° by some after supper, for entertainment sake, that Callisthenes who was an eloquent man might speak of some theme or purpose° at his own choice; which Callisthenes did; choosing the praise of the Macedonian nation for his discourse, and performing the same with so good manner as the hearers were much ravished; whereupon Alexander, nothing pleased, said, 'It was easy to be eloquent upon so good a subject': but saith he, 'Turn your style,° and let us hear what you can say against us': which Callisthenes presently undertook, and did with that sting and life, that Alexander interrupted him, and said, 'The goodness of the cause made him eloquent before, and despite° made him eloquent then again.'

Consider further, for tropes of rhetoric, that excellent use of a metaphor or translation,° wherewith he taxed° Antipater, who was an imperious and tyrannous° governor: for when one of Antipater's friends commended him to Alexander for his moderation, that he did not degenerate, as his other lieutenants did, into the Persian pride, in use of purple, but kept the ancient habit° of Macedon, of black; 'True', (saith Alexander), 'but Antipater is all purple within'.° Or that other, when Parmenio came to him in the plain of Arbella,° and shewed him the innumerable multitude of his enemies, specially as they appeared by the infinite number of lights,° as° it had been a new firmament of stars, and thereupon advised him to assail them by night: whereupon he answered,° 'that he would not steal the victory'.

For matter of policy, weigh° that significant distinction, so much in all ages embraced,° that he made between his two friends Hephaestion and Craterus, when he said, 'that the one loved Alexander, and the other loved the king'; describing the principal difference of princes' best servants, that some in affection love their person, and others in duty love their crown.

Weigh also that excellent taxation° of an error ordinary with counsellors of princes, that they counsel their masters according to the model° of their own mind and fortune, and not of their masters; when upon Darius' great offers Parmenio had said,° 'Surely I would accept these offers, were I as Alexander'; saith Alexander, 'So would I, were I as Parmenio'.

Lastly, weigh that quick and acute reply which he made when he gave so large gifts to his friends and servants, and was asked what he did reserve for himself, and he answered, 'Hope';° weigh, I say, whether he had not cast up° his account aright, because hope must be the portion of all that resolve upon great enterprises. For this was Caesar's portion° when he went first into Gaul, his estate being then utterly overthrown with largesses.° And this was likewise the portion of that noble prince, howsoever transported with ambition, Henry° duke of Guise, of whom it was usually said, that he was the greatest usurer in France, because he had turned all his estate into obligations.°

To conclude therefore: as certain critics are used to say hyperbolically, 'that if all sciences were lost, they might be found in Virgil'; so certainly this may be said truly, there are the prints and footsteps of learning in those few speeches which are reported of this prince: the admiration of whom, when I consider him not as Alexander the Great, but as Aristotle's scholar, hath carried me too far.

As for Julius Caesar,° the excellency of his learning needeth not to be argued from his education, or his company,° or his speeches; but in a further degree doth declare itself in his writings and works; whereof some are extant and permanent,° and some unfortunately perished. For first, we see there is left unto us that excellent history of his own wars, which he intitled only a Commentary,° wherein all succeeding times have admired the solid weight of matter, and the real passages° and lively images° of actions and persons, expressed in the greatest propriety° of words and perspicuity of narration that ever was; which that it was not the effect of a natural gift, but of learning and precept, is well witnessed by that work of his intitled *De Analogia*,° being a grammatical philosophy, wherein he did labour to make this same 'vox ad placitum'° to become 'vox ad licitum',° and to reduce° custom of speech to congruity of speech; and took as it were the picture of words from the life of reason.°

So we receive from him, as a monument both of his power and learning, the then reformed computation° of the year; well expressing, that he took it to be as great a glory to himself to observe and know the law of the heavens as to give law to men upon the earth.

So likewise in that book of his *Anti-Cato*,° it may easily appear that he did aspire as well to victory of wit° as victory of war; undertaking therein a conflict against the greatest champion with the pen that then lived, Cicero the orator.

So again in his book of *Apophthegms*° which he collected, we see that he esteemed it more honour to make himself but a pair of tables° to take° the wise and pithy words of others, than to have every word of his own to be made an apophthegm or an oracle;° as vain princes, by custom of flattery, pretend to do. And yet if I should enumerate divers of his speeches, as I did those of Alexander, they are truly such as Salomon noteth, when he saith, 'Verba sapientum tanquam aculei, et tanquam clavi in altum defixi': whereof I will only recite three, not so delectable° for elegancy, but admirable for vigour and efficacy.

As first,° it is reason he be thought a master of words, that could with one word appease a mutiny in his army; which was thus. The Romans, when their generals did speak to their army, did use the word *Milites*;° but when the magistrates spake to the people, they did use the word *Quirites*.° The soldiers were in tumult, and seditiously prayed to be cashiered;° not that they so meant, but by expostulation thereof° to draw Caesar to other conditions; wherein he being resolute not to give way, after some silence, he began his speech, 'Ego, Quirites'; which did admit them already cashiered; wherewith they were so surprised, crossed,° and confused, as they would not suffer° him to go on in his speech, but relinquished their demands, and made it their suit to be again called by the name of *Milites*.

The second° speech was thus: Caesar did extremely affect° the name of king; and some were set on,° as he passed by, in popular acclamation to salute him king; whereupon, finding the cry weak and poor, he put it off° thus in a kind of jest, as if they had mistaken his surname; 'Non Rex sum, sed Caesar':° a speech, that if it be searched,° the life and fullness of it can scarce be expressed: for first it was a refusal of the name, but yet not serious: again it did signify° an infinite confidence and magnanimity, as if he presumed Caesar was the greater title; as by his worthiness it is come to pass till this day. But chiefly it was a speech of great allurement towards° his own purpose; as if the state did strive with him but for a name whereof mean families were vested; for Rex was a surname with the Romans, as well as King is with us.

The last° speech which I will mention, was used to Metellus; when Caesar, after war declared,° did possess himself of the city of Rome; at which time entering into the inner treasury to take the money there accumulate, Metellus being tribune forbade him: whereto Caesar said, 'that if he did not desist, he would lay him dead in the place'; and presently taking himself up,° he added, 'Young man, it is harder for me to speak it than to do it.' 'Adolescens, durius est mihi hoc dicere quam facere.' A speech compounded of the greatest terror and greatest clemency that could proceed out of the mouth of man.

But to return and conclude with him: it is evident himself knew well his own perfection in learning, and took it upon him;° as appeared° when upon occasion that some spake what a strange resolution° it was in Lucius Sulla to resign his dictature,° he scoffing at him, to his own advantage, answered, that Sulla could not skill of° letters, and therefore knew not how to dictate.

And here it were fit to leave this point touching the concurrence of military virtue and learning; (for what example would come with any grace after those two of Alexander and Caesar?) were it not in regard of the rareness of circumstance that I find in one other particular, as that which did so suddenly pass from extreme scorn to extreme wonder; and it is of Xenophon° the philosopher, who went from Socrates' school into Asia, in the expedition of Cyrus the younger against king Artaxerxes.° This Xenophon at that time was very young, and never had seen the wars before; neither had any command in the army, but only followed the war as a voluntary,° for the love and conversation° of Proxenus° his friend. He was present when Falinus came in message° from the great king° to the Grecians, after that Cyrus was slain in the field, and they a handful of men left to themselves in the midst of the king's territories, cut off from their country by many navigable rivers, and many hundred miles. The message imported that they should deliver up their arms, and submit themselves to the king's mercy. To which message before answer was made, divers° of the army conferred° familiarly with Falinus; and amongst the rest Xenophon° happened to say, 'Why Falinus, we have now but these two things left, our arms and our virtue;° and if we yield up our arms, how shall we make use of our virtue?' Whereto Falinus smiling on him, said, 'If I be not deceived, young gentleman, you are an Athenian; and I believe you study philosophy, and it is pretty that° you say; but you are much abused° if you think your virtue can withstand the king's power.' Here was the scorn; the wonder followed: which was, that this young scholar or philosopher, after all the captains were murdered in parley° by treason, conducted those ten thousand foot through the heart of all the king's high countries° from Babylon to Graecia in safety, in despite of all the king's forces, to the astonishment of the world, and the encouragement of the Grecians in time succeeding to make invasion upon the kings of Persia; as was after° purposed by Jason the Thessalian,° attempted by Agesilaus° the Spartan, and achieved by Alexander the Macedonian;° all upon the ground° of the act of that young scholar.

To proceed now from imperial and military virtue to moral and private° virtue: first, it is an assured truth which is contained in the verses,

> Scilicet ingenuas didicisse fideliter artes
> Emollit mores, nec sinit esse feros.°

It taketh away the wildness and barbarism and fierceness of men's minds: but indeed the accent° had need be upon *fideliter*:° for a little superficial learning doth rather work a contrary effect. It taketh away all levity,° temerity, and insolency, by copious suggestion of all doubts and difficulties, and acquainting° the mind to balance reasons on both sides, and to turn back° the first offers and conceits° of the mind, and to accept of° nothing but examined and tried. It taketh away vain admiration of any thing, which is the root of all weakness. For all things are admired, either because they are new, or because they are great. For novelty, no man that wadeth in° learning or contemplation° throughly,° but will find that printed in his heart 'Nil novi super terram'.° Neither can any man marvel at the play of puppets,° that goeth behind the curtain and adviseth° well of the motion.° And for magnitude, as Alexander the Great after that he was used to great armies and the great conquests of the spacious° provinces in Asia, when he received letters out of Greece of some fights and services° there, which were commonly for a passage° or a fort or some walled town at the most, he said, 'It seemed to him that he was advertised° of the battles of the frogs and the mice,° that the old tales went of':° so certainly if a man meditate much upon the universal frame° of nature, the earth with men upon it (the divineness of souls except)° will not seem much other than an ant-hill, whereas° some ants carry corn, and some carry their young, and some go empty, and all to and fro° a little heap of dust.° It taketh away or mitigateth° fear of death or adverse fortune; which is one of the greatest impediments of virtue and imperfections of manners. For if a man's mind be deeply seasoned° with the consideration of the mortality and corruptible nature of things, he will easily concur with Epictetus,° who went forth one day and saw a woman weeping for her pitcher of earth° that was broken, and went forth the next day and saw a woman weeping for her son that was dead; and thereupon said, 'Heri vidi fragilem frangi, hodie vidi mortalem mori'.° And therefore Virgil did excellently and profoundly couple the knowledge of causes and the conquest of all fears together, as *concomitantia*:°

> Felix qui potuit rerum cognoscere causas,
> Quique metus omnes et inexorabile fatum
> Subjecit pedibus, strepitumque Acherontis avari.°

It were too long to go over the particular remedies which learning doth minister to all the diseases of the mind; sometimes purging the ill humours, sometimes opening° the obstructions, sometimes helping digestion, sometimes increasing appetite, sometimes healing the wounds and exulcerations° thereof, and the like; and therefore I will conclude with that which hath 'rationem totius';° which is, that it disposeth the constitution of the mind not to be fixed or settled in the defects thereof, but still to be capable and susceptible of growth and reformation. For the unlearned man knows not what it is to descend into himself° or to call himself to account, nor the pleasure of that 'suavissima vita, in dies sentire se fieri meliorem'.° The good parts he hath he will learn to shew to the full and use them dexterously, but not much to increase them: the faults he hath he will learn how to hide and colour° them, but not much to amend them; like an ill° mower, that mows on still° and never whets his scythe: whereas with the learned man it fares otherwise, that he doth ever intermix the correction and amendment° of his mind with the use and employment thereof. Nay further, in general and in sum, certain it is that *veritas* and *bonitas*° differ but as the seal and the print;° for truth prints° goodness, and they be the clouds of error which descend° in the storms of passions and perturbations.°

From moral virtue let us pass on to matter of power and commandment,° and consider whether in right reason there be any comparable with that wherewith knowledge investeth and crowneth man's nature.

We see the dignity of the commandment is according to the dignity of the commanded:° to have commandment over beasts, as herdsmen have, is a thing contemptible; to have commandment over children, as school-masters have, is a matter of small honour; to have commandment over galley-slaves is a disparagement° rather than an honour. Neither is the commandment of tyrants much better, over people which have put off° the generosity° of their minds: and therefore it was ever holden° that honours in free monarchies and commonwealths had a sweetness more than in tyrannies; because the commandment extendeth more over the wills of men, and not only over their deeds and services. And therefore when Virgil putteth himself forth° to attribute to Augustus Caesar the best of human honours, he doth it in these words:

victorque volentes
Per populos dat jura, viamque affectat Olympo.°

But yet the commandment of knowledge is yet higher than the commandment over the will; for it is a commandment over the reason, belief, and understanding of man, which is the highest part of the mind, and giveth law to the will itself. For there is no power on earth which setteth up a throne or chair of estate° in the spirits and souls of men, and in their cogitations, imaginations, opinions, and beliefs, but knowledge and learning. And therefore we see the detestable and extreme pleasure that arch-heretics and false prophets and impostors are transported with, when they once find in themselves that they have a superiority in the faith and conscience of men; so great, that if they have once tasted of it, it is seldom seen that any torture or persecution can make them relinquish or abandon it. But as this is that which the author of the Revelation calleth the depth or profoundness of Satan;° so by argument of contraries, the just and lawful sovereignty over men's understanding, by force of truth rightly interpreted, is that which approacheth nearest to the similitude° of the divine rule.

As for fortune and advancement, the beneficence of learning is not so confined to give fortune only to states and commonwealths, as° it doth not likewise give fortune to particular persons. For it was well noted long ago, that Homer° hath given more men their livings than either Sulla or Caesar or Augustus ever did, notwithstanding their great largesses° and donatives and distributions of lands to so many legions. And no doubt it is hard to say whether arms or learning have advanced° greater numbers. And in case° of sovereignty, we see that if arms or descent have carried away° the kingdom, yet learning hath carried the priesthood, which ever hath been in some competition with empire.

Again, for° the pleasure and delight of knowledge and learning, it far surpasseth all other in nature: for shall the pleasures of the affections° so exceed the pleasures of the senses, as much as the obtaining of desire or victory exceedeth a song or a dinner; and must not of consequence° the pleasures of the intellect or understanding exceed the pleasures of the affections? We see in all other pleasures there is satiety, and after they be used, their verdure° departeth; which sheweth well they be but deceits of pleasure,° and not pleasures; and that it was the novelty which pleased, and not the quality. And therefore we see that voluptuous men turn friars, and ambitious princes° turn melancholy. But of knowledge there is no satiety, but satisfaction and appetite are perpetually interchangeable; and therefore appeareth to be good in itself simply,° without fallacy° or accident.° Neither is that pleasure of small efficacy and contentment to the mind of man, which the poet Lucretius describeth elegantly,

Suave mari magno, turbantibus aequora ventis, &c.°

'It is a view of delight' (saith he) 'to stand or walk upon the shore side, and to see a ship tossed with tempest upon the sea; or to be in a fortified tower, and to see two battles° join upon a plain. But it is a pleasure incomparable, for the mind of man to be settled, landed, and fortified in the certainty of truth; and from thence to descry° and behold the errors, perturbations, labours, and wanderings up and down of other men.'

Lastly, leaving the vulgar° arguments, that by learning man excelleth man in that wherein man excelleth beasts; that by learning man ascendeth to the heavens and their motions,° where in body he cannot come; and the like; let us conclude with the dignity and excellency of knowledge and learning in that whereunto man's nature doth most aspire; which is immortality or continuance; for to this tendeth° generation, and raising of houses and families; to this tend buildings, foundations, and monuments; to this tendeth the desire of memory, fame, and celebration; and in effect, the strength of all other human desires. We see then how far the monuments of wit and learning are more durable than the monuments of power or of the hands. For have not the verses of Homer continued twenty-five hundred years or more, without the loss of a syllable or letter; during which time infinite° palaces, temples, castles, cities, have been decayed° and demolished? It is not possible to have the true pictures or statuaes of Cyrus, Alexander, Caesar, no nor of the kings or great personages of much later years; for the originals cannot last, and the copies cannot but leese of° the life and truth. But the images of men's wits and knowledges remain in books, exempted from the wrong of time and capable of perpetual renovation. Neither are they fitly to be called images, because they generate still,° and cast their seeds in° the minds of others, provoking and causing infinite actions and opinions in succeeding ages. So that if the invention of the ship was thought so noble, which carrieth riches and commodities from place to place, and consociateth° the most remote regions in participation of their fruits, how much more are letters° to be magnified,° which as ships pass through the vast seas of time, and make ages so distant to participate of the wisdom, illuminations, and inventions, the one of the other? Nay further, we see some of the philosophers° which were least divine and most immersed in the senses and denied generally the immortality of the soul, yet came to this point,° that whatsoever motions the spirit of man could act and perform without the organs of the body they thought might remain after death; which were only those of the understanding, and not of the affections; so immortal and incorruptible a thing did knowledge seem unto them to be. But we, that know by divine revelation that not only the understanding but the affections purified, not only the spirit but the body changed,° shall be advanced to immortality, do disclaim in° these rudiments of the senses.° But it must be remembered both in this last point, and so it may likewise be needful in other places, that in probation° of the dignity of knowledge or learning I did in the beginning separate divine testimony from human; which method I have pursued, and so handled them both apart.

Nevertheless I do not pretend, and I know it will be impossible for me by any pleading of mine, to reverse the judgment, either of Aesop's cock,° that preferred the barleycorn before the gem; or of Midas,° that being chosen judge between Apollo president of the Muses, and Pan god of the flocks, judged for plenty; or of Paris,° that judged° for beauty and love against wisdom and power; or of Agrippina, 'occidat matrem, modo imperet',° that preferred empire with any condition never so detestable; or of Ulysses, 'qui vetulam praetulit immortalitati',° being a figure° of those which prefer custom and habit before all excellency; or of a number of the like popular judgments. For these things continue as they have been: but so will that also continue whereupon learning hath ever relied, and which faileth not: 'Justificata est sapientia a filiis suis'.°

Book Two

It might seem to have more convenience,° though it come often otherwise to pass (excellent King), that those which are fruitful in their generations, and have in themselves the foresight of immortality in their descendants, should likewise be more careful of the good estate of future times; unto which they know they must transmit and commend over their dearest pledges. Queen Elizabeth was a sojourner° in the world in respect of her unmarried life; and was a blessing to her own times; and yet so as the impression of her good government, besides her happy memory, is not without some effect which doth survive her. But to your Majesty, whom God hath already blessed with so much royal issue,° worthy to continue and represent you for ever, and whose youthful and fruitful bed doth yet promise many the like renovations, it is proper and agreeable to be conversant not only in° the transitory parts of good government, but in those acts also which are in their nature permanent and perpetual. Amongst the which (if affection° do not transport° me) there is not any more worthy than the further endowment of the world with sound and fruitful knowledge. For why should a few received authors stand up like Hercules' Columns,° beyond which there should be no sailing or discovering, since we have so bright and benign a star as your Majesty to conduct and prosper us? To return therefore where we left, it remaineth to consider of what kind those acts are, which have been undertaken and performed by kings and others for the increase and advancement of learning: wherein I purpose to speak actively° without digressing or dilating.°

Let this ground° therefore be laid, that all works are overcomen° by amplitude of reward, by soundness of direction, and by the conjunction of labours. The first multiplieth endeavour, the second preventeth error, and the third supplieth° the frailty of man. But the principal of these is direction: for 'claudus in via antevertit cursorem extra viam';° and Salomon excellently setteth it down, 'If the iron be not sharp, it requireth more strength; but wisdom is that which prevaileth';° signifying that the invention or election° of the mean° is more effectual than any inforcement or accumulation of endeavours. This I am induced to speak, for that (not derogating° from the noble intention of any that have been deservers towards the state of learning) I do observe nevertheless that their works and acts are rather matters of magnificence and memory than of progression° and proficience,° and tend rather to augment the mass of learning in the multitude of learned men than to rectify or raise° the sciences themselves.

The works or acts of merit towards learning are conversant about° three objects; the places of learning, the books of learning, and the persons of the learned. For as water, whether it be the dew of heaven or the springs of the earth, doth scatter and leese itself in the ground, except it be collected into some receptacle, where it may by union comfort° and sustain itself; and for that cause the industry of man hath made and framed spring-heads,° conduits,° cisterns, and pools, which men have accustomed likewise to beautify and adorn with accomplishments° of magnificence and state, as well as of use and necessity; so this excellent liquor of knowledge, whether it descend from divine inspiration or spring from human sense, would soon perish and vanish to oblivion, if it were not preserved in books, traditions,° conferences,° and places appointed, as universities, colleges, and schools, for the receipt° and comforting° of the same.

The works which concern the seats and places of learning are four; foundations and buildings, endowments with revenues, endowments with franchises° and privileges, institutions and ordinances for government; all tending to quietness and privateness of life, and discharge of cares° and troubles; much like the stations which Virgil prescribeth for the hiving of bees:

> Principio sedes apibus statioque petenda,
> Quo neque sit ventis aditus, &c.°

The works touching books are two: first libraries, which are as the shrines where all the relics of the ancient saints, full of true virtue and that without delusion or imposture,° are preserved and reposed;° secondly, new editions of authors, with more correct impressions,° more faithful translations, more profitable glosses, more diligent annotations, and the like.

The works pertaining to the persons of learned men (besides the advancement and countenancing° of them in general) are two: the reward and designation° of readers° in sciences already extant and invented; and the reward and designation of writers and inquirers concerning any parts of learning not sufficiently laboured° and prosecuted.°

These are summarily the works and acts, wherein the merits of many excellent princes and other worthy personages have been conversant. As for any particular commemorations, I call to mind what Cicero said, when he gave general thanks; 'Difficile non aliquem, ingratum quenquam praeterire':° Let us rather, according to the Scriptures,° look unto that part of the race which is before us than look back to that which is already attained.

First therefore, amongst so many great foundations of colleges in Europe, I find it strange° that they are all dedicated to professions,° and none left free to arts and sciences at large. For if men judge that learning should be referred to action, they judge well; but in this they fall into the error described in the ancient fable;° in which the other parts of the body did suppose the stomach had been idle, because it neither performed the office of motion, as the limbs do, nor of sense, as the head doth; but yet notwithstanding it is the stomach that digesteth and distributeth to all the rest. So if any man think philosophy and universality° to be idle studies, he doth not consider that all professions are from thence served and supplied. And this I take to be a great cause that hath hindered the progression° of learning, because these fundamental knowledges have been studied but in passage.° For if you will have a tree bear more fruit than it hath used to do, it is not any thing you can do to the boughs, but it is the stirring of the earth and putting new mould about the roots that must work it. Neither is it to be forgotten that this dedicating of foundations and dotations° to professory° learning hath not only had a malign aspect and influence° upon the growth of sciences, but hath also been prejudicial to states and governments. For hence it proceedeth° that princes find a solitude° in regard of able men to serve them in causes of estate,° because there is no education collegiate which is free; where such as were so disposed might give themselves to histories, modern languages, books of policy and civil discourse, and other the like enablements unto service of estate.°

And because founders of colleges do plant and founders of lectures do water, it followeth well in order° to speak of the defect which is in public lectures;° namely, in the smallness and meanness of the salary or reward which in most places is assigned unto them; whether they be lectures of arts, or of professions. For it is necessary to the progression of sciences that readers° be of the most able and sufficient° men; as those which are ordained for generating and propagating of sciences, and not for transitory use. This cannot be, except their condition and endowment be such as may content the ablest man to appropriate his whole labour° and continue his whole age° in that function and attendance;° and therefore must have a proportion answerable to that mediocrity° or competency of advancement which may be expected from a profession or the practice of a profession. So as, if you will have sciences flourish, you must observe David's military law, which was, that 'those which staid with the carriage° should have equal part with those which were in the action';° else will the carriages be ill° attended.° So readers in sciences are indeed the guardians of the stores and provisions of sciences whence men in active courses are furnished, and therefore ought to have equal entertainment° with them; otherwise if the fathers in sciences be of the weakest sort or be ill-maintained, 'Et patrum invalidi referent jejunia nati.'°

Another defect I note, wherein I shall need some alchemist to help me, who call upon men to sell their books and to build furnaces; quitting and forsaking Minerva° and the Muses as barren virgins, and relying upon Vulcan.° But certain it is that unto the deep, fruitful, and operative° study of many sciences, specially natural philosophy and physic,° books be not only° the instrumentals;° wherein also the beneficence of men hath not been altogether wanting; for we see spheres,° globes, astrolabes, maps, and the like, have been provided as appurtenances to astronomy and cosmography, as well as books: we see likewise that some places instituted° for physic have annexed the commodity° of gardens for simples° of all sorts, and do likewise command the use of dead bodies for anatomies.° But these do respect° but a few things. In general, there will hardly be any main° proficience in the disclosing° of nature, except there be some allowance for expenses about experiments; whether they be experiments appertaining to Vulcanus or Daedalus,° furnace or engine,° or any other kind; and therefore as secretaries and spials° of princes and states bring in bills° for intelligence,° so you must allow the spials and intelligencers° of nature to bring in their bills,° or else you shall be ill advertised.°

And if Alexander° made such a liberal assignation to Aristotle of treasure for the allowance of hunters, fowlers, fishers, and the like, that he might compile an History° of nature, much better do they deserve it that travail° in Arts of nature.

Another defect which I note, is an intermission or neglect in those which are governors in universities of consultation, and in princes or superior persons of visitation;° to enter into account and consideration, whether the readings, exercises, and other customs appertaining unto learning, anciently begun and since continued, be well instituted or no; and thereupon to ground an amendment or reformation in that which shall be found inconvenient.° For it is one of your Majesty's° own most wise and princely maxims, that 'in all usages and precedents, the times be considered wherein they first began; which if they were weak or ignorant, it derogateth from the authority of the usage, and leaveth it for suspect'.° And therefore inasmuch as most of the usages and orders of the universities were derived from more obscure times, it is the more requisite they be re-examined.

In this kind I will give an instance or two for example sake, of things that are the most obvious and familiar. The one is a matter which though it be ancient and general, yet I hold to be an error; which is, that scholars° in universities come too soon and too unripe° to logic and rhetoric;° arts fitter for graduates than children and novices: for these two, rightly taken, are the gravest° of sciences; being the arts of arts, the one for judgment, the other for ornament;° and they be the rules and directions how to set forth and dispose° matter;° and therefore for minds empty and unfraught° with matter, and which have not gathered that which Cicero° calleth *sylva* and *supellex*,° stuff and variety, to begin with those arts (as if one should learn to weigh or to measure or to paint the wind), doth work but this effect, that the wisdom of those arts, which is great and universal, is almost made contemptible, and is degenerate into childish sophistry and ridiculous affectation. And further, the untimely learning of them hath drawn on by consequence the superficial and unprofitable teaching and writing of them, as fitteth° indeed to the capacity of children. Another is a lack I find in the exercises used in the universities, which do make too great a divorce between invention and memory.° For their speeches are either premeditate 'in verbis conceptis',° where nothing is left to invention, or merely extemporal,° where little is left to memory: whereas in life and action there is least use of either of these, but rather of intermixtures of premeditation and invention, notes and memory; so as the exercise fitteth not the practice, nor the image the life; and it is ever a true rule in exercises, that they be° framed as near as may be to the life of practice; for otherwise they do pervert the motions and faculties of the mind, and not prepare them. The truth whereof is not obscure, when scholars come to the practices of professions, or other actions of civil life; which when they set into,° this want is soon found by themselves, and sooner by others. But this part, touching the amendment of the institutions and orders of universities, I will conclude with the clause of Caesar's letter to Oppius and Balbus, 'Hoc quemadmodum fieri possit, nonnulla mihi in mentem veniunt, et multa reperiri possunt; de iis rebus rogo vos ut cogitationem suscipiatis'.°

Another defect which I note, ascendeth a little higher than the precedent.° For as the proficience of learning consisteth much in the orders and institutions of universities in the same states and kingdoms, so it would be yet more advanced, if there were more intelligence mutual° between the universities of Europe than now there is. We see there be many orders and foundations,° which though they be divided under several sovereignties and territories, yet they take themselves to have a kind of contract, fraternity, and correspondence one with the other, insomuch as they have Provincials and Generals. And surely as nature createth brotherhood in families, and arts mechanical contract° brotherhoods in communalties,° and the anointment° of God superinduceth° a brotherhood in kings and bishops; so in like manner there cannot but be a fraternity in learning and illumination, relating to that paternity which is attributed to God, who is called the Father of illuminations or lights.°

The last defect which I will note is, that there hath not been, or very rarely been, any public designation° of writers or inquirers concerning such parts of knowledge as may appear not to have been already sufficiently laboured or undertaken. Unto which point it is an inducement,° to enter into a view and examination what parts of learning have been prosecuted,° and what omitted; for the opinion of plenty is amongst the causes of want, and the great quantity of books maketh a shew rather of superfluity than lack; which surcharge° nevertheless is not to be remedied by making no more books, but by making more good books, which, as the serpent of Moses,° might devour the serpents of the enchanters.

The removing of all the defects formerly enumerate,° except the last, and of the active part also of the last (which is the designation of writers), are 'opera basilica',° towards which the endeavours of a private man may be but as an image in a crossway,° that may point at the way but cannot go° it. But the inducing° part of the latter (which is the survey of learning) may be set forward° by private travel. Wherefore I will now attempt to make a general and faithful perambulation° of learning, with an inquiry what parts thereof lie fresh and waste,° and not improved and converted° by the industry of man; to the end that such a plot° made and recorded to memory may both minister light to any public designation, and also serve to excite voluntary endeavours.° Wherein nevertheless my purpose is at this time to note only omissions and deficiencies, and not to make any redargution° of errors or incomplete prosecutions;° for it is one thing to set forth what ground lieth unmanured,° and another thing to correct ill husbandry in that which is manured.

In the handling and undertaking of which work I am not ignorant what it is that I do now move and attempt, nor insensible of mine own weakness to sustain my purpose. But my hope is that if my extreme love to learning carry me too far, I may obtain the excuse of affection;° for that 'it is not granted to man to love and to be wise.'° But I know well I can use no other liberty of judgment than I must leave to others; and I for my part shall be indifferently° glad either to perform myself or accept from another that duty of humanity, 'Nam qui erranti comiter monstrat viam, &c'.° I do foresee likewise that of those things which I shall enter and register as deficiencies and omissions, many will conceive and censure° that some of them are already done and extant; others to be but curiosities,° and things of no great use; and others to be of too great difficulty and almost impossibility to be compassed and effected. But for the two first, I refer myself to the particulars. For the last, touching impossibility, I take it those things are to be held possible which may be done by some person, though not by every one; and which may be done by many, though not by any one; and which may be done in succession of ages, though not within the hour-glass of one man's life; and which may be done by public designation, though not by private endeavour. But notwithstanding, if any man will take to himself rather that of Salomon, 'Dicit piger, Leo est in via',° than that of Virgil, 'Possunt quia posse videntur',° I shall be content that my labours be esteemed but as the better sort of wishes; for as it asketh° some knowledge to demand° a question not impertinent,° so it requireth some sense to make a wish not absurd.

The parts° of human learning have reference to the three parts of Man's Understanding, which is the seat of learning: History to his Memory, Poesy to his Imagination, and Philosophy to his Reason. Divine learning receiveth the same distribution; for the spirit of man is the same, though the revelation of oracle° and sense be diverse:° so as theology consisteth also of History of the Church; of Parables, which is divine poesy; and of holy Doctrine or precept. For as for that part which seemeth supernumerary,° which is Prophecy, it is but divine history; which hath that prerogative over human, as the narration may be before the fact as well as after.

HISTORIA LITERARUM.° History is Natural, Civil, Ecclesiastical, and Literary; whereof the three first I allow as extant, the fourth I note as deficient. For no man hath propounded to himself the general state of learning to be described and represented from age to age, as many have done the works of nature and the state civil and ecclesiastical; without which the history of the world seemeth to me to be as the statua of Polyphemus° with his eye out; that part being wanting which doth most shew the spirit and life of the person. And yet I am not ignorant that in divers° particular sciences, as of the jurisconsults,° the mathematicians, the rhetoricians, the philosophers, there are set down some small memorials of the schools,° authors, and books; and so likewise some barren relations° touching the invention of arts or usages. But a just story of learning,° containing the antiquities and originals of knowledges, and their sects; their inventions, their traditions; their diverse administrations and managings; their flourishings, their oppositions, decays, depressions, oblivions, removes;° with the causes and occasions of them, and all other events concerning learning, throughout the ages of the world; I may truly affirm to be wanting. The use and end of which work I do not so much design for curiosity,° or satisfaction of those that are the lovers of learning; but chiefly for a more serious and grave purpose, which is this in few words, that it will make learned men wise in the use and administration of learning. For it is not St. Augustine's nor St. Ambrose's works that will make so wise a divine, as ecclesiastical history throughly read and observed; and the same reason is of learning.

HISTORIA NATURAE ERRANTIS.° History of Nature is of three sorts; of nature in course,° of nature erring or varying,° and of nature altered or wrought;° that is, history of Creatures, history of Marvels, and history of Arts.° The first of these no doubt is extant, and that in good perfection; the two latter are handled so weakly and unprofitably, as I am moved to note them as deficient. For I find no sufficient or competent collection of the works of nature which have a digression and deflexion° from the ordinary course of generations,° productions, and motions;° whether they be singularities of place and region, or the strange events of time and chance, or the effects of yet unknown proprieties,° or the instances of exception to general kinds. It is true, I find a number of books of fabulous experiments and secrets, and frivolous impostures for pleasure and strangeness. But a substantial and severe° collection of the Heteroclites° or Irregulars of nature, well examined and described, I find not; specially not with due rejection of fables and popular errors:° for as things now are, if an untruth in nature be once on foot,° what by reason of the neglect of examination and countenance° of antiquity, and what by reason of the use of the opinion° in similitudes° and ornaments of speech, it is never called down.°

The use of this work, honoured with a precedent in Aristotle,° is nothing less than° to give contentment to the appetite of curious° and vain wits, as the manner of Mirabilaries° is to do; but for two reasons, both of great weight; the one to correct the partiality of axioms° and opinions, which are commonly framed only upon common and familiar examples; the other because from the wonders of nature is the nearest intelligence° and passage towards the wonders of art: for it is no more but by following and as it were hounding° Nature in her wanderings, to be able to lead her afterwards to the same place again. Neither am I of opinion, in this History of Marvels, that superstitious narrations of sorceries,° witchcrafts, dreams, divinations,° and the like, where there is an assurance and clear evidence of the fact, be altogether excluded. For it is not yet known in what cases, and how far, effects attributed to superstition do participate of natural causes; and therefore howsoever the practice of such things is to be condemned, yet from the speculation° and consideration of them light may be taken, not only for the discerning of the offences, but for the further disclosing of nature. Neither ought a man to make scruple of entering into these things for inquisition° of truth, as your Majesty hath shewed in your own example;° who with the two clear eyes of religion and natural philosophy have looked deeply and wisely into these shadows, and yet proved yourself to be of the nature of the sun, which passeth through pollutions and itself remains as pure as before. But this I hold fit, that these narrations which have mixture with superstition be sorted by themselves, and not to be mingled with the narrations which are merely° and sincerely° natural. But as for the narrations touching the prodigies and miracles of religions, they are either not true or not natural; and therefore impertinent° for the story of nature.

HISTORIA MECHANICA.° For History of Nature Wrought° or Mechanical, I find some collections° made of agriculture, and likewise of manual arts; but commonly with a rejection of experiments familiar and vulgar.° For it is esteemed a kind of dishonour unto learning to descend to inquiry or meditation upon matters mechanical, except they be such as may be thought secrets, rarities, and special subtleties;° which humour of vain and supercilious arrogancy is justly derided in Plato;° where he brings in Hippias, a vaunting sophist, disputing with Socrates, a true and unfeigned inquisitor° of truth; where the subject being touching beauty, Socrates, after his wandering manner of inductions,° put first an example of a fair virgin, and then of a fair horse, and then of a fair pot well glazed, whereat

Hippias was offended, and said, 'more than for courtesy's sake, he did think much° to dispute with any that did allege° such base and sordid instances': whereunto Socrates answereth, 'You have reason, and it becomes you well, being a man so trim in your vestiments, &c.' and so goeth on in an irony.° But the truth is, they be not the highest instances that give the securest information; as may be well expressed in the tale so common of the philosopher,° that while he gazed upwards to the stars fell into the water; for if he had looked down he might have seen the stars in the water, but looking aloft° he could not see the water in the stars. So it cometh often to pass that mean and small things discover great better than great can discover the small; and therefore Aristotle° noteth well, that 'the nature of every thing is best seen in his° smallest portions', and for that cause he inquireth° the nature of a commonwealth, first in a family, and the simple conjugations° of man and wife, parent and child, master and servant, which are in every cottage: even so likewise the nature of this great city of the world and the policy° thereof must be first sought in mean concordances° and small portions. So we see how that secret of nature, of the turning of iron touched with the loadstone° towards the north, was found out in needles of iron, not in bars of iron.

But if my judgment be of any weight, the use of History Mechanical is of all others the most radical and fundamental towards natural philosophy; such natural philosophy as shall not vanish in the fume° of subtile,° sublime,° or delectable° speculation, but such as shall be operative to the endowment and benefit of man's life. For it will not only minister and suggest for the present many ingenious practices in all trades, by a connexion and transferring of the observations of one art to the use of another, when the experiences of several mysteries° shall fall under the consideration of one man's mind; but further it will give a more true and real illumination concerning causes and axioms° than is hitherto attained. For like as a man's disposition is never well known till he be crossed,° nor Proteus° ever changed shapes till he was straitened° and held fast; so the passages° and variations of nature cannot appear so fully in the liberty of nature, as in the trials and vexations of art.°

For Civil History, it is of three kinds; not unfitly to be compared with the three kinds of pictures or images. For of pictures or images, we see some are unfinished, some are perfect,° and some are defaced. So of histories we may find three kinds, Memorials, Perfect Histories, and Antiquities; for Memorials are history unfinished, or the first or rough draughts of history, and Antiquities are history defaced, or some remnants of history which have casually° escaped the shipwrack of time.

Memorials, or Preparatory History, are of two sorts; whereof the one may be termed Commentaries, and the other Registers. Commentaries are they which set down a continuance of the naked events and actions, without the motives or designs, the counsels, the speeches, the pretexts, the occasions, and other passages of action: for this is the true nature of a Commentary; though Caesar,° in modesty mixed with greatness, did for his pleasure apply the name of a Commentary to the best history of the world.° Registers are collections of public acts, as decrees of council, judicial proceedings, declarations and letters of estate, orations, and the like, without a perfect continuance or contexture of the thread of the narration.

Antiquities° or Remnants of History are, as was said, 'tanquam tabulae naufragii',° when industrious persons by an exact and scrupulous diligence and observation, out of monuments, names, words, proverbs, traditions, private records and evidences, fragments of stories, passages of books that concern not story,° and the like, do save and recover somewhat from the deluge of time.

In these kinds of unperfect histories I do assign no deficience, for they are 'tanquam imperfecte mista',° and therefore any deficience in them is but their nature. As for the corruptions and moths of history, which are Epitomes,° the use of them deserveth to be banished, as all men of sound judgment have confessed; as those that have fretted° and corroded the sound bodies of many excellent histories, and wrought them into base and unprofitable dregs.

History which may be called Just and Perfect History is of three kinds, according to the object which it propoundeth, or pretendeth° to represent: for it either representeth a Time, or a Person, or an Action. The first we call Chronicles, the second Lives, and the third Narrations or Relations. Of these, although the first be the most complete and absolute kind of history and hath most estimation and glory, yet the second excelleth it in profit and use, and the third in verity° and sincerity. For History of Times representeth the magnitude of actions and the public faces and deportments of persons, and passeth over in silence the smaller passages° and motions of men and matters. But such being the workmanship of God as° he doth hang the greatest weight upon the smallest wires, 'maxima e minimis suspendens',° it comes therefore to pass, that such histories do rather set forth the pomp of business than the true and inward resorts°

thereof. But Lives, if they be well written, propounding to themselves a person to represent in whom actions both greater and smaller, public and private, have a commixture, must of necessity contain a more true, native, and lively representation. So again Narrations and Relations° of actions, as the War of Peloponnesus, the Expedition of Cyrus Minor, the Conspiracy of Catiline, cannot but be more purely and exactly true than Histories of Times, because they may choose an argument comprehensible within the notice and instructions of the writer: whereas he that undertaketh the story of a time, especially of any length,° cannot but meet with many blanks and spaces° which he must be forced to fill up out of his own wit and conjecture.°

For the History of Times, (I mean of civil history) the providence of God hath made the distribution:° for it hath pleased God to ordain and illustrate° two exemplar° states of the world, for arms, learning, moral virtue, policy, and laws; the state of Graecia, and the state of Rome; the histories whereof occupying the middle part of time, have more ancient to° them, histories which may by one common name be termed the Antiquities° of the World; and after them, histories which may be likewise called by the name of Modern History.

Now to speak of the deficiencies. As to the Heathen Antiquities of the world, it is in vain to note them for deficient. Deficient they are no doubt, consisting most of fables and fragments; but the deficience cannot be holpen;° for antiquity is like fame, 'caput inter nubila condit',° her head is muffled from our sight. For the History of the Exemplar States, it is extant in good perfection. Not but I could wish there were a perfect course of history for Graecia° from Theseus to Philopoemen (what time° the affairs of Graecia drowned° and extinguished in the affairs of Rome); and for Rome° from Romulus to Justinianus, who may be truly said to be 'ultimus Romanorum'.° In which sequences of story the text of Thucydides and Xenophon in the one, and the texts of Livius, Polybius, Sallustius, Caesar, Appianus, Tacitus, Herodianus° in the other, to be kept entire without any diminution at all, and only to be supplied° and continued. But this is matter of magnificence, rather to be commended than required: and we speak now of parts of learning supplemental, and not of supererogation.°

But for Modern Histories, whereof there are some few very worthy, but the greater part beneath mediocrity, leaving the care of foreign stories to foreign states, because I will not be 'curiosus in aliena republica',° I cannot fail to represent to your Majesty the unworthiness of the history of England in the main continuance thereof, and the partiality and obliquity° of that of Scotland in the latest and largest author° that I have seen; supposing that it would be honour for your Majesty and a work very memorable, if this island of Great Britain, as it is now joined in monarchy for the ages to come, so were joined in one history° for the times passed;° after the manner of the sacred history, which draweth down the story of the Ten Tribes and of the Two Tribes° as twins together. And if it shall seem that the greatness of this work may make it less exactly performed, there is an excellent period of a much smaller compass of time, as to the story of England; that is to say, from the Uniting of the Roses to the Uniting of the Kingdoms;° a portion of time, wherein to my understanding, there hath been the rarest varieties that in like number of successions of any hereditary monarchy hath been known. For it beginneth with the mixed adeption° of a crown, by arms and title; an entry by battle, an establishment by marriage; and therefore times answerable,° like waters after a tempest, full of working and swelling, though without extremity of storm; but well passed through by the wisdom of the pilot,° being one of the most sufficient° kings of all the number. Then followeth the reign of a king,° whose actions, howsoever conducted, had much intermixture with the affairs of Europe, balancing and inclining them variably; in whose time also began that great alteration in the state ecclesiastical,° an action which seldom cometh upon the stage; then the reign of a minor:° then an offer° of an usurpation,° though it was but as 'febris ephemera',° then the reign of a queen° matched with a foreigner: then of a queen° that lived solitary and unmarried, and yet her government so masculine as it had greater impression° and operation° upon the states abroad than it any ways received from thence: and now last, this most happy and glorious event, that this island of Britain, divided from all the world, should be united in itself; and that oracle of rest given to Aeneas, 'Antiquam exquirite matrem',° should now be performed and fulfilled upon the nations of England and Scotland, being now reunited in the ancient mother name of Britain,° as a full period° of all instability and peregrinations:° so that as it cometh to pass in massive bodies, that they have certain trepidations° and waverings° before they fix and settle; so it seemeth that by the providence of God this monarchy, before it was to settle in your Majesty and your generations (in which I hope it is now established for ever), it had these prelusive° changes and varieties.

For Lives,° I do find strange° that these times have so little esteemed the virtues of the times, as that the writing of lives should be no more frequent. For although there be not many sovereign princes or absolute commanders, and that states are most° collected into monarchies, yet are there many worthy personages that deserve better than dispersed report or barren elogies.° For herein the invention° of one of the late poets is proper, and doth well enrich the ancient fiction:° for he feigneth that at the end of the thread or web of every man's life there was a little medal containing the person's name, and that Time waited upon° the shears,° and as soon as the thread was cut, caught the medals and carried them to the river of Lethe;° and about the bank there were many birds flying up and down, that would get the medals and carry them in their beak a little while, and then let them fall into the river: only there were a few swans, which if they got a name, would carry it to a temple where it was consecrate.° And although many men more mortal in their affections° than in their bodies, do esteem desire of name and memory but as a vanity and ventosity,°

Animi nil magnae laudis egentes;°

which opinion cometh from that root,° 'non prius laudes contempsimus, quam laudanda facere desivimus';° yet that will not alter Salomon's judgment, 'Memoria justi cum laudibus, at impiorum nomen putrescet';° the one flourisheth, the other either consumeth° to present oblivion, or turneth to an ill odour. And therefore in that style° or addition, which is and hath been long well received and brought in use, 'felicis memoriae, piae memoriae, bonae memoriae',° we do acknowledge that which Cicero saith, borrowing it from Demosthenes, that 'bona fama propria possessio defunctorum';° which possession I cannot but note that in our times it lieth much waste,° and that therein there is a deficience.

For Narrations and Relations° of particular actions, there were also to be wished a greater diligence therein; for there is no great action but hath some good pen° which attends it. And because it is an ability not common to write a good history, as may well appear by the small number of them; yet if particularity° of actions memorable were but tolerably reported as they pass, the compiling of a complete History of Times might be the better expected, when a writer should arise that were fit for it: for the collection of such relations might be as a nursery garden,° whereby to plant a fair and stately garden when time should serve.

There is yet another partition of history which Cornelius Tacitus maketh, which is not to be forgotten, specially with that application which he accoupleth° it withal, Annals° and Journals:° appropriating to the former matters of estate,° and to the latter acts and accidents of a meaner nature. For giving but a touch° of certain magnificent buildings, he addeth, 'Cum ex dignitate populi Romani repertum sit, res illustres annalibus, talia diurnis urbis actis mandare'.° So as there is a kind of contemplative heraldry,° as well as civil. And as nothing doth derogate from the dignity of a state more than confusion of degrees;° so it doth not a little embase° the authority of an history, to intermingle matters of triumph or matters of ceremony or matters of novelty° with matters of state. But the use of a Journal hath not only been in the history of times, but likewise in the history of persons, and chiefly of actions; for princes in ancient time had, upon point of° honour and policy both, journals kept of what passed day by day. For we see the Chronicle which was read before Ahasuerus,° when he could not take rest,° contained matter of affairs indeed, but such as had passed in his own time, and very lately before: but the Journal° of Alexander's house expressed every small particularity, even concerning his person and court; and it is yet an use well received in enterprises memorable, as expeditions of war, navigations,° and the like, to keep diaries of that which passeth continually.

I cannot likewise be ignorant of a form of writing which some grave and wise men have used, containing a scattered history of those actions which they have thought worthy of memory, with politic discourse° and observation thereupon; not incorporate into the history, but separately, and as the more principal in their intention; which kind of Ruminated° History I think more fit to place amongst books of policy, whereof we shall hereafter speak, than amongst books of history; for it is the true office of history to represent the events themselves together with the counsels, and to leave the observations and conclusions thereupon to the liberty and faculty of every man's judgment. But mixtures are things irregular, whereof no man can define.

So also is there another kind of history manifoldly° mixed, and that is History of Cosmography:° being compounded of natural history, in respect of the regions themselves; of history civil, in respect of the habitations, regiments,° and manners of the people; and the mathematics,° in respect of the climates and configurations° towards the heavens: which part of learning of all others in this latter time hath obtained most proficience. For it may be truly affirmed to the honour of these times, and in a virtuous emulation with antiquity, that this great building of the world had never through-lights° made in it, till the age of us and our fathers; for although they had knowledge of the antipodes,

> Nosque ubi primus equis oriens afflavit anhelis,
> Illic sera rubens accendit lumina Vesper:°

yet that might be by demonstration,° and not in fact; and if by travel, it requireth the voyage but of half the globe. But to circle the earth, as the heavenly bodies do, was not done nor enterprised° till these latter times: and therefore these times may justly bear in their word,° not only *plus ultra*,° in precedence of the ancient *non ultra*,° and 'imitabile fulmen'° in precedence of the ancient 'non imitabile fulmen',

> Demens qui nimbos et non imitabile fulmen &c.°

but likewise 'imitabile coelum';° in respect of the many memorable voyages, after the manner of heaven, about the globe of the earth.°

And this proficience in navigation and discoveries may plant also an expectation of the further proficience and augmentation of all sciences; because it may seem they are ordained by God to be coevals,° that is, to meet in one age. For so the prophet Daniel speaking of the latter times foretelleth, 'Plurimi pertransibunt, et multiplex erit scientia':° as if the openness and through passage° of the world and the increase of knowledge were appointed to be in the same ages; as we see it is already performed in great part; the learning of these latter times not much giving place° to the former two periods or returns° of learning, the one of the Grecians, the other of the Romans.

History Ecclesiastical receiveth the same divisions with° History Civil: but further in the propriety thereof may be divided into History of the Church, by a general name; History of Prophecy; and History of Providence. The first describeth the times of the militant church;° whether it be fluctuant,° as the ark of Noah;° or moveable, as the ark in the wilderness;° or at rest, as the ark in the temple;° that is, the state of the church in persecution, in remove,° and in peace. This part I ought in no sort to note as deficient; only I would that the virtue and sincerity° of it were according° to the mass and quantity. But I am not now in hand° with censures, but with omissions.

HISTORIA PROPHETICA.° The second, which is History of Prophecy, consisteth of two relatives,° the prophecy and the accomplishment; and therefore the nature of such a work ought to be, that every prophecy of the scripture be sorted° with the event fulfilling the same, throughout the ages of the world; both for the better confirmation of faith, and for the better illumination of the church touching those parts of prophecies which are yet unfulfilled; allowing nevertheless that latitude° which is agreeable and familiar unto divine prophecies; being of the nature of their author, with whom a thousand years are but as one day;° and therefore are not fulfilled punctually° at once, but have springing° and germinant° accomplishment throughout many ages, though the height° or fullness of them may refer to some one age. This is a work which I find deficient, but is to be done with wisdom, sobriety, and reverence, or not at all.

The third, which is History of Providence, containeth that excellent correspondence° which is between God's revealed will and his secret will; which though it be so obscure as for the most part it is not legible to the natural man;° no, nor many times to those that behold it from the tabernacle;° yet at some times it pleaseth God, for our better establishment° and the confuting of those which are as without God in the world,° to write it in such text and capital letters that, as the prophet saith, 'he that runneth by may read it';° that is, mere sensual° persons, which hasten by° God's judgments and never bend or fix their cogitations° upon them, are nevertheless in their passage and race° urged to discern it. Such are the notable events and examples of God's judgments, chastisements, deliverances, and blessings. And this is a work which hath passed through the labour of many, and therefore I cannot present° as omitted.

There are also other parts of learning which are Appendices° to history. For all the exterior proceedings of man consist of words and deeds; whereof history doth properly receive and retain in memory the deeds, and if words, yet but as inducements° and passages° to deeds; so are there other books and writings, which are appropriate to the custody and receit° of words only; which likewise are of three sorts: Orations, Letters, and Brief Speeches or Sayings. Orations are pleadings,° speeches of counsel;° laudatives,° invectives,° apologies, reprehensions;° orations of formality or ceremony, and the like. Letters are according to all the variety of occasions; advertisements,° advices, directions,° propositions, petitions, commendatory, expostulatory, satisfactory, of compliment, of pleasure, of discourse, and all other passages° of action. And such as are written from wise men are, of all the words of man, in my judgment the best; for they are more natural than orations and public speeches, and more advised° than conferences or present° speeches. So again letters of affairs from such as manage them or are privy to them are of all others the best instructions for history, and to a diligent reader the best histories in themselves. For Apophthegms,° it is a great loss of that book of Caesar's; for as his history and those few letters of his which we have and those apophthegms which were of his own excel all men's else, so I suppose would his collection of Apophthegms have done; for as for those which are collected by others, either I have no taste in such matters, or else their choice hath not been happy. But upon these three kinds of writings I do not insist,° because I have no deficiences to propound concerning them.

Thus much therefore concerning History; which is that part of learning which answereth to one of the cells,° domiciles,° or offices° of the mind of man; which is that of the Memory.

Poesy is a part of learning in measure° of words for the most part restrained,° but in all other points extremely licensed,° and doth truly refer to the Imagination; which, being not tied to the laws of matter, may at pleasure join that which nature hath severed, and sever that which nature hath joined, and so make unlawful matches and divorces of things: 'Pictoribus atque poetis, &c.'° It is taken in two senses, in respect of words or matter. In the first sense it is but a character of style, and belongeth to arts of speech, and is not pertinent for the present. In the latter, it is (as hath been said) one of the principal portions of learning, and is nothing else but Feigned History, which may be styled° as well in prose as in verse.°

The use of this Feigned History hath been to give some shadow of satisfaction to the mind of man in those points wherein the nature of things doth deny it; the world being in proportion° inferior to the soul; by reason whereof there is agreeable to the spirit of man a more ample greatness, a more exact goodness, and a more absolute variety, than can be found in the nature of things. Therefore, because the acts or events of true history° have not that magnitude which satisfieth the mind of man, poesy feigneth acts and events greater and more heroical; because true history propoundeth the successes° and issues of actions not so agreeable to the merits of virtue and vice, therefore poesy feigns them more just in retribution, and more according to revealed providence; because true history representeth actions and events more ordinary and less interchanged, therefore poesy endueth them with more rareness, and more unexpected and alternative° variations. So as it appeareth that poesy serveth and conferreth° to magnanimity,° morality, and to delectation.° And therefore it was ever thought to have some participation of divineness, because it doth raise and erect° the mind, by submitting the shews° of things to the desires of the mind; whereas reason doth buckle° and bow the mind unto the nature of things. And we see that by these insinuations° and congruities with man's nature and pleasure, joined also with the agreement and consort° it hath with music, it hath had access° and estimation° in rude times and barbarous regions, where other learning stood excluded.

The division of poesy which is aptest in the propriety thereof, (besides those divisions which are common unto it with history, as feigned chronicles, feigned lives; and the appendices of history, as feigned epistles, feigned orations, and the rest); is into Poesy Narrative, Representative,° and Allusive.° The Narrative is a mere imitation of history, with the excesses before remembered;° choosing for subject commonly wars and love, rarely state,° and sometimes pleasure or mirth. Representative is as a visible history, and is an image of actions as if they were present, as history is of actions in nature as they are, (that is) past. Allusive or Parabolical° is a narration applied only to express some special purpose or conceit.° Which latter kind of parabolical wisdom was much more in use in the ancient times, as by the fables of Aesop and the brief sentences of the Seven° and the use of hieroglyphics° may appear. And the cause was, for that it was then of necessity to express any point of reason which was more sharp° or subtile° than the vulgar in that manner; because men in those times wanted° both variety of examples and subtilty of conceit. and as hieroglyphics were before letters, so parables were before arguments: and nevertheless now and at all times they do retain much life and vigour, because reason cannot be so sensible,° nor examples° so fit.°

But there remaineth yet another use of Poesy Parabolical, opposite to that which we last mentioned: for that tendeth to demonstrate and illustrate° that which is taught or delivered, and this other to retire° and obscure it: that is when the secrets and mysteries of religion, policy, or philosophy are involved° in fables or parables. Of this in divine poesy° we see the use is authorized. In heathen° poesy we see the exposition of fables doth fall out° sometimes with great felicity; as in the fable that the giants being overthrown in their war against the gods, the Earth their mother in revenge thereof brought forth Fame:°

Illam Terra parens, ira irritata deorum,
Extremam, ut perhibent, Coeo Enceladoque sororem
Progenuit.°

Expounded° that when princes and monarchs have suppressed actual and open rebels, then the malignity of people (which is the mother of rebellion) doth bring forth libels and slanders and taxations° of the state, which is of the same kind with rebellion, but more feminine. So in the fable that the rest of the gods having conspired to bind Jupiter, Pallas° called Briareus° with his hundred hands to his aid: expounded that monarchies need not fear any curbing of their absoluteness by mighty subjects, as long as by wisdom they keep the hearts of the people, who will be sure to come in° on their side. So in the fable that Achilles was brought up under Chiron° the Centaur, who was part a man and part a beast: expounded ingeniously but corruptly by Machiavel,° that it belongeth to the education and discipline° of princes to know as well how to play the part of the lion in violence and the fox in guile, as of the man in virtue and justice. Nevertheless in many the like encounters,° I do rather think that the fable was° first, and the exposition devised,° than that the moral was first, and thereupon the fable framed.° For I find it was an ancient vanity in Chrysippus,° that troubled himself with great contention° to fasten the assertions of the Stoics upon the fictions of the ancient poets. But yet that all the fables and fictions of the poets were but pleasure° and not figure,° I interpose no opinion.° Surely of those poets which are now extant, even Homer himself (notwithstanding he was made a kind of Scripture° by the latter schools of the Grecians), yet I should without any difficulty pronounce that his fables had no such inwardness° in his own meaning; but what they might have upon° a more original tradition, is not easy to affirm; for he was not the inventor of many of them.

In this third° part of learning, which is poesy, I can report no deficience. For being as a plant that cometh of the lust of the earth, without a formal seed, it hath sprung up and spread abroad more than any other kind. But to ascribe unto it that which is due; for the expressing of affections, passions, corruptions, and customs, we are beholding° to poets more than to the philosophers' works; and for wit and eloquence not much less than to orators' harangues. But it is not good to stay too long in the theatre. Let us now pass on to the judicial° place or palace of the mind, which we are to approach and view with more reverence and attention.

The knowledge of man is as the waters, some descending from above, and some springing from beneath; the one informed° by the light of nature, the other inspired by divine revelation. The light of nature consisteth in the notions of the mind and the reports of the senses; for as for knowledge which man receiveth by teaching, it is cumulative° and not original; as in a water° that besides his own spring-head is fed with other springs and streams. So then according to these two differing illuminations or originals, knowledge is first of all divided into Divinity and Philosophy.

PHILOSOPHIA PRIMA, SIVE DE FONTIBUS SCIENTIARUM.° In Philosophy, the contemplations of man do either penetrate unto God, or are circumferred° to Nature, or are reflected or reverted° upon Himself. Out of which several inquiries there do arise three knowledges, Divine philosophy, Natural philosophy, and Human philosophy or Humanity.° For all things are marked and stamped with this triple character, of the power of God, the difference of° nature, and the use of man. But because the distributions° and partitions of knowledge are not like several lines° that meet in one angle, and so touch but in a point; but are like branches of a tree that meet in a stem, which hath a dimension and quantity of entireness and continuance, before it come to discontinue and break itself into arms and boughs; therefore it is good, before we enter into the former distribution, to erect and constitute one universal science, by the name of 'Philosophia Prima', Primitive° or Summary° Philosophy, as the main and common° way, before we come where the ways part and divide themselves; which science whether I should report as deficient or no, I stand doubtful. For I find a certain rhapsody° of Natural Theology, and of divers parts of Logic; and of that part of Natural Philosophy which concerneth the Principles, and of that other part of Natural Philosophy which concerneth the Soul or Spirit; all these strangely commixed° and confused; but being examined, it seemeth to me rather a depredation° of other sciences, advanced and exalted unto some height of terms,° than any thing solid or substantive° of itself.

Nevertheless I cannot be ignorant of the distinction which is current, that the same things are handled but in several° respects; as for example, that logic considereth of many things as they are in notion,° and this philosophy as they are in nature; the one in appearance, the other in existence. But I find this difference better made than pursued.° For if they had considered Quantity, Similitude, Diversity, and the rest of those Extern° Characters of things, as philosophers, and in nature, their inquiries must of force° have been of a far other kind than they are. For doth any of them, in handling Quantity, speak of the force of union, how and how far it multiplieth virtue?° Doth any give the reason, why some things in nature are so common and in so great mass, and others so rare and in so small quantity? Doth any, in handling Similitude and Diversity, assign the cause why iron should not move to iron, which is more like, but move to the loadstone, which is less like? Why in all diversities of things there should be certain participles° in nature, which are almost ambiguous to which kind they should be referred? But there is a mere° and deep silence touching the nature and operation of those Common Adjuncts° of things, as in nature; and only a resuming° and repeating of the force and use of them in speech or argument.° Therefore, because in a writing of this nature I avoid all subtlety,° my meaning touching this original or universal philosophy is thus, in a plain and gross description by negative: 'that it be a receptacle for all such profitable observations and axioms° as fall not within the compass of any of the special parts of philosophy or sciences, but are more common and of a higher stage.'

Now that there are many of that kind need not be doubted. For example; is not the rule, 'Si inaequalibus aequalia addas, omnia erunt inaequalia',° an axiom as well of justice as of the mathematics? And is there not a true coincidence° between commutative° and distributive° justice, and arithmetical and geometrical proportion? Is not that other rule, 'Quae in eodem tertio conveniunt, et inter se conveniunt',° a rule taken from the mathematics, but so potent in logic as all syllogisms° are built upon it? Is not the observation, 'Omnia mutantur, nil interit',° a contemplation in philosophy thus, That the *quantum*° of nature is eternal? in natural theology thus, That it requireth the same omnipotence to make somewhat nothing, which at the first made nothing somewhat? according to the scripture, 'Didici quod omnia opera quae fecit Deus perseverent in perpetuum; non possumus eis quicquam addere nec auferre'.° Is not the ground,° which Machiavel° wisely and largely discourseth concerning governments, that the way to establish and preserve them is to reduce° them *ad principia*,° a rule in religion and nature as well as in civil administration? Was not the Persian Magic° a reduction or correspondence of the principles and architectures of nature to the rules and policy of governments? Is not the precept of a musician, to fall from a discord or harsh accord upon a concord or sweet accord, alike true in affection?° Is not the trope° of music, to avoid or slide° from the close° or cadence, common with the trope of rhetoric of deceiving expectation?° Is not the delight of the quavering° upon a stop° in music the same with the playing of light upon the water?

> Splendet tremulo sub lumine pontus . . .°

Are not the organs of the senses of one° kind with the organs of reflexion, the eye with a glass,° the ear with a cave or strait° determined and bounded? Neither are these only similitudes,° as men of narrow observation may conceive them to be, but the same footsteps of nature, treading or printing upon several subjects or matters. This science therefore (as I understand it) I may justly report as deficient; for I see sometimes the profounder sort of wits, in handling some particular argument, will now and then draw a bucket of water out of this well for their present use; but the springhead thereof seemeth to me not to have been visited, being of so excellent use both for the disclosing° of nature and the abridgment of art.°

This science being therefore first placed as a common parent, like unto Berecynthia,° which had so much heavenly issue,

> Omnes coelicolas, omnes supera alta tenentes°

we may return to the former distribution of the three philosophies; Divine, Natural, and Human. And as concerning Divine Philosophy or Natural Theology, it is that knowledge or rudiment of knowledge concerning God which may be obtained by the contemplation of his creatures; which knowledge may be truly termed divine in respect of the object, and natural in respect of the light.° The bounds of this knowledge are, that it sufficeth to convince° atheism, but not to inform° religion: and therefore there was never miracle wrought by God to convert an atheist, because the light of nature might have led him to confess° a God. But miracles have been wrought to convert idolaters and the superstitious, because no light of nature extendeth to declare the will and true worship of God. For as all works do shew forth the power and skill of the workman, and not his image; so it is of the works of God; which do shew the omnipotency and wisdom of the maker, but not his image: and therefore therein the heathen opinion differeth from the sacred truth; for they supposed the world to be the image of God,° and man to be an extract° or compendious image of the world; but the Scriptures never vouchsafe to attribute to the world that honour, as to be the image of God, but only 'the work of his hands';° neither do they speak of any other image of God, but man. Wherefore by the contemplation of nature to induce° and inforce° the acknowledgement of God, and to demonstrate his power, providence, and goodness, is an excellent argument, and hath been excellently handled by divers.°

But on the other side, out of the contemplation of nature, or ground of human knowledges, to induce any verity or persuasion concerning the points of faith, is in my judgment not safe: 'Da fidei quae fidei sunt'.° For the Heathen themselves conclude as much in that excellent and divine fable of the golden chain:° that 'men and gods were not able to draw Jupiter down to the earth; but contrariwise, Jupiter was able to draw them up to heaven.' So as we ought not to attempt to draw down or submit the mysteries of God to our reason; but contrariwise to raise and advance our reason to the divine truth. So as in this part of knowledge touching° divine philosophy, I am so far from noting any deficience, as I rather note an excess: whereunto I have digressed, because of the extreme prejudice° which both religion and philosophy hath received and may receive by being commixed° together; as that which undoubtedly will make an heretical religion, and an imaginary and fabulous philosophy.

Otherwise it is of° the nature of angels and spirits,° which is an appendix of theology both divine and natural, and is neither inscrutable nor interdicted; for although the Scripture saith, 'Let no man deceive you in sublime° discourse touching the worship of angels, pressing into that° he knoweth not,'° &c. yet notwithstanding if you observe well that precept, it may appear thereby that there be two things only forbidden, adoration of them, and opinion fantastical° of them; either to extol them further than appertaineth to the degree of a creature,° or to extol a man's knowledge of them further than he hath ground.° But the sober and grounded° inquiry° which may arise out of the passages of holy Scriptures, or out of the gradations° of nature, is not restrained. So of degenerate and revolted° spirits, the conversing with them or the employment of them is prohibited, much more° any veneration towards them. But the contemplation or science° of their nature, their power, their illusions, either by Scripture or reason, is a part of spiritual wisdom. For so the apostle saith, 'We are not ignorant of his stratagems';° and it is no more unlawful to inquire the nature of evil spirits than to inquire the force of poisons in nature, or the nature of sin and vice in morality. But this part touching angels and spirits, I cannot note as deficient, for many have occupied themselves in it; I may rather challenge° it, in many of the writers thereof, as fabulous° and fantastical.

Leaving therefore Divine Philosophy or Natural Theology (not Divinity or Inspired Theology, which we reserve for the last of all, as the haven and sabbath° of all man's contemplations), we will now proceed to Natural Philosophy. If then it be true that Democritus said, that 'the truth of nature lieth hid in certain deep mines and caves',° and if it be true likewise that the Alchemists do so much inculcate,° that Vulcan° is a second nature, and imitateth that dexterously and compendiously° which nature worketh by ambages° and length of time; it were good to divide natural philosophy into the mine and the furnace, and to make two professions or occupations of natural philosophers, some to be pioners° and some smiths; some to dig, and some to refine and hammer. And surely I do best allow of a division of that kind, though in more familiar and scholastical° terms; namely, that these be the two parts of natural philosophy,—the Inquisition of Causes, and the Production of Effects; Speculative, and Operative;° Natural Science, and Natural Prudence.° For as in civil matters there is a wisdom of discourse and a wisdom of direction;° so is it in natural. And here I will make a request, that for the latter (or at least for a part thereof) I may revive and reintegrate° the misapplied and abused name of Natural Magic;° which in the true sense is but Natural Wisdom, or Natural Prudence; taken according to the ancient acception,° purged from vanity and superstition. Now although it be true, and I know it well, that there is an intercourse between Causes and Effects, so as both these knowledges, Speculative and Operative, have a great connexion between themselves; yet because all true and fruitful Natural Philosophy hath a double scale or ladder, ascendent and descendent; ascending from experiments to the invention° of causes, and descending from causes to the invention of new experiments; therefore I judge it most requisite that these two parts be severally considered and handled.

Natural Science or Theory° is divided into Physic° and Metaphysic:° wherein I desire it may be conceived that I use the word Metaphysic in a differing sense from that that is received: and in like manner I doubt not but it will easily appear to men of judgment that in this and other particulars, wheresoever my conception and notion may differ from the ancient, yet I am studious to keep the ancient terms. For hoping well to deliver myself from mistaking° by the order and perspicuous expressing of that° I do propound, I am otherwise zealous and affectionate° to recede as little from antiquity, either in terms or opinions, as may stand with° truth and the proficience of knowledge. And herein I cannot a little marvel at the philosopher Aristotle,° that did proceed in such a spirit of difference and contradiction towards all antiquity; undertaking not only to frame new words of science at pleasure, but to confound° and extinguish all ancient wisdom; insomuch as he never nameth or mentioneth an ancient author or opinion, but to confute and reprove;° wherein for glory,° and drawing followers and disciples, he took the right course. For certainly there cometh to pass and hath place in human truth, that which was noted and pronounced in the highest truth: 'Veni in nomine Patris, nec recipitis me; si quis venerit in nomine suo, eum recipietis'.° But in this divine aphorism (considering to whom it was applied, namely to Antichrist, the highest deceiver), we may discern well that 'the coming in a man's own name,' without regard of antiquity or paternity, is no good sign of truth; although it be joined with the fortune and success of an 'Eum recipietis'.° But for this excellent person Aristotle,° I will think of him that he learned that humour of his scholar,° with whom it seemeth he did emulate,° the one to conquer all opinions,° as the other to conquer all nations. Wherein nevertheless, it may be, he may at some men's hands that are of a bitter° disposition get a like° title as his scholar did;

> Felix terrarum praedo, non utile mundo
> Editus exemplum, &c.

so 'Felix doctrinae praedo'.° But to me on the other side that do desire, as much as lieth in my pen, to ground° a sociable intercourse between antiquity° and proficience,° it seemeth best to keep way with° antiquity 'usque ad aras',° and therefore to retain the ancient terms, though I sometimes alter the uses and definitions; according to the moderate proceeding in civil government, where although there be some alteration, yet that holdeth which Tacitus wisely noteth, 'eadem magistratuum vocabula'.°

To return therefore to the use and acception° of the term Metaphysic, as I do now understand the word: It appeareth by that which hath been already said, that I intend Philosophia Prima, Summary° Philosophy, and Metaphysic, which heretofore have been confounded° as one, to be two distinct things. For the one I have made as a parent or common ancestor to all knowledge, and the other I have now brought in as a branch or descendent of Natural Science. It appeareth likewise that I have assigned to Summary Philosophy the common principles and axioms which are promiscuous° and indifferent° to several sciences. I have assigned unto it likewise the inquiry touching the operation of the relative and adventive° characters of essences,° as Quantity, Similitude, Diversity, Possibility, and the rest; with this distinction and provision; that they be handled as they have efficacy in nature, and not logically.° It appeareth likewise that Natural Theology, which heretofore hath been handled confusedly with Metaphysic, I have inclosed and bounded by itself. It is therefore now a question, what is left remaining for Metaphysic; wherein I may without prejudice preserve thus much of the conceit° of antiquity, that Physic should contemplate that which is inherent in matter and therefore transitory,° and Metaphysic that which is abstracted° and fixed.° And again that Physic should handle that which supposeth° in nature only a being and moving, and Metaphysic should handle that which supposeth further in nature a reason, understanding, and platform.° But the difference, perspicuously expressed, is most familiar and sensible.° For as we divided Natural Philosophy in general into the Inquiry of Causes and Production of Effects; so that part which concerneth the Inquiry of Causes we do subdivide, according to the received and sound division of Causes; the one part, which is Physic, enquireth and handleth the Material and Efficient Causes; and the other, which is Metaphysic, handleth the Formal and Final Causes.°

Physic (taking it according to the derivation,° and not according to our idiom for Medicine) is situate° in a middle term or distance between Natural History and Metaphysic. For Natural History describeth the 'variety of things'; Physic, the causes, but 'variable or respective° causes', and Metaphysic, the 'fixed and constant causes.'

> Limus ut hic durescit, et haec ut cera liquescit,
> Uno eodemque igni.°

Fire is the cause of induration,° but respective to clay; fire is the cause of colliquation,° but respective to wax; but fire is no constant cause either of induration or colliquation. So then the physical causes are but the efficient and the matter.°

Physic hath three parts; whereof two respect nature united or collected,° the third contemplateth nature diffused° or distributed.° Nature is collected either into one entire total, or else into the same principles or seeds.° So as the first doctrine is touching the Contexture or Configuration° of things, as 'de mundo, de universitate rerum'.° The second is the doctrine concerning the Principles or Originals of things. The third is the doctrine concerning all Variety and Particularity of things, whether it be of the differing substances, or their differing qualities and natures; whereof there needeth no enumeration, this part being but as a gloss° or paraphrase, that attendeth upon the text of Natural History. Of these three I cannot report any as deficient.° In what truth or perfection they are handled, I make not now any judgment: but they are parts of knowledge not deserted° by the labour of man.

METAPHYSICA SIVE DE FORMIS ET FINIBUS RERUM.° For Metaphysic, we have assigned unto it the inquiry of Formal and Final Causes; which assignation, as to the former of them, may seem to be nugatory° and void, because of the received and inveterate° opinion that the inquisition of man is not competent to find out essential forms° or true differences:° of which opinion we will take this hold; that the invention° of Forms° is of all other parts of knowledge the worthiest to be sought, if it be possible to be found. As for the possibility, they are ill discoverers that think there is no land when they can see nothing but sea. But it is manifest that Plato° in his opinion of Ideas, as one that had a wit of elevation° situate as upon a cliff, did descry° that forms were the true object of knowledge; but lost the real fruit of his opinion, by considering of forms as absolutely abstracted from matter, and not confined° and determined° by matter; and so turning his opinion upon Theology, wherewith all his natural philosophy is infected. But if any man shall keep a continual watchful and severe eye upon action, operation, and the use of knowledge, he may advise° and take notice what are the Forms, the disclosures whereof are fruitful and important to the state of man. For as to the Forms of substances—Man only except,° of whom it is said, 'Formavit hominem de limo terrae, et spiravit in faciem ejus spiraculum vitae',° and not as of all other creatures, 'Producant aquae, producat terra',°—the Forms of Substances I say (as they are now by compounding° and transplanting° multiplied) are so perplexed,° as they are not to be enquired;° no more than it were either possible or to purpose to seek in gross° the forms of those sounds which make words, which by composition° and transposition of letters are infinite. But on the other side, to enquire the form of those sounds or voices which make simple letters is easily comprehensible, and being known, induceth° and manifesteth the forms of all words, which consist° and are compounded of them. In the same manner to enquire the Form of a lion, of an oak, of gold, nay of water, of air, is a vain pursuit: but to enquire the Forms of sense, of voluntary motion, of vegetation, of colours, of gravity and levity,° of density, of tenuity,° of heat, of cold, and all other natures and qualities, which like an alphabet are not many, and of which the essences (upheld° by matter) of all creatures do consist; to enquire I say the true forms of these, is that part° of Metaphysic which we now define of.° Not but that Physic doth make inquiry and take consideration of the same natures: but how? Only as to the Material and Efficient Causes of them, and not as to the Forms. For example; if the cause of Whiteness° in snow or froth be enquired, and it be rendered thus, that 'the subtile° intermixture of air and water is the cause,' it is well rendered; but nevertheless, is this the Form of Whiteness? No; but it is the Efficient, which is ever but 'vehiculum formae'.° This part of Metaphysic I do not find laboured° and performed; whereat I marvel not, because I hold it not possible to be invented by that course of invention° which hath been used; in regard that men (which is the root of all error) have made too untimely a departure and too remote a recess° from particulars.

But the use of this part of Metaphysic which I report as deficient, is of the rest the most excellent in two respects; the one, because it is the duty and virtue of all knowledge to abridge the infinity of individual experience as much as the conception of truth will permit, and to remedy the complaint of 'vita brevis, ars longa',° which is performed by uniting° the notions and conceptions of sciences. For knowledges are as pyramides,° whereof history° is the basis: so of Natural Philosophy the basis is Natural History; the stage next° the basis is Physic; the stage next the vertical point° is Metaphysic. As for the vertical point, 'Opus quod operatur Deus a principio usque ad finem',° the Summary Law of Nature, we know not whether man's inquiry can attain unto it. But these three be the° true stages of knowledge; and are to them that are depraved° no better than the giants' hills,

> Ter sunt conati imponere Pelio Ossam,
> Scilicet atque Ossae frondosum involvere Olympum:°

but to those which refer all things to the glory of God, they are as the three acclamations, 'Sancte, sancte, sancte';° holy in the description or dilatation° of his works, holy in the connexion or concatenation° of them, and holy in the union of them in a perpetual and uniform law.° And therefore the speculation was excellent in Parmenides and Plato,° although but a speculation in them, That all things by scale° did ascend to unity. So then always that knowledge is worthiest, which is charged° with least multiplicity; which appeareth to be Metaphysic; as that which considereth the Simple Forms or Differences of things, which are few in number, and the degrees and co-ordinations whereof make all this variety.

The second respect which valueth° and commendeth this part of Metaphysic, is that it doth enfranchise° the power of man unto the greatest liberty and possibility of works and effects. For Physic carrieth men in narrow and restrained ways, subject to many accidents of impediments, imitating the ordinary flexuous° courses of nature; but 'latae undique sunt sapientibus viae':° to sapience° (which was anciently defined to be 'rerum divinarum et humanarum scientia')° there is ever choice of means. For physical causes give light to new invention 'in simili materia';° but whosoever knoweth any form, knoweth the utmost possibility of superinducing° that nature upon any variety of matter, and so is less restrained in operation, either to the basis of the Matter,° or the condition of the Efficient: which kind of knowledge Salomon likewise, though in a more divine sense, elegantly describeth: 'Non arctabuntur gressus tui, et currens non habebis offendiculum'.° The ways of sapience are not much liable° either to particularity or chance.

The second part of Metaphysic is the inquiry of final causes, which I am moved to report not as omitted, but as misplaced.° And yet if it were but a fault in order,° I would not speak of it; for order is matter of illustration, but pertaineth not to the substance of sciences: but this misplacing hath caused a deficience, or at least a great improficience° in the sciences themselves. For the handling of final causes mixed with the rest in physical inquiries, hath intercepted° the severe and diligent inquiry of all real and physical causes, and given men the occasion to stay upon° these satisfactory° and specious° causes, to the great arrest and prejudice of further discovery. For this I find done not only by Plato,° who ever anchoreth upon that shore, but by Aristotle, Galen, and others, which do usually likewise fall° upon these flats° of discoursing° causes. For to say° that the hairs of the eye lids are for a quickset° and fence about the sight; or that the firmness of the skins and hides of living creatures is to defend them from the extremities of heat or cold; or that the bones are for the columns or beams, whereupon the frames of the bodies of living creatures are built; or that the leaves of trees are for protecting of the fruit; or that the clouds are for watering of the earth; or that the solidness° of the earth is for the station° and mansion° of living creatures, and the like, is well enquired and collected° in Metaphysic; but in Physic they are impertinent.° Nay, they are indeed but remoras° and hinderances to stay and slug° the ship from further sailing, and have brought this to pass, that the search of the Physical Causes hath been neglected and passed in silence.

And therefore the natural philosophy of Democritus° and some others,° who did not suppose a mind or reason in the frame of things,° but attributed the form thereof able° to maintain itself to infinite essays or proofs° of nature, which they term 'fortune', seemeth to me (as far as I can judge by the recital° and fragments which remain unto us) in particularities of physical causes more real and better enquired than that of Aristotle and Plato; whereof both intermingled final causes, the one as a part of theology, and the other as a part of logic, which were the favourite studies respectively of both those persons. Not because those final causes are not true, and worthy to be enquired, being kept within their own province; but because their excursions° into the limits of physical causes hath bred a vastness° and solitude in that track.° For otherwise keeping their precincts and borders, men are extremely deceived if they think there is an enmity or repugnancy at all between them. For the cause rendered,° that the hairs about the eye-lids are for the safeguard of the sight, doth not impugn° the cause rendered, that pilosity° is incident° to orifices of moisture; 'Muscosi fontes',° &c. Nor the cause rendered, that the firmness of hides is for the armour of the body against extremities of heat or cold, doth not impugn the cause rendered, that contraction of pores is incident° to the outwardest° parts, in regard of their adjacence° to foreign or unlike bodies; and so of the rest: both causes being true and compatible, the one declaring an intention,° the other a consequence only. Neither doth this call in question or derogate from divine providence, but highly confirm and exalt it. For as in civil actions he is the greater and deeper politique,° that can make other men the instruments of his will and ends and yet never acquaint them with his purpose, so as they shall do it and yet not know what they do, than he that imparteth his meaning to those he employeth; so is the wisdom of God more admirable, when nature intendeth one thing and providence draweth forth another, than if he had communicated to particular creatures and motions the characters and impressions° of his providence. And thus much for Metaphysic; the latter part whereof I allow as extant, but wish it confined to its proper place.

Nevertheless there remaineth yet another part of Natural Philosophy, which is commonly made a principal part, and holdeth rank with Physic special° and Metaphysic; which is Mathematic; but I think it more agreeable to the nature of things and to the light of order to place it as a branch of Metaphysic; for the subject of it being Quantity; not Quantity indefinite, which is but a relative and belongeth to 'philosophia prima' (as hath been said),° but Quantity determined° or proportionable; it appeareth to be one of the Essential

Forms of things; as that that is causative in nature of a number of effects; insomuch as we see in the schools both of Democritus and of Pythagoras,° that the one did ascribe figure° to the first seeds of things, and the other did suppose numbers to be the principles and originals of things. And it is true also that of all other forms (as we understand forms) it is the most abstracted and separable from matter, and therefore most proper to Metaphysic; which hath likewise been the cause why it hath been better laboured and enquired than any of the other forms, which are more immersed into matter. For it being the nature of the mind of man (to the extreme prejudice of knowledge) to delight in the spacious liberty of generalities, as in a champain° region, and not in the inclosures of particularity;° the Mathematics of all other knowledge were the goodliest fields to satisfy that appetite. But for the placing of this science, it is not much material:° only we have endeavoured in these our partitions° to observe a kind of perspective,° that one part may cast light upon another.

The Mathematics are either Pure or Mixed. To the Pure Mathematics are those sciences belonging which handle Quantity Determinate, merely° severed from any axioms of natural philosophy; and these are two, Geometry and Arithmetic; the one handling Quantity continued,° and the other dissevered.° Mixed hath for subject some axioms or parts of natural philosophy, and considereth Quantity determined, as it is auxiliary° and incident° unto them. For many parts of nature can neither be invented° with sufficient subtlety° nor demonstrated with sufficient perspicuity nor accommodated unto use with sufficient dexterity, without the aid and intervening of the Mathematics: of which sort° are Perspective, Music, Astronomy, Cosmography, Architecture, Enginery,° and divers others. In the Mathematics I can report no deficience, except it be that men do not sufficiently understand the excellent use of the Pure Mathematics, in that they do remedy and cure many defects in the wit and faculties intellectual. For if the wit be too dull, they sharpen it; if too wandering,° they fix it; if too inherent° in the sense, they abstract it. So that as tennis is a game of no use in itself, but of great use in respect it maketh a quick eye and a body ready to put itself into all postures; so in the Mathematics, that use which is collateral and intervenient° is no less worthy than that which is principal and intended. And as for the Mixed Mathematics,° I may only make this prediction, that there cannot fail to be more kinds of them, as nature grows further disclosed.° Thus much of Natural Science, or the part of nature Speculative.

For Natural Prudence, or the part Operative° of Natural Philosophy, we will divide it into three parts, Experimental, Philosophical, and Magical; which three parts active have a correspondence and analogy with the three parts Speculative, Natural History, Physic, and Metaphysic. For many operations have been invented, sometimes by a casual incidence° and occurrence, sometimes by a purposed experiment;° and of those which have been found by an intentional experiment, some have been found out by varying or extending the same experiment, some by transferring and compounding divers° experiments the one into the other, which kind of invention an empiric° may manage. Again, by the knowledge of physical causes there cannot fail to follow many indications and designations of new particulars, if men in their speculation will keep one eye upon use and practice. But these are but coastings° along the shore, 'premendo littus iniquum':° for it seemeth to me there can hardly be discovered any radical or fundamental alterations and innovations in nature, either by the fortune and essays of experiments, or by the light and direction° of physical causes.

NATURALIS MAGIA SIVE PHYSICA OPERATIVA MAJOR.° If therefore we have reported Metaphysic deficient, it must follow that we do the like of Natural Magic, which hath relation thereunto. For as for the Natural Magic whereof now there is mention in books,° containing certain credulous and superstitious conceits and observations of Sympathies and Antipathies° and hidden proprieties,° and some frivolous experiments, strange rather by disguisement° than in themselves; it is as far differing in truth of nature from such a knowledge as we require, as the story of king Arthur of Britain,° or Hugh of Bordeaux,° differs from Caesar's commentaries in truth of story. For it is manifest that Caesar did greater things 'de vero'° than those imaginary heroes were feigned to do. But he did them not in that fabulous manner. Of this kind of learning the fable° of Ixion was a figure,° who designed to enjoy Juno, the goddess of power; and instead of her had copulation with a cloud, of which mixture were begotten centaurs and chimeras.° So whosoever shall entertain° high and vaporous° imaginations instead of a laborious and sober inquiry of truth, shall beget hopes and beliefs of strange and impossible shapes.

And therefore we may note in these sciences which hold so much of° imagination and belief, as this degenerate Natural Magic, Alchemy, Astrology, and the like, that in their propositions the description of the means is ever more monstrous than the pretence° or end. For it is a thing more probable, that he that knoweth well the natures of

Weight, of Colour, of Pliant and Fragile° in respect of the hammer, of Volatile and Fixed in respect of the fire, and the rest, may superinduce upon some metal the nature and form of gold by such mechanique° as belongeth to the production of the natures afore rehearsed, than that some grains of the medicine° projected° should in a few moments of time turn a sea of quicksilver or other material into gold. So it is more probable, that he that knoweth the nature of arefaction,° the nature of assimilation° of nourishment to the thing nourished, the manner of increase and clearing° of spirits,° the manner of the depredations° which spirits make upon the humours° and solid parts, shall by ambages° of diets, bathings, anointings, medicines, motions,° and the like, prolong life or restore some degree of youth or vivacity, than that it can be done with the use of a few drops or scruples° of a liquor or receit.°

INVENTARIUM OPUM HUMANARUM.° To conclude therefore, the true Natural Magic, which is that great liberty and latitude° of operation which dependeth upon the knowledge of Forms, I may report deficient, as the relative thereof° is. To which part, if we be serious and incline not to vanities and plausible° discourse, besides the deriving and deducing the operations themselves from Metaphysic, there are pertinent two points of much purpose, the one by way of preparation, the other by way of caution. The first is, that there be made a Calendar resembling an inventory of the estate of man, containing all the inventions (being the works or fruits of nature or art) which are now extant and whereof man is already possessed; out of which doth naturally result a note,° what things are yet held impossible, or not invented; which calendar will be the more artificial° and serviceable, if to every reputed impossibility you add what thing is extant which cometh the nearest in degree to that impossibility; to the end that by these optatives° and potentials° man's inquiry may be the more awake in deducing direction of works from the speculation° of causes. And secondly, that those experiments be not only esteemed which have an immediate and present use, but those principally which are of most universal consequence for invention of other experiments, and those which give most light to the invention of causes; for the invention of the mariner's needle,° which giveth the direction, is of no less benefit for navigation than the invention of the sails, which give the motion.

Thus have I passed through Natural Philosophy, and the deficiencies thereof; wherein if I have differed from the ancient and received doctrines, and thereby shall move° contradiction; for my part, as I affect not to dissent, so I purpose not to contend. If it be truth,

Non canimus surdis, respondent omnia sylvae:°

the voice of nature will consent, whether the voice of man do or no. And as Alexander Borgia° was wont to say of the expedition of the French for Naples, that they came with chalk in their hands to mark up their lodgings, and not with weapons to fight; so I like better that entry of truth which cometh peaceably with chalk to mark up those minds which are capable to lodge and harbour it, than that which cometh with pugnacity and contention.

CONTINUATIO PROBLEMATUM IN NATURA. CATALOGUS FALSITATUM GRASSANTIUM IN HISTORIA NATURAE.° But there remaineth a division of Natural Philosophy according to the report of the inquiry, and nothing concerning the matter or subject; and that is Positive and Considerative;° when the inquiry reporteth either an Assertion or a Doubt. These doubts or 'non liquets'° are of two sorts, Particular and Total. For the first, we see a good example thereof in Aristotle's *Problems*,° which deserved to have had a better continuance, but so nevertheless as there is one point whereof warning is to be given and taken. The registering of doubts hath two excellent uses: the one, that it saveth philosophy from errors and falsehoods; when that which is not fully appearing is not collected into assertion,° whereby error might draw error, but reserved in doubt: the other, that the entry of doubts are as so many suckers or spunges to draw° use° of knowledge; insomuch as that which, if doubts had not preceded, a man should° never have advised° but passed it over without note, by the suggestion and solicitation° of doubts is made to be attended and applied. But both these commodities do scarcely countervail° an inconvenience which will intrude itself, if it be not debarred; which is, that when a doubt is once received° men labour rather how to keep it a doubt still than how to solve it, and accordingly° bend their wits. Of this we see the familiar example in lawyers and scholars,° both which if they have once admitted a doubt, it goeth ever after authorised for a doubt. But that use of wit and knowledge is to be allowed, which laboureth to make doubtful things certain, and not those which labour to make certain things doubtful. Therefore these calendars of doubts I commend as excellent things, so that there be this caution used, that when they be throughly sifted and brought to resolution,° they be from thenceforth omitted, decarded,° and not continued to cherish and encourage men in doubting. To which calendar of doubts or problems,

I advise be annexed another calendar, as much or more material,° which is a calendar of popular errors:° I mean chiefly, in natural history such as pass° in speech and conceit, and are nevertheless apparently° detected and convicted of untruth; that man's knowledge be not weakened nor imbased° by such dross and vanity.

DE ANTIQUIS PHILOSOPHIIS.° As for the doubts or non liquets general or in total, I understand those differences of opinions touching the principles of nature and the fundamental points of the same, which have caused the diversity of sects,° schools, and philosophies; as that of Empedocles, Pythagoras, Democritus, Parmenides, and the rest. For although Aristotle, as though he had been of the race of the Ottomans,° thought he could not reign except the first thing he did he killed all his brethren; yet to those that seek truth and not magistrality,° it cannot but seem a matter of great profit to see before them the several opinions touching the foundations of nature; not for any exact truth that can be expected in those theories; for as the same phaenomena in astronomy° are satisfied by the received astronomy° of the diurnal motion and the proper motions of the planets with their eccentrics and epicycles,° and likewise by the theory of Copernicus° who supposed the earth to move; and the calculations are indifferently° agreeable to both; so the ordinary face and view of experience° is many times satisfied by several theories and philosophies; whereas to find the real truth requireth another manner of severity and attention. For as Aristotle saith° that children at the first will call every woman mother, but afterward they come to distinguish according to truth; so experience, if it be in childhood, will call every philosophy mother, but when it cometh to ripeness it will discern the true mother. So as in the mean time it is good to see the several glosses and opinions upon nature, whereof it may be every one in some one point hath seen clearer than his fellows. Therefore I wish some collection to be made painfully and understandingly° *de antiquis philosophiis*,° out of all the possible light which remaineth to us of them. Which kind of work I find deficient. But here I must give warning, that it be done distinctly and severedly;° the philosophies of every one throughout by themselves; and not by titles packed and faggoted° up together, as hath been done by Plutarch.° For it is the harmony of a philosophy in itself which giveth it light and credence; whereas if it be singled and broken, it will seem more foreign° and dissonant. For as when I read in Tacitus° the actions of Nero or Claudius, with circumstances of times, inducements,° and occasions, I find them not so strange; but when I read them in Suetonius°

Tranquillus gathered into titles° and bundles, and not in order of time, they seem more monstrous and incredible; so is it of any philosophy reported entire, and dismembered by articles.° Neither do I exclude opinions of latter times to be likewise represented in this calendar of sects of philosophy, as that of Theophrastus Paracelsus,° eloquently reduced into an harmony by the pen of Severinus° the Dane; and that of Telesius,° and his scholar Donius,° being as a pastoral° philosophy, full of sense but of no great depth; and that of Fracastorius,° who though he pretended not to make any new philosophy, yet did use the absoluteness° of his own sense upon the old; and that of Gilbertus° our countryman, who revived, with some alterations and demonstrations,° the opinions of Xenophanes;° and any other worthy to be admitted.

Thus have we now dealt with two of the three beams of man's knowledge; that is *Radius Directus*,° which is referred to nature, *Radius Refractus*,° which is referred to God, and cannot report truly because of the inequality° of the medium. There resteth° *Radius Reflexus*° whereby Man beholdeth and contemplateth himself.

We come therefore now to that knowledge whereunto the ancient oracle° directeth us, which is the knowledge of ourselves; which deserveth the more accurate handling, by how much° it toucheth us more nearly. This knowledge, as it is the end and term° of natural philosophy in the intention° of man, so notwithstanding it is but a portion of natural philosophy in the continent of nature. And generally let this be a rule, that all partitions of knowledges be accepted rather for lines and veins,° than for sections and separations; and that the continuance and entireness of knowledge be preserved. For the contrary hereof hath made particular sciences to become barren, shallow, and erroneous, while they have not been nourished and maintained from the common fountain. So we see Cicero° the orator complained of Socrates and his school, that he was the first that separated philosophy and rhetoric; whereupon rhetoric became an empty and verbal art. So we may see that the opinion of Copernicus touching the rotation of the earth, which astronomy itself cannot correct because it is not repugnant to° any of the phaenomena, yet natural philosophy may correct. So we see also that the science of medicine, if it be destituted° and forsaken by natural philosophy, it is not much better than an empirical practice. With this reservation therefore we proceed to Human Philosophy or Humanity,° which hath two parts: the one considereth man segregate,° or distributively; the other congregate,° or in society. So as Human Philosophy is either

Simple and Particular, or Conjugate° and Civil.° Humanity Particular consisteth of the same parts whereof man consisteth; that is, of knowledges which respect the Body, and of knowledges that respect the Mind. But before we distribute so far, it is good to constitute.° For I do take the consideration in general and at large of Human Nature to be fit to be emancipate° and made a knowledge by itself; not so much in regard of those delightful and elegant discourses which have been made of the dignity of man, of his miseries,° of his state and life, and the like adjuncts° of his common and undivided nature; but chiefly in regard of the knowledge concerning the sympathies and concordances° between the mind and body, which, being mixed, cannot be properly assigned to the sciences of either.

PARS PHYSIOGNOMIAE, DE GESTU SIVE MOTU CORPORIS.° This knowledge hath two branches: for as all leagues and amities consist of mutual Intelligence° and mutual Offices,° so this league of mind and body hath these two parts; how the one discloseth the other, and how the one worketh upon the other; Discovery, and Impression.° The former of these hath begotten two arts, both of Prediction or Prenotion;° whereof the one is honoured with the inquiry of Aristotle,° and the other of Hippocrates.° And although they have of latter time been used to be coupled with superstitious and fantastical arts,° yet being purged and restored to their true state, they have both of them a solid ground in nature, and a profitable use in life. The first is Physiognomy,° which discovereth the disposition of the mind by the lineaments of the body. The second is the Exposition of Natural Dreams,° which discovereth the state of the body by the imaginations of the mind. In the former of these I note a deficience. For Aristotle hath very ingeniously and diligently handled the factures° of the body, but not the gestures of the body, which are no less comprehensible by art,° and of greater use and advantage. For the Lineaments of the body do disclose the disposition and inclination of the mind in general; but the Motions of the countenance and parts do not only so, but do further disclose the present humour and state of the mind and will. For as your Majesty saith most aptly and elegantly, 'As the tongue speaketh to the ear, so the gesture speaketh to the eye'.° And therefore a number of subtile° persons, whose eyes do dwell upon the faces and fashions of men, do well know the advantage of this observation, as being most° part of their ability; neither can it be denied but that it is a great discovery of dissimulations,° and a great direction° in business.°

The latter branch, touching Impression, hath not been collected into art, but hath been handled dispersedly; and it hath the same relation or antistrophe° that the former hath. For the consideration is double: Either how, and how far the humours and affects° of the body do alter or work upon the mind; or again, how and how far the passions or apprehensions of the mind do alter or work upon the body. The former of these hath been inquired and considered as a part and appendix of Medicine, but much more as a part of Religion or Superstition.° For the physician prescribeth cures of the mind in phrensies° and melancholy passions; and pretendeth° also to exhibit° medicines to exhilarate° the mind, to confirm the courage, to clarify the wits, to corroborate° the memory, and the like; but the scruples° and superstitions of diet and other regiment° of the body in the sect of the Pythagoreans,° in the heresy of the Manicheans,° and in the law of Mahomet,° do exceed.° So likewise the ordinances in the Ceremonial Law,° interdicting the eating of the blood and the fat, distinguishing between beasts clean and unclean for meat, are many and strict. Nay the faith° itself being clear and serene° from all clouds of Ceremony, yet retaineth the use of fastings, abstinences, and other macerations° and humiliations of the body, as things real, and not figurative.° The root and life of all which prescripts° is, (besides the ceremony), the consideration of that dependency which the affections of the mind are submitted unto upon the state and disposition of the body. And if any man of weak judgment do conceive° that this suffering of the mind from the body doth either question° the immortality or derogate from the sovereignty of the soul, he may be taught in easy instances, that the infant in the mother's womb is compatible with° the mother and yet separable; and the most absolute monarch is sometimes led by his servants and yet without subjection.

As for the reciprocal knowledge, which is the operation of the conceits and passions of the mind upon the body, we see all wise physicians in the prescriptions of their regiments to their patients do ever consider 'accidentia animi',° as of great force to further or hinder remedies or recoveries; and more specially it is an inquiry of great depth and worth concerning Imagination,° how and how far it altereth the body proper of the imaginant.° For although it hath a manifest power to hurt, it followeth not it hath the same degree of power to help; no more than a man can conclude, that because there be pestilent airs,° able suddenly to kill a man in health, therefore there should be sovereign° airs, able suddenly to cure a man in sickness. But the inquisition of this part is of great use, though it needeth, as

Socrates said, 'a Delian diver',° being difficult and profound. But unto all this knowledge 'de communi vinculo',° of the concordances between the mind and the body, that part of inquiry is most necessary, which considereth of the seats and domiciles° which the several faculties of the mind do take and occupate° in the organs of the body; which knowledge hath been attempted, and is controverted,° and deserveth to be much better enquired. For the opinion of Plato,° who placed the understanding in the brain, animosity (which he did unfitly call anger, having a greater mixture with pride) in the heart, and concupiscence or sensuality in the liver, deserveth not to be despised; but much less to be allowed.° So then we have constituted (as in our own wish and advice) the inquiry touching human nature entire, as a just portion of knowledge to be handled apart.

The knowledge that concerneth man's body is divided as the good of man's body is divided, unto which° it referreth. The good of man's body is of four kinds,° Health, Beauty, Strength, and Pleasure: so the knowledges are Medicine, or art of Cure;° art of Decoration, which is called Cosmetic;° art of Activity, which is called Athletic; and art Voluptuary,° which Tacitus truly calleth 'eruditus luxus'.° This subject of man's body is of all other things in nature most susceptible of remedy; but then that remedy is most susceptible of error. For the same subtility° of the subject doth cause large possibility and easy failing; and therefore the inquiry ought to be the more exact.

To speak therefore of Medicine, and to resume that we have said, ascending a little higher:° The ancient opinion that man was Microcosmus,° an abstract or model of the world, hath been fantastically strained by Paracelsus° and the alchemists,° as if there were to be found in man's body certain correspondences and parallels, which should have respect to all varieties of things, as stars, planets, minerals, which are extant in the great world. But thus much is evidently true, that of all substances which nature hath produced, man's body is the most extremely compounded.° For we see herbs and plants are nourished by earth and water; beasts for the most part by herbs and fruits; man by the flesh of beasts, birds, fishes, herbs, grains, fruits, water, and the manifold alterations, dressings,° and preparations of these several bodies, before they come to be his food and aliment. Add hereunto that beasts have a more simple order of life, and less change of affections to work upon their bodies; whereas man in his mansion,° sleep, exercise, passions, hath infinite variations; and it cannot be denied but that the Body of man of all other things is of the most compounded° mass. The Soul on the other side is the simplest of substances, as is well expressed,

> Purumque reliquit
> Aethereum sensum atque auraï simplicis ignem.°

So that it is no marvel though the soul so placed enjoy no rest, if that principle be true that 'Motus rerum est rapidus extra locum, placidus in loco'.°

But to the purpose. This variable composition of man's body hath made it as an instrument easy to distemper;° and therefore the poets did well to conjoin Music and Medicine in Apollo:° because the office of medicine is but to tune this curious° harp of man's body and to reduce it to harmony. So then the subject being so variable hath made the art by consequent° more conjectural; and the art being conjectural hath made so much the more place to be left for imposture. For almost all other arts and sciences are judged by acts or masterpieces,° as I may term them, and not by the successes° and events. The lawyer is judged by the virtue of his pleading, and not by the issue of the cause.° The master in the ship is judged by the directing his course aright, and not by the fortune of the voyage. But the physician, and perhaps the politique, hath no particular acts demonstrative of his ability, but is judged most by the event; which is ever but as it is taken. For who can tell, if a patient die or recover, or if a state be preserved or ruined, whether it be art or accident? And therefore many times the impostor is prized, and the man of virtue taxed.° Nay, we see the weakness and credulity of men is such, as they will often prefer a mountebank° or witch before a learned physician. And therefore the poets were clear-sighted in discerning this extreme folly, when they made Aesculapius° and Circe° brother and sister, both children of the sun, as in the verses,

> Ipse repertorem medicinae talis et artis
> Fulmine *Phoebigenam* Stygias detrusit ad undas.°

And again,

> Dives inaccessos ubi *Solis filia* lucos,° &c.

For in all times, in the opinion of the multitude, witches and old women and impostors have had a competition with physicians. And what followeth? Even this, that physicians say to themselves, as Salomon expresseth it upon an higher occasion; 'If it befall to me as befalleth to the fools, why should I labour to be more wise?'° And therefore I cannot much blame physicians, that they use commonly to intend° some other art or practice, which they fancy, more than their profession. For you shall have of° them antiquaries, poets, humanists,° statesmen, merchants, divines, and in every° of these better seen° than in their profession; and no doubt upon this ground, that they find that mediocrity° and excellency in their art maketh no difference in profit or reputation towards their fortune; for the weakness of patients and sweetness of life and nature of hope maketh men depend upon physicians with all their defects.

But nevertheless these things which we have spoken of are courses° begotten between a little occasion° and a great deal of sloth and default; for if we will excite and awake our observation,° we shall see in familiar instances what a predominant faculty° the subtilty of spirit° hath over the variety of matter or form. Nothing more variable than faces and countenances; yet men can bear in memory the infinite distinctions of them; nay, a painter with a few shells° of colours, and the benefit of his eye and habit of his imagination, can imitate them all that ever have been, are, or may be, if they were brought before him. Nothing more variable than voices; yet men can likewise discern° them personally; nay, you shall have a buffon° or pantomimus° will express as many as he pleaseth. Nothing more variable than the differing sounds of words; yet men have found the way to reduce them to a few simple letters. So that it is not the insufficiency or incapacity of man's mind, but it is the remote standing or placing thereof, that breedeth these mazes and incomprehensions:° for as the sense° afar off is full of mistaking but is exact at hand, so is it of the understanding; the remedy whereof is not to quicken or strengthen the organ, but to go nearer to the object; and therefore there is no doubt but if the physicians will learn and use the true approaches and avenues of nature, they may assume as much as the poet saith:

> Et quoniam variant morbi, variabimus artes;
> Mille mali species, mille salutis erunt.°

Which that they should do, the nobleness of their art doth deserve; well shadowed° by the poets, in that they made Aesculapius to be the son° of the Sun, the one being the fountain of life, the other as the second stream; but infinitely more honoured by the example of our Saviour, who made the body of man the object of his miracles,° as the soul was the object of his doctrine. For we read not that ever he vouchsafed to do any miracle about honour, or money (except that one for giving tribute to Caesar),° but only about the preserving, sustaining, and healing the body of man.

Medicine is a science which hath been (as we have said) more professed° than laboured, and yet more laboured than advanced; the labour having been, in my judgment, rather in circle than in progression. For I find much iteration,° but small addition. It considereth causes of diseases, with the occasions or impulsions;° the diseases themselves, with the accidents;° and the cures, with the preservations.° The deficiencies which I think good to note, being a few of many, and those such as are of a more open and manifest nature, I will enumerate, and not place.°

NARRATIONES MEDICINALES.° The first is the discontinuance of the ancient and serious diligence of Hippocrates,° which used to set down a narrative of the special cases of his patients, and how they proceeded, and how they were judged° by recovery or death. Therefore having an example proper in the father of the art, I shall not need to allege an example foreign,° of the wisdom of the lawyers, who are careful to report new cases and decisions for the direction of future judgments. This continuance of Medicinal History I find deficient; which I understand neither to be so infinite as to extend to every common case, nor so reserved° as to admit none but wonders: for many things are new in the manner, which are not new in the kind; and if men will intend° to observe, they shall find much worthy to observe.

ANATOMIA COMPARATA.° In the inquiry which is made by Anatomy° I find much deficience: for they inquire of the parts, and their substances, figures,° and collocations;° but they inquire not of the diversities of the parts, the secrecies of the passages,° and the seats or nestling° of the humours, nor much of the footsteps° and impressions° of diseases: the reason of which omission I suppose to be, because the first inquiry may be satisfied in the view of one or a few anatomies;° but the latter, being comparative and casual,° must arise from the view of many. And as to the diversity of parts, there is no doubt but the facture° or framing of the inward parts is as full of difference as the outward, and in that is the cause continent° of many diseases; which not being observed, they quarrel many times with the humours, which are not in fault; the fault being in the very frame and mechanic° of the part, which cannot be removed by medicine alterative,° but must be accommodate° and palliate° by diets and medicines familiar.° And for the passages and pores, it is true which was anciently noted, that the more subtile° of them appear not in anatomies, because they are shut and latent in dead bodies, though they be open and manifest in live: which being supposed, though the inhumanity of *anatomia vivorum*° was by Celsus° justly reproved;° yet in regard of the great use of this observation, the inquiry needed not by him so slightly° to have been relinquished altogether, or referred to the casual practices of surgery; but might have been well diverted upon the dissection of beasts alive, which notwithstanding the dissimilitude of their parts, may sufficiently satisfy this inquiry.

And for the humours, they are commonly passed over in anatomies as purgaments;° whereas it is most necessary to observe what cavities, nests, and receptacles the humours do find in the parts, with the differing kind of the humour so lodged and received. And as for the footsteps of diseases, and their devastations° of the inward parts, imposthumations,° exulcerations,° discontinuations,° putrefactions, consumptions,° contractions, extensions, convulsions, dislocations,° obstructions,° repletions,° together with all preternatural° substances, as stones,° carnosities,° excrescences, worms, and the like; they ought to have been exactly observed by multitude of anatomies and the contribution of men's several experiences,° and carefully set down both historically° according to the appearances, and artificially° with a reference to the diseases and symptoms which resulted from them, in case where the anatomy is of a defunct patient; whereas now upon opening of bodies they are passed over slightly and in silence.

INQUISITIO ULTERIOR DE MORBIS INSANABILIBUS.° In the inquiry of diseases, they do abandon the cures of many, some as in their nature incurable, and others as past the period of cure; so that Sulla° and the triumvirs° never proscribed so many men to die, as they do by their ignorant edicts; whereof numbers do escape with less difficulty than they did in the Roman proscriptions.° Therefore I will not doubt to note as a deficience, that they inquire not the perfect° cures of many diseases, or extremities of diseases, but pronouncing them incurable do enact a law of neglect,° and exempt ignorance from discredit.

DE EUTHANASIA EXTERIORE.° Nay further, I esteem it the office of a physician not only to restore health, but to mitigate pain and dolors;° and not only when such mitigation may conduce to recovery, but when it may serve to make a fair and easy passage:° for it is no small felicity which Augustus° Caesar was wont to wish to himself, that same *Euthanasia*; and which was specially noted in the death of Antoninus° Pius, whose death was after the fashion and semblance of a kindly° and pleasant sleep. So it is written of Epicurus,° that after his disease was judged desperate, he drowned his stomach and senses with a large draught and ingurgitation° of wine; whereupon the epigram° was made, 'Hinc stygias ebrius hausit aquas'; 'he was not sober enough to taste any bitterness of the Stygian° water'. But the physicians contrariwise do make a kind of scruple and religion to stay with the patient after the disease is deplored;° whereas, in my judgment, they ought both to enquire the skill and to give the attendances° for the facilitating and assuaging of the pains and agonies of death.

MEDICINAE EXPERIMENTALES.° In the consideration of the Cures of diseases, I find a deficience in the receipts of propriety° respecting the particular cures of diseases: for the physicians have frustrated the fruit of tradition and experience by their magistralities,° in adding and taking out and changing *quid pro quo*° in their receipts, at their pleasures; commanding° so over the medicine as the medicine cannot command over the disease. For except it be treacle° and mithridatum,° and of late diascordium,° and a few more, they tie themselves to no receipts severely° and religiously:° for as to the confections of sale° which are in the shops, they are for readiness° and not for propriety; for they are upon general intentions of purging,° opening,° comforting, altering, and not much appropriate to particular diseases: and this is the cause why empirics° and old women are more happy° many times in their cures than learned° physicians, because they are more religious in holding° their medicines. Therefore here is the deficience which I find, that physicians have not, partly out of their own practice, partly out of the constant probations° reported in books, and partly out of the traditions of empirics, set down and delivered over certain experimental medicines for the cure of particular diseases, besides their own conjectural and magistral° descriptions. For as they were the men of the best composition° in the state of Rome, which either being consuls° inclined to the people, or being tribunes° inclined to the senate; so in the matter we now handle, they be the best° physicians, which being learned incline to the traditions of experience, or being empirics incline to the methods of learning.

IMITATIO NATURAE IN BALNEIS, ET AQUIS MEDICINALIBUS.° In preparation of Medicines, I do find strange,° specially considering how mineral medicines° have been extolled, and that they are safer for the outward than inward parts, that no man hath sought to make an imitation by art of Natural Baths and Medicinable° Fountains; which nevertheless are confessed to receive their virtues° from minerals: and not so only, but discerned and distinguished from what particular mineral they receive tincture,° as sulphur, vitriol,° steel, or the like; which nature if it may° be reduced° to compositions of art,° both the variety of them will be increased, and the temper° of them will be more commanded.°

FILUM MEDICINALE, SIVE DE VICIBUS MEDICINARUM.° But lest I grow to be more particular° than is agreeable either to my intention or to proportion, I will conclude this part with the note of one deficience more, which seemeth to me of greatest consequence; which is, that the prescripts° in use are too compendious° to attain their end: for, to my understanding, it is a vain and flattering opinion to think any medicine can be so sovereign or so happy, as that the receit or use of it can work any great effect upon the body of man. It were a strange speech which spoken, or spoken oft, should reclaim a man from a vice to which he were by nature subject. It is order,° pursuit,° sequence, and interchange of application, which is mighty in nature; which although it require more exact knowledge in prescribing and more precise obedience in observing, yet is recompensed with the magnitude of effects. And although a man would think, by the daily visitations of the physicians, that there were a pursuance° in the cure; yet let a man look into their prescripts and ministrations, and he shall find them but inconstancies and every day's devices,° without any settled providence° or project.° Not that every scrupulous or superstitious prescript° is effectual, no more than every straight way° is the way to heaven; but the truth of the direction must precede severity of observance.

For Cosmetic, it hath parts civil,° and parts effeminate:° for cleanness of body was ever esteemed to proceed from a due reverence to God, to society, and to ourselves.° As for artificial decoration,° it is well worthy of the deficiencies which it hath; being neither fine enough to deceive, nor handsome to use, nor wholesome to please.°

For Athletic, I take the subject of it largely;° that is to say, for any point of ability whereunto the body of man may be brought, whether it be of activity or of patience;° whereof activity hath two parts, strength and swiftness; and patience likewise hath two parts, hardness° against wants and extremities, and endurance of pain or torment: whereof we see the practices° in tumblers,° in savages,° and in those that suffer punishment. Nay, if there be any other faculty which falls not within any of the former divisions, as in those that dive, that obtain a strange power of containing respiration,° and the like, I refer it to this part. Of these things the practices are known, but the philosophy that concerneth them is not much enquired; the rather, I think, because they are supposed to be obtained either by an aptness of nature, which cannot be taught, or only by continual custom, which is soon prescribed; which though it be not true, yet I forbear to note any deficiencies; for the Olympian Games are down°

long since, and the mediocrity of these things is for use;° as for the excellency of them, it serveth for the most part but for mercenary ostentation.

For Arts° of Pleasure Sensual, the chief deficience in them is of laws to repress them. For as it hath been well observed that the arts which flourish in times while virtue is in growth, are military; and while virtue is in state,° are liberal;° and while virtue is in declination,° are voluptuary; so I doubt° that this age of the world is somewhat upon the descent of the wheel. With arts voluptuary I couple practices joculary;° for the deceiving of the senses is one of the pleasures of the senses. As for games of recreation, I hold them to belong to civil life and education. And thus much of that particular Human Philosophy which concerns the Body, which is but the tabernacle° of the mind.

For Human Knowledge which concerns the Mind,° it hath two parts; the one that enquireth of the substance or nature of the soul or mind, the other that enquireth of the faculties or functions thereof. Unto the first of these, the considerations of the original of the soul, whether it be native or adventive,° and how far it is exempted from laws of matter, and of the immortality thereof, and many other points, do appertain: which have been not more laboriously enquired than variously reported; so as the travail therein taken seemeth to have been rather in a maze than in a way.° But although I am of opinion that this knowledge may be more really° and soundly enquired, even in nature, than it hath been; yet I hold that in the end it must be bounded by religion, or else it will be subject to deceit and delusion; for as the substance of the soul in the creation was not extracted out of the mass of heaven and earth by the benediction of a 'producat',° but was immediately inspired° from God; so it is not possible that it should be (otherwise than by accident) subject to the laws of heaven and earth, which are the subject of philosophy; and therefore the true knowledge of the nature and state of the soul, must come by the same inspiration that gave the substance. Unto this part of knowledge touching the soul there be two appendices; which, as they have been handled, have rather vapoured forth fables than kindled truth; Divination° and Fascination.°

Divination hath been anciently and fitly divided into artificial and natural; whereof artificial is when the mind maketh a prediction by argument,° concluding upon° signs and tokens;° natural is when the mind hath a presention° by an internal power, without the inducement° of a sign. Artificial is of two sorts; either when the argument is coupled with a derivation of causes, which is rational; or when it is only grounded upon a coincidence of the effect, which is experimental:° whereof the latter for the most part is superstitious; such as were the heathen observations upon the inspection of sacrifices, the flights of birds, the swarming of bees; and such as was the Chaldean° Astrology, and the like. For artificial divination, the several kinds thereof are distributed amongst particular knowledges. The Astronomer° hath his predictions, as of conjunctions,° aspects,° eclipses, and the like. The Physician hath his predictions, of death, of recovery, of the accidents° and issues of diseases. The Politique hath his predictions; 'O urbem venalem, et cito perituram, si emptorem invenerit!'° which stayed not long to be performed, in Sulla first, and after in Cæsar. So as these predictions are now impertinent,° and to be referred over.°

But the divination which springeth from the internal nature of the soul, is that which we now speak of; which hath been made to be of two sorts, primitive° and by influxion.° Primitive is grounded upon the supposition that the mind, when it is withdrawn and collected into itself and not diffused into the organs of the body, hath some extent and latitude of prenotion;° which therefore appeareth most in sleep, in extasies,° and near death; and more rarely in waking apprehensions; and is induced and furthered by those abstinences and observances° which make the mind most to consist in° itself. By influxion is grounded upon the conceit that the mind, as a mirror or glass, should take illumination from the foreknowledge of God and spirits; unto which the same regiment° doth likewise conduce. For the retiring of the mind within itself° is the state which is most susceptible of divine influxions; save° that it is accompanied in this case with a fervency and elevation° (which the ancients noted° by fury),° and not with a repose and quiet, as it is in the other.

Fascination is the power and act of imagination, intensive° upon other bodies than the body of the imaginant:° for of that we spake° in the proper place: wherein the school of Paracelsus and the disciples of pretended° Natural Magic have been so intemperate, as they have exalted the power of the imagination to be much one with° the power of miracle-working faith.° Others that draw nearer to probability, calling to their view the secret° passages of things, and especially of the contagion° that passeth from body to body, do conceive it should likewise be agreeable to nature that there should be some transmissions and operations from spirit to spirit, without the mediation of the senses; whence the conceits have grown (now almost made civil)° of the Mastering Spirit, and the force of confidence, and the like. Incident unto this is the inquiry how to raise and fortify° the imagination; for if the imagination fortified have power, then it is material° to know how to fortify and exalt it. And herein comes in crookedly and dangerously a palliation° of a great part of Ceremonial° Magic. For it may be pretended that Ceremonies, Characters,° and Charms, do work not by any tacit or sacramental° contract with evil spirits, but serve only to strengthen the imagination of him that useth it; as images are said by the Roman church to fix° the cogitations° and raise° the devotions of them that pray before them. But for mine own judgment, if it be admitted that imagination hath power, and that Ceremonies fortify imagination, and that they be used sincerely and intentionally for that purpose; yet I should hold them unlawful, as opposing° to that first edict which God gave unto man, 'In sudore vultus comedes panem tuum'.° For they propound° those noble effects which God hath set forth unto man to be bought at the price of labour, to be attained by a few easy and slothful observances. Deficiencies in these knowledges I will report none, other than the general deficience, that it is not known how much of them is verity and how much vanity.

The knowledge which respecteth the Faculties° of the Mind of man is of two kinds; the one respecting his Understanding and Reason, and the other his Will, Appetite,° and Affection;° whereof the former produceth Position° or Decree, the latter Action or Execution. It is true that the Imagination is an agent or *nuncius*° in both provinces, both the judicial and the ministerial. For Sense sendeth over° to Imagination before Reason have judged: and Reason sendeth over to Imagination before the Decree can be acted; for Imagination ever precedeth Voluntary Motion:° saving that this Janus° of Imagination hath differing faces; for the face towards Reason hath the print of Truth, but the face towards Action hath the print of Good; which nevertheless are faces, 'Quales decet esse sororum'.° Neither is the Imagination simply and only a messenger; but is invested with or at leastwise usurpeth° no small authority in itself, besides the duty of the message. For it was well said by Aristotle,° that 'the mind hath over the body that commandment, which the lord hath over a bondman; but that reason hath over the imagination that commandment which a magistrate hath over a free citizen'; who may come also to rule in his turn. For we see that in matters of Faith and Religion we raise° our Imagination above our Reason; which is the cause why Religion sought ever access to the mind by similitudes,° types,° parables, visions, dreams. And again in all persuasions that are wrought by eloquence and other impression° of like nature, which do paint° and disguise the true appearance of things, the chief recommendation° unto Reason is from the Imagination.° Nevertheless, because I find not any science that doth properly or fitly pertain to the Imagination, I see no cause to alter the former division.° For as for Poesy, it is rather a pleasure or play of imagination, than a work or duty thereof. And if it be a work, we speak not now of such parts of learning as the Imagination produceth, but of such sciences as handle and consider of the Imagination; no more than we shall speak now of such knowledges as Reason produceth (for that extendeth to all philosophy), but of such knowledges as do handle and inquire of the faculty of Reason: so as Poesy had his° true place. As for the power of the Imagination in nature, and the manner of fortifying the same, we have mentioned° it in the doctrine *De Anima*, whereunto most fitly it belongeth. And lastly, for Imaginative or Insinuative° Reason, which is the subject of Rhetoric, we think it best to refer it to the Arts of Reason. So therefore we content ourselves with the former division, that Human Philosophy which respecteth the faculties of the mind of man hath two parts, Rational and Moral.

The part of Human Philosophy which is rational,° is of all knowledges, to the most wits, the least delightful; and seemeth but a net of subtility° and spinosity.° For as it was truly said, that knowledge is 'pabulum animi',° so in the nature of men's appetite to this food, most men are of the taste and stomach of the Israelites in the desert, that would fain have returned 'ad ollas carnium',° and were weary of manna;° which, though it were celestial, yet seemed less nutritive and comfortable.° So generally men taste well° knowledges that are drenched in flesh and blood, Civil History, Morality, Policy, about the which men's affections, praises, fortunes, do turn and are conversant; but this same 'lumen siccum'° doth parch and offend° most men's watery and soft natures. But to speak truly of things as they are in worth, Rational Knowledges are the keys of all other arts; for as Aristotle° saith aptly and elegantly, that 'the hand is the Instrument of Instruments, and the mind is the Form of Forms': so these be truly said to be the Art of Arts: neither do they only direct, but likewise confirm and strengthen; even as the habit of shooting doth not only enable to shoot a nearer shoot,° but also to draw a stronger bow.

The Arts Intellectual° are four in number; divided according to the ends whereunto they are referred: for man's labour is to invent° that which is sought or propounded; or to judge that which is invented; or to retain that which is judged; or to deliver over that which is retained. So as the arts must be four: Art of Inquiry or Invention; Art of Examination or Judgment; Art of Custody° or Memory; and Art of Elocution or Tradition.°

Invention is of two kinds,° much differing; the one, of Arts and Sciences; and the other, of Speech and Arguments. The former of these I do report deficient; which seemeth to me to be such a deficience as if in the making of an inventory touching the estate of a defunct° it should be set down that 'there is no ready money'. For as money will fetch° all other commodities, so this knowledge is that which should purchase all the rest. And like as the West-Indies had never been discovered if the use of the mariner's needle had not been first discovered, though the one be vast regions and the other a small motion; so it cannot be found strange if sciences be no further discovered, if the art itself of invention and discovery hath been passed over.

That this part of knowledge is wanting, to my judgment standeth plainly confessed: for first, Logic doth not pretend° to invent Sciences or the Axioms of Sciences, but passeth it over with a 'cuique in sua arte credendum'.° And Celsus° acknowledgeth it gravely, speaking of the empirical and dogmatical sects of physicians, that 'medicines and cures were first found out, and then after° the reasons and causes were discoursed;° and not the causes first found out, and by light from them the medicines and cures discovered.' And Plato in his *Theaetetus*° noteth well, that 'particulars are infinite, and the higher generalities give no sufficient direction; and that the pith of all sciences, which maketh the arts-man° differ from the inexpert, is in the middle propositions, which in every particular knowledge are taken from tradition and experience.' And therefore we see that they which discourse of the inventions and originals° of things, refer them rather to chance than to art, and rather to beasts, birds, fishes, serpents, than to men.

Dictamnum genetrix Cretaea carpit ab Ida,
Puberibus caulem foliis et flore comantem
Purpureo: non illa feris incognita capris
Gramina, cum tergo volucres haesere sagittae.°

So that it was no marvel (the manner of antiquity being to consecrate inventors) that the Egyptians had so few human idols in their temples, but almost all brute:

> Omnigenumque Deum monstra, et latrator Anubis,
> Contra Neptunum et Venerem, contraque Minervam, &c.°

And if you like better the tradition of the Grecians, and ascribe the first inventions to men, yet you will rather believe that Prometheus° first struck the flints, and marvelled at the spark, than that when he first struck the flints he expected the spark; and therefore we see the West-Indian Prometheus had no intelligence° with the European,° because of the rareness with them of flint, that gave the first occasion.° So as it should seem that hitherto men are rather beholden° to a wild goat for surgery, or to a nightingale for music, or to the Ibis for some part of physic, or to the pot lid that flew open for artillery, or generally to chance° or any thing else, than to Logic, for the invention of arts and sciences. Neither is the form of invention which Virgil describeth much other:°

> Ut varias usus meditando extunderet artes
> Paulatim.°

For if you observe the words well, it is no other method than that which brute beasts are capable of, and do put in ure;° which is a perpetual intending° or practising some one thing, urged and imposed by an absolute necessity of conservation of being: for so Cicero saith very truly, 'Usus uni rei deditus et naturam et artem saepe vincit'.° And therefore if it be said of men,

> Labor omnia vincit
> Improbus, et duris urgens in rebus egestas,°

it is likewise said of beasts, 'Quis psittaco docuit suum χαῖρε?'° Who taught the raven in a drowth° to throw pebbles into an hollow tree where she spied water, that the water might rise so as she might come to it? Who taught the bee to sail through such a vast sea of air, and to find the way from a field in flower a great way off to her hive? Who taught the ant to bite every grain of corn that she burieth in her hill, lest it should take root and grow? Add then the word *extundere*,° which importeth° the extreme difficulty, and the word *paulatim*,° which importeth the extreme slowness, and we are where we were, even amongst the Egyptians' gods; there being little left to the faculty of Reason, and nothing to the duty of Art, for matter of invention.

Secondly, the induction° which the logicians speak of, and which seemeth familiar with Plato,° whereby the Principles of sciences may be pretended° to be invented, and so the middle propositions by derivation from the principles,—their form of induction, I say, is utterly vicious° and incompetent: wherein their error is the fouler, because it is the duty of Art to perfect and exalt Nature; but they contrariwise have wronged, abused, and traduced° nature. For he that shall attentively observe how the mind doth gather this excellent dew of knowledge, like unto that which the poet speaketh of, 'Aërei mellis coelestia dona,'° distilling and contriving it out of particulars natural and artificial, as the flowers of the field and garden, shall find that the mind of herself by nature doth manage and act° an induction much better than they describe it. For to conclude upon an enumeration of particulars without instance contradictory° is no conclusion, but a conjecture; for who can assure° (in many subjects) upon those particulars which appear of a side,° that there are not other on the contrary side which appear not? As if Samuel should have rested upon° those sons of Issay° which were brought before him, and failed of° David, which was in the field. And this form (to say truth) is so gross, as it had not been possible for wits so subtile as have managed these things to have offered it to the world, but that they hasted to their theories and dogmaticals,° and were imperious and scornful toward particulars; which their manner was to use but as *lictores*° and *viatores*,° for sergeants and whifflers, 'ad summovendam turbam',° to make way and make room for their opinions, rather than in their true use and service. Certainly it is a thing may touch a man with a religious wonder, to see how the footsteps of seducement° are the very same in divine and human truth: for as in divine truth man cannot endure to become as a child;° so in human, they reputed the attending° the Inductions (whereof we speak) as if it were a second infancy or childhood.

Thirdly, allow some Principles or Axioms were rightly induced, yet nevertheless certain it is that Middle Propositions cannot be deduced from them in subject of nature° by Syllogism, that is, by touch° and reduction° of them to principles in a middle term. It is true that in sciences popular, as moralities,° laws, and the like, yea and divinity (because it pleaseth God to apply° himself to the capacity of the simplest), that form may have use; and in natural philosophy° likewise, by way of argument° or satisfactory° reason, 'quae assensum parit, operis effoeta est',° but the subtilty of nature and operations will not be enchained in those bonds: for Arguments consist of

Propositions, and Propositions of Words; and Words are but the current tokens or marks of Popular Notions of things;° which notions, if they be grossly° and variably° collected out of particulars, it is not the laborious examination either of consequences of arguments or of the truth of propositions, that can ever correct that error; being (as the physicians speak) in the first digestion:° and therefore it was not without cause, that so many excellent philosophers became Sceptics° and Academics,° and denied any certainty of knowledge or comprehension, and held opinion that the knowledge of man extended only to appearances and probabilities. It is true that in Socrates° it was supposed to be but a form of irony, 'Scientiam dissimulando simulavit',° for he used to disable° his knowledge, to the end to enhance his knowledge; like the humour of Tiberius° in his beginnings, that would reign, but would not acknowledge so much; and in the latter Academy, which Cicero embraced, this opinion also of *acatalepsia*° (I doubt)° was not held sincerely: for that all those which excelled in copie° of speech seem to have chosen that sect, as that which was fittest to give glory to their eloquence and variable° discourses; being rather like progresses° of pleasure than journeys to an end. But assuredly many scattered in both° Academies did hold it in subtilty° and integrity. But here was their chief error; they charged the deceit upon° the Senses; which in my judgment (notwithstanding all their cavillations)° are very sufficient to certify° and report truth, though not always immediately,° yet by comparison, by help of instruments,° and by producing and urging such things as are too subtile for the sense to some effect comprehensible by the sense, and other like assistance. But they ought to have charged the deceit upon the weakness of the intellectual powers, and upon the manner of collecting and concluding upon the reports of the senses. This I speak not to disable the mind of man, but to stir it up to seek help:° for no man, be he never so cunning or practised, can make a straight line or perfect circle by steadiness of hand, which may be easily done by help of a ruler or compass.°

EXPERIENTIA LITERATA, ET INTERPRETATIO NATURAE.° This part of invention, concerning the invention of sciences, I purpose (if God give me leave) hereafter to propound; having digested° it into two parts; whereof the one I term *Experientia literata*,° and the other *Interpretatio Naturae*,° the former being but a degree and rudiment° of the latter. But I will not dwell too long, nor speak too great upon a promise.

The invention of speech or argument is not properly an invention: for to invent is to discover that° we know not, and not to recover or resummon° that which we already know; and the use of this invention is no other but out of the knowledge whereof our mind is already possessed, to draw forth or call before us that which may be pertinent to the purpose which we take into our consideration. So as, to speak truly, it is no Invention, but a Remembrance or Suggestion,° with an application; which is the cause why the schools° do place it after judgment, as subsequent and not precedent. Nevertheless, because we do account it a Chase° as well of deer in an inclosed park as in a forest at large, and that it hath already obtained the name, let it be called invention: so as° it be perceived and discerned,° that the scope and end of this invention is readiness and present use of our knowledge, and not addition or amplification thereof.

To procure this ready use of knowledge there are two courses, Preparation and Suggestion. The former of these seemeth scarcely a part of Knowledge, consisting rather of diligence than of any artificial erudition.° And herein Aristotle° wittily, but hurtfully, doth deride the sophists near his time, saying, 'they did as if one that professed the art of shoe-making should not teach how to make up a shoe, but only exhibit in a readiness a number of shoes of all fashions and sizes'. But yet a man might reply, that if a shoe-maker should° have no shoes in his shop, but only work as he is bespoken,° he should be weakly customed.° But our Saviour, speaking of Divine Knowledge, saith that 'the kingdom of heaven is like a good householder, that bringeth forth both new and old store';° and we see the ancient writers of rhetoric do give it in precept, that pleaders should have the Places° whereof they have most continual use ready handled in all the variety that may be; as that, to speak for the literal interpretation of the law against equity,° and contrary; and to speak for presumptions and inferences against testimony,° and contrary. And Cicero himself, being broken unto° it by great experience, delivereth it plainly, that whatsoever a man shall have occasion to speak of, (if he will take the pains) he may have it in effect premeditate,° and handled *in thesi*;° so that when he cometh to a particular,° he shall have nothing to do but to put to° names and times and places, and such other circumstances of individuals. We see likewise the exact diligence of Demosthenes;° who, in regard of the great force that the entrance and access into causes hath to make a good impression, had ready framed a number of prefaces for orations and speeches. All which authorities and precedents may overweigh Aristotle's opinion, that would have us change a rich wardrobe for a pair of shears.°

But the nature of the collection of this provision or preparatory store, though it be common both to logic and rhetoric, yet having made an entry of it here, where it came first to be spoken of, I think fit to refer over the further handling° of it to rhetoric.

The other part of Invention, which I term° Suggestion, doth assign and direct us to certain marks or Places,° which may excite° our mind to return and produce such knowledge as it hath formerly collected, to the end we may make use thereof. Neither is this use (truly taken) only to furnish argument to dispute probably° with others, but likewise to minister° unto our judgment to conclude aright within ourselves.° Neither may these Places serve only to apprompt° our invention, but also to direct our inquiry. For as faculty of wise interrogating is half a knowledge. For as Plato° saith, 'Whosoever seeketh, knoweth that which he seeketh for in a general notion; else how shall he know it when he hath found it?' And therefore the larger your Anticipation is, the more direct and compendious is your search. But the same Places which will help us what to produce of that which we know already, will also help us, if a man of experience were before° us, what questions to ask; or if we have books and authors to instruct us, what points to search and revolve:° so as I cannot report that this part of invention, which is that which the schools call Topics,° is deficient.

Nevertheless Topics are of two sorts, general and special. The general we have spoken to;° but the particular hath been touched by some, but rejected generally as inartificial° and variable.° But leaving the humour which hath reigned too much in the schools (which is to be vainly subtile° in a few things which are within their command, and to reject the rest), I do receive° particular Topics, that is places or directions of invention and inquiry in every particular knowledge, as things of great use; being mixtures of Logic with the matter of sciences; for in these it holdeth, 'Ars inveniendi adolescit cum inventis',° for as in going of a way° we do not only gain that part of the way which is passed, but we gain the better sight of that part of the way which remaineth; so every degree of proceeding in a science giveth a light to that which followeth; which light if we strengthen, by drawing it forth into questions or places of inquiry, we do greatly advance our pursuit.

Now we pass unto the arts of Judgment,° which° handle the natures of Proofs and Demonstrations; which° as to Induction hath a coincidence° with Invention; for in all inductions, whether in good or vicious form, the same action of the mind which inventeth, judgeth;° all one° as in the sense;° but otherwise° it is in proof by syllogism; for the proof being not immediate but by mean,° the invention of the mean is one thing, and the judgment of the consequence is another; the one exciting° only, the other examining. Therefore for the real and exact form of judgment we refer ourselves to that which we have spoken° of Interpretation of Nature.

For the other judgment by Syllogism, as it is a thing most agreeable° to the mind of man, so it hath been vehemently and excellently laboured. For the nature of man doth extremely covet to have somewhat in his understanding fixed and immoveable, and as a rest and support of the mind. And therefore as Aristotle° endeavoureth to prove that in all motion there is some point quiescent;° and as he elegantly expoundeth the ancient fable of Atlas° (that stood fixed and bare up° the heaven from falling) to be meant of the poles or axle-tree° of heaven, whereupon the conversion° is accomplished; so assuredly men have a desire to have an Atlas or axle-tree within to keep them from fluctuation, which is like to a perpetual peril of falling;° therefore men did hasten to set down some Principles about which the variety of their disputations might turn.

So then this art of Judgment is but the reduction° of propositions to principles in a middle term: the Principles to be agreed° by all and exempted from argument; the Middle Term to be elected° at the liberty of every man's invention; the Reduction to be of two kinds, direct and inverted; the one when the proposition is reduced to the principle, which they term a 'Probation ostensive';° the other when the contradictory of the proposition is reduced to the contradictory of the principle, which is that which they call 'per incommodum',° or 'pressing° an absurdity'; the number of middle terms to be° as the proposition standeth degrees° more or less removed from the principle.

But this art hath two several methods of doctrine; the one by way of direction,° the other by way of caution: the former frameth and setteth down a true form of consequence,° by the variations and deflexions from which errors and inconsequences may be exactly judged; toward the composition and structure of which form, it is incident° to handle the parts thereof, which are propositions, and the parts of propositions, which are simple words; and this is that part of logic which is comprehended in the Analytics.°

The second method of doctrine was introduced for expedite° use and assurance sake; discovering the more subtile forms of sophisms° and illaqueations° with their redargutions,° which is that which is termed *Elenches.*° For although in the more gross sorts of fallacies it happeneth (as Seneca° maketh the comparison well) as in juggling feats, which though we know not how they are done, yet we know well it is not as it seemeth to be; yet the more subtile sort of them doth not only put a man besides his answer,° but doth many times abuse° his judgment.

This part concerning *Elenches* is excellently handled by Aristotle° in precept, but more excellently by Plato in example, not only in the persons° of the Sophists,° but even in Socrates° himself; who professing to affirm nothing, but to infirm° that which was affirmed by another, hath exactly expressed all the forms of objection, fallace,° and redargution. And although we have said that the use of this doctrine is for redargution, yet it is manifest the° degenerate and corrupt use is for caption° and contradiction; which passeth for a great faculty, and no doubt is of very great advantage: though the difference° be° good which was made between orators and sophisters,° that the one is as the greyhound, which hath his advantage in the race, and the other as the hare, which hath her advantage in the turn,° so as it is the advantage of the weaker creature.

But yet further, this doctrine of *Elenches* hath a more ample latitude° and extent than is perceived; namely, unto divers parts of knowledge; whereof some are laboured° and other omitted. For first, I conceive (though it may seem at first somewhat strange) that that part which is variably referred, sometimes to Logic sometimes to Metaphysic, touching the common adjuncts of essences,° is but an elenche; for the great sophism of all sophisms being equivocation or ambiguity of words and phrase, specially of such words as are most general and intervene° in every inquiry, it seemeth to me that the true and fruitful use (leaving vain subtilties and speculations) of the inquiry of majority, minority, priority, posteriority, identity, diversity, possibility, act, totality, parts, existence, privation, and the like,° are but wise cautions against ambiguities of speech. So again the distribution of things into certain tribes,° which we call categories or predicaments,° are but cautions against the confusion of definitions and divisions.

Secondly, there is a seducement that worketh by the strength of the impression° and not by the subtilty of the illaqueation; not so much perplexing the reason as overruling it by power of the imagination. But this part I think more proper to handle when I shall speak of Rhetoric.°

But lastly, there is yet a much more important and profound kind of fallacies° in the mind of man, which I find not observed or enquired at all, and think good to place here, as that which of all others appertaineth° most to rectify° judgment: the force whereof is such, as it doth not dazzle or snare the understanding in some particulars, but doth more generally and inwardly infect and corrupt the state thereof. For the mind of man is far from the nature of a clear and equal glass,° wherein the beams of things should reflect° according to their true incidence; nay, it is rather like an enchanted° glass, full of superstition and imposture,° if it be not delivered° and reduced.° For this purpose, let us consider the false appearances that are imposed upon us by the general nature of the mind,° beholding them in an example or two; as first, in that instance which is the root of all superstition, namely, that to the nature of the mind of all men it is consonant° for the affirmative or active to affect more than the negative or privative:° so that a few times hitting° or presence, countervails° oft-times failing or absence; as was well answered by Diagoras° to him that shewed him in Neptune's temple the great number of pictures of such as had scaped° shipwrack and had paid their vows to Neptune, saying, 'Advise° now, you that think it folly to invocate° Neptune in tempest': 'Yea but' (saith Diagoras) 'where are they painted that are drowned?' Let us behold it in another instance, namely, that the spirit of man, being of an equal and uniform substance, doth usually suppose and feign in nature a greater equality and uniformity than is in truth. Hence it cometh, that the mathematicians cannot satisfy themselves, except they reduce the motions of the celestial bodies to perfect circles,° rejecting spiral lines, and labouring to be discharged of eccentrics.° Hence it cometh, that whereas there are many things in nature as it were *monodica*,° 'sui juris',° yet the cogitations of man do feign unto them relatives, parallels, and conjugates,° whereas no such thing is; as they have feigned an element of Fire,° to keep square with° Earth, Water, and Air, and the like: nay, it is not credible, till it be opened,° what a number of fictions and fancies the similitude° of human actions and arts, together with the making of man *communis mensura*,° have brought into Natural Philosophy; not much better than the heresy of the Anthropomorphites,° bred in the cells of gross° and solitary monks, and the opinion of Epicurus,° answerable° to the same in heathenism, who supposed the gods to be of human shape. And therefore Velleius° the Epicurean needed not to have asked, why God should have° adorned the heavens with stars, as if he had been an Aedilis,° one that should have° set forth some magnificent shews or plays. For if that great work-master° had been of an human disposition, he would have cast the stars into some pleasant and beautiful works° and orders,° like the frets° in the roofs of houses; whereas one can scarce find a posture° in square or triangle or straight line amongst such an infinite number; so differing an harmony there is between the spirit of Man and the spirit of Nature.

Let us consider again° the false appearances imposed upon us by every man's own individual nature and custom,° in that feigned supposition that Plato° maketh of the cave: for certainly if a child were continued° in a grot° or cave under the earth until maturity of age, and came suddenly abroad, he would have strange and absurd° imaginations; so in like manner, although our persons° live in the view of heaven, yet our spirits are included° in the caves of our own complexions° and customs; which minister unto us infinite errors and vain opinions, if they be not recalled to examination.° But hereof we have given many examples in one of the errors, or peccant humours, which we ran briefly over° in our first book.

ELENCHI MAGNI, SIVE DE IDOLIS ANIMI HUMANI NATIVIS ET ADVENTITIIS.° And lastly, let us consider the false appearances that are imposed upon us by words,° which are framed and applied according to the conceit and capacities of the vulgar sort:° and although we think we govern our words, and prescribe it well, 'Loquendum ut vulgus, sentiendum ut sapientes',° yet certain it is that words, as a Tartar's bow,° do shoot back upon the understanding of the wisest, and mightily entangle and pervert° the judgment; so as it is almost° necessary in all controversies and disputations to imitate the wisdom of the Mathematicians, in setting down in the very beginning the definitions of our words and terms, that others may know how we accept and understand them, and whether they concur with us or no. For it cometh to pass for want of this, that we are sure to end there where we ought to have begun, which is in questions and differences about words. To conclude therefore, it must be confessed that it is not possible to divorce ourselves from these fallacies and false appearances, because they are inseparable from our nature and condition of life; so yet nevertheless the caution° of them (for all elenches, as was said, are but cautions) doth extremely import° the true conduct of human judgment. The particular elenches or cautions against these three false appearances I find altogether deficient.

DE ANALOGIA DEMONSTRATIONUM.° There remaineth one part of judgment of great excellency, which to mine understanding is so slightly touched,° as I may report that also deficient; which is the application of the differing kinds of proofs to the differing kinds of subjects. For there being but four kinds of demonstrations, that is, by the immediate consent of the mind or sense; by induction; by sophism; and by congruity,° which is that which Aristotle calleth 'demonstration in orb or circle', and not *a notioribus*;° every° of these hath certain subjects in the matter of sciences, in which respectively they have chiefest use; and certain other, from which respectively they ought to be excluded: and the rigour and curiosity in requiring the more severe proofs in some things, and chiefly the facility° in contenting ourselves with the more remiss° proofs in others, hath been amongst the greatest causes of detriment and hindrance to knowledge. The distributions and assignations of demonstrations, according to the analogy of sciences, I note as deficient.

The custody or retaining of knowledge is either in Writing or Memory; whereof Writing hath two parts, the nature of the character,° and the order of the entry. For° the art of characters, or other visible notes of words or things, it hath nearest conjugation° with grammar, and therefore I refer it to the due place.° For the disposition° and collocation° of that knowledge which we preserve in writing, it consisteth in a good digest of commonplaces;° wherein I am not ignorant of the prejudice imputed to the use of common-place books, as causing a retardation° of reading, and some sloth or relaxation° of memory. But because it is but a counterfeit thing in knowledges to be forward° and pregnant,° except a man be deep and full, I hold the entry° of common-places to be a matter of great use and essence° in studying; as that which assureth copie° of invention, and contracteth° judgment to a strength. But this is true, that of the methods of common-places that I have seen, there is none of any sufficient worth; all of them carrying merely the face of a school, and not of a world; and referring to vulgar matters and pedantical divisions without all life or respect to action.

For the other principal part of the custody of knowledge, which is Memory, I find that faculty in my judgment weakly enquired of. An art° there is extant of it; but it seemeth to me that there are better precepts than that art, and better practices of that art than those received. It is certain the art (as it is) may be raised to points° of ostentation prodigious: but in use (as it is now managed) it is barren; not burdensome nor dangerous to natural memory, as is imagined, but barren; that is, not dexterous° to be applied to the serious use of business and occasions. And therefore I make no more estimation of repeating a great number of names or words upon once hearing, or the pouring forth of a number of verses or rhymes *ex tempore*, or the making of a satirical simile of every thing, or the turning of every thing to a jest, or the falsifying or contradicting of every thing by cavil,° or the like (whereof in the faculties of the mind there is great copie, and such as by device and practice may be exalted to an extreme degree of wonder), than I do of the tricks of tumblers, funambuloes,° baladines;° the one being the same in the mind that the other is in the body; matters of strangeness without worthiness.°

This art of Memory is but built upon two intentions;° the one Prenotion,° the other Emblem.° Prenotion dischargeth° the indefinite seeking of that we would remember, and directeth us to seek in a narrow compass; that is, somewhat that hath congruity with our place of memory. Emblem reduceth conceits intellectual to images sensible, which strike the memory more; out of which axioms may be drawn much better practique° than that in use; and besides which axioms, there are divers more touching help of memory, not inferior to them. But I did in the beginning distinguish,° not to report those things deficient, which are but only ill managed.

There remaineth the fourth kind of Rational Knowledge, which is transitive,° concerning the expressing or transferring our knowledge to others; which I will term by the general name of Tradition° or Delivery. Tradition hath three parts; the first concerning the organ of tradition; the second concerning the method of tradition; and the third concerning the illustration of tradition.

For the organ of tradition, it is either Speech or Writing: for Aristotle° saith well, 'Words are the images of cogitations, and letters are the images of words'; but yet it is not of necessity that cogitations be expressed by the medium of words. For whatsoever is capable of sufficient differences,° and those perceptible by the sense, is in nature competent to express° cogitations. And therefore we see in the commerce° of barbarous people that understand not one another's language, and in the practice of divers° that are dumb and deaf, that men's minds are expressed in gestures, though not exactly, yet to serve the turn. And we understand further that it is the use° of China and the kingdoms of the high Levant° to write in Characters Real,° which express neither letters nor words in gross, but Things or Notions; insomuch as countries and provinces, which understand not one another's language, can nevertheless read one another's writings, because the characters are accepted more generally than the languages do extend; and therefore they have a vast multitude of characters; as many, I suppose, as radical° words.

DE NOTIS RERUM.° These Notes of Cogitations are of two sorts; the one when the note° hath some similitude or congruity with the notion; the other *ad placitum*,° having force only by contract° or acceptation. Of the former sort are Hieroglyphics° and Gestures.° For as to Hieroglyphics (things of ancient use, and embraced chiefly by the Egyptians, one of the most ancient nations), they are but as continued° impresses° and emblems.° And as for Gestures, they are as transitory° Hieroglyphics, and are to Hieroglyphics as words spoken are to words written, in that they abide not;° but they have evermore, as well as the other, an affinity with the things signified: as Periander,° being consulted with how to preserve a tyranny newly usurped, bid the messenger attend and report what he saw him do; and went into his garden and topped all the highest flowers; signifying, that it consisted in the cutting off and keeping low of the nobility and grandees. *Ad placitum* are the Characters Real before mentioned, and Words: although some have been willing by curious inquiry, or rather by apt feigning,° to have derived imposition of names from reason° and intendment;° a speculation elegant, and, by reason° it searcheth into antiquity, reverent;° but sparingly mixed with truth, and of small fruit.

This portion of knowledge, touching the Notes of Things and cogitations in general, I find not enquired, but deficient. And although it may seem of no great use, considering that words and writings by letters do far excel all the other ways; yet because this part concerneth as it were the mint° of knowledge (for words are the tokens current and accepted for conceits,° as moneys° are for values, and that it is fit° men be not ignorant that moneys may be of another kind° than gold and silver), I thought good to propound it to better enquiry.

Concerning Speech and Words, the consideration of them hath produced the science of Grammar: for man still striveth to reintegrate° himself in° those benedictions, from which by his fault° he hath been deprived; and as he hath striven against the first general curse° by the invention of all other arts, so hath he sought to come forth of the second general curse° (which was the confusion of tongues) by the art of Grammar: whereof the use in a mother tongue is small; in a foreign tongue more; but most in such foreign tongues as have ceased to be vulgar° tongues, and are turned° only to learned tongues.° The duty of it is of two natures;° the one popular,° which is for the speedy and perfect attaining languages, as well° for intercourse of speech as for understanding of authors; the other philosophical, examining the power and nature of words as they are the footsteps and prints of reason: which kind of analogy between words and reason is handled *sparsim*,° brokenly, though not entirely; and therefore I cannot report it deficient, though I think it very worthy to be reduced° into a science by itself.

Unto Grammar also belongeth, as an appendix, the consideration of the Accidents° of Words; which are measure,° sound, and elevation° or accent,° and the sweetness and harshness of them; whence hath issued some curious° observations in Rhetoric, but chiefly Poesy, as we consider it in respect of the verse and not of the argument: wherein though men in learned tongues do tie themselves to the ancient measures,° yet in modern languages it seemeth to me as free to make new measures of verses as of dances; for a dance is a measured pace, as a verse is a measured speech. In these things the sense is better judge than the art;

> Coenae fercula nostrae
> Mallem convivis quam placuisse cocis.°

And of the servile expressing° antiquity° in an unlike and an unfit subject, it is well said, 'Quod tempore antiquum videtur, id incongruitate est maxime novum'.°

For Ciphers,° they are commonly in letters or alphabets, but may be in words. The kinds of Ciphers (besides the simple ciphers with changes and intermixtures of nulls and non-significants)° are many, according to the nature or rule of the infolding;° Wheel-ciphers, Key-ciphers, Doubles, &c. But the virtues of them, whereby they are to be preferred,° are three; that they be not laborious to write and read; that they be impossible to decipher; and, in some cases, that they be without suspicion. The highest degree whereof is to write 'omnia per omnia';° which is undoubtedly possible, with a proportion quintuple° at most° of the writing infolding to the writing infolded, and no other restraint whatsoever. This art of Ciphering,° hath for relative° an art of Disciphering; by supposition° unprofitable; but, as things are, of great use. For suppose that ciphers were well managed, there be multitudes of them which exclude the discipherer. But in regard of the rawness and unskilfulness of the hands through which they pass, the greatest matters are many times carried in the weakest ciphers.

In the enumeration of these private and retired arts, it may be thought I seek to make a great muster-roll° of sciences; naming them for shew and ostentation, and to little other purpose. But let those which are skilful in them judge whether I bring them in only for appearance, or whether in that which I speak of them (though in few marks)° there be not some seed of proficience. And this must be remembered, that as there be many of great account in their countries and provinces, which when they come up to the Seat of the Estate° are but of mean rank and scarcely regarded; so these arts being here placed with the principal and supreme sciences, seem petty things; yet to such as have chosen them to spend their studies° in them, they seem great matters.

For the Method of Tradition, I see it hath moved a controversy° in our time. But as in civil business, if there be a meeting and men fall at words° there is commonly an end of the matter for that time and no proceeding° at all; so in learning, where there is much controversy there is many times little inquiry. For this part of knowledge of Method seemeth to me so weakly enquired as I shall report it deficient.

Method hath been placed,° and that not amiss, in Logic, as a part of Judgment: for as the doctrine of Syllogisms comprehendeth the rules of judgment upon that which is invented, so the doctrine of Method containeth the rules of judgment upon that which is to be delivered;° for judgment precedeth Delivery, as it followeth Invention. Neither is the method or the nature of the tradition material° only to the use of knowledge, but likewise to the progression° of knowledge: for since the labour and life of one man cannot attain to perfection° of knowledge, the wisdom of the Tradition is that which inspireth the felicity of continuance and proceeding. And therefore the most real diversity of method is of method referred to Use, and method referred to Progression; whereof the one may be termed Magistral,° and the other of Probation.°

The latter whereof seemeth to be 'via deserta et interclusa.'° For as knowledges are now delivered, there is a kind of contract of error between the deliverer and the receiver: for he that delivereth knowledge desireth to deliver it in such form as may be best believed, and not as may be best examined; and he that receiveth knowledge desireth rather present satisfaction° than expectant inquiry; and so rather not to doubt than not to err: glory° making the author not to lay open his weakness, and sloth making the disciple not to know his strength.

DE METHODO SINCERA, SIVE AD FILIOS SCIENTIARUM.° But knowledge that is delivered as a thread to be spun on,° ought to be delivered and intimated,° if it were possible, in the same method wherein it was invented, and so is it possible of knowledge induced.° But in this same anticipated and prevented° knowledge, no man knoweth how he came to the knowledge which he hath obtained. But yet nevertheless, 'secundum majus et minus',° a man may revisit and descend unto the foundations of his knowledge and consent;° and so transplant it into another as it grew in his own mind. For it is in knowledges as it is in plants: if you mean to use the plant, it is no matter for the roots; but if you mean to remove it to grow, then it is more assured° to rest upon roots than slips.° So the delivery of knowledges (as it is now used) is as of° fair bodies of trees without the roots; good for the carpenter, but not for the planter; but if you will have sciences grow, it is less matter for the shaft or body of the tree, so you look well to the taking up of the roots. Of which kind of delivery the method of the mathematiques,° in that subject, hath some shadow;° but generally I see it neither put in ure° nor put in inquisition,° and therefore note it for deficient.

Another diversity of Method there is, which hath some affinity with the former, used in some cases by the discretion° of the ancients, but disgraced since by the impostures of many vain persons, who have made it as a false light for their counterfeit merchandises; and that is, Enigmatical° and Disclosed.° The pretence whereof° is to remove° the vulgar capacities from being admitted to the secrets of knowledges, and to reserve them to selected auditors, or wits of such sharpness as can pierce the veil.

Another diversity of Method, whereof the consequence is great, is the delivery of knowledge in° Aphorisms,° or in Methods; wherein we may observe that it hath been too much taken into custom, out of a few Axioms or observations upon any subject to make a solemn and formal° art; filling it with some discourses,° and illustrating it with examples, and digesting it into a sensible° Method; but the writing in Aphorisms hath many excellent virtues, whereto the writing in Method° doth not approach.

For first, it trieth° the writer, whether he be superficial or solid: for Aphorisms, except they should be ridiculous, cannot be made but of the pith and heart of sciences; for discourse of illustration is cut off; recitals of examples are cut off; discourse of connexion and order is cut off; descriptions of practice are cut off; so there remaineth nothing to fill the Aphorisms but some good quantity of observation: and therefore no man can suffice,° nor in reason will attempt to write Aphorisms, but he that is sound and grounded. But in Methods,

Tantum series juncturaque pollet,
Tantum de medio sumptis accedit honoris,°

as a man shall make a great shew° of an art, which if it were disjointed would come to little. Secondly, Methods are more fit to win consent or belief, but less fit to point to action; for they carry a kind of demonstration in orb or circle,° one part illuminating another, and therefore satisfy;° but particulars, being dispersed,° do best agree with dispersed directions. And lastly, Aphorisms, representing a knowledge broken,° do invite° men to enquire farther; whereas Methods, carrying the shew of a total, do secure° men, as if they were at furthest.°

Another diversity of Method, which is likewise of great weight, is the handling of knowledge by Assertions and their Proofs, or by Questions and their Determinations;° the latter kind whereof, if it be immoderately followed, is as prejudicial to the proceeding° of learning, as it is to the proceeding of an army to go about° to besiege every little fort or hold.° For if the field be kept and the sum° of the enterprise pursued, those smaller things will come in° of themselves: indeed a man would° not leave some important piece enemy° at his back. In like manner, the use of confutation in the delivery° of sciences ought to be very sparing; and to serve to remove strong preoccupations and prejudgments,° and not to minister and excite disputations and doubts.

Another diversity of Methods is according to the subject or matter which is handled; for there is a great difference in delivery of the Mathematics, which are the most abstracted of knowledges, and Policy,° which is the most immersed: and howsoever contention hath been moved touching an uniformity of method° in multiformity of matter, yet we see how that opinion, besides the weakness of it, hath been of ill desert towards learning, as that which taketh the way° to reduce learning to certain empty and barren generalities; being but the very husks and shells of sciences, all the kernel being forced out and expulsed° with the torture and press° of the method; and therefore as I did allow well° of particular° Topics for invention, so I do allow likewise of particular Methods of tradition.°

Another diversity of judgment° in the delivery and teaching of knowledge is according unto the light and presuppositions of that which is delivered; for that knowledge which is new and foreign from opinions received, is to be delivered in another form than that that is agreeable° and familiar; and therefore Aristotle,° when he thinks to tax° Democritus, doth in truth commend him, where he saith, 'If we shall indeed dispute, and not follow after similitudes,' &c. For those whose conceits° are seated in popular opinions, need only but to prove or dispute; but those whose conceits are beyond popular opinions, have a double labour; the one to make themselves conceived,° and the other to prove and demonstrate; so that it is of necessity with them to have recourse to similitudes and translations° to express themselves. And therefore in the infancy of learning, and in rude times, when those conceits which are now trivial° were then new, the world was full of Parables and Similitudes; for else would men either have passed over without mark° or else rejected for paradoxes that which was offered, before they had understood or judged. So in divine learning we see how frequent Parables and Tropes are: for it is a rule, that 'whatsoever science is not consonant° to presuppositions, must pray in aid of° similitudes'.

DE PRUDENTIA TRADITIONIS.° There be also other diversities of Methods, vulgar° and received; as that of Resolution or Analysis,° of Constitution or Systasis,° of Concealment or Cryptic,° &c. which I do allow well of; though I have stood upon° those which are least handled and observed. All which I have remembered° to this purpose, because I would erect and constitute one general inquiry, which seems to me deficient, touching the Wisdom of Tradition.

But unto this part of knowledge concerning Method doth further belong not only the Architecture of the whole frame of a work, but also the several beams and columns thereof; not as to their stuff, but as to their quantity and figure;° and therefore Method considereth not only the disposition of the Argument or Subject, but likewise the Propositions; not as to their truth or matter, but as to their limitation and manner.° For herein Ramus merited better a great deal in reviving the good rules of Propositions,° καθόλου, πρῶτον κατα παντόζ &c. than he did in introducing the canker of Epitomes;° and yet (as it is the condition of human things that, according to the ancient fables,° 'The most precious things have the most pernicious keepers') it was so, that the attempt of the one made him fall upon° the other. For he had need be well conducted° that should design to make Axioms convertible,° if he make them not withal circular, and non-promovent,° or 'incurring° into themselves': but yet the intention was excellent.

DE PRODUCTIONE AXIOMATUM.° The other considerations of Method concerning Propositions are chiefly touching the utmost° propositions, which limit° the dimensions of sciences; for every knowledge may be fitly said, besides the profundity (which is the truth and substance of it, that makes it solid), to have a longitude and a latitude; accounting the latitude towards other sciences,° and the longitude towards action;° that is, from the greatest generality to the most particular precept. The one giveth rule how far one knowledge ought to intermeddle within the province of another, which is the rule they call καθ αντὸ;° the other giveth rule unto what degree of particularity a knowledge should descend: which latter° I find passed over in silence, being in my judgment the more material;° for certainly there must be somewhat left to practice; but how much is worthy the inquiry. We see remote and superficial generalities do but offer knowledge to scorn° of practical men; and are no more aiding to practice, than an Ortelius' universal map° is to direct the way between London and York. The better sort of rules have been not unfitly compared to glasses° of steel unpolished, where you may see the images of things, but first they must be filed:° so the rules will help, if they be laboured and polished by practice. But how chrystalline° they may be made at the first, and how far forth° they may be polished aforehand, is the question; the inquiry whereof seemeth to me deficient.

There hath been also laboured and put in practice a method, which is not a lawful method, but a method of imposture; which is to deliver knowledges in such manner, as men may speedily come to make a shew of learning who have it not: such was the travail of Raymundus Lullius,° in making that art which bears his name; not unlike to some books of Typocosmy° which have been made since; being nothing but a mass of words of all arts, to give men countenance° that those which use the terms might be thought to understand the art; which collections are much like a fripper's° or broker's° shop, that hath ends° of every thing, but nothing of worth.

Now we descend to that part which concerneth the Illustration of Tradition,° comprehended in that science which we call Rhetoric, or Art of Eloquence; a science excellent, and excellently well laboured. For although in true value it is inferior to wisdom, as it is said by God to Moses,° when he disabled° himself for want of this faculty, 'Aaron shall be thy speaker, and thou shalt be to him as God'; yet with people it is the more mighty: for so Salomon saith, 'Sapiens corde appellabitur prudens, sed dulcis eloquio majora reperiet,'° signifying that profoundness of wisdom will help a man to a name or admiration, but that it is eloquence that prevaileth in an active life.

And as to the labouring of it, the emulation of Aristotle° with the rhetoricians of his time, and the experience of Cicero, hath made them in their works of Rhetorics exceed themselves. Again, the excellency of examples of eloquence in the orations of Demosthenes and Cicero, added to the perfection of the precepts of eloquence, hath doubled the progression in this art; and therefore the deficiencies which I shall note will rather be in some collections which may as handmaids attend the art, than in the rules or use of the art itself.

Notwithstanding, to stir the earth a little about the roots of this science, as we have done of the rest: The duty and office of Rhetoric is to apply Reason to Imagination° for the better moving of the will. For we see Reason is disturbed in the administration thereof by three means; by Illaqueation° or Sophism, which pertains to Logic; by Imagination or Impression,° which pertains to Rhetoric; and by Passion or Affection, which pertains to Morality.° And as in negotiation° with others men are wrought° by cunning, by importunity,° and by vehemency; so in this negotiation within ourselves men are undermined° by Inconsequences,° solicited and importuned by Impressions or Observations,° and transported by Passions. Neither is the nature of man so unfortunately built, as that those powers and arts should have force to disturb reason, and not to establish and advance it: for the end of Logic is to teach a form of argument to secure reason, and not to entrap it; the end of Morality is to procure the affections to obey reason, and not to invade it; the end of Rhetoric is to fill the imagination° to second° reason, and not to oppress it: for these abuses of arts come in but *ex obliquo*,° for caution.

And therefore it was great injustice in Plato,° though springing out of a just hatred of the rhetoricians of his time, to esteem of° Rhetoric but as a voluptuary art, resembling° it to cookery, that did mar wholesome meats, and help unwholesome by variety of sauces to the pleasure of the taste. For we see that speech is much more conversant in adorning that which is good than in colouring° that which is evil;° for there is no man but speaketh more honestly than he can do or think: and it was excellently noted by Thucydides° in Cleon, that because he used to hold on° the bad side in causes of estate,° therefore he was ever inveighing against eloquence and good speech; knowing that no man can speak fair of courses sordid and base. And therefore as Plato° said elegantly, that 'virtue, if she could be seen, would move great love and affection'; so seeing that she cannot be shewed to the Sense by corporal shape, the next degree is to shew her to the Imagination in lively representation: for to shew her to Reason only in subtilty of argument, was a thing ever derided in Chrysippus° and many of the Stoics; who thought to thrust virtue upon men by sharp disputations and conclusions, which have no sympathy with the will of man.

Again, if the affections in themselves were pliant and obedient to reason, it were true there should be no great use of persuasions and insinuations° to the will, more than of naked proposition and proofs; but in regard of the continual mutinies and seditions of the affections,

> Video meliora, proboque;
> Deteriora sequor:°

reason would become captive and servile, if Eloquence of Persuasions did not practise° and win the Imagination from the Affection's part, and contract a confederacy° between the Reason and Imagination against the Affections. For the affections themselves carry ever an appetite to good, as reason doth; the difference is, that 'the affection beholdeth merely the present;° reason beholdeth the future and sum of time';° and therefore the present filling the imagination more, reason is commonly vanquished; but after that force of eloquence and persuasion hath made things future and remote appear as present, then upon the revolt of the imagination reason prevaileth.

DE PRUDENTIA SERMONIS PRIVATI.° We conclude therefore, that Rhetoric can be no more charged with the colouring of the worse part,° than Logic with Sophistry, or Morality with Vice. For we know the doctrines of contraries° are the same, though the use be opposite. It appeareth also that Logic differeth from Rhetoric, not only as the fist from the palm,° the one close the other at large;° but much more in this, that Logic handleth reason exact and in truth, and Rhetoric handleth it as it is planted in popular opinions and manners. And therefore Aristotle° doth wisely place Rhetoric as between Logic on the one side and moral or civil knowledge on the other, as participating of both: for the proofs and demonstrations of Logic are toward all men indifferent° and the same; but the proofs and persuasions of Rhetoric ought to differ according to the auditors:°

> Orpheus in sylvis, inter delphinas Arion:°

which application, in perfection of idea, ought to extend so far, that if a man should speak of the same thing to several persons, he should speak to them all respectively° and several ways: though this politic part of eloquence in private speech it is easy for the greatest orators to want,° whilst by the observing their well-graced forms of speech they leese° the volubility° of application: and therefore it shall not be amiss to recommend this to better inquiry; not being curious° whether we place it here, or in that part which concerneth policy.

COLORES BONI ET MALI, SIMPLICIS ET COMPARATI.° Now therefore will I descend to the deficiencies, which (as I said) are but attendances.° And first, I do not find the wisdom and diligence of Aristotle° well pursued, who began to make a collection of the popular signs and colours of good and evil, both simple and comparative, which are as the Sophisms of Rhetoric (as I touched before). For example:

SOPHISMA.

Quod laudatur, bonum: quod vituperatur, malum.°

REDARGUTIO.

Laudat venales qui vult extrudere merces.°

Malum est, malum est, inquit emptor: sed cum recesserit, tum gloriabitur.°

The defects in the labour of Aristotle are three: one, that there be but a few of many; another, that their Elenches° are not annexed:° and the third, that he conceived but a part of the use of them: for their use is not only in probation,° but much more in impression.° For many forms are equal in signification which are differing in impression; as the difference is great in the piercing of that which is sharp and that which is flat, though the strength of the percussion° be the same; for there is no man but will be a little more raised° by hearing it said, 'Your enemies will be glad of this':

Hoc Ithacus velit, et magno mercentur Atridae:°

than by hearing it said only, 'This is evil for you.'

Secondly, I do resume° also that which I mentioned before° touching Provision or Preparatory store for the furniture° of speech and readiness° of invention; which appeareth to be of two sorts; the one in resemblance to a shop° of pieces unmade up, the other to a shop of things ready made up; both to be applied to that which is frequent and most in request: the former of these I will call *Antitheta*,° and the latter *Formulae.*°

ANTITHETA RERUM.° *Antitheta* are Theses argued *pro et contra*;° wherein men may be more large and laborious:° but (in such as are able to do it) to avoid prolixity of entry,° I wish° the seeds of the several arguments to be cast up° into some brief and acute° sentences; not to be cited, but to be as skeins or bottoms° of thread, to be unwinded° at large when they come to be used; supplying authorities and examples by reference.

PRO VERBIS LEGIS.

Non est interpretatio, sed divinatio, quae recedit a litera.
Cum receditur a litera, judex transit in legislatorem.°

PRO SENTENTIA LEGIS.

Ex omnibus verbis est eliciendus sensus qui interpretatur singula.°

FORMULAE. Formulae are but decent° and apt° passages or conveyances of speech, which may serve indifferently° for differing subjects; as of preface, conclusion, digression, transition, excusation,° &c. For as in buildings there is great pleasure and use in the well° casting° of the stair-cases, entries, doors, windows, and the like; so in speech the conveyances and passages are of special ornament and effect.

A CONCLUSION IN A DELIBERATIVE.°

So may we redeem the faults passed, and prevent the inconveniences future.°

There remain two appendices touching the tradition of knowledge, the one Critical, the other Pedantical.° For all knowledge is either delivered by teachers, or attained by men's proper° endeavours: and therefore as the principal part of tradition of knowledge concerneth chiefly writing of books,° so the relative° part thereof concerneth reading of books. Whereunto appertain incidently° these considerations. The first is concerning the true correction and edition of authors; wherein nevertheless rash diligence hath done great prejudice. For these critics° have often presumed that that which they understand not is false° set down: as the Priest that where he found it written of St. Paul, 'Demissus est per sportam'° mended his book,° and made it 'Demissus est per portam',° because *sporta* was an hard word, and out of° his reading; and surely their errors, though they be not so palpable and ridiculous, are yet of the same kind. And therefore as it hath been wisely noted, the most corrected copies° are commonly the least correct.

The second is concerning the exposition and explication of authors, which resteth° in annotations and commentaries; wherein it is over usual° to blanch° the obscure places, and discourse upon the plain.°

The third is concerning the times,° which in many cases give great light to true interpretations.

The fourth is concerning some brief censure° and judgment of the authors; that men thereby may make some election° unto themselves what books to read.

And the fifth is concerning the syntax° and disposition of studies; that men may know in what order or pursuit° to read.

For Pedantical° knowledge, it containeth that difference° of Tradition which is proper for youth; whereunto appertain divers considerations of great fruit.

As first, the timing and seasoning of knowledges; as with what to initiate them, and from what for a time to refrain° them.

Secondly, the consideration where to begin with the easiest and so proceed to the more difficult; and in what courses° to press° the more difficult and then to turn them to the more easy: for it is one method to practise swimming with bladders, and another to practise dancing with heavy shoes.

A third is the application of learning according unto the propriety° of the wits; for there is no defect in the faculties intellectual but seemeth to have a proper cure contained in some studies: as for example, if a child be bird-witted,° that is, hath not the faculty of attention, the Mathematics giveth a remedy thereunto; for in them, if the wit be caught away but a moment, one is new to begin.° And as sciences have a propriety towards faculties for cure and help, so faculties or powers have a sympathy towards sciences for excellency or speedy profiting; and therefore it is an inquiry of great wisdom, what kinds of wits and natures are most apt° and proper for what sciences.

Fourthly, the ordering of exercises is matter of great consequence to hurt or help; for as is well observed by Cicero,° men in exercising their faculties, if they be not well advised, do exercise their faults and get ill habits as well as good; so as there is a great judgment to be had in the continuance and intermission° of exercises. It were too long to particularise a number of other considerations of this nature, things but of mean appearance, but of singular efficacy. For as the wronging or cherishing° of seeds or young plants is that that is most important to their thriving; and as it was noted° that the first six kings being in truth as tutors° of the state of Rome in the infancy thereof, was the principal cause of the immense greatness of that state which followed: so the culture and manurance° of minds in youth hath such a forcible (though unseen) operation, as hardly any length of time or contention° of labour can countervail it afterwards. And it is not amiss to observe also how small and mean faculties gotten by education, yet when they fall into great men or great matters, do work great and important effects; whereof we see a notable example in Tacitus° of two stage-players, Percennius and Vibulenus, who by their faculty of playing° put the Pannonian armies into an extreme tumult and combustion.° For there arising a mutiny amongst them upon the death of Augustus Caesar, Blaesus the lieutenant had committed° some of the mutineers; which were suddenly° rescued; whereupon Vibulenus got to be heard speak, which he did in this manner: 'These poor innocent wretches, appointed to cruel death, you have restored to behold the light. But who shall restore my brother to me, or life unto my brother? that was sent hither in message from the legions of Germany to treat of the common cause, and he hath murdered him this last night by some of his fencers and ruffians, that he hath about him for his executioners upon soldiers. Answer, Blaesus, what is done with his body? The mortalest° enemies do not deny burial. When I have performed my last duties to the corpse with kisses, with tears, command me to be slain beside him; so that these my fellows, for our good meaning and our true hearts to the legions, may have leave to bury us.' With which speech he put the army into an infinite fury and uproar; whereas truth was he had no brother, neither was there any such matter, but he played it merely as if he had been upon the stage.°

But to return: we are now come to a period° of Rational Knowledges; wherein if I have made the divisions° other than those that are received, yet would I not be thought to disallow all those divisions which I do not use. For there is a double necessity imposed upon me of altering the divisions. The one, because it differeth in end and purpose, to sort together those things which are next° in nature, and those things which are next in use. For if a secretary of state should sort his papers, it is like° in his study or general cabinet° he would sort together things of a° nature, as treaties, instructions, &c. but in his boxes or particular° cabinet he would sort together those that he were like° to use together, though of several° natures; so in this general cabinet of knowledge it was necessary for me to follow the divisions of the nature of things; whereas if myself had been° to handle any particular knowledge, I would have respected the divisions fittest for use. The other, because the bringing in of the deficiencies did by consequence alter the partitions of the rest: for let the knowledge extant (for demonstration sake) be fifteen; let the knowledge with the deficiencies be twenty; the parts° of fifteen are not the parts of twenty; for the parts of fifteen are three and five; the parts of twenty are two, four, five, and ten. So as these things are without° contradiction, and could not otherwise be.

We proceed now to that knowledge which considereth of the Appetite and Will° of Man; whereof Salomon saith, 'Ante omnia, fili, custodi cor tuum; nam inde procedunt actiones vitae.'° In the handling of this science, those which have written seem to me to have done as if a man that professeth to teach to write did only exhibit fair copies of alphabets and letters joined, without giving any precepts or directions for the carriage° of the hand and framing° of the letters. So have they made good and fair exemplars° and copies, carrying the draughts and portraitures of Good, Virtue, Duty, Felicity; propounding them well described as the true objects and scopes of man's will and desires; but how to attain these excellent marks,° and how to frame° and subdue the will of man to become true° and conformable° to these pursuits, they pass it over altogether, or slightly and unprofitably. For it is not the disputing that moral virtues are in the mind of man by habit and not by nature, or the distinguishing that generous spirits are won by doctrines and persuasions, and the vulgar sort by reward and punishment,° and the like scattered glances and touches,° that can excuse the absence of this part.

The reason of this omission I suppose to be that hidden rock whereupon both this and many other barks° of knowledge have been cast away; which is, that men have despised to be conversant in ordinary and common matters;° the judicious direction whereof nevertheless is the wisest doctrine (for life consisteth not in novelties or subtilities);° but contrariwise they have compounded° sciences chiefly of a certain resplendent or lustrous mass of matter, chosen to give glory either to the subtility of disputations or to the eloquence of discourses. But Seneca giveth an excellent check° to eloquence; 'Nocet illis eloquentia, quibus non rerum cupiditatem facit, sed sui'.° Doctrines should be such as should make men in love with the lesson, and not with the teacher; being directed to the auditor's benefit, and not to the author's commendation: and therefore those are of the right kind which may be concluded as Demosthenes concludes his counsel, 'Quae si feceritis, non oratorem duntaxat in praesentia laudabitis, sed vosmetipsos etiam non ita multo post statu rerum vestrarum meliore'.°

Neither needed men of so excellent parts° to have despaired of a fortune which the poet Virgil promised himself (and indeed obtained), who got as much glory of eloquence, wit, and learning in the expressing of the observations of husbandry,° as of the heroical acts of Aeneas:—

> Nec sum animi dubius, verbis ea vincere magnum
> Quam sit, et angustis his addere rebus honorem.°

And surely if the purpose be in good earnest not to write at leisure that which men may read at leisure, but really to instruct and suborn° action and active life, these Georgics° of the mind, concerning the husbandry and tillage° thereof, are no less worthy than the heroical° descriptions of Virtue, Duty, and Felicity. Wherefore the main and primitive° division of moral knowledge seemeth to be into the Exemplar or Platform° of Good, and the Regiment or Culture of the Mind; the one describing the nature of good, the other prescribing rules how to subdue, apply, and accommodate° the will of man thereunto.

The doctrine touching the Platform or Nature of Good considereth it either Simple° or Compared; either the kinds of good, or the degrees of good: in the latter whereof those infinite disputations° which were touching the supreme degree thereof, which they term felicity, beatitude, or the highest good, the doctrines concerning which were as° the heathen divinity,° are by the Christian faith discharged.° And as Aristotle° saith, that 'young men may be happy, but not otherwise but by hope'; so we° must all acknowledge our minority,° and embrace the felicity which is by hope of the future world.

Freed therefore and delivered from this doctrine of the philosophers' heaven, whereby they feigned an higher elevation of man's nature than was, (for we see in what an height of style° Seneca writeth, 'Vere magnum, habere fragilitatem hominis, securitatem Dei',° we may with more sobriety° and truth receive° the rest of their inquiries and labours.° Wherein for the Nature of Good Positive° or Simple, they have set it down excellently, in describing the forms of Virtue and Duty, with their situations and postures,° in distributing them into their kinds, parts, provinces, actions, and administrations, and the like: nay farther, they have commended them to man's nature and spirit with great quickness of argument and beauty of persuasions; yea, and fortified and intrenched° them (as much as discourse can do) against corrupt and popular opinions. Again, for the Degrees and Comparative Nature of Good, they have also excellently handled it in their triplicity of Good,° in the comparisons between a contemplative and an active life,° in the distinction between virtue with reluctation° and virtue secured,° in their encounters between honesty and profit,° in their balancing of virtue with virtue,° and the like; so as this part deserveth to be reported for excellently laboured.°

Notwithstanding, if before they had comen to the popular and received notions of virtue and vice, pleasure and pain, and the rest, they had stayed a little longer upon the inquiry concerning the roots of good and evil, and the strings° of those roots, they had given, in my opinion, a great light to that which followed; and specially if they had consulted with nature, they had made their doctrines less prolix and more profound; which being by them in part omitted and in part handled with much confusion, we will endeavour to resume° and open° in a more clear manner.

There is formed in every thing a double nature of good: the one, as every thing is a total° or substantive° in itself; the other, as it is a part or member of a greater body; whereof the latter° is in degree the greater and the worthier, because it tendeth to the conservation of a more general form. Therefore we see the iron in particular° sympathy° moveth to the loadstone;° but yet if it exceed a certain quantity, it forsaketh the affection° to the loadstone, and like a good patriot° moveth to the earth, which is the region and country of massy° bodies. So may we go forward,° and see that water and massy bodies move to the centre of the earth; but rather than to suffer a divulsion° in the continuance° of nature, they will move upwards from the centre of the earth, forsaking their duty to the earth in regard of their duty to the world. This double nature of good, and the comparative thereof, is much more engraven° upon man, if he degenerate not; unto whom the conservation of duty to the public ought to be much more precious than the conservation of life and being: according to that memorable speech of Pompeius Magnus, when being in commission of purveyance for a famine° at Rome, and being dissuaded with great vehemency and instance° by his friends about him that he should not hazard himself to sea in an extremity of weather, he said only to them, 'Necesse est ut eam, non ut vivam'.° But it may be truly affirmed that there was never any philosophy, religion, or other discipline,° which did so plainly and highly exalt the good which is communicative, and depress° the good which is private and particular, as the Holy Faith; well declaring, that it was the same God that gave the Christian law to men, who gave those laws of nature to inanimate creatures that we spake of before; for we read that the elected saints of God° have wished themselves anathematized° and razed out° of the book of life, in an ecstasy of charity and infinite feeling of communion.

This being set down and strongly planted, doth judge and determine most of the controversies wherein Moral Philosophy is conversant. For first it decideth the question touching the preferment of the contemplative or active life, and decideth it against Aristotle.° For all the reasons which he bringeth for the contemplative are private, and respecting the pleasure and dignity of a man's self (in which respects no question the contemplative life hath the pre-eminence): not much unlike to that comparison which Pythagoras° made for the gracing and magnifying of philosophy and contemplation; who being asked what he was, answered, 'that if Hiero were ever at the Olympian games, he knew the manner, that some came to try their fortune for the prizes, and some came as merchants to utter° their commodities, and some came to make good cheer and meet their friends, and some came to look on; and that he was one of them that came to look on.' But men must know, that in this theatre° of man's life it is reserved only for God° and Angels to be lookers on. Neither could the like question ever have been received in the church, notwithstanding their 'Pretiosa in oculis Domini mors sanctorum ejus',° by which place° they would exalt their civil death° and regular° professions,° but upon this defence, that the monastical life is not simple° contemplative, but performeth the duty either of incessant prayers and supplications, which hath been truly esteemed as an office° in the church, or else of writing or taking instructions for writing concerning the law of God, as Moses did when he abode so long in the mount.° And so we see Henoch° the seventh from Adam, who was the first Contemplative° and walked with God, yet did also endow the church with prophecy, which St. Jude citeth. But for contemplation which should be finished° in itself without casting beams° upon society, assuredly divinity knoweth it not.

It decideth also the controversies between Zeno and Socrates and their schools and successions on the one side, who placed felicity in virtue simply or attended;° the actions and exercises° whereof do chiefly embrace and concern society;° and on the other side,° the Cyrenaics° and Epicureans, who placed it in pleasure,° and made virtue (as it is used in some comedies of errors,° wherein the mistress and the maid change habits),° to be but as a servant, without which pleasure cannot be served and attended; and the reformed school of the Epicureans, which placed it in serenity of mind and freedom from perturbation; as if they would have deposed Jupiter again, and restored Saturn and the first age,° when there was no summer nor winter, spring nor autumn, but all after° one air and season; and Herillus,° which placed felicity in extinguishment of the disputes of the mind, making no fixed nature of good and evil, esteeming things according to the clearness° of the desires, or the reluctation;° which opinion was revived in the heresy of the Anabaptists,° measuring things according to the motions° of the spirit, and the constancy or wavering of belief: all which are manifest to tend to private repose and contentment, and not to point of° society.

It censureth° also the philosophy of Epictetus,° which presupposeth that felicity must be placed in those things which are in our power, lest we be liable° to fortune and disturbance: as if it were not a thing much more happy to fail in good and virtuous ends for the public, than to obtain all that we can wish to ourselves in our proper fortune; as Consalvo° said to his soldiers, shewing them Naples, and protesting he had rather die one foot forwards than to have his life secured for long by one foot of retreat; whereunto the wisdom of that heavenly leader hath signed,° who hath affirmed that 'a good conscience is a continual feast':° shewing plainly that the conscience of good intentions, howsoever succeeding, is a more continual joy to nature than all the provision which can be made for security and repose.

It censureth likewise that abuse of philosophy which grew general about the time of Epictetus,° in converting it into an occupation or profession; as if the purpose had been, not to resist and extinguish perturbations, but to fly and avoid the causes of them, and to shape a particular kind and course of life to that end; introducing such an health of mind, as was that health of body of which Aristotle° speaketh of Herodicus, who did nothing all his life long but intend° his health: whereas if men refer themselves to duties of society, as° that health of body is best which is ablest to endure all alterations and extremities, so likewise that health of mind is most proper° which can go through the greatest temptations and perturbations. So as Diogenes'° opinion is to be accepted, who commended not them which abstained, but them which sustained,° and could refrain° their mind *in praecipitio*,° and could give unto the mind (as is used in horsemanship) the shortest stop or turn.

Lastly, it censureth the tenderness° and want of application° in some of the most ancient and reverend philosophers and philosophical men, that did retire too easily from civil business, for avoiding° of indignities and perturbations; whereas the resolution of men truly moral ought to be such as the same Consalvo said the honour of a soldier should be, 'e tela crassiore',° and not so fine as that every thing should catch in it and endanger it.

To resume Private or Particular Good, it falleth into the division of Good Active and Passive: for this difference of Good (not unlike to that which amongst the Romans was expressed in the familiar or household terms of *Promus* and *Condus*)° is formed also in all things;

and is best disclosed in the two several appetites in creatures, the one to preserve or continue themselves, and the other to dilate° or multiply themselves; whereof the latter seemeth to be the worthier. For in nature, the heavens, which are the more worthy, are the agent; and the earth, which is the less worthy, is the patient. In the pleasures of living creatures, that of generation° is greater than that of food. In divine doctrine, 'Beatius est dare quam accipere'.° And in life, there is no man's spirit so soft,° but esteemeth° the effecting° of somewhat that he hath fixed in his desire more than sensuality. Which priority of the Active Good is much upheld by the consideration of our estate to be mortal and exposed to fortune; for if we might have a perpetuity and certainty in our pleasures, the state° of them would advance their price; but when we see it is but 'Magni aestimamus mori tardius',° and 'Ne glorieris de crastino, nescis partum diei',° it maketh us to desire to have somewhat secured and exempted from time; which are only our deeds and works; as it is said 'Opera eorum sequuntur eos'.° The pre-eminence likewise of this Active Good is upheld by the affection which is natural in man towards variety and proceeding;° which in the pleasures of the sense (which is the principal part of Passive Good) can have no great latitude:° 'Cogita quamdiu eadem feceris; cibus, somnus, ludus; per hunc circulum curritur; mori velle non tantum fortis, aut miser, aut prudens, sed etiam fastidiosus potest'.° But in enterprises, pursuits, and purposes of life, there is much variety; whereof men are sensible° with pleasure in their inceptions,° progressions, recoils,° reintegrations,° approaches, and attainings to their ends: so as it was well said, 'Vita sine proposito languida et vaga est.'°

Neither hath this Active Good any identity with the good of society, though in some case it hath an incidence into° it: for although it do many times bring forth acts of beneficence, yet it is with a respect° private° to a man's own power, glory, amplification, continuance; as appeareth plainly when it findeth a contrary° subject. For that gigantine° state of mind which possesseth the troublers of the world, such as was Lucius Sulla,° and infinite° other in smaller model,° who would have all men happy or unhappy as° they were their friends or enemies, and would give form to the world according to their own humours, (which is the true Theomachy),° pretendeth° and aspireth to active good, though it recedeth furthest from good of society, which we have determined to be the greater.

To resume Passive Good, it receiveth a subdivision of Conservative° and Perfective.° For let us take a brief review of that which we have said: we have spoken first of the Good of Society, the intention° whereof embraceth the form of Human Nature, whereof we are members and portions, and not our own proper and individual form. We have spoken of Active Good, and supposed° it as a part of Private and Particular Good; and rightly; for there is impressed upon all things a triple desire or appetite proceeding from love to themselves; one of preserving and continuing their form; another of advancing and perfecting their form; and a third of multiplying and extending their form upon other things; whereof the multiplying or signature° of it upon other things is that which we handled by the name of Active Good. So as there remaineth the conserving° of it, and perfecting or raising of it; which latter is the highest degree of Passive Good. For to preserve in state° is the less, to preserve with advancement is the greater.° So in man,

Igneus est ollis vigor, et coelestis origo.°

His approach or assumption° to divine or angelical nature is the perfection of his form; the error or false imitation of which good is that which is the tempest of human life; while man, upon the instinct of an advancement formal and essential, is carried to seek an advancement local. For as those which are sick, and find no remedy, do tumble° up and down and change place, as if by a remove local° they could obtain a remove internal; so is it with men in ambition, when failing of the mean° to exalt their nature, they are in a perpetual estuation° to exalt their place. So then Passive Good is, as was said, either Conservative or Perfective.

To resume the good of Conservation or Comfort, which consisteth in the fruition of that which is agreeable to our natures; it seemeth to be the most pure and natural of pleasures,° but yet the softest° and the lowest.° And this also receiveth a difference,° which hath neither been well judged of nor well enquired. For the good of fruition or contentment is placed either in the sincereness° of the fruition, or in the quickness and vigour of it; the one superinduced° by the equality,° the other by vicissitude;° the one having less mixture of evil, the other more impression° of good. Whether of° these is the greater good, is a question controverted; but whether man's nature may not be capable of both, is a question not enquired.

The former question being debated between Socrates and a Sophist, Socrates° placing felicity in an equal and constant peace of mind, and the Sophist in much desiring and much enjoying, they fell from argument to ill words:° the Sophist saying that Socrates' felicity was the felicity of a block or stone; and Socrates saying that the Sophist's felicity was the felicity of one that had the itch, who did nothing but itch and scratch. And both these opinions do not want their supports.° For the opinion of Socrates is much upheld by the general consent even of the Epicures° themselves, that virtue beareth a great part in felicity; and if so, certain it is that virtue hath more use in clearing° perturbations than in compassing° desires. The Sophist's opinion is much favoured by the assertion we last spake of, that good of advancement is greater than good of simple preservation; because every obtaining a desire hath a shew of advancement,° as motion though in a circle° hath a shew of progression.

But the second question,° decided the true way, maketh the former superfluous. For can it be doubted but that there are some who take more pleasure in enjoying pleasures than some other, and yet nevertheless are less troubled with the loss or leaving of them? So as this same 'Non uti ut non appetas, non appetere ut non metuas, sunt animi pusilli et diffidentis'.° And it seemeth to me, that most of the doctrines of the philosophers are more fearful and cautionary° than the nature of things requireth. So have they increased the fear of death in offering to cure it. For when they would have a man's whole life to be but a discipline° or preparation to die,° they must needs make men think that it is a terrible enemy against whom there is no end of preparing.° Better saith the poet:°

> Qui finem vitae extremum inter munera ponat
> Naturae.°

So have they sought to make men's minds too uniform and harmonical,° by not breaking° them sufficiently to contrary motions: the reason whereof I suppose to be, because they themselves were men dedicated to a private, free, and unapplied° course of life. For as we see, upon the lute or like instrument, a ground,° though it be sweet and have shew of many changes,° yet breaketh° not the hand to such strange and hard stops and passages° as a set song° or voluntary;° much after the same manner was the diversity between a philosophical and a civil° life. And therefore men are to imitate the wisdom of jewellers; who, if there be a grain or a cloud or an ice° which may be ground forth without taking too much of the stone, they help it; but if it should lessen and abate° the stone too much, they will not meddle with it: so ought men so to procure serenity as they destroy not magnanimity.°

Having therefore deduced the Good of Man which is Private and Particular as far as seemeth fit, we will now return to that good of man which respecteth and beholdeth society, which we may term Duty; because the term of Duty is more proper to a mind well framed and disposed towards others, as the term of Virtue is applied to a mind well formed and composed in itself; though neither can a man understand Virtue without some relation to society, nor Duty without an inward disposition.° This part may seem at first to pertain to science° civil and politic; but not if it be well observed. For it concerneth the regiment and government of every man over himself, and not over others. And as in architecture the direction of framing° the posts, beams, and other parts of building, is not the same with the manner of joining them and erecting the building; and in mechanicals,° the direction how to frame an instrument or engine, is not the same with the manner of setting it on work° and employing it; and yet nevertheless in expressing of the one you incidently express the aptness towards the other; so the doctrine of conjugation° of men in society differeth from that of their conformity° thereunto.

This part of Duty is subdivided into two parts; the common duty of every man, as a man or member of a state; the other, the respective° or special duty of every man, in his profession, vocation, and place. The first of these is extant and well laboured, as hath been said. The second likewise I may report rather dispersed° than deficient; which manner of dispersed writing in this kind of argument I acknowledge to be best. For who can take upon him to write of the proper duty, virtue, challenge,° and right of every several vocation, profession and place?° For although sometimes a looker-on° may see more than a gamester,° and there be a proverb more arrogant than sound, that 'the vale best discovereth the hill';° yet there is small doubt but that men can write best and most really and materially° in their own professions; and that the writing of speculative° men of° active matter for the most part doth seem to men of experience, as Phormio's argument° of the wars seemed to Hannibal, to be but dreams and dotage.° Only there is one vice which accompanieth them that write in their own professions, that they magnify them in excess. But generally it were to be wished (as that which would make learning indeed solid and fruitful) that active men would or could become writers.

In which kind I cannot but mention, *honoris causa*,° your Majesty's excellent book° touching the duty of a king: a work richly compounded of divinity, morality, and policy, with great aspersion° of all other arts; and being in mine opinion one of the most sound and healthful writings that I have read; not distempered° in the heat of invention, nor in the coldness of negligence; not sick of dizziness,° as those are who leese themselves in their order;° nor of convulsions,° as those which cramp in matters impertinent;° not savouring° of perfumes and paintings, as those do who seek to please the reader more than nature beareth;° and chiefly well disposed in the spirits° thereof, being agreeable to truth and apt for action; and far removed from that natural infirmity, whereunto I noted those that write in their own professions to be subject, which is, that they exalt it above measure. For your Majesty hath truly described, not a king of Assyria or Persia in their extern° glory, but a Moses or a David, pastors° of their people. Neither can I ever leese° out of my remembrance what I heard your Majesty in the same sacred spirit of government deliver in a great cause of judicature,° which was, that 'Kings ruled by their laws as God did by the laws of nature, and ought as rarely to put in use their supreme prerogative as God doth his power of working miracles'. And yet notwithstanding, in your book of a free monarchy,° you do well give men to understand, that you know the plenitude of the power and right of a King, as well as the circle° of his office and duty. Thus have I presumed to allege° this excellent writing of your Majesty, as a prime° or eminent example of tractates concerning special and respective duties; wherein I should have said as much, if it had been written a thousand years since. Neither am I moved with certain courtly decencies,° which esteem it flattery to praise in presence. No, it is flattery to praise in absence; that is, when either the virtue is absent, or the occasion is absent; and so the praise is not natural, but forced, either in truth or in time. But let Cicero be read in his oration *pro Marcello*,° which is nothing but an excellent table° of Caesar's virtue, and made to his face;° besides the example of many other excellent persons, wiser a great deal than such observers; and we will never doubt,° upon a full occasion, to give just praises to present or absent.

DE CAUTELIS ET MALIS ARTIBUS.° But to return: there belongeth further to the handling of this part touching the duties of professions and vocations, a Relative° or opposite, touching the frauds, cautels,° impostures, and vices of every profession; which hath been likewise handled: but how? rather in a satire and cynically,° than seriously and wisely: for men have rather sought by wit to deride and traduce much of that which is good in professions, than with judgment to discover and sever° that which is corrupt. For, as Salomon saith, He that cometh to seek after knowledge with a mind to scorn and censure, shall be sure to find matter° for his humour, but no matter for his instruction: 'Quaerenti derisori scientiam ipsa se abscondit; sed studioso fit obviam'.° But the managing of this argument with integrity and truth, which I note as deficient, seemeth to me to be one of the best fortifications for honesty and virtue that can be planted. For as the fable goeth of the Basilisk,° that if he see you first you die for it, but if you see him first he dieth; so is it with deceits and evil arts; which if they be first espied they leese their life, but if they prevent° they endanger. So that we are much beholden to Machiavel° and others, that write what men do and not what they ought to do. For it is not possible to join serpentine wisdom with the columbine innocency,° except men know exactly all the conditions of the serpent; his baseness° and going upon his belly, his volubility° and lubricity,° his envy° and sting, and the rest; that is, all forms and natures of evil. For without this, virtue lieth open and unfenced.° Nay an honest man can do no good upon° those that are wicked to reclaim them, without the help of the knowledge of evil. For men of corrupted minds presuppose that honesty groweth out of simplicity° of manners, and believing of preachers, school-masters, and men's exterior language: so as, except you can make them perceive that you know the utmost reaches° of their own corrupt opinions, they despise all morality. 'Non recipit stultus verba prudentiae, nisi ea dixeris quae versantur in corde ejus'.°

Unto this part touching Respective Duty doth also appertain the duties between husband and wife, parent and child, master and servant: so likewise the laws of friendship and gratitude, the civil bond of companies,° colleges, and politic bodies,° of neighbourhood, and all other proportionate duties; not as they are parts of government and society,° but as to the framing° of the mind of particular persons.°

The knowledge concerning good respecting Society doth handle it also not simply alone, but comparatively; whereunto belongeth the weighing of duties between person and person, case and case, particular and public: as we see in the proceeding of Lucius Brutus° against his own sons, which was so much extolled; yet what was said?

Infelix, utcunque ferent ea facta minores.°

So the case was doubtful,° and had opinion on both sides. Again, we see when M. Brutus and Cassius° invited to a supper certain whose opinions they meant to feel,° whether they were fit to be made their associates, and cast forth° the question touching the killing of a tyrant being an usurper, they were divided in opinion; some holding that servitude was the extreme of evils, and others that tyranny° was better than a civil war:° and a number of the like cases there are of comparative duty. Amongst which that of all others is the most frequent, where the question is of a great deal of good to ensue of° a small injustice. Which Jason of Thessalia determined against the truth: 'Aliqua sunt injuste facienda, ut multa juste fieri possint':° But the reply is good, 'Authorem praesentis justitiae habes, sponsorem futurae non habes'.° Men must pursue things which are just in present,° and leave the future to the divine Providence. So then we pass on from this general part touching the exemplar and description of good.

DE CULTURA ANIMI.° Now therefore that we have spoken of this fruit of life, it remaineth to speak of the husbandry° that belongeth thereunto; without which part the former seemeth to be no better than a fair image or statua,° which is beautiful to contemplate, but is without life and motion: whereunto Aristotle himself subscribeth in these words: 'Necesse est scilicet de virtute dicere, et quid sit, et ex quibus gignatur. Inutile enim fere fuerit virtutem quidem nosse, acquirendae autem ejus modos et vias ignorare. Non enim de virtute tantum, qua specie sit, quaerendum est, sed et quomodo sui copiam faciat: utrumque enim volumus, et rem ipsam nosse, et ejus compotes fieri: hoc autem ex voto non succedet, nisi sciamus et ex quibus et quomodo'.° In such full words and with such iteration° doth he inculcate° this part. So saith Cicero in great commendation of Cato the second,° that he had applied himself to philosophy 'non ita disputandi causa, sed ita vivendi'.° And although the neglect of our times, wherein few men do hold any consultations touching the reformation of their life (as Seneca excellently saith, 'De partibus vitae quisque deliberat, de summa nemo')° may make this part seem superfluous; yet I must conclude with that aphorism of Hippocrates, 'Qui gravi morbo correpti dolores non sentiunt, iis mens aegrotat';° they need medicine not only to assuage the disease but to awake the sense.° And if it be said that the cure of men's minds° belongeth to sacred Divinity, it is most true: but yet Moral Philosophy may be preferred unto° her as a wise servant and humble handmaid. For as the Psalm° saith, that 'the eyes of the handmaid look perpetually towards the mistress,' and yet no doubt many things are left to the discretion of the handmaid to discern° of the mistress' will; so ought

Moral Philosophy to give a constant attention to the doctrines of Divinity, and yet so as it may yield of herself (within due limits) many sound and profitable directions.

This part therefore, because of the excellency thereof, I cannot but find exceeding strange that it is not reduced to written inquiry;° the rather because it consisteth of much matter wherein both speech and action is often conversant, and such wherein the common talk of men (which is rare, but yet cometh sometimes to pass) is wiser than their books. It is reasonable therefore that we propound° it in the more particularity,° both for the worthiness, and because we may acquit ourselves° for reporting it deficient; which seemeth almost incredible, and is otherwise conceived and presupposed by those themselves that have written. We will therefore enumerate some heads or points thereof, that it may appear the better what it is, and whether it be extant.

First therefore, in this, as in all things which are practical, we ought to cast up our account, what is in our power and what not; for the one may be dealt with by way of alteration, but the other by way of application° only. The husbandman° cannot command neither the nature of the earth nor the seasons of the weather; no more can the physician the constitution of the patient nor the variety of accidents.° So in the culture and cure° of the mind of man, two things are without° our command; points of nature, and points of fortune; for to the basis of the one, and the conditions of the other, our work is limited and tied. In these things therefore it is left unto us to proceed by application:

Vincenda est omnis fortuna ferendo

and so likewise,

Vincenda est omnis natura ferendo.°

But when that we speak of suffering, we do not speak of a dull° and neglected suffering, but of a wise and industrious suffering, which draweth and contriveth use and advantage out of that which seemeth adverse and contrary; which is that property which we call Accommodating or Applying.° Now the wisdom of application resteth principally in the exact and distinct knowledge of the precedent state or disposition unto which we do apply: for we cannot fit a garment, except we first take measure of the body.

So then the first article° of this knowledge is to set down sound and true distributions and descriptions of the several characters and tempers of men's natures and dispositions, specially having regard to those differences which are most radical° in being the fountains and causes of the rest, or most frequent in concurrence or commixture;° wherein it is not the handling of a few of them in passage,° the better to describe the mediocrities° of virtues, that can satisfy this intention; for if it deserve to be considered, that there are minds which are proportioned to great matters, and others to small (which Aristotle° handleth or ought to have handled by the name of Magnanimity), doth it not deserve as well to be considered, that there are minds proportioned to intend° many matters, and others to° few? so that some can divide themselves,° others can perchance do exactly well,° but it must be but in few things at once; and so there cometh to be a narrowness of mind, as well as a pusillanimity.° And again, that some minds are proportioned to that which may be dispatched at once, or within a short return of time;° others to that which begins afar off, and is to be won with length of pursuit;

Jam tum tenditque fovetque:°

so that there may be fitly said to be a longanimity;° which is commonly also ascribed to God as a magnanimity. So further deserved it to be considered by Aristotle,° that there is a disposition in conversation° (supposing it in things which do in no sort touch or concern a man's self) to soothe° and please, and a disposition contrary to contradict and cross;° and deserveth it not much better to be considered, that there is a disposition, not in conversation or talk but in matter of more serious nature (and supposing it still in things merely indifferent),° to take pleasure in the good of another, and a disposition contrariwise to take distaste at the good of another; which is that property which we call good-nature or ill-nature, benignity or malignity? And therefore I cannot sufficiently marvel that this part of knowledge touching the several characters of natures and dispositions should be omitted both in morality and policy, considering it is of so great ministery° and suppeditation° to them both. A man shall find in the traditions of astrology some pretty and apt divisions of men's natures, according to the predominances of the planets;° lovers of quiet, lovers of action, lovers of victory, lovers of honour, lovers of pleasure, lovers of arts, lovers of change, and so forth. A man shall find in the wisest sort of these Relations° which the Italians make touching Conclaves, the natures of the several Cardinals handsomely° and lively° painted forth. A man shall meet with in every day's conference° the denominations of sensitive, dry,° formal,° real,°

humorous,° certain,° 'huomo di prima impressione,'° 'huomo di ultima impressione',° and the like: and yet nevertheless this kind of observations wandereth° in words, but is not fixed° in inquiry. For the distinctions are found° (many of them), but we conclude° no precepts upon them; wherein our fault is the greater, because both history, poesy, and daily experience are as goodly fields where these observations grow; whereof we make a few posies° to hold in our hands, but no man bringeth them to the confectionary,° that receits° might be made of them for use of life.

Of much like kind are those impressions of nature,° which are imposed upon the mind by the sex, by the age, by the region,° by health and sickness, by beauty and deformity, and the like, which are inherent and not extern; and again those which are caused by extern fortune; as sovereignty, nobility, obscure birth, riches, want, magistracy,° privateness, prosperity, adversity, constant fortune, variable fortune, rising *per saltum, per gradus*,° and the like. And therefore we see that Plautus maketh it a wonder to see an old man beneficent; 'benignitas hujus ut adolescentuli est':° St. Paul concludeth that severity of discipline was to be used to the Cretans, 'Increpa eos dure',° upon the disposition of their country; 'Cretenses semper mendaces, malae bestiae, ventres pigri'.° Sallust noteth that it is usual with Kings to desire contradictories; 'Sed plerumque regiae voluntates, ut vehementes sunt, sic mobiles, saepeque ipsae sibi adversae'.° Tacitus observeth how rarely raising of the fortune mendeth° the disposition; 'Solus Vespasianus mutatus in melius'.° Pindarus maketh an observation that great and sudden fortune for the most part defeateth° men; 'Qui magnam felicitatem concoquere non possunt':° so the Psalm sheweth it is more easy to keep a measure in the enjoying of fortune than in the increase of fortune; 'Divitiae si affluant, nolite cor apponere'.° These observations and the like I deny not but are touched a little by Aristotle° as in passage in his Rhetorics, and are handled in some scattered discourses; but they were never incorporate into Moral Philosophy, to which they do essentially appertain; as the knowledge of the diversity of grounds and moulds° doth to agriculture, and the knowledge of the diversity of complexions and constitutions° doth to the physician; except we mean to follow the indiscretion° of empirics,° which minister the same medicines to all patients.

Another article of this knowledge is the inquiry touching the affections;° for as in medicining° of the body it is in order first° to know the divers complexions and constitutions, secondly the diseases, and lastly the cures; so in medicining of the mind, after knowledge of the divers characters of men's natures, it followeth in order° to know the diseases and infirmities of the mind, which are no other than the perturbations and distempers° of the affections. For as the ancient politiques° in popular estates° were wont to compare the people to the sea and the orators to the winds, because as the sea would of itself be calm and quiet if the winds did not move and trouble it, so the people would be peaceable and tractable if the seditious orators did not set them in working and agitation; so it may be fitly said, that the mind in the nature thereof would be temperate and stayed, if the affections, as winds, did not put it into tumult and perturbation. And here again I find strange, as before, that Aristotle should have written divers volumes of Ethics,° and never handled the affections,° which is the principal subject thereof; and yet in his Rhetorics, where they are considered but collaterally and in a second degree (as they may be moved by speech), he findeth place for them, and handleth them well for the quantity; but where their true place is, he pretermitteth° them. For it is not his disputations about pleasure and pain° that can satisfy this inquiry, no more than he that should generally handle the nature of light can be said to handle the nature of colours; for pleasure and pain are to the particular affections as light is to particular colours. Better travails° I suppose had the Stoics taken in this argument, as far as I can gather by that which we have at second hand:° but yet it is like it was after their manner, rather in subtilty° of definitions (which in a subject of this nature are but curiosities)° than in active° and ample descriptions and observations. So likewise I find some particular writings° of an elegant nature touching some of the affections; as of anger, of comfort upon adverse accidents, of tenderness of countenance, and other. But the poets and writers of histories are the best doctors° of this knowledge; where we may find painted forth° with great life, how affections are kindled and incited; and how pacified and refrained; and how again contained from act and further degree;° how they disclose themselves, how they work, how they vary, how they gather and fortify,° how they are inwrapped° one within another, and how they do fight and encounter one with another, and other the like particularities: amongst the which this last is of special use in moral and civil matters; how (I say) to set affection against affection, and to master one by another; even as we use to hunt beast with beast and fly° bird with bird, which otherwise percase° we could not so easily recover:° upon which foundation is erected that excellent use of *praemium* and *poena*,° whereby civil states consist;°

employing the predominant affections of fear and hope, for the suppressing and bridling the rest. For as in the government of states it is sometimes necessary to bridle one faction with another, so it is in the government within.

Now come we to those points° which are within our own command, and have force and operation upon the mind to affect the will and appetite and to alter manners: wherein they ought to have handled custom, exercise, habit, education, example, imitation, emulation, company, friends, praise, reproof, exhortation, fame, laws, books, studies: these as° they have determinate° use in moralities,° from these the mind suffereth, and of these are such receipts and regiments° compounded° and described,° as may seem to recover or preserve the health and good estate of the mind, as far as pertaineth to human medicine: of which number we will visit upon some one or two as an example of the rest, because it were too long to prosecute all; and therefore we do resume Custom and Habit to speak of.

The opinion of Aristotle° seemeth to me a negligent opinion, that of those things which consist° by nature nothing can be changed by custom; using for example, that if a stone be thrown ten thousand times up, it will not learn to ascend; and that by often seeing or hearing, we do not learn to see or hear the better. For though this principle be true in things wherein nature is peremptory° (the reason whereof we cannot now stand° to discuss), yet it is otherwise in things wherein nature admitteth a latitude.° For he might see that a strait° glove will come more easily on with use, and that a wand° will by use° bend otherwise than it grew, and that by use of the voice we speak louder and stronger, and that by use of enduring heat or cold we endure it the better, and the like: which latter sort have a nearer resemblance unto that subject of manners he handleth than those instances which he allegeth.° But allowing his conclusion, that virtues and vices consist in habit, he ought so much the more to have taught the manner of superinducing° that habit: for there be many precepts of the wise ordering the exercises of the mind, as there is of ordering the exercises of the body; whereof we will recite a few.

The first shall be, that we beware we take not at the first either too high a strain° or too weak: for if too high, in a diffident nature you discourage; in a confident nature you breed an opinion of facility,° and so a sloth; and in all natures you breed a further expectation than can hold out, and so an insatisfaction° on the end: if too weak, of the other side, you may not look° to perform and overcome any great task.

Another precept is, to practise all things chiefly at two several° times, the one when the mind is best disposed, the other when it is worst disposed; that by the one you may gain a great step,° by the other you may work out° the knots° and stonds° of the mind, and make the middle times the more easy and pleasant.

Another precept is, that which Aristotle mentioneth by the way, which is to bear ever towards the contrary extreme of that whereunto we are by nature inclined:° like unto the rowing against the stream, or making a wand straight by bending him contrary to his natural crookedness.

Another precept is that the mind is brought to any thing better, and with more sweetness and happiness, if that whereunto you pretend° be not first in the intention, but 'tanquam aliud agendo',° because of the natural hatred of the mind against necessity and constraint. Many other axioms there are touching the managing of Exercise and Custom; which being so conducted,° doth prove indeed another nature;° but being governed by chance, doth commonly prove but an ape° of nature, and bringeth forth that which is lame and counterfeit.°

So if we should handle books and studies, and what influence and operation° they have upon manners, are there not divers precepts of great caution° and direction° appertaining thereunto? Did not one of the fathers in great indignation call Poesy 'vinum daemonum',° because it increaseth temptations, perturbations, and vain opinions? Is not the opinion of Aristotle° worthy to be regarded, wherein he saith that young men are no fit auditors of moral philosophy, because they are not settled from° the boiling heat of their affections, nor attempered° with time and experience? And doth it not hereof come, that those excellent books and discourses of the ancient writers (whereby they have persuaded unto virtue most effectually, by representing her in state and majesty, and popular° opinions against virtue in their parasites' coats,° fit to be scorned and derided), are of so little effect towards honesty of life, because they are not read and revolved° by men in their mature and settled years, but confined almost to boys and beginners? But is it not true also, that much less young men are fit auditors of matters of policy, till they have been throughly seasoned° in religion and morality; lest their judgments be corrupted, and made apt to think that there are no true differences° of things, but according to utility° and fortune; as the verse describes it, 'Prosperum et felix scelus virtus vocatur';° and again, 'Ille crucem pretium sceleris tulit, hic diadema';° which the poets do speak satirically, and in indignation on virtue's behalf; but books of policy do speak it seriously and positively;° for so it pleaseth Machiavel° to say, that 'if Caesar had been overthrown he would have been more odious° than ever was Catiline'; as if there had been no difference but in fortune, between a very fury of lust and blood, and the most excellent spirit (his ambition reserved)° of the world?° Again, is there not a caution likewise to be given of the doctrines of moralities themselves (some kinds of them), lest they make men too precise,° arrogant, incompatible;° as Cicero saith of Cato, 'In Marco Catone haec bona quae videmus divina et egregia, ipsius scitote esse propria; quae nonnunquam requirimus, ea sunt omnia non a natura, sed a magistro'?° Many other axioms and advices there are touching those proprieties° and effects which studies do infuse and instil into manners. And so likewise is there touching the use of all those other points, of company, fame, laws, and the rest, which we recited in the beginning in the doctrine of morality.

But there is a kind of Culture of the Mind that seemeth yet more accurate° and elaborate than the rest, and is built upon this ground;° that the minds of all men are at some times in a state° more perfect,° and at other times in a state more depraved. The purpose therefore of this practice is to fix and cherish the good hours of the mind, and to obliterate and take forth the evil.° The fixing of the good hath been practised by two means; vows or constant resolutions; and observances or exercises; which are not to be regarded so much in themselves, as because they keep the mind in continual obedience. The obliteration of the evil hath been practised by two means; some kind of redemption° or expiation of that which is past; and an inception° or account *de novo*° for the time to come. But this part seemeth sacred and religious, and justly; for all good Moral Philosophy (as was said) is but an handmaid to religion.

Wherefore we will conclude with that last point which is of all other means the most compendious° and summary,° and again the most noble and effectual, to the reducing° of the mind unto virtue and good estate; which is the electing and propounding unto a man's self good and virtuous ends of his life, such as may be in a reasonable sort within his compass to attain. For if these two things be supposed, that a man set before him honest and good ends, and again that he be resolute, constant, and true° unto them, it will follow that he shall mould himself into all virtue at once.° And this is indeed like the work of nature; whereas the other course is like the work of the hand. For as when a carver° makes an image, he shapes only that part whereupon he worketh; as if he be upon the face, that part which shall be the body is but a rude° stone still, till such times as he comes to it; but contrariwise when nature makes a flower or living creature, she formeth rudiments of all the parts at one time;° so in obtaining virtue by habit, while a man practiseth temperance, he doth not profit much to fortitude, nor the like; but when he dedicateth and applieth himself to good ends, look what virtue soever the pursuit and passage° towards those ends doth commend unto him, he is invested of° a precedent° disposition to conform himself thereunto; which° state of mind Aristotle doth excellently express himself, that it ought not to be called virtuous, but divine: his words are these: 'Immanitati autem consentaneum est opponere eam, quae supra humanitatem est, heroicam sive divinam virtutem':° and a little after, 'Nam ut ferae neque vitium neque virtus est, sic neque Dei: sed hic quidem status altius quiddam virtute est, ille aliud quiddam a vitio'.°

And therefore we may see what celsitude° of honour Plinius Secundus attributeth to Trajan in his funeral oration,° where he said, that 'men needed to make no other prayers to the gods, but that they would continue as good lords to them as Trajan had been'; as if he had not been only an imitation of divine nature, but a pattern° of it. But these be heathen and profane passages,° having but a shadow of that divine state of mind which religion and the holy faith doth conduct men unto, by imprinting upon their souls Charity,° which is excellently called the bond of Perfection,° because it comprehendeth° and fasteneth all virtues together. And as it is elegantly said by Menander of vain° love, which is but a false imitation of divine love, 'Amor melior sophista laevo ad humanam vitam,'° that love teacheth a man to carry° himself better than the sophist or preceptor, which he calleth left-handed, because with all his rules and preceptions° he cannot form a man so dexterously,° nor with that facility to prize° himself and govern himself, as love can do; so certainly if a man's mind be truly inflamed with charity, it doth work him suddenly into greater perfection than all the doctrine of morality can do, which is but a sophist° in comparison of the other. Nay further, as Xenophon° observed truly that all other affections, though they raise the mind, yet they do it by distorting and uncomeliness of ecstasies or excesses; but only love doth exalt the mind, and nevertheless at the same instant doth settle and compose it; so in all other excellencies, though they advance° nature, yet they are subject to excess; only charity admitteth no excess: for so we see, aspiring to be like God in power, the angels transgressed° and fell; 'Ascendam, et ero similis Altissimo';°

by aspiring to be like God in knowledge, man transgressed and fell; 'Eritis sicut Dii, scientes bonum et malum';° but by aspiring to a similitude of° God in goodness or love, neither man nor angel ever transgressed or shall transgress. For unto that imitation we are called: 'Diligite inimicos vestros, benefacite eis qui oderunt vos, et orate pro persequentibus et calumniantibus vos, ut sitis filii Patris vestri qui in coelis est, qui solem suum oriri facit super bonos et malos, et pluit super justos et injustos'.° So in the first platform° of the divine nature itself, the heathen religion speaketh thus, 'Optimus Maximus',° and the sacred Scriptures thus, 'Misericordia ejus super omnia opera ejus'.°

Wherefore° I do conclude this part of moral knowledge, concerning the Culture and Regiment of the Mind; wherein if any man, considering the parts thereof which I have enumerated, do judge that my labour is but to collect into an Art or Science that which hath been pretermitted° by others as matter of common sense and experience, he judgeth well.° But as Philocrates sported with Demosthenes,° 'You may not marvel (Athenians), that Demosthenes and I do differ, for he drinketh water, and I drink wine'; and like as we read of an ancient parable of the two gates of sleep,

> Sunt geminae somni portae: quarum altera fertur
> Cornea, qua veris facilis datur exitus umbris:
> Altera candenti perfecta nitens elephanto,
> Sed falsa ad coelum mittunt insomnia manes:°

so if we put on sobriety° and attention, we shall find it a sure maxim in knowledge, that the more pleasant liquor (of wine) is the more vaporous,° and the braver° gate (of ivory) sendeth forth the falser dreams.

But we have now concluded that general part of Human Philosophy, which contemplateth man segregate,° and as he consisteth of body and spirit. Wherein we may further note, that there seemeth to be a relation or conformity between the good of the mind and the good of the body. For as we divided the good of the body into health, beauty, strength, and pleasure; so the good of the mind,° inquired in rational and moral knowledges,° tendeth to this, to make the mind sound,° and without perturbation; beautiful, and graced with decency;° and strong and agile for all duties of life. These three, as in the body so in the mind, seldom meet, and commonly sever.° For it is easy to observe that many have strength of wit and courage, but have neither health from° perturbations, nor any beauty or decency in their doings: some again have an elegancy and fineness of carriage,° which have neither soundness of honesty, nor substance of sufficiency:° and some again have honest and reformed° minds, that can neither become themselves° nor manage business: and sometimes° two of them meet, and rarely all three. As for pleasure, we have likewise determined that the mind ought not to be reduced to stupid,° but to retain° pleasure; confined rather in the subject of it, than in the strength and vigour of it.

Civil° Knowledge is conversant about a subject which of all others is most immersed in matter,° and hardliest° reduced to axiom.° Nevertheless, as Cato° the censor said, that 'the Romans were like sheep, for that a man might better drive a flock of them, than one of them; for in a flock, if you could get but some few go° right, the rest would follow': so in that respect moral philosophy is more difficile° than policy. Again, moral philosophy propoundeth to itself the framing° of internal goodness; but civil knowledge requireth only an external goodness; for that as to society sufficeth; and therefore it cometh oft to pass that there be evil times in good governments: for so we find in the holy story, when the kings were good, yet it is added, 'Sed adhuc populus non direxerat cor suum ad Dominum Deum patrum suorum'.° Again, States, as° great engines,° move slowly, and are not so soon put out of frame:° for as in Egypt the seven good years° sustained the seven bad, so governments for a time well grounded° do bear out° errors following: but the resolution of particular persons is more suddenly subverted.° These respects° do somewhat qualify° the extreme difficulty of civil knowledge.

This knowledge hath three parts, according to the three summary actions of society; which are Conversation,° Negotiation,° and Government. For man seeketh in society comfort, use, and protection:° and they be three wisdoms of divers natures, which do often sever;° wisdom of the behaviour, wisdom of business, and wisdom of state.

The wisdom of Conversation ought not to be over-much affected,° but much less despised; for it hath not only an honour in itself, but an influence also into° business and government.° The poet saith,

Nec vultu destrue verba tuo:°

a man may destroy the force of his words with his countenance: so may he of his deeds, saith Cicero; recommending to his brother affability and easy access; 'Nil interest habere ostium apertum, vultum clausum';° it is nothing won to admit men with an open door, and to receive them with a shut and reserved countenance. So we see Atticus, before the first interview between Caesar and Cicero,° the war depending,° did seriously advise Cicero touching the composing and ordering of his countenance and gesture. And if the government of the countenance be of such effect, much more is that of the speech, and other carriage° appertaining to conversation; the true model whereof seemeth to me well expressed by Livy, though not meant for this purpose; 'Ne aut arrogans videar, aut obnoxius; quorum alterum est alienae libertatis obliti, alterum suae':° the sum of behaviour is to retain a man's own dignity, without intruding upon the liberty of others.

On the other side, if behaviour and outward carriage be intended° too much, first it may pass into affectation, and then 'quid deformius quam scenam in vitam transferre',° to act a man's life? But although it proceed not to that extreme, yet it consumeth time, and employeth the mind too much. And therefore as we use to advise young students from° company keeping, by saying, 'Amici fures temporis',° so certainly the intending of° the discretion° of behaviour is a great thief of meditation. Again, such as are accomplished in that honour° of urbanity° please themselves in name, and seldom aspire to higher virtue; whereas those that have defect in it do seek comeliness by reputation: for where reputation is, almost every thing becometh;° but where that is not, it must be supplied by *puntos*° and compliments. Again, there is no greater impediment of action than an over-curious observance of decency,° and the guide of decency, which is time and season. For as Salomon sayeth, 'Qui respicit ad ventos, non seminat; et qui respicit ad nubes, non metet';° a man must make his opportunity, as oft as find it. To conclude; Behaviour seemeth to me as a garment of the mind, and to have the conditions of a garment. For it ought to be made in fashion; it ought not to be too curious;° it ought to be shaped so as to set forth any good making° of the mind, and hide any deformity; and above all, it ought not to be too strait° or restrained for exercise or motion. But this part of civil knowledge hath been elegantly° handled, and therefore I cannot report it for deficient.

DE NEGOTIIS GERENDIS.° The wisdom touching Negotiation° or Business hath not been hitherto collected into writing, to the great derogation° of learning and the professors of learning. For from this root springeth chiefly that note° or opinion, which by us is expressed in adage° to this effect, that there is no great concurrence between learning and wisdom. For of the three wisdoms which we have set down to pertain to civil life, for° wisdom of Behaviour, it is by learned men for the most part despised, as an inferior to virtue and an enemy to meditation;° for wisdom of Government, they acquit themselves well when they are called to it, but that happeneth to few; but for the wisdom of Business, wherein man's life is most conversant, there be no books of it, except some few scattered advertisements,° that have no proportion to the magnitude of this subject. For if books were written of this as the other,° I doubt not but learned men with mean° experience would far excel men of long experience without learning, and outshoot° them in their own bow.

Neither needeth it at all to be doubted that this knowledge should be so variable as it falleth not under precept; for it is much less infinite than science of Government, which we see is laboured° and in some part reduced.° Of this wisdom it seemeth some of the ancient Romans in the saddest° and wisest times were professors;° for Cicero° reporteth that it was then° in use for senators that had name and opinion° for general wise men, as Coruncanius, Curius, Laelius, and many others, to walk at certain hours in the Place,° and to give audience to those that would use° their advice; and that the particular° citizens would resort unto them, and consult with them of the marriage of a daughter, or of the employing of a son, or of a purchase or bargain, or of an accusation, and every other occasion incident to man's life; so as there is a wisdom of counsel and advice even in private causes, arising out of an universal insight into the affairs of the world; which is used indeed upon particular cases° propounded, but is gathered by general observation of causes of like nature. For so we see in the book which Q. Cicero° writeth to his brother *De petitione consulatus* (being the only book of business that I know written by the ancients), although it concerned a particular action then on foot,° yet the substance thereof consisteth of many wise and politic axioms, which contain not a temporary but a perpetual direction in the case of popular elections.°

But chiefly we may see in those aphorisms which have place amongst divine writings, composed by Salomon the king, of whom the Scriptures° testify that his heart was as the sands of the sea, encompassing the world and all worldly matters; we see, I say, not a few profound and excellent cautions, precepts, positions,° extending to much variety of occasions; whereupon we will stay awhile, offering to consideration some number of examples.

'Sed et cunctis sermonibus qui dicuntur ne accommodes aurem tuam, ne forte audias servum tuum maledicentem tibi'.° Here is concluded° the provident stay° of inquiry of that which we would be loath to find: as it was judged° great wisdom in Pompeius Magnus that he burned Sertorius' papers unperused.

'Vir sapiens si cum stulto contenderit, sive irascatur sive rideat, non inveniet requiem'.° Here is described the great disadvantage which a wise man hath in undertaking° a lighter° person than himself; which is such an engagement° as whether a man turn the matter to jest, or turn it to heat,° or howsoever he change copy,° he can no ways quit himself well of it.

'Qui delicate a pueritia nutrit servum suum, postea sentiet eum contumacem'.° Here is signified, that if a man begin too high a pitch° in his favours, it doth commonly end in unkindness° and unthankfulness.

'Vidisti virum velocem in opere suo? Coram regibus stabit, nec erit inter ignobiles'.° Here is observed that, of all virtues for rising to honour, quickness of dispatch° is the best; for superiors many times love not to have those they employ too deep or too sufficient,° but ready and diligent.

'Vidi cunctos viventes qui ambulant sub sole, cum adolescente secundo qui consurgit pro eo'.° Here is expressed that which was noted by Sulla first, and after him by Tiberius: 'Plures adorant solem orientem quam occidentem vel meridianum'.°

'Si spiritus potestatem habentis ascenderit super te, locum tuum ne dimiseris; quia curatio faciet cessare peccata maxima.'° Here caution is given that upon° displeasure, retiring° is of all courses the unfittest; for a man leaveth things at worst, and depriveth himself of means to make them better.

'Erat civitas parva, et pauci in ea viri: venit contra eam rex magnus, et vadavit eam, intruxitque munitiones per gyrum, et perfecta est obsidio: inventusque est in ea vir pauper et sapiens, et liberavit eam per sapientiam suam; et nullus deinceps recordatus est hominis illius pauperis.'° Here the corruption of states is set forth, that esteem not virtue or merit longer than they have use of it.

'Mollis responsio frangit iram'.° Here is noted that silence or rough answer exasperateth; but an answer present° and temperate pacifieth.

'Iter pigrorum quasi sepes spinarum'.° Here is lively represented how laborious sloth proveth in the end; for when things are deferred till the last instant and nothing prepared beforehand, every step findeth a briar or an impediment, which catcheth or stoppeth.

'Melior est finis orationis quam principium'.° Here is taxed° the vanity of formal speakers, that study more about prefaces and inducements° than upon the conclusions and issues of speech.

'Qui cognoscit in judicio faciem, non bene facit; iste et pro buccella panis deseret veritatem'.° Here is noted, that a judge were° better be a briber° than a respecter of persons; for a corrupt judge offendeth not so lightly° as a facile.°

'Vir pauper calumnians pauperes similis est imbri vehementi, in quo paratur fames.'° Here is expressed the extremity° of necessitous extortions,° figured in the ancient fable° of the full and hungry horse-leech.

'Fons turbatus pede, et vena corrupta, est justus cadens coram impio.'° Here is noted, that one judicial° and exemplar° iniquity in the face of the world,° doth trouble the fountains of justice more than many particular injuries passed over by connivance.

'Qui subtrahit aliquid a patre et a matre, et dicit hoc non esse peccatum, particeps est homicidii.'° Here is noted, that whereas men in wronging their best friends use to° extenuate° their fault, as if they might presume or be bold upon° them, it doth contrariwise indeed aggravate their fault, and turneth it from injury to impiety.

'Noli esse amicus homini iracundo, nec ambulato cum homine furioso.'° Here caution is given, that in the election° of our friends we do principally avoid those which are impatient, as those that will espouse° us to many factions and quarrels.

'Qui conturbat domum suam, possidebit ventum.'° Here is noted, that in domestical° separations and breaches men do promise to themselves quieting of their mind and contentment; but still they are deceived° of their expectation, and it turneth to wind.

'Filius sapiens laetificat patrem: filius vero stultus maestitia est matri suae.'° Here is distinguished, that fathers have most comfort of the good proof of their sons; but mothers have most discomfort of their ill proof,° because women have little discerning of virtue, but of fortune.

'Qui celat delictum, quaerit amicitiam; sed qui altero sermone repetit, separat foederatos.'° Here caution is given, that reconcilement° is better managed by an amnesty, and passing over that which is past, than by apologies and excusations.

'In omni opere bono erit abundantia; ubi autem verba sunt plurima, ibi frequenter egestas.'° Here is noted that words and discourse abound most where there is idleness and want.

'Primus in sua causa justus; sed venit altera pars, et inquirit in eum.'° Here is observed, that in all causes° the first tale possesseth° much; in sort that the prejudice thereby wrought will be hardly° removed, except some abuse° or falsity in the information be detected.

'Verba bilinguis quasi simplicia, et ipsa perveniunt ad interiora ventris.'° Here is distinguished, that flattery and insinuation which seemeth set° and artificial° sinketh not far; but that entereth deep which hath shew of nature, liberty,° and simplicity.°

'Qui erudit derisorem, ipse sibi injuriam facit; et qui arguit impium, sibi maculam generat.'° Here caution is given how we tender° reprehension to arrogant and scornful natures, whose manner is to esteem it for contumely,° and accordingly to return it.

'Da sapienti occasionem, et addetur ei sapientia.'° Here is distinguished the wisdom brought into habit,° and that which is but verbal and swimming only in conceit;° for the one upon the occasion presented is quickened° and redoubled,° the other is amazed° and confused.

'Quomodo in aquis resplendent vultus prospicientium, sic corda hominum manifesta sunt prudentibus.'° Here the mind of a wise man is compared to a glass,° wherein the images of all diversity of natures and customs are represented; from which representation proceedeth that application,

Qui sapit, innumeris moribus aptus erit.°

Thus have I staid somewhat longer upon these sentences politic of Salomon than is agreeable to the proportion of an example; led with a desire to give authority° to this part of knowledge, which I noted as deficient, by so excellent a precedent; and have also attended° them with brief observations, such as to my understanding offer no violence to the sense, though I know they may be applied to a more divine use. But it is allowed even in divinity, that some interpretations, yea and some writings, have more of the Eagle° than others. But taking them as instructions for life, they might have received large discourse,° if I would have broken° them and illustrated them by deducements° and examples.

Neither was this in use only with the Hebrews; but it is generally to be found in the wisdom of the more ancient times, that as men found out any observation that they thought was good for life, they would gather it and express it in parable or aphorism or fable.° But for° fables, they were vicegerents° and supplies° where examples failed: now that the times abound with history, the aim is better when the mark° is alive. And therefore the form of writing which of all others is fittest for this variable° argument of negotiation° and occasions is that which Machiavel° chose wisely and aptly for government; namely, discourse upon histories or examples. For knowledge drawn freshly and in our view° out of particulars, knoweth the way best to particulars again. And it hath much greater life for practice when the discourse attendeth upon° the example, than when the example attendeth upon the discourse. For this is no point of order,° as it seemeth at first, but of substance. For when the example is the ground,° being set down in an history° at large, it is set down with all circumstances, which may sometimes control° the discourse thereupon made and sometimes supply° it, as a very pattern for action; whereas the examples alleged for the discourse's sake are cited succinctly and without particularity, and carry a servile° aspect toward the discourse which they are brought in to make good.

But this difference is not amiss to be remembered, that as history of Times is the best ground for discourse of government, such as Machiavel handleth, so histories° of Lives is the most proper for discourse of business, as° more conversant in private actions. Nay there is a ground of discourse for this purpose fitter than them both, which is discourse upon letters, such as are wise and weighty, as many are of Cicero° *ad Atticum* and others. For letters have a great° and more particular° representation of business than either Chronicles or Lives. Thus have we spoken both of the matter and form of this part of civil knowledge touching Negotiation, which we note to be deficient.

But yet there is another part of this part, which differeth as much from that whereof we have spoken as *sapere* and *sibi sapere*,° the one moving as it were to the circumference, the other to the centre. For there is a wisdom of counsel,° and again there is a wisdom of pressing° a man's own fortune; and they do sometimes meet, and often sever.° For many are wise in their own ways that are weak for government or counsel; like an ant, which is a wise creature for itself, but very hurtful for the garden. This wisdom the Romans did take much knowledge of:° 'Nam pol sapiens' (saith the comical° poet) 'fingit fortunam sibi',° and it grew to an adage, 'Faber quisque fortunae propriae,'° and Livy attributeth it to Cato the first, 'In hoc viro tanta vis animi et ingenii inerat, ut quocunque loco natus esset, sibi ipse fortunam facturus videretur'.°

This conceit or position° if it be too much declared and professed, hath been thought a thing impolitic and unlucky; as was observed in Timotheus the Athenian; who having done many great services to the estate° in his government, and giving an account thereof to the people as the manner was, did conclude every particular° with this clause, 'and in this fortune had no part'. And it came so to pass that he never prospered in any thing he took in hand afterward: for this is too high° and too arrogant, savouring of that which Ezekiel saith of Pharaoh, 'Dicis, Fluvius est meus, et ego feci memet ipsum',° or of that which another prophet° speaketh, that men offer sacrifices to their nets and snares; and that which the poet expresseth,

> Dextra mihi Deus, et telum quod missile libro,
> Nunc adsint!°

For these confidences° were ever unhallowed, and unblessed. And therefore those that were great politiques indeed ever ascribed their successes to their felicity,° and not to their skill or virtue. For so Sulla surnamed himself *Felix*, not *Magnus*.° So Caesar said to the master of the ship, 'Caesarem portas et fortunam ejus'.°

FABER FORTUNAE, SIVE DE AMBITU VITAE.° But yet nevertheless these positions, 'Faber quisque fortunae suae'; 'Sapiens dominabitur astris';° 'Invia virtuti nulla est via';° and the like, being taken and used as spurs° to industry, and not as stirrups° to insolency, rather for resolution° than for presumption° or outward declaration, have been ever thought sound and good, and are no question imprinted in the greatest minds; who are so sensible of° this opinion as they can scarce contain it within. As we see in Augustus Caesar (who was rather diverse from his uncle than inferior in virtue), how when he died,° he desired his friends about him to give him a *Plaudite*;° as if he were conscient° to himself that he had played his part well upon the stage. This part of knowledge we do report also as deficient: not but that it is practised too much, but it hath not been reduced to writing. And therefore lest it should seem to any that it is not comprehensible by axiom, it is requisite, as we did in the former, that we set down some heads° or passages of it.

Wherein it may appear at the first a new and unwonted argument to teach men how to raise and make their fortune; a doctrine wherein every man perchance will be ready to yield° himself a disciple, till he see the difficulty: for Fortune layeth as heavy impositions° as Virtue; and it is as hard and severe a thing to be a true politique,° as to be truly moral. But the handling hereof concerneth learning greatly, both in honour and in substance: in honour, because° pragmatical° men may not go away with an opinion that learning is like a lark, that can mount and sing and please herself, and nothing else; but may know that she holdeth as well of° the hawk, that can soar aloft, and can also descend and strike upon the prey: in substance, because it is the perfect law of inquiry of truth, that nothing be in the globe of matter, which should° not be likewise in the globe of crystal,° or form; that is that there be not any thing in being and action, which should not be drawn and collected into contemplation and doctrine. Neither doth learning admire or esteem of° this architecture of fortune otherwise than as of an inferior work: for no man's fortune can be an end° worthy of his being, and many times the worthiest men do abandon their fortune willingly for better respects:° but nevertheless fortune as an organ° of virtue and merit deserveth the consideration.

First therefore, the precept which I conceive to be most summary° towards the prevailing in fortune, is to obtain that window which Momus° did require, who seeing in the frame of man's heart such angles and recesses, found fault there was not a window to look into them; that is, to procure good informations of particulars touching persons, their natures, their desires and ends, their customs and fashions, their helps and advantages, and whereby they chiefly stand; so again their weaknesses and disadvantages, and where they lie most open and obnoxious;° their friends, factions, dependances;° and again their opposites,° enviers, competitors, their moods and times, 'Sola viri molles aditus et tempora noras';° their principles, rules, and observations, and the like: and this not only of persons, but of actions; what° are on foot° from time to time, and how they are conducted, favoured, opposed; and how they import,° and the like. For the knowledge of present actions is not only material in itself, but without it also the knowledge of persons is very erroneous: for men change with the actions; and whiles they are in pursuit they are one, and when they return to their nature they are another.° These informations of particulars touching persons and actions are as the minor propositions in every active syllogism; for no excellency of observations (which are as the major propositions) can suffice to ground a conclusion, if there be error and mistaking in the minors.

That this knowledge is possible, Salomon is our surety;° who saith, 'Consilium in corde viri tanquam aqua profunda; sed vir prudens exhauriet illud'.° And although the knowledge itself falleth not under precept, because it is of individuals,° yet the instructions for the obtaining of it may.

We will begin therefore with this precept, according to the ancient opinion, that the sinews of wisdom° are slowness of belief and distrust; that more trust be given to countenances and deeds than to words; and in words, rather to sudden passages and surprised° words, than to set° and purposed° words. Neither let that be feared which is said, 'fronti nulla fides',° which is meant of a general outward behaviour, and not of the private and subtile motions and labours of the countenance and gesture; which as Q. Cicero° elegantly saith, is *animi janua*, the gate of the mind. None more close° than Tiberius, and yet Tacitus saith of Gallus, 'Etenim vultu offensionem conjectaverat'.° So again, noting the differing character and manner of his commending Germanicus and Drusus in the senate, he saith touching his fashion wherein he carried° his speech of Germanicus, thus; 'Magis in speciem adornatis verbis, quam ut penitus sentire videretur',° but of Drusus thus; 'Paucioribus, sed intentior, et fida oratione',° and in another place, speaking of his character of speech when he did any thing that was gracious and popular, he saith that in other things he was 'velut eluctantium verborum',° but then again, 'solutius loquebatur quando subveniret'.° So that there is no such artificer of dissimulation, nor no such commanded° countenance ('vultus jussus')° that can sever° from a feigned tale some of these fashions,° either a more slight and careless fashion, or more set and formal, or more tedious and wandering,° or coming from a man more drily and hardly.°

Neither are deeds such assured pledges,° as that they may be trusted without a judicious consideration of their magnitude and nature: 'Fraus sibi in parvis fidem praestruit, ut majore emolumento fallat',° and the Italian° thinketh himself upon the point to be bought and sold, when he is better used than he was wont to be without manifest cause. For° small favours,° they do but lull men asleep, both as to caution and as to industry, and are as Demosthenes calleth them, 'Alimenta socordiae'.° So again we see how false the nature of some deeds are, in that particular which Mutianus practised upon Antonius Primus, upon that hollow° and unfaithful reconcilement° which was made between them; whereupon Mutianus advanced° many of the friends of Antonius: 'simul amicis ejus praefecturas et tribunatus largitur',° wherein under pretence to strengthen him, he did desolate° him, and won from him his dependances.°

As for Words (though they be like waters° to physicians, full of flattery° and uncertainty), yet they are not to be despised, specially with the advantage° of passion and affection. For so we see Tiberius upon a stinging and incensing° speech of Agrippina came a step forth of° his dissimulation, when he said, 'You are hurt because you do not reign'; of which Tacitus saith, 'Audita haec raram occulti pectoris vocem elicuere; correptamque Graeco versu admonuit, ideo laedi quia non regnaret'.° And therefore the poet doth elegantly call passions tortures, that urge men to confess their secrets:

Vino tortus et ira.°

And experience sheweth, there are few men so true to themselves° and so settled,° but that, sometimes upon° heat,° sometimes upon bravery,° sometimes upon kindness, sometimes upon trouble of mind and weakness, they open themselves; specially if they be put to it with a counter-dissimulation, according to the proverb of Spain, 'Di mentira, y sacaras verdad', 'Tell a lie and find a truth'.°

As for the knowing of men which is at second hand from reports; men's weaknesses and faults are best known from their enemies, their virtues and abilities from their friends, their customs and times from their servants, their conceits and opinions from their familiar friends with whom they discourse most. General fame° is light,° and the opinions conceived by superiors or equals are deceitful; for to such men are more masked: 'Verior fama e domesticis emanat'.°

But the soundest disclosing and expounding of men is by their natures and ends; wherein the weakest sort of men are best interpreted by their natures, and the wisest° by their ends. For it was both pleasantly and wisely said (though I think very untruly) by a nuncio° of the pope, returning from a certain nation where he served as lieger;° whose opinion being asked touching the appointment of one to go in his place, he wished that in any case they did° not send one that was too wise; because no very wise man would ever imagine what they in that country were like to do. And certainly it is an error frequent for men to shoot over,° and to suppose deeper ends and more compass reaches° than are: the Italian proverb being elegant, and for the most part true:

> Di danari, di senno, e di fede,
> Cè nè manco che non credi.°

There is commonly less money, less wisdom, and less good faith, than men do account° upon.

But Princes upon a far other° reason are best interpreted by their natures, and private persons by their ends; for princes being at the top of human desires, they have for the most part no particular ends whereto they aspire, by distance from which a man might take measure and scale° of the rest of their actions and desires; which is one of the causes that maketh their hearts more inscrutable.° Neither is it sufficient to inform ourselves in° men's ends and natures of the variety of them only, but also of the predominancy, what humour° reigneth most, and what end is principally sought. For so we see, when Tigellinus saw himself outstripped by Petronius Turpilianus in Nero's humours of pleasures, 'metus ejus rimatur',° he wrought upon Nero's fears, whereby he brake the other's neck.

But to all this part of inquiry the most compendious way resteth in three things. The first, to have general acquaintance and inwardness° with those which have general acquaintance and look° most into the world; and specially according to the diversity of business and the diversity of persons, to have privacy and conversation with some one friend at least which° is perfect and well intelligenced° in every several kind. The second is to keep a good mediocrity in liberty of speech° and secrecy; in most things liberty; secrecy where it importeth;° for liberty of speech inviteth and provoketh liberty to be used again, and so bringeth much to a man's knowledge; and secrecy, on the other side, induceth trust and inwardness.° The last is the reducing° of a man's self to this watchful and serene° habit, as to make account and purpose, in every conference and action, as well to observe as to act. For as Epictetus would have a philosopher in every particular action to say to himself, 'Et hoc volo, et etiam institutum servare',° so a politic man in every thing should say to himself, 'Et hoc volo, ac etiam aliquid addiscere'.°

I have stayed° the longer upon this precept of obtaining good information, because it is a main part by itself, which answereth° to all the rest. But, above all things, caution must be taken that men have a good stay° and hold of themselves, and that this much knowledge do not draw on° much meddling; for nothing is more unfortunate than light and rash intermeddling in many matters; so that this variety of knowledge tendeth in conclusion but only° to this, to make a better and freer choice of those actions which may concern us, and to conduct them with the less error and the more dexterity.

The second precept concerning this knowledge is, for men to take good information touching their own person, and well to understand themselves: knowing that, as St. James° saith, though men look oft in a glass, yet they do suddenly° forget themselves; wherein as the divine glass is the word of God, so the politic glass is the state of the world or times wherein we live; in the which we are to behold ourselves.

For men ought to take an unpartial view of their own abilities and virtues; and again of their wants and impediments; accounting these with the most, and those other with the least; and from this view and examination to frame the considerations following.

First, to consider how the constitution of their nature sorteth with° the general state of the times; which if they find agreeable and fit,° then in all things to give themselves more scope and liberty; but if differing and dissonant,° then in the whole course of their life to be more close,° retired, and reserved. As we see in Tiberius, who was never seen at a play and came not into the senate in twelve of his last years; whereas Augustus Caesar lived ever in men's eyes, which Tacitus observeth: 'Alia Tiberio morum via'.°

Secondly, to consider how their nature sorteth with professions and courses of life, and accordingly to make election,° if they be free; and, if engaged,° to make the departure at the first opportunity. As we see was done by duke Valentine,° that was designed by his father to° a sacerdotal profession, but quitted it soon after in regard of his parts° and inclination; being such nevertheless, as a man cannot tell well whether they were worse for a prince or for a priest.

Thirdly, to consider how they sort with those whom they are like to have competitors and concurrents,° and to take that course wherein there is most solitude,° and themselves like° to be most eminent: as Caesar° Julius did, who at first was an orator or pleader; but when he saw the excellency of Cicero, Hortensius, Catulus, and others, for eloquence, and saw there was no man of reputation for the wars but Pompeius, upon whom the state was forced to rely, he forsook his course begun toward a civil° and popular° greatness, and transferred his designs to a martial greatness.

Fourthly, in the choice of their friends and dependances,° to proceed according to the composition of their own nature; as we may see in Caesar,° all whose friends and followers were men active and effectual,° but not solemn° or of reputation.

Fifthly, to take special heed how they guide themselves by examples, in thinking they can do as they see others do; whereas perhaps their natures and carriages° are far differing; in which error it seemeth Pompey was, of whom Cicero° saith, that he was wont often to say, 'Sulla potuit, ego non potero?', wherein he was much abused,° the natures and proceedings of himself and his example being the unlikest° in the world; the one being fierce, violent, and pressing the fact;° the other solemn, and full of majesty and circumstance, and therefore the less effectual.

But this precept touching the politic knowledge of ourselves hath many other branches whereupon we cannot insist.°

Next to the well understanding and discerning of a man's self, there followeth the well opening° and revealing a man's self; wherein we see nothing more usual than for the more able man to make the less shew. For there is a great advantage in the well setting forth of a man's virtues, fortunes, merits; and again in the artificial° covering of a man's weaknesses, defects, disgraces; staying upon° the one, sliding from° the other; cherishing° the one by circumstances,° gracing° the other by exposition,° and the like: wherein we see what Tacitus saith of Mutianus, who was the greatest politique of his time, 'Omnium quae dixerat feceratque arte quadam ostentator',° which requireth indeed some art, lest it turn tedious and arrogant; but yet so as ostentation (though° it be to the first degree of vanity) seemeth to me rather a vice in manners° than in policy:° for as it is said, 'Audacter calumniare, semper aliquid haeret',° so, except it be in a ridiculous degree of deformity, 'Audacter te vendita, semper aliquid haeret'.° For it will stick with the more ignorant and inferior sort of men, though men of wisdom and rank do smile at it and despise it; and yet the authority won with° many doth countervail the disdain of a few. But if it be carried with decency and government,° as with a natural, pleasant, and ingenious° fashion; or at times when it is mixed with some peril and unsafety° (as in military persons); or at times when others are most envied; or with easy and careless° passage to it and from it, without dwelling too long or being too serious; or with an equal freedom of taxing° a man's self as well as gracing° himself; or by occasion of repelling or putting down° others' injury° or insolency; it doth greatly add to reputation: and surely not a few solid° natures, that want° this ventosity° and cannot sail in the height of the winds,° are not without some prejudice and disadvantage by their moderation.

But for these flourishes° and enhancements of virtue, as they are not perchance unnecessary, so it is at least necessary that virtue be not disvalued° and imbased under the just price; which is done in three manners: by offering and obtruding° a man's self; wherein men think he is rewarded, when he is accepted: by doing too much; which will not give that which is well done leave to settle, and in the end induceth satiety: and by finding too soon the fruit of a man's virtue, in commendation, applause, honour, favour; wherein if a man be pleased with a little, let him hear what is truly said, 'Cave ne insuetus rebus majoribus videaris, si haec te res parva sicuti magna delectat'.°

But the covering of defects is of no less importance than the valuing of good parts; which may be done likewise in three manners; by Caution, by Colour,° and by Confidence.° Caution is when men do ingeniously and discreetly avoid to be put into those things for which they are not proper: whereas contrariwise bold and unquiet spirits will thrust themselves into matters without difference,° and so publish and proclaim all their wants.° Colour is when men make a way for themselves to have a construction° made of their faults or wants as proceeding from a better cause, or intended for some other purpose: for of the one it is well said, 'Saepe latet vitium proximitate boni',°

and therefore whatsoever want a man hath, he must see that he pretend° the virtue that shadoweth° it; as if he be dull, he must affect gravity;° if a coward, mildness; and so the rest. For the second, a man must frame some probable cause why he should not do his best, and why he should dissemble his abilities; and for that purpose must use° to dissemble those abilities which are notorious in him, to give colour that his true wants are but industries° and dissimulations. For Confidence, it is the last but the surest remedy; namely, to depress° and seem to despise whatsoever a man cannot attain; observing the good principle of the merchants, who endeavour to raise the price of their own commodities, and to beat down the price of others. But there is a confidence that passeth° this other; which is, to face out° a man's own defects, in seeming to conceive that he is best in those things wherein he is failing;° and, to help that again, to seem on the other side that he hath least opinion of himself in those things wherein he is best: like as we shall see it commonly in poets, that if they shew their verses, and you except° to any, they will say 'that that line cost them more labour than any of the rest'; and presently will seem to disable and suspect rather some other line, which they know well enough to be the best in the number.° But above all, in this righting and helping of a man's self in his own carriage,° he must take heed he shew not himself dismantled° and exposed to scorn and injury,° by too much dulceness,° goodness, and facility° of nature, but shew some sparkles° of liberty, spirit, and edge: which kind of fortified carriage, with a ready rescussing° of a man's self from scorns, is sometimes of necessity imposed upon men by somewhat in their person° or fortune;° but it ever succeedeth with good felicity.

Another precept of this knowledge is, by all possible endeavour to frame the mind to be pliant and obedient to occasion; for nothing hindereth men's fortunes so much as this 'Idem manebat neque idem decebat',° men are where they were, when occasions turn:° and therefore to Cato, whom Livy maketh such an architect of fortune, he addeth that he had 'versatile ingenium'.° And thereof it cometh that these grave solemn wits, which must be like themselves and cannot make departures,° have more dignity than felicity.° But in some it is nature° to be somewhat viscous° and inwrapped,° and not easy to turn. In some it is a conceit° that is almost a nature, which is, that men can hardly make themselves believe that they ought to change their course, when they have found good by it in former experience. For Machiavel° noteth wisely, how Fabius Maximus would have been temporizing still, according to his old bias,° when the nature of the war was altered and required hot pursuit. In some other it is want of point and penetration in their judgment, that they do not discern when things have a period,° but come in too late after the occasion; as Demosthenes° compareth the people of Athens to country fellows when they play in a fence school, that if they have° a blow, then they remove their weapon to that ward,° and not before. In some other it is a lothness° to leese° labours passed,° and a conceit that they can bring about occasions to their ply;° and yet in the end, when they see no other remedy, then they come to it with disadvantage; as Tarquinius,° that gave for the third part of Sibylla's books the treble price, when he might at first have had all three for the simple.° But from whatsoever root or cause this restiveness of mind proceedeth, it is a thing most prejudicial; and nothing is more politic° than to make the wheels of our mind concentric° and voluble° with the wheels of fortune.°

Another precept of this knowledge, which hath some affinity with that we last spake of, but with difference, is that which is well expressed, 'Fatis accede Deisque',° that men do not only turn with the occasions but also run with the occasions, and not strain their credit or strength to over-hard or extreme points, but choose in their actions that which is most passable:° for this will preserve men from foil,° not occupy them too much about one matter, win opinion of moderation, please the most,° and make a shew of a perpetual felicity in all they undertake; which cannot but mightily increase reputation.

Another part of this knowledge seemeth to have some repugnancy° with the former two, but not as I understand it; and it is that which Demosthenes° uttereth in high° terms; 'Et quemadmodum receptum est, ut exercitum ducat imperator, sic et a cordatis viris res ipsae ducendae; ut quae ipsis videntur, ea gerantur, et non ipsi eventus persequi cogantur'.° For if we observe, we shall find two differing kinds of sufficiency in managing of business: some can make use of occasions aptly and dexterously, but plot° little; some can urge and pursue their own plots well, but cannot accommodate° nor take in;° either of which is very unperfect without the other.

Another part of this knowledge is the observing a good mediocrity° in the declaring° or not declaring a man's self: for although depth of secrecy, and making way 'qualis est via navis in mari',° (which the French calleth *sourdes menées*,° when men set things in work° without opening° themselves at all), be sometimes both prosperous and admirable; yet many times 'Dissimulatio errores parit qui dissimulatorem ipsum illaqueant'.° And therefore we see the greatest politiques have in a natural and free° manner professed° their desires, rather than been reserved and disguised in them. For so we see that Lucius Sulla° made a kind of profession, that 'he wished all men happy or unhappy as they stood his friends or enemies.' So Caesar,° when he went first into Gaul, made no scruple to profess 'that he had rather be first in a village than second at Rome.' So again as soon as he had begun the war, we see what Cicero saith of him; 'Alter' (meaning of Caesar) 'non recusat, sed quodammodo postulat, ut (ut est) sic appelletur tyrannus.'° So we may see in a letter of Cicero to Atticus, that Augustus Caesar in his very entrance into affairs, when he was a dearling of the senate, yet in his harangues to the people would swear 'Ita parentis honores consequi liceat',° which was no less than the tyranny,° save that, to help° it he would stretch forth his hand towards a statua of Caesar's that was erected in the place: whereat many men° laughed and wondered and said: Is it possible? or Did you ever hear the like? and yet thought he meant no hurt, he did it° so handsomely and ingenuously. And all these were prosperous: whereas Pompey,° who tended to the same end but in a more dark and dissembling manner, as Tacitus saith of him, 'Occultior non melior',° wherein Sallust concurreth, 'ore probo, animo inverecundo',° made it his design by infinite secret engines° to cast the state into an absolute anarchy and confusion, that the state might cast itself into his arms for necessity and protection, and so the sovereign power be put upon him, and he never seen in it: and when he had brought it (as he thought) to that point, when he was chosen consul alone, as never any was, yet he could make no great matter of it, because men understood him not; but was fain° in the end to go the beaten track of getting arms into his hands, by colour° of the doubt° of Caesar's designs: so tedious,° casual,° and unfortunate are these deep dissimulations; whereof it seemeth Tacitus made this judgment, that they were a cunning of an inferior form in regard of true policy; attributing the one to Augustus, the other to Tiberius, where speaking of Livia he saith, 'Et cum artibus mariti simulatione filii bene composita',° for surely the continual habit of dissimulation is but a weak and sluggish cunning, and not greatly politic.

Another precept of this Architecture of Fortune is to accustom our minds to judge of the proportion or value of things as they conduce and are material to our particular ends; and that to do substantially,° and not superficially. For we shall find the logical part (as I may term it) of some men's minds good, but the mathematical part erroneous; that is, they can well judge of consequences, but not of proportions° and comparison; preferring things of shew° and sense° before things of substance and effect. So some fall in love with access° to princes, others with popular fame and applause, supposing they are things of great purchase;° when in many cases they are but matters of envy, peril, and impediment. So some measure things according to the labour and difficulty or assiduity which are spent about them; and think if they be ever° moving, that they must needs advance and proceed; as Caesar saith in a despising manner of Cato the second, when he describeth how laborious and indefatigable he was to no great purpose; 'Haec omnia magno studio agebat'.° So in most things men are ready to abuse themselves in thinking the greatest° means to be best, when it should be the fittest.

As for the true marshalling° of men's pursuits towards their fortune as they are more or less material, I hold them to stand thus. First the amendment° of their own minds; for the remove° of the impediments of the mind will sooner clear the passages of fortune, than the obtaining fortune will remove the impediments of the mind. In the second place I set down wealth and means; which I know most men would have placed first, because of the general use which it beareth towards all variety of occasions. But that opinion I may condemn with like reason as Machiavel° doth that other, that moneys were the sinews of the wars; whereas (saith he) the true sinews of the wars are the sinews of men's arms, that is, a valiant, populous,° and military nation; and he voucheth° aptly the authority of Solon,° who when Croesus shewed him his treasury of gold said to him, that if another came that had better iron he would be master of his gold. In like manner it may be truly affirmed that it is not moneys that are the sinews of fortune, but it is the sinews and steel of men's minds, wit, courage, audacity, resolution, temper, industry, and the like. In third place I set down reputation, because of the peremptory° tides and currents it hath; which if they be not taken in their due time are seldom recovered, it being extreme° hard to play an after-game° of reputation. And lastly I place honour, which is more easily won by any of the other three, much more by all, than any of them can be purchased by honour. To conclude this precept, as there is order and priority in matter,° so is there in time, the preposterous° placing whereof is one of the commonest errors; while men fly to their ends when they should intend° their beginnings, and do not take things in order of time as they come on, but marshal them according to greatness and not according to instance;° not observing the good precept, 'Quod nunc instat agamus.'°

Another precept of this knowledge is, not to embrace any matters which do occupy too great a quantity of time, but to have that sounding in a man's ears, 'Sed fugit interea, fugit irreparabile tempus',° and that is the cause why those which take their course of rising by professions of burden,° as lawyers, orators, painful° divines,° and the like, are not commonly so politic° for their own fortune, otherwise than in their ordinary way, because they want time to learn particulars, to wait occasions, and to devise plots.°

Another precept of this knowledge is to imitate nature which doth nothing in vain; which surely a man may do, if he do well interlace° his business, and bend not his mind too much upon that which he principally intendeth. For a man ought in every particular action so to carry the motions of his mind, and so to have one thing under° another, as if he cannot have that he seeketh in the best degree, yet to have it in a second, or so in a third; and if he can have no part of that which he purposed, yet to turn the use of it to somewhat else;° and if he cannot make anything of it for the present, yet to make it as a seed of somewhat in time to come; and if he can contrive no effect or substance from it, yet to win some good opinion by it, or the like; so that he should exact an account° of himself, of every action to reap° somewhat, and not to stand amazed and confused if he fail of that he chiefly meant: for nothing is more impolitic than to mind° actions wholly one by one; for he that doth so looseth infinite occasions which intervene, and are many times more proper and propitious for somewhat that he shall need afterwards, than for that which he urgeth for the present; and therefore men must be perfect in that rule, 'Haec oportet facere, et illa non omittere'.°

Another precept of this knowledge is, not to engage a man's self peremptorily in any thing, though it seem not liable to accident; but ever to have a window to fly out at, or a way° to retire;° following the wisdom in the ancient fable of the two frogs, which consulted when their plash° was dry whither they should go; and the one moved° to go down into a pit, because it was not likely the water would dry° there; but the other answered, 'True, but if it do, how shall we get out again?'

Another precept of this knowledge is that ancient precept of Bias, construed° not to any point of perfidiousness but only to caution and moderation, 'Et ama tanquam inimicus futurus, et odi tanquam amaturus',° for it utterly betrayeth° all utility° for men to embark themselves too far in unfortunate friendships, troublesome spleens,° and childish and humorous° envies or emulations.

But I continue this beyond the measure of an example; led, because I would not° have such knowledges which I note as deficient to be thought things imaginative° or in the air,° or an observation or two much made of; but things of bulk and mass, whereof an end is hardlier° made than a beginning. It must be likewise conceived, that in these points which I mention and set down, they are far from complete tractates° of them, but only as small pieces for patterns.° And lastly, no man I suppose will think that I mean fortunes are not obtained without all this ado; for I know they come tumbling into some men's laps; and a number obtain good fortunes by diligence in a plain° way, little intermeddling, and keeping themselves from gross errors.

But as Cicero,° when he setteth down an Idea of a perfect Orator, doth not mean that every pleader° should be such; and so likewise, when a Prince or a Courtier° hath been described by such as have handled those subjects, the mould° hath used to be made according to the perfection of the art, and not according to common practice: so I understand it that it ought to be done in the description of a Politic man; I mean politic for his own fortune.

But it must be remembered all this while, that the precepts which we have set down are of that kind which may be counted and called 'bonae artes'.° As for evil arts, if a man would set down for himself that principle of Machiavel,° that 'a man seek not to attain virtue itself, but the appearance only thereof; because the credit of virtue is a help, but the use of it is cumber';° or that other° of his principles, that 'he presuppose that men are not fitly to be wrought otherwise but by fear, and therefore that he seek to have every man obnoxious,° low,° and in strait',° which the Italians call 'seminar spine', to sow thorns; or that other principle contained in the verse which Cicero citeth, 'Cadant amici, dummodo inimici intercidant',° as the Triumvirs,° which sold every° one to other the lives of their friends for the deaths of their enemies; or that other protestation° of L. Catilina,° to set on fire and trouble° states, to the end° to fish in droumy° waters, and to unwrap° their fortunes; 'Ego si quid in fortunis meis excitatum sit incendium, id non aqua sed ruina restinguam',° or that other principle of Lysander° 'that children are to be deceived with comfits,° and men with oaths': and the like evil and corrupt positions,° whereof (as in all things) there are more in number than of the good: certainly with these dispensations from the laws of charity and integrity the pressing° of a man's fortune may be more hasty and compendious.° But it is in life as it is in ways;° the shortest way is commonly the foulest, and surely the fairer way is not much about.°

But men if they be in their own power and do bear and sustain themselves, and be not carried away with a whirlwind or tempest of ambition, ought in the pursuit of their own fortune to set before their eyes not only that general map° of the world, that 'all things are vanity and vexation of spirit',° but many other more particular cards° and directions:° chiefly that, that being without well-being° is a curse and the greater being the greater curse, and that all virtue is most rewarded and all wickedness most punished in itself: according as the poet saith excellently:

> Quae vobis, quae digna, viri, pro laudibus istis
> Praemia posse rear solvi? pulcherrima primum
> Dii *moresque* dabunt vestri:°

and so of the contrary. And secondly they ought to look up to the eternal providence and divine judgment, which often subverteth the wisdom of evil plots and imaginations, according to that Scripture, 'He hath conceived mischief, and shall bring forth a vain thing'.° And although men should refrain themselves from injury and evil arts, yet this incessant and sabbathless° pursuit of a man's fortune leaveth not tribute° which we owe to God of our time; who (we see) demandeth a tenth° of our substance, and a seventh,° which is more strict, of our time: and it is to small purpose to have an erected face towards heaven, and a perpetual grovelling spirit upon earth,° eating dust as doth the serpent; 'Atque affigit humo divinae particulam aurae'.° And if any man flatter° himself that he will employ his fortune well though he should obtain it ill,° as was said concerning Augustus Caesar, and after of Septimius Severus, that 'either they should never have been born or else they should never have died',° they did so much mischief° in the pursuit and ascent of their greatness, and so much good when they were established; yet these compensations and satisfactions are good to be used, but never good to be purposed.° And lastly, it is not amiss for men in their race toward their fortune to cool themselves a little with that conceit which is elegantly expressed by the emperor Charles the fifth° in his instructions to the king his son, that 'fortune hath somewhat of the nature of a woman, that if she be too much wooed she is the farther off.' But this last is but a remedy for those whose tastes° are corrupted: let men rather build upon that foundation which is as a corner-stone of divinity and philosophy, wherein they join close, namely that same *Primum quaerite.*° For divinity saith, 'Primum quaerite regnum Dei, et ista omnia adjicientur vobis',° and philosophy saith, 'Primum quaerite bona animi, caetera aut aderunt aut non oberunt'.° And although the human foundation hath somewhat of the sand,° as we see in M. Brutus when he brake forth into that speech,

> Te colui, Virtus, ut rem; at tu nomen inane es;°

yet the divine foundation is upon the rock. But this may serve for a taste of that knowledge which I noted as deficient.

Concerning Government,° it is a part of knowledge secret and retired,° in both these respects in which things are deemed secret; for some things are secret because they are hard to know, and some because they are not fit to utter. We see all governments are obscure and invisible.

> Totamque infusa per artus
> Mens agitat molem, et magno se corpore miscet.°

Such is the description of governments. We see the government of God over the world is hidden, insomuch as it seemeth to participate of° much irregularity and confusion. The government of the Soul in moving the Body is inward° and profound,° and the passages° thereof hardly° to be reduced to demonstration.° Again, the wisdom of antiquity° (the shadows° whereof are in the poets) in the description of torments and pains, next unto the crime of rebellion which was the Giants'° offence, doth detest the offence of futility,° as in Sisyphus° and Tantalus.° But this was meant of particulars:° nevertheless even unto the general rules and discourses° of policy and government there is due a reverent and reserved° handling.

But contrariwise in the governors toward° the governed all things ought, as far as the frailty of man permitteth, to be manifest and revealed. For so it is expressed in the Scriptures touching the government of God, that this globe, which seemeth to us a dark and shady° body, is in the view of God as crystal: 'Et in conspectu sedis tanquam mare vitreum simile crystallo'.° So unto princes and states, and specially towards wise senates and councils, the natures and dispositions of the people, their conditions and necessities,° their factions and combinations,° their animosities and discontents, ought to be, in regard of the variety of their intelligences,° the wisdom of their observations, and the height of their station where they keep sentinel, in great part clear and transparent. Wherefore, considering that I write to a king that is a master of this science, and is so well assisted,° I think it decent to pass over this part in silence, as willing to obtain the certificate° which one of the ancient philosophers aspired unto; who being silent, when others contended° to make demonstration of their abilities by speech, desired it might be certified for his part, that 'there was one that knew how to hold his peace.'°

DE PRUDENTIA LEGISLATORIA, SIVE, DE FONTIBUS JURIS.° Notwithstanding, for the more public part of government, which is Laws, I think good to note only one deficience; which is, that all those which have written of laws, have written either as philosophers or as lawyers, and none as statesmen. As for the philosophers, they make imaginary laws° for imaginary commonwealths; and their discourses are as the stars, which give little light because they are so high. For the lawyers, they write according to the states where they live, what is received° law, and not what ought to be law: for the wisdom of a lawmaker is one, and of a lawyer is another. For there are in nature certain fountains of justice,° whence all civil laws are derived but as streams; and like as waters do take tinctures° and tastes from the soils through which they run, so do civil laws vary according to the regions and governments where they are planted, though they proceed from the same fountains. Again, the wisdom of a lawmaker consisteth not only in a platform° of justice, but in the application thereof; taking into consideration by what means laws may be made certain, and what are the causes and remedies of the doubtfulness° and incertainty of law; by what means laws may be made apt and easy to be executed, and what are the impediments and remedies in the execution of laws; what influence laws touching private right of *meum* and *tuum*° have into° the public state, and how they may be made apt° and agreeable; how laws are to be penned° and delivered, whether in Texts or in Acts;° brief or large; with preambles or without; how they are to be pruned and reformed from time to time; and what is the best means to keep them from being too vast° in volumes or too full of multiplicity and crossness;° how they are to be expounded, when upon causes° emergent° and judicially° discussed, and when upon responses° and conferences° touching general points or questions; how they are to be pressed,° rigorously or tenderly; how they are to be mitigated° by equity and good conscience; and whether discretion° and strict° law are to be mingled in the same courts or kept apart in several courts; again, how the practice, profession, and erudition of law is to be censured° and governed; and many other points touching the administration, and (as I may term it) animation° of laws. Upon which I insist the less, because I purpose (if God give me leave), having begun a work° of this nature in aphorisms, to propound it hereafter noting it in the mean time for deficient.

And for your Majesty's laws° of England, I could say much of their dignity, and somewhat of their defect; but they cannot but excel the civil laws° in fitness for the government: for the civil law was 'non hos quaesitum munus in usus';° it was not made for the countries which it governeth. Hereof I cease to speak, because I will not intermingle matter of action° with matter of general learning.

Thus have I concluded this portion of learning touching Civil Knowledge; and with civil knowledge have concluded Human Philosophy; and with human philosophy, Philosophy in General. And being now at some pause, looking back into that I have passed through, this writing seemeth to me, ('si nunquam fallit imago')° as far as a man can judge of his own work, not much better than that noise or sound which musicians make while they are tuning their instruments; which is nothing° pleasant to hear, but yet is a cause why the music is sweeter afterwards. So have I been content to tune the instruments of the muses, that they may play that have better hands. And surely, when I set before me the condition of these times, in which learning hath made her third visitation° or circuit, in all the qualities thereof; as the excellency and vivacity of the wits° of this age; the noble helps and lights° which we have by° the travails of ancient writers; the art of printing, which communicateth books to men of all fortunes; the openness of the world by navigation, which hath disclosed multitudes of experiments,° and a mass of natural history; the leisure wherewith these times abound, not employing men so generally in civil business, as the states of Graecia did in respect of their popularity,° and the state of Rome in respect of the greatness of their monarchy; the present disposition of these times at this instant to peace; the consumption° of all that ever can be said in controversies of religion, which have so much diverted° men from other sciences; the perfection of your Majesty's learning, which as a phoenix may call whole vollies° of wits to follow you; and the inseparable propriety° of time, which is ever more and more to disclose truth;° I cannot but be raised to this persuasion,° that this third period of time will far surpass that of the Graecian and Roman learning: only if° men will know their own strength and their own weakness both; and take one from the other° light of invention, and not fire of contradiction; and esteem of the inquisition of truth as of an enterprise,° and not as of a quality° or ornament; and employ wit and magnificence° to° things of worth and excellency, and not to things vulgar and of popular estimation. As for my labours, if any man shall please himself or others in the reprehension° of them, they° shall make that ancient and patient request, 'Verbera sed audi',° let men reprehend them, so° they observe and weigh them. For the appeal is (lawful though it may be it shall not be needful) from the first cogitations of men to their second, and from the nearer times to the times further off. Now let us come to that learning, which both the former times were not so blessed as to know, sacred and inspired Divinity, the Sabbath° and port of all men's labours and peregrinations.

The prerogative° of God extendeth as well to the reason as to the will of man; so that as we are to obey his law though we find a reluctation° in our will, so we are to believe his word though we find a reluctation in our reason. For if we believe only that which is agreeable to our sense, we give consent to the matter and not to the author; which is no more than we would do towards a suspected and discredited witness; but that faith which was accounted to Abraham for righteousness° was of such a point as whereat Sarah° laughed, who therein was an image of natural reason.°

Howbeit (if we will truly consider it) more worthy it is to believe than to know as we now know. For in knowledge man's mind suffereth from sense,° but in belief it suffereth from spirit, such one° as it holdeth for more authorised° than itself, and so suffereth from the worthier agent. Otherwise it is° of the state of man glorified; for then faith shall cease,° and we shall know as we are known.

Wherefore we conclude that sacred Theology (which in our idiom we call Divinity) is grounded° only upon the word and oracle of God, and not upon the light of nature: for it is written, 'Coeli enarrant gloriam Dei',° but it is not written, 'Coeli enarrant voluntatem Dei',° but of that it is said, 'Ad legem et testimonium: si non fecerint secundum verbum istud, &c.'° This holdeth not only in those points of faith which concern the great mysteries of the Deity, of the Creation, of the Redemption, but likewise those which concern the law moral truly interpreted: 'Love your enemies: do good to them that hate you: be like to your heavenly Father, that suffereth his rain to fall upon the just and unjust'.° To this it ought to be applauded,° 'Nec vox hominem sonat':° it is a voice beyond the light of nature.° So° we see the heathen poets, when they fall upon° a libertine° passion, do still expostulate with° laws and moralities, as if they were opposite° and malignant to nature: 'Et quod natura remittit, invida jura negant'.° So said Dendamis° the Indian unto

Alexander's messengers, that he had heard somewhat of Pythagoras and some other of the wise men of Graecia, and that he held them for excellent men: but that they had a fault, which was that they had in too great reverence and veneration a thing they called law and manners. So it must be confessed that a great part of the law moral is of that perfection, whereunto the light of nature cannot aspire. How then is it that man is said to have by the light and law of nature some notions and conceits of virtue and vice, justice and wrong, good and evil? Thus; because the light of nature is used in two several° senses; the one, that which springeth from reason, sense, induction, argument, according to the laws of heaven and earth; the other, that which is imprinted upon the spirit of man by an inward° instinct, according to the law of conscience,° which is a sparkle° of the purity of his first estate:° in which latter sense only he is participant° of some light and discerning° touching the perfection of the moral law: but how? sufficient to check the vice, but not to inform° the duty. So then the doctrine of religion, as well moral as mystical, is not to be attained but by inspiration and revelation from God.

The use notwithstanding of reason in spiritual things, and the latitude° thereof, is very great and general: for it is not for nothing that the apostle calleth religion 'our reasonable service of God';° insomuch as the very ceremonies and figures of the old law° were full of reason and signification,° much more than the ceremonies of idolatry and magic, that are full of non-significants and surd characters.° But most specially the Christian Faith, as in all things so in this, deserveth to be highly magnified; holding and preserving the golden mediocrity° in this point between the law of the Heathen and the law of Mahumet,° which have embraced the two extremes. For the religion of the Heathen had no constant belief or confession, but left all to the liberty of argument; and the religion of Mahumet on the other side interdicteth° argument altogether: the one having the very face of error, and the other of imposture:° whereas the Faith doth both admit and reject disputation with difference.°

The use of human reason in religion is of two sorts: the former, in the conception and apprehension of the mysteries of God to us revealed; the other, in the inferring and deriving of doctrine and direction° thereupon. The former extendeth to the mysteries themselves; but how? by way of illustration,° and not by way of argument. The latter consisteth indeed of probation° and argument. In the former we see God vouchsafeth° to descend to our capacity, in the expressing of his mysteries in sort as may be sensible° unto us; and doth grift° his revelations and holy doctrine upon the notions of our reason, and applieth his inspirations to open our understanding, as the form of the key to the ward of the lock: for the latter, there is allowed us an use of reason and argument secondary° and respective,° although not original and absolute. For after the articles and principles of religion are placed,° and exempted from examination of reason, it is then permitted unto us to make derivations and inferences from and according to the analogy° of them, for our better direction. In nature° this holdeth not; for both the principles are examinable° by induction, though not by a medium or syllogism;° and besides, those principles or first positions have no discordance with that reason which draweth down and deduceth the inferior° positions. But yet it holdeth not in religion alone, but in many knowledges both of greater and smaller nature, namely wherein there are not only *posita*° but *placita*;° for in such there can be no use of absolute reason. We see it familiarly in games of wit,° as chess, or the like; the draughts and first laws of the game are positive,° but how? merely *ad placitum*,° and not examinable by reason; but then how to direct our play thereupon with best advantage to win the game, is artificial° and rational. So in human laws there be many grounds and maxims which are *placita juris*,° positive upon authority and not upon reason, and therefore not to be disputed: but what is most just, not absolutely, but relatively and according to those maxims, that affordeth a long field° of disputation. Such therefore is that secondary reason which hath place in divinity, which is grounded upon the *placets*° of God.

DE USU LEGITIMO RATIONIS HUMANAE IN DIVINIS.° Here therefore I note this deficience, that there hath not been to my understanding sufficiently enquired and handled the true limits and use of reason in spiritual things, as a kind of divine dialectic:° which for that it is not done, it seemeth to me a thing usual, by pretext of true conceiving that which is revealed, to search and mine° into that which is not revealed; and by pretext of enucleating° inferences° and contradictories,° to examine that which is positive; the one sort falling into the error of Nicodemus, demanding to have things made more sensible° than it pleaseth God to reveal them; 'Quomodo possit homo nasci cum sit senex?'° the other sort into the error of the disciples, which were scandalized° at a show° of contradiction; 'Quid est hoc quod dicit nobis? Modicum, et non videbitis me; et iterum, modicum, et videbitis me',° &c.

Upon this I have insisted the more in regard of the great and blessed use thereof; for this point well laboured and defined of° would in my judgment be an opiate° to stay and bridle not only the vanity of curious° speculations, wherewith the schools labour, but the fury of controversies, wherewith the church laboureth. For it cannot but open men's eyes, to see that many controversies do merely pertain to that which is either not revealed or positive;° and that many others do grow upon° weak and obscure inferences or derivations: which latter sort, if men would revive the blessed style of that great doctor° of the Gentiles, would be carried thus, 'Ego, non Dominus',° and again, 'Secundum consilium meum',° in opinions and counsels, and not in positions and oppositions.° But men are now over-ready to usurp the style° 'Non ego, sed Dominus',° and not so only, but to bind it with the thunder and denunciation of curses and anathemas, to the terror of those which have not sufficiently learned out of Salomon° that 'the causeless curse shall not come'.

Divinity° hath two principal parts; the matter informed° or revealed,° and the nature of the information or revelation: and with the latter we will begin, because it hath most coherence with that which we have now last handled. The nature of the information consisteth of three branches; the limits of the information, the sufficiency° of the information, and the acquiring or obtaining the information. Unto the limits of the information belong these considerations; how far forth particular persons continue to be inspired; how far forth the church is inspired; and how far forth reason may be used: the last point whereof I have noted as deficient. Unto the sufficiency of the information belong two considerations; what points of religion are fundamental, and what perfective,° being matter of further building and perfection upon one and the same foundation; and again, how the gradations of light° according to the dispensation° of times° are material° to the sufficiency of belief.

DE GRADIBUS UNITATIS IN CIVITATE DEI.° Here again I may rather give it in advice° than note it as deficient, that the points fundamental, and the points of further perfection only, ought to be with piety and wisdom distinguished: a subject tending to much like end as that I noted before; for as that other were likely to abate the number of controversies, so this is like to abate the heat of many of them. We see Moses when he saw the Israelite and the Egyptian fight, he did not say, 'Why strive you?'° but drew his sword and slew the Egyptian: but when he saw the two Israelites fight, he said, 'You are brethren, why strive you?'° If the point of doctrine be an Egyptian, it must be slain by the sword of the Spirit, and not reconciled; but if it be an Israelite, though in the wrong, then, 'Why strive you?' We see of the fundamental points, our Saviour penneth° the league° thus, 'He that is not with us, is against us';° but of points not fundamental, thus, 'He that is not against us, is with us'.° So we see the coat of our Saviour was entire without seam,° and so is the doctrine of the Scriptures in itself; but the garment of the Church was of divers colours, and yet not divided. We see the chaff° may and ought to be severed from the corn in the ear, but the tares° may not be pulled up from the corn in the field: so as it is a thing of great use well to define what and of what latitude° those points are, which do make men merely° aliens and disincorporate° from the Church of God.

For the obtaining of the information, it resteth upon the true and sound interpretation of the Scriptures, which are the fountains of the water of life. The interpretations of the Scriptures are of two sorts; methodical,° and solute° or at large. For this divine water, which excelleth so much that of Jacob's well,° is drawn forth much in the same kind as natural water useth to be out of wells and fountains; either it is first forced up into a cistern,° and from thence fetched and derived° for use; or else it is drawn and received in buckets and vessels immediately where it springeth. The former sort whereof, though it seem to be the more ready, yet in my judgment is more subject° to corrupt.° This is that method which hath exhibited unto us the scholastical divinity;° whereby divinity hath been reduced° into an art, as into a cistern, and the streams of doctrine or positions fetched and derived from thence.

In this men have sought three things, a summary° brevity, a compacted° strength, and a complete perfection; whereof the two first they fail to find, and the last they ought not to seek. For as to brevity, we see in all summary methods, while men purpose to abridge they give cause to dilate.° For the sum or abridgment by contraction becometh obscure, the obscurity requireth exposition, and the exposition is deduced° into large commentaries, or into common places and titles,° which grow to be more vast than the original writings whence the sum was at first extracted. So we see the volumes of the schoolmen are greater much than the first writings of the fathers,° whence the Master of the Sentences° made his sum or collection. So in like manner the volumes of the modern doctors of the civil law° exceed those of the ancient jurisconsults,° of which Tribonian° compiled the digest. So as this course of sums and commentaries is that which doth infallibly make the body of sciences more immense in quantity, and more base in substance.

And for strength, it is true that knowledges reduced into exact methods° have a shew of strength, in that each part seemeth to support and sustain the other; but this is more satisfactory° than substantial; like unto buildings which stand by architecture and compaction,° which are more subject to ruin than those which are built more strong in their several° parts, though less compacted. But it is plain that the more you recede from your grounds° the weaker do you conclude;° and as in nature the more you remove yourself from particulars the greater peril of error you do incur, so much more in divinity the more you recede from the Scriptures by inferences and consequences, the more weak and dilute° are your positions.

And as for perfection or completeness in divinity, it is not to be sought; which makes this course of artificial° divinity the more suspect.° For he that will reduce a knowledge into an art, will make it round° and uniform: but in divinity many things must be left abrupt° and concluded with this: 'O altitudo sapientiae et scientiae Dei! quam incomprehensibilia sunt judicia ejus, et non investigabiles viae ejus!'° So again the apostle saith, 'Ex parte scimus',° and to have the form of a total where there is but matter for a part, cannot be without supplies° by supposition° and presumption.° And therefore I conclude, that the true use of these Sums and Methods hath place in institutions° or introductions preparatory unto knowledge; but in them, or by deducement° from them, to handle the main body and substance of a knowledge, is in all sciences prejudicial,° and in divinity dangerous.

As to the interpretation of the Scriptures solute and at large, there have been divers kinds introduced and devised; some of them rather curious° and unsafe, than sober and warranted.° Notwithstanding thus much must be confessed, that the Scriptures, being given by inspiration and not by human reason, do differ from all other books in the author;° which by consequence doth draw on° some difference to be used by the expositor. For the inditer° of them did know four things which no man attains to know; which are, the mysteries of the kingdom of glory; the perfection of the laws of nature; the secrets of the heart of man; and the future succession of all ages. For as to the first, it is said, 'He that presseth into the light, shall be oppressed of the glory':° and again, 'No man shall see my face and live'.° To the second, 'When he prepared the heavens I was present, when by law and compass he inclosed the deep'.° To the third, 'Neither was it needful that any should bear witness to him of Man, for he knew well what was in Man'.° And to the last, 'From the beginning are known to the Lord all his works'.°

From the former two° of these have been drawn certain senses and expositions of Scriptures, which had need be contained° within the bounds of sobriety;° the one anagogical,° and the other philosophical. But as to the former, man is not to prevent° his time: 'Videmus nunc per speculum in aenigmate, tunc autem facie ad faciem':° wherein nevertheless there seemeth to be a liberty granted, as far forth as° the polishing of this glass,° or some moderate explication of this aenigma.° But to press too far into it, cannot but cause a dissolution° and overthrow of the spirit of man. For in the body there are three degrees of that we receive into it; Aliment,° Medicine, and Poison; whereof aliment is that which the nature of man can perfectly alter and overcome:° medicine is that which is partly converted by nature, and partly converteth nature; and poison is that which worketh wholly upon° nature, without that that nature can in any part work upon it. So in the mind whatsoever knowledge reason cannot at all work upon and convert, is a mere intoxication,° and endangereth a dissolution of the mind and understanding.

But for the latter,° it hath been extremely set on foot of late time by the school of Paracelsus,° and some others, that have pretended to find the truth of all natural philosophy in the Scriptures; scandalizing° and traducing° all other philosophy as heathenish and profane. But there is no such enmity between God's word and his works. Neither do they give honour to the Scriptures, as they suppose, but much imbase them. For to seek heaven and earth in the word of God, whereof it is said, 'Heaven and earth shall pass, but my word shall not pass',° is to seek temporary things amongst eternal: and as to seek divinity in philosophy is to seek the living amongst the dead, so to seek philosophy in divinity is to seek the dead amongst the living:° neither are the pots or lavers° whose place was in the outward part of the temple to be sought in the holiest place of all, where the ark of the testimony° was seated. And again, the scope or purpose of the Spirit of God is not to express matters of nature in the Scriptures, otherwise than in passage,° and for application to man's capacity and to matters moral or divine. And it is a true rule, 'Authoris aliud agentis parva authoritas':° for it were a strange conclusion, if a man should use a similitude for ornament or illustration sake, borrowed from nature or history according to vulgar conceit, as of a Basilisk, an Unicorn, a Centaur, a Briareus, an Hydra, or the like, that therefore he must needs be thought to affirm the matter thereof positively to be true. To conclude therefore, these two interpretations, the one by reduction or aenigmatical, the other philosophical or physical, which have been received and pursued in imitation of the rabbins and cabalists,° are to be confined with a 'Noli altum sapere, sed time'.°

But the two latter points, known to God and unknown to man, touching the secrets of the heart, and the successions of time, doth make a just and sound difference between the manner of the exposition of the Scriptures, and all other books. For it is an excellent observation which hath been made upon the answers° of our Saviour Christ to many of the questions which were propounded to him, how that they are impertinent° to the state of the question demanded; the reason whereof is, because not being like man, which knows man's thoughts by his words, but knowing man's thoughts immediately, he never answered their words, but their thoughts: much in the like manner it is with the Scriptures, which being written to the thoughts of men, and to the succession of all ages, with a foresight of all heresies, contradictions, differing estates of the church, yea and particularly of the elect,° are not to be interpreted only according to the latitude° of the proper sense of the place, and respectively° towards that present occasion whereupon the words were uttered; or in precise congruity or contexture with the words before or after; or in contemplation of the principal scope of the place; but have in themselves, not only totally or collectively, but distributively in clauses and words, infinite springs and streams of doctrine to water the church in every part; and therefore as the literal sense° is as it were the main stream or river; so the moral sense chiefly, and sometimes the allegorical or typical,° are they whereof the church hath most use: not that I wish men to be bold in allegories, or indulgent° or light° in allusions; but that I do much condemn that interpretation of the Scripture which is only after the manner as men use to interpret a profane book.

EMANATIONES SCRIPTURARUM IN DOCTRINAS POSITIVAS.° In this part touching the exposition of the Scriptures, I can report no deficience; but by way of remembrance this I will add: In perusing books of divinity, I find many books of controversies; and many of common places° and treatises; a mass of positive divinity, as it is made an art;° a number of sermons and lectures, and many prolix commentaries upon the Scriptures, with harmonies° and concordances. But that form of writing in divinity, which in my judgment is of all others most rich and precious, is positive divinity collected upon particular texts of Scriptures in brief observations; not dilated into common places, not chasing after controversies, not reduced into method of art; a thing abounding in sermons, which will vanish,° but defective in books, which will remain; and a thing wherein this age excelleth. For I am persuaded, and I may speak it with an 'Absit invidia verbo',°

and no ways in derogation of antiquity, but as in a good emulation between the vine and the olive, that if the choice and best of those observations upon texts of Scriptures which have been made dispersedly in sermons within this your Majesty's island of Britain by the space of these forty years° and more (leaving out the largeness° of exhortations° and applications thereupon) had been set down in a continuance,° it had been the best work in divinity which had been written since the apostles' times.

The matter informed° by divinity is of two kinds; matter of belief and truth of opinion, and matter of service and adoration; which is also judged and directed by the former; the one being as the internal soul of religion, and the other as the external body thereof. And therefore the heathen religion was not only a worship of idols, but the whole religion was an idol in itself; for it had no soul, that is, no certainty of belief or confession; as a man may well think, considering the chief doctors of their church were the poets; and the reason was, because the heathen gods were no jealous° gods, but were glad to be admitted into part,° as they had reason. Neither did they respect° the pureness of heart, so° they might have external honour and rites.

But out of these two do result and issue four main branches of divinity; Faith, Manners, Liturgy,° and Government.° Faith containeth the doctrine of the nature of God, of the attributes of God, and of the works of God. The nature of God consisteth of three persons in unity of Godhead. The attributes of God are either common to the Deity, or respective° to the persons. The works of God summary are two, that of the Creation, and that of the Redemption; and both these works, as in total they appertain to the unity of the Godhead, so in their parts they refer to the three persons: that of the Creation, in the mass of the matter to the Father; in the disposition of the form to the Son; and in the continuance and conservation of the being to the Holy Spirit: so that of the Redemption, in the election° and counsel to the Father; in the whole act and consummation° to the Son; and in the application to the Holy Spirit; for by the Holy Ghost was Christ conceived in flesh, and by the Holy Ghost are the elect regenerate in spirit. This work likewise we consider either effectually in the elect; or privatively° in the reprobate;° or according to appearance in the visible church.

For Manners, the doctrine thereof is contained in the law, which discloseth° sin. The law itself is divided, according to the edition° thereof, into the law of Nature, the law Moral, and the law Positive; and according to the style, into Negative and Affirmative, Prohibitions and Commandments. Sin, in the matter and subject thereof, is divided according to the commandments; in the form thereof, it referreth to the three persons in Deity: sins of Infirmity° against the Father, whose more special attribute is Power; sins of Ignorance against the Son, whose attribute is Wisdom; and sins of Malice against the Holy Ghost, whose attribute is Grace or Love. In the motions of it,° it either moveth to the right hand or to the left; either to blind devotion, or to profane and libertine° transgression; either in imposing restraint where God granteth liberty, or in taking liberty where God imposeth restraint. In the degrees and progress of it, it divideth itself into thought, word, or act.° And in this part I commend much the deducing° of the law of God to cases of conscience;° for that I take indeed to be a breaking,° and not exhibiting whole, of the bread of life. But that which quickeneth both these doctrines of faith and manners, is the elevation and consent° of the heart; whereunto appertain books of exhortation, holy meditation, Christian resolution, and the like.

For the Liturgy or service, it consisteth of the reciprocal acts between God and man; which, on the part of God, are the preaching of the word and the sacraments,° which are seals to the covenant,° or as the visible word;° and on the part of man, invocation of the name of God, and under the law, sacrifices, which were as visible° prayers or confessions: but now the adoration being 'in spiritu et veritate',° there remaineth only 'vituli labiorum',° although the use of holy vows of thankfulness and retribution° may be accounted also as sealed petitions.°

And for the Government of the church, it consisteth of the patrimony° of the church, the franchises° of the church, and the offices and jurisdictions of the church, and the laws of the church directing the whole; all which have two considerations, the one in themselves, the other how they stand compatible and agreeable to the civil estate.

This matter of divinity is handled either in form of instruction of truth, or in form of confutation of falsehood. The declinations° from religion, besides the privative,° which is atheism and the branches thereof, are three; Heresies, Idolatry, and Witchcraft; Heresies, when we serve the true God with a false worship; Idolatry, when we worship false gods, supposing° them to be true; and Witchcraft, when we adore false gods, knowing them to be wicked and false. For so your Majesty° doth excellently well observe, that Witchcraft is the height of Idolatry. And yet we see though these be true degrees,

Samuel teacheth us that they are all of a° nature, when there is once a receding from the word of God; for so he saith, 'Quasi peccatum ariolandi est repugnare, et quasi scelus idololatriae nolle acquiescere'.°

These things I have passed over so briefly because I can report no deficience concerning them: for I can find no space or ground that lieth vacant and unsown in the matter of divinity; so diligent have men been, either in sowing of good seed or in sowing of tares.°

Thus have I made as it were a small Globe of the Intellectual World, as truly and faithfully as I could discover; with a note and description of those parts which seem to me not constantly occupate,° or not well converted° by the labour of man. In which, if I have in any point receded from that which is commonly received, it hath been with a purpose of proceeding *in melius*,° and not *in aliud*;° a mind of amendment° and proficience,° and not of change and difference.° For I could not be true and constant to the argument I handle, if I were not willing to go beyond others; but yet not more willing than to have others go beyond me again: which may the better appear by this, that I have propounded my opinions naked and unarmed, not seeking to preoccupate° the liberty of men's judgments by confutations. For in any thing which is well set down, I am in good hope that if the first reading move an objection,° the second reading will make an answer. And in those things wherein I have erred, I am sure I have not prejudiced° the right by litigious° arguments; which certainly have this contrary effect and operation, that they add authority to error, and destroy the authority of that which is well invented:° for question° is an honour and preferment° to falsehood, as on the other side it is a repulse° to truth. But the errors I claim and challenge to myself as mine own. The good, if any be, is due 'tanquam adeps sacrificii', to be incensed° to the honour, first of the Divine Majesty, and next of your Majesty,° to whom on earth I am most bounden.°

Essays (1612) [*Selections*]

Of Religion

THE quarrels and divisions for religion were evils unknown to the heathen: and no marvel, for it is the true God that is the jealous God,° and the gods of the heathen were good fellows.° But yet the bonds of religious unity are so° to be strengthened, as the bonds of human society be not dissolved. Lucretius° the poet, when he beheld the act of Agamemnon, enduring and assisting at the sacrifice of his daughter, concludes with this verse:

Tantum religio potuit suadere malorum.°

But what would he have done if he had known the massacre° of France, or the powder treason° of England? Certainly he would have been seven times more epicure and atheist° than he was. Nay, he would rather have chosen to be one of the madmen of Münster° than to have been a partaker of those counsels. For it is better that religion should deface men's understanding than their piety and charity, retaining reason only but as an engine° and chariot driver of cruelty and malice.

It was a great blasphemy when the devil said, 'I will ascend, and be like the highest'.° But it is a greater blasphemy if they make God to say, 'I will descend, and be like the prince of darkness'; and it is no better when they make the cause of religion descend to the execrable actions of murdering of princes, butchery of people, and firing° of states. Neither is there such a sin against the person of the holy Ghost (if one should take it literally) as, instead of the likeness of a dove,° to bring him down in the likeness of a vulture or raven; nor such a scandal to their church, as out of the bark° of Saint Peter to set forth the flag of a bark of pirates and assassins.° Therefore, since these things are the common enemies of human society, princes by their power, churches by their decrees, and all learning° (Christian, moral, or whatsoever sect or opinion) by their Mercury rod,° ought to join in the damning to hell for ever these facts,° and their supports.° And in all counsels concerning religion that counsel of the apostle would° be prefixed, 'Ira hominis non implet iustitiam Dei'.°

OF FRIENDSHIP

THERE is no greater desert or wilderness than to be without true friends. For without friendship, society is but meeting.° And as it is certain that in bodies inanimate, union° strengtheneth any natural° motion, and weakeneth any violent° motion; so amongst men, friendship multiplieth joys, and divideth griefs. Therefore, whosoever wanteth° fortitude,° let him worship friendship, for the yoke of friendship maketh the yoke of fortune more light.

There be some whose lives are as if they perpetually played upon a stage, disguised to all others, open only to themselves. But perpetual dissimulation is painful,° and he that is all fortune and no nature is an exquisite° hireling.° Live not in continual smother,° but take some friends with whom to communicate.° It will unfold thy understanding; it will evaporate thy affections;° it will prepare thy business.

A man may keep a corner of his mind from his friend, and it be but° to witness to himself that it is not upon facility,° but upon true use of friendship that he imparteth himself. Want of true friends, as it is the reward of perfidious natures, so it is an imposition° upon great fortunes. The one deserveth it, the other cannot escape it. And therefore it is good to retain sincerity,° and to put it into the reckoning° of ambition that the higher one goeth, the fewer friends he shall have.

Perfection of friendship is but a speculation.° It is friendship, when a man can say to himself, I love this man without respect of utility; I am open-hearted to him; I single him from the generality of those with whom I live; I make him a portion of my own wishes.

OF THE TRUE GREATNESS° OF KINGDOMS

THE speech° of Themistocles, which was arrogant in challenge,° is profitable in censure.° Desired at a banquet to touch° a lute, he said, 'He could not fiddle;° but he could make a small town to become a great city'. This speech at a time of solace,° and not serious, was uncivil, and at no time could be decent of a man's self. But it may have a pretty° application. For to speak truly of politiques and statesmen, there are sometimes, though rarely, those that can make a small estate° great, and cannot fiddle; and there be many that can fiddle very cunningly, and yet the procedure° of their art is to make a flourishing estate ruinous and distressed.° For certainly, those degenerate arts whereby divers politiques and governors do gain both satisfaction with their masters, and admiration with the vulgar,° deserve no better name than fiddling, if they add nothing to the safety, strength, and amplitude° of the states they govern.

The greatness of a state in bulk or territory doth fall under measure;° and the greatness of finances and revenue doth fall under computation.° The population may appear° by musters,° and the number of cities and towns by charts° and maps. But yet there is nothing among civil affairs more subject to error, than the right valuation and true judgment concerning the greatness of an estate. Certainly there is a kind of resemblance between the kingdom of heaven and the kingdoms upon the earth.° The kingdom of heaven is compared° not to any great kernel or nut, but to a grain of mustard, which is one of the least° of grains, but hath in it a property and spirit hastily to get up° and spread. So are there states that are great in territory, and yet not apt° to conquer or enlarge; and others that have but a small dimension or stem,° and yet apt to be the foundation of great monarchies.

Walled towns, stored arsenals and armories, goodly stables, elephants (if you will), mass of treasure, number in armies, ordnance° and artillery, they are all but a sheep in a lion's skin,° except the breed and disposition of the people be military. The help is mercenary aids.° But a prince or state that resteth upon waged° companies of foreign arms, and not of his own natives, may spread° his feathers for a time, but he will mew° them soon after. The blessing° of Judah and Issachar will never meet,° to be both the lion's whelp and the ass laid between burdens.° Neither will a people overcharged with tributes° be ever fit for empire.

Nobility and gentlemen, multiplying in too great a proportion, maketh the common subject grow to be a peasant and base swain,° driven out of heart,° and but the gentleman's labourer. Like as it is in coppices,° where if you leave your staddles° too thick you shall never have clean underwood, but shrubs and bushes. And take away the middle people, you take away the infantry, which is the nerve of an army:° and you bring it to this, that not the hundredth poll° will be fit for a helmet,° and so, great population and little strength. Certainly Virgil coupled arms and the plough together well in the constitution of ancient Italy:

Terra potens armis atque ubere glebae.°

For it is the plough that yieldeth the best soldier. But how? Maintained in plenty and in the hand of owners, and not of mere labourers.°

Sedentary and within-doors arts,° and nice manufactures,° that require° rather the finger than the hand or arm, have in their nature a contrariety to a disposition military; and generally, all warlike people are a little idle, and love danger better than pain.° Neither must they be too much broken of it,° if they shall° be preserved in vigour.

No body can be healthful without exercise, neither natural body nor politic; and to the politic body of a kingdom or state, a civil war is as the heat of a fever: but an honourable foreign war is like the heat of exercise. At least discoveries, navigations, honourable succours° of other states may keep health: for in a slothful peace° both courages will effeminate, and manners corrupt.

States liberal of naturalization° are capable of greatness;° and the jealous states that rest upon° the first tribe° and stirp,° quickly want° body to carry the boughs and branches. Many are the ingredients into the receipt of° greatness.° No man can by care-taking add a cubit to his stature,° in the little model° of a man's body. But certainly in the great frame of kingdoms and commonwealths, it is in the power of princes or estates by ordinances° and constitutions, and manners° which they may introduce, to sow greatness to° their posterity and succession. But these things are commonly left to chance.°

The Charge° touching duels

My Lords,

I thought it fit for my place, and for these times, to bring to hearing before your Lordships some cause° touching private duels, to see if this Court can do any good to tame and reclaim° that evil which seems unbridled. And I could have wished that I had met with some greater persons,° as a subject for your censure, both because it had been more worthy of this presence,° and also the better to have shewed the resolution myself hath to proceed without respect of persons in this business. But finding this cause on foot° in my predecessor's time, and published° and ready for hearing, I thought to lose no time, in a mischief° that groweth every day; and besides it passeth not amiss sometimes in government, that the greater sort be admonished by an example made in the meaner, and the dog to be beaten before the lion.° Nay I should think (my Lords) that men of birth and quality° will leave° the practice, when it begins to be vilified,° and come so low as to barbers' surgeons° and butchers, and such base mechanical° persons.

And for the greatness of this presence, in which I take much comfort, both as I consider it in itself, and much more in respect it is by his Majesty's direction, I will supply° the meanness of the particular cause, by handling of the general point; to the end that by occasion of this present cause, both my purpose of prosecution against duels and the opinion of the Court (without which I am nothing) for the censure of them may appear, and thereby offenders in that kind may read their own case, and know what they are to expect; which may serve for a warning until example may be made in some greater person, which I doubt° the times will but too soon afford.

Therefore before I come to the particular° whereof your Lordships are now to judge, I think it time best spent to speak somewhat,

First, of the nature and greatness of this mischief.

Secondly, of the causes and remedies.

Thirdly, of the justice of the law of England, which some stick not° to think defective in this matter.

Fourthly, of the capacity of this Court, where certainly the remedy of this mischief is best to be found.

And fifthly, touching mine own purpose and resolution,° wherein I shall humbly crave your Lordships' aid and assistance.

For the mischief itself, it may please your Lordships to take into your consideration that when revenge is once extorted° out of the magistrate's hand contrary to God's ordinance, 'Mihi vindicta, ego retribuam',° and every man shall bear the sword° not to defend but to assail,° and private men begin once to presume to give law to themselves, and to right their own wrongs, no man can foresee the dangers and inconveniencies that may arise and multiply thereupon. It may cause sudden storms in Court, to the disturbance of his Majesty, and unsafety of his person. It may grow from quarrels to banding,° and from banding to trooping,° and so to tumult and commotion, from particular persons to dissension of families and alliances,° yea to national quarrels, according to the infinite variety of accidents,° which fall not under foresight: so that the state by this means shall be like to a distempered° and unperfect° body, continually subject to inflammations° and convulsions.°

Besides, certainly both in divinity and in policy,° offences of presumption° are the greatest. Other offences yield and consent to the law that it° is good, not daring to make defence, or to justify themselves; but this offence expressly gives the law an affront, as if there were two laws, one a kind of gown-law,° and the other a law of reputation,° as they term it; so that Paul's° and Westminster,° the pulpit and the courts of justice, must give place to the law (as the King speaketh in his proclamation) of Ordinary tables,° and such reverend assemblies; the year-books° and statute-books° must give place to some French and Italian pamphlets,° which handle the doctrine of duels, which if they be in the right, *transeamus ad illa*, let's receive° them, and not keep the people in conflict and distraction between two laws.

Again (my Lords) it is a miserable effect, when young men full of towardness° and hope, such as the poets call 'Aurorae filii',° sons of the morning, in whom the expectation and comfort of their friends consisteth, shall be cast away and destroyed in such a vain° manner. But much more it is to be deplored when so much noble and gentle° blood shall be spilt upon such follies, as, if it were adventured° in the field in service of the King and realm, were able to make the fortune of a day,° and to change the fortune of a kingdom. So as your Lordships see what a desperate evil this is; it troubleth peace, it disfurnisheth° war, it bringeth calamity upon private men, peril upon the State, and contempt upon the law.

Touching the causes of it; the first motive no doubt is a false and erroneous imagination of honour and credit;° and therefore the King, in his last proclamation,° doth most amply and excellently call them 'bewitching° duels'. For, if one judge of it truly, it is no better than a sorcery that enchanteth the spirits of young men that bear great minds,° with a false shew, 'species° falsa'; and a kind of satanical illusion and apparition° of honour; against religion, against law, against moral virtue, and against the precedents and examples of the best times and valiantest nations, as I shall tell you by and by, when I shall shew you that the law of England is not alone in this point.

But then the seed of this mischief being such, it is nourished by vain discourses,° and green° and unripe conceits, which nevertheless have so prevailed, as though a man were staid° and sober-minded, and a right believer touching the vanity and unlawfulness of these duels, yet the stream of vulgar opinion is such, as it imposeth a necessity upon men of value° to conform themselves; or else there is no living or looking upon men's faces. So that we have not to do, in this case, so much with particular persons, as with unsound and depraved opinions, like the dominations° and spirits of the air which the Scripture° speaketh of.

Hereunto may be added, that men have almost lost the true notion and understanding of Fortitude° and Valour. For Fortitude distinguisheth of the grounds° of quarrels, whether they be just; and not only so, but whether they be worthy; and setteth a better price upon men's lives than to bestow them idly. Nay it is weakness and dis-esteem of a man's self, to put a man's life upon such ledgier° performances. A man's life is not to be trifled away, it is to be offered up and sacrificed to honourable services, public merits, good causes, and noble adventures. It is in expense of blood as it is in° expense of money. It is no liberality° to make a profusion of money upon every vain occasion, nor no more it is fortitude to make effusion of blood, except the cause be of worth. And thus much for the causes of this evil.

For the remedies, I hope some great and noble person will put his hand to this plough, and I wish° that my labours of this day may be but forerunners to the work of a higher and better hand. But yet to deliver my opinion, as may be proper for this time and place, there be four things that I have thought on, as the most effectual° for the repressing of this depraved custom of particular° combats.

The first is, that there do appear and be declared a constant and settled resolution in the State to abolish it. For this is a thing (my Lords) must go down° at once, or not at all. For then every particular man will think himself acquitted° in his reputation, when he sees that the State takes it° to heart, as an insult against the King's power and authority, and thereupon hath absolutely resolved to master it; like unto that which was set down in express words in the edict of Charles the ninth° of France touching duels, that 'the King himself took upon him the honour of all that took themselves grieved or interessed° for not having performed the combat'. So must the State do in this business: and in my conscience there is none° that is but of a reasonable sober° disposition, be he never so valiant, (except it be some furious person that is like a firework) but will be glad of it, when he shall see the law and rule of State disinterest° him of a vain and unnecessary hazard.

Secondly, care must be taken that this evil be no more cockered,° nor the humour° of it fed; wherein I humbly pray your Lordships that I may speak my mind freely, and yet be understood aright. The proceedings of the great and noble Commissioners Marshall° I honour and reverence much, and of them I speak not in any sort. But I say the compounding° of quarrels, which is otherwise° in use, by private noblemen and gentlemen, it is so punctual,° and hath such reference and respect unto the received conceits,° what's before-hand, and what's behind-hand,° and I cannot tell what, as without all question it doth in a fashion countenance and authorize this practice of duels, as if it had in it somewhat of right.

Thirdly, I must acknowledge that I learned out of the King's last proclamation the most prudent and best applied remedy for this offence (if it shall please his Majesty to use it) that the wit of man can devise. This offence (my Lords) is grounded upon a false conceit of honour, and therefore it would° be punished in the same kind. 'In eo quis rectissime plectitur, in quo peccat'.° The fountain of honour is the King, and his aspect° and the access to his person continueth° honour in life, and to be banished from his presence° is one of the greatest eclipses of honour that can be. If his Majesty shall be pleased that when this Court shall censure any of these offences in persons of eminent quality, to add this out of his own power and discipline,° that these persons shall be banished and excluded from his Court for certain years, and the Courts of his Queen and Prince,° I think there is no man that hath any good blood in him will commit an act that shall cast him into that darkness, that he may not behold his Sovereign's face.

Lastly, and that which more properly concerneth this Court, we see (my Lords) the root of this offence is stubborn; for it despiseth death, which is the utmost of punishments, and it were a just but a miserable severity,° to execute the law without all remission or mercy, where the case proveth capital.° And yet the late severity in France was more, where by a kind of martial law established by ordinance of the King and Parliament, the party that had slain another was presently° had° to the gibbet; insomuch as gentlemen of great quality were hanged, their wounds bleeding, lest a natural death should prevent° the example of justice. But (my Lords) the course which we shall take is of far greater lenity, and yet of no less efficacy; which is to punish, in this Court, all the middle° acts and proceedings which tend to the duel (which I will enumerate to you anon), and so to hew and vex the root in the branches;° which, no doubt, in the end will kill the root, and yet prevent the extremity of law.

Now for the law of England, I see it excepted° to, though ignorantly, in two points:

The one, that it should make no difference between an insidious° and foul murther, and the killing of a man upon fair terms, as they now call it.

The other, that the law hath not provided sufficient punishment and reparations for contumely° of words, as the Lie,° and the like.

But these are no better than childish novelties, against the divine law, and against all laws in effect, and against the examples of all the bravest and most virtuous nations of the world.

For first for the law of God, there is never to be found any difference° made in homicide, but between homicide voluntary and involuntary, which we term misadventure.° And for the case of misadventure itself, there were cities of refuge;° so that the offender was put to his flight,° and that flight was subject to accident,° whether the revenger of blood° should overtake him before he had gotten° sanctuary or no. It is true that our law hath made a more subtle distinction between the will inflamed° and the will advised,° between manslaughter in heat and murther upon prepensed malice,° or cold blood, as the soldiers call it; an indulgence° not unfit for a choleric and warlike nation; for it is true, 'ira furor brevis';° a man in fury is not himself. This privilege of passion the ancient Roman law restrained,° but to a case:° that was, if the husband took° the adulterer in the manner;° to that rage and provocation only it gave way, that it was an homicide was justifiable. But for a difference to be made in case of killing and destroying man, upon a fore-thought purpose, between foul and fair, and as it were between single° murther and vied° murther, it is but a monstrous child of this latter age, and there is no shadow of it in any law divine or human. Only it is true, I find in the Scripture that Cain° inticed his brother into the field and slew him treacherously, but Lamech° vaunted of his manhood that 'he would kill a young man and if it were in his hurt'. So as I see no difference between an insidious murther and a braving° or presumptuous° murther, but the difference between Cain and Lamech.

As for examples in civil states, all memory doth consent that Graecia and Rome were the most valiant and generous° nations of the world; and, that which is more to be noted, they were free estates, and not under a monarchy, whereby a man would think it a great deal the more reason that particular persons should have righted° themselves; and yet they had not this practice of duels, nor anything that bare show° thereof: and sure they would have had it, if there had been any virtue in it. Nay as he saith, 'Fas est et ab hoste doceri',° it is memorable, that° is reported by a counsellor and ambassador of the Emperor's,° touching the censure of the Turks of these duels. There was a combat of this kind performed by two persons of quality of the Turks, wherein one of them was slain, the other party was convented before the council of Bassaes;° the manner of the reprehension was in these words: 'How durst you undertake to fight one with the other? Are there not Christians enough to kill? Did you not know that whether° of you should be slain, the loss would be the Great Seigneur's?'° So as we may see that the most warlike nations, whether generous or barbarous,° have ever despised this wherein now men glory.

It is true (my Lords) that I find combats of two natures authorized, how justly I will not dispute as to the latter° of them.

The one, when upon the approaches of armies in the face one of the other, particular persons have made challenges for trial of valours in the field, upon the public quarrel. This the Romans called 'pugna per provocationem.'° And this was never, but either between the generals themselves, who were absolute,° or between particulars° by licence° of the generals; never upon private authority. So you see David asked leave when he fought with Goliath,° and Joab, when the armies were met, gave leave, and said, 'Let the young men play before us';° and of this kind was that famous example in the wars of Naples,° between twelve Spaniards and twelve Italians, where the Italians bare away the victory; besides other infinite like examples worthy and laudable, sometimes by singles,° sometimes by numbers.

The second combat is a judicial° trial of right, where the right is obscure, introduced by the Goths and the Northern nation, but more anciently entertained° in Spain; and this yet remains in some cases as a divine lot° of battle, though controverted° by divines touching the lawfulness of it. So that a wise writer saith, 'Taliter pugnantes videntur tentare Deum, quia hoc volunt ut Deus ostendat et faciat miraculum, ut justam causam habens victor efficiatur, quod saepe contra accidit.'° But howsoever it be, this kind of fight taketh his warrant° from law. Nay, the French themselves, whence° this folly seemeth chiefly to have flown, never had it but only in practice and toleration,° but never as authorized by law. And yet now of late they have been fain to purge their folly with extreme rigour, in so much as many gentlemen left between death and life in the duels (as I spake before) were hastened to hanging with their wounds bleeding. For the State found it had been neglected so long, as nothing could be thought cruelty which tended to the putting of it down.

As for the second defect pretended° in our law, that it hath provided no remedy for lies and fillips,° it may receive like answer. It would have been thought a madness amongst the ancient lawgivers, to have set a punishment upon the lie given,° which in effect is but a word of denial, a negative of another's saying. Any lawgiver, if he had been asked the question, would have made Solon's answer°: that 'he had not ordained any punishment for it, because he never imagined the world would have been so fantastical° as to take it so highly.' The civilians,° they dispute whether an action of injury lie° for it, and rather resolve the contrary. And Francis the first° of France, who first set on° and stamped this disgrace so deep,° is taxed° by the judgment of all wise writers for beginning the vanity of it; for it was he, that when he had himself given the lie and defy° to the Emperor, to make it current° in the world, said in a solemn assembly, that 'he was no honest man that would bear° the lie': which was the fountain of this new learning.°

As for words of reproach and contumely (whereof the lie was esteemed none) it is not credible (but° that the orations themselves are extant) what extreme and exquisite° reproaches were tossed up and down in the senate of Rome and the places of assembly, and the like in Graecia, and yet no man took himself fouled° by them, but took them but for breath° and the stile of an enemy, and either despised them or returned them, but no blood spilt about them.

So of every touch° or light° blow of the person,° they are not in themselves considerable, save that they have got upon them the stamp of a disgrace, which maketh these light things pass for great matter. The law of England, and all laws, hold these degrees of injury to the person, slander, battery,° maim,° and death; and if there be extraordinary circumstances of despite and contumely, as in case of libels and bastinadoes,° and the like, this Court taketh them in hand, and punisheth them exemplarly. But for this apprehension° of a disgrace, that a fillip to the person should be a mortal wound to the reputation, it were good that men did hearken unto the saying of Consalvo,° the great and famous commander, that was wont to say, 'a gentleman's honour should be *de tela crassiore*', of a good strong warp or web, that every little thing should not catch in it; when as now it seems they are but of cobweb-lawn° or such light stuff, which certainly is weakness, and not true greatness of mind,° but like a sick man's body, that is so tender° that it feels every thing. And so much in maintenance and demonstration of the wisdom and justice of the law of the land.

For the capacity° of this Court, I take this to be a ground° infallible that wheresoever an offence is capital, or matter of felony,° if it be acted, there the combination° or practice° tending to that offence is punishable in this Court as a high misdemeanour. So practice to impoison, though it took no effect; waylaying to murther, though it took no effect, and the like; have been adjudged heinous° misdemeanours punishable in this Court. Nay inceptions° and preparations in inferior crimes (that are not capital), as suborning° and preparing of witnesses that were never deposed,° or deposed nothing material,° have likewise been censured in this Court, as appeareth by the decree in Garnon's case.

Why then, the major proposition being such, the minor° cannot be denied: for every appointment of the field° is but combination and plotting of murther; let them gild° it how they list, they shall never have fairer terms° of° me in place of justice. Then the conclusion° followeth, that it is a case fit for the censure° of this Court. And of this there be precedents in the very point of challenge.

It was the case of Wharton, plaintiff, against Ellekar and Acklam defendants, where Acklam being a follower of Ellekar's, was censured for carrying a challenge from Ellekar to Wharton, though the challenge was not put in writing, but delivered only by word of message; and there are words in the decree, that such challenges are to° the subversion of government.

These things are well known, and therefore I needed not so much to have insisted upon them, but that in this case I would be thought not° to innovate any thing of° mine own head, but to follow the former precedents of the Court, though I mean to do it more thoroughly, because the time requires it more.

Therefore now to come to that which concerneth my part;° I say, that by the favour of the King and the Court, I will prosecute in this Court in the cases following.

If any man shall appoint the field, though the fight be not acted or performed.

If any man shall send any challenge in writing, or any message of challenge.

If any man carry or deliver any writing or message of challenge.

If any man shall accept or return a challenge.

If any man shall accept to be a second in a challenge of either side.

If any man shall depart the realm, with intention and agreement to perform the fight beyond the seas.

If any man shall revive a quarrel by any scandalous bruits or writings, contrary to a former proclamation published by his Majesty in that behalf.

Nay I hear there be some Counsel learned° of duels, that tell young men when they are before-hand, and when they are otherwise, and thereby incense and incite them to the duel, and make an art° of it. I hope I shall meet with some of them too: and I am sure (my Lords) this course of preventing duels in nipping them in the bud, is fuller of clemency and providence than the suffering them to go on, and hanging men with their wounds bleeding, as they did in France.

To conclude, I have some petitions to make, first to your Lordship, my Lord Chancellor, that in case I be advertised° of a purpose in any° to go beyond the sea to fight, I may have granted his Majesty's writ of 'Ne exeat regnum'° to stop him, for this giant bestrideth the sea,° and I would take and snare him by the foot on this side; for the combination and plotting is on this side, though it should be acted beyond sea. And your Lordship said notably the last time I made a motion° in this business, that a man may be as well 'fur de se', as 'felo de se',° if he steal out of the realm for a bad purpose; and for the satisfying° of the words of the writ, no man will doubt but he doth 'machinari contra coronam' (as the words of the writ be) that seeketh to murther a subject; for that is ever 'contra coronam et dignitatem'.° I have also a suit° to your Lordships all in general, that for justice sake, and for true honour's sake, honour of religion, law, and the King our master, against this fond° and false disguise or puppetry° of honour, I may in my prosecution, which it is like enough may sometimes stir coals° (which I esteem not for my particular, but as it may hinder the good service), I may (I say) be countenanced° and assisted from your Lordships. Lastly, I have a petition to the noblesse° and gentlemen of England, that they would learn to esteem themselves at a just price. 'Non hos quaesitum munus in usus,'° their blood is not to be spilt like water or a vile thing; therefore that they would rest persuaded there cannot be a form of honour, except it be upon a worthy matter. But for this, 'ipsi viderint,'° I am resolved.

The Charge against Somerset for poisoning Overbury°

It may please your Grace, my Lord High Steward of England, and you my Lords the Peers; you have here before you Robert Earl of Somerset, to be tried for his life, concerning the procuring and consenting to the impoisonment of Sir Thomas Overbury, then the King's prisoner in the Tower of London, as an accessary° before the fact.°

I know your Lordships cannot behold this nobleman, but you must remember his great favour with the King,° and the great place° that he hath had and borne, and must be sensible° that he is yet of your number and body, a Peer as you are; so as you cannot cut him off from your body but with grief; and therefore that you will expect from us that give in the King's evidence sound and sufficient matter of proof to satisfy your honours and consciences.

And for the manner of the evidence also, the King our master (who among his other virtues excelleth in that virtue of the imperial throne, which is justice)° hath given us commandment that we should not expatiate° nor make invectives,° but materially° pursue the evidence, as it conduceth to the point in question; a matter that (though we are glad of so good a warrant) yet we should have done of ourselves: for far be it from us, by any strains° of wit or art, to seek to play prizes,° or to blazon° our names in blood, or to carry the day otherwise than upon just grounds. We shall carry the lanthorn° of justice (which is the evidence) before your eyes upright: to be able° to save it from being put out with any winds of evasions or vain defences, that is our part, and within that we shall contain ourselves; not doubting at all but that this evidence in itself will carry that force as it shall little need vantages° or aggravations.

My Lords, the course which I shall hold in delivering that which I shall say, (for I love order),° is this:

First, I will speak somewhat of the nature and greatness of the offence which is now to be tried; not to weigh down my Lord with the greatness of it, but rather contrariwise to shew that a great offence needs a great proof, and that the King, however he might use this gentleman heretofore as the signet upon his finger (to use the

Scripture-phrase)° yet in this case could not but put him off,° and deliver him into the hands of justice.

Secondly, I will use some few words touching the nature of the proofs which in such a case are competent.

Thirdly, I will state the proofs.

And lastly, I will produce the proofs, either out of the examinations and matters in writing, or witnesses *viva voce*.

For the offence itself; it is of crimes (next unto high treason) the greatest; it is the foulest of felonies. And take this offence with the circumstances, it hath three degrees or stages; that it is murder; that it is murder by impoisonment; that it is murder committed upon the King's prisoner in the Tower: I might say, that it is murder under the colour° of friendship; but that is a circumstance moral;° I leave that to the evidence itself.

For murder, my Lords, the first record of justice which was in the world was a judgment upon murder, in the person of Adam's first-born, Cain;° and though it were not punished by death, but with banishment and mark of ignominy, in respect of the primogeniture,° or of the population° of the world, or other points of God's secret will, yet it was adjudged, and was (as I said) the first record of justice. So it appeareth likewise in Scripture, that the murder of Abner by Joab,° though it were by David respited° in respect of great services past, or reason of state, yet it was not forgotten. But of this I will say no more, because I will not discourse.° It was ever admitted, and so ranked in God's own tables,° that murder is of offences between man and man (next to treason and disobedience of authority, which some divines have referred to the first table,° because of the lieutenancy° of God in princes and fathers) the greatest.

For impoisonment, I am sorry it should be heard of in this kingdom: it is not 'nostri generis nec sanguinis':° it is an Italian° crime, fit for the court of Rome,° where that person that intoxicateth the Kings of the earth with his cup of poison in heretical doctrine, is many times really and materially intoxicated and impoisoned himself.

But it hath three circumstances, which make it grievous° beyond other murders. Whereof the first is, that it takes a man in full peace; in God's and the King's peace: He thinks no harm, but is comforting nature with refection° and food; so that (as the Scripture saith) 'his table is made a snare.'°

The second is, that it is easily committed, and easily concealed; and on the other side, hardly° prevented, and hardly discovered. For murder by violence princes have guards, and private men have houses, attendants, and arms: neither can such murders be committed but 'cum sonitu',° and with some overt and apparent act that may discover° and trace the offender. But for poison, the say-cup° itself of princes will scarce serve, in regard of many poisons that neither discolour nor distaste;° and so passeth without noise or observation.

And the last is, because it containeth not only the destruction of the maliced man, but of any other; 'Quis modo tutus erit?', for many times the poison is prepared for one and is taken by another: so that men die other men's deaths, 'concidit infelix alieno vulnere':° and it is, as the Psalm calleth it, 'sagitta nocte volans';° 'the arrow that flies by night',° it hath no aim or certainty. And therefore if any man shall say to himself here is great talk of impoisonment but I am sure I am safe, for I have no enemies, neither have I anything that another man should long for, why that is all one:° he may sit next him at the table that is meant to be poisoned and pledge° him of his cup; as we may see by an example in the 21st year of King Henry the eighth,° that where the purpose was to poison one man there was poison put into barm° or yeast, and with that barm pottage° or gruel was made, whereby sixteen of the Bishop of Rochester's servants were impoisoned; nay it went into the alms basket° likewise, and the poor at the gate were poisoned. And therefore with great judgment did the statute made that year touching this accident° make this impoisoning high treason, because it tendeth to the dissolution of human society; for whatsoever offence doth so, is in the nature thereof high treason.

Now for the third degree of this particular offence, which is that it was committed upon the King's prisoner, who was out of° his own defence and merely in the King's protection, and for whom the King and state was a kind of respondent,° it is a thing that aggravates the fault much. For certainly (my Lord of Somerset) let me tell you this, that Sir Thomas Overbury is the first man that was murdered in the Tower of London, since the murder of the two young princes.° Thus much of the offence, now to the proofs.

For the nature of the proofs, your Lordships must consider that impoisonment of offences is the most secret; so secret, as if in all cases of impoisonment you should require testimony, you were as good proclaim impunity.° I will put book-examples.°

Who could have impeached Livia,° by testimony, of the impoisoning of the figs upon the tree, which her husband was wont for his pleasure to gather with his own hands?

Who could have impeached Parysatis° for the poisoning of one side of the knife that she carved with, and keeping the other side clean; so that herself did eat of the same piece of meat that the lady did that she did impoison?

The cases are infinite (and indeed not fit to be spoken of) of the secrecy of impoisonments; but wise triers° must take upon them, in these secret cases, Salomon's spirit,° that, where there could be no witnesses, collected° the act by the affection.°

But yet we are not at our case: for that which your Lordships are to try is not the act of impoisonment, (for that is done to your hand): all the world by law is concluded° to say that Overbury was impoisoned by Weston. But the question before you is of the procurement° only, and of the abetting° (as the law termeth it), as accessary before the fact: which abetting is no more but to do or use any act or means which may aid or conduce unto the impoisonment.

So that it is not the buying or making of the poison, or the preparing, or confecting,° or commixing° of it, or the giving or sending or laying the poison, that are the only acts that do amount unto abetment. But if there be any other act or means done or used to give the opportunity of impoisonment, or to facilitate the execution of it, or to stop or divert any impediments that might hinder it, and this be with an intention to accomplish and achieve the impoisonment; all these are abetments, and accessaries before the fact. I will put you a familiar° example. Allow there be a conspiracy to murder a man as he journeys by the ways,° and it be one man's part to draw him forth to that journey by invitation, or by colour° of some business; and another takes upon him to dissuade some friend of his, whom he had a purpose to take in his company, that° he be not too strong to make his defence; and another hath the part to go along with him, and to hold him in talk till the first blow be given: all these (my Lords) without scruple are abettors to this murder, though none of them give the blow, nor assist to give the blow.

My Lords, he is not the hunter alone that lets slip° the dog upon the deer, but he that lodges° the deer, or raises° him, or puts him out,° or he that sets a toil° that he cannot escape, or the like.

But this (my Lords) little needeth in this present case, where there is such a chain of acts of impoisonment as hath been seldom seen, and could hardly have been expected, but that greatness of fortune maketh commonly grossness° in offending.

To descend to the proofs themselves, I shall keep this course.

First, I will make a narrative or declaration of the fact itself.

Secondly, I will break° and distribute the proofs, as they concern the prisoner.

And thirdly, according to that distribution, I will produce them, and read them, or use them.

So that there is nothing that I shall say, but your Lordship (my Lord of Somerset) shall have three thoughts or cogitations to answer it: First, when I open° it, you may take your aim.° Secondly, when I distribute it, you may prepare your answers without confusion.° And lastly, when I produce the witnesses or examinations themselves, you may again ruminate and re-advise how to make your defence. And this I do the rather, because your memory or understanding may not be oppressed or overladen with length of evidence, or with confusion of order. Nay more, when your Lordship shall make your answers in your time, I will put you in mind (when cause shall be) of your omissions.

First, therefore, for the simple narrative of the fact. Sir Thomas Overbury for a time was known to have had great interest° and great friendship with my Lord of Somerset, both in his meaner fortunes and after; insomuch as he was a kind of oracle of direction° unto him; and if you will believe his own vaunts (being of an insolent Thrasonical° disposition), he took upon him that the fortune, reputation, and understanding of this gentleman (who is well known to have had a better teacher)° proceeded from his company and counsel.

And this friendship rested not only in conversation and business of court, but likewise in communication of secrets of estate.° For my Lord of Somerset, at that time exercising (by his Majesty's special favour and trust) the office of the Secretary° provisionally, did not forbear to acquaint Overbury with the King's packets of dispatches from all parts, Spain, France, the Low Countries, etc. And this not by glimpses, or now and then rounding° in the ear for a favour, but in a settled° manner: packets were sent, sometimes opened by my Lord, sometimes unbroken, unto Overbury, who perused them, copied, registered them, made tables° of them as he thought good: so that I will undertake the time was when Overbury knew more of the secrets of state than the Council-table° did. Nay, they were grown to such an inwardness, as they made a play° of all the world besides themselves: so as they had ciphers° and jargons° for the King, the Queen, and all the great men; things seldom used, but either by princes and their ambassadors and ministers, or by such as work° and practise° against, or at least upon princes.

But understand me (my Lord) I shall not charge you this day with any disloyalty; only I lay this for a foundation, that there was a great communication of secrets between you and Overbury, and that it had relation to matters of estate, and the greatest causes of this kingdom.

But (my Lords) as it is a principle in nature, that the best things are in their corruption the worst, and the sweetest wine makes the sharpest vinegar;° so fell it out with them, that this excess (as I may term it) of friendship ended in mortal hatred on my Lord of Somerset's part.

For it fell out, some twelve months before Overbury's imprisonment in the Tower, that my Lord of Somerset was entered into an unlawful love towards his° unfortunate Lady, then Countess of Essex: which went so far, as it was then secretly projected° (chiefly between my Lord Privy Seal° and my Lord of Somerset) to effect a nullity in the marriage with my Lord of Essex, and so to proceed to a marriage with Somerset.

This marriage and purpose did Overbury mainly oppugn,° under pretence to do the true part of a friend (for that he counted° her an unworthy woman); but the truth was that Overbury, who (to speak plainly) had little that was solid for religion or moral virtue, but was a man possessed with ambition and vain-glory, was loth to have any partners in the favour of my Lord of Somerset, and especially not the house of the Howards,° against whom he had always professed hatred and opposition. So all was but miserable bargains° of ambition.

And (my Lords) that this is no sinister construction, will well appear unto you, when you shall hear that Overbury makes his brags to my Lord of Somerset, that he had won him the love of the lady by his letters and industry: so far was he from cases of conscience in this matter. And certainly (my Lords) howsoever the tragical misery of that poor gentleman Overbury ought somewhat to obliterate his faults; yet because we are not now upon point of civility,° but to discover the face of truth to the face of justice; and that it is material to the true understanding of the state of this cause; Overbury was naught° and corrupt, the ballads° must be amended for that point.

But to proceed; when Overbury saw that he was like to be dispossessed of my Lord here, whom he had possessed so long, and by whose greatness he had promised himself to do wonders; and being a man of an unbounded and impetuous spirit, he began not only to dissuade, but to deter him from that love and marriage; and finding him fixed, thought to try stronger remedies, supposing that he had my Lord's head under his girdle,° in respect of communication of secrets of estate (or, as he calls them himself in his letters, secrets of all natures); and therefore dealt violently with him to make him desist, with menaces of discovery of secrets,° and the like.

Hereupon grew two streams of hatred upon Overbury; the one from the Lady, in respect that he crossed° her love and abused her name, which are furies° to women; the other of a deeper and more mineral° nature, from my Lord of Somerset himself; who was afraid of Overbury's nature, and that if he did break from him and fly out, he would mine into° him and trouble his whole fortunes.

I might add a third stream from the Earl of Northampton's ambition, who desires to be first in favour with my Lord of Somerset; and knowing Overbury's malice to himself and his house, thought that man must be removed and cut off. So it was amongst them resolved and decreed that Overbury must die.

Hereupon they had variety of devices. To send him beyond sea, upon occasion of employment, that was too weak; and they were so far from giving way to it, as they crossed° it. There rested but two ways, quarrel or assault, and poison. For that of assault, after some proposition and attempt, they passed from it; it was a thing too open, and subject to more variety of chances. That of poison likewise was a hazardous thing, and subject to many preventions° and cautions, especially to such a jealous and working brain as Overbury had, except he were first fast in their hand.

Therefore the way was first to get him into a trap, and lay him up,° and then they could not miss the mark. Therefore in execution of this plot it was devised, that Overbury should be designed to some honourable imployment in foreign parts, and should under-hand° by the Lord of Somerset be incouraged to refuse it; and so upon that contempt° he should be laid prisoner in the Tower, and then they would look he should be close° enough, and death should be his bail. Yet were they not at their end. For they considered that if there was not a fit lieutenant of the Tower for their purpose, and likewise a fit under-keeper of Overbury; first, they should meet with many impediments in the giving and exhibiting° the poison: secondly, they should be exposed to note° and observation, that might discover them: and thirdly, Overbury in the meantime might write clamorous and furious letters to other his friends, and so all might be disappointed. And therefore the next link of the chain was to displace the then lieutenant Waade, and to place Helwisse, a principal abettor in the impoisonment: again, to displace Cary, that was the under-keeper in Waade's time, and to place Weston, who was the principal actor in the impoisonment: and this was done in such a while, that it may appear to be done as it were with one breath; as there were but fifteen days° between the commitment of Overbury, the displacing of Waade, the placing of Helwisse, the displacing of Cary the under-keeper, the placing of Weston, and the first poison given two days after.

Then when they had this poor gentleman in the Tower close prisoner, where he could not escape nor stir, where he could not feed but by their hands, where he could not speak nor write but through their trunks;° then was the time to execute the last act of this tragedy.

Then must Franklin be purveyor of the poisons, and procure five, six, seven several° potions, to be sure to hit his complexion.° Then must Mistress Turner be the say-mistress° of the poisons to try upon poor beasts, what's present,° and what works at distance of time. Then must Weston be the tormentor, and chase° him with poison after poison; poison in salts, poison in meats, poison in sweetmeats, poison in medicines and vomits, until at last his body was almost come, by use of poisons, to the state that Mithridates'° body was by the use of treacle° and preservatives,° that the force of the poisons were blunted upon him: Weston confessing, when he was chid° for not dispatching him, that he had given him enough to poison twenty men. Lastly, because all this asked time, courses were taken by Somerset both to divert all means of Overbury's delivery, and to entertain Overbury by continual letters, partly of hopes and projects for his delivery, and partly of other fables and negotiations; somewhat like some kind of persons° (which I will not name) which keep men in talk of fortune-telling, when they have a felonious meaning.

And this is the true narrative of this act of impoisonment, which I have summarily recited.

Now for the distribution of the proofs, there are four heads of proofs to prove you guilty (my Lord of Somerset) of this impoisonment; whereof two are precedent to the impoisonment, the third is present, and the fourth is following or subsequent. For it is in proofs as it is in lights,° there is a direct light, and there is a reflexion of light, or back-light.

The first head or proof thereof is, That there was a root of bitterness, a mortal malice or hatred, mixed with deep and bottomless fears, that you had towards Sir Thomas Overbury.

The second is, That you were the principal actor, and had your hand in all those acts which did conduce to the impoisonment, and which gave opportunity and means to effect it; and without which the impoisonment could never have been, and which could serve or tend to no other end but to the impoisonment.

The third is, That your hand was in the very impoisonment itself, which is more than needs to be proved; that you did direct poison, that you did deliver poison, that you did continually hearken to the success of the impoisonment and that you spurred it on and called for dispatch° when you thought it lingered.

And lastly, That you did all the things after the impoisonment, which may detect° a guilty conscience, for the smothering of it and avoiding punishment for it: which can be but of three kinds: That you suppressed, as much as in you was, testimony: That you did deface, and destroy, and clip° and misdate° all writings that might give light to the impoisonment; and that you did fly to the altar of guiltiness, which is a pardon,° and a pardon of murder, and a pardon for yourself, and not for your Lady.

In this (my Lord) I convert° my speech to you, because I would have you attend the points of your charge, and so of your defence, the better. And two of these heads° I have taken to myself, and left the other two to the King's two Serjeants.°

For the first main part, which is the mortal hatred, coupled with fear, that was in my Lord of Somerset towards Overbury, although he did palliate it with a great deal of hypocrisy and dissimulation even to the end; I shall prove it (my Lord Steward, and you my Lords and Peers) manifestly, by matter both of oath and writing. The root of this hatred was that that hath cost many a man's life; that is, fear of discovering secrets: secrets (I say) of a high and dangerous nature. Wherein the course that I will hold shall be this.

First, I will shew that such a breach and malice was between my Lord and Overbury, and that it burst forth into violent menaces and threats on both sides.

Secondly, That these secrets were not light,° but of a high nature; for I will give you the elevation of the pole.° They were such as my Lord of Somerset for his part had made a vow, that Overbury should neither live in court nor country. That he had likewise opened himself and his own fears so far, that if Overbury ever came forth of the Tower, either Overbury or himself must die for it. And of Overbury's part, he had threatened my Lord, That whether he did live or die, my Lord's shame should never die, but he would leave him the most odious man of the world. And further that my Lord was like enough to repent it in the place where Overbury wrote, which was the Tower of London. He was a true prophet in that. So here is the height of the secrets.

Thirdly, I will shew you that all the King's business was by my Lord put into Overbury's hands: So as there is work enough for secrets, whatsoever they were. And like Princes' confederates, they had their ciphers and jargons.

And lastly I will shew you that it is but a toy° to say that the malice was only in respect he spake dishonourably of the Lady; or for doubt° of breaking the marriage: for that Overbury was a coadjutor° to that love, and the Lord of Somerset was as deep in speaking ill of the Lady as Overbury. And again, it was too late for that matter, for the bargain of the match was then made and past. And if it had been no more but to remove Overbury from disturbing of the match, it had been an easy matter to have banded over Overbury beyond seas, for which they had a fair way; but that would not serve their turn.

And lastly, 'periculum periculo vincitur',° to go so far as an impoisonment must have a deeper malice than flashes:° for the cause must bear a proportion to the effect.

For the next general head of proofs, which consists in acts preparatory to the middle acts,° they are in eight several points of the compass, as I may term it.

First, That there were devices and projects to despatch Overbury, or to overthrow him, plotted between the Countess of Somerset, the Earl of Somerset, and the Earl of Northampton, before they fell upon the impoisonment: for always before men fix upon a course of mischief there be some rejections; but die he must one way or other.

Secondly, That my Lord of Somerset was principal practiser (I must speak it) in a most perfidious manner to set a train° or trap for Overbury to get him into the Tower; without which they never durst have attempted the impoisonment.

Thirdly, That the placing of the Lieutenant Helwisse, one of the impoisoners, and the displacing of Waade, was by the means of my Lord of Somerset.

Fourthly, That the placing of Weston the under-keeper, who was the principal impoisoner, and the displacing of Cary, and the doing of all this within fifteen days after Overbury's commitment, was by the means and countenance of my Lord of Somerset. And these two were the active instruments of the impoisonment: and this was a business that the Lady's power could not reach unto.

Fifthly, That because there must be a time for the tragedy to be acted, and chiefly because they would not have the poisons work upon the sudden; and for that the strength of Overbury's nature, or the very custom of receiving poison into his body, did overcome the poisons that they wrought° not so fast; therefore Overbury must be held in the Tower. And as my Lord of Somerset got him into the trap, so he kept him in, and abused him with continual hopes of liberty; and diverted all the true and effectual means of his liberty, and made light of his sickness and extremities.°

Sixthly, That not only the plot of getting Overbury into the Tower, and the devices to hold him and keep him there, but the strange manner of his close keeping (being in but for a contempt) was by the device and means of my Lord of Somerset, who denied° his father to see him, denied his servants that offered to be shut up close prisoners with him, and in effect handled it so that he was close prisoner to all his friends, and open and exposed to all his enemies.

Seventhly, That the advertisements° which my Lady received from time to time from the Lieutenant or Weston touching Overbury's state of body or health, were ever sent up to the Court, though it were in progress,° and that from my Lady: such a thirst and listening they had to hear that he was despatched.

Lastly, there was a continual negotiation to set Overbury's head on work,° that he should make some recognition° to clear the honour of the Lady, and that he should become a good instrument towards her and her friends: all which was but entertainment:° for your Lordships shall plainly see divers of my Lord of Northampton's letters (whose hand was deep in this business) written (I must say it) in dark° words and clauses; that there was one thing pretended, another intended; that there was a real charge and there was somewhat not real; a main drift and a dissimulation. Nay further, there are some passages which the Peers in their wisdom will discern to point directly at the impoisonment.

[After this introduction followed the evidence itself. That concluded, the petition which Somerset had sent to the King in February was read, and with the following remarks on it Bacon closed the case for the prosecution.]

Mr. Attorney. You see, my Lords, in this declaration of my Lord Somerset there is a brink of confession; I would to God it had a bottom.° He urges that in respect he hath formerly been so great in the King's favour, and had never committed any treason, neither against his person nor state, that he should never have been called to an account for this fault, though he had been guilty: That grace timely given is a benefit; and that it is not only enough to give life, but to save reputation. But if he must be urged,° then he desires his wife might be pardoned, having confessed the fact: And that if he must be put upon the hazard of a trial, the King will before give him leave to dispose of his lands and goods to the use of his wife and child; and that in the meantime he will give my Lord Hay and Sir R. Carr leave to come to him.

Mr. Serjeant Crew. This declaration is an implicative° confession.

Mr. Attorney. I think there is none° here but wonders, seeing that all poisons be works of darkness, how this should so clearly appear. But it seems his greatness in fortune caused this grossness in offending.

[Somerset made his defence but was unable to explain away much incriminating evidence. Bacon then addressed the court.]

It hath, my Lord, formerly at arraignments been a custom after the King's counsel and the prisoner's defence hath been heard, briefly to sum up what hath been said: but in this we have been so formal° in the distribution that I do not think it necessary. And therefore now there is no more to be done, but that the Peers will be pleased to confer, and the prisoner to withdraw until the censures be past.

[The Peers withdrew to consider their verdict, and returned after an hour with the unanimous judgment, 'Guilty'. Bacon then concluded the trial.]

My Lord High Steward, Robert Earl of Somerset hath been indicted and arraigned, and put himself upon his Peers, who all without the difference of one voice have found him Guilty. I pray judgment.

Letter to King James I, March 1621

It may please your most excellent Majesty,

Time hath been when I have brought unto you *gemitum columbae*° from others. Now I bring it from myself. I fly unto your Majesty with the wings of a dove, which once within these seven days I thought would have carried me a higher flight.°

When I enter into myself,° I find not the materials° of such a tempest as is comen upon me. I have been (as your Majesty knoweth best) never author of any immoderate° counsel, but always desired to have things carried *suavibus modis.*° I have been no avaricious oppressor of the people. I have been no haughty or intolerable or hateful man, in my conversation° or carriage.° I have inherited no hatred from my father, but am a good patriot born. Whence should this be? For these are the things that use to° raise dislikes abroad.

For the House of Commons, I began my credit° there, and now it must be the place of the sepulture° thereof; and yet this Parliament, upon the message touching religion, the old love revived, and they said I was the same man still, only honesty was turned into honour.

For the Upper House, even within these days before these troubles, they seemed as to take me into their arms, finding in me ingenuity° which they took to be the true straight line of nobleness, without crooks or angles.

And for the briberies and gifts wherewith I am charged, when the books of hearts shall be opened, I hope I shall not be found to have the troubled fountain of a corrupt heart in a depraved habit of taking rewards to pervert justice; howsoever I may be frail, and partake of the abuse of the times.

And therefore I am resolved when I come to my answer, not to trick up° my innocency (as I writ to the Lords) by cavillations° or voidances,° but to speak to them the language that my heart speaketh to me, in excusing, extenuating, or ingenuous confessing; praying to God to give me the grace to see to the bottom of my faults, and that no hardness of heart° do steal upon me, under shew of more neatness of conscience than is cause.

But not to trouble your Majesty longer, craving pardon for this long mourning letter; That which I thirst after as the hart after the streams,° is that I may know by my matchless friend that presenteth to you this letter, your Majesty's heart (which is an *abyssus*° of goodness, as I am an *abyssus* of misery) towards me. I have been ever your man, and counted myself but an usufructuary° of myself, the property being yours: and now making myself an oblation° to do with me as may best conduce to the honour of your justice, the honour of your mercy, and the use of your service, resting as

clay in your Majesty's gracious hands.

FR. ST. ALBAN, Canc.

March 25, 1621.

Letter to Lancelot Andrewes, Bishop of Winchester, 1622

My Lord,

Amongst consolations, it is not the least to represent to a man's self like examples of calamity in others.° For examples give a quicker° impression than arguments; and besides they certify us that which the Scripture also tendereth for satisfaction,° that 'no new thing is happened unto us'.° This they do the better, by how much the examples are liker° in circumstances to our own case; and more especially if they fall upon persons that are greater and worthier than ourselves. For as it savoureth° of vanity to match ourselves highly in our own conceit; so on the other side it is a good sound conclusion, that if our betters have sustained the like events, we have the less cause to be grieved.

In this kind of consolation I have not been wanting to myself, though as a Christian I have tasted (through God's great goodness) of higher remedies. Having therefore, through the variety of my reading, set before me many examples both of ancient and latter times, my thoughts (I confess) have chiefly stayed upon three particulars,° as the most eminent and the most resembling. All three, persons that had held chief place of authority in their countries; all three ruined, not by war, or by any other disaster, but by justice and sentence,° as delinquents and criminals; all three famous writers, insomuch as the remembrance of their calamity is now as to posterity but as a little picture of night-work,° remaining amongst the fair and excellent tables° of their acts and works: and all three (if that were any thing to the matter) fit examples to quench any man's ambition of rising again; for that they were every one of them restored with great glory, but to their further ruin and destruction, ending in a violent death.

The men were Demosthenes,° Cicero,° and Seneca;° persons that I durst° not claim affinity with, except the similitude of our fortunes had contracted° it. When I had cast mine eyes upon these examples, I was carried on further to observe how they did bear their fortunes, and principally how they did employ their times, being banished and disabled for public business: to the end that I might learn by them; and that they might be as well my counsellors as my comforters.

Whereupon I happened to note how diversely their fortunes wrought upon° them; especially in that point at which I did most aim, which was the employing of their times and pens. In Cicero, I saw that during his banishment (which was almost two years) he was so softened° and dejected, as he wrote nothing but a few womanish epistles.° And yet, in mine opinion, he had least reason of the three to be discouraged: for that although it was judged, and judged by the highest kind of judgment, in form of a statute or law, that he should be banished, and his whole estate confiscated and seized, and his houses pulled down, and that it should be highly penal° for any man to propound his repeal; yet his case even then had no great blot of ignominy; but it was thought but a tempest of popularity° which overthrew him. Demosthenes contrariwise, though his case was foul, being condemned for bribery, and not simple bribery, but bribery in the nature of treason and disloyalty;° yet nevertheless took so little knowledge of his fortune, as during his banishment he did much busy himself, and intermeddle with matters of state; and took upon him to counsel the state (as if he had been still at the helm) by letters; as appears by some epistles of his which are extant.° Seneca indeed, who was condemned for many corruptions and crimes, and banished into a solitary island,° kept a mean;° and though his pen did not freeze, yet he abstained from intruding into matters of business;° but spent his time in writing books of excellent argument and use for all ages;° though he might have made better choice (sometimes) of his dedications.°

These examples confirmed me much in a resolution (whereunto I was otherwise inclined) to spend my time wholly in writing; and to put forth that poor talent, or half talent, or what it is, that God hath given me,° not as heretofore to particular exchanges, but to banks or mounts° of perpetuity,° which will not break.° Therefore having not long since set forth a part of my *Instauration*;° which is the work that in mine own judgment ('si nunquam fallit imago')° I do most esteem; I think to proceed in some new parts thereof. And although I have received from many parts beyond the seas, testimonies° touching that work, such as beyond which I could not expect at the first in so abstruse an argument; yet nevertheless I have just cause to doubt that it flies too high over men's heads. I have a purpose therefore (though I break the order of time)° to draw it down to the sense,° by some patterns of a *Natural Story*° and *Inquisition*.° And again, for that my book of *Advancement of Learning* may be some preparative, or key, for the better opening of the *Instauration*; because it exhibits a mixture of new conceits and old; whereas the *Instauration* gives the new unmixed, otherwise than with some little aspersion° of the old for taste's sake; I have thought good to procure a translation of that book into the general language, not without great and ample additions, and enrichment thereof; especially in the second book, which handleth the *Partition of Sciences*;° in such sort, as I hold° it may serve in lieu of the first part of the *Instauration*, and acquit my promise in that part. Again, because I cannot altogether desert the civil person° that I have borne, which if I should forget, enough would remember; I have also entered into a work touching *Laws*,° propounding a character of justice in a middle term,° between the speculative and reverend° discourses of philosophers, and the writings of lawyers, which are tied and obnoxious° to their particular laws. And although it be true that I had a purpose to make a particular digest or recompilement of the laws of mine own nation;° yet because it is a work of assistance,° and that that I cannot master by my own forces and pen, I have laid it aside.

Now having in the work of my *Instauration* had in contemplation° the general good of men in their very being, and the dowries° of nature; and in my work of laws, the general good of men likewise in society, and the dowries of government; I thought in duty I owed somewhat unto mine own country, which I ever loved; insomuch as although my place° hath been far above my desert, yet my thoughts and cares concerning the good thereof were beyond, and over, and above my place: so now being (as I am) no more able to do my country service, it remained unto me [to] do it honour: which I have endeavoured to do in my work of *The Reign of King Henry the Seventh*.° As for my *Essays*, and some other particulars of that nature, I count them but as the recreations of my other studies, and in that sort purpose to continue them; though I am not ignorant that those kind of writings would, with less pains and embracement° (perhaps), yield more lustre and reputation to my name, than those other which I have in hand. But I account the use that a man should seek of the publishing of his own writings before his death, to be but an untimely anticipation of that which is proper to follow a man, and not to go along with him.

But revolving° with myself my writings, as well those which I have published, as those which I had in hand, methought they went all into the City, and none into the Temple;° where, because I have found so great consolation, I desire likewise to make some poor oblation.° Therefore I have chosen an argument, mixt of religious and civil considerations; and likewise mixt between contemplative and active. For who can tell whether there may not be an *exoriare aliquis*?° Great matters (especially if they be religious) have (many times) small beginnings: and the platform° may draw on° the building. This work,° because I was ever an enemy to flattering dedications, I have dedicated to your Lordship, in respect of our ancient and private acquaintance; and because amongst the men of our times I hold you in especial reverence.

Your Lordship's loving friend,

FR. ST. ALBAN.

Poems and Psalms

'The world's a bubble'

THE world's a bubble, and the life of man
less than a span;
In his conception wretched, from the womb
so to the tomb:
Curst from the cradle, and brought up to years
with cares and fears.
Who then to frail mortality shall trust,
But limns° on° water, or but writes in dust.

Yet since with sorrow here we live opprest,
what life is best?
Courts are but only superficial schools
to dandle fools.
The rural parts are turned into a den
of savage men.
And where's a city from all vice so free,
But may be term'd the worst of all the three?

Domestic cares afflict the husband's bed,
or pains his head.
Those that live single take it for a curse,
or do things worse.
These° would have children; those that have them moan,
or wish them gone.
What is it then to have or have no wife,
But single thraldom, or a double strife?

Our own affections still at home to please
is a disease:
To cross the seas to any foreign soil,
perils and toil.
Wars with their noise affright us: when they cease,
we are worse in peace.
What then remains, but that we still should cry
Not to be born, or being born to die?

The Translation of Certain Psalms
[*Selection*]

Psalm 90

O Lord, thou art our home, to whom we fly,
  And so hast always been from age to age:
Before the hills did intercept° the eye,
  Or that the frame was up of earthly stage,°
    One God thou wert, and art, and still shall be;
    The line of Time, it doth not measure thee.

Both death and life obey thy holy lore,
  And visit in their turns,° as they are sent;
A thousand years with thee they are no more
  Than yesterday, which, ere it is, is spent:
    Or as a watch by night, that course doth keep,
    And goes, and comes, unwares to them that sleep.°

Thou carriest man away as with a tide:
  Then down swim all his thoughts that mounted high:
Much like a mocking dream, that will not bide,
  But flies before the sight of waking eye;
    Or as the grass, that cannot term° obtain
    To see the summer come about again.°

At morning, fair it musters on the ground;
  At even, it is cut down and laid along:
And though it spared were and favour found,
  The weather would perform the mower's wrong:
    Thus hast thou hang'd our life on brittle pins,°
    To let us know it will not bear our sins.°

Thou buriest not within oblivion's tomb
  Our trespasses, but ent'rest them aright;
Ev'n those that are conceiv'd in darkness' womb,
  To thee appear as done at broad day-light.
    As a tale told, which sometimes men attend,°
    And sometimes not, our life steals to an end.°

The life of man is threescore years and ten,
  Or, that if he be strong, perhaps fourscore;

Yet all things are but labour to him then,
New sorrows still come on, pleasures no more.
Why should there be such turmoil and such strife,
To spin in length this feeble line of life?°

But who considers duly of thine ire?
Or doth the thoughts thereof wisely embrace?
For thou, O God, art a consuming fire:
Frail man, how can he stand before thy face?
If thy displeasure thou dost not refrain,
A moment brings all back to dust again.°

Teach us, O Lord, to number well our days,
Thereby our hearts to wisdom to apply;
For that which guides man best in all his ways,
Is meditation of mortality.
This bubble light, this vapour of our breath,
Teach us to consecrate to hour of death.°

Return unto us, Lord, and balance now
With days of joy our days of misery;
Help us right soon, our knees to thee we bow,
Depending wholly on thy clemency;
Then shall thy servants both with heart and voice,
All the days of their life in thee rejoice.°

Begin thy work, O Lord, in this our age,
Shew it unto thy servants that now live;
But to our children raise it many a stage,
That all the world to thee may glory give.
Our handy-work likewise, as fruitful tree,
Let it, O Lord, blessed, not blasted be.°

Psalm 104

FATHER and King of pow'rs, both high and low,
Whose sounding° fame all creatures serve to blow;
My soul shall with the rest strike up thy praise,
And carol of thy works and wondrous ways.°
But who can blaze° thy beauties, Lord, aright?
They turn the brittle beams of mortal sight.°
Upon thy head thou wear'st a glorious crown,

All set with virtues, polish'd with renown:
Thence round about a silver veil doth fall
Of crystal light, mother of colours all.
The compass° heaven, smooth without grain or fold,
All set with spangs of glitt'ring stars untold,
And strip'd with golden beams of power unpent,°
Is raised up for a removing tent.°
Vaulted and arched are his chamber beams
Upon the seas, the waters, and the streams:
The clouds as chariots swift do scour the sky;
The stormy winds upon their wings do fly.
His angels spirits are, that wait his will,
As flames of fire his anger they fulfil.
In the beginning, with a mighty hand,
He made the earth by counterpoise° to stand;
Never to move, but to be fixèd still;
Yet hath no pillars but his sacred will.
This earth, as with a veil, once cover'd was,
The waters over-flowed all the mass:
But upon his rebuke away they fled,
And then the hills began to shew their head;
The vales their hollow bosoms open'd plain,
The streams ran trembling down the vales again:
And that the earth no more might drownèd be,
He set the sea his bounds of liberty;
And though his waves resound, and beat the shore,
Yet it is bridled by his holy lore.
Then did the rivers seek their proper places,
And found their heads, their issues, and their races;
The springs do feed the rivers all the way,
And so the tribute to the sea repay:
Running along through many a pleasant field,
Much fruitfulness unto the earth they yield:
That know the beasts and cattle feeding by,
Which for to slake their thirst do thither hie.
Nay desert grounds the streams do not forsake,
But through the unknown ways their journey take:
The asses wild, that hide in wilderness,
Do thither come, their thirst for to refresh.
The shady trees along their banks do spring,
In which the birds do build, and sit, and sing;

Stroking the gentle air with pleasant notes,
Plaining° or chirping through their warbling throats.
The higher grounds, where waters cannot rise,
By rain and dews are water'd from the skies;
Causing the earth put forth the grass for beasts,
And garden herbs, serv'd at the greatest feasts;
And bread, that is all viands' firmament,
And gives a firm and solid nourishment;
And wine, man's spirits for to recreate;
And oil, his face for to exhilarate.°
The sappy cedars, tall like stately tow'rs,
High-flying birds do harbour in their bow'rs:
The holy storks, that are the travellers,°
Choose for to dwell and build within the firs;
The climbing goats hang on steep mountain's side;
The digging conies° in the rocks do bide.
The moon, so constant in inconstancy,
Doth rule the monthly seasons orderly;
The sun, eye of the world, doth know his race,
And when to shew, and when to hide his face.
Thou makest darkness, that it may be night,
When as the savage beasts, that fly the light,
(As conscious of man's hatred) leave their den,
And range abroad, secur'd from sight of men.
Then do the forests ring of lions roaring,
That ask their meat of God,° their strength restoring;
But when the day appears, they back do fly,
And in their dens again do lurking lie.
Then man goes forth° to labour in the field,
Whereby his grounds more rich increase may yield.
O Lord, thy providence sufficeth all;
Thy goodness, not restrained, but general
Over thy creatures: the whole earth doth flow
With thy great largeness pour'd forth here below.
Nor is it earth alone exalts thy name,
But seas and streams likewise do spread the same.
The rolling seas unto the lot doth fall°
Of beasts innumerable, great and small;
There do the stately ships plough up the floods,
The greater navies look like walking woods;
The fishes there far voyages do make,

To divers shores their journey they do take.
There hast thou set the great Leviathan,
That makes the seas to seethe like boiling pan.°
All these do ask of thee their meat to live,
Which in due season thou to them dost give.
Ope thou thy hand, and then they have good fare;
Shut thou thy hand, and then they troubled are.
All life and spirit from thy breath proceed,
Thy word doth all things generate and feed.
If thou withdraw'st it, then they cease to be,
And straight return to dust and vanity;°
But when thy breath° thou dost send forth again,
Then all things do renew and spring amain;°
So that the earth, but lately desolate,
Doth now return unto the former state.
The glorious majesty of God above
Shall ever reign in mercy and in love:
God shall rejoice all his fair works to see,
For as they come from him all perfect be.
The earth shall quake, if aught his wrath provoke;
Let him but touch the mountains, they shall smoke.°
As long as life doth last I hymns will sing,
With cheerful voice, to the eternal King;
As long as I have being, I will praise
The works of God, and all his wondrous ways.
I know that he my words will not despise,
Thanksgiving is to him a sacrifice.
But as for sinners, they shall be destroy'd
From off the earth; their places shall be void.
Let all his works praise him with one accord;
O praise the Lord, my soul; praise ye the Lord!

Psalm 126

WHEN God return'd us graciously
  Unto our native land,
We seem'd as in a dream to be,
  And in a maze to stand.°

The heathen likewise they could say:
  'The God, that these men serve,

Hath done great things for them this day,°
 Their nation to preserve.'

'Tis true, God hath pour'd out his grace
 On us abundantly;
For which we yield him psalms and praise,
 And thanks with jubilee.

O Lord, turn our captivity,
 As winds, that blow at south,
Do pour the tides with violence
 Back to the river's mouth.°

Who sows in tears shall reap in joy,
 The Lord doth so ordain;
So° that his seed be pure and good,
 His harvest shall be gain.

Psalm 137

When as we sat all sad and desolate,
 By Babylon upon the river's side,
Eas'd from the tasks which in our captive state
 We were enforced daily to abide,
  Our harps we had brought with us to the field,
  Some solace to our heavy souls to yield.

But soon we found we fail'd of our account,
 For when our minds some freedom did obtain,
Straightways the memory of Sion Mount
 Did cause afresh our wounds to bleed again;
  So that with present griefs, and future fears,
  Our eyes burst forth into a stream of tears.°

As for our harps, since sorrow struck them dumb,
 We hang'd them on the willow-trees were near;
Yet did our cruel masters to us come,
 Asking of us some Hebrew songs to hear:
  Taunting us rather in our misery,°
  Than much delighting in our melody.

Alas (said we), who can once force or frame
 His grieved and oppressed heart to sing

The praises of Jehovah's glorious name,
  In banishment, under a foreign king?
    In Sion is his seat and dwelling place,
    Thence doth he shew the brightness of his face.

Hierusalem, where God his throne hath set,
  Shall any hour absent thee from my mind?
Then let my right hand quite her skill forget,
  Then let my voice and words no passage find;
    Nay, if I do not thee prefer in all,
    That in the compass of my thoughts can fall.

Remember thou, O Lord, the cruel cry
  Of Edom's children, which did ring and sound,
Inciting the Chaldean's cruelty,
  'Down with it, down with it, even unto the ground.'°
    In that good day repay it unto them,
    When thou shalt visit thy Hierusalem.

And thou, O Babylon, shalt have thy turn
  By just revenge, and happy shall he be,
That thy proud walls and tow'rs shall waste and burn,
  And as thou didst by us, so do by thee.
    Yea, happy he, that takes thy children's bones,
    And dasheth them against the pavement stones.°

Psalm 149

O SING a new song to our God above;
  Avoid profane ones, 'tis for holy quire:
Let Israel sing songs of holy love
  To him that made them, with their hearts on fire:
    Let Sion's sons lift up their voice, and sing
    Carols and anthems to their heavenly King.

Let not your voice alone his praise forth tell,
  But move withal and praise him in the dance;
Cymbals and harps let them be tunèd well:
  'Tis he that doth the poor's estate advance:
    Do this not only on the solemn days,
    But on your secret beds your spirits raise.°

O let the saints bear in their mouth his praise;
  And a two-edged sword drawn in their hand;
Therewith for to revenge° the former days
  Upon all nations that their zeal withstand;
    To bind their kings in chains of iron strong,
    And manacle their nobles for their wrong.

Expect° the time, for 'tis decreed in Heaven,
Such honour shall unto his saints be given.

The Essays or Counsels Civil and Moral (1625)

NEWLY ENLARGED

1. OF TRUTH

'WHAT is Truth?'° said jesting° Pilate; and would not stay for an answer. Certainly there be that° delight in giddiness,° and count° it a bondage to fix a belief;° affecting° free-will in thinking, as well as in acting. And though the sects of philosophers of that kind° be gone, yet there remain certain discoursing wits° which are of the same veins,° though there be not so much blood in them as was in those of the ancients. But it is not only the difficulty and labour which men take in finding out of truth; nor again that when it is found it imposeth° upon men's thoughts; that doth bring lies in favour; but a natural though corrupt love of the lie itself. One of the latter school of the Grecians° examineth the matter, and is at a stand° to think what should be in it, that men should love lies, where neither they make for° pleasure, as with poets, nor for advantage, as with the merchant; but for the lie's sake. But I cannot tell: this same truth is a naked and open day-light,° that doth not shew the masks and mummeries° and triumphs° of the world, half so stately and daintily° as candle-lights. Truth may perhaps come to the price of a pearl, that sheweth° best by day; but it will not rise to the price of a diamond or carbuncle,° that sheweth best in varied lights. A mixture of a lie doth ever add pleasure.

Doth any man doubt, that if there were taken out of men's minds vain opinions, flattering hopes, false valuations, imaginations as one would,° and the like, but it would leave the minds of a number of men poor shrunken things, full of melancholy and indisposition,° and unpleasing to themselves? One of the Fathers,° in great severity, called poesy 'vinum daemonum'° because it filleth° the imagination; and yet it is but with the shadow of a lie. But it is not the lie that passeth through the mind, but the lie that sinketh in and settleth in it, that doth the hurt; such as we spake of before. But howsoever° these things are thus in men's depraved° judgments° and affections,° yet truth,° which only doth judge itself, teacheth that the inquiry of truth, which is the love-making or wooing of it, the knowledge of truth, which is the presence of it, and the belief° of truth, which is the enjoying of it, is the sovereign good of human nature.

The first creature° of God, in the works of the days,° was the light of the sense;° the last was the light of reason;° and his sabbath° work ever since, is the illumination of his Spirit.° First he breathed light upon the face of the matter or chaos; then he breathed light into the face of man; and still° he breatheth and inspireth light into the face of his chosen. The poet that beautified° the sect° that was otherwise inferior to the rest, saith yet° excellently well: 'It is a pleasure to stand upon the shore, and to see ships tossed upon the sea; a pleasure to stand in the window of a castle, and to see a battle and the adventures° thereof below: but no pleasure is comparable to the standing upon the vantage ground of Truth' (a hill not to be commanded,° and where the air is always clear and serene),° 'and to see the errors, and wanderings, and mists, and tempests, in the vale below'; so° always that this prospect° be with pity,° and not with swelling° or pride. Certainly, it is heaven upon earth, to have a man's mind move° in charity, rest° in providence, and turn upon the poles° of truth.

To pass from theological and philosophical truth, to the truth of civil business;° it will be acknowledged even by those that practise it not, that clear and round° dealing is the honour of man's nature; and that mixture of falsehood is like allay° in coin of gold and silver, which may make the metal work the better, but it embaseth° it. For these winding° and crooked courses are the goings° of the serpent; which goeth basely upon the belly, and not upon the feet. There is no vice that doth so cover a man with shame as to be found false and perfidious. And therefore Montaigne° saith prettily, when he inquired the reason, why the word of the lie should be such a disgrace and such an odious charge? Saith he, 'If it be well weighed,° to say that a man lieth, is as much to say, as that he is brave towards God and a coward towards men'. For a lie faces God, and shrinks from man. Surely the wickedness of falsehood and breach of faith cannot possibly be so highly° expressed, as in that it shall be the last peal° to call the judgments of God upon the generations of men; it being foretold,° that when Christ cometh, 'he shall not find faith upon the earth'.

2. Of Death°

Men fear Death, as children fear to go° in the dark;° and as that natural fear in children is increased with tales, so is the other. Certainly, the contemplation° of death, as the wages of sin° and passage to another world, is holy and religious; but the fear of it, as a tribute° due unto nature, is weak.° Yet in religious meditations there is sometimes mixture of vanity° and of superstition. You shall read in some of the friars' books of mortification,° that a man should think° with himself what the pain is if he have but his finger's end pressed or tortured, and thereby imagine what the pains of death are, when the whole body is corrupted° and dissolved;° when many times death passeth with less pain than the torture of a limb: for the most vital parts° are not the quickest of sense.° And by him° that spake only as a philosopher and natural° man, it was well said, 'Pompa mortis magis terret, quam mors ipsa'.° Groans and convulsions, and a discoloured face, and friends weeping, and blacks,° and obsequies,° and the like, shew° death terrible.

It is worthy the observing, that there is no passion in the mind of man so weak, but it mates° and masters the fear of death; and therefore death is no such terrible enemy when a man hath so many attendants about him that can win the combat of him.° Revenge triumphs over death; Love slights it; Honour aspireth to it;° Grief flieth to it; Fear pre-occupateth° it. Nay we read, after Otho° the emperor had slain himself, Pity (which is the tenderest° of affections) provoked many to die, out of mere compassion to their sovereign, and as the truest sort of followers. Nay Seneca adds niceness° and satiety: 'Cogita quamdiu eadem feceris; mori velle, non tantum fortis, aut miser, sed etiam fastidiosus potest'.° A man would° die, though he were neither valiant nor miserable, only upon° a weariness to do° the same thing so oft over and over.

It is no less worthy to observe, how little alteration in good spirits° the approaches of death make; for they appear to be the same men till the last instant. Augustus Caesar° died in a compliment; 'Livia, conjugii nostri memor, vive et vale'.° Tiberius° in dissimulation; as Tacitus saith of him, 'Jam Tiberium vires et corpus, non dissimulatio, deserebant'.° Vespasian° in a jest; sitting upon the stool,° 'Ut puto Deus fio'.° Galba° with a sentence;° 'Feri, si ex re sit populi Romani,'° holding forth his neck. Septimius Severus° in despatch;° 'Adeste si quid mihi restat agendum'.° And the like. Certainly the Stoics bestowed too much cost° upon death, and by their great preparations made it appear more fearful. Better saith he,° 'qui finem vitae extremum inter munera ponat naturae'.° It is as natural to die as to be born; and to a little infant, perhaps, the one is as painful as the other. He that dies in an earnest pursuit° is like one that is wounded in hot blood;° who, for the time, scarce feels the hurt; and therefore a mind fixed and bent upon somewhat° that is good doth avert the dolours° of death. But above all, believe it, the sweetest canticle° is, 'Nunc dimittis';° when a man hath obtained° worthy ends and expectations. Death hath° this also; that it openeth the gate to good fame, and extinguisheth envy. 'Extinctus amabitur idem'.°

3. Of Unity° in Religion

Religion being the chief band° of human society, it is a happy thing when itself is well contained° within the true band of Unity. The quarrels and divisions about religion were evils unknown to the heathen. The reason was, because the religion of the heathen consisted rather in rites and ceremonies, than in any constant° belief.° For you may imagine what kind of faith theirs was, when the chief doctors° and fathers of their church were the poets. But the true God hath this attribute, that he is a 'jealous God';° and therefore his worship and religion will endure no mixture° nor partner. We shall therefore speak a few words concerning the Unity of the Church; what are the Fruits thereof; what the Bounds;° and what the Means.°

The Fruits of Unity (next unto the well pleasing of God, which is all in all)° are two; the one towards those that are without° the church, the other towards those that are within. For the former; it is certain that heresies and schisms° are of all others the greatest scandals;° yea, more than corruption of manners.° For as in the natural body a wound or solution of continuity° is worse than a corrupt humour;° so in the spiritual. So that nothing doth so much keep men out of the church, and drive men out of the church, as breach of unity. And therefore, whensoever it cometh to that pass,° that one saith 'Ecce in deserto',° another saith 'Ecce in penetralibus';° that is, when some men seek Christ in the conventicles° of heretics, and others in an outward face of a church, that voice° had need continually to sound in men's ears, 'Nolite exire',—'Go not out'. The Doctor° of the Gentiles° (the propriety° of whose vocation drew him to have a special care of those without) saith, 'If an heathen come in, and hear you speak with several tongues,° will he not say that you are mad?'° And certainly it is little better, when atheists and profane persons do hear of so many discordant and contrary opinions in religion; it doth avert° them from the church, and maketh them 'to sit down in the chair of the scorners'.° It is but a light° thing to be vouched° in so serious a matter, but yet it expresseth well the deformity: there is a master of scoffing,° that in his catalogue of books of a feigned library sets down° this title of a book, 'The morris-dance° of Heretics'. For indeed every sect of them° hath a diverse posture° or cringe° by themselves,° which cannot but move° derision in worldlings° and depraved politiques,° who are apt to contemn° holy things.

As for the fruit towards those that are within; it is peace; which containeth infinite blessings. It establisheth° faith. It kindleth charity. The outward peace of the church distilleth° into peace of conscience. And it turneth the labours of writing and reading of controversies into treatises of mortification and devotion.

Concerning the Bounds of Unity; the true placing of them importeth exceedingly.° There appear to be two extremes. For to certain zelants° all speech of pacification is odious. 'Is it peace, Jehu? What hast thou to do with peace? turn thee behind me'.° Peace is not the matter, but following and party.° Contrariwise,° certain Laodiceans° and lukewarm persons think they may accommodate° points° of religion by middle ways, and taking part* of both, and witty° reconcilements;° as if they would make an arbitrement° between God and man. Both these extremes are to be avoided; which will be done, if the league of Christians° penned° by our Saviour himself were in the two cross° clauses thereof soundly and plainly expounded: 'He that is not with us is against us';° and again, 'He that is not against us is with us';° that is, if the points fundamental and of substance in religion were truly discerned° and distinguished from points not merely° of faith, but of opinion, order,° or good intention.° This is a thing may seem to many a matter trivial, and done already.° But if it were done less partially,° it would be embraced more generally.

Of this I may give only this advice, according to my small model.° Men ought to take heed of rending God's church by two kinds of controversies. The one is, when the matter of the point controverted is too small and light, not worth the heat and strife about it, kindled only by contradiction. For as it is noted by one of the fathers,° 'Christ's coat indeed had no seam,° but the church's vesture° was of divers colours';° whereupon he saith, 'In veste varietas sit, scissura non sit':° they be two° things, Unity and Uniformity. The other is, when the matter of the point controverted is great, but it is driven to an over-great subtilty° and obscurity; so that it becometh a thing rather ingenious° than substantial. A man that is of judgment and understanding shall sometimes hear ignorant men differ,° and know well within himself that those which so differ mean one thing,° and yet they themselves would never agree. And if it come so to pass in that distance of judgment which is between man and man, shall we not° think that God above, that knows the heart, doth not discern that frail men° in some of their contradictions intend° the same thing; and accepteth° of both? The nature of such controversies is excellently expressed by St. Paul in the warning and precept that he giveth concerning the same, 'Devita profanas vocum novitates, et oppositiones falsi nominis scientiae.'° Men create oppositions which are not;° and put them into new terms° so fixed, as° whereas the meaning ought to govern the term, the term in effect governeth the meaning. There be also two false peaces or unities: the one, when the peace is grounded° but upon an implicit° ignorance; for all colours will agree in the dark:° the other, when it is pieced up° upon a direct admission of contraries in fundamental points. For truth and falsehood, in such things, are like the iron and clay in the toes of Nebuchadnezzar's image;° they may cleave,° but they will not incorporate.°

Concerning the Means of procuring Unity; men must beware, that in the procuring or muniting° of religious unity they do not dissolve° and deface° the laws of charity and of human society. There be two swords° amongst Christians, the spiritual and temporal; and both have their due office and place in the maintenance of religion. But we may not take up the third sword, which is Mahomet's° sword, or like unto it; that is, to propagate religion by wars or by sanguinary° persecutions to force consciences; except it be in cases of overt scandal, blasphemy, or intermixture of practice° against the state; much less to nourish seditions; to authorise° conspiracies and rebellions; to put the sword into the people's hands; and the like; tending to the subversion of all government, which is the ordinance° of God. For this is but to dash the first table against the second;° and so to consider men as Christians, as we forget that they are men. Lucretius the poet, when he beheld° the act of Agamemnon, that could endure the sacrificing of his own daughter, exclaimed:

Tantum Religio potuit suadere malorum.°

What would he have said, if he had known of the massacre in France,° or the powder treason of England?° He would have been seven times more Epicure and atheist° than he was. For as the temporal sword is to be drawn with great circumspection in cases of religion; so it is a thing monstrous to put it into the hands of the common people. Let that be left unto the Anabaptists,° and other furies.° It was great blasphemy when the devil said, 'I will ascend and be like the Highest';° but it is greater blasphemy to personate° God, and bring him in° saying, 'I will descend, and be like the prince of darkness':° and what is it better, to make the cause of religion to descend to the cruel and execrable actions of murthering princes, butchery of people, and subversion of states and governments? Surely this is to bring down the Holy Ghost, instead of the likeness of a dove,° in the shape of a vulture or raven; and set out of the bark° of a Christian church a flag of a bark of pirates and Assassins.° Therefore it is most necessary that the church by doctrine° and decree, princes by their sword, and all learnings,° both Christian and moral, as° by their Mercury rod,° do damn and send to hell for ever those facts° and opinions tending to the support of the same;° as hath been already in good part done. Surely in counsels° concerning religion, that counsel of the apostle would° be prefixed, 'Ira hominis non implet justitiam Dei'.° And it was a notable observation of a wise father,° and no less ingenuously° confessed; that 'those which held and persuaded° pressure of consciences, were commonly interested° therein themselves for their own ends'.

4. Of Revenge°

Revenge is a kind of wild° justice; which the more man's nature runs to, the more ought law to weed it out. For as for the first wrong, it doth but° offend the law; but the revenge of that wrong putteth the law out of office.° Certainly, in taking revenge, a man is but even with his enemy; but in passing it over, he is superior; for it is a prince's part° to pardon. And Salomon, I am sure, saith, 'It is the glory of a man to pass by an offence'.° That which is past is gone, and irrevocable; and wise men have enough to do with things present and to come; therefore they do but trifle with themselves, that labour in past matters. There is no man doth a wrong for the wrong's sake; but thereby to purchase° himself profit, or pleasure, or honour, or the like. Therefore why should I be angry with a man for loving himself better than me? And if any man should do wrong merely° out of ill-nature, why, yet it is but like the thorn° or briar, which prick and scratch, because they can do no other. The most tolerable sort of revenge is for those wrongs° which there is no law to remedy; but then let a man take heed° the revenge be such as there is no law to punish; else a man's enemy is still before hand,° and it is two for one.°

Some, when they take revenge, are desirous the party° should know whence it cometh.° This the more generous. For the delight seemeth to be not so much in doing the hurt as in making the party repent. But base and crafty cowards are like the arrow that flieth in the dark.° Cosmus,° duke of Florence, had a desperate° saying against perfidious or neglecting° friends, as if those wrongs were unpardonable; 'You shall read' (saith he) 'that we are commanded to forgive our enemies; but you never read that we are commanded to forgive our friends'. But yet the spirit of Job was in a better tune:° 'Shall we' (saith he) 'take good at God's hands, and not be content to take evil also?'° And so of friends in a proportion.° This is certain, that a man that studieth° revenge keeps his own wounds green,° which otherwise would heal and do well. Public revenges are for the most part fortunate;° as that for the death of Caesar;° for the death of Pertinax;° for the death of Henry the Third° of France; and many more. But in private revenges it is not so. Nay rather, vindictive persons live the life of witches; who, as they are mischievous,° so end they infortunate.

5. Of Adversity

It was a high° speech of Seneca° (after° the manner of the Stoics), that 'the good things which belong to prosperity are to be wished;° but the good things that belong to adversity are to be admired'.° 'Bona rerum secundarum optabilia; adversarum mirabilia.' Certainly if miracles be° the command over nature,° they appear most in adversity. It is yet a higher° speech of his than the other (much too high° for a heathen), 'It is true greatness to have in one° the frailty of a man, and the security° of a God'. 'Vere magnum habere fragilitatem hominis, securitatem Dei'.° This would have done better in poesy, where transcendences° are more allowed. And the poets indeed have been busy with it; for it is in effect the thing which is figured° in that strange fiction° of the ancient poets, which seemeth not to be without mystery;° nay, and to have some approach to the state of a Christian; that 'Hercules, when he went to unbind Prometheus' (by whom human nature is represented), 'sailed the length of the great ocean in an earthen pot or pitcher';° lively° describing Christian resolution, that saileth in the frail bark of the flesh thorough° the waves of the world.

But to speak in a mean.° The virtue of Prosperity is temperance, the virtue of Adversity is fortitude; which in morals° is the more heroical virtue. Prosperity is the blessing of the Old Testament; Adversity is the blessing of the New; which carrieth the greater benediction,° and the clearer revelation of God's favour. Yet even in the Old Testament, if you listen to David's harp,° you shall hear as many hearse-like airs° as carols;° and the pencil of the Holy Ghost hath laboured more in describing the afflictions of Job than the felicities° of Salomon. Prosperity is not without many fears and distastes;° and Adversity is not without comforts and hopes. We see in needle-works and embroideries, it is more pleasing to have a lively work° upon a sad° and solemn ground,° than to have a dark and melancholy work upon a lightsome° ground: judge therefore of the pleasure of the heart by the pleasure of the eye. Certainly virtue is like precious odours, most fragrant when they are incensed° or crushed:° for Prosperity doth best discover° vice, but Adversity doth best discover virtue.

6. Of Simulation° and Dissimulation°

Dissimulation is but a faint° kind of policy° or wisdom; for it asketh° a strong wit° and a strong heart to know when to tell truth, and to do it. Therefore it is the weaker sort of politiques° that are the great dissemblers.

Tacitus° saith, 'Livia sorted° well with the arts° of her husband and dissimulation of her son'; attributing arts or policy to Augustus, and dissimulation to Tiberius. And again, when Mucianus° encourageth Vespasian to take arms against Vitellius, he saith, 'We rise not against the piercing judgment of Augustus, nor the extreme caution or closeness° of Tiberius'. These properties,° of arts or policy and dissimulation or closeness, are indeed habits and faculties several,° and to be distinguished. For if a man have that penetration of judgment as° he can discern what things are to be laid open, and what to be secreted,° and what to be shewed at half lights,° and to whom and when (which indeed are arts of state° and arts of life, as Tacitus° well calleth them), to him a habit of dissimulation is a hinderance and a poorness.° But if a man cannot obtain to° that judgment, then it is left to him generally to be close,° and a dissembler. For where a man cannot choose or vary in particulars,° there it is good to take the safest and wariest way in general; like the going softly,° by one that cannot well° see. Certainly the ablest men that ever were have had all an openness° and frankness of dealing; and a name of certainty° and veracity; but then they were like horses well managed;° for they could tell passing° well when to stop or turn; and at such times when they thought the case indeed required dissimulation, if then they used it, it came to pass that the former opinion° spread abroad of their good faith and clearness of dealing made them almost invisible.

There be three degrees of this hiding and veiling of a man's self. The first, Closeness,° Reservation,° and Secrecy; when a man leaveth himself without observation, or without hold to be taken, what he is.° The second, Dissimulation, in the negative; when a man lets fall signs and arguments,° that he is not that° he is. And the third, Simulation,° in the affirmative; when a man industriously and expressly° feigns and pretends to be that he is not.

For the first° of these, Secrecy; it is indeed the virtue of a confessor. And assuredly the secret man° heareth many confessions. For who will open himself to a blab° or babbler?° But if a man be thought secret, it inviteth discovery;° as the more close air sucketh in the more open;° and as in confession the revealing is not for worldly use, but for the ease of a man's heart, so secret men come to the knowledge of many things in that kind;° while° men rather discharge° their minds than impart° their minds. In few words, mysteries are due to secrecy.° Besides (to say truth) nakedness is uncomely,° as well in mind as body; and it addeth no small reverence° to men's manners and actions, if they be not altogether open. As for talkers° and futile° persons, they are commonly vain° and credulous withal.° For he that talketh what he knoweth, will also talk what he knoweth not. Therefore set it down,° that an habit of secrecy is both politic° and moral. And in this part,° it is good that a man's face give his tongue leave° to speak. For the discovery° of a man's self by the tracts° of his countenance is a great weakness and betraying; by how much° it is many times more marked° and believed than a man's words.

For the second, which is Dissimulation; it followeth many times upon° secrecy by a necessity; so that he that will be secret must be a dissembler in some degree. For men are too cunning to suffer° a man to keep an indifferent carriage° between both, and to be secret, without swaying the balance on either side. They will so beset° a man with questions, and draw him on, and pick it out of him, that, without an absurd° silence, he must shew an inclination one way; or if he do not, they will gather as much by his silence as by his speech. As for equivocations,° or oraculous speeches,° they cannot hold out long. So that no man can be secret, except he give himself a little scope° of dissimulation; which is, as it were, but the skirts or train° of secrecy.

But for the third degree, which is Simulation and false profession;° that I hold more culpable, and less politic; except it be in great and rare matters. And therefore a general custom of simulation (which is this last degree) is a vice, rising° either of° a natural falseness or fearfulness,° or of a mind that hath some main° faults, which because a man must needs disguise, it maketh him practise simulation in other things, lest his hand should be out of ure.°

The great advantages of simulation and dissimulation are three. First, to lay asleep opposition, and to surprise.° For where a man's intentions are published,° it is an alarum° to call up all that are against them. The second is, to reserve to a man's self a fair° retreat. For if a man engage° himself by a manifest° declaration, he must go through° or take a fall.° The third is, the better to discover the mind of another. For to him that opens himself men will hardly shew themselves adverse; but will (fair)° let him go on, and turn their freedom of speech to freedom of thought.° And therefore it is a good shrewd proverb° of the Spaniard, 'Tell a lie and find a truth.' As if there were no way of discovery but by simulation. There be also three disadvantages, to set it even.° The first, that simulation and dissimulation commonly carry with them a shew° of fearfulness, which in any business doth spoil the feathers of round flying up to the mark.° The second, that it puzzleth and perplexeth° the conceits° of many, that perhaps would otherwise co-operate with him; and makes a man walk almost alone to his own ends. The third and greatest, is, that it depriveth a man of one of the most principal instruments for action; which is trust and belief.° The best composition° and temperature° is to have openness in fame and opinion;° secrecy in habit;° dissimulation in seasonable use;° and a power to feign, if there be no remedy.

7. Of Parents and Children

The joys of parents are secret; and so are their griefs and fears. They cannot utter° the one; nor they will not utter the other. Children sweeten labours; but they make misfortunes more bitter. They increase the cares of life; but they mitigate° the remembrance° of death. The perpetuity° by generation° is common to beasts; but memory,° merit, and noble works, are proper° to men. And surely a man shall see the noblest works and foundations° have proceeded from childless men; which° have sought to express° the images of their minds, where those of their bodies have failed.° So the care of° posterity is most° in them that have no posterity. They that are the first raisers of their houses° are most indulgent towards their children; beholding them as the continuance not only of their kind° but of their work;° and so both children and creatures.°

The difference in affection° of parents towards their several children is many times unequal;° and sometimes unworthy; especially in the mother; as Salomon saith, 'A wise son rejoiceth the father, but an ungracious son shames the mother'.° A man shall see, where there is a house full of children, one or two of the eldest respected,° and the youngest made wantons;° but in the midst some that are as it were forgotten, who many times nevertheless prove the best. The illiberality of parents in allowance° towards their children is an harmful° error; makes them base;° acquaints them with shifts;° makes them sort with° mean company; and makes them surfeit° more when they come to plenty. And therefore the proof° is best, when men keep their authority towards their children, but not their purse.° Men have a foolish manner (both parents and schoolmasters and servants) in creating and breeding an emulation° between brothers during childhood, which many times sorteth to° discord when they are men, and disturbeth families. The Italians make little difference between children and nephews or near kinsfolks;° but so they be of the lump,° they care not though they pass not° through their own body. And, to say truth, in nature it is much a like matter;° insomuch that we see a nephew sometimes resembleth an uncle or a kinsman more than his own parent; as the blood happens.°

Let parents choose betimes° the vocations and courses° they mean their children should take; for then they are most flexible; and let them not too much apply° themselves to the disposition of their children, as thinking they will take best to that which they have most mind to. It is true, that if the affection or aptness° of the children be extraordinary, then it is good not to cross° it; but generally the precept° is good, 'optimum elige, suave et facile illud faciet consuetudo'.° Younger brothers are commonly fortunate, but seldom or never where the elder are disinherited.

8. OF MARRIAGE AND SINGLE LIFE

HE that hath wife and children hath given hostages° to fortune; for they are impediments° to great enterprises, either of virtue or mischief. Certainly the best works, and of greatest merit for the public, have proceeded from the unmarried or childless men; which° both in affection° and means° have married and endowed the public. Yet it were great reason° that those that have children should have greatest care of future times; unto which they know they must transmit their dearest pledges. Some there are, who though they lead a single life, yet their thoughts° do end with° themselves, and account future times impertinences.° Nay, there are some other that account wife and children but as bills of charges.° Nay more, there are some foolish rich covetous men, that take a pride in having no children, because° they may be thought so much the richer. For perhaps they have heard some talk, 'Such an one is a great rich man,' and another except° to it, 'Yea, but he hath a great charge° of children'; as if it were an abatement° to his riches. But the most ordinary cause of a single life is liberty, especially in certain self-pleasing and humorous° minds, which are so sensible of° every restraint, as° they will go near to° think their girdles° and garters to be bonds and shackles.°

Unmarried men are best friends, best masters, best servants; but not always best subjects; for they are light to° run away; and almost all fugitives° are of that condition. A single life doth well with° churchmen;° for charity will hardly water the ground° where it must first fill a pool.° It is indifferent° for judges and magistrates; for if they be facile° and corrupt, you shall have a servant five times worse than a wife. For soldiers, I find the generals commonly in their hortatives° put men in mind of their wives and children; and I think the despising of marriage amongst the Turks maketh the vulgar° soldier more base. Certainly wife and children are a kind of discipline° of humanity; and single men, though they may be many times more charitable, because their means are less exhaust,° yet, on the other side, they are more cruel and hardhearted (good to make severe inquisitors),° because their tenderness is not so oft called upon. Grave natures,° led by custom, and therefore constant, are commonly loving husbands; as was said of Ulysses, 'vetulam suam praetulit immortalitati'.° Chaste women are often proud and froward,° as presuming upon the merit of their chastity. It is one of the best bonds° both of chastity and obedience in the wife, if she think her husband wise;° which she will never do if she find him jealous. Wives are young men's mistresses;° companions for middle age; and old men's nurses. So as° a man may have a quarrel° to marry when he will. But yet he was reputed one of the wise men, that made answer° to the question, when a man should marry?—'A young man not yet, an elder man not at all'. It is often seen that bad husbands have very good wives; whether it be that it raiseth the price° of their husband's kindness when it comes; or that the wives take a pride in their patience. But this never fails, if the bad husbands were of their own choosing, against their friends' consent; for then they will be sure to make good° their own folly.

9. Of Envy

There be none of the affections° which have been noted to fascinate° or bewitch, but love and envy. They both have vehement° wishes; they frame themselves readily into imaginations° and suggestions;° and they come easily into the eye,° especially upon the presence of the objects; which are the points° that conduce to fascination, if any such thing there be. We see likewise the scripture° calleth envy an 'evil eye'; and the astrologers call the evil influences° of the stars 'evil aspects';° so that still° there seemeth to be acknowledged, in the act of envy, an ejaculation or irradiation of the eye.° Nay some have been so curious° as to note, that the times when the stroke or percussion° of an envious eye doth most hurt, are when the party envied is beheld in glory° or triumph; for that sets an edge° upon envy: and besides, at such times the spirits° of the person envied do come forth most into the outward parts,° and so meet the blow.

But leaving these curiosities° (though not unworthy to be thought on in fit place), we will handle,° what persons are apt° to envy others; what persons are most subject to be envied themselves; and what is the difference between public and private envy.

A man that hath no virtue° in himself, ever envieth virtue in others. For men's minds will either feed upon their own good or upon others' evil;° and who° wanteth° the one will prey upon the other; and whoso° is out of hope to attain to another's virtue, will seek to come at even hand° by depressing° another's fortune.

A man that is busy° and inquisitive° is commonly envious. For to know much of other men's matters cannot be because all that ado° may concern his own estate;° therefore it must needs° be that he taketh a kind of play-pleasure° in looking upon the fortunes of others.

Neither can he that mindeth but° his own business find much matter for envy. For envy is a gadding° passion, and walketh the streets, and doth not keep home: 'Non est curiosus, quin idem sit malevolus'.°

Men of noble birth are noted to be envious towards new men° when they rise. For the distance is altered; and it is like a deceit of the eye,° that when others come on° they think themselves go back.

Deformed persons, and eunuchs, and old men,° and bastards, are envious. For he that cannot possibly mend his own case° will do what he can to impair another's; except these defects light upon° a very brave and heroical nature, which thinketh to° make his natural wants° part of his honour; in that it should be said that an eunuch, or a lame man, did such great matters; affecting° the honour of a miracle; as it was in Narses° the eunuch, and Agesilaus° and Tamberlanes,° that were lame men.

The same is the case of men that rise° after calamities and misfortunes. For they are as men fallen out with the times; and think other men's harms a redemption of their own sufferings.

They that desire to excel in too many matters, out of levity° and vain-glory,° are ever envious. For they cannot want work;° it being impossible but many in some one of those things should surpass them. Which was the character of Hadrian° the Emperor; that mortally envied poets and painters and artificers,° in works wherein he had a vein° to excel.

Lastly, near kinsfolks,° and fellows in office,° and those that have been bred together, are more apt to envy their equals when they are raised.° For it doth upbraid° unto them their own fortunes, and pointeth at° them, and cometh oftener into their remembrance, and incurreth° likewise more into the note° of others; and envy ever redoubleth° from speech and fame. Cain's envy was the more vile and malignant towards his brother Abel, because when his sacrifice° was better accepted there was nobody to look on.° Thus much for those that are apt to envy.

Concerning those that are more or less subject to envy: First, persons of eminent virtue, when they are advanced,° are less envied. For their fortune seemeth but due unto them; and no man envieth the payment of a debt, but rewards and liberality rather. Again, envy is ever joined with the comparing of a man's self; and where there is no comparison, no envy; and therefore kings are not envied but by kings.° Nevertheless it is to be noted that unworthy° persons are most envied at their first coming in,° and afterwards overcome it better; whereas contrariwise,° persons of worth and merit are most envied when their fortune continueth long. For by that time, though their virtue be the same, yet it hath not the same lustre; for fresh men grow up that darken it.

Persons of noble blood are less envied in their rising. For it seemeth but right done to their birth. Besides, there seemeth not much added to their fortune; and envy is as the sunbeams, that beat hotter upon a bank or steep rising ground, than upon a flat. And for the same reason those that are advanced by degrees are less envied than those that are advanced suddenly and *per saltum*.°

Those that have joined with their honour great travels,° cares, or perils, are less subject to envy. For men think that they earn their honours hardly,° and pity them sometimes; and pity ever healeth° envy. Wherefore you shall observe that the more deep° and sober° sort of politique persons,° in their greatness, are ever bemoaning° themselves, what a life they lead; chanting a 'quanta patimur'.° Not that they feel it so, but only to abate° the edge of envy. But this is to be understood of business that is laid upon men, and not such as they call unto themselves. For nothing increaseth envy more than an unnecessary and ambitious engrossing° of business. And nothing doth extinguish envy more than for a great person to preserve all other inferior officers in their full rights and pre-eminences° of their places. For by that means there be so many screens between him and envy.

Above all, those are most subject to envy, which carry the greatness of their fortunes in an insolent and proud manner; being never well° but while they are shewing how great they are, either by outward pomp, or by triumphing over all opposition or competition; whereas wise men will rather do sacrifice° to envy, in suffering themselves sometimes of purpose° to be crossed° and overborne° in things that do not much concern them. Notwithstanding so much is truc, that the carriage° of greatness in a plain and open manner (so it be without arrogancy and vain-glory) doth draw less envy than if it be in a more crafty and cunning fashion. For in that° course a man doth but disavow fortune;° and seemeth to be conscious of his own want in° worth; and doth but teach others to envy him.

Lastly, to conclude this part; as we said in the beginning that the act of envy had somewhat in it of witchcraft, so there is no other cure of envy but the cure of witchcraft; and that is, to remove the 'lot'° (as they call it) and to lay it upon another. For which purpose, the wiser sort of great persons bring in ever upon the stage somebody upon whom to derive° the envy that would come upon themselves; sometimes upon ministers and servants; sometimes upon colleagues and associates; and the like; and for that turn° there are never wanting some persons of violent and undertaking° natures, who, so° they may have power and business, will take it at any cost.

Now, to speak of public envy. There is yet° some good in public envy, whereas in private there is none. For public envy is as an ostracism,° that eclipseth° men when they grow too great. And therefore it is a bridle also to great ones, to keep them within bounds.

This° envy, being in the Latin word *invidia*, goeth in the modern languages by the name of 'discontentment'; of which we shall speak in handling Sedition.° It is a disease in a state like to infection. For as infection spreadeth upon that which is sound,° and tainteth° it; so when envy is gotten° once into a state, it traduceth° even the best actions thereof, and turneth them into an ill odour.° And therefore there is little won° by intermingling of plausible° actions. For that doth argue but a weakness and fear of envy, which hurteth so much the more; as it is likewise usual in infections; which if you fear them, you call them upon you.

This public envy seemeth to beat chiefly upon° principal officers or ministers, rather than upon kings and estates° themselves. But this is a sure rule, that if the envy upon the minister be great, when the cause of it in him is small; or if the envy be general° in a manner upon all the ministers of an estate;° then the envy (though hidden) is truly upon the state itself. And so much of public envy or discontentment, and the difference thereof from private envy, which was handled° in the first place.

We will add this in general, touching° the affection of envy; that of all other affections it is the most importune° and continual. For of other affections there is occasion given but now and then; and therefore it was well said, 'Invidia festos dies non agit'°: for it is ever working upon some° or other. And it is also noted that love and envy do make a man pine,° which other affections do not, because they are not so continual. It is also the vilest affection, and the most depraved; for which cause it is the proper attribute of the devil,° who is called 'The envious man, that soweth tares° amongst the wheat by night';° as it always cometh to pass, that envy worketh subtilly,° and in the dark; and to the prejudice° of good things, such as is the wheat.

10. Of Love

THE stage is more beholding° to Love, than the life of man. For as to the stage, love is ever matter° of comedies, and now and then of tragedies; but in life it doth much mischief; sometimes like a syren,° sometimes like a fury.° You may observe, that amongst all the great and worthy persons whereof the memory remaineth, either ancient or recent, there is not one that hath been transported to the mad degree of love: which shews that great spirits° and great business° do keep out this weak° passion. You must except nevertheless Marcus Antonius,° the half-partner° of the empire of Rome, and Appius Claudius,° the decemvir° and lawgiver; whereof the former was indeed a voluptuous man, and inordinate;° but the latter was an austere and wise man: and therefore it seems (though rarely) that love can find entrance not only into an open heart, but also into a heart well fortified, if watch be not well kept. It is a poor° saying of Epicurus, 'Satis magnum alter alteri theatrum sumus';° as if man, made for the contemplation of heaven and all noble objects, should do nothing but kneel before a little idol,° and make himself a subject, though not of the mouth (as beasts are), yet of the eye; which was given him for higher purposes.°

It is a strange° thing to note the excess of this passion, and how it braves° the nature and value of things, by this; that the speaking in a perpetual hyperbole° is comely° in nothing but in love. Neither is it merely in the phrase;° for whereas it hath been well said° that the arch-flatterer, with whom all the petty flatterers have intelligence,° is a man's self; certainly the lover is more. For there was never proud man thought so absurdly well of himself as the lover doth of the person loved; and therefore it was well said, that 'it is impossible to love and to be wise'.° Neither doth this weakness appear to others only, and not to the party loved; but to the loved most of all, except the love be reciproque.° For it is a true rule, that love is ever rewarded either with the reciproque° or with an inward and secret contempt. By how much the more men ought to beware of this passion, which loseth not only other things, but itself. As for the other losses, the poet's relation° doth well figure° them; That he that preferred Helena, quitted the gifts of Juno and Pallas. For whosoever esteemeth° too much of amorous affection quitteth both riches and wisdom.

This passion hath his floods° in the very times of weakness; which are great prosperity and great adversity; though this latter hath been less observed: both which times kindle love, and make it more fervent,° and therefore shew it to be the child of folly. They do best, who if they cannot but admit° love, yet make it keep quarter;° and sever° it wholly from their serious affairs and actions of life; for if it check° once with business, it troubleth men's fortunes, and maketh men that they can no ways° be true to their own ends.° I know not how, but martial men° are given° to love: I think it is but as they are given to wine; for perils commonly ask° to be paid in pleasures. There is in man's nature a secret inclination and motion° towards love of others, which if it be not spent upon some one or a few, doth naturally spread itself towards many, and maketh men become humane and charitable; as it is seen sometime in friars. Nuptial love maketh° mankind; friendly love perfecteth it; but wanton love corrupteth and embaseth° it.

11. Of Great Place°

Men in great place are thrice servants: servants of the sovereign or state;° servants of fame;° and servants of business. So as they have no freedom; neither in their persons,° nor in their actions, nor in their times. It is a strange desire, to seek power and to lose liberty:° or to seek power over others and to lose power over a man's self. The rising unto place° is laborious; and by pains men come to greater pains; and it is sometimes base;° and by indignities° men come to dignities. The standing is slippery, and the regress is either a downfall, or at least an eclipse, which is a melancholy thing. 'Cum non sis qui fueris, non esse cur velis vivere'.° Nay, retire men cannot when they would, neither will they when it were reason;° but are impatient of° privateness,° even in age and sickness, which require the shadow;° like old townsmen,° that will be still sitting at their street door, though thereby they offer age to scorn.

Certainly great persons had need to° borrow other men's opinions, to think themselves happy; for if they judge by their own feeling, they cannot find it: but if they think with themselves what other men think of them, and that other men would fain° be as they are, then they are happy as it were by report; when perhaps they find the contrary within. For they are the first that find their own griefs, though they be the last that find their own faults. Certainly men in great fortunes are strangers to themselves, and while they are in the puzzle of business° they have no time to tend° their health either of body or mind. 'Illi mors gravis incubat, qui notus nimis omnibus, ignotus moritur sibi'.°

In place° there is licence° to do good and evil; whereof the latter is a curse: for in evil the best condition is not to will;° the second not to can.° But power to do good is the true and lawful end of aspiring.° For good thoughts° (though God accept them) yet towards men are little better than good dreams, except they be put in act;° and that cannot be without power and place, as the vantage and commanding ground.° Merit and good works is the end° of man's motion;° and conscience° of the same is the accomplishment of man's rest. For if a man can be partaker of God's theatre,° he shall likewise be partaker of God's rest.° 'Et conversus Deus, ut aspiceret opera quae fecerunt manus suae, vidit quod omnia essent bona nimis';° and then the sabbath.

In the discharge° of thy place set before thee the best examples; for imitation is a globe° of precepts. And after a time set before thee thine own example; and examine thyself strictly whether thou didst not best at first. Neglect not also the examples of those that have carried themselves ill° in the same place; not to set off thyself° by taxing° their memory, but to direct° thyself what to avoid. Reform therefore, without bravery° or scandal of former times and persons; but yet set it down° to thyself as well to create good precedents as to follow them. Reduce° things to the first institution,° and observe wherein and how they have degenerate;° but yet ask counsel of both times; of the ancient time, what is best; and of the latter time, what is fittest. Seek to make thy course regular, that men may know beforehand what they may expect; but be not too positive° and peremptory;° and express° thyself well when thou digressest from thy rule. Preserve the right° of thy place; but stir° not questions of jurisdiction: and rather assume thy right in silence and *de facto*,° than voice° it with claims and challenges. Preserve likewise the rights of inferior places; and think it more honour to direct in chief° than to be busy in all. Embrace and invite° helps° and advices touching° the execution° of thy place; and do not drive away such as bring thee information, as meddlers; but accept of° them in good part.

The vices of authority are chiefly four: delays, corruption, roughness, and facility.° For delays: give easy access; keep times appointed; go through° with that which is in hand, and interlace° not business but of necessity. For corruption: do not only bind thine own hands or thy servants' hands from taking, but bind the hands of suitors also from offering.° For integrity used° doth the one; but integrity professed,° and with a manifest detestation of bribery, doth the other. And avoid not only the fault, but the suspicion. Whosoever is found variable,° and changeth manifestly without manifest cause, giveth suspicion of corruption. Therefore always when thou changest thine opinion or course, profess it plainly, and declare it, together with the reasons that move thee to change; and do not think to steal° it. A servant or a favourite, if he be inward,° and no other apparent cause of esteem, is commonly thought but a by-way° to close° corruption. For roughness:° it is a needless° cause of discontent; severity breedeth fear, but roughness breedeth hate. Even reproofs from authority ought to be grave, and not taunting.° As for facility: it is worse than bribery. For bribes come but now and then; but if importunity° or idle respects° lead a man, he shall never be without. As Salomon saith, 'To respect persons is not good; for such a man will transgress for a piece of bread'.° It is most true that was anciently° spoken, 'A place sheweth the man'. And it sheweth some to the better, and some to the worse. 'Omnium consensu capax imperii, nisi imperasset,'° saith Tacitus of Galba; but of Vespasian he saith, 'Solus imperantium, Vespasianus mutatus in melius'°: though the one was meant of sufficiency,° the other of manners and affection.° It is an assured° sign of a worthy and generous spirit, whom honour° amends.° For honour is, or should be, the place of virtue; and as in nature things move violently to their place and calmly in their place,° so virtue in ambition is violent, in authority settled and calm.

All rising to great place is by a winding stair;° and if there be factions, it is good to side a man's self° whilst he is in the rising, and to balance himself when he is placed.° Use the memory of thy predecessor fairly and tenderly;° for if thou dost not, it is a debt° will sure be paid when thou art gone. If thou have colleagues, respect them, and rather call them when they look not for it, than exclude them when they have reason to look to be called.° Be not too sensible° or too remembering of thy place° in conversation and private answers to suitors; but let it rather be said, 'When he sits in place he is another man'.

12. Of Boldness

It is a trivial grammar-school text,° but yet worthy a wise man's consideration. Question was asked of Demosthenes,° 'what was the chief part of an orator?' he answered, 'action': what next? 'action': what next again? 'action'.° He said it that knew it best, and had by nature himself no advantage in that° he commended. A strange thing, that that part of an orator which is but superficial, and rather the virtue of a player,° should be placed so high, above those other noble°

parts of invention, elocution, and the rest; nay almost alone, as if it were all in all. But the reason is plain. There is in human nature generally more of the fool than of the wise; and therefore those faculties by which the foolish part of men's minds is taken° are most potent. Wonderful like° is the case of Boldness, in civil business;° what first? Boldness: what second and third? Boldness. And yet boldness is a child of ignorance and baseness,° far inferior to other parts. But nevertheless it doth fascinate and bind hand and foot those that are either shallow in judgment or weak in courage, which are the greatest part; yea and prevaileth with wise men at weak times.° Therefore we see it hath done wonders in popular° states; but with senates and princes less; and more ever upon the first entrance° of bold persons into action than soon after; for boldness is an ill keeper of promise. Surely as there are mountebanks° for the natural body, so are there mountebanks for the politic body;° men that undertake° great cures, and perhaps have been lucky in two or three experiments, but want the grounds of science,° and therefore cannot hold out.°

Nay you shall see a bold fellow many times do Mahomet's miracle.° Mahomet made the people believe that he would call an hill to him, and from the top of it offer up his prayers for the observers of his law. The people assembled; Mahomet called the hill to come to him, again and again; and when the hill stood still, he was never a whit° abashed, but said, 'If the hill will not come to Mahomet, Mahomet will go to the hill.' So these men, when they have promised great matters and failed most shamefully, yet (if they have the perfection of boldness) they will but slight it over,° and make a turn,° and no more ado.° Certainly to men of great judgment, bold persons are a sport° to behold; nay and to the vulgar also, boldness has somewhat of the ridiculous. For if absurdity be the subject of laughter, doubt you not but great boldness is seldom without some absurdity. Especially it is a sport to see, when a bold fellow is out of countenance; for that puts his face into a most shrunken and wooden° posture; as needs it must; for in bashfulness the spirits do a little go and come; but with bold men, upon like occasion, they stand at a stay;° like a stale° at chess, where it is no mate, but yet the game cannot stir.° But this last were fitter for a satire than for a serious observation.

This is well to be weighed; that boldness is ever blind; for it seeth° not dangers and inconveniences. Therefore it is ill in counsel,° good in execution; so that the right use° of bold persons is, that they never command in chief, but be seconds, and under the direction of others.

For in counsel it is good to see dangers; and in execution not to see them, except they be very great.

13. Of Goodness and Goodness of Nature

I TAKE Goodness in this sense, the affecting° of the weal° of men, which is that° the Grecians call *Philanthropia*;° and the word 'humanity'° (as it is used) is a little too light to express it. Goodness I call the habit, and Goodness of Nature the inclination.° This of all virtues and dignities° of the mind is the greatest; being the character° of the Deity: and without it man is a busy,° mischievous, wretched° thing; no better than a kind of vermin. Goodness answers° to the theological virtue Charity, and admits no excess,° but° error. The desire of power in excess caused the angels to fall; the desire of knowledge in excess caused man to fall: but in charity there is no excess; neither can angel or man come in danger° by it. The inclination to goodness is imprinted deeply in the nature of man; insomuch that if it issue° not towards men, it will take unto° other living creatures; as it is seen in the Turks,° a cruel people, who nevertheless are kind to beasts, and give alms to dogs and birds; insomuch as° Busbechius° reporteth, a Christian boy in Constantinople had like° to have been stoned for gagging in a waggishness° a long-billed fowl.°

Errors° indeed in this virtue of goodness or charity may be committed. The Italians have an ungracious° proverb, *Tanto buon che val niente*; 'So good, that he is good for nothing'. And one of the doctors° of Italy, Nicholas Machiavel,° had the confidence° to put in writing, almost in plain terms, that 'the Christian faith had given up good men in prey to those that are tyrannical and unjust'. Which he spake,° because indeed there was never law, or sect, or opinion, did so much magnify° goodness, as the Christian religion doth. Therefore, to avoid the scandal° and the danger both, it is good to take knowledge of° the errors of an habit so excellent.

Seek the good of other men, but be not in bondage to their faces or fancies;° for that is but facility or softness; which taketh an honest mind prisoner. Neither give thou Aesop's cock° a gem, who would be better pleased and happier if he had a barley-corn. The example of God teacheth the lesson truly; 'He sendeth his rain, and maketh his sun to shine, upon the just and unjust';° but he doth not rain wealth, nor shine honour and virtues, upon men equally. Common°

benefits are to be communicate° with all; but peculiar° benefits with choice. And beware how in making the portraiture° thou breakest the pattern.° For divinity maketh the love of ourselves the pattern; the love of our neighbours but the portraiture. 'Sell all thou hast, and give it to the poor, and follow me':° but sell not all thou hast, except thou come and follow me;° that is, except thou have a vocation wherein thou mayest do as much good with little means as with great; for otherwise in feeding the streams thou driest the fountain.

Neither is there only a habit of goodness,° directed by right reason;° but there is in some men, even in nature, a disposition° towards it; as on the other side there is a natural malignity. For there be° that in their nature do not affect° the good of others. The lighter sort of malignity turneth but to a crossness,° or frowardness,° or aptness to oppose, or difficilness,° or the like; but the deeper sort to envy and mere° mischief. Such men in other men's calamities are, as it were, in season,° and are ever on the loading part:° not so good as the dogs that licked Lazarus' sores;° but like flies that are still° buzzing upon any thing that is raw; *misanthropi*, that make it their practice to bring men to the bough, and yet have never a tree for the purpose in their gardens, as Timon° had. Such dispositions are the very errours° of human nature; and yet they are the fittest timber° to make great politiques° of; like to knee timber,° that is good for ships, that are ordained to be tossed; but not for building houses, that shall stand firm.

The parts and signs of goodness are many. If a man be gracious and courteous to strangers, it shews he is a citizen of the world,° and that his heart is no island cut off from other lands,° but a continent that joins to them. If he be compassionate towards the afflictions of others, it shews that his heart is like the noble tree° that is wounded itself when it gives the balm. If he easily pardons and remits offences, it shews that his mind is planted above° injuries; so that he cannot be shot.° If he be thankful for small benefits, it shews that he weighs men's minds,° and not their trash.° But above all, if he have St. Paul's perfection,° that he would wish to be an *anathema*° from Christ for the salvation of his brethren, it shews much of a divine nature, and a kind of conformity with Christ himself.

14. Of Nobility

We will speak of Nobility first as a portion of an estate;° then as a condition of particular° persons. A monarchy where there is no nobility at all, is ever a pure° and absolute tyranny; as that of the Turks. For nobility attempers° sovereignty, and draws the eyes of the people somewhat aside° from the line royal.° But for° democracies, they need it not; and they are commonly more quiet and less subject to sedition, than where there are stirps° of nobles. For men's eyes are upon the business,° and not upon the persons; or if upon the persons, it is for the business' sake, as fittest,° and not for flags° and pedigree.° We see the Switzers° last well,° notwithstanding their diversity of religion and of cantons. For utility° is their bond, and not respects.° The united provinces° of the Low Countries in their government excel; for where there is an equality, the consultations° are more indifferent,° and the payments and tributes more cheerful. A great and potent nobility addeth majesty to a monarch, but diminisheth power; and putteth life and spirit into the people, but presseth° their fortune. It is well when nobles are not too great for sovereignty nor for justice;° and yet maintained in that height, as the insolency of inferiors may be broken upon them before it come on too fast° upon the majesty of kings. A numerous nobility causeth poverty and inconvenience in a state; for it is a surcharge° of expense; and besides, it being of necessity that many of the nobility fall° in time° to be weak in fortune, it maketh a kind of disproportion between honour and means.°

As for nobility in particular persons; it is a reverend° thing to see an ancient castle or building not in decay; or to see a fair° timber tree° sound and perfect. How much more to behold an ancient noble family, which hath stood against the waves and weathers° of time. For new nobility is but the act of power,° but ancient nobility is the act of time. Those that are first raised to nobility are commonly more virtuous,° but less innocent, than their descendants; for there is rarely any rising° but by a commixture° of good and evil arts. But it is reason° the memory of their virtues remain to their posterity, and their faults die with themselves. Nobility of birth commonly abateth industry;° and he that is not industrious, envieth him that is. Besides, noble persons cannot go much higher: and he that standeth at a stay° when others rise, can hardly avoid motions° of envy. On the other side, nobility extinguisheth the passive envy° from others towards them; because they are in possession of honour. Certainly, kings that have able men of° their nobility shall find ease in employing them, and a better slide° into their business; for people naturally bend to them, as born in some sort to command.

15. Of Seditions and Troubles

Shepherds° of people had need know the calendars° of tempests in state; which are commonly greatest when things grow to equality;° as natural tempests are greatest about the *Equinoctia*.° And as there are certain hollow° blasts of wind and secret° swellings of seas before a tempest, so are there in states:

> ———Ille etiam caecos instare tumultus
> Saepe monet, fraudesque et operta tumescere bella.°

Libels° and licentious° discourses against the state, when they are frequent and open; and in like sort, false news° often running up and down to the disadvantage of the state, and hastily embraced; are amongst the signs of troubles. Virgil giving the pedigree of Fame, saith she was sister to the Giants:

> Illam Terra parens, ira irritata Deorum,
> Extremam (ut perhibent) Coeo Enceladoque sororem
> Progenuit.°

As if fames° were the relics of seditions past; but they are no less indeed the preludes of seditions to come. Howsoever he noteth it right, that seditious tumults and seditious fames differ no more but as brother and sister, masculine and feminine; especially if it come to that, that the best actions of a state, and the most plausible,° and which ought to give greatest contentment, are taken in ill sense, and traduced:° for that shews the envy° great, as Tacitus saith, 'conflata magna invidia, seu bene seu male gesta premunt'.° Neither doth it follow, that because these fames are a sign of troubles, that the suppressing of them with too much severity should be a remedy of troubles. For the despising of them many times checks them best; and the going about° to stop them doth but make a wonder long-lived.° Also that kind of obedience which Tacitus speaketh of, is to be held suspected: 'Erant in officio, sed tamen qui mallent mandata imperantium interpretari quam exequi';° disputing, excusing, cavilling° upon mandates° and directions,° is a kind of shaking off the yoke, and assay° of disobedience; especially if in those disputings they which are for the direction speak fearfully° and tenderly,° and those that are against it audaciously.

Also, as Machiavel° noteth well, when princes, that ought to be common parents,° make themselves as a party,° and lean to a side, it is as a boat that is overthrown by uneven weight on the one side; as was well seen in the time of Henry the Third of France; for first himself entered league° for the extirpation of the Protestants; and presently after° the same league was turned upon himself. For when the authority of princes is made but an accessary to a cause,° and that there be other bands° that tie faster° than the band of sovereignty, kings begin to be put almost out of possession.°

Also, when discords, and quarrels, and factions, are carried° openly and audaciously, it is a sign the reverence of government is lost. For the motions of the greatest persons in a government ought to be as the motions of the planets under *primum mobile*° (according to the old opinion),° which is, that every° of them is carried swiftly by the highest motion, and softly° in their own motion. And therefore, when great ones in their own particular° motion move violently, and, as Tacitus expresseth it well, 'liberius quam ut imperantium meminissent',° it is a sign the orbs are out of frame.° For reverence is that wherewith princes are girt° from God; who threateneth the dissolving thereof: 'Solvam cingula regum'.°

So when any of the four pillars of government are mainly° shaken or weakened (which are Religion, Justice, Counsel,° and Treasure), men had need to pray for fair weather. But let us pass from this part° of predictions (concerning which, nevertheless, more light may be taken from that which followeth); and let us speak first of the Materials° of seditions; then of the Motives° of them; and thirdly of the Remedies.

Concerning the Materials of seditions. It is a thing well to be considered; for the surest way to prevent seditions (if the times do bear° it) is to take away the matter of them. For if there be fuel° prepared, it is hard to tell whence the spark shall come that shall set it on fire. The matter° of seditions is of two kinds; much poverty and much discontentment.° It is certain, so many overthrown estates,° so many votes for troubles. Lucan noteth well the state° of Rome before the civil war,

> Hinc usura vorax, rapidumque in tempore foenus,
> Hinc concussa fides, et multis utile bellum.°

This same *multis utile bellum*, is an assured and infallible sign of a state disposed to seditions and troubles. And if this poverty and broken estate in the better sort° be joined with a want and necessity in the mean people, the danger is imminent and great. For the rebellions of the belly° are the worst. As for discontentments, they are in the politic body° like to humours° in the natural, which are apt to gather a preternatural° heat and to inflame. And let no prince measure the danger of them by this, whether they be just or unjust: for that were to imagine people° to be too reasonable; who do often spurn at their own good: nor yet by this, whether the griefs° whereupon they rise° be in fact great or small: for they are the most dangerous discontentments where the fear is greater than the feeling: 'Dolendi modus, timendi non item'.° Besides, in great oppressions, the same things that provoke the patience, do withal mate° the courage; but in fears° it is not so. Neither let any prince or state be secure° concerning discontentments, because they have been often,° or have been long, and yet no peril hath ensued: for as it is true that every vapour or fume° doth not turn into a storm; so it is nevertheless true that storms, though they blow over° divers times, yet may fall at last; and, as the Spanish proverb noteth well, 'The cord breaketh at the last by the weakest pull'.

The Causes and Motives° of seditions are, innovation° in religion; taxes; alteration of laws and customs; breaking° of privileges; general oppression; advancement of unworthy persons; strangers;° dearths;° disbanded soldiers; factions grown desperate; and whatsoever, in offending people, joineth and knitteth° them in a common cause.

For the Remedies; there may be some general preservatives,° whereof we will speak: as for the just° cure, it must answer to the particular disease; and so be left to counsel rather than rule.°

The first remedy or prevention is to remove by all means possible that material cause° of sedition whereof we spake; which is, want and poverty in the estate.° To which purpose serveth, the opening and well-balancing of trade; the cherishing° of manufactures; the banishing of idleness;° the repressing° of waste and excess by sumptuary laws;° the improvement and husbanding° of the soil; the regulating of prices of things vendible;° the moderating of taxes and tributes,° and the like. Generally, it is to be foreseen° that the population of a kingdom (especially if it be not mown down by wars) do not exceed the stock° of the kingdom which should maintain them. Neither is the population to be reckoned only by number; for a smaller number that spend more and earn less, do wear out° an estate° sooner than a greater number that live lower and gather more.° Therefore the multiplying of nobility° and other degrees of quality° in an over-proportion to the common people, doth speedily bring a state to necessity; and so doth likewise an overgrown clergy; for they bring nothing to the stock; and in like manner, when more are bred° scholars than preferments° can take off.°

It is likewise to be remembered, that forasmuch as the increase of any estate must be upon the foreigner° (for whatsoever is somewhere gotten is somewhere lost),° there be but three things which one nation selleth unto another: the commodity° as nature yieldeth it; the manufacture; and the vecture,° or carriage. So that if these three wheels° go, wealth will flow as in a spring tide.° And it cometh many times to pass, that 'materiam superabit opus';° that the work and carriage is more worth than the material, and enricheth a state more; as is notably° seen in the Low-Countrymen, who have the best mines above ground° in the world.

Above all things, good policy is to be used that the treasure and monies in a state be not gathered into few hands. For otherwise a state may have a great stock, and yet starve. And money is like muck,° not good except it be spread. This is done chiefly by suppressing, or at the least keeping a strait° hand° upon the devouring° trades of usury, ingrossing,° great pasturages,° and the like.

For removing discontentments, or at least the danger of them; there is in every state (as we know) two portions of subjects; the nobless° and the commonalty. When one of these is discontent,° the danger is not great; for common people are of slow motion, if they be not excited° by the greater sort;° and the greater sort are of small strength, except the multitude be apt and ready to move of themselves.° Then is the danger, when the greater sort do but wait for the troubling of the waters° amongst the meaner, that then they may declare themselves.° The poets° feign,° that the rest of the gods would have bound Jupiter; which he hearing of, by the counsel of Pallas, sent for Briareus, with his hundred hands, to come in to his aid. An emblem, no doubt, to show how safe° it is for monarchs to make sure of the good will of common people.

To give moderate liberty for griefs and discontentments to evaporate (so it be without too great insolency or bravery),° is a safe way. For he that turneth the humours back,° and maketh the wound bleed inwards,° endangereth° malign ulcers and pernicious imposthumations.°

The part° of Epimetheus° might well become° Prometheus,° in the case of discontentments; for there is not a better provision° against them. Epimetheus, when griefs and evils flew abroad, at last shut the lid, and kept hope in the bottom of the vessel. Certainly, the politic° and artificial° nourishing and entertaining° of hopes, and carrying men from hopes to hopes, is one of the best antidotes against the poison of discontentments. And it is a certain sign of a wise government and proceeding,° when it can hold men's hearts by hopes, when it cannot by satisfaction; and when it can handle things in such manner, as no evil shall appear so peremptory° but that it hath some outlet of hope: which is the less hard to do, because both particular persons° and factions are apt° enough to flatter themselves, or at least to brave° that, which° they believe not.

Also the foresight and prevention, that there be no likely or fit° head° whereunto discontented persons may resort, and under whom they may join, is a known,° but an excellent point of caution. I understand a fit head to be one that hath greatness° and reputation; that hath confidence with° the discontented party, and upon whom they turn their eyes; and that is thought discontented in his own particular:° which kind of persons are either to be won° and reconciled to the state, and that in a fast° and true manner; or to be fronted° with some other of the same party, that may oppose them, and so divide the reputation. Generally, the dividing and breaking of all factions and combinations° that are adverse to the state, and setting them at distance,° or at least distrust, amongst themselves, is not one of the worst remedies. For it is a desperate case, if those that hold with° the proceeding of the state be full of discord and faction, and those that are against it be entire and united.

I have noted that some witty and sharp° speeches which have fallen from° princes have given fire to seditions. Caesar did himself infinite hurt in that speech, 'Sulla nescivit literas, non potuit dictare':° for it did utterly cut off that hope which men had entertained, that he would at one time or other give over his dictatorship. Galba undid himself by that speech, 'legi a se militem, non emi';° for it put the soldiers out of hope of the donative.° Probus likewise, by that speech, 'si vixero, non opus erit amplius Romano imperio militibus';° a speech of great despair° for the soldiers. And many the like. Surely princes had need, in tender° matters and ticklish° times, to beware what they say; especially in these short speeches, which fly abroad like darts,° and are thought to be shot out of their secret intentions. For as for large discourses,° they are flat° things, and not so much noted.

Lastly, let princes, against all events,° not be without some great person, one or rather more, of military valour, near unto them, for the repressing of seditions in their beginnings. For without that, there useth° to be more trepidation° in court upon the first breaking out of troubles than were fit. And the state runneth the danger of that which Tacitus saith; 'Atque is habitus animorum fuit, ut pessimum facinus auderent pauci, plures vellent, omnes paterentur'.° But let such military persons be assured,° and well reputed of,° rather than factious and popular;° holding also good correspondence° with the other great men in the state; or else the remedy is worse than the disease.

16. Of Atheism

I HAD rather believe all the fables in the Legend,° and the Talmud,° and the Alcoran,° than that this universal frame° is without a mind. And therefore God never wrought miracle to convince° atheism, because his ordinary works convince it. It is true, that a little philosophy inclineth man's mind to atheism; but depth in philosophy bringeth men's minds about° to religion. For while the mind of man looketh upon second causes scattered,° it may sometimes rest° in them, and go no further; but when it beholdeth the chain° of them, confederate° and linked together, it must needs fly to° Providence and Deity. Nay, even that school which is most accused of atheism doth most demonstrate religion; that is, the school° of Leucippus° and Democritus° and Epicurus.° For it is a thousand times more credible, that four mutable elements,° and one immutable fifth essence,° duly and eternally placed,° need no God, than that an army of infinite small portions° or seeds unplaced,° should have produced this order and beauty without a divine marshal.° The scripture saith, 'The fool hath said in his heart, there is no God';° it is not said, 'The fool hath thought in his heart'; so as he rather saith it by rote° to himself as that he would have,° than that he can throughly° believe it or be persuaded of it. For none deny there is a God, but° those for whom it maketh° that there were no God.

It appeareth° in nothing more,° that atheism is rather in the lip than in the heart of man, than by this; that atheists will ever be talking of that their opinion, as if they fainted° in it within themselves, and would be glad to be strengthened by the consent° of others. Nay more, you shall have° atheists strive to get disciples, as it fareth° with other sects. And, which is most of all,° you shall have of them° that will suffer for atheism, and not recant; whereas if they did truly think that there were no such thing as God, why should they trouble themselves? Epicurus is charged° that he did but dissemble for his credit's sake, when he affirmed there were blessed natures,° but such as enjoyed themselves without having respect to° the government of the world. Wherein they say he did temporize;° though in secret he thought there was no God. But certainly he is traduced;° for his words are noble and divine: 'Non Deos vulgi negare profanum; sed vulgi opiniones Diis applicare profanum'.° Plato could have said no more. And although he° had the confidence to deny the administration,° he had not the power to deny the nature.

The Indians of the west° have names for their particular gods, though they have no name for God: as if the heathens should have had the names Jupiter, Apollo, Mars, &c. but not the word *Deus*; which shews that even those barbarous people have the notion, though they have not the latitude° and extent of it. So that against atheists the very savages° take part° with the very subtlest° philosophers. The contemplative° atheist is rare: a Diagoras,° a Bion,° a Lucian° perhaps, and some others; and yet they seem to be more than they are; for that all that impugn° a received° religion or superstition are by the adverse part branded with the name of atheists. But the great atheists indeed are hypocrites; which are ever handling° holy things, but without feeling; so as they must needs be° cauterized° in the end.

The causes of atheism are: divisions in religion, if they be many; for any one main division addeth zeal to both sides; but many divisions introduce atheism. Another is, scandal° of priests; when it is come to that which St. Bernard saith, 'Non est jam dicere, ut populus sic sacerdos; quia nec sic populus ut sacerdos'.° A third is, custom of profane scoffing° in holy matters; which doth by little and little deface° the reverence of religion. And lastly, learned times, specially with peace and prosperity; for troubles and adversities do more bow men's minds to religion. They that deny a God destroy man's nobility; for certainly man is of kin° to the beasts by his body; and, if he be not of kin to God by his spirit, he is a base and ignoble creature. It destroys likewise magnanimity,° and the raising° of human nature; for take an example of a dog, and mark° what a generosity and courage he will put on when he finds himself maintained° by a man; who to him is instead of a god, or *melior natura*;° which courage is manifestly such as that creature, without that confidence of a better nature than his own, could never attain. So man, when he resteth° and assureth° himself upon divine protection and favour, gathereth° a force and faith which human nature in itself could not obtain.

Therefore, as atheism is in all respects hateful, so in this, that it depriveth human nature of the means to exalt itself above human frailty. As it is in particular persons, so it is in nations. Never was there such a state for magnanimity as Rome. Of this state hear what Cicero saith: 'Quam volumus licet, patres conscripti, nos amemus, tamen nec numero Hispanos, nec robore Gallos, nec calliditate Poenos, nec artibus Graecos, nec denique hoc ipso hujus gentis et terrae domestico nativoque sensu Italos ipsos et Latinos; sed pietate, ac religione, atque hac una sapientia, quod Deorum immortalium numine omnia regi gubernarique perspeximus, omnes gentes nationesque superavimus'.°

17. Of Superstition

It were better to have no opinion° of God at all, than such an opinion as is unworthy of him. For the one is unbelief, the other is contumely:° and certainly superstition is the reproach of the Deity.° Plutarch° saith well to that purpose: 'Surely' (saith he) 'I had rather a great deal men should say there was no such man at all as Plutarch, than that they should say that there was one Plutarch that would eat his children as soon as they were born'; as the poets° speak of Saturn.° And as the contumely is greater towards God, so the danger is greater towards men. Atheism leaves° a man to sense,° to philosophy, to natural piety,° to laws, to reputation;° all which may be guides to an outward moral virtue, though religion were not;° but superstition dismounts° all these, and erecteth an absolute monarchy in the minds of men. Therefore atheism did never perturb states; for it makes men wary of themselves, as looking no further:° and we see the times inclined to atheism (as the time of Augustus Caesar)° were civil times. But superstition hath been the confusion° of many states, and bringeth in a new *primum mobile*,° that ravisheth° all the spheres° of government. The master of superstition is the people; and in all superstition wise men follow fools; and arguments are fitted to practice,° in a reversed° order. It was gravely° said by some of the prelates in the council of Trent,° where the doctrine of the schoolmen° bare great sway,° that 'the schoolmen were like astronomers, which did feign eccentrics and epicycles, and such engines of orbs,° to save the phaenomena;° though they knew there were no such things'; and in like manner, that the schoolmen had framed a number of subtle and intricate axioms and theorems,° to save the practice of the church.

The causes of superstition are: pleasing and sensual° rites and ceremonies; excess of outward and pharisaical° holiness; over-great reverence of traditions, which cannot but load° the church; the stratagems of prelates for their own ambition and lucre;° the favouring too much of good intentions, which openeth the gate° to conceits° and novelties;° the taking an aim at divine matters by human,° which cannot but breed mixture of imaginations;° and, lastly, barbarous times, especially joined with calamities and disasters. Superstition, without a veil,° is a deformed thing; for as it addeth deformity to an ape to be so like a man, so the similitude° of superstition to religion makes it the more deformed. And as wholesome° meat corrupteth° to little worms, so good forms and orders corrupt into a number of petty° observances. There is a superstition in avoiding superstition, when men think to do best if they go furthest° from the superstition formerly received; therefore care would be had° that (as it fareth° in ill purgings)° the good be not taken away with the bad; which commonly is done when the people is the reformer.

18. OF TRAVEL°

TRAVEL, in the younger sort, is a part of education; in the elder, a part of experience. He that travelleth into a country before he hath some entrance into° the language, goeth to school,° and not to travel. That young men travel under some tutor, or grave° servant, I allow° well; so° he be such a one that hath° the language, and hath been in the country before; whereby he may be able to tell them what things are worthy to be seen in the country where they go; what acquaintances they are to seek; what exercises or discipline° the place yieldeth.° For else young men shall go hooded,° and look abroad° little. It is a strange thing, that in sea voyages, where there is nothing to be seen but sky and sea, men should make° diaries; but in land-travel, wherein so much is to be observed, for the most part they omit it; as if chance were fitter to be registered than observation. Let diaries therefore be brought in use.

The things to be seen and observed are: the courts of princes, specially when they give audience to ambassadors; the courts of justice, while they sit° and hear causes;° and so of consistories° ecclesiastic; the churches and monasteries, with the monuments which are therein extant; the walls and fortifications of cities and towns, and so the havens and harbours; antiquities and ruins; libraries; colleges, disputations,° and lectures, where any are; shipping and navies; houses and gardens of state and pleasure,° near great cities; armories;° arsenals; magazines;° exchanges; burses;° warehouses; exercises of horsemanship, fencing, training of soldiers, and the like; comedies, such whereunto the better sort of persons do resort; treasuries of jewels and robes; cabinets and rarities;° and, to conclude, whatsoever is memorable in the places where they go. After all which the tutors or servants ought to make diligent inquiry. As for triumphs,° masks, feasts, weddings, funerals, capital executions, and such shows,° men need not to be put in mind of them; yet are they not to be neglected.

If you will have a young man to put his travel into a little room,° and in short time to gather much, this you must do. First as was said, he must have some entrance into the language before he goeth. Then he must have such a servant or tutor as knoweth the country, as was likewise said. Let him carry with him also some card° or book describing the country where he travelleth; which will be a good key° to his inquiry. Let him keep also a diary. Let him not stay long in one city or town; more or less as the place deserveth, but not long; nay, when he stayeth in one city or town, let him change his lodging from one end and part of the town to another; which is a great adamant° of acquaintance. Let him sequester himself from the company of his countrymen, and diet° in such places where there is good company of the nation where he travelleth. Let him upon his removes° from one place to another, procure recommendation° to some person of quality° residing in the place whither he removeth; that he may use his favour in those things he desireth to see or know. Thus he may abridge° his travel with much profit.

As for the acquaintance which is to be sought in travel; that which is most of all profitable, is acquaintance with the secretaries and employed men° of ambassadors: for so in travelling in one country he shall suck° the experience of many. Let him also see and visit eminent persons in all kinds, which are of great name abroad; that he may be able to tell° how the life° agreeth° with the fame. For quarrels, they are with care and discretion° to be avoided. They are commonly for mistresses, healths,° place,° and words.° And let a man beware how he keepeth company with choleric and quarrelsome persons; for they will engage him into° their own quarrels. When a traveller returneth home, let him not leave the countries where he hath travelled altogether behind him; but maintain a correspondence by letters with those of his acquaintance which are of most worth. And let his travel appear rather in his discourse than in his apparel or gesture;° and in his discourse let him be rather advised° in his answers, than forward°

to tell stories; and let it appear that he doth not change his country manners° for those of foreign parts; but only prick in° some flowers° of that he hath learned abroad into the customs of his own country.

19. Of Empire°

It is a miserable state of mind to have few things to desire, and many things to fear; and yet that commonly is the case of° kings; who, being at the highest, want° matter of desire,° which makes their minds more languishing;° and have many representations° of perils and shadows, which makes their minds the less clear. And this is one reason also of that effect° which the Scripture speaketh of, that 'the king's heart is inscrutable'.° For multitude of jealousies,° and lack of some predominant desire that should marshal and put in order all the rest, maketh any man's heart hard to find or sound.° Hence it comes likewise, that princes many times make themselves° desires, and set their hearts upon toys;° sometimes upon a building; sometimes upon erecting° of an order;° sometimes upon the advancing° of a person; sometimes upon obtaining excellency in some art or feat° of the hand; as Nero° for playing on the harp, Domitian° for certainty of the hand with the arrow,° Commodus° for playing at fence, Caracalla° for driving chariots, and the like. This seemeth incredible unto those that know not the principle that 'the mind of man is more cheered and refreshed by profiting° in small things, than by standing at a stay° in great'. We see also that kings that have been fortunate conquerors in their first years, it being not possible for them to go forward infinitely, but that they must have some check° or arrest in their fortunes, turn in their latter years to be superstitious and melancholy; as did Alexander° the Great; Diocletian;° and in our memory, Charles the Fifth;° and others. For he that is used to go forward, and findeth a stop,° falleth out of his own favour,° and is not the thing he was.

To speak now of the true temper° of empire; it is a thing rare and hard to keep; for both temper and distemper° consist of contraries. But it is one thing to mingle contraries, another to interchange them. The answer° of Apollonius to Vespasian is full of excellent instruction. Vespasian asked him, 'what was Nero's overthrow?'° He answered, 'Nero could touch and tune the harp well; but in government sometimes he used to wind the pins° too high, sometimes to let them down too low'. And certain it is that nothing destroyeth authority so much as the unequal° and untimely interchange of power pressed° too far, and relaxed too much.

This is true, that the wisdom° of all these latter times in princes' affairs is rather fine° deliveries° and shiftings° of dangers and mischiefs when they are near, than solid and grounded° courses° to keep them aloof.° But this is but to try masteries° with fortune. And let men beware how they neglect and suffer° matter° of trouble to be prepared; for no man can forbid° the spark, nor tell whence it may come. The difficulties in princes' business are many and great; but the greatest difficulty is often in their own mind. For it is common with princes (saith Tacitus) to will° contradictories: 'Sunt plerumque regum voluntates vehementes, et inter se contrariae'.° For it is the solecism° of power, to think to command the end, and yet not to endure the mean.°

Kings have to deal with their neighbours, their wives, their children, their prelates° or clergy, their nobles, their second-nobles° or gentlemen, their merchants, their commons, and their men of war;° and from all these arise dangers, if care and circumspection be not used.

First for° their neighbours: there can no general rule be given (the occasions are so variable), save one, which ever holdeth;° which is, that princes do keep due sentinel,° that none of their neighbours do overgrow so (by increase of territory, by embracing° of trade, by approaches,° or the like), as they become more able to annoy them than they were. And this is generally the work of standing counsels° to foresee and to hinder it. During° that triumvirate of kings, King Henry the Eighth of England, Francis the First King of France, and Charles the Fifth Emperor, there was such a watch kept, that none of the three could win a palm° of ground, but the other two would straightways° balance it, either by confederation, or, if need were,° by a war; and would not in any wise take up peace at interest.° And the like was done by that league° (which Guicciardine saith was the security of Italy) made between Ferdinando King of Naples, Lorenzius Medices, and Ludovicus Sforza, potentates, the one of Florence, the other of Milan. Neither is the opinion of some of the schoolmen° to be received,° that 'a war cannot justly be made but upon a precedent° injury or provocation'. For there is no question but a just° fear of an imminent danger, though there be no blow given,° is a lawful cause of a war.

For their wives: there are cruel examples of them. Livia is infamed° for the poisoning of her husband;° Roxalana, Solyman's wife, was the destruction° of that renowned prince Sultan Mustapha, and otherwise° troubled his house and succession; Edward the Second of England his queen° had the principal hand° in the deposing and murther of her husband. This kind of danger is then to be feared chiefly, when the wives have° plots for the raising° of their own children; or else that they be advoutresses.°

For their children: the tragedies° likewise of dangers from them have been many. And generally, the entering of fathers into suspicion° of their children hath been ever unfortunate. The destruction of Mustapha (that° we named before) was so fatal to Solyman's line,° as the succession° of the Turks from Solyman until this day is suspected to be untrue, and of strange° blood; for that Selymus the Second was thought to be suppositious.° The destruction of Crispus,° a young prince of rare towardness,° by Constantinus the Great, his father, was in like manner fatal to his house;° for both Constantinus and Constance, his sons, died violent deaths; and Constantius, his other son, did little better; who died indeed of sickness, but after that Julianus had taken arms against him. The destruction of Demetrius,° son to Philip the Second° of Macedon, turned upon the father, who died of repentance. And many like examples there are; but few or none where the fathers had good° by such distrust; except it were where the sons were up in open arms° against them; as was Selymus° the First against Bajazet; and the three sons° of Henry the Second, King of England.

For their prelates: when they are proud and great, there is also danger from them; as it was in the times of Anselmus° and Thomas Becket,° Archbishops of Canterbury; who with their crosiers° did almost try it° with the king's sword; and yet they had to deal with stout° and haughty kings; William Rufus, Henry the First, and Henry the Second. The danger is not from that state,° but° where it hath a dependance of foreign authority;° or where the churchmen come in° and are elected, not by the collation° of the king, or particular patrons, but by the people.

For their nobles: to keep them at a distance, it is not amiss; but to depress° them, may make a king more absolute, but less safe; and less able to perform any thing that he desires. I have noted° it in my *History of King Henry the Seventh of England*, who depressed his nobility; whereupon it came to pass° that his times were full of difficulties and troubles; for the nobility, though they continued° loyal unto him, yet did they not co-operate with him in his business. So that in effect he was fain° to do all things himself.

For their second° nobles: there is not much danger from them, being a body dispersed.° They may sometimes discourse high,° but that doth little hurt; besides, they are a counterpoise to the higher nobility, that they grow not too potent; and, lastly, being the most immediate in authority with the common people, they do best temper° popular commotions.°

For their merchants: they are *vena porta*;° and if they flourish not, a kingdom may have good limbs, but will have empty veins, and nourish° little. Taxes and imposts° upon them do seldom good to the king's revenue; for that that he wins in the hundred° he leeseth in the shire;° the particular rates° being increased, but the total bulk∪ of trading rather decreased.

For their commons:° there is little danger from them, except it be where they have great and potent heads;° or where you meddle with the point of religion, or their customs,° or means of life.

For their men of war: it is a dangerous state° where they live and remain in a body,° and are used to donatives;° whereof we see examples in the janizaries,° and pretorian bands° of Rome; but trainings of men, and arming them in several° places, and under several commanders, and without donatives, are things of defence, and no danger.

Princes are like to heavenly bodies, which cause good or evil times; and which have much veneration, but no rest.∪ All precepts concerning kings are in effect comprehended in those two remembrances; 'memento quod es homo'; and 'memento quod es Deus', or 'vice Dei';° the one bridleth their power, and the other their will.°

20. Of Counsel

The greatest trust between man and man is the trust of giving counsel. For in other° confidences men commit the parts of life; their lands, their goods, their child, their credit, some particular affair; but to such as° they make their counsellors, they commit the whole: by how much the more they° are obliged° to all faith and integrity. The wisest princes need not think it any diminution to their greatness, or derogation° to their sufficiency,° to rely upon counsel. God himself is not without,° but hath made it one of the great names° of his blessed Son; *The Counsellor*.° Salomon° hath pronounced that 'in counsel is stability'. Things will have their first or second agitation:° if they be not tossed upon the arguments of counsel, they will be tossed upon the waves of fortune; and be full of inconstancy, doing and undoing, like the reeling° of a drunken man. Salomon's son° found° the force of counsel, as his father saw the necessity of it. For the beloved kingdom of God was first rent° and broken by ill counsel;° upon which counsel there are set for our instruction the two marks° whereby bad counsel is for ever best discerned; that it was young counsel, for° the persons; and violent counsel, for the matter.°

The ancient times do set forth in figure° both the incorporation and inseparable conjunction of counsel with kings, and the wise and politic use of counsel by kings: the one, in that they say Jupiter did marry Metis, which signifieth counsel; whereby they intend° that Sovereignty is married to Counsel: the other in that which followeth, which was thus: They say, after Jupiter was married to Metis, she conceived by him and was with child, but Jupiter suffered her not° to stay till she brought forth,° but ate her up; whereby he became himself with child, and was delivered of Pallas armed,° out of his head.° Which monstrous° fable containeth a secret of empire;° how kings are to make use of their council of state.° That first they ought to refer matters unto them, which is the first begetting or impregnation; but when they are elaborate,° moulded, and shaped in the womb of their counsel, and grow ripe and ready to be brought forth, that then they suffer not their council to go through with° the resolution and direction,° as if it depended on them; but take the matter back° into their own hands, and make it appear to the world that the decrees and final directions (which, because they come forth with prudence and power, are resembled to° Pallas armed) proceeded from themselves; and not only from their authority, but (the more to add reputation to themselves) from their head and device.°

Let us now speak of the inconveniences of counsel, and of the remedies. The inconveniences that have been noted in calling° and using counsel, are three. First, the revealing of affairs, whereby they become less secret. Secondly, the weakening of the authority of princes, as if they were less of themselves.° Thirdly, the danger of being unfaithfully counselled, and more for the good of them that counsel than of him that is counselled. For which inconveniences, the doctrine of Italy, and practice of France, in some kings' times, hath introduced *cabinet* councils;° a remedy worse than the disease.

As to secrecy; princes are not bound to communicate all matters with° all counsellors; but may extract and select. Neither is it necessary that he that consulteth what he should do, should declare° what he will do. But let princes beware that the unsecreting° of their affairs comes not from themselves. And as for cabinet councils, it may be their motto, 'plenus rimarum sum':° one futile° person that maketh it his glory to tell, will do more hurt than many that know it their duty to conceal. It is true there be some affairs which require extreme secrecy, which will hardly go beyond one or two persons° besides the king: neither are those counsels unprosperous; for, besides the secrecy, they commonly go on constantly° in one spirit of direction,° without distraction.° But then it must be a prudent king, such as is able to grind with a hand-mill;° and those inward° counsellors had need also be wise men, and especially true and trusty to the king's ends; as it was with King Henry the Seventh° of England, who in his greatest business imparted° himself to none, except it were to Morton and Fox.°

For weakening of authority; the fable° showeth the remedy. Nay, the majesty of kings is rather exalted than diminished when they are in the chair of council; neither was there ever prince bereaved of his dependances° by his council; except where there hath been either an over-greatness° in one counsellor or an over-strict combination° in divers; which are things soon found and holpen.°

For the last inconvenience, that men will counsel with an eye to themselves; certainly, 'non inveniet fidem super terram'° is meant of the nature of times,° and not of all particular persons. There be° that are in nature faithful, and sincere, and plain, and direct; not crafty and involved;° let princes, above all, draw to themselves such natures. Besides, counsellors are not commonly so united, but that one counsellor keepeth sentinel over another; so that if any do counsel out of faction° or private ends, it commonly comes to the king's ear. But the best remedy is, if princes know their counsellors, as well as their counsellors know them:

Principis est virtus maxima nosse suos.°

And on the other side,° counsellors should not be too speculative° into their sovereign's person. The true composition of a counsellor is rather to be skilful° in their master's business, than in his nature; for then he is like to advise him, and not feed his humour.° It is of singular° use to princes if they take the opinions of their council both separately° and together. For private opinion is more free; but opinion before others is more reverent.° In private, men are more bold in their own humours; and in consort,° men are more obnoxious° to others' humours; therefore it is good to take both;° and of the inferior sort° rather in private, to preserve freedom; of the greater rather in consort, to preserve respect. It is in vain for princes to take counsel concerning matters,° if they take no counsel likewise concerning persons; for all matters are as dead images; and the life° of the execution of affairs resteth in the good choice of persons. Neither is it enough to consult concerning persons *secundum genera*,° as in an idea,° or mathematical description, what the kind and character of the person should be; for the greatest errors are committed, and the most° judgment is shown, in the choice of individuals. It was truly said, 'optimi consiliarii mortui';° books will speak plain when counsellors blanch.° Therefore it is good to be conversant° in them, specially the books of such as themselves have been actors upon the stage.°

The councils at this day in most places are but familiar meetings, where matters are rather talked on° than debated. And they run too swift to the order or act of council.° It were better that in causes of weight, the matter° were propounded one day and not spoken to° till the next day; 'in nocte consilium'.° So was it done in the Commission of Union° between England and Scotland; which was a grave° and orderly assembly. I commend set° days for petitions; for both it gives the suitors more certainty for their attendance, and it frees the meetings for matters of estate,° that they may 'hoc agere'.° In choice of committees for ripening° business for the council, it is better to choose indifferent° persons, than to make an indifferency° by putting in those that are strong on both sides. I commend also standing commissions;° as for trade, for treasure, for war, for suits,° for some provinces;° for where there be divers° particular councils and but one council of estate (as it is in Spain), they are, in effect, no more than standing commissions: save° that they have greater authority. Let such as are to inform councils out of their particular professions, (as lawyers, seamen, mintmen,° and the like), be first heard before committees; and then, as occasion serves, before the council. And let them not come in multitudes, or in a tribunitious° manner; for that is to clamour° councils, not to inform them. A long table and a square table, or seats about the walls, seem things of form,° but are things of substance; for at a long table a few at the upper end,° in effect, sway all the business; but in the other form° there is more use of the counsellors' opinions that sit lower. A king, when he presides in council, let him beware how he opens° his own inclination too much in that which he propoundeth; for else counsellors will but take the wind of him,° and instead of giving free counsel, sing him a song of *placebo*.°

21. Of Delays

FORTUNE is like the market; where many times, if you can stay° a little, the price will fall. And again, it is sometimes like Sibylla's offer;° which at first offereth the commodity at full,° then consumeth part and part, and still holdeth up° the price. For 'occasion' (as it is in the common verse)° 'turneth a bald noddle,° after she hath presented her locks in front, and no hold taken';° or at least turneth the handle of the bottle first to be received, and after° the belly, which is hard to clasp. There is surely no greater wisdom than well to time the beginnings° and onsets of things. Dangers are no more light,° if they once seem light; and more dangers have deceived men than forced° them. Nay, it were better to meet some dangers half way, though they come nothing° near, than to keep too long a watch° upon their approaches; for if a man watch too long, it is odds he will fall asleep. On the other side, to be deceived with too long shadows (as some have been when the moon was low and shone on their enemies' back),° and so to shoot off before the time; or to teach dangers to come on, by over early buckling° towards them; is another extreme.

The ripeness° or unripeness of the occasion (as we said) must ever be well weighed; and generally it is good to commit the beginnings of all great actions to Argos° with his hundred eyes, and the ends to Briareus° with his hundred hands; first to watch, and then to speed. For the helmet of Pluto,° which maketh the politique° man go invisible, is secrecy in the counsel and celerity in the execution. For when things are once come to the execution, there is no secrecy comparable to celerity; like the motion of a bullet in the air, which flieth so swift as it outruns the eye.

22. Of Cunning

WE take° Cunning for a sinister° or crooked° wisdom. And certainly there is a great difference between a cunning man and a wise man; not only in point of honesty, but in point of ability. There be that can pack the cards,° and yet cannot play well; so there are some that are good in canvasses° and factions, that are otherwise weak men. Again, it is one thing to understand persons, and another thing to understand matters; for many are perfect° in men's humours,° that are not greatly capable° of the real part of business; which is the constitution° of one that hath studied men more than books. Such men are fitter for practice° than for counsel; and they are good but in their own alley:° turn them to new men, and they have lost their aim; so as the old rule to know a fool from a wise man, 'Mitte ambos nudos ad ignotos, et videbis',° doth scarce hold° for them. And because these cunning men are like haberdashers° of small wares, it is not amiss to set forth their shop.°

It is a point of cunning, to wait upon° him with whom you speak, with your eye; as the Jesuits° give it in precept: for there be many wise men that have secret hearts and transparent countenances. Yet this would° be done with a demure° abasing° of your eye sometimes, as the Jesuits also do use.°

Another is, that when you have anything to obtain of present despatch,° you entertain and amuse the party with whom you deal with some other discourse; that he be not too much awake° to make objections. I knew a counsellor° and secretary, that never came to Queen Elizabeth of England with bills° to sign, but he would always first put her into° some discourse of estate,° that she might the less mind the bills.

The like surprise may be made by moving° things when the party is in haste, and cannot stay to consider advisedly of that° is moved.

If a man would cross° a business that he doubts° some other° would handsomely and effectually move, let him pretend to wish it well, and move it himself in such sort as may foil° it.

The breaking off in the midst of that one was about to say, as if he took himself up,° breeds a greater appetite in him with whom you confer to know more.

And because it works° better when anything seemeth to be gotten from you by question,° than if you offer it of yourself,° you may lay a bait for a question, by showing another° visage and countenance than you are wont; to the end to give occasion for the party to ask what the matter is of the change? As Nehemias did; 'And I had not before that time been sad before the king'.°

In things that are tender° and unpleasing, it is good to break the ice by some whose words are of less weight, and to reserve the more weighty voice to come in as by chance, so that he may be asked the question upon° the other's speech; as Narcissus° did, in relating to Claudius the marriage of Messalina and Silius.

In things that a man would not be seen in himself, it is a point of cunning to borrow the name of the world; as to say, 'The world says', or 'There is a speech abroad'.

I knew one that, when he wrote a letter, he would put that which was most material° in the postscript, as if it had been a bye-matter.°

I knew another that, when he came to have speech,° he would pass over that that he intended most; and go forth, and come back again, and speak of it as of a thing that he had almost forgot.

Some procure themselves to be surprised at such times as it is like° the party that they work upon° will suddenly come upon° them; and to be found with a letter in their hand, or doing somewhat which they are not accustomed; to the end° they may be apposed of° those things which of themselves they are desirous to utter.

It is a point of cunning, to let fall those words in a man's own name, which he would have another man learn and use, and thereupon take advantage. I knew two that were competitors for the secretary's place° in queen Elizabeth's time, and yet kept good quarter° between themselves; and would confer one with another upon the business; and the one of them said, that to be a secretary 'in the declination° of a monarchy' was a ticklish° thing, and that he did not affect° it: the other straight° caught up those words, and discoursed with divers of his friends, that he had no reason to desire to be secretary in the declination of a monarchy. The first man took hold of it, and found means it was told the Queen; who hearing of 'a declination of a monarchy', took it so ill, as she would never after hear of the other's suit.

There is a cunning, which we in England call 'The turning of the cat in the pan';° which is, when that which a man says to another, he lays° it as if another had said it to him. And to say truth, it is not easy, when such a matter passed° between two, to make it appear from which of them it first moved and began.

It is a way that some men have, to glance and dart° at others by justifying themselves by negatives; as to say 'This I do not'; as Tigellinus° did towards Burrhus, 'Se non diversas spes, sed incolumitatem imperatoris simpliciter spectare'.°

Some have in readiness so many tales and stories, as there is nothing they would insinuate, but they can wrap it into a tale; which serveth both to keep themselves more in guard,° and to make others carry it° with more pleasure.

It is a good point of cunning, for a man to shape the answer he would have° in his own words and propositions; for it makes the other party stick° the less.

It is strange how long some men will lie in wait° to speak somewhat they desire to say; and how far about they will fetch;° and how many other matters they will beat over,° to come near it. It is a thing of great patience, but yet of much use.

A sudden, bold, and unexpected question doth many times surprise a man, and lay him open.° Like to him that, having changed his name and walking in Paul's,° another suddenly came behind him and called him by his true name, whereat straightways° he looked back.

But these small wares and petty points° of cunning are infinite; and it were a good deed to make a list of them; for that° nothing doth more hurt in a state than that cunning men pass for wise.

But certainly some there are that know the resorts and falls° of business, that cannot sink into the main° of it; like a house that hath convenient stairs and entries, but never a fair° room. Therefore you shall see them find out pretty looses° in the conclusion, but are no ways able to examine or debate matters. And yet commonly they take advantage of their inability, and would be thought wits of direction.° Some build rather upon the abusing of others, and (as we now say) 'putting tricks upon° them', than upon soundness of their own proceedings. But Salomon saith, 'Prudens advertit ad gressus suos: stultus divertit ad dolos'.°

23. Of Wisdom for a Man's Self°

An ant is a wise° creature for itself, but it is a shrewd° thing in an orchard or garden. And certainly men that are great lovers of themselves waste° the public. Divide° with reason between self-love° and society;° and be so true° to thyself, as thou be not false to others; specially to thy king and country. It is a poor centre of a man's actions, *himself*. It is right earth.° For that only stands fast° upon his° own centre; whereas all things that have affinity with the heavens, move upon the centre of another, which they benefit. The referring of all to a man's self is more tolerable in a sovereign prince; because themselves are not only themselves, but their good and evil is at the peril° of the public fortune. But it is a desperate evil in a servant to a prince, or a citizen in a republic. For whatsoever affairs° pass such a man's hands, he crooketh° them to his own ends; which must needs be often eccentric° to the ends of his master or state. Therefore let princes, or states, choose such servants as have not this mark;° except they mean° their service should be made but the accessary.° That which maketh the effect more pernicious is that all proportion° is lost. It were disproportion enough for the servant's good to be preferred before° the master's; but yet it is a greater extreme, when a little good of the servant shall carry things° against a great good of the master's. And yet that is the case of bad officers, treasurers, ambassadors, generals, and other false and corrupt servants; which set a bias upon their bowl,° of their own petty ends and envies,° to the overthrow° of their master's great and important affairs. And for the most part, the good such servants receive is after the model° of their own fortune; but the hurt they sell° for that good is after the model of their master's fortune. And certainly it is the nature of extreme self-lovers, as° they will set an house on fire, and° it were but to roast their eggs; and yet these men many times hold credit° with their masters, because their study is but to please them and profit themselves; and for either respect° they will abandon the good of their affairs.

Wisdom for a man's self is, in many branches thereof, a depraved thing. It is the wisdom of rats,° that will be sure to leave a house somewhat before it fall. It is the wisdom of the fox, that thrusts out the badger, who digged° and made room for him. It is the wisdom of crocodiles, that shed tears° when they would° devour. But that which is specially to be noted is, that those which (as Cicero says of Pompey) are 'sui amantes, sine rivali',° are many times unfortunate. And whereas they have all their times sacrificed to themselves, they become in the end themselves sacrifices to the inconstancy of fortune; whose wings they thought by their self-wisdom to have pinioned.°

24. Of Innovations°

As the births° of living creatures at first are ill-shapen, so are all Innovations, which are the births of time. Yet notwithstanding, as those that first bring honour into their family are commonly more worthy° than most that succeed,° so the first precedent (if it be good) is seldom attained° by imitation. For Ill,° to man's nature as it stands perverted,° hath a natural motion,° strongest in continuance;° but Good, as a forced motion,° strongest at first. Surely every medicine is an innovation;° and he that will not apply new remedies must expect new evils; for time is the greatest innovator; and if time of course° alter things to the worse, and wisdom and counsel shall not alter them to the better, what shall be the end? It is true, that what is settled by custom, though it be not good,° yet at least it is fit;° and those things which have long gone together, are as it were confederate° within themselves; whereas new things piece° not so well; but though they help by their utility,° yet they trouble° by their inconformity.° Besides, they are like strangers;° more admired° and less favoured.

All this is true, if time stood still; which contrariwise moveth so round,° that a froward° retention of custom is as turbulent° a thing as an innovation; and they that reverence too much old times, are but a scorn° to the new. It were good therefore that men in their innovations would follow the example of time itself; which indeed innovateth greatly, but quietly, and by degrees scarce to be perceived. For otherwise,° whatsoever is new is unlooked for; and ever° it mends° some, and pairs° other;° and he that is holpen° takes it for a fortune, and thanks the time; and he that is hurt, for a wrong, and imputeth it to the author.° It is good also not to try experiments in states, except the necessity be urgent, or the utility evident; and well to beware° that it be the reformation that draweth on° the change, and not the desire of change that pretendeth° the reformation. And lastly, that the novelty, though it be not rejected, yet be held for a suspect;° and, as the Scripture° saith, that 'we make a stand upon the ancient way, and then look about us, and discover what is the straight and right way, and so to walk in it'.

25. Of Dispatch°

Affected dispatch° is one of the most dangerous things to business that can be. It is like that which the physicians call *predigestion*,° or hasty digestion; which is sure to fill the body full of crudities° and secret seeds of diseases. Therefore measure not dispatch by the times of sitting,° but by the advancement of the business. And as in races it is not the large stride or high lift° that makes the speed; so in business, the keeping close to the matter, and not taking of it° too much at once, procureth dispatch. It is the care of some only to come off speedily° for the time;° or to contrive some false periods° of business, because° they may seem men of dispatch. But it is one thing to abbreviate by contracting,° another by cutting off. And business so handled at several sittings or meetings goeth commonly backward and forward in an unsteady manner. I knew a wise man° that had it for a by-word,° when he saw men hasten to a conclusion, 'Stay° a little, that° we may make an end the sooner'.

On the other side, true dispatch is a rich° thing. For time is the measure of business, as money is of wares; and business is bought at a dear hand° where there is small dispatch. The Spartans and Spaniards° have been noted to be of small dispatch; 'Mi venga la muerte de Spagna'; 'Let my death come from Spain'; for then it will be sure to be long in coming.

Give good hearing to those that give the first information in business; and rather direct° them in the beginning, than interrupt them in the continuance of their speeches; for he that is put out of his own order will go forward and backward, and be more tedious while he waits upon his memory, than he could have been if he had gone on in his own course. But sometimes it is seen that the moderator° is more troublesome than the actor.°

Iterations° are commonly loss of time. But there is no such gain of time as to iterate often the state of the question; for it chaseth away many a frivolous speech as it is coming forth. Long and curious° speeches are as fit for dispatch, as a robe or mantle with a long train° is for race. Prefaces and passages,° and excusations,° and other speeches of reference to the person,° are great wastes of time; and though they seem to proceed of° modesty, they are bravery.° Yet beware of being too material° when there is any impediment or obstruction in men's wills; for pre-occupation of mind° ever requireth preface of speech, like a fomentation° to make the unguent° enter.

Above all things, order, and distribution,° and singling out of parts, is the life of dispatch; so as the distribution be not too subtle:° for he that doth not divide° will never enter well into business; and he that divideth too much will never come out of it clearly. To choose time is to save time; and an unseasonable motion is but beating the air. There be three parts of business; the preparation, the debate or examination, and the perfection.° Whereof, if you look for dispatch, let the middle° only be the work of many, and the first and last the work of few. The proceeding° upon somewhat conceived° in writing doth for the most part facilitate dispatch: for though it should be wholly rejected, yet that negative is more pregnant of direction° than an indefinite; as ashes are more generative than dust.°

26. Of Seeming Wise

It hath been an opinion, that the French are wiser than they seem, and the Spaniards seem wiser than they are. But howsoever it be between nations, certainly it is so between man and man. For as the Apostle saith of godliness, 'Having a shew of godliness, but denying the power° thereof';° so certainly there are in point of° wisdom and sufficiency,° that do nothing or little very solemnly:° 'magno conatu nugas'.°

It is a ridiculous thing and fit for a satire to persons of judgment, to see what shifts° these formalists° have, and what prospectives° to make *superficies*° to seem body that hath depth and bulk. Some are so close and reserved,° as they will not shew their wares but by a dark light;° and seem always to keep back somewhat;° and when they know within themselves they speak of that they do not well know, would nevertheless seem to others to know of that which they may not well speak. Some help themselves with countenance and gesture, and are wise by signs;° as Cicero saith of Piso, that when he answered him, he fetched one of his brows up to his forehead, and bent the other down to his chin; 'Respondes, altero ad frontem sublato, altero ad mentum depresso supercilio, crudelitatem tibi non placere'.° Some think to bear it° by speaking a great word,° and being peremptory; and go on, and take by admittance that which they cannot make good.° Some, whatsoever is beyond their reach, will seem to despise or make light of it as impertinent° or curious;° and so would have their ignorance seem judgment. Some are never without a difference,° and commonly by amusing men with a subtilty,° blanch° the matter; of whom A. Gellius saith, 'Hominem delirum, qui verborum minutiis rerum frangit pondera'.° Of which kind also, Plato° in his *Protagoras* bringeth in Prodicus in scorn,° and maketh him make a speech that consisteth of distinctions from the beginning to the end. Generally, such men in all deliberations find ease° to be° of the negative side, and affect a credit to object° and foretell difficulties; for when propositions° are denied, there is an end of them; but if they be allowed,° it requireth a new work; which false point of wisdom is the bane° of business.

To conclude, there is no decaying merchant, or inward° beggar, hath so many tricks to uphold the credit of their wealth, as these empty persons have to maintain the credit of their sufficiency. Seeming wise men may make shift° to get opinion;° but let no man choose them for employment; for certainly you were better° take for business a man somewhat absurd° than over-formal.

27. Of Friendship

It had been hard° for him that spake it to have put more truth and untruth together in few words, than in that speech, 'Whosoever is delighted in solitude is either a wild beast or a god'.° For it is most true that a natural° and secret hatred and aversation° towards society in any man, hath somewhat of the savage beast; but it is most untrue that it should have any character° at all of the divine nature; except it proceed, not out of a pleasure in solitude, but out of a love and desire to sequester° a man's self for a higher conversation:° such as is found to have been falsely and feignedly° in some of the heathen; as Epimenides° the Candian, Numa° the Roman, Empedocles° the Sicilian, and Apollonius of Tyana;° and truly and really in divers of the ancient hermits and holy fathers of the church. But little do men perceive what solitude is, and how far it extendeth. For a crowd is not company; and faces are but a gallery° of pictures; and talk but a tinkling cymbal,° where there is no love. The Latin adage meeteth with° it a little: 'Magna civitas, magna solitudo';° because in a great town friends are scattered; so that there is not that fellowship, for the most part, which is in less° neighbourhoods. But we may go further, and affirm most truly that it is a mere° and miserable solitude to want° true friends; without which the world is but a wilderness; and even in this sense also of solitude, whosoever in the frame° of his nature and affections is unfit for friendship, he taketh it° of the beast, and not from humanity.°

A principal fruit of friendship is the ease° and discharge of the fullness and swellings of the heart,° which passions of all kinds do cause and induce. We know diseases of stoppings° and suffocations are the most dangerous in the body; and it is not much otherwise in the mind; you may take sarza° to open° the liver, steel° to open the spleen,° flower of sulphur° for the lungs, castoreum° for the brain; but no receipt° openeth the heart, but a true friend; to whom you may impart griefs, joys, fears, hopes, suspicions, counsels, and whatsoever lieth upon the heart to oppress it, in a kind of civil° shrift° or confession.

It is a strange thing to observe how high a rate° great kings and monarchs do set upon this fruit of friendship whereof we speak: so great, as° they purchase it many times at the hazard of their own safety and greatness. For princes, in regard° of the distance of their fortune from that of their subjects and servants, cannot gather this fruit, except (to make themselves capable thereof) they raise some persons to be as it were companions and almost equals to themselves, which many times sorteth to° inconvenience. The modern languages give unto such persons the name of favourites, or privadoes;° as if it were matter of grace,° or conversation.° But the Roman name attaineth° the true use and cause thereof, naming them *participes curarum*;° for it is that which tieth the knot.° And we see plainly that this hath been done, not by weak and passionate° princes only, but by the wisest and most politique that ever reigned; who have oftentimes joined to themselves some of their servants; whom both themselves have called friends, and allowed others likewise to call them in the same manner; using the word which is received° between private men.

L. Sulla, when he commanded Rome, raised Pompey (after surnamed the Great) to that° height, that Pompey vaunted himself for Sulla's over-match. For when he had carried° the consulship for a friend of his, against the pursuit° of Sulla, and that Sulla did a little resent thereat, and began to speak great,° Pompey turned upon him again, and in effect bade him be quiet; 'for that more men adored the sun rising than the sun setting'.° With Julius Caesar,° Decimus Brutus had obtained that interest,° as he set him down in his testament for heir in remainder after his nephew.° And this was the man that had power with° him to draw him forth to his death. For when Caesar would have discharged° the senate, in regard of some ill presages, and specially a dream of Calpurnia; this man lifted him gently by the arm out of his chair, telling him he hoped he would not dismiss the senate till his wife had dreamt a better dream. And it seemeth his favour° was so great, as° Antonius, in a letter which is recited verbatim in one of Cicero's° *Philippics*, calleth him *venefica*, 'witch'; as if he had enchanted Caesar.

Augustus raised Agrippa (though of mean birth) to that height, as when he consulted with Maecenas° about the marriage of his daughter Julia, Maecenas took the liberty to tell him, that 'he must either marry his daughter to Agrippa, or take away his life: there was no third way, he had made him so great'. With Tiberius Caesar, Sejanus had ascended to that height, as they two were termed and reckoned as a pair of friends.° Tiberius in a letter to him saith, 'haec pro amicitia nostra non occultavi';° and the whole senate dedicated an altar to Friendship, as to a goddess, in respect of the great dearness of friendship between them two. The like or more was between Septimius Severus° and Plautianus. For he forced his eldest son to marry the daughter of Plautianus; and would often maintain° Plautianus in doing affronts° to his son; and did write also in a letter to the senate, by these words:° 'I love the man so well, as I wish he may over-live° me'. Now if these princes had been as a Trajan° or a Marcus Aurelius,° a man might have thought that this had proceeded of an abundant goodness of nature; but being men so wise,° of such strength and severity of mind, and so° extreme lovers of themselves, as all these were, it proveth most plainly that they found their own felicity (though as great as ever happened to mortal men) but as an half piece,° except they might have a friend to make it entire; and yet, which is more, they were princes that had wives, sons, nephews;° and yet all these could not supply the comfort of friendship.

It is not to be forgotten what Comineus° observeth of his first master, Duke Charles the Hardy; namely, that he would communicate° his secrets with none; and least of all, those secrets which troubled him most. Whereupon he goeth on and saith that towards his latter time that 'closeness° did impair and a little perish° his understanding'. Surely Comineus might have made the same judgment also, if it had pleased him, of his second master Lewis the Eleventh, whose closeness was indeed his tormentor. The parable° of Pythagoras° is dark,° but true; 'Cor ne edito'; 'Eat not the heart'. Certainly, if a man would give it a hard phrase,° those that want friends to open° themselves unto are cannibals of their own hearts. But one thing is most admirable° (wherewith I will conclude this first fruit of friendship), which is, that this communicating of a man's self to his friend works° two contrary effects; for it redoubleth joys, and cutteth griefs in halfs. For there is no man that imparteth° his joys to his friend, but he joyeth° the more: and no man that imparteth his griefs to his friend, but he grieveth the less. So that it is in truth of operation° upon a man's mind, of like virtue° as the alchymists use° to attribute to their stone° for man's body; that it worketh all contrary effects, but still to the good and benefit of nature. But yet without praying in aid of° alchymists, there is a manifest image° of this in the ordinary course of nature. For in bodies,° union° strengtheneth and cherisheth° any natural action; and on the other side weakeneth and dulleth any violent impression:° and even so it is of minds.

The second fruit of friendship is healthful and sovereign° for the understanding, as the first is for the affections.° For friendship maketh indeed a fair day° in the affections, from° storm and tempests; but it maketh daylight in the understanding,° out of darkness and confusion of thoughts. Neither is this to be understood only of faithful counsel, which a man receiveth from his friend; but before you come to that, certain it is that whosoever hath his mind fraught° with many thoughts, his wits and understanding do clarify° and break up,° in the communicating and discoursing with another; he tosseth° his thoughts more easily; he marshalleth° them more orderly; he seeth how they look when they are turned into words: finally, he waxeth° wiser than himself; and that more by an hour's discourse° than by a day's meditation.° It was well said by Themistocles to the king of Persia, that 'speech was like cloth of Arras,° opened and put abroad;° whereby the imagery° doth appear in figure;° whereas in thoughts they lie but as in packs'.° Neither is this second fruit of friendship, in opening the understanding, restrained° only to such friends as are able to give a man counsel (they indeed are best); but even without that, a man learneth of himself, and bringeth his own thoughts to light, and whetteth° his wits as against a stone, which itself cuts not. In a word, a man were better° relate himself° to a statua or picture, than to suffer his thoughts to pass in smother.°

Add now, to make this second fruit of friendship complete, that other point which lieth more open° and falleth within vulgar° observation; which is faithful counsel from a friend. Heraclitus° saith well in one of his enigmas, 'Dry light is ever the best'. And certain it is, that the light that a man receiveth by counsel from another, is drier and purer than that which cometh from his own understanding and judgment; which is ever infused and drenched° in his affections and customs. So as there is as much difference between the counsel that a friend giveth, and that a man giveth himself, as there is between the counsel of a friend and of a flatterer.° For there is no such flatterer as is a man's self; and there is no such remedy against flattery of a man's self, as the liberty of a friend. Counsel is of two sorts; the one concerning manners,° the other concerning business.° For the first, the best preservative° to keep the mind in health is the faithful admonition° of a friend. The calling of a man's self to a strict account° is a medicine, sometime, too piercing and corrosive. Reading good books of morality is a little flat° and dead. Observing our faults in others is sometimes improper° for our case. But the best receipt (best, I say, to work, and best to take)° is the admonition of a friend. It is a strange thing to behold what gross errors and extreme absurdities many (especially of the greater sort)° do commit, for want of a friend to tell them of them;° to the great damage both of their fame and fortune: for, as St. James saith,° they are as men that 'look sometimes into a glass, and presently° forget their own shape and favour'.°

As for business, a man may think, if he will, that two eyes see no more than one; or that a gamester° seeth always more than a looker-on; or that a man in anger is as wise as he that hath said over°

the four and twenty letters;° or that a musket may be shot off as well upon the arm as upon a rest;° and such other fond and high° imaginations, to think himself all in all.° But when all is done, the help of good counsel is that which setteth business straight. And if any man think that he will take counsel, but it shall be by pieces;° asking counsel in one business of one man, and in another business of another man; it is well (that is to say, better perhaps than if he asked none at all); but he runneth° two dangers; one, that he shall not be faithfully counselled; for it is a rare thing, except it be from a perfect and entire friend, to have counsel given, but such as shall be bowed and crooked° to some ends which he hath that giveth it. The other, that he shall have counsel given, hurtful and unsafe° (though with good meaning),° and mixed partly of mischief° and partly of remedy; even as if you would call a physician that is thought good for the cure of the disease you complain of, but is unacquainted with your body; and therefore may put you in way for° a present° cure, but overthroweth your health in some other kind;° and so cure the disease and kill the patient. But a friend that is wholly acquainted with a man's estate° will beware, by furthering any present business, how he dasheth upon° other inconvenience. And therefore rest not upon scattered° counsels; they will rather distract and mislead, than settle and direct.

After these two noble fruits of friendship (peace in the affections, and support of the judgment), followeth the last fruit; which is like the pomegranate, full of many kernels;° I mean aid and bearing a part in° all actions and occasions. Here the best way to represent to life the manifold use of friendship, is to cast° and see how many things there are which a man cannot do himself; and then it will appear that it was a sparing speech° of the ancients, to say,° that 'a friend is another himself'; for that a friend is far more than himself. Men have their time,° and die many times in desire° of some things which they principally take to heart;° the bestowing° of a child, the finishing of a work, or the like. If a man have a true friend, he may rest almost secure that the care of those things will continue after him. So that a man hath, as it were, two lives in his desires.° A man hath a body, and that body is confined to a place; but where friendship is, all offices° of life are as it were granted to him and his deputy. For he may exercise° them by his friend. How many things are there which a man cannot, with any face° or comeliness,° say or do himself?° A man can scarce allege his own merits with modesty, much less extol them; a man cannot sometimes brook° to supplicate° or beg; and a number of the like.° But all these things are graceful in a friend's mouth, which are blushing° in a man's own. So again, a man's person° hath many proper° relations which he cannot put off. A man cannot speak to his son but as a father; to his wife but as a husband; to his enemy but upon terms:° whereas a friend may speak as the case requires, and not as it sorteth with° the person. But to enumerate these things were endless; I have given the rule, where a man cannot fitly play° his own part: if he have not a friend, he may quit the stage.

28. Of Expense°

RICHES are for spending, and spending for honour° and good actions. Therefore extraordinary expense must be limited° by the worth of the occasion; for voluntary undoing° may be as well for a man's country as for the kingdom of heaven.° But ordinary expense ought to be limited by a man's estate; and governed with such regard, as it be within his compass;° and not subject to deceit and abuse° of servants; and ordered to the best shew,° that the bills may be less than the estimation abroad.° Certainly, if a man will keep but of even hand,° his ordinary expenses ought to be but to the half of his receipts;° and if he think to wax° rich, but to the third part. It is no baseness° for the greatest to descend and look into their own estate. Some forbear it, not upon negligence alone, but doubting° to bring themselves into melancholy, in respect° they shall find it broken.° But wounds cannot be cured without searching.° He that cannot look into his own estate at all, had need° both choose well those whom he employeth, and change them often; for new° are more timorous and less subtle.° He that can look into his estate but seldom, it behoveth him to turn all to certainties.° A man had need, if he be plentiful° in some° kind of expense, to be as saving again° in some other. As if he be plentiful in diet,° to be saving in apparel; if he be plentiful in the hall,° to be saving in the stable; and the like. For he that is plentiful in expenses of all kinds will hardly be preserved from decay.° In clearing° of a man's° estate, he may as well° hurt° himself in being too sudden,° as in letting it run on too long. For hasty selling is commonly as disadvantageable as interest.° Besides, he that clears° at once° will relapse; for finding himself out of straits,° he will revert to his customs:° but he that cleareth by degrees induceth a habit of frugality,° and gaineth as well upon° his mind as upon his estate. Certainly, who° hath a state° to repair, may not despise small things;

and commonly it is less dishonourable to abridge° petty° charges,° than to stoop to petty gettings. A man ought warily to begin charges which once begun will continue: but in matters that return not° he may be more magnificent.

29. Of the True Greatness of Kingdoms and Estates°

The speech° of Themistocles the Athenian, which was haughty and arrogant in taking so much to himself, had° been a grave and wise observation and censure,° applied at large° to others. Desired at a feast to touch a lute, he said, 'He could not fiddle,° but yet he could make a small town a great city'. These words (holpen° a little with a metaphor)° may express two differing abilities in those that deal in business of estate.° For if a true survey be taken of counsellors and statesmen, there may be found (though rarely) those which can make a small state great, and yet cannot fiddle: as on the other side, there will be found a great many that can fiddle very cunningly, but yet are so far from being able to make a small state great, as their gift° lieth the other way; to bring a great and flourishing estate to ruin and decay. And, certainly those degenerate arts and shifts,° whereby many counsellors and governors gain both favour with their masters and estimation with the vulgar, deserve no better name than fiddling; being things rather pleasing for the time, and graceful° to themselves only, than tending to the weal° and advancement of the state which they serve. There are also (no doubt) counsellors and governors which may be held sufficient° (*negotiis pares*),° able to manage° affairs, and to keep them from precipices° and manifest inconveniences;° which° nevertheless are far from the ability to raise and amplify an estate in power, means,° and fortune. But be the workmen what they may be, let us speak of the work; that is, the true Greatness° of Kingdoms and Estates, and the means° thereof. An argument° fit for great and mighty princes to have in their hand; to the end that neither by over-measuring° their forces, they leese themselves in vain enterprises; nor on the other side, by undervaluing them, they descend to fearful° and pusillanimous counsels.°

The greatness of an estate in bulk and territory, doth fall under measure;° and the greatness of finances and revenue doth fall under computation. The population may appear° by musters;° and the number and greatness of cities and towns by cards° and maps. But yet there is not anything amongst° civil affairs more subject to error, than the right valuation and true judgment concerning the power and forces of an estate. The kingdom of heaven is compared,° not to any great kernel or nut, but to a grain of mustard-seed; which is one of the least° grains, but hath in it a property and spirit hastily to get up° and spread. So are there states great in territory, and yet not apt° to enlarge or command; and some that have but a small dimension of stem,° and yet apt to be the foundations of great monarchies.

Walled towns, stored° arsenals and armories,° goodly races° of horse, chariots of war, elephants, ordnance,° artillery,° and the like; all this is but a sheep in a lion's skin,° except the breed and disposition of the people be stout° and warlike. Nay, number (itself) in armies importeth° not much, where the people is of weak courage; for (as Virgil° saith) 'It never troubles a wolf how many the sheep be'. The army of the Persians in the plains of Arbela was such a vast sea of people, as it did somewhat astonish the commanders in Alexander's army; who came to him therefore, and wished° him to set upon° them by night; but he answered,° 'He would not pilfer the victory'. And the defeat was easy. When Tigranes the Armenian, being encamped upon a hill with four hundred thousand men, discovered the army of the Romans, being not above fourteen thousand, marching towards him, he made himself merry with it, and said,° 'Yonder men are too many for an ambassage,° and too few for a fight'. But, before the sun set, he found them enough to give him the chase° with infinite slaughter. Many are the examples of the great odds° between number and courage: so that a man may truly make a judgment, that the principal point of greatness in any state is to have a race of military men.° Neither is money the sinews of war (as it is trivially said),° where the sinews of men's arms, in base and effeminate people, are failing. For Solon° said well to Croesus (when in ostentation he shewed him his gold), 'Sir, if any other come that hath better iron than you, he will be master of all this gold'. Therefore let any prince or state think soberly° of his forces, except his militia° of natives be of good and valiant soldiers. And let princes, on the other side, that have subjects of martial disposition, know their own strength; unless they be otherwise wanting unto themselves.° As for mercenary forces (which is the help° in this case), all examples show that whatsoever estate or prince doth rest° upon them, he may spread° his feathers for a time, but he will mew° them soon after.

The blessing° of Judah and Issachar will never meet; that 'the same people or nation should be both the lion's whelp and the ass between burthens';° neither will it be, that a people overlaid° with taxes should ever become valiant and martial. It is true that taxes levied by consent of the estate° do abate° men's courage less: as it hath been seen notably° in the excises° of the Low Countries; and, in some degree, in the subsidies° of England. For you must note that we speak now of the heart and not of the purse. So that although the same tribute° and tax, laid by consent or by imposing, be all one° to the purse, yet it works diversely° upon the courage. So that you may conclude, that no people over-charged with tribute is fit for empire.

Let states that aim at greatness, take heed how° their nobility and gentlemen do multiply too fast. For that maketh the common subject grow to be a peasant° and base swain,° driven out of heart,° and in effect but the gentleman's labourer. Even as you may see in coppice° woods; if you leave your staddles° too thick, you shall never have clean underwood, but shrubs and bushes. So in countries, if the gentlemen be too many, the commons will be base;° and you will bring it to that, that not the hundred° poll° will be fit for an helmet;° especially as to the infantry, which is the nerve of an army;° and so there will be great population and little strength. This which I speak of hath been no where better seen than by comparing of England and France; whereof England, though far less in territory and population, hath been (nevertheless) an over-match;° in regard° the middle people° of England make good soldiers, which the peasants of France do not.° And herein the device° of king Henry the Seventh (whereof I have spoken° largely in the history of his life) was profound and admirable; in making farms and houses of husbandry° of a standard;° that is, maintained with such a proportion of land unto them, as may breed a subject to live in convenient° plenty and no servile condition; and to keep the plough in the hands of the owners, and not mere hirelings.° And thus indeed you shall attain to Virgil's character° which he gives to ancient Italy:

Terra potens armis atque ubere glebae.°

Neither is that state° (which, for anything I know, is almost peculiar to England, and hardly to be found anywhere else, except it be perhaps in Poland) to be passed over; I mean the state of free servants° and attendants upon noblemen and gentlemen; which are no ways° inferior unto the yeomanry° for arms. And therefore out of all question, the splendour and magnificence and great retinues and hospitality° of noblemen and gentlemen, received into custom,° doth much conduce° unto martial greatness. Whereas, contrariwise, the close and reserved° living of noblemen and gentlemen causeth a penury° of military forces.

By all means it is to be procured, that the trunk of Nebuchadnezzar's tree° of monarchy be great enough to bear the branches and the boughs;° that is, that the natural° subjects of the crown or state bear a sufficient proportion to the stranger° subjects that they govern. Therefore all states that are liberal of° naturalisation towards strangers are fit for empire.° For to think that an handful of people can, with the greatest courage and policy in the world, embrace too large extent of dominion, it° may hold for a time, but it will fail suddenly. The Spartans were a nice° people in point of° naturalisation; whereby, while they kept their compass,° they stood firm; but when they did spread, and their boughs were becomen too great for their stem, they became a windfall° upon the sudden.° Never any state was in this point so open to receive strangers into their body as were the Romans. Therefore it sorted with° them accordingly; for they grew to the greatest monarchy.° Their manner was to grant naturalisation (which they called *jus civitatis*),° and to grant it in the highest degree; that is, not only *jus commercii, jus connubii, jus haereditatis*; but also *jus suffragii*, and *jus honorum*.° And this not to singular° persons alone, but likewise to whole families; yea to cities, and sometimes to nations. Add to this their custom of plantation of colonies;° whereby the Roman plant was removed° into the soil of other nations. And putting both constitutions° together, you will say that it was not the Romans that spread upon the world, but it was the world that spread upon the Romans; and that was the sure way of greatness. I have marvelled° sometimes at Spain, how they clasp and contain° so large dominions with so few natural° Spaniards; but sure the whole compass of Spain is a very great body° of a tree; far above Rome and Sparta at the first.° And besides, though they have not had that usage to naturalise liberally, yet they have that which is next to it; that is, to employ almost indifferently all nations in their militia° of ordinary soldiers; yea and sometimes in their highest° commands. Nay it seemeth at this instant they are sensible° of this want of natives; as by the Pragmatical Sanction,° now published, appeareth.

It is certain, that sedentary and within-door arts,° and delicate manufactures° (that require° rather the finger than the arm), have in their nature a contrariety to a military disposition. And generally, all warlike people are a little idle, and love danger better than travail.° Neither must they be too much broken of° it, if they shall° be preserved in vigour. Therefore it was great advantage in the ancient states of Sparta, Athens, Rome, and others, that they had the use of slaves, which commonly did rid° those manufactures.° But that is abolished,° in greatest part, by the Christian law. That which cometh nearest to it, is to leave those arts chiefly to strangers (which for that purpose are the more easily to be received),° and to contain° the principal bulk of the vulgar natives° within those three kinds,—tillers° of the ground; free servants; and handicraftsmen° of strong and manly arts, as smiths, masons, carpenters, &c: not reckoning professed° soldiers.

But above all, for empire and greatness, it importeth° most that a nation do profess arms as their principal honour, study,° and occupation. For the things which we formerly have spoken of are but habilitations° towards arms; and what is habilitation without intention and act?° Romulus,° after his death (as they report or feign), sent a present° to the Romans, that above all they should intend° arms; and then they should prove the greatest empire of the world. The fabric° of the state of Sparta was wholly (though not wisely) framed° and composed to that scope° and end. The Persians and Macedonians had it for a flash.° The Gauls, Germans, Goths, Saxons, Normans, and others, had it for a time. The Turks have it at this day, though in great declination.° Of Christian Europe, they that have it are, in effect, only the Spaniards. But it is so plain that every man profiteth in that° he most intendeth, that it needeth not to be stood upon.° It is enough to point at° it; that no nation which doth not directly° profess arms, may look to° have greatness fall into their mouths. And on the other side, it is a most certain oracle of time,° that those states that continue long in that profession (as the Romans and Turks principally have done) do wonders. And those that have professed arms but for an age, have notwithstanding° commonly attained that greatness in that age which maintained them long after, when their profession and exercise of arms hath grown to decay.

Incident to this point° is, for a state to have those laws or customs which may reach forth° unto them just occasions (as may be pretended)° of war. For there is that justice° imprinted in the nature of men, that they enter not upon° wars (whereof so many calamities do ensue) but upon some, at the least specious,° grounds and quarrels.° The Turk hath at hand, for cause of war, the propagation of his law or sect;° a quarrel that he may always command.° The Romans, though they esteemed the extending the limits of their empire to be great honour to their generals when it was done, yet they never rested upon that alone to begin a war. First therefore, let nations that pretend to° greatness have this; that they be sensible of° wrongs, either upon borderers,° merchants, or politique ministers;° and that they sit° not too long upon a provocation. Secondly, let them be prest° and ready to give aids and succours to their confederates; as it ever was with the Romans; insomuch as, if the confederates had leagues defensive with divers other states, and, upon invasion offered,° did implore their aids severally,° yet the Romans would ever be the foremost, and leave it to none other to have the honour. As for the wars which were anciently made on the behalf of a kind of party,° or tacit conformity° of estate, I do not see how they may be well justified: as when the Romans° made a war for the liberty of Graecia; or when the Lacedaemonians and Athenians° made wars to set up or pull down democracies and oligarchies; or when wars were made by foreigners, under the pretence of justice or protection, to deliver the subjects of others from tyranny and oppression; and the like. Let it suffice, that no estate expect to be great, that is not awake° upon any just occasion of arming.

No body can be healthful without exercise, neither natural body nor politic; and certainly to a kingdom or estate, a just and honourable war is the true exercise. A civil war indeed is like the heat of a fever; but a foreign war is like the heat of exercise, and serveth to keep the body in health; for in a slothful peace,° both courages° will effeminate° and manners corrupt. But howsoever it be for happiness, without all question, for greatness it maketh,° to be still for the most part° in arms; and the strength of a veteran° army (though it be a chargeable° business) always on foot,° is that which commonly giveth the law,° or at least the reputation, amongst all neighbour states; as may well be seen in° Spain, which hath had, in one part or other, a veteran army almost continually, now by the space of six score° years.

To be master of the sea is an abridgment° of a monarchy. Cicero, writing to Atticus of Pompey his° preparation against Caesar, saith, 'Consilium Pompeii plane Themistocleum est; putat enim, qui mari potitur, eum rerum potiri'.° And, without doubt, Pompey had° tired out Caesar, if upon vain confidence° he had not left° that way.° We see the great effects of battles by sea. The battle of Actium° decided the empire of the world. The battle of Lepanto° arrested the greatness of the Turk. There be many examples where sea-fights have been final° to the war; but this is when princes or states have set up their rest° upon the battles. But thus much is certain, that he that commands the sea is at great liberty, and may take as much and as little° of the war as he will. Whereas those that be strongest by land are many times nevertheless in great straits.° Surely, at this day, with us of Europe,° the vantage° of strength at sea (which is one of the principal dowries of this kingdom of Great Britain) is great; both because most of the kingdoms of Europe are not merely° inland, but girt° with the sea most part of their compass;° and because the wealth of both Indies seems in great part but an accessary° to the command of the seas.

The wars of latter ages seem to be made in the dark, in respect of the glory and honour which reflected° upon men from the wars in ancient time. There be now, for martial encouragement, some degrees° and orders of chivalry; which nevertheless are conferred promiscuously° upon soldiers and no soldiers;° and some remembrance perhaps upon the scutcheon;° and some hospitals° for maimed soldiers; and such like things. But in ancient times, the trophies° erected upon the place of the victory; the funeral laudatives° and monuments for those that died in the wars; the crowns and garlands personal;° the style of Emperor,° which the great kings of the world after° borrowed; the triumphs of the generals upon their return; the great donatives° and largesses° upon the disbanding° of the armies; were things able° to inflame all men's courages. But above all, that of the Triumph,° amongst the Romans, was not pageants or gaudery,° but one of the wisest and noblest institutions that ever was. For it contained three things; honour to the general; riches to the treasury out of the spoils; and donatives to the army. But that honour perhaps were° not fit for monarchies; except it be in the person of the monarch himself, or his sons, as it came to pass in the times of the Roman emperors, who did impropriate° the actual triumphs to themselves and their sons, for such wars as they did achieve° in person; and left only, for wars achieved by subjects, some triumphal garments and ensigns° to the general.

To conclude: no man can 'by care taking' (as the Scripture° saith) 'add a cubit to his stature', in this little model° of a man's body; but in the great frame of kingdoms and commonwealths, it is in the power of princes or estates to add amplitude and greatness° to their kingdoms; for by introducing such ordinances, constitutions, and customs, as we have now touched,° they may sow greatness to° their posterity and succession. But these things are commonly not observed,° but left to take their chance.

30. Of Regiment° of Health

THERE is a wisdom in this° beyond the rules of physic:° a man's own observation, what he finds good of, and what he finds hurt of,° is the best physic to preserve health. But it is a safer conclusion to say, 'This agreeth not well with me, therefore I will not continue it'; than this, 'I find no offence of this, therefore I may use it'. For strength of nature in youth passeth over many excesses, which are owing° a man till his age. Discern of the coming on of years, and think not to do the same things still;° for age will not be defied. Beware of sudden change in any great point of diet, and if necessity inforce it, fit the rest to it. For it is a secret both in nature and state, that it is safer to change many things than one.° Examine thy customs of diet, sleep, exercise, apparel, and the like; and try, in any thing thou shalt judge hurtful, to discontinue it by little and little; but so, as° if thou dost find any inconvenience by° the change, thou come back to it again: for it is hard to distinguish that which is generally held good and wholesome, from that which is good particularly,° and fit for thine own body. To be free-minded° and cheerfully disposed at hours of meat and of sleep and of exercise, is one of the best precepts of long lasting. As for the passions and studies° of the mind: avoid envy; anxious fears; anger fretting inwards;° subtle and knotty° inquisitions;° joys and exhilarations in excess; sadness not communicated.° Entertain° hopes; mirth rather than joy; variety of delights, rather than surfeit of them; wonder and admiration, and therefore novelties; studies that fill the mind with splendid and illustrious objects, as histories, fables, and contemplations of nature.

If you fly physic in health altogether, it will be too strange° for your body when you shall need it. If you make it too familiar, it will work no extraordinary effect when sickness cometh. I commend° rather some diet for certain seasons, than frequent use of physic, except it be grown into a custom.° For those diets alter the body more, and trouble it less. Despise no new accident° in your body, but ask opinion of it. In sickness, respect° health principally; and in health, action.° For those that put their bodies to endure° in health, may in most sicknesses, which are not very sharp,° be cured only with diet and tendering.° Celsus° could never have spoken° it as a physician, had he not been a wise man withal, when he giveth it for one of the great precepts of health and lasting,° that a man° do vary and interchange contraries,° but with an inclination to the more benign extreme: use fasting and full eating, but rather full eating; watching° and sleep, but rather sleep; sitting° and exercise, but rather exercise; and the like. So shall nature be cherished, and yet taught masteries.°

Physicians are some of them so pleasing° and conformable to the humour° of the patient, as they press° not the true cure of the disease; and some other are so regular° in proceeding according to art° for the disease, as they respect not sufficiently the condition of the patient. Take one of a middle temper; or if it may not be found in one man, combine two of either° sort; and forget not to call as well the best acquainted with your body, as the best reputed of for his faculty.°

31. Of Suspicion

Suspicions amongst thoughts are like bats amongst birds, they ever fly by twilight. Certainly they are to be repressed, or at the least well guarded:° for they cloud° the mind; they leese° friends; and they check with° business, whereby° business cannot go on currently° and constantly. They dispose kings to tyranny, husbands to jealousy, wise men to irresolution and melancholy. They are defects, not in the heart,° but in the brain; for they take place° in the stoutest° natures; as in the example of Henry the Seventh of England. There was not a more suspicious° man, nor a more stout.° And in such a composition° they do small hurt. For commonly they are not admitted, but with examination, whether they be likely or no? But in fearful° natures they gain ground too fast. There is nothing makes a man suspect much, more than to know little; and therefore men should remedy° suspicion by procuring to know more, and not to keep their suspicions in smother.° What would men have? Do they think those they employ and deal with are saints? Do they not think they will have their own ends, and be truer to themselves than to them? Therefore there is no better way to moderate° suspicions, than to account upon° such suspicions as true and yet to bridle° them as false. For so far a man ought to make use of suspicions, as to provide,° as if that should be true that he suspects, yet it may do him no hurt.

Suspicions that the mind of itself gathers are but buzzes;° but suspicions that are artificially° nourished, and put into men's heads by the tales and whisperings of others, have stings. Certainly, the best mean° to clear the way in this same wood° of suspicions, is frankly to communicate° them with the party° that he suspects; for thereby he shall be sure to know more of the truth° of them than he did before; and withal° shall make that party more circumspect not to give further cause of suspicion. But this would° not be done to men of base° natures; for they, if they find themselves once suspected, will never be true. The Italian says, 'Sospetto licentia fede';° as if suspicion did give a passport° to faith; but it ought rather to kindle it to discharge itself.°

32. Of Discourse°

Some in their discourse desire rather commendation of wit,° in being able to hold° all arguments, than of judgment, in discerning° what is true; as if it were a praise to know what might be said, and not what should be thought.° Some have certain common places and themes° wherein they are good, and want variety;° which kind of poverty is for the most part tedious, and when it is once perceived, ridiculous. The honourablest part° of talk is to give the occasion;° and again to moderate° and pass to somewhat else; for then a man leads the dance. It is good, in discourse° and speech of conversation, to vary and intermingle speech of the present occasion° with arguments,° tales with reasons,° asking of questions with telling of opinions, and jest with earnest: for it is a dull thing to tire, and, as we say now, to jade,° any thing too far.

As for jest,° there be certain things which ought to be privileged° from it; namely, religion, matters of state, great persons,° any man's present° business of importance, and any case that deserveth pity. Yet there be some that think their wits have been asleep, except they dart out° somewhat that is piquant,° and to the quick.° That is a vein° which would° be bridled;

Parce, puer, stimulis, et fortius utere loris.°

And generally, men ought to find the difference between saltness° and bitterness. Certainly, he that hath a satirical vein, as he maketh others afraid of his wit, so he had need be afraid of others'° memory.

He that questioneth much, shall learn much, and content° much; but especially if he apply° his questions to the skill of the persons whom he asketh; for he shall give them occasion to please themselves in speaking, and himself shall continually gather knowledge. But let his questions not be troublesome; for that is fit for a poser.° And let him be sure to leave other men their turns to speak. Nay, if there be any that would reign° and take up all the time, let him find means to take them off,° and to bring others on; as musicians use° to do with those that dance too long galliards.°

If you dissemble sometimes your knowledge of that° you are thought to know, you shall be thought another time to know that you know not. Speech of a man's self° ought to be seldom, and well chosen. I knew one was wont to say in scorn, 'He must needs be a wise man, he speaks so much of himself': and there is but one case wherein a man may commend himself with good grace,° and that is in commending virtue in another; especially if it be such a virtue whereunto himself pretendeth.°

Speech of touch° towards others should be sparingly used; for discourse ought to be as a field,° without coming home° to any man. I knew two noblemen, of the west part of England, whereof the one was given to scoff,° but kept ever royal cheer° in his house; the other would ask of those that had been at the other's table, 'Tell truly, was there never a flout° or dry blow° given?' To which the guest would answer, 'Such and such a thing passed'.° The lord would say, 'I thought he would mar° a good dinner'.

Discretion of speech is more than eloquence; and to speak agreeably° to him with whom we deal, is more than to speak in good words or in good order. A good continued° speech, without a good speech of interlocution, shews slowness; and a good reply or second speech, without a good settled speech, sheweth shallowness and weakness. As we see in beasts, that those that are weakest in the course, are yet nimblest in the turn;° as it is betwixt the greyhound and the hare. To use too many circumstances° ere° one come to the matter, is wearisome; to use none at all, is blunt.

33. Of Plantations°

Plantations are amongst ancient, primitive, and heroical° works. When the world was young° it begat more children; but now it is old it begets fewer: for I may justly account new plantations to be the children of former° kingdoms. I like a plantation in a pure soil;° that is, where people° are not displanted° to the end to plant in others. For else it is rather an extirpation° than a plantation. Planting of countries is like planting of woods; for you must make account° to leese° almost twenty years profit, and expect your recompense° in the end. For the principal thing° that hath been the destruction of most plantations, hath been the base° and hasty drawing° of profit in the first years. It is true, speedy profit is not to be neglected, as far as may stand with° the good of the plantation, but no further.

It is a shameful and unblessed° thing to take the scum of people,° and wicked condemned men,° to be the people with whom you plant;° and not only so, but it spoileth the plantation; for they will ever live like rogues, and not fall to° work, but be lazy, and do mischief, and spend° victuals,° and be quickly weary, and then certify° over to their country to the discredit of the plantation. The people wherewith you plant ought to be gardeners, ploughmen, labourers, smiths, carpenters, joiners, fishermen, fowlers,° with some few apothecaries, surgeons, cooks, and bakers.

In a country of plantation, first look about what kind of victual the country yields of itself° to hand; as chestnuts, wallnuts, pine-apples,° olives, dates, plums, cherries, wild honey, and the like; and make use of them. Then consider what victual or esculent° things there are, which grow speedily, and within the year; as parsnips, carrots, turnips, onions, radish, artichokes of Hierusalem,° maize, and the like. For° wheat, barley, and oats, they ask° too much labour; but with pease and beans you may begin, both because they ask less labour, and because they serve for meat° as well as for bread. And of rice likewise cometh a great increase,° and it is a kind of meat. Above all, there ought to be brought store° of biscuit,° oat-meal, flour, meal,° and the like, in the beginning, till bread may be had. For beasts, or birds, take chiefly such as are least subject to diseases, and multiply fastest; as swine, goats, cocks, hens, turkeys, geese, house-doves, and the like. The victual in plantations ought to be expended° almost as in a besieged town; that is, with certain allowance.° And let the main part of the ground employed to° gardens or corn, be to a common stock;° and to be laid in,° and stored up, and then delivered out in proportion;° besides some spots of ground that any particular person will manure° for his own private.°

Consider likewise what commodities the soil where the plantation is doth naturally yield, that they may some way help to defray the charge° of the plantation, (so° it be not, as was said, to the untimely prejudice° of the main business), as it hath fared with tobacco° in Virginia. Wood commonly aboundeth but too much; and therefore timber is fit to be one.° If there be iron ore, and streams whereupon to set° the mills, iron is a brave° commodity where wood aboundeth. Making of bay-salt,° if the climate be proper for it, would be put in experience.° Growing silk° likewise, if any be, is a likely° commodity. Pitch and tar, where store of firs and pines are, will not fail. So drugs and sweet woods,° where they are, cannot but yield great profit. Soap-ashes° likewise, and other things that may be thought of. But moil° not too much under ground; for the hope of mines° is very uncertain, and useth to make the planters lazy in other things.

For government, let it be in the hands of one, assisted with some counsel;° and let them have commission° to exercise martial laws,° with some limitation.° And above all, let men make that° profit of being in the wilderness, as they have God always, and his service, before their eyes. Let not the government of the plantation depend upon too many counsellors and undertakers° in the country that planteth, but upon a temperate° number; and let those be rather noblemen and gentlemen, than merchants; for they look ever to the present° gain. Let there be freedoms from custom,° till the plantation be of strength; and not only freedom from custom, but freedom to carry their commodities where they may make their best° of them, except there be some special cause of caution.

Cram not in people, by sending too fast company° after company; but rather hearken° how they waste,° and send supplies° proportionably; but so as the number may live well in the plantation, and not by surcharge° be in penury. It hath been a great endangering to the health of some plantations, that they have built along the sea and rivers, in marish° and unwholesome grounds. Therefore, though you begin there, to avoid carriage° and other like discommodities,° yet build still° rather upwards° from the streams, than along. It concerneth likewise the health of the plantation that they have good store of salt with them, that they may use it in their victuals when it shall be necessary.°

If you plant where savages are, do not only entertain° them with trifles and gingles;° but use them justly and graciously, with sufficient guard° nevertheless; and do not win their favour by helping them to invade their enemies, but for their defence it is not amiss,° and send oft of them° over to the country that plants, that they may see a better condition° than their own, and commend it when they return. When the plantation grows to strength, then it is time to plant with women as well as with men; that the plantation may spread into generations,° and not be ever pieced° from without. It is the sinfullest thing in the world to forsake or destitute° a plantation once in forwardness;° for besides the dishonour, it is the guiltiness of blood° of many commiserable° persons.

34. Of Riches

I CANNOT call Riches better than the baggage° of virtue. The Roman word° is better, *impedimenta*. For as the baggage is to an army, so is riches to virtue. It cannot be spared nor left behind, but it hindereth the march; yea and the care of it sometimes loseth° or disturbeth the victory. Of great riches there is no real use, except it be in the distribution;° the rest is but conceit.° So saith Salomon, 'Where much is, there are many to consume it; and what hath the owner but the sight of it with his eyes?'° The personal fruition° in any man cannot reach° to feel° great riches: there is a custody of them; or a power of dole and donative° of them; or a fame of them; but no solid use to the owner. Do you not see what feigned° prices are set upon little stones and rarities? and what works of ostentation are undertaken, because° there might seem to be some use of great riches? But then you will say, they may be of use to buy men° out of dangers or troubles. As Salomon saith, 'Riches are as a stronghold, in the imagination of the rich man'.° But this is excellently expressed that it is in imagination, and not always in fact. For certainly great riches have sold° more men than they have bought out.° Seek not proud° riches, but such as thou mayest get justly,° use soberly, distribute cheerfully, and leave contentedly. Yet have no abstract° nor friarly° contempt of them. But distinguish, as Cicero saith well of Rabirius Posthumus, 'In studio rei amplificandae apparebat, non avaritiae praedam, sed instrumentum bonitati quaeri'.° Hearken also to Salomon, and beware of hasty gathering of riches; 'Qui festinat ad divitias, non erit insons'.° The poets feign,° that when Plutus° (which is Riches) is sent from Jupiter, he limps and goes slowly; but when he is sent from Pluto,° he runs and is swift of foot. Meaning that riches gotten by good means and just labour pace° slowly; but when they come by the death of others (as by the course of° inheritance, testaments, and the like), they come tumbling upon a man. But it might be applied likewise to Pluto, taking him for the devil. For when riches come from the devil (as by fraud and oppression and unjust means), they come upon speed.°

The ways to enrich° are many, and most of them foul. Parsimony is one of the best, and yet is not innocent; for it withholdeth men from works of liberality and charity. The improvement of the ground is the most natural obtaining° of riches; for it is our great mother's blessing, the earth's; but it is slow. And yet where men of great wealth do stoop° to husbandry,° it multiplieth riches exceedingly. I knew a nobleman° in England, that had the greatest° audits° of any man in my time; a great grazier,° a great sheep-master,° a great timber man,° a great collier,° a great corn-master,° a great lead-man,° and so of iron, and a number of the like points° of husbandry.° So as the earth seemed a sea to him, in respect of the perpetual importation. It was truly observed by one,° that himself came very hardly° to a little riches, and very easily to great riches. For when a man's stock is come to that, that he can expect the prime° of markets, and overcome° those bargains° which for their greatness° are few men's money,° and be partner in the industries° of younger men, he cannot but increase mainly.°

The gains of ordinary trades° and vocations° are honest; and furthered by two things chiefly; by diligence, and by a good name for good and fair dealing. But the gains of bargains° are of a more doubtful nature; when men shall wait upon° others' necessity, broke° by servants and instruments° to draw them on, put off° others cunningly that would be better chapmen,° and the like practices,° which are crafty and naught.° As for the chopping° of bargains, when a man buys not to hold° but to sell over again, that commonly grindeth double,° both upon the seller and upon the buyer. Sharings° do greatly enrich, if the hands be well chosen that are trusted. Usury is the certainest° means of gain, though one of the worst; as that whereby a man doth eat his bread 'in sudore vultus alieni';° and besides, doth plough upon Sundays.° But yet certain though it be, it hath flaws;° for that the scriveners and brokers° do value° unsound° men to serve their own turn.

The fortune in being the first in an invention or in a privilege,° doth cause sometimes a wonderful overgrowth in riches, as it was with the first sugar man° in the Canaries.° Therefore if a man can play the true logician,° to have as well judgment as invention, he may do great matters; especially if the times be fit. He that resteth upon gains certain,° shall hardly grow to great riches; and he that puts all upon adventures,° doth oftentimes break° and come to poverty: it is good therefore to guard° adventures with certainties,° that may uphold° losses. Monopolies, and coemption of wares° for re-sale, where they are not restrained,° are great means to enrich; especially if the party° have intelligence° what things are like to come into request, and so store° himself beforehand. Riches gotten by service, though it be of the best rise,° yet when they are gotten by flattery, feeding humours,° and other servile conditions, they may be placed amongst the worst. As for fishing for testaments and executorships° (as Tacitus saith of Seneca, 'testamenta et orbos tamquam indagine capi'),° it is yet worse; by how much° men submit themselves to meaner persons than in service.

Believe not much them that° seem to despise riches; for they despise them that despair of them; and none worse° when they come to them. Be not penny-wise;° riches have wings, and sometimes they fly away of themselves,° sometimes they must be set flying to bring in more. Men leave their riches either to their kindred, or to the public; and moderate portions° prosper best in both. A great state° left to an heir, is as a lure° to all the birds of prey round about to seize on him, if he be not the better stablished° in years and judgment. Likewise glorious° gifts and foundations are like 'sacrifices without salt';° and but the painted sepulchres° of alms, which soon will putrefy° and corrupt inwardly. Therefore measure not thine advancements° by quantity, but frame them by measure:° and defer not charities till death; for, certainly, if a man weigh it rightly, he that doth so is rather liberal of another man's than of his own.°

35. Of Prophecies

I MEAN not to speak of divine prophecies;° nor of heathen oracles; nor of natural° predictions; but only of prophecies that have been of certain memory, and from hidden causes. Saith the Pythonissa° to Saul,° 'To-morrow thou and thy son shall be with me'. Homer hath these verses:

> At domus Aeneae cunctis dominabitur oris,
> Et nati natorum, et qui nascentur ab illis.°

A prophecy, as it seems, of the Roman empire. Seneca the tragedian hath these verses:

> ———Venient annis
> Saecula seris, quibus Oceanus
> Vincula rerum laxet, et ingens
> Pateat Tellus, Tiphysque novos
> Detegat orbes; nec sit terris
> Ultima Thule.°

—a prophecy of the discovery of America. The daughter of Polycrates° dreamed that Jupiter bathed her father, and Apollo anointed him; and it came to pass that he was crucified in an open place, where the sun made his body run with sweat, and the rain washed it. Philip of Macedon° dreamed he sealed up his wife's belly; whereby he did expound it,° that his wife should be barren; but Aristander the soothsayer° told him his wife was with child, because men do not use to seal vessels that are empty. A phantasm° that appeared to M. Brutus in his tent, said to him, 'Philippis iterum me videbis'.° Tiberius said to Galba, 'Tu quoque, Galba, degustabis imperium'.° In Vespasian's time,° there went° a prophecy in the East, that those that should come forth of Judea should reign over the world: which though it may be was meant of our Saviour, yet Tacitus expounds it of° Vespasian. Domitian° dreamed, the night before he was slain, that a golden head was growing out of the nape of his neck: and indeed the succession° that followed him, for many years, made golden times. Henry the Sixth° of England said of Henry the Seventh, when he was a lad, and gave him water, 'This is the lad that shall enjoy the crown for which we strive'.

When I was in France,° I heard from one Dr. Pena, that the Queen Mother,° who was given° to curious° arts, caused the King her husband's nativity° to be calculated, under a false name; and the astrologer gave a judgment, that he should be killed in a duel; at which the Queen laughed, thinking her husband to be above challenges and duels:° but he was slain upon a course at tilt,° the splinters of the staff of Montgomery going in at his beaver.° The trivial° prophecy, which I heard when I was a child, and queen Elizabeth was in the flower of her years, was,

> When hempe° is sponne
> England's done:

whereby it was generally conceived, that after the princes had reigned which had the principial° letters of that word *hempe* (which were Henry, Edward, Mary, Philip, and Elizabeth), England should come to utter confusion;° which, thanks be to God, is verified only in the change of the name; for that the King's style is now no more of England, but of Britain.° There was also another prophecy, before the year of eighty-eight,° which I do not well understand.

> There shall be seen upon a day,
> Between the Baugh° and the May,°
> The black fleet of Norway.
> When that that is come and gone,
> England build houses of lime and stone,
> For after wars shall you have none.

It was generally conceived to be meant of the Spanish fleet that came in eighty-eight: for that the king of Spain's surname, as they say, is Norway. The prediction of Regiomontanus,°

> Octogesimus octavus mirabilis annus,°

was thought likewise accomplished in the sending of that great fleet, being the greatest in strength, though not in number, of all that ever swam upon the sea. As for Cleon's dream,° I think it was a jest. It was, that he was devoured of a long dragon; and it was expounded of a maker of sausages, that troubled him exceedingly. There are numbers of the like kind; especially if you include dreams, and predictions of astrology. But I have set down these few only of certain credit,° for example.

My judgment is, that they ought all to be despised;° and ought to serve but for winter talk° by the fireside. Though when I say *despised*, I mean it as for belief; for otherwise, the spreading or publishing of them is in no sort° to be despised. For they have done much mischief; and I see many severe laws° made to suppress them. That that hath given them grace,° and some credit, consisteth in three things. First, that men mark° when they hit, and never mark when they miss; as they do generally also of dreams. The second is, that probable conjectures, or obscure traditions, many times turn themselves into prophecies; while the nature of man, which coveteth divination,° thinks it no peril to foretell that which indeed they do but collect.° As that of Seneca's verse.° For so much was then° subject to demonstration, that the globe of the earth had great parts° beyond the Atlantic, which might be probably conceived not to be all sea: and adding thereto the tradition in Plato's *Timaeus*, and his *Atlanticus*,° it might encourage one° to turn it to a prediction. The third and last° (which is the great° one) is, that almost all of them, being infinite in number, have been impostures, and by idle and crafty° brains merely° contrived and feigned after the event past.

36. Of Ambition°

Ambition is like choler;° which is an humour that maketh men active, earnest, full of alacrity, and stirring,° if it be not stopped.° But if it be stopped, and cannot have his° way, it becometh adust,° and thereby malign and venomous.° So ambitious men, if they find the way open for their rising, and still° get forward, they are rather busy° than dangerous; but if they be checked° in their desires, they become secretly discontent, and look upon men and matters with an evil eye,° and are best pleased when things go backward;° which is the worst property° in a servant of a prince or state. Therefore it is good for princes, if they use ambitious men, to handle it° so as they be still progressive° and not retrograde;° which because it cannot be without inconvenience, it is good not to use such natures at all. For if they rise not with their service,° they will take order° to make their service fall with them.

But since we have said it were° good not to use men of ambitious natures, except it be upon necessity, it is fit we speak in what cases they are of necessity. Good commanders in the wars must be taken, be they never so ambitious;° for the use of their service dispenseth with° the rest; and to take a soldier without ambition is to pull off his spurs. There is also great use of ambitious men in being screens° to princes in matters of danger and envy;° for no man will take that part, except he be like a seeled° dove, that mounts and mounts because he cannot see about him. There is use also of ambitious men in pulling down the greatness of any subject that overtops;° as Tiberius used Macro in the pulling down of Sejanus.°

Since therefore they must be used in such cases, there resteth° to speak how they are to be bridled,° that they may be less dangerous. There is less danger of them if they be of mean birth, than if they be noble; and if they be rather harsh of nature,° than gracious° and popular:° and if they be rather new° raised, than grown cunning and fortified in their greatness. It is counted by some a weakness in princes to have favourites; but it is of all others the best° remedy against ambitious great-ones. For when the way of pleasuring and displeasuring° lieth by the favourite,° it is impossible any other° should be over-great. Another means to curb them, is to balance them by others as proud as they. But then there must be some middle° counsellors, to keep things steady; for without that ballast° the ship will roll too much. At the least, a prince may animate and inure° some meaner° persons, to be as it were scourges° to ambitious men. As for the having of them obnoxious° to ruin; if they be of fearful° natures, it may do well;° but if they be stout° and daring, it may precipitate° their designs, and prove dangerous. As for the pulling of them down, if the affairs require it, and that it may not be done with safety suddenly, the only way is, the interchange continually of favours and disgraces; whereby they may not know what to expect, and be as it were in a wood.°

Of ambitions, it is less harmful,° the ambition to prevail in great things, than that other to appear in every thing; for that breeds confusion,° and mars business.° But yet it is less danger to have an ambitious man stirring in business, than great in dependances.° He that seeketh to be eminent amongst able men hath a great task; but that is ever good for the public.° But he that plots to be the only figure amongst ciphers° is the decay of a whole age.

Honour° hath three things in it; the vantage° ground to do good; the approach to kings and principal persons; and the raising of a man's own fortunes. He that hath the best of these intentions, when he aspireth,° is an honest man; and that prince that can discern of these intentions in another° that aspireth, is a wise prince. Generally, let princes and states choose such ministers as are more sensible° of duty than of rising; and such as love business rather upon conscience° than upon bravery;° and let them discern° a busy nature from a willing mind.

37. Of Masques° and Triumphs°

THESE things are but toys,° to come amongst such serious observations.° But yet, since princes will have such things, it is better they should be graced with elegancy than daubed with cost.°

Dancing to song,° is a thing of great state° and pleasure. I understand it, that the song be in quire,° placed aloft,° and accompanied with some broken music;° and the ditty fitted to the device.° Acting in song,° especially in dialogues, hath an extreme good grace; I say acting, not dancing° (for that is a mean and vulgar thing); and the voices of the dialogue would° be strong and manly, (a bass and a tenor; no treble);° and the ditty high and tragical;° not nice or dainty.° Several° quires, placed one over against another,° and taking the voice by catches,° anthem-wise, give great pleasure. Turning dances into figure° is a childish curiosity.° And generally let it be noted, that those things which I here set down are such as do naturally take the sense,° and not respect° petty wonderments.° It is true, the alterations° of scenes, so it be quietly and without noise, are things of great beauty and pleasure; for they feed and relieve the eye, before it be full° of the same object.

Let the scenes abound with light, specially coloured and varied; and let the masquers,° or any other,° that are to come down from the scene,° have some motions° upon the scene itself before their coming down; for it draws the eye strangely,° and makes it with great pleasure to desire to see that° it cannot perfectly discern. Let the songs be loud and cheerful, and not chirpings or pulings.° Let the music likewise be sharp° and loud, and well placed. The colours that shew° best by candle-light, are white, carnation, and a kind of sea-water-green; and oes,° or spangs,° as they are of no great cost, so they are of most glory.° As for rich embroidery, it is lost° and not discerned. Let the suits° of the masquers be graceful, and such as become° the person when the vizards° are off; not after examples of known attires;° Turks, soldiers, mariners, and the like.

Let anti-masques° not be long; they have been commonly of fools, satyrs, baboons, wild-men, antics,° beasts, sprites,° witches, Ethiops,° pigmies, turquets,° nymphs, rustics, Cupids, statua's moving,° and the like. As for angels, it is not comical enough° to put them in anti-masques; and any thing that is hideous, as devils, giants, is on the other side as unfit.° But chiefly, let the music of them be recreative,° and with some strange changes.° Some sweet odours° suddenly coming forth, without any drops falling, are, in such a company as there is steam° and heat, things of great pleasure and refreshment. Double masques, one of men, another of ladies, addeth state° and variety. But all is nothing except the room be kept clear and neat.

For jousts, and tourneys,° and barriers;° the glories of them are chiefly in the chariots, wherein the challengers make their entry; especially if they be drawn with strange beasts: as lions, bears, camels, and the like; or in the devices° of their entrance; or in the bravery° of their liveries; or in the goodly furniture° of their horses and armour. But enough of these toys.

38. Of Nature° in Men

Nature is often hidden; sometimes overcome; seldom extinguished. Force maketh nature more violent in the return;° doctrine° and discourse° maketh nature less importune;° but custom° only doth alter and subdue nature. He that seeketh victory over his nature, let him not set himself too great nor too small tasks; for the first will make him dejected by often failings; and the second will make him a small proceeder, though by often prevailings. And at the first let him practise° with helps, as swimmers do with bladders° or rushes;° but after a time let him practise with disadvantages,° as dancers do with thick° shoes. For it breeds great perfection, if the practice be harder than the use.° Where nature is mighty, and therefore the victory hard, the degrees° had need be, first to stay° and arrest nature in time;° like to him that would say over the four and twenty letters° when he was angry; then to go less in quantity; as if one should, in forbearing° wine, come from drinking healths° to a draught° at a meal; and lastly, to discontinue altogether. But if a man have the fortitude and resolution to enfranchise° himself at once,° that is the best:

> Optimus ille animi vindex laedentia pectus
> Vincula qui rupit, dedoluitque semel.°

Neither is the ancient rule° amiss,° to bend nature as a wand° to a contrary extreme, whereby to set it right;° understanding it,° where the contrary extreme is no vice. Let not a man force a habit upon himself with a perpetual continuance, but with some intermission. For both the pause reinforceth the new onset; and if a man that is not perfect be ever in practice, he shall as well practise his errors as his abilities,° and induce one habit of both;° and there is no means to help this but by seasonable intermissions. But let not a man trust his victory over his nature too far; for nature will lay° buried a great time, and yet revive upon the occasion or temptation. Like as it was with Aesop's damsel,° turned from a cat to a woman, who sat very demurely at the board's end, till a mouse ran before her. Therefore let a man either avoid the occasion altogether; or put° himself often to it, that° he may be little moved with it.

A man's nature is best perceived in privateness, for there is no affectation; in passion,° for that putteth a man out of his precepts; and in a new case° or experiment,° for there custom leaveth him. They are happy men whose natures sort with° their vocations; otherwise they may say, 'multum incola fuit anima mea',° when they converse° in those things they do not affect.° In studies,° whatsoever a man commandeth° upon himself, let him set hours° for it; but whatsoever is agreeable to his nature, let him take no care for any set times; for his thoughts will fly to it of themselves; so as° the spaces° of other business or studies will suffice. A man's nature runs either to herbs or weeds; therefore let him seasonably water the one, and destroy the other.

39. Of Custom and Education

Men's thoughts are much according to their inclination;° their discourse and speeches according to their learning and infused° opinions; but their deeds are after as° they have been accustomed. And therefore as Machiavel° well noteth (though in an evil-favoured° instance), there is no trusting to the force of nature nor to the bravery° of words, except it be corroborate° by custom. His instance is, that for the achieving° of a desperate conspiracy, a man should not rest° upon the fierceness of any man's nature, or his resolute undertakings;° but take such an one as hath had his hands formerly° in blood. But Machiavel knew not of a friar Clement, nor a Ravillac, nor a Jaureguy, nor a Baltazar Gerard;° yet his rule holdeth still, that nature, nor the engagement° of words, are not so forcible° as custom. Only° superstition is now so well advanced,° that men of the first blood° are as firm as butchers by occupation;° and votary resolution° is made equipollent° to custom even in matter of blood. In other things the predominancy of custom is everywhere visible; insomuch as a man would wonder to hear men profess, protest, engage, give great words, and then do just as they have done before; as if they were dead images,° and engines° moved only by the wheels of custom.

We see also the reign or tyranny of custom, what it is.° The Indians° (I mean the sect of their wise men) lay themselves quietly upon a stack of wood, and so sacrifice themselves by fire. Nay the wives strive to be burned with the corpses of their husbands. The lads of Sparta,° of ancient time, were wont to be scourged upon the altar of Diana, without so much as queching.° I remember, in the beginning of Queen Elizabeth's time of England, an Irish rebel° condemned, put up a petition to the Deputy° that he might be hanged in a with,° and not in an halter;° because it had been so used with former rebels. There be monks in Russia, for penance,° that will sit a whole night in a vessel of water, till they be engaged° with hard ice. Many examples may be put° of the force of custom, both upon mind and body.

Therefore, since custom is the principal magistrate° of man's life, let men by all means endeavour to obtain good customs. Certainly custom is most perfect when it beginneth in young years: this we call education; which is, in effect, but an early custom. So we see, in languages the tongue is more pliant° to all expressions and sounds, the joints are more supple to all feats of activity and motions,° in youth than afterwards. For it is true that late learners cannot so well take the ply;° except it be in some minds that have not suffered themselves to fix,° but have kept themselves open and prepared to receive continual amendment,° which is exceeding rare.

But if the force of custom simple and separate° be great, the force of custom copulate° and conjoined and collegiate° is far greater. For there° example teacheth, company comforteth,° emulation quickeneth,° glory raiseth: so as° in such places the force of custom is in his exaltation.° Certainly the great multiplication° of virtues upon human nature resteth upon° societies° well ordained and disciplined. For commonwealths and good governments do nourish virtue grown, but do not much mend° the seeds. But the misery is, that the most effectual means are now applied to the ends least to be desired.°

40. Of Fortune

It cannot be denied, but° outward accidents° conduce° much to fortune; favour,° opportunity, death of others, occasion fitting virtue.° But chiefly, the mould° of a man's fortune is in his own hands. 'Faber quisque fortunae suae',° saith the poet. And the most frequent of external causes is, that the folly of one man is the fortune of another. For no man prospers so suddenly as by others' errors. 'Serpens nisi serpentem comederit non fit draco'.° Overt and apparent° virtues bring forth praise; but there be secret and hidden virtues that bring forth fortune; certain deliveries° of a man's self, which have no name. The Spanish name, *desemboltura*,° partly expresseth them; when there be not stonds° nor restiveness° in a man's nature; but that° the wheels of his mind keep way° with the wheels of his fortune. For so Livy (after he had described Cato Major° in these words, 'In illo viro tantum robur corporis et animi fuit, ut quocunque loco natus esset, fortunam sibi facturus videretur'),° falleth upon° that, that he had 'versatile ingenium'.° Therefore if a man look sharply and attentively, he shall see Fortune: for though she be blind,° yet she is not invisible. The way of fortune is like the milken way in the sky; which is a meeting or knot of a number of small stars; not seen asunder,° but giving light together. So are there a number of little and scarce° discerned virtues, or rather faculties and customs,° that make men fortunate. The Italians note some of them, such as a man would little think.° When they speak of one that cannot do amiss, they will throw in into his other conditions, that he hath 'Poco di matto'.° And certainly there be not two more fortunate properties,° than to have a little of the fool, and not too much of the honest. Therefore extreme° lovers of their country or masters were never fortunate, neither can they be. For when a man placeth his thoughts without° himself, he goeth not his own way. An hasty fortune° maketh an enterpriser° and remover° (the French hath it better, *entreprenant*, or *remuant*); but the exercised° fortune maketh the able man.

Fortune is to be honoured and respected, and it be but° for her daughters, Confidence° and Reputation. For those two felicity°

breedeth; the first within a man's self, the latter in others towards him. All wise men, to decline° the envy of their own virtues,° use° to ascribe them to Providence and Fortune; for so they may the better assume° them: and, besides, it is greatness in° a man to be the care° of the higher powers. So Caesar said to the pilot in the tempest, 'Caesarem portas, et fortunam ejus'.° So Sulla chose the name of *Felix*,° and not of *Magnus*.° And it hath been noted, that those who ascribe openly too much to their own wisdom and policy, end infortunate. It is written that Timotheus° the Athenian, after he had, in the account he gave to the state of his government, often interlaced° this speech, 'and in this Fortune had no part', never prospered in anything he undertook afterwards. Certainly there be,° whose fortunes are like Homer's verses, that have a slide° and easiness more than the verses of other poets; as Plutarch saith of Timoleon's fortune, in respect of that of Agesilaus or Epaminondas. And that this should be, no doubt it is much° in a man's self.

41. Of Usury°

Many have made witty invectives against Usury. They say that it is a pity° the devil should have God's part, which is the tithe.° That the usurer is the greatest sabbath-breaker,° because his plough goeth every Sunday. That the usurer is the drone that Virgil speaketh of;

> Ignavum fucos pecus a praesepibus arcent.°

That the usurer breaketh the first law° that was made for mankind after the fall, which was, 'in sudore vultus tui comedes panem tuum', not, 'in sudore vultus alieni'.° That usurers should have orange-tawny bonnets,° because they do judaize.° That it is against nature° for money to beget money; and the like. I say this only, that usury is a 'concessum propter duritiem cordis':° for since there must be borrowing and lending, and men are so hard of heart as they will not lend freely,° usury must be permitted. Some others have made suspicious° and cunning propositions of° banks, discovery° of men's estates, and other inventions. But few have spoken of usury usefully. It is good to set before us° the incommodities and commodities° of usury, that the good may be either weighed out° or culled° out; and warily to provide, that while we make forth° to that which is better, we meet not with that which is worse.

The discommodities of usury are, First, that it makes fewer° merchants. For were it not for this lazy trade of usury, money would not lie still, but would in great part be employed upon merchandizing; which is the *vena porta*° of wealth in a state. The second, that it makes poor merchants. For as a farmer cannot husband° his ground so well if he sit at a great rent;° so the merchant cannot drive° his trade so well, if he sit at great usury.° The third is incident° to the other two; and that is the decay of customs° of kings or states, which ebb or flow with merchandizing. The fourth, that it bringeth the treasure of a realm or state° into a few hands. For the usurer being at certainties,° and others at uncertainties, at the end of the game most of the money will be in the box;° and ever a state flourisheth when wealth is more equally spread.° The fifth, that it beats down° the price of land; for the employment of money is chiefly either merchandizing or purchasing;° and usury waylays° both. The sixth, that it doth dull and damp° all industries, improvements, and new inventions, wherein money would be stirring, if it were not for this slug.° The last, that it is the canker° and ruin of many men's estates; which in process of time breeds a public poverty.°

On the other side,° the commodities of usury are, first, that howsoever usury in some respect hindereth merchandizing,° yet in some other it advanceth° it; for it is certain that the greatest part of trade is driven by young merchants, upon borrowing at interest; so as if the usurer either call in° or keep back his money, there will ensue presently° a great stand° of trade. The second is, that were it not for this easy borrowing upon interest, men's necessities° would draw upon them a most sudden undoing;° in that they would be forced to sell their means° (be it lands or goods) far under foot;° and so, whereas usury doth but gnaw upon° them, bad markets° would swallow them quite up. As for mortgaging° or pawning,° it will little mend the matter: for either men will not take pawns without use;° or if they do, they will look precisely° for the forfeiture.° I remember a cruel monied man in the country, that would say, 'The devil take this usury, it keeps us from° forfeitures of mortgages and bonds'. The third and last is, that it is a vanity° to conceive that there would be ordinary borrowing without profit; and it is impossible to conceive the number of inconveniences that will ensue, if borrowing be cramped.° Therefore to speak of the abolishing of usury is idle.° All states have ever had it, in one kind or rate,° or other. So as that opinion must be sent to Utopia.°

To speak now of the reformation and reiglement° of usury; how the discommodities of it may be best avoided, and the commodities retained. It appears by the balance of commodities and discommodities of usury, two things are to be reconciled. The one, that the tooth of usury be grinded,° that it bite not too much; the other, that there be left open a means to invite° monied men to lend to the merchants, for the continuing and quickening of trade.° This cannot be done, except you introduce two several° sorts of usury, a less° and a greater. For if you reduce usury to one low rate, it will ease the common borrower, but the merchant will be to seek for° money. And it is to be noted, that the trade of merchandize, being the most lucrative, may bear usury at a good rate: other contracts° not so.

To serve both intentions,° the way would be briefly thus. That there be two rates of usury; the one free, and general° for all; the other under licence only, to certain persons and in certain places of merchandizing. First therefore, let usury in general be reduced to five in the hundred;° and let that rate be proclaimed to be free and current;° and let the state shut itself out° to take° any penalty° for the same. This will preserve borrowing from any general stop or dryness.° This will ease infinite borrowers in the country. This will, in good part, raise the price of land, because land purchased at sixteen years' purchase° will yield six in the hundred, and somewhat more; whereas this rate of interest yields but five. This by like reason will encourage and edge° industrious and profitable improvements; because many will rather venture in that kind° than take five in the hundred, especially having been used to greater profit. Secondly, let there be certain persons licensed to lend to known merchants upon usury at a higher rate; and let it be with the cautions° following. Let the rate be, even with the merchant himself, somewhat more easy° than that he used formerly to pay; for by that means all borrowers shall have some ease° by this reformation, be he merchant, or whosoever. Let it be no bank or common° stock, but every man be master of his own money. Not that I altogether mislike° banks, but they will hardly be brooked,° in regard of certain suspicions. Let the state be answered some small matter° for the licence,° and the rest left to the lender; for if the abatement° be but small, it will no whit° discourage the lender. For he, for example, that took before ten or nine in the hundred, will sooner descend to eight in the hundred, than give over° his trade of usury, and go from certain gains to gains of hazard.° Let these licensed lenders be in number indefinite, but restrained to certain principal cities and towns of merchandizing; for then they will be hardly able to colour° other men's monies in the country: so as the licence of nine will not suck away the current rate of five; for no man will send° his monies far off, nor put them into unknown hands.

If it be objected that this doth in a sort° authorize usury, which before was in some places but permissive; the answer is, that it is better to mitigate usury by declaration,° than to suffer it to rage by connivance.

42. Of Youth and Age°

A man that is young in years may be old in hours,° if he have lost no time.° But that happeneth rarely. Generally, youth is like the first cogitations,° not so wise as the second. For there is a youth in thoughts, as well as in ages. And yet the invention° of young men is more lively than that of old; and imaginations° stream into their minds better, and as it were more divinely. Natures that have much heat° and great and violent desires and perturbations, are not ripe for action till they have passed the meridian° of their years; as it was with Julius Caesar,° and Septimius Severus.° Of the latter of whom it is said, 'Juventutem egit erroribus, imo furoribus, plenam'.° And yet he was the ablest emperor, almost, of all the list. But reposed° natures may do well in youth. As it is seen in Augustus Caesar,° Cosmus° Duke of Florence, Gaston de Fois,° and others. On the other side, heat and vivacity in age° is an excellent composition° for business.°

Young men are fitter to invent than to judge;° fitter for execution than for counsel;° and fitter for new projects than for settled° business. For the experience of age,° in things that fall within the compass° of it, directeth° them; but in new things, abuseth° them. The errors of young men are the ruin of business; but the errors of aged men amount but to this, that more might have been done, or sooner. Young men, in the conduct and manage° of actions, embrace more than they can hold; stir° more than they can quiet;° fly to the end, without consideration of the means and degrees;° pursue some few principles which they have chanced upon absurdly;° care not° to innovate, which draws unknown inconveniences;° use extreme remedies at first; and that which doubleth° all errors, will not acknowledge or retract them; like an unready° horse, that will neither stop nor turn. Men of age object too much, consult too long, adventure too little,° repent too soon, and seldom drive business home to the full period,° but content themselves with a mediocrity° of success.

Certainly it is good to compound° employments of both; for that will be good for the present, because the virtues of either age may correct the defects of both; and good for succession,° that young men may be learners, while men in age are actors;° and, lastly, good for extern accidents,° because authority followeth old men, and favour and popularity youth. But for the moral° part, perhaps youth will have the pre-eminence, as age hath for the politic.° A certain rabbin,° upon the text, 'Your young men shall see visions, and your old men shall dream dreams',° inferreth that young men are admitted nearer to God than old, because vision is a clearer revelation than a dream. And certainly, the more a man drinketh of the world, the more it intoxicateth:° and age doth profit rather in° the powers of understanding, than in the virtues of the will and affections.° There be some have an over-early ripeness in their years, which fadeth betimes.° These are, first, such as have brittle° wits, the edge° whereof is soon turned; such as was Hermogenes° the rhetorician, whose books are exceeding subtle; who afterwards waxed° stupid. A second sort is of those that have some natural dispositions which have better grace° in youth than in age; such as is a fluent and luxuriant° speech; which becomes youth well, but not age: so Tully saith of Hortensius, 'Idem manebat, neque idem decebat'.° The third° is of such as° take too high a strain° at the first, and are magnanimous° more than tract° of years can uphold. As was Scipio Africanus, of whom Livy saith in effect, 'Ultima primis cedebant'.°

43. OF BEAUTY°

VIRTUE is like a rich stone, best plain set;° and surely virtue is best in a body that is comely,° though not of delicate° features; and that hath rather dignity of presence,° than beauty of aspect.° Neither is it almost° seen, that very beautiful persons are otherwise of great virtue; as if nature were rather busy° not to err, than in labour to produce excellency.° And therefore they prove accomplished,° but not of great spirit; and study rather° behaviour than virtue.° But this holds° not always: for Augustus Caesar,° Titus Vespasianus,° Philip le Bel° of France, Edward the Fourth° of England, Alcibiades° of Athens, Ismael the Sophy° of Persia, were all high and great spirits;° and yet the most beautiful men of their times. In beauty, that of favour° is more than that of colour;° and that of decent° and gracious° motion° more than that of favour. That is the best part of beauty, which a picture cannot express; no nor the first sight of the life.°

There is no excellent beauty that hath not some strangeness° in the proportion.° A man cannot tell whether Apelles° or Albert Durer°

were the more trifler;° whereof the one° would make a personage° by geometrical proportions; the other, by taking the best parts out of divers faces, to make one excellent. Such personages, I think, would please nobody but the painter that made them. Not but° I think a painter may make a better face than ever was; but he must do it by a kind of felicity° (as a musician that maketh an excellent air° in music), and not by rule. A man shall see faces, that if you examine them part by part, you shall find never a° good; and yet altogether° do well. If it be true that the principal part of beauty is in decent motion, certainly it is no marvel° though° persons in years seem many times° more amiable;° 'pulchrorum autumnus pulcher';° for no youth can be comely° but by pardon,° and considering the youth as to make up° the comeliness. Beauty is as summer fruits, which are easy to corrupt, and cannot last; and for the most part it makes a dissolute youth,° and an age° a little out of countenance;° but yet certainly again, if it light well,° it maketh virtue shine, and vices blush.°

44. Of Deformity

Deformed persons° are commonly even° with nature; for as nature hath done ill by° them, so do they by nature; being for the most part (as the Scripture° saith) 'void of natural affection'; and so they have their revenge of nature. Certainly there is a consent° between the body and the mind; and where nature erreth° in the one, she ventureth° in the other: 'Ubi peccat in uno, periclitatur in altero'. But because there is in man an election° touching the frame° of his mind, and a necessity in the frame of his body, the stars of natural inclination are sometimes obscured by the sun of discipline° and virtue. Therefore it is good to consider of deformity, not as a sign, which is more deceivable;° but as a cause, which seldom faileth of° the effect.

Whosoever hath any thing fixed° in his person° that doth induce° contempt, hath also a perpetual spur in himself to rescue and deliver himself from scorn. Therefore all deformed persons are extreme bold. First, as° in their own defence, as being exposed to scorn; but in process of time by a general habit. Also it stirreth in them industry, and especially of this kind, to watch and observe the weakness of others, that they may have somewhat to repay.° Again, in their superiors, it quencheth jealousy towards them, as persons that they think they may at pleasure despise: and it layeth their competitors and emulators asleep;° as never believing they should be in possibility of advancement, till they see them in possession.° So that upon the matter,° in a great wit,° deformity is an advantage to rising.°

Kings in ancient times (and at this present° in some countries) were wont° to put great trust in eunuchs; because they that are envious towards all are more obnoxious° and officious° towards one. But yet their trust towards them hath rather been as to good spials° and good whisperers,° than good magistrates and officers. And much like is the reason° of deformed persons. Still° the ground is, they will, if they be of spirit,° seek to free themselves from scorn; which must be either by virtue or malice; and therefore let it not be marvelled° if sometimes they prove excellent persons; as was Agesilaus,° Zanger° the son of Solyman, Aesop,° Gasca° President of Peru; and Socrates° may go likewise amongst them; with others.

45. Of Building°

Houses are built to live in, and not to look on; therefore let use be preferred before uniformity,° except where both may be had. Leave the goodly fabrics of houses, for beauty only, to the enchanted palaces* of the poets; who build them with small cost. He that builds a fair house upon an ill seat,° committeth himself to prison. Neither do I reckon it an ill seat only where the air° is unwholesome; but likewise where the air is unequal;° as you shall see many fine seats° set upon a knap° of ground, environed with higher hills round about it; whereby the heat of the sun is pent in,° and the wind gathereth as in troughs;° so as° you shall have, and that suddenly, as great diversity of heat and cold as if you dwelt in several° places. Neither is it ill air only that maketh an ill seat, but ill ways,° ill markets: and, if you will consult with Momus,° ill neighbours. I speak not of many more:° want of water; want of wood, shade, and shelter; want of fruitfulness, and mixture° of grounds of several natures; want of prospect;° want of level grounds; want of places at some near distance for sports of hunting, hawking, and races; too near the sea, too remote; having [not] the commodity of navigable rivers, or the discommodity of their overflowing; too far off from great cities, which may hinder business, or too near them, which lurcheth° all provisions,° and maketh every thing dear; where a man hath a great living° laid together, and where he is scanted:° all which, as it is impossible perhaps to find together, so it is good to know them, and think of them, that a man may take° as many as he can; and if he have several dwellings, that he sort° them so, that what he wanteth in the one he may find in the other. Lucullus° answered Pompey well; who, when he saw his stately galleries,° and rooms so large and lightsome,° in one of his houses, said, 'Surely an excellent place for summer, but how do you in winter?' Lucullus answered, 'Why, do you not think me as wise as some fowl° are, that ever change their abode towards the winter?'

To pass from the seat to the house itself; we will do as Cicero doth in the orator's art; who writes books *De Oratore*, and a book he entitles *Orator*; whereof the former delivers° the precepts° of the art, and the latter the perfection.° We will therefore describe a princely palace, making a brief model thereof. For it is strange to see, now in Europe, such huge buildings as the Vatican° and Escurial° and some others be, and yet scarce a very fair° room in them.

First therefore, I say you cannot have a perfect palace, except you have two several sides;° a side for the banquet,° as is spoken of in the book of Hester,° and a side for the household;° the one for feasts and triumphs,° and the other for dwelling. I understand both these sides to be not only returns,° but parts of the front; and to be uniform without, though severally partitioned° within; and to be on both sides of a great and stately tower in the midst of the front, that, as it were, joineth them together on either hand. I would have on the side of the banquet, in front, one only goodly room above stairs, of some forty foot high; and under it a room for a dressing or preparing place° at times of triumphs. On the other side, which is the household side, I wish it divided at the first° into a hall and a chapel (with a partition° between); both of good state° and bigness;° and those not to go all the length, but to have at the further end a winter and a summer parlour,° both fair.° And under these rooms, a fair and large cellar sunk under ground; and likewise some privy° kitchens, with butteries° and pantries,° and the like. As for the tower, I would have it two stories, of eighteen foot high a piece,° above the two wings; and a goodly leads° upon the top, railed° with statua's interposed; and the same tower to be divided into rooms, as shall be thought fit. The stairs likewise to the upper rooms, let them be upon a fair open newel,° and finely railed in with images of wood,° cast into a brass colour; and a very fair landing-place° at the top. But this to be, if you do not point° any of the lower rooms for a dining place of servants. For otherwise you shall have the servants' dinner after your own: for the steam of it will come up as in a tunnel.° And so much for the front. Only I understand the height of the first stairs to be sixteen foot, which is the height of the lower room.°

Beyond this front is there to be a fair court, but three sides of it, of a far lower building than the front. And in all the four corners of that court fair stair-cases, cast into turrets, on the outside,° and not within the row of buildings themselves. But those towers are not to be of the height of the front, but rather proportionable to the lower building. Let the court not be paved, for that striketh up° a great heat in summer, and much cold in winter. But° only some side alleys, with a cross, and the quarters to graze,° being kept shorn, but not too near° shorn. The row of return° on the banquet side, let it be all stately galleries: in which galleries let there be three, or five fine cupolas in the length of it, placed at equal distance; and fine coloured windows of several° works.° On the household side, chambers of presence° and ordinary entertainments,° with some bed-chambers; and let all three sides be a double house,° without thorough lights° on the sides, that you may have rooms from° the sun, both for forenoon and afternoon. Cast° it also, that you may have rooms both for summer and winter; shady for summer, and warm for winter. You shall have sometimes fair houses so full of glass, that one cannot tell where to become° to be out of the sun or cold. For inbowed° windows, I hold them of good use (in cities, indeed, upright° do better, in respect of the uniformity° towards the street); for they be pretty° retiring places for conference;° and besides, they keep both the wind and sun off; for that which would strike° almost thorough the room doth scarce pass the window. But let them be but few, four in the court, on the sides° only.

Beyond this court, let there be an inward° court, of the same square° and height; which is to be environed with the garden on all sides; and in the inside, cloistered° on all° sides, upon decent° and beautiful arches, as high as the first story.° On the under story, towards the garden, let it be turned to a grotta,° or place of shade, or estivation.° And only have opening and windows towards the garden; and be level upon the floor, no whit° sunken under ground, to avoid all dampishness. And let there be a fountain, or some fair work of statua's° in the midst of this court; and to be paved as the other court was. These buildings to be for privy lodgings° on both sides; and the end for privy galleries. Whereof you must foresee° that one of them° be for an infirmary,° if the prince or any special person should be sick, with chambers, bed-chamber, antecamera,° and recamera,°

joining to it. This upon the second story. Upon the ground story, a fair gallery,° open, upon pillars; and upon the third story likewise, an open gallery,° upon pillars, to take° the prospect and freshness of the garden. At both corners of the further side, by way of return, let there be two delicate or rich cabinets,° daintily° paved, richly hanged,° glazed with crystalline° glass, and a rich cupola in the midst; and all other elegancy° that may be thought upon. In the upper° gallery too, I wish that there may be, if the place will yield° it, some fountains running in divers places from the wall, with some fine avoidances.°

And thus much for the model of the palace; save° that you must have, before you come to the front, three courts. A green court plain,° with a wall about it; a second court of the same,° but more garnished,° with little turrets, or rather embellishments, upon the wall; and a third court, to make a square with the front, but not to be built,° nor yet enclosed with a naked wall, but enclosed with tarrasses,° leaded° aloft, and fairly garnished, on the three sides; and cloistered on the inside, with pillars, and not with arches below. As for offices,° let them stand at distance, with some low galleries,° to pass from them to the palace itself.

46. Of Gardens°

God Almighty first planted a Garden.° And indeed it is the purest of human pleasures. It is the greatest refreshment to the spirits of man; without which, buildings and palaces are but gross handy-works:° and a man shall ever see that when ages grow to civility° and elegancy,° men come to build stately° sooner than to garden finely;° as if gardening were the greater perfection. I do hold° it, in the royal ordering° of gardens, there ought to be gardens for all the months in the year; in which severally° things of beauty may be then in season. For December, and January, and the latter part of November, you must take such things as are green all winter: holly; ivy; bays; juniper; cypress-trees; yew; pine-apple trees;° fir-trees; rosemary; lavender; periwinkle, the white, the purple, and the blue; germander;° flags;° orange-trees; lemon-trees; and myrtles, if they be stooved;° and sweet marjoram, warm set.° There followeth, for the latter part of January and February, the mezereon-tree,° which then blossoms; crocus vernus,° both the yellow and the grey; primroses; anemones; the early tulippa; hyacinthus orientalis; chamaïris;° fritellaria. For March, there come violets, specially the single blue, which are the earliest; the yellow daffodil; the daisy; the almond-tree in blossom; the peach-tree in blossom; the cornelian-tree° in blossom; sweet-briar. In April follow the double white violet; the wall-flower; the stock-gilliflower; the cowslip; flower-de-lices, and lilies of all natures;° rosemary-flowers; the tulippa; the double piony; the pale daffodil; the French honeysuckle; the cherry-tree in blossom; the dammasin° and plum-trees in blossom; the white thorn° in leaf; the lilac-tree. In May and June come pinks° of all sorts, specially the blush-pink; roses of all kinds, except the musk, which comes later; honeysuckles; strawberries; bugloss;° columbine; the French marigold; flos Africanus;° cherry-tree in fruit; ribes;° figs in fruit; rasps;° vine-flowers; lavender in flowers; the sweet satyrian,° with the white flower; herba muscaria;° lilium convallium;° the apple-tree in blossom. In July come gilliflowers of all varieties; musk-roses; the lime-tree in blossom; early pears and plums in fruit; genitings,° quadlins.° In August come plums of all sorts in fruit; pears; apricocks; barberries;° filberds;° musk-melons;° monks-hoods, of all colours. In September come grapes; apples; poppies of all colours; peaches; melocotones;° nectarines; cornelians;° wardens;° quinces. In October and the beginning of November come° services;° medlars;° bullaces;° roses cut or removed to come late; holly-oaks; and such like. These particulars are for the climate of London; but my meaning is perceived, that you may have *ver perpetuum*,° as the place° affords.

And because the breath° of flowers is far sweeter in the air (where it comes and goes like the warbling° of music) than in the hand, therefore nothing is more fit for that delight, than to know what be the flowers and plants that do best perfume the air. Roses, damask and red,° are fast° flowers of their smells; so that you may walk by a whole row of them, and find nothing of° their sweetness; yea though it be in a morning's dew. Bays likewise yield no smell as they grow. Rosemary little; nor sweet marjoram. That which above all others yields the sweetest smell in the air, is the violet, specially the white double violet, which comes twice a year; about the middle of April, and about Bartholomew-tide.° Next to that is the musk-rose. Then the strawberry-leaves dying, which [yield]° a most excellent cordial° smell. Then the flower of the vines; it is a little dust,° like the dust of a bent,° which grows upon the cluster in the first coming forth. Then sweet-briar. Then wall-flowers, which are very delightful to be set under a parlour or lower chamber window. Then pinks and gilliflowers, specially the matted pink and clove gilliflower. Then the flowers of the lime-tree. Then the honeysuckles, so° they be somewhat afar off. Of bean-flowers I speak not, because they are field flowers. But those which perfume the air most delightfully, not passed by° as the rest, but being trodden upon and crushed, are three; that is, burnet,° wild-thyme, and watermints. Therefore you are to set whole alleys of them, to have the pleasure when you walk or tread.

For gardens (speaking of those which are indeed prince-like, as we have done of buildings), the contents ought not well to be under thirty acres° of ground; and to be divided into three parts; a green in the entrance; a heath or desert° in the going forth;° and the main garden in the midst; besides alleys on both sides.° And I like well that four acres of ground be assigned to the green; six to the heath; four and four to either side; and twelve to the main garden. The green hath two pleasures: the one, because nothing is more pleasant to the eye than green grass kept finely shorn; the other, because it will give you a fair alley in the midst, by which you may go in front upon° a stately hedge,° which is to enclose the garden. But because the alley will be long, and, in great heat of the year or day, you ought not to buy° the shade in the garden by going in the sun thorough° the green, therefore you are, of either side the green, to plant a covert alley,° upon carpenter's work,° about twelve foot in height, by which you may go in shade into the garden. As for the making of knots° or figures with divers coloured earths,° that they may lie under the windows of the house on that side which the garden stands, they be but toys: you may see as good sights many times in tarts.°

The garden is best to be square, encompassed on all the four sides with a stately arched hedge.° The arches to be upon pillars of carpenter's work, of some ten foot high, and six foot broad; and the spaces between of the same dimension with the breadth of the arch. Over the arches let there be an entire° hedge of some four foot high, framed also upon carpenter's work; and upon the upper hedge, over every arch, a little turret, with a belly,° enough to receive a cage of birds; and over every space between the arches some other little figure, with broad plates of round coloured glass gilt, for the sun to play upon. But this hedge I intend to be raised upon a bank, not steep, but gently slope, of some six foot, set all with flowers. Also I understand, that this square of the garden should not be the whole breadth of the ground, but to leave on either side ground enough for diversity of side alleys; unto which the two covert alleys of the green may deliver° you. But there must be no alleys with hedges at either end of this great enclosure; not at the hither end, for letting your prospect° upon this fair hedge from the green; nor at the further end, for letting your prospect° from the hedge through the arches upon the heath.

For the ordering° of the ground within the great hedge, I leave it to variety of device;° advising nevertheless that whatsoever form you cast it into, first, it be not too busy,° or full of work. Wherein I, for my part, do not like images cut out° in juniper or other garden stuff; they be for children. Little low hedges, round, like welts,° with some pretty pyramides,° I like well; and in some places, fair columns upon frames of carpenter's work. I would also have the alleys spacious and fair. You may have closer° alleys upon the side grounds, but none in the main garden. I wish also, in the very middle, a fair mount,° with three ascents,° and alleys, enough for four to walk abreast; which I would have to be perfect circles, without any bulwarks° or embossments;° and the whole mount to be thirty foot high; and some fine banqueting-house,° with some chimneys neatly cast,° and without too much glass.

For fountains, they are a great beauty and refreshment; but pools° mar° all, and make the garden unwholesome, and full of flies and frogs. Fountains I intend to be of two natures: the one that sprinkleth or spouteth water; the other a fair receipt° of water, of some thirty or forty foot square, but without fish, or slime, or mud. For the first, the ornaments of images gilt, or of marble, which are in use, do well:° but the main matter is so to convey the water, as it never stay,° either in the bowls or in the cistern;° that the water be never by rest discoloured, green or red or the like; or gather any mossiness or putrefaction. Besides that, it is to be cleansed every day by the hand. Also some steps up to it, and some fine pavement about it, doth well. As for the other kind of fountain, which we may call a bathing pool, it may admit much curiosity° and beauty; wherewith we will not trouble ourselves: as, that the bottom be finely paved, and with images;° the sides likewise; and withal embellished with coloured glass, and such things of lustre;° encompassed also with fine rails of low statua's. But the main point is the same which we mentioned in the former kind of fountain; which is, that the water be in perpetual motion, fed by a water higher than the pool, and delivered° into it by fair spouts,° and then discharged away under ground, by some equality of bores,° that it stay little. And for fine devices, of arching water without spilling, and making it rise in several forms (of feathers, drinking glasses, canopies,° and the like), they be pretty things to look on, but nothing to health and sweetness.

For the heath, which was the third part of our plot,° I wish it to be framed,° as much as may be, to a natural wildness. Trees I would have none in it, but some thickets made only of sweet-briar and honeysuckle, and some wild vine amongst;° and the ground set° with violets, strawberries, and primroses. For these are sweet, and prosper in the shade. And these to be in the heath, here and there, not in any order. I like also little heaps, in the nature of mole-hills (such as are in wild heaths), to be set, some with wild thyme; some with pinks; some with germander, that gives a good flower to the eye; some with periwinkle; some with violets; some with strawberries; some with cowslips; some with daisies; some with red roses; some with lilium convallium; some with sweet-williams red; some with bear's-foot:° and the like low flowers, being withal sweet° and sightly.° Part of which heaps are to be with standards° of little bushes pricked° upon their top, and part without. The standards to be roses; juniper; holly; barberries (but° here and there, because of the smell of their blossom); red currants; gooseberries; rosemary; bays; sweet-briar; and such like. But these standards to be kept with cutting,° that they grow not out of course.°

For the side grounds, you are to fill them with variety of alleys, private,° to give a full shade, some of them, wheresoever the sun be. You are to frame° some of them likewise for shelter, that when the wind blows sharp, you may walk as in a gallery. And those alleys must be likewise hedged at both ends, to keep out the wind; and these closer alleys must be ever finely gravelled, and no grass, because of going wet.° In many of these alleys likewise, you are to set fruit-trees of all sorts; as well upon the walls as in ranges.° And this would° be generally observed, that the borders wherein you plant your fruit-trees be fair and large, and low, and not steep; and set with fine flowers, but thin and sparingly, lest they deceive° the trees. At the end of both the side grounds, I would have a mount of some pretty height, leaving the wall of the enclosure breast high, to look abroad into the fields.

For the main garden, I do not deny but° there should be some fair alleys ranged on both sides, with fruit trees; and some pretty tufts of fruit trees, and arbours with seats, set in some decent order; but these to be by no means set too thick;° but to leave the main garden so as it be not close, but the air open and free. For as for shade, I would have you rest upon° the alleys of the side grounds, there to walk, if you be disposed, in the heat of the year or day; but to make account° that the main garden is for the more temperate parts of the year; and in the heat of summer, for the morning and the evening, or overcast° days.

For aviaries, I like them not, except they be of that largeness as they may be turfed,° and have living plants and bushes set in them; that the birds may have more scope,° and natural nestling,° and that no foulness° appear in the floor of the aviary.

So I have made a platform° of a princely garden, partly by precept, partly by drawing, not a model,° but some general lines of it; and in this I have spared for no cost. But it is nothing for great princes, that for the most part taking advice with workmen,° with no less cost set their things together; and sometimes add statua's, and such things, for state and magnificence,° but nothing° to the true pleasure of a garden.

47. Of Negotiating°

It is generally better to deal by speech than by letter; and by the mediation of a third° than by a man's self.° Letters are good, when a man would draw° an answer by letter back again; or when it may serve for a man's justification afterwards to produce his own letter; or where it° may be danger° to be interrupted, or heard by pieces.° To deal in person is good, when a man's face breedeth regard,° as commonly with inferiors; or in tender° cases, where a man's eye upon the countenance° of him with whom he speaketh may give him a direction how far to go; and generally, where a man will reserve° to himself liberty either to disavow° or to expound.° In choice of instruments,° it is better to choose men of a plainer° sort, that are like° to do that that is committed to them, and to report back again faithfully the success,° than those that are cunning to contrive out of other men's business somewhat° to grace° themselves, and will help the matter in report for satisfaction sake.° Use also such persons as affect° the business wherein they are employed; for that quickeneth much;° and such as are fit for the matter; as bold men for expostulation,° fair-spoken° men for persuasion, crafty° men for inquiry and observation, froward° and absurd° men for business that doth not well bear out° itself. Use also such as have been lucky, and prevailed° before in things wherein you have employed them; for that breeds confidence, and they will strive to maintain their prescription.° It is better to sound° a person with whom one deals afar off,° than to fall upon° the point at first; except you mean to surprise him by some short° question. It is better dealing with men in appetite,° than with those that are where they would be. If a man deal with another upon conditions,° the start or first performance° is all;° which a man cannot reasonably demand,° except either the nature of the thing be such, which° must go before;° or else a man can persuade the other party° that he shall still need him in some other thing; or else that he° be counted° the honester° man. All practice° is to discover,° or to work.° Men discover themselves in trust,° in passion,° at unawares,° and of necessity, when they would have somewhat done and cannot find an apt pretext. If you would work any man, you must either know his nature and fashions,° and so lead him; or his ends, and so persuade him; or his weakness and disadvantages,° and so awe° him; or those that have interest in° him, and so govern him. In dealing with cunning persons, we must ever consider their ends, to interpret their speeches; and it is good to say little to them, and that which they least look for.° In all negotiations of difficulty, a man may not look° to sow and reap at once; but must prepare business, and so ripen it by degrees.

48. Of Followers and Friends

Costly° followers are not to be liked; lest while a man maketh his train° longer, he make his wings shorter. I reckon to be costly, not them alone which° charge° the purse, but which are wearisome and importune° in suits.° Ordinary followers ought to challenge° no higher conditions° than countenance,° recommendation, and protection from wrongs. Factious° followers are worse to be liked, which follow not upon affection to him with whom they range° themselves, but upon discontentment conceived against some other; whereupon commonly ensueth° that ill intelligence° that we many times see between great personages. Likewise glorious° followers, who make themselves as trumpets of the commendation of those they follow, are full of inconvenience; for they taint° business through want of secrecy; and they export° honour° from a man, and make him a return° in envy. There is a kind of followers likewise which are dangerous, being indeed espials;° which inquire the secrets of the house,° and bear tales of them° to others. Yet such men, many times, are in great favour; for they are officious,° and commonly exchange tales.° The following by certain estates° of men, answerable° to that° which a great person himself professeth (as of soldiers to him° that hath been employed in the wars, and the like), hath ever been a thing civil,° and well taken even in monarchies; so° it be without too much pomp or popularity.° But the most honourable kind of following is to be followed as one that apprehendeth° to advance° virtue and desert° in all sorts of persons. And yet, where there is no eminent odds in sufficiency,° it is better to take with° the more passable,° than with the more able. And besides, to speak truth, in base° times active men are of more use than virtuous.° It is true that in government° it is good to use men of one rank° equally:° for to countenance some extraordinarily, is to make them insolent, and the rest discontent;° because they may claim a due.° But contrariwise,° in favour,° to use men with much difference and election° is good; for it maketh the persons preferred more thankful, and the rest more officious:° because all is of favour.° It is good discretion not to make too much° of any man at the first; because one cannot hold out° that proportion.° To be governed (as we call it) by one,° is not safe; for it shews softness,° and gives a freedom° to scandal and disreputation;° for those that would not censure or speak ill of a man immediately,° will talk more boldly of those that are so great with° them, and thereby wound their° honour. Yet to be distracted° with many is worse; for it makes men to be of the last impression,° and full of change. To take advice of some few friends is ever honourable; for 'lookers-on many times see more than gamesters';° and 'the vale best discovereth the hill'.° There is little friendship in the world, and least of all between equals, which was wont to be magnified.° That that is, is between superior and inferior, whose fortunes may comprehend° the one the other.

49. Of Suitors°

Many ill matters° and projects° are undertaken;° and private suits do putrefy° the public good. Many good matters are undertaken with bad minds; I mean not only corrupt minds, but crafty minds, that intend not performance.° Some embrace° suits, which° never mean to deal° effectually in them; but if they see there may be life in the matter by some other mean,° they will be content to win a thank, or take a second° reward, or at least to make use in the mean time of the suitor's hopes. Some take hold of suits only for an occasion to cross° some other; or to make° an information whereof they could not otherwise have apt pretext; without care what become of the suit when that turn is served; or, generally, to make other men's business a kind of entertainment° to bring in their own. Nay some undertake suits, with a full purpose to let them fall; to the end° to gratify the adverse party or competitor. Surely there is in some sort° a right in every suit; either a right in equity,° if it be a suit of controversy;° or a right of desert,° if it be a suit of petition.° If affection° lead a man to favour the wrong side in justice,° let him rather use his countenance° to compound° the matter than to carry° it. If affection lead a man to favour the less worthy in desert, let him do it without depraving° or disabling° the better deserver. In suits which a man doth not well understand, it is good to refer them to some friend of trust and judgment, that may report whether he may deal° in them with honour:° but let him choose well his referendaries,° for else he may be led by the nose.° Suitors are so distasted° with delays and abuses,° that plain dealing in denying° to deal in suits at first,° and reporting the success barely,° and in challenging° no more thanks than one hath deserved, is grown not only honourable but also gracious.° In suits of favour,° the first coming° ought to take little place:° so far forth° consideration may be had of his° trust, that if intelligence° of the matter could not otherwise have been had but by him, advantage be not taken of the note,° but the party° left to his other means; and in some sort recompensed for his discovery.° To be ignorant of the value° of a suit is simplicity;° as well as to be ignorant of the right° thereof is want of conscience. Secrecy in suits is a great mean° of obtaining; for voicing° them to be in forwardness° may discourage some kind of suitors, but doth quicken and awake others. But timing° of the suit is the principal. Timing, I say, not only in respect of the person that should grant it, but in respect of those which° are like° to cross it. Let a man, in the choice of his mean,° rather choose the fittest° mean than the greatest° mean; and rather them that deal in certain° things, than those that are general. The reparation° of a denial° is sometimes equal to the first grant;° if a man shew himself neither dejected nor discontented. 'Iniquum petas ut aequum feras',° is a good rule, where a man hath strength of favour:° but otherwise a man were° better rise in his suit;° for he that would have ventured at first to have lost the suitor, will not in the conclusion lose both the suitor and his own former favour. Nothing is thought so easy° a request to a great person, as his letter; and yet, if it be not in° a good cause, it is so much out of his reputation. There are no worse instruments than these general° contrivers° of suits; for they are but a kind of poison and infection to public proceedings.

50. Of Studies

Studies serve for delight, for ornament, and for ability.° Their chief use for delight, is in privateness and retiring;° for ornament, is in discourse;° and for ability, is in the judgment and disposition of business.° For expert° men can execute, and perhaps judge of particulars, one by one; but the general counsels,° and the plots and marshalling° of affairs, come best from those that are learned. To spend too much time in studies is sloth;° to use them too much for ornament, is affectation; to make judgment wholly by their rules, is the humour° of a scholar.° They perfect nature, and are perfected by experience: for natural abilities are like natural plants, that need proyning° by study; and° studies themselves do give forth directions too much at large,° except they be bounded in by experience. Crafty° men contemn° studies, simple° men admire° them, and wise° men use them; for they teach not their own use;° but that is a wisdom without them, and above them,° won by observation.

Read not to contradict and confute; nor to believe and take for granted; nor to find talk and discourse; but to weigh and consider.° Some books are to be tasted, others to be swallowed, and some few to be chewed and digested; that is, some books are to be read only in parts; others to be read, but not curiously;° and some few to be read wholly, and with diligence and attention. Some books also may be read by deputy,° and extracts° made of them by others; but that would° be only in the less important arguments,° and the meaner sort of books; else distilled books° are like common distilled waters,° flashy° things.

Reading maketh a full man; conference° a ready man; and writing an exact man. And therefore, if a man write little, he had need have a great memory; if he confer° little, he had need have a present wit;° and if he read little, he had need have much cunning, to seem to know that he doth not. Histories make men wise; poets witty;° the mathematics subtile; natural philosophy deep; moral grave;° logic and rhetoric able to contend. 'Abeunt studia in mores'.° Nay there is no stond° or impediment in the wit,° but may be wrought out° by fit studies: like as diseases of the body may have appropriate exercises. Bowling° is good for the stone and reins;° shooting° for the lungs and breast; gentle walking for the stomach; riding for the head; and the like. So if a man's wit be wandering,° let him study the mathematics; for in demonstrations,° if his wit be called away° never so little,° he must begin again. If his wit be not apt to distinguish or find differences,° let him study the schoolmen;° for they are *cymini sectores.*° If he be not apt to beat over° matters, and to call up one thing to prove° and illustrate° another, let him study the lawyers' cases.° So every defect of the mind may have a special receipt.°

51. Of Faction°

Many have an opinion not wise,° that for a prince to govern his estate,° or for a great person to govern his proceedings,° according to the respect of° factions, is a principal part of policy;° whereas contrariwise,° the chiefest wisdom is either in ordering° those things which are general,° and wherein men of several° factions do nevertheless agree; or in dealing with correspondence to° particular persons, one by one. But I say not that the consideration of factions is to be neglected. Mean° men, in their rising,° must adhere;° but great° men, that have strength in themselves,° were better to maintain themselves indifferent° and neutral. Yet even in beginners, to adhere so moderately, as he be a man of the one faction which is most passable° with the other, commonly giveth best way.° The lower and weaker faction is the firmer in conjunction;° and it is often seen that a few that are stiff° do tire out a greater number that are more moderate.° When one of the factions is extinguished,° the remaining subdivideth; as the faction between Lucullus° and the rest of the nobles of the senate (which they called *Optimates*)° held out awhile against the faction of Pompey and Caesar; but when the senate's authority was pulled down, Caesar and Pompey soon after brake.° The faction or party of Antonius and Octavianus Caesar against Brutus and Cassius, held out likewise for a time; but when Brutus and Cassius were overthrown, then soon after Antonius and Octavianus brake and subdivided. These examples are of wars,° but the same holdeth in private° factions. And therefore those that are seconds° in factions do many times, when the faction subdivideth, prove principals; but many times also they prove cyphers° and cashiered;° for many a man's strength is in opposition; and when that faileth he groweth out of use.° It is commonly seen that men once placed° take in° with the contrary faction to that by which they enter:° thinking belike° that they have the first sure,° and now are ready for a new purchase.° The traitor in faction lightly goeth away with it;° for when matters have stuck long in balancing,° the winning of some one man casteth° them, and he getteth all the thanks. The even carriage°

between two factions proceedeth not always of° moderation, but of a trueness to a man's self,° with end° to make use of both.° Certainly in Italy they hold it a little suspect° in popes, when they have often in their mouth *Padre commune*:° and take it to be a sign of one that meaneth to refer° all to the greatness of his own house. Kings had need beware how they side° themselves, and make themselves as of a faction or party; for leagues within the state are ever pernicious to monarchies: for they raise° an obligation° paramount° to obligation of sovereignty, and make the king 'tanquam unus ex nobis';° as was to be seen in the League of France.° When factions are carried too high and too violently,° it is a sign of weakness in princes;° and much to the prejudice both of their authority and business. The motions of factions under kings ought to be like the motions (as the astronomers speak) of the inferior orbs, which may have their proper° motions, but yet still are quietly° carried by the higher motion of *primum mobile*.°

52. Of Ceremonies and Respects°

He that is only real,° had need have exceeding great parts of virtue;° as the stone had need to be rich that is set without foil.° But if a man mark it well, it is in° praise and commendation of men as it is in gettings and gains: for the proverb is true, that 'light gains° make heavy purses';° for light gains come thick,° whereas great come but now and then. So it is true that small matters° win great commendation, because they are continually in use and in note:° whereas the occasion of any great virtue cometh but on festivals.° Therefore it doth much add to a man's reputation, and is (as queen Isabella° said) 'like perpetual letters commendatory',° to have good forms.° To attain° them it almost sufficeth not to despise them; for so° shall a man observe them in others; and° let him trust himself with the rest.° For if he labour too much to express them, he shall lose their grace; which is to be natural and unaffected.° Some men's behaviour is like a verse, wherein every syllable is measured;° how can a man comprehend° great matters, that breaketh° his mind too much to small observations?

Not to use ceremonies at all, is to teach others not to use them again;° and so diminisheth respect to himself; especially they be° not to be omitted to strangers and formal° natures; but the dwelling upon° them, and exalting them above the moon, is not only tedious, but doth diminish the faith and credit° of him that speaks. And certainly there is a kind of conveying° of effectual and imprinting° passages° amongst compliments,° which is of singular use, if a man can hit upon it. Amongst a man's peers a man shall be sure of familiarity; and therefore it is good a little to keep state.° Amongst a man's inferiors one shall be sure of reverence; and therefore it is good a little to be familiar. He that is too much° in anything, so that he giveth another occasion of satiety, maketh himself cheap. To apply° one's self to others is good; so it be with demonstration that a man doth it upon regard,° and not upon facility.° It is a good precept generally in seconding° another, yet to add somewhat of one's own: as if you will grant his opinion, let it be with some distinction; if you will follow his motion,° let it be with condition; if you allow his counsel,° let it be with alleging° further reason.

Men had need beware how they be too perfect° in compliments; for be they never so sufficient° otherwise, their enviers will be sure to give them that attribute, to the disadvantage of their greater virtues. It is loss also in business° to be too full of respects,° or to be curious° in observing times and opportunities.° Salomon saith, 'He that considereth the wind shall not sow, and he that looketh to the clouds shall not reap'.° A wise man will make more opportunities than he finds. Men's behaviour should be like their apparel, not too strait° or point device,° but free for exercise or motion.

53. Of Praise

Praise is the reflection of virtue.° But it is as° the glass or body which giveth the reflection. If it be from the common people, it is commonly false and naught;° and rather followeth vain° persons than virtuous. For the common people understand not many excellent virtues. The lowest virtues draw praise from them; the middle virtues work° in them astonishment or admiration; but of the highest virtues they have no sense° or perceiving° at all. But shews,° and 'species virtutibus similes',° serve best° with them. Certainly fame is like a river, that beareth up things light and swoln,° and drowns things weighty and solid. But if persons of quality° and judgment° concur, then it is (as the Scripture saith), 'Nomen bonum instar unguenti fragrantis'.° It filleth all round about, and will not easily away.° For the odours of ointments° are more durable than those of flowers.

There be so many false points° of praise, that a man may justly hold it a suspect.° Some praises proceed merely of° flattery; and if he be an ordinary flatterer, he will have° certain common attributes, which may serve every man; if he be a cunning flatterer, he will follow the arch-flatterer,° which is a man's self; and wherein a man thinketh best of himself, therein the flatterer will uphold° him most: but if he be an impudent flatterer, look° wherein a man is conscious to himself that he is most defective, and is most out of countenance° in himself, that will the flatterer entitle° him to perforce,° 'spreta conscientia'.° Some praises come of good wishes and respects,° which is a form° due in civility to kings and great persons, 'laudando praecipere';° when by telling men what they are, they represent to them what they should be.° Some men are praised maliciously to their hurt, thereby to stir° envy and jealousy towards them; 'pessimum genus inimicorum laudantium';° insomuch as it was a proverb amongst the Grecians,° that 'he that was praised to his hurt, should have a push rise upon his nose'; as we say, that 'a blister will rise upon one's tongue° that tells a lie'. Certainly moderate praise, used with opportunity,° and not vulgar,° is that which doth the good. Salomon saith, 'He that praiseth his friend aloud, rising early, it shall be to him no better than a curse'.° Too much magnifying of man or matter doth irritate° contradiction, and procure envy and scorn.

To praise a man's° self cannot be decent,° except it be in rare cases; but to praise a man's office or profession, he° may do it with good grace, and with a kind of magnanimity.° The Cardinals of Rome, which° are theologues, and friars, and schoolmen, have a phrase of notable° contempt and scorn towards civil business:° for they call all temporal business of wars, embassages,° judicature, and other employments, *sbirrerie*,° which is 'under-sheriffries';° as if they were but matters for under-sheriffs and catch-poles:° though many times those under-sheriffries do more good than their high speculations. St. Paul, when he boasts of himself, he doth oft interlace,° 'I speak like a fool';° but speaking of his calling, he saith, 'magnificabo apostolatum meum'.°

54. Of Vain-Glory°

It was prettily devised° of Aesop; 'the fly sat upon the axle-tree of the chariot wheel, and said, "What a dust do I raise!"' So are there some vain° persons, that whatsoever goeth alone° or moveth upon greater means,° if they have never so little hand in it, they think it is they that carry° it. They that are glorious° must needs be factious;°

for all bravery stands upon comparisons.° They must needs be violent, to make good their own vaunts.° Neither can they be secret, and therefore not effectual;° but according to the French proverb, 'Beaucoup de bruit, peu de fruit'; 'Much bruit,° little fruit'. Yet certainly there is use of this quality° in civil affairs. Where there is an opinion and fame to be created either of virtue or greatness, these men are good trumpeters. Again, as Titus Livius noteth in the case of Antiochus and the Aetolians, 'There are sometimes great effects of cross° lies';° as if a man that negotiates between two princes, to draw them to join in a war against the third, doth extol the forces of either° of them above measure, the one to the other: and sometimes he that deals between man and man, raiseth his own credit with both, by pretending greater interest° than he hath in either. And in these and the like kinds, it often falls out that somewhat is produced of nothing;° for lies are sufficient to breed opinion,° and opinion brings on substance.°

In military commanders and soldiers, vain-glory is an essential point;° for as iron sharpens iron,° so by glory one courage sharpeneth another. In cases of great enterprise upon charge and adventure,° a composition° of glorious natures doth put life into business; and° those that are of solid° and sober natures have more of the ballast than of the sail.° In fame of learning, the flight° will be slow without some feathers of ostentation. 'Qui de contemnenda gloria libros scribunt, nomen suum inscribunt'.° Socrates, Aristotle, Galen,° were men full of ostentation. Certainly vain-glory helpeth to perpetuate a man's memory; and virtue was never so beholding° to human nature, as° it received his due at the second hand.° Neither had° the fame of Cicero, Seneca, Plinius Secundus,° borne her age so well, if it had not been joined with some vanity in themselves; like unto varnish, that makes ceilings° not only shine but last.

But all this while, when I speak of vain-glory, I mean not of that property° that Tacitus doth attribute to Mucianus; 'Omnium, quae dixerat feceratque, arte quadam ostentator':° for that proceeds not of vanity, but of natural magnanimity° and discretion;° and in some persons is not only comely,° but gracious.° For excusations,° cessions,° modesty itself well governed,° are but arts of ostentation. And amongst those arts there is none better than that which Plinius Secundus speaketh of, which is to be liberal° of praise and commendation to others, in that wherein a man's self hath any perfection. For saith Pliny very wittily,° 'In commending another you do yourself right; for he that you commend is either superior to you in that you commend, or inferior. If he be inferior, if he be to be° commended, you much more; if he be superior, if he be not to be commended, you much less'.°

Glorious men are the scorn° of wise men, the admiration of fools, the idols of parasites,° and the slaves of their own vaunts.

55. OF HONOUR AND REPUTATION

THE winning° of Honour is but the revealing° of a man's virtue and worth without disadvantage.° For some in their actions do woo and affect° honour and reputation; which sort of men are commonly much talked of, but inwardly° little admired. And some, contrariwise, darken° their virtue in the shew of it; so as they be undervalued in opinion. If a man perform that which hath not been attempted before; or attempted and given over;° or hath been achieved, but not with so good circumstance;° he shall purchase° more honour, than by effecting a matter of greater difficulty or virtue,° wherein he is but a follower.° If a man so temper° his actions, as in some one of them he doth content every faction or combination° of people, the music will be the fuller. A man is an ill husband° of his honour, that entereth into° any action, the failing wherein may disgrace him more than the carrying of it through can honour him.° Honour that is gained and broken° upon another° hath the quickest° reflexion, like diamonds cut with facets. And therefore let a man contend° to excel any competitors of his in honour, in outshooting them, if he can, in their own bow.° Discreet followers and servants help much to reputation. 'Omnis fama a domesticis emanat'.° Envy, which is the canker° of honour, is best extinguished by declaring° a man's° self in his ends rather to seek merit than fame; and by attributing a man's successes rather to divine Providence and felicity,° than to his own virtue or policy.° The true marshalling° of the degrees of sovereign honour are these. In the first place are *conditores imperiorum*,° founders of states and commonwealths;° such as were Romulus,° Cyrus,° Caesar,° Ottoman,° Ismael.° In the second place are *legislatores*, lawgivers; which are also called 'second founders', or *perpetui principes*,° because they govern by their ordinances after they are gone; such were Lycurgus,° Solon,° Justinian,° Eadgar,° Alphonsus of Castile,° the wise, that made the *Siete partidas*.° In the third place are *liberatores*, or *salvatores*,° such as compound° the long miseries of civil wars, or deliver their countries from servitude of strangers or tyrants; as

Augustus Caesar,° Vespasianus,° Aurelianus,° Theodoricus,° King Henry the Seventh° of England, King Henry the Fourth° of France. In the fourth place are *propagatores* or *propugnatores imperii*;° such as in honourable wars enlarge their territories, or make noble defence against invaders. And in the last place are *patres patriae*,° which° reign justly, and make the times good° wherein they live. Both which last kinds need no examples, they are in such number. Degrees of honour in subjects are, first *participes curarum*,° those upon whom princes do discharge the greatest weight of their affairs; their 'right hands', as we call them. The next are *duces belli*,° great leaders; such as are princes' lieutenants,° and do them notable services in the wars. The third are *gratiosi*, favourites; such as exceed not this scantling,° to be solace to the sovereign, and harmless to the people. And the fourth, *negotiis pares*;° such as have great places under princes, and execute their places° with sufficiency.° There is an honour, likewise, which may be ranked amongst the greatest, which happeneth rarely; that is, of° such as sacrifice themselves to death or danger for the good of their country; as was M. Regulus,° and the two Decii.°

56. Of Judicature°

JUDGES ought to remember that their office is *jus dicere*, and not *jus dare*;° to interpret law, and not to make law, or give law. Else will it be like the authority claimed by the church of Rome, which under pretext of exposition of Scripture doth not stick° to add and alter; and to pronounce° that which they do not find; and by shew of antiquity to introduce novelty.° Judges ought to be more learned than witty,° more reverend° than plausible,° and more advised° than confident. Above all things, integrity is their portion° and proper virtue. 'Cursed' (saith the law) 'is he that removeth the landmark'.° The mislayer° of a mere-stone° is to blame.° But it is the unjust judge that is the capital° remover of landmarks, when he defineth amiss° of lands and property. One foul° sentence° doth more hurt than many foul examples. For these do but corrupt the stream, the other corrupteth the fountain. So saith Salomon, 'Fons turbatus, et vena corrupta, est justus cadens in causa sua coram adversario'.° The office of judges may have reference° unto the parties that sue,° unto the advocates that plead, unto the clerks and ministers of justice underneath them, and to the sovereign or state° above them.

First, for the causes° or parties that sue. 'There be'° (saith the Scripture) 'that turn judgment into wormwood';° and surely there be also that turn it into vinegar; for injustice maketh it bitter, and delays make it sour. The principal duty of a judge is to suppress force and fraud;° whereof force is the more pernicious when it is open, and fraud when it is close° and disguised. Add thereto contentious suits,° which ought to be spewed out,° as the surfeit° of courts. A judge ought to prepare his way to a just sentence, as God useth to prepare his way, by raising valleys and taking down hills:° so when there appeareth on either side an high hand,° violent prosecution,° cunning advantages taken, combination,° power, great counsel,° then is the virtue of a judge seen, to make° inequality equal; that he may plant his judgment as upon an even ground. 'Qui fortiter emungit, elicit sanguinem';° and where° the wine-press is hard wrought,° it yields a harsh wine, that tastes of the grape-stone. Judges must beware of hard constructions° and strained inferences; for there is no worse torture° than the torture of laws. Specially in case of laws penal, they ought to have care that that which was meant for terror be not turned into rigour;° and that they bring not upon the people that shower whereof the Scripture speaketh, 'Pluet super eos laqueos';° for penal laws pressed° are a 'shower of snares'° upon the people. Therefore let penal laws, if they have been sleepers of long,° or if they be grown unfit for the present time, be by wise judges confined° in the execution: 'Judicis officium est, ut res, ita tempora rerum, &c.'° In causes of life and death, judges ought (as far as the law permitteth) in justice° to remember mercy; and to cast a severe eye upon the example,° but a merciful eye upon the person.

Secondly, for the advocates and counsel° that plead. Patience and gravity of hearing is an essential part of justice; and an overspeaking° judge is no well-tuned cymbal.° It is no grace° to a judge first to find° that which he might have heard in due time from the bar;° or to show quickness of conceit° in cutting off evidence or counsel too short;° or to prevent° information by questions, though pertinent. The parts° of a judge in hearing are four: to direct the evidence;° to moderate° length, repetition, or impertinency° of speech; to recapitulate, select, and collate the material points of that which hath been said; and to give the rule° or sentence. Whatsoever is above° these is too much; and proceedeth either of glory° and willingness° to speak, or of impatience to hear,° or of shortness of memory, or of want of a staid and equal° attention. It is a strange thing to see that the boldness of advocates should prevail with judges; whereas they° should imitate God, in whose seat they sit; who 'represseth the presumptuous, and giveth grace to the modest'.° But it is more strange, that judges should have noted° favourites; which cannot but cause multiplication of fees,° and suspicion of bye-ways.° There is due from the judge to the advocate some commendation and gracing,° where causes are well handled and fair pleaded; especially towards the side which obtaineth° not; for that upholds in the client the reputation of his counsel, and beats down in him the conceit° of his cause. There is likewise due to the public° a civil° reprehension of advocates, where there appeareth cunning counsel, gross° neglect, slight° information, indiscreet pressing,° or an over-bold defence. And let not the counsel at the bar chop° with the judge, nor wind himself into° the handling of the cause anew after the judge hath declared his sentence; but on the other side, let not the judge meet the cause half way,° nor give occasion for the party to say his counsel or proofs° were not heard.

Thirdly, for that that concerns clerks and ministers.° The place of justice is an hallowed° place; and therefore not only the bench, but the foot-pace° and precincts and purprise° thereof, ought to be preserved without scandal and corruption. For certainly 'Grapes' (as the Scripture saith) 'will not be gathered of thorns or thistles';° neither can justice yield her fruit with sweetness amongst the briars and brambles of catching° and polling° clerks and ministers. The attendance° of courts is subject to four bad instruments. First, certain persons that are sowers of suits;° which make the court swell,° and the country pine.° The second sort is of those that engage courts in quarrels of jurisdiction,° and are not truly *amici curiae*,° but *parasiti curiae*, in puffing° a court up beyond her bounds, for their own scraps° and advantage. The third sort is of those that may be accounted the left hands of courts; persons that are full of nimble and sinister° tricks and shifts, whereby they pervert° the plain and direct courses of courts, and bring justice into oblique lines and labyrinths. And the fourth is the poller° and exacter of fees; which justifies the common resemblance° of the courts of justice to the bush° whereunto while the sheep flies for defence in weather,° he is sure to lose part of his fleece. On the other side, an ancient° clerk, skilful in precedents, wary in proceeding, and understanding° in the business° of the court, is an excellent finger° of a court; and doth many times point the way to the judge himself.

Fourthly, for that which may concern the sovereign and estate. Judges ought above all to remember the conclusion of the Roman Twelve Tables;° 'Salus populi suprema lex';° and to know that laws, except they be in order to that end,° are but things captious,° and oracles° not well inspired. Therefore it is an happy thing in a state when kings and states° do often consult with judges; and again when judges do often consult with the king and state: the one, when there is matter of law intervenient° in business of state;° the other, when there is some consideration of state intervenient in matter of law. For many times the things deduced° to judgment may be *meum* and *tuum*,° when the reason° and consequence thereof may trench to° point of° estate:° I call matter of estate, not only the parts° of sovereignty, but whatsoever introduceth any great alteration° or dangerous precedent; or concerneth manifestly any great portion of people. And let no man weakly° conceive that just laws and true policy° have any antipathy;° for they are like the spirits° and sinews, that° one moves with the other. Let judges also remember, that Salomon's throne was supported by lions° on both sides: let them be lions, but yet lions under the throne; being circumspect° that they do not check or oppose any points of sovereignty. Let not judges also be so ignorant of their own right, as to think there is not left to them, as a principal part of their office, a wise use and application of laws. For they may remember what the apostle saith of a greater law than theirs; 'Nos scimus quia lex bona est, modo quis ea utatur legitime'.°

57. Of Anger°

To seek to extinguish Anger utterly is but a bravery° of the Stoics.° We have better oracles: 'Be angry, but sin not. Let not the sun go down upon your anger'.° Anger must be limited and confined both in race° and in time. We will first speak how the natural inclination and habit° to be angry may be attempered° and calmed. Secondly, how the particular motions of anger may be repressed, or at least refrained° from doing mischief. Thirdly, how to raise anger or appease anger in another.

For the first; there is no other way but to meditate and ruminate well upon the effects of anger, how it troubles man's life. And the best time to do this, is to look back upon anger when the fit is throughly° over. Seneca saith well, that 'anger is like ruin,° which breaks itself upon that it° falls'.° The Scripture exhorteth us 'To possess our souls in patience'.° Whosoever is out of patience, is out of possession of his soul.° Men must not turn bees;

. . . animasque in vulnere ponunt.°

Anger is certainly a kind of baseness;° as it appears well in the weakness of those subjects in whom it reigns; children, women, old folks, sick folks.° Only° men must beware° that they carry their anger rather with scorn° than with fear; so that they may seem rather to be above the injury than below it; which is a thing easily done, if a man will give law to himself in it.°

For the second point; the causes and motives of anger are chiefly three.° First, to be too sensible° of hurt; for no man is angry that feels not himself hurt; and therefore tender and delicate persons must needs be oft angry; they have so many things to trouble them, which more robust natures have little sense of. The next is, the apprehension and construction° of the injury offered to be, in the circumstances thereof, full of contempt: for contempt is that which putteth an edge upon° anger, as much or more than the hurt itself.° And therefore when men are ingenious in° picking out circumstances° of contempt, they do kindle their anger much. Lastly, opinion of the touch° of a man's reputation doth multiply and sharpen anger. Wherein the remedy is, that a man should have, as Consalvo° was wont to say, 'telam honoris crassiorem'.° But in all refrainings° of anger, it is the best remedy to win time;° and to make a man's self believe° that the opportunity of his revenge is not yet come, but that he foresees a time° for it; and so to still° himself in the mean time, and reserve° it.

To contain° anger from mischief,° though it take hold of a man, there be two things whereof you must have special caution.° The one, of extreme bitterness of words, especially if they be aculeate° and proper;° for *communia maledicta*° are nothing so much;° and again, that in anger a man reveal no secrets; for that makes him not fit for society. The other, that you do not peremptorily break off, in any business,° in a fit of anger; but howsoever you shew bitterness, do not act anything that is not revocable.

For raising and appeasing anger in another; it is done chiefly by choosing of times, when men are frowardest° and worst disposed, to incense them. Again, by gathering (as was touched° before) all that you can find out to aggravate° the contempt. And the two remedies are by the contraries. The former to take good times,° when first to relate° to a man an angry business;° for the first impression is much; and the other is, to sever, as much as may be, the construction° of the injury from the point of contempt;° imputing it to misunderstanding, fear, passion, or what you will.

58. Of Vicissitude° of Things

SALOMON saith, 'There is no new thing upon the earth'.° So that as Plato had an imagination,° that 'all knowledge was but remembrance';° so Salomon giveth his sentence,° that 'all novelty is but oblivion'.° Whereby you may see that the river of Lethe° runneth as well above ground as below. There is an abstruse° astrologer° that saith, 'if it were not for two things that are constant (the one is, that the fixed stars ever stand at like distance one from another, and never come nearer together, nor go further asunder; the other, that the diurnal motion perpetually keepeth time), no individual would last one moment'. Certain it is, that the matter° is in a perpetual flux,° and never at a stay.° The great winding-sheets,° that bury all things in oblivion, are two: deluges and earthquakes. As for conflagrations and great droughts, they do not merely° dispeople° and destroy. Phaëton's car° went but a day. And the three years' drought in the time of Elias° was but particular,° and left people alive. As for the great burnings by lightnings, which are often in the West Indies,° they are but narrow.° But in the other two destructions, by deluge and earthquake, it is further to be noted, that the remnant of people which hap to be reserved,° are commonly ignorant and mountainous° people, that can give no account of the time past; so that the oblivion is all one° as if none had been left. If you consider well of the people of the West Indies, it is very probable that they are a newer or a younger people than the people of the old world.° And it is much more likely that the destruction that hath heretofore° been there, was not by earthquakes (as the Egyptian priest told Solon concerning the island of Atlantis, that 'it was swallowed by an earthquake'),° but rather that it was desolated by a particular deluge. For earthquakes are seldom° in those parts. But on the other side, they have such pouring° rivers, as the rivers of Asia and Africk and Europe are but brooks to them. Their Andes° likewise, or mountains, are far higher than those with us; whereby° it seems that the remnants of generation of men° were in such a particular deluge saved. As for the observation that Machiavel° hath, that the jealousy of sects doth much extinguish the memory of things; traducing Gregory the Great,° that° he did what in him lay to extinguish all heathen antiquities; I do not find that those zeals° do° any great effects, nor last long; as it appeared in the succession of Sabinian,° who did revive the former antiquities.

The vicissitude or mutations in the Superior Globe° are no fit matter for this present argument. It may be, Plato's great year,° if the world should last so long, would have some effect; not in renewing the state° of like° individuals (for that is the fume° of those that conceive the celestial bodies have more accurate° influences° upon these things below than indeed they have), but in gross.° Comets, out of question, have likewise power and effect over the gross and mass of things; but they are rather gazed upon, and waited upon° in their journey, than wisely observed in their effects; specially in their respective° effects; that is, what kind of comet, for magnitude, colour, version° of the beams, placing° in the region° of heaven, or lasting, produceth what kind of effects.

There is a toy° which I have heard, and I would not have it given over,° but waited upon° a little. They say it is observed in the Low Countries (I know not in what part) that every five and thirty years the same kind and suit° of years and weathers comes about again; as great frosts, great wet, great droughts, warm winters, summers with little heat, and the like; and they call it the *Prime*.° It is a thing I do the rather mention, because, computing backwards, I have found some concurrence.°

But to leave these points° of nature, and to come to men. The greatest vicissitude of things amongst men, is the vicissitude of sects and religions. For those orbs° rule in men's minds most. The true religion is 'built upon the rock';° the rest are tossed upon the waves of time. To speak° therefore of the causes of new sects; and to give some counsel concerning them, as far as the weakness of human judgment can give stay° to so great revolutions.°

When the religion formerly received° is rent by discords; and when the holiness of the professors° of religion is decayed and full of scandal; and withal° the times be stupid, ignorant, and barbarous; you may doubt° the springing up of a new sect; if then also there should arise any extravagant° and strange spirit to make° himself author thereof. All which points held° when Mahomet° published his law. If a new sect have not two properties,° fear it not; for it will not spread. The one is, the supplanting or the opposing of authority° established; for nothing is more popular° than that. The other is, the giving licence to pleasures and a voluptuous life. For as for speculative° heresies, (such as were in ancient times the Arians,° and now the Arminians),° though they work mightily upon men's wits, yet they do not produce any great alterations in states; except it be by the help of civil° occasions. There be three manner of plantations° of new sects. By the power of signs° and miracles; by the eloquence and wisdom of speech and persuasion; and by the sword. For martyrdoms, I reckon them amongst miracles; because they seem to exceed the strength of human nature: and I may do the like of° superlative and admirable holiness of life. Surely there is no better way to stop the rising of new sects and schisms, than to reform abuses; to compound° the smaller differences; to proceed mildly, and not with sanguinary persecutions; and rather to take off° the principal authors by winning° and advancing° them, than to enrage them by violence and bitterness.

The changes and vicissitude in wars are many; but chiefly in three things; in the seats or stages° of the war; in the weapons; and in the manner of the conduct. Wars, in ancient time, seemed more to move from east to west; for the Persians, Assyrians, Arabians, Tartars° (which were the invaders), were all eastern people.° It is true, the Gauls were western; but we read but of two incursions of theirs: the one to Gallo-Graecia,° the other to Rome. But East and West have no certain° points of heaven; and no more have the wars, either from the east or west, any certainty of observation. But North and South are fixed;° and it hath seldom or never been seen that the far southern people have invaded the northern, but contrariwise.° Whereby it is manifest that the northern tract of the world is in nature the more martial region:° be it in respect° of the stars° of that hemisphere; or of the great continents that are upon the north, whereas the south part, for aught that is known, is almost all sea; or (which is most apparent) of the cold of the northern parts, which is that which, without aid of discipline,° doth make the bodies hardest,° and the courages warmest.

Upon the breaking and shivering° of a great state and empire, you may be sure to have wars. For great empires, while they stand, do enervate° and destroy the forces of the natives which they have subdued, resting° upon their own protecting forces; and then when they fail also, all goes to ruin, and they become a prey.° So was it in the decay of the Roman empire; and likewise in the empire of Almaigne,° after Charles the Great,° every bird taking a feather; and were° not unlike° to befall° to Spain, if it should break.° The great accessions° and unions of kingdoms do likewise stir up wars: for when a state grows to an over-power,° it is like a great flood, that will be sure to overflow. As it hath been seen in the states of Rome, Turkey, Spain, and others. Look when the world hath fewest barbarous peoples,° but such as commonly will not marry or generate, except they know means° to live (as it is almost every where at this day, except Tartary), there is no danger of inundations of people:° but when there be great shoals° of people, which go on to populate, without foreseeing° means of life and sustentation,° it is of necessity that once in an age or two they discharge a portion of their people upon other nations; which the ancient northern people were wont to do by lot; casting lots what part should stay at home, and what should seek their fortunes. When a warlike state grows soft and effeminate, they may be sure of a war. For commonly such states are grown rich in the time of their degenerating; and so the prey inviteth, and their decay in valour encourageth a war.

As for the weapons, it° hardly falleth under rule and observation:° yet we see even they have returns and vicissitudes. For certain it is, that ordnance° was known in the city of the Oxidrakes° in India; and was that which the Macedonians° called thunder and lightning, and magic. And it is well known° that the use of ordnance hath been° in China above two thousand years. The conditions of weapons, and their improvement, are: First, the fetching° afar off; for that outruns the danger; as it is seen in ordnance and muskets. Secondly, the strength of the percussion; wherein likewise ordnance do exceed all arietations° and ancient inventions. The third is, the commodious° use of them; as that they may serve in all weathers; that the carriage may be light and manageable; and the like.

For the conduct of the war: at the first, men rested° extremely° upon number: they did put the wars likewise upon main force° and valour; pointing° days for pitched fields,° and so trying° it out upon an even match:° and they were more ignorant° in ranging and arraying° their battles.° After° they grew° to rest upon number rather competent than vast; they grew to advantages of place, cunning diversions, and the like: and they grew more skilful in the ordering° of their battles.

In the youth of a state, arms do flourish; in the middle age of a state, learning,° and then both of them together for a time; in the declining age of a state, mechanical arts and merchandise. Learning hath his° infancy, when it is but beginning and almost childish: then his youth, when it is luxuriant° and juvenile: then his strength of years, when it is solid and reduced:° and lastly, his old age, when it waxeth dry and exhaust.° But it is not good to look too long upon these turning wheels of vicissitude, lest we become giddy. As for the philology° of them, that is but a circle of tales,° and therefore not fit for this writing.

59. A Fragment of an Essay on Fame°

The poets° make Fame a monster. They describe her in part finely and elegantly; and in part gravely° and sententiously.° They say, look how many feathers she hath, so many eyes she hath underneath; so many tongues;° so many voices; she pricks up so many ears.

This is a flourish.° There follow excellent parables;° as that she gathereth strength in going:° that she goeth upon the ground, and yet hideth her head in the clouds:° that in the day-time she sitteth in a watch tower,° and flieth most by night:° that she mingleth things done with things not done:° and that she is a terror to great cities.° But that which passeth° all the rest is; they° do recount that the Earth, mother of the Giants that° made war against Jupiter and were by him destroyed, thereupon in an anger° brought forth Fame; for certain it is that rebels, figured° by the giants, and seditious fames and libels, are but brothers and sisters; masculine° and feminine. But now, if a man can tame this monster, and bring her to feed at the hand, and govern her, and with her fly° other ravening fowl° and kill them, it is somewhat worth.°

But we are infected with the stile of the poets. To speak now in a sad° and a serious manner. There is not in all the politics° a place° less handled,° and more worthy to be handled, than this of fame. We will therefore speak of these points. What are false fames; and what are true fames; and how they may be best discerned;° how fames may be sown° and raised;° how they may be spread and multiplied; and how they may be checked and laid dead.° And other things concerning the nature of fame. Fame is of that force, as there is scarcely any great action wherein it hath not a great part; especially in the war. Mucianus° undid Vitellius, by a fame that he scattered, that Vitellius had in purpose° to remove° the legions of Syria into Germany, and the legions of Germany into Syria; whereupon the legions of Syria were infinitely inflamed. Julius Caesar took Pompey unprovided,° and laid asleep° his industry and preparations, by a fame that he cunningly° gave out, how Caesar's own soldiers loved him not; and being wearied with wars, and laden with the spoils of Gaul, would forsake him as soon as he came into Italy. Livia° settled all things for the succession of her son Tiberius, by continual giving out that her husband Augustus was upon recovery and amendment.° And it is an usual thing with the Bashaws,° to conceal° the death of the great Turk from the Janizaries° and men of war, to save° the sacking° of Constantinople and other towns, as their manner is. Themistocles° made Xerxes King of Persia post apace° out of Graecia, by giving out that the Grecians had a purpose to break his bridge of ships which he had made athwart° Hellespont. There be a thousand such-like examples, and the more they are, the less they need to be repeated;

because a man meeteth with them every where. Therefore let all wise governors° have as great a watch and care over fames, as they have of the actions° and designs themselves.

New Atlantis

WE sailed from Peru, (where we had continued° by the space of one whole year), for China and Japan, by the South Sea;° taking with us victuals for twelve months; and had good winds from the east,° though soft and weak, for five months' space and more. But then the wind came about,°, and settled in the west for many days, so as° we could make little or no way,° and were sometimes in purpose to turn back. But then again there arose strong and great winds from the south, with a point° east; which carried us up (for all that we could do) towards the north: by which time our victuals failed us, though we had made good spare° of them. So that finding ourselves in the midst of the greatest wilderness of waters in the world, without victual, we gave ourselves for lost men, and prepared for death. Yet we did lift up our hearts and voices to God above, who 'showeth his wonders in the deep';° beseeching him of his mercy, that as in the beginning he discovered° the face of the deep, and brought forth dry land, so he would now discover land to us, that we might not perish.

And it came to pass that the next day about evening, we saw within a kenning° before us, towards the north, as it were thick clouds, which did put us in some hope of land; knowing how that part of the South Sea was utterly unknown; and might° have islands or continents, that hitherto were not come to light. Wherefore we bent° our course thither, where we saw the appearance of land, all that night; and in the dawning of the next day, we might plainly discern that it was a land; flat to our sight, and full of boscage;° which made it shew the more dark. And after an hour and a half's sailing, we entered into a good haven, being the port of a fair city; not great° indeed, but well built, and that gave° a pleasant view from the sea: and we thinking every minute long till we were on land, came close to the shore, and offered° to land. But straightways we saw divers° of the people, with bastons° in their hands, as it were forbidding us to land; yet without any cries or fierceness, but only as warning us off by signs that they made. Whereupon being not a little discomforted,° we were advising° with ourselves what we should do. During which time there made forth to us a small boat, with about eight persons in it; whereof one of them had in his hand a tipstaff° of a yellow cane, tipped at both ends with blue,° who came aboard our ship, without any show of distrust at all. And when he saw one of our number present himself somewhat afore the rest, he drew forth a little scroll of parchment (somewhat yellower than our parchment, and shining like the leaves of writing tables,° but otherwise soft and flexible), and delivered it to our foremost man.° In which scroll were written in ancient Hebrew, and in ancient Greek, and in good Latin of the School,° and in Spanish, these words; 'Land ye not, none of you; and provide° to be gone from this coast within sixteen days, except° you have further time given you. Meanwhile, if you want fresh water, or victual, or help for your sick, or that your ship needeth repair, write down your wants, and you shall have that which belongeth to mercy.' This scroll was signed with a stamp° of cherubins' wings,° not spread but hanging downwards, and by them a cross. This being delivered, the officer returned, and left only a servant with us to receive our answer.

Consulting hereupon amongst ourselves, we were much perplexed. The denial° of landing and hasty warning us away troubled us much; on the other side, to find that the people had° languages and were so full of humanity,° did comfort us not a little. And above all, the sign of the cross to that instrument° was to us a great rejoicing, and as it were a certain presage of good. Our answer was in the Spanish tongue; 'That for° our ship, it was well; for we had rather met with calms and contrary winds than any tempests. For our sick, they were many, and in very ill case;° so that if they were not permitted to land, they ran danger of their lives.' Our other wants we set down in particular;° adding, 'that we had some little store of merchandise, which if it pleased them to deal° for, it might supply our wants without being chargeable° unto them.' We offered some reward in pistolets° unto the servant, and a piece of crimson velvet to be presented to the officer; but the servant took them not, nor would scarce look upon them; and so left us, and went back in another little boat which was sent for him.

About three hours after we had dispatched our answer, there came towards us a person (as it seemed) of place.° He had on him a gown with wide sleeves, of a kind of water chamolet,° of an excellent azure colour, far more glossy than ours; his under-apparel was green; and so was his hat, being in the form of a turban, daintily made, and not so huge as the Turkish turbans; and the locks of his hair came down below the brims of it. A reverend° man was he to behold. He came in a boat, gilt° in some part of it, with four persons more only in that boat; and was followed by another boat, wherein were some twenty. When he was come within a flight-shot° of our ship, signs were made to us that we should send forth some to meet him upon the water; which we presently° did in our ship-boat, sending the principal man amongst us save° one, and four of our number with him.

When we were come within six yards of their boat, they called to us to stay, and not to approach farther; which we did. And thereupon the man whom I before described stood up, and with a loud voice in Spanish, asked, 'Are ye Christians?' We answered, 'We were;' fearing the less, because of the cross we had seen in the subscription.° At which answer the said person lifted up his right hand towards heaven, and drew it softly to his mouth (which is the gesture they use when they thank God), and then said: 'If ye will swear (all of you) by the merits of the Saviour that ye are no pirates,° nor have shed blood° lawfully nor unlawfully within forty days past, you may have licence° to come on land.' We said, 'We were all ready to take that oath.' Whereupon one of those that were with him, being (as it seemed) a notary, made an entry of this act.° Which done, another of the attendants of the great person, which was with him in the same boat, after his lord had spoken a little to him, said aloud; 'My lord would have you know, that it is not of° pride or greatness° that he cometh not aboard your ship; but for that° in your answer you declare that you have many sick amongst you, he was warned by the Conservator of Health° of the city that he should keep a distance.' We bowed ourselves towards him, and answered, 'We were his humble servants; and accounted° for great honour and singular humanity towards us that which was already done; but hoped well° that the nature of the sickness of our men was not infectious.' So he returned; and a while after came the notary to us aboard our ship; holding in his hand a fruit of that country, like an orange, but of colour between orange-tawney° and scarlet, which cast° a most excellent odour. He used it (as it seemeth) for a preservative against infection. He gave us our oath; 'By the name of Jesus and his merits:' and after° told us that the next day by six of the clock in the morning we should be sent to, and brought to the Strangers' House (so he called it), where we should be accommodated° of things both for our whole° and for our sick. So he left us; and when we offered him some pistolets, he smiling said, 'He must not be twice paid for one labour:' meaning (as I take it) that he had salary sufficient of the state for his service. For (as I after learned) they call an officer that taketh rewards, 'twice paid'.°

The next morning early, there came to us the same officer that came to us at first with his cane, and told us, he came to conduct us to the Strangers' House; and that he had prevented° the hour,° because° we might have the whole day before us for our business. 'For,' said he, 'if you will° follow my advice, there shall first go with me some few of you, and see the place, and how it may be made convenient for you; and then you may send for your sick, and the rest of your number which ye will bring on land.' We thanked him, and said, 'That this care which he took of desolate strangers God would reward.' And so six of us went on land with him: and when we were on land, he went before us, and turned to us, and said, 'He was but our servant, and our guide.' He led us through three fair° streets; and all the way we went there were gathered some people on both sides standing in a row; but in so civil a fashion, as if it had been not to wonder at us but to welcome us; and divers of them, as we passed by them, put their arms a little abroad;° which is their gesture when they bid any° welcome.

The Strangers' House is a fair and spacious house, built of brick, of somewhat a bluer colour than our brick; and with handsome windows, some of glass, some of a kind of cambric oiled.° He brought us first into a fair parlour above stairs,° and then asked us, 'What number of persons we were? And how many sick?' We answered, 'We were in all (sick and whole) one and fifty persons, whereof our sick were seventeen.' He desired us to have patience a little, and to stay till he came back to us; which was about an hour after; and then he led us to see the chambers° which were provided for us, being in number nineteen: they having cast° it (as it seemeth) that four of those chambers, which were better than the rest, might receive four of the principal men of our company, and lodge them alone by themselves; and the other fifteen chambers were to lodge us two and two together. The chambers were handsome and cheerful chambers, and furnished civilly.° Then he led us to a long gallery, like a dorture,° where he showed us all along the one side (for the other side was but wall and window) seventeen cells,° very neat ones, having partitions of cedar wood. Which gallery and cells, being in all forty (many more than we needed), were instituted° as an infirmary for sick persons. And he told us withal, that as any of our sick waxed well,° he might be removed from his cell to a chamber; for which purpose there were set forth° ten spare chambers, besides the number we spake of before. This done, he brought us back to the parlour, and lifting up his cane a little, (as they do when they give any charge° or command), said to us, 'Ye are to know that the custom of the land requireth, that after this day and to-morrow, (which we give you for removing of your people from your ship), you are to keep within doors for three days. But let it not trouble you, nor do not think yourselves restrained, but rather left to your rest and ease. You shall want nothing, and there are six of our people appointed to attend° you, for any business you may have abroad.'° We gave him thanks with all affection° and respect, and said, 'God surely is manifested in this land.' We offered him also twenty pistolets; but he smiled, and only said; 'What? twice paid!' And so he left us.

Soon after our dinner was served in; which was right° good viands,° both for bread and meat: better than any collegiate° diet that I have known in Europe. We had also drink of three sorts, all wholesome and good; wine of the grape; a drink of grain, such as is with us our ale, but more clear; and a kind of cider made of a fruit of that country; a wonderful pleasing and refreshing drink. Besides, there were brought in to us great store° of those scarlet oranges for our sick; which (they said) were an assured° remedy for sickness taken at sea. There was given us also a box of small grey or whitish pills, which they wished our sick should take, one of the pills every night before sleep; which (they said) would hasten their recovery.

The next day, after that our trouble of carriage° and removing of our men and goods out of our ship was somewhat settled and quiet,° I thought good to call our company together; and when they were assembled said unto them; 'My dear friends, let us know ourselves,° and how it standeth with us. We are men cast on land, as Jonas° was out of the whale's belly, when we were as° buried in the deep: and now we are on land, we are but° between death and life; for we are beyond both the old world and the new; and whether ever we shall see Europe, God only knoweth. It is a kind of miracle hath brought us hither: and it must be little less that shall bring us hence. Therefore in regard of° our deliverance past, and our danger present and to come, let us look up to God, and every man reform his own ways. Besides we are come here amongst a Christian people, full of piety and humanity: let us not bring that confusion of face° upon ourselves, as to show our vices or unworthiness before them. Yet there is more. For they have by commandment (though in° form of courtesy) cloistered us within these walls for three days: who knoweth whether it be not to take some taste° of our manners° and conditions?° and if they find them bad, to banish us straightways; if good, to give us further time. For these men that they have given us for attendance° may withal° have an eye upon us. Therefore for God's love, and as we love the weal° of our souls and bodies, let us so behave ourselves as we may be at peace with God, and may find grace° in the eyes of this people.' Our company with one voice thanked me for my good admonition, and promised me to live soberly and civilly, and without giving any the least occasion of offence. So we spent our three days joyfully and without care, in expectation what would be done with us when they were expired. During which time, we had every hour joy of the amendment° of our sick; who thought themselves cast into some divine pool of healing,° they mended° so kindly° and so fast.

The morrow° after our three days were past, there came to us a new man that we had not seen before, clothed in blue as the former was, save that his turban was white, with a small red cross on the top. He had also a tippet° of fine linen. At his coming in, he did bend to us a little, and put his arms abroad. We of° our parts saluted him in a very lowly and submissive manner; as looking° that from him we should receive sentence of life or death. He desired to speak with some few of us: whereupon six of us only stayed, and the rest avoided° the room. He said, 'I am by office governor of this House of Strangers, and by vocation I am a Christian priest; and therefore am come to you to offer you my service, both as strangers and chiefly as Christians. Some things I may tell you, which I think you will not be unwilling to hear. The state hath given you licence to stay on land for the space of six weeks: and let it not trouble you if your occasions° ask° further time, for the law in this point is not precise; and I do not doubt but myself shall be able to obtain for you such further time as may be convenient. Ye shall also understand, that the Strangers' House is at this time° rich, and much aforehand;° for it hath laid up° revenue these thirty-seven years; for so long it is since any stranger arrived in this part: and therefore take ye no care; the state will defray° you all the time you stay; neither shall you stay one day the less for that. As for any merchandise ye have brought, ye shall be well used,° and have your return° either in merchandise or in gold and silver: for to us it is all one.° And if you have any other request to make, hide it not. For ye shall find we will not make your countenance to fall° by the answer ye shall receive. Only this I must tell you, that none of you must go above a *karan*' (that is with° them a mile and an half) 'from the walls of the city, without especial leave.'

We answered, after we had looked awhile one upon another, admiring° this gracious and parent-like usage;° 'That we could not tell what to say: for we wanted words to express our thanks; and his noble free° offers left us nothing to ask. It seemed to us that we had before° us a picture of our salvation in heaven; for we that were awhile since in the jaws of death, were now brought into a place where we found nothing but consolations. For the commandment laid upon us, we would not fail to obey it, though it was impossible but our hearts should be inflamed to tread further upon this happy and holy ground.' We added; 'That our tongues should first cleave to the roofs of our mouths, ere we should forget° either his reverend person or this whole nation in our prayers.' We also most humbly besought him to accept of us as his true servants, by as just a right as ever men on earth were bounden;° laying and presenting both our persons and all we had at his feet. He said; 'He was a priest, and looked for a priest's reward: which was our brotherly love and the good of our souls and bodies.' So he went from us, not without tears of tenderness in his eyes; and left us also confused with joy and kindness, saying amongst ourselves, 'That we were come into a land of angels, which° did appear to us daily and prevent us° with comforts, which we thought not of, much less expected.'

The next day, about ten of the clock, the governor came to us again, and after salutations said familiarly,° 'That he was come to visit us': and called for a chair, and sat him down: and we, being some ten of us (the rest were of the meaner sort,° or else gone abroad), sat down with him. And when we were set,° he began thus: 'We of this island of Bensalem'° (for so they call it in their language), 'have this;° that by means° of our solitary situation, and of the laws of secrecy which we have for our travellers, and our rare admission of strangers, we know well most part of the habitable world, and are ourselves unknown. Therefore because he that knoweth least is fittest to ask questions, it is more reason, for the entertainment of the time,° that ye ask me questions, than that I ask you.'

We answered; 'That we humbly thanked him that he would give us leave so to do: and that we conceived by the taste we had already, that there was no worldly thing on earth more worthy to be known than the state° of that happy land. But above all' (we said), 'since that we were met° from the several° ends of the world, and hoped assuredly that we should meet one day in the kingdom of heaven (for that we were both parts° Christians), we desired to know (in respect° that land was so remote, and so divided by vast and unknown seas, from the land where our Saviour walked on earth), who was the apostle of that nation, and how it was converted to the faith?' It appeared in his face that he took great contentment in this our question: he said, 'Ye knit my heart to you, by asking this question in the first place; for it sheweth that you "first seek the kingdom of heaven";° and I shall gladly and briefly satisfy your demand.

'About twenty years after the ascension of our Saviour, it came to pass that there was seen by the people of Renfusa (a city upon the eastern coast of our island), within night° (the night was cloudy and calm), as it might be some° mile into° the sea, a great pillar of light;° not sharp,° but in form of a column or cylinder, rising from the sea a great way up towards heaven: and on the top of it was seen a large cross of light, more bright and resplendent than the body of the pillar. Upon° which so strange a spectacle, the people of the city gathered apace° together upon the sands, to wonder; and so after put themselves° into a number of small boats, to go nearer to this marvellous sight. But when the boats were come within about sixty yards of the pillar, they found themselves all bound,° and could go no further; yet so as they might° move to go about,° but might not approach nearer: so as the boats stood all as in a theatre,° beholding this light as an heavenly sign. It so fell out, that there was in one of the boats one of the wise men of the society of Salomon's House; which house or college° (my good brethren) is the very eye° of this kingdom; who having awhile attentively and devoutly viewed and contemplated this pillar and cross, fell down upon his face; and then raised himself upon his knees, and lifting up his hands to heaven, made his prayers in this manner:

' "Lord God of heaven and earth, thou hast vouchsafed° of thy grace to those of our order,° to know thy works of creation, and the secrets of them; and to discern° (as far as appertaineth to the generations of men) between divine miracles, works of nature, works of art, and impostures and illusions° of all sorts. I do here acknowledge and testify before this people, that the thing which we now see before our eyes is thy Finger° and a true Miracle; and forasmuch as° we learn in our books that thou never workest miracles but to a divine and excellent end (for the laws of nature are thine own laws,° and thou exceedest° them not but upon great cause),° we most humbly beseech thee to prosper° this great sign, and to give us the interpretation° and use of it in mercy; which thou dost in some part° secretly promise by sending it unto us."

'When he had made his prayer, he presently° found the boat he was in moveable and unbound; whereas all the rest remained still fast; and taking that for an assurance of leave to approach, he caused the boat to be softly° and with silence rowed towards the pillar. But ere he came near it, the pillar and cross of light brake up, and cast itself abroad,° as it were, into a firmament° of many stars; which also vanished soon after, and there was nothing left to be seen but a small ark or chest of cedar, dry, and not wet at all with water, though it swam.° And in the fore-end° of it, which was towards him, grew a small green branch of palm; and when the wise man had taken it with all reverence into his boat, it opened of itself, and there were found in it a Book and a Letter; both written in fine parchment, and wrapped in sindons° of linen. The Book contained all the canonical° books of the Old and New Testament, according as you have them, (for we know well what the Churches with you receive);° and the Apocalypse° itself, and some other books of the New Testament which were not at that time written,° were nevertheless in the Book. And for the Letter, it was in these words:

' "I Bartholomew,° a servant of the Highest, and Apostle of Jesus Christ, was warned by an angel that appeared to me in a vision of glory, that I should commit this ark to the floods° of the sea. Therefore I do testify and declare unto that people where God shall ordain this ark to come to land, that in° the same day is come unto them salvation and peace and good-will, from the Father, and from the Lord Jesus."

'There was also in both these writings,° as well the Book as° the Letter, wrought° a great miracle, conform to° that of the Apostles in the original Gift of Tongues.° For there being at that time in this land Hebrews, Persians, and Indians, besides the natives, every one read upon° the Book and Letter, as if they had been written in his own language. And thus was this land saved from infidelity° (as the remain° of the old world was from water) by an ark, through the apostolical and miraculous evangelism° of St. Bartholomew.' And here he paused, and a messenger came, and called him from us. So this was all that passed in that conference.

The next day, the same governor came again to us immediately after dinner, and excused himself, saying, 'That the day before he was called from us somewhat abruptly, but now he would make us amends, and spend time with us, if we held° his company and conference° agreeable.' We answered, 'That we held it so agreeable and pleasing to us, as we forgot both dangers past and fears to come, for the time we heard him speak; and that we thought an hour spent with him, was worth years of our former life.' He bowed himself a little to us, and after we were set again, he said; 'Well, the questions are on your part.'°

One of our number said, after a little pause, 'That there was a matter we were no less desirous to know, than fearful to ask, lest we might presume too far. But encouraged by his rare humanity towards us (that could scarce think ourselves strangers, being his vowed and professed servants), we would take the hardiness° to propound it: humbly beseeching him, if he thought it not fit to be answered, that he would pardon it, though he rejected it.' We said, 'We well observed° those his words, which he formerly° spake, that this happy island where we now stood was known to few, and yet knew most of the nations of the world; which we found to be true, considering they had the languages of Europe, and knew much of our state and business;° and yet we in Europe (notwithstanding all the remote discoveries and navigations of this last age), never heard any of° the least inkling or glimpse of this island. This we found wonderful strange; for that° all nations have inter-knowledge° one of another either by voyage into foreign parts, or by strangers that come to them: and though the traveller into a foreign country doth commonly know more by the eye, than he that stayeth at home can by relation° of the traveller; yet both ways suffice to make a mutual knowledge, in some degree, on both parts.° But for° this island, we never heard tell of any ship of theirs that had been seen to arrive upon any shore of Europe; no, nor of either the East or West Indies; nor yet of any ship of any other part of the world that had made return from them. And yet the marvel rested° not in this. For the situation of it (as his lordship said) in the secret conclave° of such a vast sea might cause it. But then that they should have knowledge of the languages, books, affairs, of those that lie such a distance from them, it was a thing we could not tell what to make of; for that it seemed to us a condition° and propriety° of divine powers and beings, to be hidden and unseen to others, and yet to have others open° and as° in a light to them.'

At this speech the governor gave a gracious smile, and said; 'That we did well to ask pardon for this question we now asked; for that it imported° as if we thought this land a land of magicians, that sent forth spirits of the air into all parts, to bring them news and intelligence of other countries.' It was answered by us all, in all possible humbleness,° but yet with a countenance taking knowledge° that we knew that he spake it but merrily,° 'That we were apt° enough to think there was somewhat supernatural in this island; but yet rather as angelical than magical.° But to let his lordship know truly what it was that made us tender° and doubtful° to ask this question, it was not any such conceit,° but because we remembered he had given a touch° in his former speech, that this land had laws of secrecy touching° strangers.' To this he said; 'You remember it aright; and therefore in that I shall say to you I must reserve° some particulars, which it is not lawful for me to reveal; but there will be enough left to give you satisfaction.

'You shall understand (that which perhaps you will scarce think credible) that about three thousand years ago, or somewhat more,° the navigation° of the world (specially for remote voyages) was greater than at this day. Do not think with yourselves that I know not how much it is increased with you within these six-score years:° I know it well: and yet I say greater then than now; whether it was, that the example of the ark, that saved the remnant of men from the universal deluge, gave men confidence to adventure upon the waters; or what it was;° but such is the truth. The Phoenicians, and especially the Tyrians, had great fleets.° So had the Carthaginians, their colony,° which is yet further west. Toward the east, the shipping of Egypt and of Palestina was likewise great. China also, and the great Atlantis (that you call America), which have now but junks° and canoes,° abounded then in tall ships. This island (as appeareth by faithful registers° of those times) had then fifteen hundred strong ships, of great content.° Of all this there is with you sparing° memory, or none; but we have large° knowledge thereof.

'At that time, this land was known and frequented by the ships and vessels of all the nations before named. And (as it cometh to pass) they had many times men of other countries, that were no sailors, that came with them; as Persians, Chaldeans, Arabians; so as almost all nations of might° and fame resorted hither; of whom we have some stirps° and little tribes with us at this day. And for our own ships, they went sundry voyages, as well to your Straits, which you call the Pillars of Hercules,° as to other parts in the Atlantic and Mediterranean Seas; as to Paguin° (which is the same with° Cambaline)° and Quinzy,° upon the Oriental Seas, as far as to the borders of the East Tartary.°

'At the same time, and an age° after, or more, the inhabitants of the great Atlantis did flourish. For though the narration and description° which is made by a great man° with° you, that the descendants of Neptune planted° there; and of the magnificent temple, palace, city, and hill; and the manifold streams of goodly navigable rivers, (which, as so many chains, environed° the same site and temple); and the several degrees° of ascent whereby men did climb up to the same, as if it had been a *scala coeli*;° be all poetical and fabulous: yet so much is true, that the said country of Atlantis, as well that of Peru, then called Coya, as that° of Mexico, then named Tyrambel,° were mighty and proud kingdoms in arms, shipping, and riches: so mighty, as at one time (or at least within the space of ten years) they both° made two great expeditions; they of Tyrambel through° the Atlantic to the Mediterranean Sea; and they of Coya through the South Sea upon this our island. And for the former of these, which was into Europe, the same author amongst you (as it seemeth) had some relation from the Egyptian priest° whom he citeth. For assuredly such a thing there was. But whether it were the ancient Athenians that had the glory of the repulse° and resistance of those forces, I can say nothing: but certain it is, there never came back either ship or man from that voyage. Neither had° the other voyage of those of Coya upon us had better fortune, if they had not met with enemies of greater clemency. For the king of this island (by name Altabin), a wise man and a great warrior, knowing well both his own strength and that of his enemies, handled° the matter so, as he cut off their land-forces from their ships; and entoiled° both their navy and their camp with a greater power than theirs, both by sea and land; and compelled them to render° themselves without striking stroke:° and after they were at his mercy, contenting himself only with their oath that they should no more bear arms against him, dismissed them all in safety.

'But the Divine Revenge° overtook not long after those proud enterprises. For within less than the space of one hundred years, the great Atlantis was utterly lost and destroyed: not by a great earthquake, as your man saith° (for that whole tract° is little subject to earthquakes), but by a particular deluge or inundation;° those countries having, at this day, far greater rivers and far higher mountains to pour down waters, than any part of the old world. But it is true that the same inundation was not deep; not past° forty foot, in most places, from the ground: so that although it destroyed man and beast generally, yet some few wild inhabitants of the wood escaped. Birds also were saved by flying to the high trees and woods. For as for men, although they had buildings in many places higher than the depth of the water, yet that inundation, though it were shallow, had a long continuance; whereby they of the vale° that were not drowned, perished for want of food and other things necessary.

'So as° marvel you not at the thin population of America, nor at the rudeness and ignorance of the people; for you must account° your° inhabitants of America as a young people; younger a thousand years, at the least, than the rest of the world; for that there was so much time between the universal flood and their particular° inundation. For the poor remnant of human seed which remained in their mountains peopled the country again slowly, by little and little; and being simple and savage people, (not like Noah and his sons, which was the chief family of the earth), they were not able to leave letters,° arts,° and civility° to their posterity; and having likewise in their mountainous habitations been used (in respect° of the extreme cold of those regions) to clothe themselves with the skins of tigers,° bears, and great hairy goats,° that they have in those parts; when after they came down into the valley, and found the intolerable heats° which are there, and knew no means° of lighter apparel, they were forced to begin the custom of going naked, which continueth at this day. Only they take great pride and delight in the feathers of birds,° and this also they took° from those their ancestors of the mountains, who were invited° unto it by the infinite° flights° of birds that came up to the high grounds, while the waters stood below. So you see, by this main° accident of time,° we lost our traffic° with the Americans, with whom of all others, in regard they lay nearest to us, we had most commerce.

'As for the other parts of the world, it is most manifest that in the ages following (whether it were in respect° of wars, or by a natural revolution of time),° navigation did every where greatly decay; and specially far voyages (the rather° by the use of galleys,° and such vessels as could hardly brook° the ocean), were altogether left and omitted.° So then, that part of intercourse which could be from other nations to sail to us, you see how it hath long since ceased; except it were by some rare accident, as this of yours. But now of the cessation of that other part of intercourse, which might be by our sailing to other nations, I must yield° you some other cause. For I cannot say (if I shall say truly), but° our shipping, for number, strength, mariners, pilots, and all things that appertain° to navigation, is as great as ever: and therefore why we should sit at home, I shall now give you an account by itself: and it will draw° nearer to give you satisfaction to your principal question.

'There reigned in this island, about nineteen hundred years ago, a King, whose memory of all others we most adore; not superstitiously, but as a divine instrument, though a mortal man; his name was Solamona: and we esteem him as the lawgiver of our nation. This king had a "large heart",° inscrutable for good;° and was wholly bent to make his kingdom and people happy. He therefore, taking into consideration how sufficient and substantive° this land was to maintain itself without any aid at all of the foreigner; being five thousand six hundred miles in circuit, and of rare fertility of soil in the greatest part thereof; and finding also the shipping of this country might be plentifully set on work,° both by fishing and by transportations from port to port,° and likewise by sailing unto some small islands that are not far from us, and are under the crown and laws of this state; and recalling into his memory the happy and flourishing estate° wherein this land then was, so as it might be a thousand ways altered to the worse, but scarce any one way to the better; thought nothing wanted° to his noble and heroical° intentions, but only (as far as human foresight might reach) to give perpetuity° to that which was in his time so happily established. Therefore amongst his other fundamental laws of this kingdom, he did ordain the interdicts and prohibitions which we have touching entrance of strangers; which at that time (though it was after the calamity of America) was frequent; doubting novelties,° and commixture° of manners.° It is true, the like° law against the admission of strangers without licence is an ancient law in the kingdom of China,° and yet° continued in use. But there it is a poor thing; and hath made them a curious,° ignorant, fearful,° foolish nation. But our lawgiver made his law of another temper.° For first, he hath preserved all points of humanity, in taking order° and making provision for the relief of strangers distressed; whereof you have tasted.'

At which speech (as reason was)° we all rose up, and bowed ourselves. He went on. 'That king also, still desiring to join humanity and policy° together; and thinking it against humanity to detain strangers here against their wills, and against policy that they should return and discover° their knowledge of this estate,° he took this course: he did ordain that of the strangers that should be permitted to land, as many (at all times) might depart as would; but as many as would stay should have very good conditions° and means to live from the state. Wherein he saw so far, that now in so many ages since the prohibition, we have memory not of one ship that ever returned; and but of thirteen persons only, at several° times, that chose to return in our bottoms.° What those few that returned may have reported abroad I know not. But you must think,° whatsoever they have said could be taken where they came but for a dream. Now for our travelling from hence into parts abroad, our Lawgiver thought fit altogether to restrain° it. So is it not in China. For the Chineses sail where they will or can; which sheweth that their law of keeping out strangers is a law of pusillanimity and fear. But this restraint of ours hath one only exception, which is admirable; preserving the good which cometh by communicating with strangers, and avoiding the hurt; and I will now open° it to you. And here I shall seem a little to digress, but you will by and by find it pertinent.

'Ye shall understand (my dear friends) that amongst the excellent acts of that king, one above all hath the preeminence. It was the erection and institution of an Order or Society which we call "Salomon's House"; the noblest foundation (as we think) that ever was upon the earth; and the lanthorn° of this kingdom. It is dedicated to the study of the Works and Creatures° of God. Some think it beareth the founder's name a little corrupted, as if it should be Solamona's House. But the records write it as it is spoken. So as I take it to be denominate° of the King of the Hebrews, which is famous with° you, and no stranger to us. For we have some parts of his works which with you are lost; namely, that Natural History which he wrote, of all plants, from the "cedar of Libanus"° to the "moss that groweth out of the wall", and of all "things that have life and motion".° This maketh me think that our king, finding himself to symbolize° in many things with that king of the Hebrews (which lived many years before him), honoured him with the title of this foundation. And I am the rather induced to be of this opinion, for that I find in ancient records this Order or Society is sometimes called Salomon's House and sometimes the College of the Six Days Works; whereby I am satisfied that our excellent king had learned from the Hebrews that God had created the world and all that therein is within six days; and therefore he instituting that House for the finding out of the true nature of all things° (whereby God might have the more glory in the workmanship° of them, and men the more fruit in the use° of them) did give it also that second name.

'But now to come to our present purpose. When the king had forbidden to all his people navigation into any part that was not under his crown, he made nevertheless this ordinance;° That every twelve years there should be set forth° out of this kingdom two ships, appointed to° several° voyages; That in either° of these ships there should be a mission° of three of the Fellows or Brethren of Salomon's House; whose errand was only° to give us knowledge of the affairs and state of those countries to which they were designed,° and especially of the sciences, arts, manufactures, and inventions° of all the world; and withal to bring unto us books, instruments, and patterns° in every kind; That the ships, after they had landed the brethren, should return; and that the brethren should stay abroad till the new mission. These ships are not otherwise fraught,° than with store of victuals, and good quantity of treasure to remain with the brethren, for the buying of such things and rewarding of such persons as they should think fit. Now for me to tell you how the vulgar° sort of mariners are contained° from being discovered at land;° and how they that must be put on shore for any time, colour° themselves under the names of other nations; and to what places these voyages have been designed; and what places of "rendez-vous" are appointed for the new missions; and the like circumstances of the practique;° I may not do it: neither is it much to your desire. But thus you see we maintain a trade, not for gold, silver, or jewels; nor for silks; nor for spices; nor any other commodity of matter;° but only for God's first creature, which was *Light*:° to have *light* (I say) of the growth° of all parts of the world.'°

And when he had said this, he was silent; and so° were we all. For indeed we were all astonished to hear so strange things so probably° told. And he, perceiving that we were willing to say somewhat but had it not ready, in great courtesy took us off,° and descended° to ask us questions of our voyage and fortunes; and in the end concluded, that we might do well to think with° ourselves what time of stay we would demand of the state; and bade us not to scant° ourselves; for he would procure such time as we desired. Whereupon we all rose up, and presented ourselves to kiss the skirt of his tippet; but he would not suffer° us; and so took his leave. But when it came once° amongst our people that the state used° to offer conditions° to strangers that would stay, we had work enough to get any of our men to look to our ship, and to keep them from going presently° to the governor to crave° conditions. But with much ado we refrained them, till we might agree what course to take.

We took° ourselves now for free men, seeing there was no danger of our utter perdition;° and lived most joyfully, going abroad and seeing what was to be seen in the city and places adjacent within our tedder;° and obtaining° acquaintance with many of the city, not of the meanest quality;° at whose hands we found such humanity, and such a freedom° and desire to take strangers as it were into their bosom, as was enough to make us forget all that was dear to us in our own countries: and continually we met with many things right worthy of observation and relation;° as indeed, if there be a mirror in the world worthy to hold° men's eyes, it is that country.

One day there were two of our company bidden to a Feast of the Family, as they call it. A most natural, pious, and reverend custom it is, shewing that nation to be compounded° of all goodness. This is the manner of it. It is granted to any man that shall live to see thirty persons descended of his body alive together, and all above three years old, to make this feast; which is done at the cost of the state. The Father of the Family, whom they call the *Tirsan*, two days before the feast, taketh to° him three of such friends as he liketh to choose; and is assisted° also by the governor of the city or place where the feast is celebrated; and all the persons of the family, of both sexes, are summoned to attend him. These two days the Tirsan sitteth in consultation° concerning the good estate° of the family. There, if there be any discord or suits° between any of the family, they are compounded° and appeased. There, if any of the family be distressed or decayed,° order is taken° for their relief and competent means to live. There, if any be subject to vice, or take ill courses,° they are reproved and censured. So likewise direction° is given touching marriages, and the courses of life which any of them should take, with divers other the like orders and advices.° The governor assisteth,° to the end to put in execution by his public authority the decrees and orders of the Tirsan, if they should be disobeyed; though that seldom needeth;° such reverence and obedience they give to the order of nature. The Tirsan doth also then ever choose one man from amongst his sons, to live in house with him: who is called ever after the Son of the Vine. The reason will hereafter appear.

On the feast-day, the Father or Tirsan cometh forth after divine service into a large room where the feast is celebrated; which room hath an half-pace° at the upper end. Against the wall, in the middle of the half-pace, is a chair placed for him, with a table and carpet before it. Over the chair is a state,° made round or oval, and it is of ivy;° an ivy somewhat whiter than ours, like the leaf of a silver asp,° but more shining; for it is green all winter. And the state is curiously wrought° with silver and silk of divers colours, broiding° or binding in the ivy; and is ever of the work of some of the daughters of the family; and veiled over at the top with a fine net of silk and silver. But the substance of it is true° ivy; whereof, after it is taken down, the friends of the family are desirous to have some leaf or sprig to keep.

The Tirsan cometh forth with all his generation or lineage, the males before him, and the females following him; and if there be a mother from whose body the whole lineage is descended, there is a traverse° placed in a loft above on the right hand of the chair, with a privy° door, and a carved° window of glass, leaded° with gold and blue; where she sitteth, but is not seen. When the Tirsan is come forth, he sitteth down in the chair; and all the lineage place themselves against the wall, both at his back and upon the return° of the half-pace, in order of their years without difference of sex; and stand upon their feet. When he is set; the room being always full of company, but well kept and without disorder; after some pause there cometh in from the lower end of the room a *Taratan*° (which is as much as an herald) and on either side of him two young lads; whereof one carrieth a scroll of their shining yellow parchment; and the other a cluster of grapes of gold, with a long foot or stalk. The herald and children are clothed with mantles of sea-water green satin; but the herald's mantle is streamed with gold,° and hath a train.

Then the herald with three curtesies,° or rather inclinations,° cometh up as far as the half-pace; and there first taketh into his hand the scroll. This scroll is the King's Charter, containing gift of revenew,° and many privileges,° exemptions,° and points of honour,° granted to the Father of the Family; and is ever styled and directed,° 'To such an one our well-beloved friend and creditor': which is a title proper only to this case. For they say the king is debtor to no man, but° for propagation of his subjects. The seal set° to the king's charter is the king's image, imbossed° or moulded in gold; and though such charters be expedited° of course,° and as of right, yet they are varied by discretion,° according to the number and dignity of the family. This charter the herald readeth aloud; and while it is read, the father or Tirsan standeth up, supported° by two of his sons, such as he chooseth. Then the herald mounteth the half-pace, and delivereth the charter into his hand: and with that there is an acclamation by all that are present in their language, which is thus much:° 'Happy are the people of Bensalem'.

Then the herald taketh into his hand from the other child the cluster of grapes, which is of gold, both the stalk and the grapes. But the grapes are daintily° enamelled; and if the males of the family be the greater number, the grapes are enamelled purple, with a little sun set on the top; if the females, then they are enamelled into a greenish yellow, with a crescent on the top. The grapes are in number as many as there are descendants of the family. This golden cluster the herald delivereth also to the Tirsan; who presently delivereth it over to that son that he had formerly chosen to be in house with him: who beareth it before his father as an ensign° of honour when he goeth in public, ever after; and is thereupon called the Son of the Vine.

After this ceremony ended, the father or Tirsan retireth; and after some time cometh forth again to dinner, where he sitteth alone under the state, as before; and none of his descendants sit with him, of what degree or dignity soever, except he hap° to be of Salomon's House. He is served only by his own children, such as are male; who perform unto him all service of° the table upon the knee;° and the women only stand about him, leaning against the wall. The room below the half-pace hath tables on the sides for the guests that are bidden;° who are served with great and comely order;° and towards the end of dinner (which in the greatest feasts with them lasteth never above an hour and an half) there is an hymn sung, varied according to the invention° of him that composeth it (for they have excellent poesy),° but the subject of it is (always) the praises° of Adam and Noah and Abraham; whereof the former two peopled the world, and the last was the Father of the Faithful:° concluding ever with a thanksgiving for the nativity of our Saviour, in whose birth the births of all are only blessed.°

Dinner being done, the Tirsan retireth again; and having withdrawn himself alone into a place where he maketh some private prayers, he cometh forth the third time, to give the blessing; with all his descendants, who stand about him as at the first. Then he calleth them forth by one and by one, by name, as he pleaseth, though seldom the order of age be inverted. The person that is called (the table being before removed) kneeleth down before the chair, and the father layeth his hand upon his head, or her head, and giveth the blessing in these words: 'Son of Bensalem (or Daughter of Bensalem), thy father saith it; the man by whom thou hast breath and life speaketh the word; The blessing of the everlasting Father, the Prince of Peace, and the Holy Dove be upon thee, and make the days of thy pilgrimage° good and many.' This he saith to every° of them; and that done, if there be any of his sons of eminent merit and virtue, (so° they be not above° two), he calleth for them again; and saith, laying his arm over their shoulders, they standing; 'Sons, it is well ye are born, give God the praise, and persevere to the end.' And withal delivereth to either° of them a jewel, made in the figure of an ear of wheat, which they ever after wear in the front of their turban or hat. This done, they fall to° music and dances, and other recreations, after their manner, for the rest of the day. This is the full order of that feast.

By that time° six or seven days were spent, I was fallen into strait° acquaintance with a merchant of that city, whose name was Joabin. He was a Jew, and circumcised: for they have some few stirps° of Jews yet remaining among them, whom they leave to their own religion. Which they may the better do, because they are of a far differing disposition from the Jews in other parts. For whereas they hate the name of Christ, and have a secret inbred rancour against the people amongst whom they live: these (contrariwise) give unto our Saviour many high attributes, and love the nation of Bensalem extremely. Surely° this man of whom I speak would ever acknowledge that Christ was born of a Virgin, and that he was more than a man; and he would tell how God made him ruler of the Seraphims which guard his throne; and they call him also the 'Milken Way',° and the *Eliah*° of the *Messiah*; and many other high names; which though they be inferior to his divine Majesty, yet they are far from the language of other Jews. And for the country of Bensalem, this man would make no end of commending it: being desirous, by tradition among the Jews there, to have it believed that the people thereof were of the generations° of Abraham, by another son, whom they call Nachoran;° and that Moses by a secret cabala° ordained the laws of Bensalem which they now use; and that when the Messiah should come, and sit in his throne at Hierusalem, the king of Bensalem should sit at his feet, whereas other kings should keep a great distance. But yet setting aside these Jewish dreams,° the man was a wise man, and learned, and of great policy,° and excellently seen° in the laws and customs of that nation.

Amongst other discourses, one day I told him I was much affected° with the relation° I had from some of the company, of their custom in holding the Feast of the Family; for that (methought) I had never heard of a solemnity wherein nature° did so much preside. And because propagation of families proceedeth from the nuptial copulation, I desired to know of him what laws and customs they had concerning marriage; and whether they kept° marriage well; and whether they were tied to one wife? For that where population° is so much affected,° and such as with them it seemed to be, there is commonly permission of plurality of wives. To this he said, 'You have reason for to commend that excellent institution of the Feast of the Family. And indeed we have experience, that those families that are partakers of the blessing of that feast do flourish and prosper ever after in an extraordinary manner. But hear me now, and I will tell you what I know. You shall understand that there is not under the heavens so chaste a nation as this of Bensalem; nor so free from all pollution or foulness. It is the virgin of the world. I remember I have read in one of your European books,° of an holy hermit amongst you that desired to see the Spirit of Fornication; and there appeared to him a little foul ugly Ethiop. But if he had desired to see the Spirit of Chastity of Bensalem, it would have appeared to him in the likeness of a fair beautiful Cherubin. For there is nothing amongst mortal men more fair and admirable, than the chaste minds of this people.

'Know therefore, that with them there are no stews,° no dissolute houses,° no courtesans, nor any thing of that kind. Nay they wonder (with detestation) at you in Europe, which permit such things. They say ye have put marriage out of office: for marriage is ordained a remedy for unlawful concupiscence;° and natural concupiscence seemeth as a spur to marriage. But when men have at hand a remedy more agreeable to their corrupt will, marriage is almost expulsed.° And therefore there are with you seen° infinite° men that marry not, but choose rather a libertine and impure single life, than to be yoked in marriage; and many that do marry, marry late, when the prime and strength of their years is past. And when they do marry, what is marriage to them but a very bargain;° wherein is sought alliance,° or portion,° or reputation, with some desire (almost indifferent) of issue;° and not the faithful nuptial union of man and wife, that was first instituted. Neither is it possible that those that have cast away so basely so much of their strength, should greatly esteem children (being of the same matter),° as chaste men do. So likewise during marriage, is the case much amended, as it ought to be if those things were tolerated only for necessity? No, but they remain still as a very affront to marriage. The haunting of those dissolute places, or resort to courtesans, are no more punished in married men than in bachelors. And the depraved custom of change,° and the delight in meretricious° embracements (where sin is turned into art), maketh marriage a dull thing, and a kind of imposition or tax.

'They hear you defend these things, as done to avoid greater evils; as advoutries,° deflowering of virgins, unnatural lust, and the like. But they say this is a preposterous° wisdom; and they call it "Lot's offer",° who to save his guests from abusing,° offered his daughters: nay they say farther that there is little gained in this; for that the same vices and appetites do still remain and abound; unlawful lust being like a furnace, that if you stop the flames altogether, it will quench; but if you give it any vent,° it will rage. As for masculine love, they have no touch° of it; and yet there are not so faithful and inviolate friendships in the world again as are there; and to speak generally (as I said before), I have not read of any such chastity in any people as theirs. And their usual saying is, that "whosoever is unchaste cannot reverence himself"; and they say, that "the reverence of a man's self is, next religion, the chiefest bridle of all vices".'

And when he had said this, the good Jew paused a little; whereupon I, far more willing to hear him speak on than to speak myself, yet thinking it decent° that upon his pause° of speech I should not be altogether silent, said only this; 'That I would say to him, as the widow of Sarepta° said to Elias;° that he was come to bring to memory our sins; and that I confess the righteousness of Bensalem was greater than the righteousness of Europe.' At which speech he bowed his head, and went on in this manner: 'They have also many wise and excellent laws touching marriage. They allow no polygamy. They have ordained that none do intermarry° or contract,° until a month be passed from their first interview. Marriage without consent of parents° they do not make void, but they mulct° it in the inheritors: for the children of such marriages are not admitted° to inherit above° a third part of their parents' inheritance. I have read in a book° of one of your men, of a Feigned Commonwealth, where the married couple are permitted, before they contract, to see one another naked. This they dislike; for they think it a scorn to give a refusal after so familiar° knowledge. But because of many hidden defects in men and women's bodies, they have a more civil way; for they have near every town a couple of pools, (which they call "Adam and Eve's pools"), where it is permitted to one of the friends of the man, and another of the friends of the woman, to see them severally° bathe naked.'

And as we were thus in conference,° there came one that seemed to be a messenger, in a rich huke,° that spake with the Jew: whereupon he turned to me and said; 'You will pardon me, for I am commanded away in haste.' The next morning he came to me again, joyful as it seemed, and said, 'There is word come to the governor of the city, that one of the Fathers of Salomon's House will be here this day seven-night: wc have seen none of them this dozen years. His coming is in state;° but the cause of his coming is secret. I will provide you and your fellows of a good standing° to see his entry.' I thanked him, and told him, 'I was most glad of the news.'

The day being come, he made his entry. He was a man of middle stature and age, comely of person, and had an aspect° as if he pitied men.° He was clothed in a robe of fine black cloth, with wide sleeves and a cape. His undergarment was of excellent white linen down to the foot, girt with a girdle of the same; and a sindon or tippet of the same about his neck. He had gloves that were curious,° and set with stone;° and shoes of peach-coloured velvet. His neck was bare to the shoulders. His hat was like a helmet, or Spanish Montera;° and his locks curled below it decently:° they were of colour brown. His beard was cut round, and of the same colour with his hair, somewhat lighter. He was carried in a rich chariot° without wheels, litter-wise;° with two horses at either° end, richly trapped° in blue velvet embroidered; and two footmen on each side in the like attire. The chariot was all of cedar, gilt, and adorned with crystal;° save that the fore-end had panels of sapphires, set in borders of gold, and the hinder-end the like of emeralds of the Peru° colour. There was also a sun of gold, radiant,° upon the top, in the midst; and on the top before, a small cherub of gold, with wings displayed.° The chariot was covered with cloth of gold tissued upon blue.° He had before him fifty attendants, young men all, in white satin loose coats to° the mid-leg; and stockings of white silk; and shoes of blue velvet; and hats of blue velvet; with fine plumes of divers colours, set round like hat-bands. Next° before the chariot went° two men, bare-headed, in linen garments down to the foot, girt,° and shoes of blue velvet; who carried the one a crosier,° the other a pastoral staff like a sheep-hook; neither of them of metal, but the crosier of balm-wood,° the pastoral staff of cedar. Horsemen he had none, neither before nor behind his chariot: as it seemeth, to avoid all tumult and trouble. Behind his chariot went all the officers and principals of the Companies of the City.° He sat alone, upon cushions of a kind of excellent plush, blue; and under his foot curious carpets of silk of divers colours, like the Persian, but far finer. He held up his bare hand as he went, as° blessing the people, but in silence. The street was wonderfully well kept:° so that there was never any army had their men stand in better battle array, than the people stood. The windows likewise were not crowded, but every one stood in them as if they had been placed.°

When the shew° was past, the Jew said to me; 'I shall not be able to attend° you as I would,° in regard of some charge the city hath laid° upon me, for the entertaining of this great person.' Three days after, the Jew came to me again, and said, 'Ye are happy men; for the Father of Salomon's House taketh knowledge of your being here, and commanded me to tell you that he will admit all your company to his presence, and have private conference with one of you that ye shall choose: and for this hath appointed the next day after to-morrow. And because he meaneth to give you his blessing, he hath appointed it in the forenoon.'

We came at our day and hour, and I was chosen by my fellows for the private access.° We found him in a fair chamber, richly hanged,° and carpeted under foot, without any degrees° to the state.° He was set upon a low throne richly adorned, and a rich cloth of state over his head, of blue satin embroidered. He was alone, save that he had two pages of honour, on either hand one, finely attired in white. His under-garments were the like that we saw him wear in the chariot; but instead of his gown, he had on him a mantle with a cape, of the same fine black, fastened about him. When we came in, as we were taught, we bowed low at our first entrance; and when we were come near his chair, he stood up, holding forth his hand ungloved, and in posture° of blessing; and we every one of us stooped down, and kissed the hem of his tippet. That done, the rest departed, and I remained. Then he warned the pages forth° of the room, and caused me to sit down beside him, and spake to me thus in the Spanish tongue:

'God bless thee, my son; I will give thee the greatest jewel I have. For I will impart unto thee, for the love of God and men, a relation° of the true state of Salomon's House. Son, to make you know the true state of Salomon's House, I will keep° this order. First, I will set forth unto you the end° of our foundation. Secondly, the preparations° and instruments we have for our works.° Thirdly, the several employments and functions whereto our fellows are assigned. And fourthly, the ordinances and rites which we observe.

'The End of our Foundation is the knowledge of Causes, and secret motions° of things; and the enlarging of the bounds of Human Empire,° to the effecting of all things possible.

'The Preparations and Instruments are these. We have large and deep caves of several depths: the deepest are sunk six hundred fathom;° and some of them are digged and made under great hills and mountains: so that if you reckon together the depth of the hill and the depth of the cave, they are (some of them) above three miles deep. For we find that the depth of a hill, and the depth of a cave from the flat, is the same thing;° both remote alike from the sun and heaven's beams, and from the open air. These caves we call the Lower Region. And we use them for all coagulations,° indurations,° refrigerations, and conservations of bodies.° We use them likewise for the imitation of natural mines;° and the producing also of new artificial metals, by compositions° and materials which we use, and lay there for many years. We use them also sometimes (which may seem strange) for curing of some diseases, and for prolongation of life in some hermits that choose to live there, well accommodated of° all things necessary; and indeed live very long; by° whom also we learn many things.

'We have burials° in several° earths,° where we put divers cements, as the Chineses do their porcelain.° But we have them in greater variety, and some of them more fine. We have also great variety of composts,° and soils for the making of the earth fruitful.

'We have high towers; the highest about half a mile° in height; and some of them likewise set upon high mountains; so that the vantage° of the hill with the tower is in the highest of them three miles at least. And these places we call the Upper Region: accounting the air between the high places and the low, as a Middle Region. We use these towers, according to their several heights and situations, for insolation,° refrigeration, conservation; and for the view of divers meteors;° as winds, rain, snow, hail; and some of the fiery meteors° also. And upon them, in some places, are dwellings of hermits, whom we visit sometimes, and instruct what to observe.

'We have great lakes both salt and fresh, whereof we have use for the fish and fowl. We use them also for burials of some natural bodies: for we find a difference in things buried in earth or in air below the earth, and things buried in water. We have also pools, of which some do strain° fresh water out of salt; and others by art do turn fresh water into salt. We have also some rocks in the midst of the sea, and some bays upon the shore, for some works wherein is required the air and vapour of the sea. We have likewise violent streams and cataracts, which serve us for many motions°, and likewise engines° for multiplying and enforcing° of winds, to set also on going° divers motions.

'We have also a number of artificial wells and fountains, made in imitation of the natural sources and baths; as° tincted° upon vitriol,° sulphur, steel,° brass,° lead, nitre, and other minerals. And again we have little wells for infusions of many things, where the waters take the virtue° quicker and better than in vessels or basins. And amongst them we have a water which we call Water of Paradise, being, by that we do to it, made very sovereign° for health, and prolongation of life.°

'We have also great and spacious houses, where we imitate and demonstrate meteors;° as snow, hail, rain, some artificial rains of bodies° and not of water, thunders, lightnings; also generations of bodies in air;° as frogs, flies, and divers others.

'We have also certain chambers, which we call Chambers of Health, where we qualify° the air as we think good and proper for the cure of divers diseases, and preservation of health.

'We have also fair and large baths, of several mixtures, for the cure of diseases, and the restoring of man's body from arefaction:° and others for the confirming° of it in strength of sinews, vital parts, and the very juice and substance of the body.

'We have also large and various orchards and gardens, wherein we do not so much respect° beauty, as variety of ground and soil, proper for divers trees and herbs: and some very spacious, where trees and berries are set whereof we make divers kinds of drinks, besides the vineyards. In these we practise likewise all conclusions° of grafting and inoculating,° as well of wild-trees as fruit-trees, which produceth many effects.° And we make (by art) in the same orchards and gardens, trees and flowers to come earlier or later° than their seasons; and to come up and bear more speedily than by their natural course they do. We make them also by art greater° much than their nature; and their fruit greater and sweeter and of differing taste, smell, colour, and figure,° from their nature. And many of them we so order,° as they become of medicinal use.

'We have also means to make divers plants rise by mixtures of earths without seeds;° and likewise to make divers new plants, differing from the vulgar;° and to make one tree or plant turn into another.°

'We have also parks and inclosures of all sorts of beasts and birds, which we use not only for view° or rareness, but likewise for dissections and trials;° that thereby we may take light° what may be wrought° upon the body of man. Wherein we find° many strange effects; as continuing life in them, though divers parts, which you account vital, be perished and taken forth;° resuscitating of some that seem dead in appearance; and the like. We try also all poisons and other medicines upon them, as well of chirurgery° as physic.° By art likewise, we make them greater or taller than their kind° is; and contrariwise dwarf them, and stay° their growth: we make them more fruitful and bearing° than their kind is; and contrariwise barren and not generative. Also we make them differ in colour, shape, activity, many ways. We find means to make commixtures° and copulations of different kinds; which have produced many new kinds, and them not barren,° as the general opinion is. We make a number of kinds of serpents,° worms, flies, fishes, of putrefaction;° whereof some are advanced° (in effect) to be perfect° creatures, like beasts or birds; and have sexes, and do propagate. Neither do we this by chance, but we know beforehand of what matter° and commixture what kind of those creatures will arise.

'We have also particular pools, where we make trials upon fishes, as we have said before of beasts and birds.

'We have also places for breed° and generation of those kinds of worms and flies which are of special use; such as are with you your silk-worms and bees.

'I will not hold you long with recounting of our brewhouses, bake-houses, and kitchens, where are made divers drinks, breads, and meats,° rare and of special effects. Wines we have of grapes; and drinks of other juice of fruits, of grains, and of roots: and of mixtures with honey, sugar, manna,° and fruits dried and decocted.° Also of the tears° or woundings of trees, and of the pulp of canes. And these drinks are of several ages, some to the age or last° of forty years. We have drinks also brewed with° several herbs, and roots, and spices; yea with several fleshes,° and white meats;° whereof some of the drinks are such, as they are in effect meat and drink both:° so that divers, especially in age,° do desire to live with° them, with little or no meat or bread. And above all, we strive to have drinks of extreme thin parts,° to insinuate° into the body, and yet without all° biting, sharpness, or fretting;° insomuch as some of them put upon the back of your hand will, with a little stay,° pass through to the palm, and yet taste mild to the mouth. We have also waters which we ripen° in that fashion, as they become nourishing; so that they are indeed excellent drink; and many will use no other. Breads we have of several grains, roots, and kernels: yea and some of flesh° and fish dried; with divers kinds of leavenings° and seasonings: so that some do extremely move° appetites; some do nourish so, as divers do live of them, without any other meat; who live very long. So for meats, we have some of them so beaten and made tender and mortified,° yet without all corrupting, as a weak heat of the stomach will turn them into good chylus,° as well as a strong heat would meat otherwise prepared. We have some meats also and breads and drinks, which taken by men enable them to fast long after;° and some other, that used make the very flesh of men's bodies sensibly° more hard and tough, and their strength far greater than otherwise it would be.

'We have dispensatories,° or shops of medicines. Wherein you may easily think, if we have such variety of plants and living creatures more than you have in Europe (for we know what you have), the simples,° drugs, and ingredients of medicines, must likewise be in so much the greater variety. We have them likewise of divers ages, and long fermentations.° And for their preparations, we have not only all manner of exquisite° distillations and separations,° and especially by gentle heats and percolations through divers strainers, yea and substances; but also exact forms of composition,° whereby they incorporate° almost, as° they were natural simples.

'We have also divers mechanical arts, which you have not; and stuffs made by them; as papers, linen, silks, tissues;° dainty works of feathers of wonderful lustre; excellent dyes, and many others; and shops° likewise, as well for such as are not brought into vulgar use amongst us as for those that are. For you must know that of the things before recited, many of them are grown into use throughout the kingdom; but yet° if they did flow from our invention, we have of them also for patterns and principals.°

'We have also furnaces of great diversities,° and that keep great diversity of heats;° fierce and quick; strong and constant; soft and mild; blown,° quiet; dry, moist; and the like. But above all, we have heats in imitation of the sun's and heavenly bodies' heats, that pass divers inequalities and (as it were) orbs,° progresses,° and returns,° whereby we produce admirable effects. Besides, we have heats of dungs, and of bellies and maws° of living creatures, and of their bloods and bodies; and of hays and herbs laid up° moist; of lime unquenched;° and such like. Instruments also which generate heat only by motion. And farther, places for strong insolations; and again, places under the earth, which by nature or art yield heat. These divers heats we use, as the nature of the operation which we intend requireth.

'We have also perspective-houses,° where we make demonstrations° of all lights and radiations; and of all colours; and out of things uncoloured and transparent, we can represent° unto you all several colours; not in rain-bows, as it is in gems and prisms, but of themselves single.° We represent also all multiplications° of light, which we carry to great distance, and make so sharp° as to discern small points° and lines; also all colorations of light: all delusions and deceits° of the sight, in figures, magnitudes, motions, colours: all demonstrations of shadows. We find also divers means, yet unknown to you, of producing of light originally° from divers bodies. We procure° means of seeing objects afar off;° as in the heaven and remote places; and represent things near as afar off, and things afar off as near; making feigned° distances. We have also helps for the sight, far above spectacles and glasses° in use. We have also glasses and means to see small and minute bodies perfectly and distinctly;° as the shapes and colours of small flies and worms, grains° and flaws in gems, which cannot otherwise be seen; observations in urine° and blood, not otherwise to be seen. We make artificial rain-bows, halos, and circles about light. We represent also all manner of reflexions, refractions, and multiplications of visual beams° of objects.

'We have also precious stones of all kinds, many of them of great beauty, and to you unknown; crystals° likewise; and glasses of divers kinds; and amongst them some of metals vitrificated,° and other materials besides those of which you make glass. Also a number of fossils,° and imperfect° minerals, which you have not. Likewise loadstones of prodigious virtue;° and other rare stones, both natural and artificial.

'We have also sound-houses,° where we practise and demonstrate all sounds, and their generation.° We have harmonies which you have not, of quarter-sounds, and lesser slides° of sounds. Divers instruments of music likewise to you unknown, some sweeter than any you have; together with bells and rings° that are dainty and sweet. We represent small sounds as great° and deep; likewise great sounds extenuate° and sharp; we make divers tremblings° and warblings of sounds, which in their original° are entire.° We represent and imitate all articulate sounds and letters, and the voices and notes of beasts and birds. We have certain helps which set° to the ear do further the hearing greatly. We have also divers strange and artificial echos, reflecting the voice many times, and as it were tossing it: and some that give back the voice louder than it came; some shriller, and some deeper; yea, some rendering the voice differing in the letters or articulate sound from that they receive. We have also means to convey sounds in trunks° and pipes, in strange° lines and distances.

'We have also perfume-houses; wherewith we join also practices of taste.° We multiply° smells, which may seem strange. We imitate smells,° making all smells to breathe out of other mixtures than those that give them. We make divers imitations of taste likewise, so that they will deceive any man's taste. And in this house we contain° also a confiture-house;° where we make all sweet-meats, dry and moist, and divers pleasant wines, milks, broths,° and sallets,° far in greater variety than you have.

'We have also engine-houses,° where are prepared engines° and instruments for all sorts of motions. There we imitate and practise to make swifter motions° than any you have, either out of your muskets or any engine that you have; and to make them and multiply them more easily, and with small force,° by wheels and other means: and to make them stronger, and more violent than yours are; exceeding your greatest cannons and basilisks.° We represent also ordnance° and instruments of war, and engines of all kinds: and likewise new mixtures and compositions of gun-powder, wildfires° burning in water, and unquenchable. Also fire-works of all variety both for pleasure and use. We imitate also flights of birds; we have some degrees of° flying in the air; we have ships and boats for going under water,° and brooking of seas;° also swimming-girdles° and supporters.° We have divers curious clocks, and other like motions of return,° and some perpetual motions.° We imitate also motions of living creatures, by images° of men, beasts, birds, fishes, and serpents. We have also a great number of other various motions, strange for equality, fineness, and subtilty.°

'We have also a mathematical house, where are represented all instruments,° as well of geometry as astronomy, exquisitely made.

'We have also houses of deceits of the senses; where we represent all manner of feats of juggling,° false apparitions, impostures, and illusions; and their fallacies.° And surely you will easily believe that we that have so many things truly natural which induce° admiration,° could in a world of particulars° deceive the senses, if we would disguise those things and labour° to make them seem more miraculous. But we do hate all impostures° and lies: insomuch as° we have severely forbidden it to all our fellows, under pain of ignominy and fines, that they do not shew any natural work° or thing, adorned or swelling;° but only pure° as it is, and without all affectation of strangeness.

'These are (my son) the riches of Salomon's House.

'For the several employments and offices° of our fellows; we have twelve that sail into foreign countries, under the names of other nations (for our own we conceal); who bring us the books, and abstracts, and patterns° of experiments of all other parts. These we call Merchants of Light.

'We have three that collect° the experiments which are in all books. These we call Depredators.°

'We have three that collect the experiments of all mechanical arts; and also of liberal sciences;° and also of practices which are not brought into° arts. These we call Mystery-men.°

'We have three that try new experiments, such as themselves think good. These we call Pioners° or Miners.

'We have three that draw° the experiments of the former four into titles° and tables,° to give the better light for the drawing of observations and axioms° out of them. These we call Compilers.°

'We have three that bend° themselves, looking into the experiments of their fellows, and cast about° how to draw out of them things of use and practice° for man's life, and knowledge as well for works as for plain demonstration of causes,° means of natural divinations,° and the easy and clear discovery of the virtues and parts of bodies.° These we call Dowry-men° or Benefactors.°

'Then after divers meetings and consults° of our whole number, to consider of the former labours and collections,° we have three that take care, out of them, to direct new experiments, of a higher light,° more penetrating into nature than the former. These we call Lamps.°

'We have three others that do execute the experiments so directed, and report them. These we call Inoculators.°

'Lastly, we have three that raise the former discoveries by experiments into greater observations,° axioms, and aphorisms.° These we call Interpreters of Nature.°

'We have also, as you must think, novices and apprentices, that the succession of the former employed men do not fail; besides a great number of servants and attendants, men and women. And this we do also: we have consultations, which of the inventions and experiences° which we have discovered shall be published, and which not: and take all an oath of secrecy, for the concealing of those which we think fit to keep secret:° though some of those we do reveal sometimes to the state, and some not.

'For our ordinances and rites: we have two very long and fair galleries: in one of these we place patterns and samples of all manner of the more rare and excellent inventions: in the other we place the statua's of all principal inventors.° There we have the statua of your Columbus, that discovered the West Indies:° also the inventor of ships: your monk° that was the inventor of ordnance and of gunpowder: the inventor of music: the inventor of letters: the inventor of printing: the inventor of observations of astronomy:° the inventor of works in metal: the inventor of glass: the inventor of silk of the worm: the inventor of wine: the inventor of corn and bread: the inventor of sugars: and all these by more certain tradition° than you have. Then have we divers inventors of our own, of excellent works; which since you have not seen, it were too long to make descriptions of them; and besides, in the right understanding of those descriptions you might easily err. For upon every invention of value, we erect a statua to the inventor, and give him a liberal and honourable reward. These statua's are some of brass; some of marble and touch-stone;° some of cedar and other special woods gilt° and adorned: some of iron; some of silver; some of gold.

'We have certain hymns and services, which we say daily, of laud° and thanks to God for his marvellous works: and forms of prayers, imploring his aid and blessing for the illumination of our labours, and the turning of them into good and holy uses.

'Lastly, we have circuits° or visits of divers principal cities of the kingdom; where, as it cometh to pass, we do publish° such new profitable inventions as we think good. And we do also declare° natural divinations° of diseases, plagues, swarms of hurtful creatures, scarcity, tempests, earthquakes, great inundations, comets,° temperature° of the year, and divers other things; and we give counsel thereupon what the people shall do for the prevention° and remedy of them.'

And when he had said this, he stood up; and I, as I had been taught, kneeled down; and he laid his right hand upon my head, and said; 'God bless thee, my son, and God bless this relation which I have made. I give thee leave to publish it for the good of other nations; for we here are in God's bosom, a land unknown.' And so he left me; having assigned a value° of about two thousand ducats,° for a bounty to me and my fellows. For they give great largesses° where they come upon all occasions.

[The rest was not perfected.]°

Magnalia Naturae,

praecipue quoad usus humanos.°

The prolongation of life.
The restitution of youth in some degree.
The retardation of age.
The curing of diseases counted incurable.
The mitigation of pain.
More easy and less loathsome purgings.°
The increasing of strength and activity.
The increasing of ability to suffer torture or pain.
The altering of complexions,° and fatness and leanness.
The altering of statures.
The altering of features.

The increasing and exalting of the intellectual parts.
Versions° of bodies into other bodies.
Making of new species.
Transplanting of one species into another.
Instruments of destruction, as of war and poison.
Exhilaration° of the spirits, and putting them in good disposition.
Force of the imagination, either upon another body, or upon the body itself.
Acceleration of time in maturations.
Acceleration of time in clarifications.°
Acceleration of putrefaction.
Acceleration of decoction.°
Acceleration of germination.
Making rich composts for the earth.
Impressions° of the air, and raising of tempests.
Great alteration; as in induration, emollition,° &c.
Turning crude and watery substances into oily and unctuous substances.
Drawing of new foods out of substances not now in use.
Making new threads for apparel;° and new stuffs; such as paper, glass, &c.
Natural divinations.
Deceptions of the senses.
Greater pleasures of the senses.
Artificial minerals and cements.

ABBREVIATIONS

Classical works are usually quoted from the Loeb Library translations. Where an author is known by only one work, the name alone is given; e.g. Dio Cassius (*Roman History*); Diogenes Laertius (*Lives of the Eminent Philosophers*); Juvenal (*Satires*); Livy (*History*); Lucretius (*De rerum natura*); Quintilian (*Institutes of Oratory*). Other abbreviations as used in the *Oxford Classical Dictionary* (*OCD*).

Adv.L.	Bacon, *The Advancement of Learning*
Aristotle	*The Complete Works of Aristotle: The Revised Oxford Translation*, ed. Jonathan Barnes, 2 vols. (Princeton, 1984): the Bollingen series
Baconiana	*Baconiana, Or certain Genuine Remains of Sr. Francis Bacon.... In Arguments Civil and Moral, Natural, Medical, Theological, and Bibliographical: Now for the First time faithfully Published*, by Thomas Tenison (London, 1679)
Beal	Peter Beal, *Index of English Literary Manuscripts*, i/1 *1450–1625: Andrewes–Donne* (London and New York, 1980)
Cambridge Companion	Markku Peltonen (ed.), *The Cambridge Companion to Francis Bacon* (Cambridge, 1996)
De Mas	Enrico de Mas (ed.), *Scritti politici, giuridici e storici di Francesco Bacone*, 2 vols. (Turin, 1971)
Essential Articles	Brian Vickers (ed.), *Essential Articles for the Study of Francis Bacon* (Hamden, Conn., 1968; London, 1972)
Gibson	R. W. Gibson, *Francis Bacon: A Bibliography of his Works and of Baconiana to the year 1750* (Oxford, 1950); *Supplement* (Oxford, 1959)
Guicciardini	Francesco Guicciardini, *Selected Writings*, ed. C. Grayson, tr. M. Grayson (London, 1965)
Kiernan	*Sir Francis Bacon: 'The Essayes or Counsels, Civill and Morall'*, ed. Michael Kiernan (Oxford, 1985)
Le Doeuff	Michèle Le Doeuff, *Du progrès et de la promotion des savoirs* (Paris, 1991), ed. and tr. *Adv.L.*
Machiavelli	Niccolò Machiavelli, *The Prince*, tr. L. Ricci, rev. E. R. P. Vincent; and *Discourses*, tr. C. E. Detmold; ed. Max Lerner (New York, 1950)
Melchionda	*Gli 'Essayes' di Francis Bacon: Studio introduttivo, testo critico e commento*, ed. Mario Melchionda (Florence, 1979)

Montaigne	*Essais*, in *OEuvres complètes*, ed. A. Thibaudet and M. Rat (Paris, 1965): Pléiade edition
OCD	*Oxford Classical Dictionary*
OED	*Oxford English Dictionary*
Promus	*The Promus of Formularies and Elegancies by Francis Bacon*, ed. Mrs Henry Pott (London, 1883)
Resuscitatio	*Resuscitatio or, Bringing into Publick Light Several Pieces of the Works, Civil, Historical, Philosophical, & Theological, Hitherto Sleeping; of the Right Honourable Francis Bacon.... According to the best Corrected Coppies*, ed. William Rawley (London, 1657)
Reynolds	*The Essays or Counsels, Civil and Moral, of Francis Bacon*, ed. Samuel Harvey Reynolds (Oxford, 1890)
Rossi	*Scritti Filosofici di Francesco Bacone*, ed. Paolo Rossi (Turin, 1975, 1986)
SOED	*Shorter Oxford English Dictionary*
Wolff	Emil Wolff, *Francis Bacon und seine Quellen*, 2 vols. (Berlin, 1910, 1913; facs. repr. Nendeln, Liechtenstein, 1977)
Works	*The Works of Francis Bacon*, ed. James Spedding *et al.*, 14 vols. (London, 1857–74)
Wright, *Adv.L.*	W. Aldis Wright (ed.), *The Advancement of Learning* (2nd edn., Oxford, 1873)
Wright, *Essays*	W. Aldis Wright (ed.), *Bacon's Essays and Colours of Good and Evil* (3rd edn., London, 1865)

NOTES

In annotating these selections I have attempted to document Bacon's biblical quotations and allusions and his references to classical and contemporary authors, and to provide translations from Greek and Latin. The constraints of space made it impossible to comment on individual allusions in the detail found in the best editions of the *Essays*, by S. H. Reynolds (1890), Mario Melchionda (1979), and Michael Kiernan (1985). I chose to devote most energy to defining the meanings and (where relevant) specific connotations of the words Bacon used. Given that the past recedes constantly from us, that the amount of books to be read increases daily, that our language is continually changing, and that at least 10 per cent of Bacon's vocabulary is no longer current, I thought it important to concentrate on this fundamental area of primary meaning. To do so I have relied on the great, but now somewhat out-of-date records of the *Oxford English Dictionary*, partly revised and updated in the splendid *New Shorter Oxford English Dictionary* (1993).

For readers knowledgeable in early modern English, the amount of linguistic annotation I have provided may seem excessive; for many others, I hope, it will be welcome. We can all recognize the English words that Bacon uses which have either disappeared or become extremely rare, such as munite, 'to fortify'; principial, 'initial'; celsitude, 'loftiness'; droumy, 'muddy'. Far more deceptive, however, are the many words that have survived in the same written form but with great changes of meaning. Such are, for instance, from the *Essays* alone, the words accommodate, meaning 'to adapt oneself'; decline, 'to turn aside'; deliver, 'to describe', but also 'to let in, admit'; inordinate, 'irregular, ungovernable'; merely, 'absolutely'; obnoxious, 'exposed to, submissive', peremptory, 'deadly, destructive'; plausible, 'praiseworthy, deserving applause'; success, 'the outcome of an action'—good or bad; suit, 'order, succession'; towardness, 'docility'. In the *Advancement of Learning*, for instance, we find abused, for 'deceived'; application, for 'adaptation'; casualty, 'uncertainty'; conjugations, 'relationships of man and wife'; combustion, 'feverish excitement'; continent, 'container'; curious, 'made with care'; depress and disable, both meaning 'disparage'; to illustrate, 'to render illustrious, enhance'; inveterate, 'firmly established'; jealousy, 'vehement feeling'; magistral, 'dogmatic', but also 'a specific (*ad hoc*) remedy'; positive, 'dogmatic', but also 'practical'; respective, 'appropriate'; retribution, 'recompense'; to reduce, 'to convert, call or turn back'; satisfactory, 'superficially convincing'. Bacon's words, many of which were then new to our language (quite a number he introduced himself), are often close to their original Latin meaning, such as futile, meaning 'idly talkative' (Lat. *futilis*, leaky); adventive, 'coming from without; inessential'; incurring, 'running into themselves'; volubility, in the sense of 'rolling or twisting motion'; officious, 'ready to serve' (Lat. *officiosus*), not in a bad sense; to suborn, meaning 'to furnish, equip'; etc. Readers referring to these notes will discover, as I did myself, many surprises.

Yet annotation, although necessary, can be distracting from the reading experience. As Dr Johnson so memorably put it, introducing his edition of Shakespeare in 1765, 'Particular passages are cleared by notes, but the general effect of the work is weakened. The mind is refrigerated by interruption; the thoughts are diverted from the principal subject; the reader is weary, he suspects not why; and at last throws away the book which he has too diligently studied.' I can think of no better advice to readers than Johnson gave: 'Notes are often necessary, but they are necessary evils. Let him that is yet unacquainted with the powers of *Shakespeare*, and who desires to feel the highest pleasure that the drama can give, read every play from the first scene to the last, with utter negligence of all his commentators. When his fancy is once on the wing, let it not stoop at correction or explanation.'

AN ADVERTISEMENT TOUCHING THE CONTROVERSIES OF THE CHURCH OF ENGLAND

The reign of Queen Elizabeth I, fortunate in so many other respects, was continually disturbed by religious controversy. One unavoidable cause of these disturbances was the situation that Elizabeth inherited on her accession to the throne in 1558. Henry VIII's break with Rome had initiated a Protestant Reformation in England, which was continued during the reign (1547–53) of his son Edward VI. Under the rule of Queen Mary (1553–8), however, England reverted to Catholicism, and many leading churchmen went into exile in the main Protestant centres in Europe, notably Calvin's Geneva, Zwingli's Zurich, Strasbourg (where the future episcopal leadership was concentrated), and Emden. Their return to England on Elizabeth's accession strengthened the movement for a more drastic reform of the English Church. During this whole period England's major enemy in religious affairs continued to be the Roman Catholic Church, and a sufficient number of conspiracies (some led by Jesuit emissaries) were revealed to justify the continuing fear of Rome as a dangerous opponent. In addition, the Papal bull of excommunication in 1570 created a stubborn enclave of Popish recusants within England. But the Protestant reformers, most of them part and parcel of the English Church, constituted an even more troublesome nuisance, continually agitating for far-reaching changes in the Church's constitution and government.

The other main cause exacerbating controversy, in the second half of the Queen's reign, was the repressive policies adopted by the leading figure in the Church of England, John Whitgift, Archbishop of Canterbury. His predecessor Edmund Grindal, during his tenure of the post (1575–83), had pursued a moderate line, tolerating much Nonconformist behaviour in religious services and in public discussion, not provoking confrontation. Whitgift, however, from the moment of his accession to the office in late 1583, pursued a line of strict orthodoxy, making several innovations that were calculated to provoke all Nonconformists, whether moderate or extremists, to bitter opposition. In the judgement of a modern authority 'the Whitgiftian policy, continued in their generations by Bancroft and the Laudians, was as much responsible as any Puritan excess for destroying the comprehensiveness of the Church of England and its fully national character' (Patrick Collinson, *The Elizabethan Puritan Movement* (London, 1967; Oxford, 1990), 245–7, 273). Whitgift introduced three articles which all ministers of the Church had to swear to accept, the first and third of which were unexceptionable (affirming the Queen's supremacy, and subscribing to the Thirty-nine Articles of faith). The second article, by contrast, which demanded total endorsement of the liturgy enshrined in the Book of Common Prayer and of the ceremonies for ordaining bishops, priests, and deacons, was totally unacceptable to the Puritans. They believed that the Church of England had retained too many elements from Catholicism, both in its forms of service and in church government. Their preferences were for sermons expounding and applying the word of God, rather than a ritualized order of service, and for a democratic church organization, resembling that of Christ's first disciples, rather than the hierarchy of archbishops, bishops, deacons, and so on.

Whitgift enforced his policies by persuading the Queen to grant new powers to the Court of High Commission for Causes Ecclesiastical, over which he presided. Anyone summoned before this body had to swear an oath *ex officio mero*, that he would answer truly any question put to him. Unlike proper legal usage, where specific charges are brought, evidence produced, and witnesses called, Whitgift's court merely confronted its victims with a set of questions ('interrogatories') concerning their religious opinions and practices, which could be artfully varied so as to make the examinee incriminate himself and his associates. Those who refused to swear the initial oath involving them in these proceedings were sent to prison, suffering appalling conditions, and could remain there for months without any legal right of appeal. Whitgift's seizure of powers belonging to the state, at the same time suspending the constraints safeguarding the subject's rights (the court could also impose heavy fines, and excommunicate whomsoever it chose), aroused much anger at many levels of the English society, from the highly competent Puritan lawyers in the House of Commons down to the ministers and their loyal congregations (Collinson, ibid. 266–73, 312–15). These autocratic proceedings were pursued successfully by Whitgift, with the Queen's support, despite the misgivings of highly placed courtiers with Puritan leanings (such as Burghley) and despite a co-ordinated Puritan campaign in the parliaments of 1583–4 and 1587. Whitgift stepped up his campaign in 1590, and again in 1593, introducing a savage piece of legislation against Puritan sectaries which justified his proceeding to the extreme: Henry Barrow and John Greenwood were hanged in April 1593, John Penry in May. Other ministers were pardoned, released from prison (their health ruined), but not fully cleared for years (Collinson, ibid. 409–21, 428–31).

The continuous confrontations produced by Whitgift's intransigent policies sparked off much activity among Puritans, both in their clandestine meetings and in (largely anonymous) publications. The Church of England had long found it necessary to issue formal self-defences against published attacks by both Catholic and Protestant opponents, and many writers became involved in controversy across the inevitable divide of attack and defence. Most of these writings are worthy but wordy, solid expositions of their own case with continuous refutation of their opponents'. But a series of publications in 1588–9, said to be written by one Martin Marprelate, suddenly raised the literary level of controversy to new heights. Skilful in logic and rhetoric, well-read in the Church's self-defences, these authors lampooned the bishops and their apologists in an irreverent, highly amusing, and extremely damaging series of pamphlets, chaotically printed on clandestine presses as their publishers tried to evade the legal officers sent to silence them. This tactical victory provoked the bishops to counter-measures, the leading figure in their campaign being Richard Bancroft (1544–1610). Bancroft was for twelve years chaplain to Sir Christopher Hatton (subsequently Lord Chancellor, and one of Whitgift's few supporters at court), and he occupied the same post with Whitgift for five years before going on to become Bishop of London in 1597 and Archbishop of Canterbury in November 1604.

Bancroft had been collecting incriminating evidence against the Puritans for several years, and his policy of answering Martin in his own style took the form of hiring university wits, such as Lyly and Nashe, to write mocking pamphlets in the same irreverent manner, and paying London theatres to stage some crude pieces satirizing Martin. Some twenty separate publications of this anti-Martinist literature appeared in 1589–90, all of a lamentable quality, while the theatrical performances created so much disturbance that Burghley temporarily closed the theatres. Martin Marprelate had used satiric means to expose genuine abuses in Church and State: the hacks hired by Bancroft lowered the tone of controversy to gutter abuse.

It is against this background of a fundamental opposition between the Church of England and the Nonconformist party, both equally intransigent in their attitudes, with a rapidly degenerating level of discussion, that Bacon produced his 'Advertisement touching the Controversies of the Church of England', an individual's call for a cessation to hostilities. It is carefully composed, and shows signs of special reading in patristic and other literature concerning church history, quoting Latin texts which Bacon subsequently never used, and for which he took the unusual trouble of giving an English translation, as if striving to reach a wider audience. (Some of these quotations have eluded identification.) That he never published the piece suggests that he designed it as a private memorandum for people in power who might share his independent stance, which found fault with both sides. In particular Bacon deplored the 'immodest and deformed manner of writing lately entertained, whereby matters of religion are handled in the style of the stage', and explicitly declared: 'I dislike the invention of him who (as it seemeth) pleased himself in it as in no mean policy, that these men are to be dealt withal at their own weapons, and pledged in their own cup' (above, pp. 3–4). As Patrick Collinson argued in a paper ('Hooker and the Elizabethan Establishment') given to the Folger Institute conference on Richard Hooker in September 1993, in this essay Bacon produced 'a genuinely detached, truly irenical critique' which reserved its 'harsher criticism for those who had used these kinds of weapon against Martin', selecting 'for particular censure' the inspiring force behind them: 'Bancroft was not named, but Bancroft was meant'.

This identification seems to me certain, although it raises the problem of dating Bacon's treatise, which Spedding tentatively assigned to 'the Summer of 1589' (*Works*, viii. 73). Bancroft had only just emerged into the public eye that year, with the publication of his lengthy (106 pages) *Sermon Preached at Paules Cross* in February 1589 (STC 1346; entered in the Stationers' Register on 3 March 1589), and if we accept Spedding's dating Bacon would have been answering him almost immediately. Since the first known allusion to Bacon's treatise dates from 1591 (see below), until more information emerges we can only date it approximately, to the period 1589–91. Bacon was, however, familiar with Bancroft's diatribe, which soon achieved notoriety. Bancroft used the quasi-official context of a Paul's Cross sermon to launch a violent attack on the 'many false prophets' who 'now remaine amongst us: *Arrians, Donatists, Papists, Anabaptists*, the *Familie* of love, and sundrie other' (p. 3). Referring approvingly (pp. 24, 51) to Thomas Cooper's *An Admonition to the People of England* (1589, entered 10 January) and to John Jewel's *An Apologie, or aunswer in defence of the Church of England* (1562, 1564, 1599; translated by Lady Ann Bacon), Bancroft went on to denounce Reformers throughout Europe for maintaining 'most strange and rebellious propositions' (p. 78), likely to cause 'the overthrow of all government' (p. 89). Once the Reformers had gained power by such illegal means, he warned, they 'would no doubt tyrannise by their censures over both princes and people at their pleasure, in most intolerable and popelike maner' (p. 94). Bancroft's violence, and his prophecies of doom, aroused great resentment among foreign churches, and he was answered in two pamphlets published in Edinburgh in 1590, one by the Scottish presbyterian John Davidson, the other by John Penry.

Bancroft's sermon used for the first time a tactic he was to employ in his later publications (*A Survay of the pretended Holy Discipline* and *Dangerous Positions and Proceedings*, both issued anonymously in 1593), of smearing the Reformers, including such a serious and responsible apologist as Thomas Cartwright, by coupling them (*Sermon*, pp. 3, 21, etc.; *Survay*, p. 1, etc.) with the Family of Love. This was an extreme radical sect founded by Henry Nicholas and formally banned by the Queen in 1580, to which the Puritans were deeply opposed. As Patrick Collinson has suggested, Bacon was undoubtedly referring to Bancroft in complaining at the Church of England apologists for libelling the Nonconformists: 'They have ever sorted and coupled them with the Family of love, whose heresies they have laboured to descry and confute' (above, p. 14). This tactic evidently formed part of the instructions Bancroft gave to the anti-Martinist writers in 1589–90 (William Pierce, *An Historical Introduction to the Marprelate Tracts* (London, 1908), 173–4, 219–20). Bancroft, with his constant warnings of impending disaster, is again one of the targets in Bacon's denouncing as mere alarmism some Church of England apologists' claim that the Puritans represented a genuine threat to national security, 'which is as unwisely acknowledged [accepted] as untruly affirmed'. The truth is, he declared, that 'the wound is no way dangerous, except we poison it with our remedies'. (Metaphors of wounds and their proper treatment form a running motif in this tract.) As for the 'bishops and governors of the church', Bacon identifies their imperfections in behaviour and government as the principal cause of 'schisms and divisions', the bishops having grown 'worldly, "lovers of themselves, and pleasers of men"' (above, p. 6).

Bacon's criticism of the bishops' greed and corruption naturally endeared him to the Reformers, although that was certainly not his aim. In *A petition directed to her most excellent majestie* (Middleburg, 1591), some anonymous writer noted that 'a learned man, and friend to the Bishops noteth as abuses, their urging of Subscription, their oath ex officio, their excommunication for trifles, and easy silencing of Ministers', adding the identifying marginal reference to 'Advert[isement] To[uching] the Church of England, not printed' (p. 2; cit. J. Martin, *Francis Bacon, the State, and the Reform of Natural Philosophy* (Cambridge, 1992), 193 n. 55). Bacon was certainly no 'friend to the Bishops', but his tactic of attacking extremism on both sides meant that the Church of England's apologists could also invoke his voice, as Bancroft himself did in the preface to his *Survay of the pretended Holy Discipline*: 'There are many in *England* I perceive, that are so addicted unto their own opinions, as concerning the pretended holy Discipline, and such a reformation, as they themselves have devised: that they cannot with anie patience endure, to heare either contradiction or argument to the contrarie. *I knowe some of them* (saith a certaine advertiser touching the controversies of the Church of England) *that would think it a tempting of God, to heare or read what might bee saide against them: as if they could be at "quod bonum est tenete", without an "omnia probate" going before*. Which maner of persons, the Prophet *David* resembleth unto the deaf Adder, *that stoppeth his eare, and will not heare the voyce of the inchaunter, though he be expert in charming*' (Sig. *2^r). (The fact that Bacon's tract was still circulating in 1592–3 is proof that it had aroused considerable interest.) By his selective quotation Bancroft avoided reporting the treatise's criticisms of the Established Church, although he must surely have recognized that he was one of the targets. The uncomfortable fact for the ecclesiastical Establishment, however, was that Bacon's denunciation of the bishops' greed and other abuses was completely justified, and was indeed shared by one of the Church's most distinguished and respected apologists, Richard Hooker, in his *The Lawes of Ecclesiasticall Politie* (the first four books of which were published in 1593). As A. S. McGrade has described it, 'the last and longest chapter of Book VII' (drafted, but not yet published) includes 'an extensive indictment of the bishops', a 'cutting account of episcopal shortcomings' which displays a 'bitterness' and personal 'anguish' far removed from the conventional picture of Hooker as 'serene . . . advocate . . . of the Elizabethan settlement' (*Works*, Folger Library edn. vi/1 (Binghamton, NY, 1993), 333–4).

Giving, as he puts it, 'the opinion of an indifferent [impartial] person', Bacon also objects to the intransigence adopted by Whitgift and his fellow bishops 'in standing so precisely upon altering nothing', a stubborn attitude which opposed even the legitimate reform of abuses. Among these abuses Bacon lists the power that Whitgift had acquired to excommunicate Nonconformists, which he dismisses as 'a base process to lackey up and down for duties and fees' (above, p. 13). Here again, Bacon's diagnosis is historically accurate. When Whitgift submitted to the Queen the new articles for his Court of High Commission in 1583, they included the clause '"that the Bishops upon their *significavit* be allowed to issue writs *De Excommunicato Capiendo*, without incurring any charges"'; and, as William Pierce reports, '*more suo* [Whitgift] tells the Queen that this would increase the number of writs and increase her income' (Pierce, *Historical Introduction*, 70). As a lawyer, with undoubted Puritan leanings, Bacon expresses a widely shared resentment at Whitgift's High Commission with its incriminating interrogatories: 'Swearing men to blanks and generalities (not included within a compass of matter certain, which the party that is to take the oath may comprehend) is a thing captious and strainable' (above, p. 14). His judgement that this measure forced moderate Nonconformists into the same exposed position as extremists has been confirmed by one of the leading modern authorities (Collinson, *Elizabethan Puritan Movement*, op. cit., 247, 273).

Yet Bacon has equally severe criticisms to make of the Reformers, showing an impartiality all the more impressive considering his political alignment with Burghley, a life-long Puritan, and the fact that his mother, Lady Ann Bacon, continued to be a patron and financial supporter of the Puritan cause (ibid. 439–41, 443). Bacon criticizes the Puritans for their purely adversarial belief that the degree to which ceremonies or modes of government are 'opposite to the institutions of the Church of Rome' is a mark of their being 'good and holy', and he argues that in the 'general demolition of the Church of Rome' there was '(as men's actions are imperfect) some good purged with the bad' (above, pp. 9–10). Bacon then reproves the Puritans for modelling themselves too closely on foreign churches, not considering what is suitable to the specifically English situation. He also rejects one of the main nonconformist principles, the 'parity and equality of ministers', as a source of 'confusion' or chaos (pp. 10–11). Surveying the history of the claims for church reformation in the sixteenth century, he notes a progression from legitimate outrage against ecclesiastical abuses to a series of increasingly violent and extreme positions, culminating in the Reformers' claim to have produced 'an only and perpetual form of policy in the church' (p. 12). This absoluteness, Bacon shows, goes along with a preference for negative positions in ethical and social issues, the Reformers failing to teach the people 'their lawful liberty, as well as their restraints and prohibitions'. But, Bacon urges, it is just 'as unlawful to shut where God hath opened, as to open where God hath shut', or 'to bind where God hath loosed, as to loose where God hath bound' (above, pp. 16–17).

Bacon's criticism of both the bishops and the Puritans, as he was well aware, was 'not like to be grateful to either part'. Yet, while showing much sympathy for the Puritans' complaints, he displayed his own loyalty to the Crown and to the national Church. Some modern readers might share the opinion expressed by the distinguished editor and bibliographer R. B. McKerrow, that 'even in such a careful and temperate balancing of the opposing views as Bacon's *Advertisement*... the only kind of toleration which the writer seems to have in his mind is the allowing of certain external matters of form and ceremony to be non-essential, and so modifying the Church that all could belong to it. Not once does the idea of allowing separate Churches within the same state seem to occur to him' (*The Works of Thomas Nashe*, 5 vols. (Oxford, 1910, 1966), v. 40). With all respect, that is an anachronistic comment, written in a period when plurality of faith had been gradually accepted. In Bacon's day one European nation had, for a while, two religions, but the dangers and instabilities involved were terribly illustrated by the St Bartholomew's massacres of 23–6 August 1572, in which some thirteen thousand Huguenots (French Calvinist Protestants) perished. If we consider the pressures that the English government was under from quarrelsome religious creeds without and within, we may well agree that Bacon's irenic stance was an admirably responsible one.

An ironic footnote to this treatise can be added, concerning events in 1603–4. Bacon addressed to King James 'at his first coming in'—according to a note in the title of one of the manuscript copies (*Works*, x. 102)—a tract called *Certain Considerations touching the better Pacification and Edification of the Church of England*, which was printed in 1604 (ibid. 103–27). In it he reviews the present state of the Church, expressing surprise that 'the civil state' should make laws every three or four years to keep itself healthy while 'contrariwise the ecclesiastical state should still continue upon the dregs of time, and receive no alteration now for these five and forty years and more' (105). He then proceeds to repeat many of his criticisms from the earlier tract: but since the Reformers are no longer a cause of dissension, the Church Establishment receives the brunt of his critique. The bishops are reproved for 'the sole exercise of their authority' (for 'the Bishop giveth orders alone; excommunicateth alone; judgeth alone', an autonomy not known in government otherwise, and likely 'to have crept in in the degenerate and corrupt times'), and for exerting their authority through deputies, equally unknown in 'all laws of the world' (108–14). Bacon again attacks 'the oath *ex officio* whereby men are enforced to accuse themselves' (114), and again denounces the bishops' facile resort to excommunication, which 'is the greatest judgment upon the earth, being that which is ratified in heaven, and being a precursory or prelusory judgment to the judgment of Christ in the end of the world', which has been grossly misused to raise fees (121–2). He denounces non-residence as 'an abuse drawn out of covetousness and sloth' (122–4), and concludes by suggesting that 'the Papists themselves should not need so much the severity of penal law if the sword of the spirit were better edged, by strengthening the authority and repressing the abuses in the Church' (126).

To this outspoken criticism of the ecclesiastical Establishment Bacon added other independent observations, defending the liturgy as the central element in church-worship, but suggesting amendments (114–18); proposing that the practice of 'prophesying' (see below, pp. 508–9) be revived, but made more of a pedagogic exercise by having experienced ministers coach younger ones

(119–20); and urging that church stipends be made adequate (124–6). Although siding with those who wanted reforms, Bacon does not endorse the Reformers' programme, continuing to take the same moderate and independent line as in his earlier treatise. His position, however, hardly endeared him to the bishops, and his book soon suffered the effects of their displeasure. Rawley recorded that the printed edition of 1604 was 'called in', and Spedding had never found 'any perfect copy of this edition', the most complete one he knew having 'sheet E printed only on pages 1, 4, 5, and 8 (as if only one side had been completed): the blank pages being supplied in MS' (102 note). His belief that 'the printing was stopped before it was completed' has been confirmed by modern bibliographers, for the revised *Short-Title Catalogue* identifies two editions in 1604, both only printed in part (1118 has sheets A–D4 only, 1118.5 has E outer forme only, with the text completed in MS). The Folger Shakespeare Library copy has a contemporary MS note on the title-page which records what happened: 'This booke is not [in] print, only four s[heets] was printed and the bishop of Lon[don] called it in an[d] would not suf[fer more] to be printed. [That] which was not p[rinted] I got writt[en] by hand as you see' (*National Union Catalog Pre-1956 Imprints*, lviii. 434). The Bishop of London, under whose jurisdiction fell the duties of ecclesiastical censorship, was none other than Richard Bancroft.

My text is from Spedding (*Works*, viii. 74–95), who chose as his main authority the Bodleian Library manuscript (MS e. Mus. 55: Beal BcF 63), which he collated with two MSS in the British Library (Harl. MS 3795: Beal BcF 68, and Add. MS 4263: Beal BcF 66), and with the version printed by William Rawley in his *Resuscitatio* (1657). The work was first published as *A Wise and Moderate Discourse concerning Church-Affaires* (London, 1641), and reissued in 1663 as *True peace: or a moderate discourse to compose the unsettled consciences, and greatest differences in ecclesiastical affaires.*

1. *Advertisement*. Admonition. *exercised*. Distressed. *church militant*. 'The body of Christians still on earth, as distinct from those in Purgatory (expectant) and those in Heaven (triumphant)' (*The Oxford Dictionary of the Christian Church*). *winnow*. Separate. *proveth*. Tests. *discovereth*. Reveals. '*That... Christ*'. Matt. 24: 23; Mark 13: 21. *person*. body. *assumed*. Simulated, adopted. '*Ecce... penetralibus*'. Matt. 24: 26. Cf. Essay 3, 'Of Unity in Religion', above, p. 344. *conventicles*. Clandestine or unlawful religious meetings. *conciliables*. Small or secret assemblies. *heretics*. Persons who hold opinions or doctrines contrary to Christian orthodoxy. *sectaries*. Adherents of a schismatical or heretical sect. *extern*. Outward, superficial. *been seduced*. Led astray. *unswathe... bonds*. Undo uniting or cementing elements. *hath used*. Has been wont. *canon*. Rule, edict. '*Erratis... Dei*'. Matt. 22: 29; Mark 12: 24. Cf. *Novum Organum*, i. 89 (*Works*, iv. 88–9) for the particular sense that Bacon gave to this text. '*Let... wrath*'. Jas. 1: 19. Also noted in *Promus*, no. 259. *make... discord*. Rejoice at our disagreements.

2. '*Vos... ye*'. Exod. 2: 13. *the party... wrong*. Ibid. *enter into*. Discuss in detail. *touching*. Concerning. *detained*. Occupied. *first peace*. After the

Emperor Constantine the Great conquered the Eastern provinces in AD 325, he summoned the first Council of Nicaea, primarily to settle the Arian dispute about the person of Christ. *what time.* At which time. *moved.* Raised, propounded. *curious.* Intricate, abstruse. *anatomies.* Analyses. *person.* Personality of Christ, as uniting his divine and human natures (*SOED*). *Catholic fathers.* Early Christian writers, especially of the first five centuries. *subtility.* Intricacy. *determinations.* Authoritative decisions settling disputes. *evasions.* Prevarications. *take.* Catch. '*illis . . . Christianum*'. Untraced. *great.* Major. '*non . . . colendo*'. Untraced. *east and west churches.* Those of the Eastern Roman or Byzantine Empire (which developed its own identity until, in the schism of 1054, it became the Greek Orthodox Church) and the Latin church of Western Europe. *images.* The worship of images was violently opposed by the 16th-century Reformers, especially Zwingli and Calvin, who were followed by the Puritans. *adoration of the Sacrament.* In the Roman Catholic Mass the 'host' (or consecrated bread of the Eucharist) which is elevated by the priest. *contend.* Differ, dispute. *ceremonies.* Outward rituals, observances. *things indifferent.* Non-essential issues. Such were often called '*adiaphora*', or 'matters of indifference', namely 'beliefs or practices which the Reformers regarded as being tolerable, in that they were neither explicitly rejected nor stipulated by scripture', such as what ministers wore at church services. This concept 'allowed the Reformers to adopt a pragmatic approach to many beliefs and practices, thus avoiding unnecessary confrontation': Alister E. McGrath, *Reformation Thought: An Introduction* (Oxford, 1988), 155. *policy.* Politics. *one faith, one baptism.* Eph. 4: 5. *league.* Covenant. *penned.* Composed, formulated. '*he . . . us*'. Matt. 12: 30; Mark 9: 40; Luke 9: 50. Cf. Essay 3, 'Of Unity in Religion', p. 345 above. '*differentia . . . doctrinae*'. Untraced. '*habet . . . temporis*'. Untraced. *St. James.* Jas. 1: 19 (as quoted above). *close up and grow together.* Like a wound healing. *over-weening.* Conceited, arrogant. *turbulent.* Violently disturbed. *humours.* Excited states of public feeling (*SOED*). *primitive church.* The Christian Church in its earliest and (supposedly) purest era. *positions.* Positing propositions. '*Si . . . affirmanti*'. Untraced.

3. '*Ego . . . Dominus*'. 1 Cor. 7: 12. '*Et . . . meum*'. 1 Cor. 7: 40. *bind.* Reinforce, secure. *denunciations.* Threatening announcements. *simple.* Uneducated. *the . . . come.* Prov. 26: 2. Cf. *Adv.L.*, above, p. 292. *accidents.* Unfavourable symptoms. *opening.* Cutting open, discovering. *green.* Unhealed. *formalizeth.* Fixes in opposition. *indisposition.* Unwillingness. *pretended.* Intended. *in reason.* As a general principle. *repetition.* Review. *de concordia ordinum.* 'Upon the harmony of the two orders' (Cicero, *Epist. ad Att.* I. xvii. 9). *reducing to memory.* Recalling, remembering. *extremities.* Extreme behaviour, opinions, utterances. '*Qui . . . componit*'. 'Whoever wants to establish peace without locating the cause of discord, rather than dealing with people honestly, deceives them with the sweet hope of peace'; untraced. *surseance.* Conclusion. *immodest.* Indecorous, offensive. *style of the stage.* Referring to the Martin Marprelate controversy and the 'anti-Martinist' response. *affection.* Passion. *politic.* Calculating, governed by

expediency. *touch and sense*. Feeling and sensation. *speculation*. Consideration. *pertaineth*. Applies. *warranted*. Authenticated. *leave*. Abandon, omit. *search*. Probe. *rip up*. Reopen (a wound or sore) roughly. *scant beseeming*. Hardly suitable to. '*Non . . . joci*'. Perhaps derived from Horace, *Epist*. I. xvi. 54, 'miscebis sacra profanis': 'you will make no difference between sacred and profane'. *ridiculous*. Deserving derisive laughter. *Atheism*. Cf. Bacon's Essay 16, 'Of Atheism', above, pp. 371–3, where one of the four causes of atheism is said to be the 'custom of profane scoffing in holy matters: which doth by little and little deface the reverence of religion'.

4. *scoffing*. Contemptuous mocking. *sect*. The atheists. *that bishop*. Thomas Cooper, Bishop of Winchester, author of *An Admonition to the people of England* (1589). '*a fool . . . him*'. Prov. 26: 4. *matter*. Subject, topic. '*If . . . not*'. Job 29: 24. *diverted*. Digressed. *glanced*. Alluded. *receive*. Attend to; take in. *invention*. Plan, idea. *him*. Richard Bancroft. *pledged . . . cup*. Answered in kind. *Cardinal Sansovino . . . Lateran*. Julius II (1443–1513), Pope from 1503, turned against his former ally France in 1511, founding the Holy League for the purpose of defeating it. Louis XII called a Council at Pisa in 1511, with the intention of deposing the Pope, but Julius responded by calling a much larger Council at the Lateran in 1512, which confirmed his authority and won the support of the Emperor Maximilian, so isolating France. Bacon's source is Guicciardini, *Storia d'Italia*, x. 2, describing how the Pope accepted the advice 'proposto da Antonio del Monte a San Sovino, uno de'cardinali creati ultimamente a Ravenna . . . [e] intimò il concilio universale . . . nella città di Roma nella chiesa di San Giovanni Laterano: per la quale convocazione pretendeva avere dissoluto il concilio [di Pisa] convocato dagli avversari'. Cf. *Storia d'Italia*, ed. C. Panigada (Bari, 1929), iii. 110, and V. Luciani, 'Bacon and Guicciardini', *PMLA* 62 (1947), 96–113, at p. 100. *encounter*. Oppose. *Mr. Jewel*. John Jewel (1522–77), Bishop of Salisbury, one of the intellectual leaders of the Reforming party, whose *Apologia Ecclesiae Anglicanae* (1562) was the first outstanding defence of the Church of England's claims to legitimacy and autonomy. See head-note above. *pretended*. Falsely claimed to be such; spurious. *by the Fathers*. Using arguments derived from patristic authors. '*Nil . . . mei*'. Cicero, *Epist. ad Att*. ix. 16: 'There is nothing I like better than that I should be true to myself and they to themselves.' Also noted in *Promus*, no. 1142. '*Dum . . . consentimus*'. Untraced. *of*. Which of. *percase*. Perhaps. '*the . . . fray*'. *The Oxford Dictionary of English Proverbs* (B 475) cites Bacon's use of this proverb together with its appearance in *Mar Mar-Martin* (1590), 1: 'If all be true that the Lawyers say, The second blowe doth make the fray.' In some versions it reads: 'The first blow makes the wrong, but the second makes the fray.' See also Bacon's *Colours of Good and Evil* (*Works*, vii. 91). *obscure*. Little known. '*Qui . . . multiplicat*'. Untraced. *determine the question*. Settle the issue. '*Alter . . . abstulit*'. Untraced. *join*. Lump together. *scoffers*. Mockers (especially of religion and morality). *their*. That their. *solicited*. Urged, pleaded. *have . . . intelligence*. Communicate. *this other libeller*. Untraced. *credit*. Trustworthiness, authority. *glosses*. Comments.

ordinary. Usual. *pressing*. Persistent in solicitation. *fawning*. Showing servile deference. *humours*. Moods, predisposition. '*Veneri . . . suem*'. Apostolius, 4.59, via Erasmus, *Adagia*, III. i. 30 (no. 2030). *satisfy*. Be satisfied.

5. *one*. Bishop Cooper. *with honour*. Honourably, favourably. *indifferent*. Impartial. *hand carried*. Policy exerted. *flieth in the dark*. Is circulated covertly. *uttered*. Published. *side*. Group, party. '*punitis . . . auctoritas*'. Tacitus, *Annals*, iv. 35: 'Genius chastised grows in authority'. *ever*. Always. *falleth out*. Happens. *meet*. Suitable. *deface*. Discredit. *the other*. The writings of the 'anti-Martinists', Nashe, Lyly, and Co. *exercises*. Religious observances, acts of preaching. *so as*. So that. *meaner*. Less eminent. *amplifications*. Exaggerations. *near*. Nearly. *force . . . sinews*. Power of movement. *convulsion*. Violent agitation, disturbance. *acknowledged*. Admitted. *his*. Unidentified. *impugning*. Finding fault with, accusing. *sympathy*. Correspondence, affinity. *civil policy*. Civic state. *taxation*. Accusation, censure. *comen abroad*. Published. *censured*. Reproved. *extravagancies*. Immoderate, unrestrained utterances. *light*. Frivolous, trivial. *pave*. Lay paving stones (and so, harden). *highway*. Road. *vein*. Style of writing or behaviour. *short madness*. Passing fit. Cf. Horace's famous phrase, *ira furor brevis est* (*Epist*. I. ii. 62) and Essay 57, 'Of Anger', above, pp. 449–50. *disport and solace*. Entertainment and relaxation. *of*. One of. *blanch*. Palliate by misrepresentation. *sincere*. Not falsified; unbiassed. *accidents*. Events, occurrences.

6. *imputation*. Accusation. *partial*. Prejudiced, biased. *courses*. Policies. *holden and entertained*. Favoured, adopted. *partizants*. Zealous supporters. *strait*. Strict, close. *importeth*. Implies, involves. *distraction*. Disorder, disruption. *propounding*. Proposing, devising. *palpable*. Obvious. *conceits*. Conceptions. *conversation*. Behaviour. *secular states*. Groups within society (nobles, parliament, common people). '*is situate . . . hill*'. Matt. 5: 14. *maketh question*. Expresses doubt. *wax*. Grow, become. '*lovers of themselves*'. 2 Tim. 3: 2. '*pleasers of men*'. Col. 3: 22. *grope*. Seek, search for. *Pharisees*. A Jewish religious party, according to the Gospels the chief opponents of Christ, who denounced their formalism and self-righteousness. *howsoever*. Notwithstanding. *Moses' chair*. Moses was the founder and lawgiver of Israel; his chair is the symbol of authority, from which he would pass judgement, as in Exod. 18: 13. '*tamquam . . . habentes*'. Untraced. *declining*. Turning aside. *trace out*. Lay down, devise. '*Nolite exire*'. Matt. 24: 26. *voice*. Opinion uttered. *natural*. With a secular, layman's realistic view. *freres*. Friars.

7. '*that . . . prelates*'. Machiavelli, *Discourses*, iii. 1, describing how, in Italy, religion was 'brought back to its pristine principles and purity by Saint Francis and Saint Dominic', who, 'by their voluntary poverty and the example of the life of Christ', managed to save religion 'from being destroyed by the licentiousness of the prelates and heads of the Church'. The same passage is quoted in *Adv.L.*, above, p. 132. *bear out*. Corroborate. *policy*. Craftiness. *spiritual enemy*. All enemies of God, including the devil:

cf. Archimago in Spenser's *Faerie Queene*, i–ii. *authorize*. Endow with authority. *affected*. Favourably disposed. *ought*. Ought to be. *contradiction*. Inconsistency. *detractation*. Defamation. *embased*. Lowered. *pierced*. Damaged, assaulted. *indisposition*. Disinclination. *want*. Lack. *correspondence*. Sympathetic response. *edifying*. Moral improvement, instruction. *two witnesses*. Matt. 18: 16 (an injunction to obtain an adequate number of witnesses before laying any accusation of wrongdoing). '*Pariter . . . canebat*'. Virgil, *Aeneid*, iv. 190: the description of *Fama* (Rumour), who 'sang alike of fact and falsehood'. Cf. *Promus*, no. 1082, and the fragmentary Essay 'Of Fame', above, p. 455. *taxations*. Censure. *coast*. Side, source. '*Magnes . . . credulitas*'. Untraced. *if any be*. If there is anyone. '*hath . . . things*'. Rev. 2: 20. '*lost . . . love*'. Rev. 2: 4. '*be . . . cold*'. Rev. 3: 15 f. *foully*. Disgracefully. *sit well*. Hold office. '*they . . . remain*'. Rev. 2: 5. V. Luciani, 'Bacon and Machiavelli', *Italica*, 24 (1947), 26–40, suggests an allusion to *Discourses*, iii. 1, which argues that 'mixed bodies, such as republics or religious sects', should periodically make those 'beneficial changes that bring them back to their original principles', and so renew them by purging the inevitable corruptions that happen 'in the process of time'. '*et . . . Domini*'. 1 Sam. 2: 17. '*that . . . thought*'. Eccles. 10: 20. *conceit*. Conception. '*Increpet te Dominus*'. Jude 8–9: 'Likewise notwithstanding these dreamers also defile the flesh, and despise government, and speake evill of them that are in authoritie. Yet Michael the Archangel, when he strove against the devill, and disputed about the body of Moyses, durst not blame him with cursed speaking, but sayd, The Lord rebuke thee': Geneva Bible, adding a cross-reference to Zaccharia 3: 2, where 'Christ under the name of the Angel rebuked Satan, as knowing that he went about to hinder the Church: but here we are admonished not to seeke to revenge ourselves by evill speaking, but to referre the thing to God'. *he . . . dealbate*. Spedding records the reading of BL Harl. MS 3795: 'in saying *Percutiet te Dominus*, the word will strike thee, he offended not, yet in saying *paries dealbate* . . .'. The words 'offended not' are a marginal insertion, and Spedding conjectures that they may have dropped out of the MS, causing a later copyist to insert the words 'he did justly denounce the judgment of God', to complete the sense (*Works*, viii. 81). '*Percutiet te Dominus*'. Acts 23: 2, where Paul rebukes the high priest Ananias: 'God shall smite thee, thou whited wall'. '*paries dealbate*'. Ibid. ('whited wall'). *retracted*. Ibid. 4–5: 'And they that stood by said, Revilest thou God's high priest? Then said Paul, I wist not, brethren, that he was the high priest'. '*ipsum . . . expavit*'. Untraced. *synods*. Councils of bishops or representatives of several churches convened for the purpose of regulating doctrine or discipline. *story*. History. *deprived*. Deposed. *Ham*. See Gen. 9: 20–5, the drunkenness of Noah, where 'Ham . . . saw the nakedness of his father, and told his two brethren', Shem and Japheth, who covered 'their father's nakedness' with their faces averted, so avoiding his curse. Bacon may also be thinking of that late medieval work of popular ethics, Dame Juliana Berners, *The Boke of Saint Albans*, frequently reprinted between 1486 and 1610, for 'the tradition which divided mankind into the descendants of Japheth, ennobled by the paternal blessing conferred on their ancestor by

Noah; and the descendants of Ham, rendered ignoble by Noah's curse' (M. James, *English Politics and the Concept of Honour 1485–1642*, *Past and Present*, suppl. 3 (Oxford, 1978), 3).

8. *calling*. Christian vocation. *presently*. immediately. *cease*. Put a stop to, cancel. '*Sacerdotes . . . sumus*'. Gregorius I, *Epistolae*, in Migne, *Patrologia Latina*, lxxvii, col. 745a. '*Nisi . . . potes*'. Anon., *De Dignitate Sacerdotali*, ibid. xvii, col. 527a. *make doubt of*. Call in question. *humour*. Temperament. *wanteth*. Lacks. '*the salutation . . . master*'. Matt. 23: 7. *compliment*. Deferential praise. '*to seek . . . lips*'. Untraced in this form; but the connection of 'knowledge' with 'lips' is frequent in the Old Testament: cf. e.g. Job 33: 3; Prov. 5: 2, 14: 7, 15: 7, 20: 15. '*Diotrephes . . . pre-eminence*'. See 3 John 9, denouncing Diotrephes, the leader of an opposition party within this community, for his ambition and inhospitable behaviour in refusing to receive the apostles. *light upon*. Descend on. *adhere*. Attach themselves, as followers. '*quorum . . . obsequio*'. Untraced. '*Pauci . . . magistrorum*'. Untraced. *wreathed*. Interwoven. *accidental*. Incidental. *sobriety*. Soundness of judgement. *These*. The following. *generalities*. General rules. *continent*. Container. *derived*. Conveyed. *side*. Take sides. '*they . . . left*'. Matt. 6: 3. '*transeunt . . . praejudicium*'. Untraced. *take . . . judgment*. Reach a sensible position. *in their way*. Along the way. '*inter . . . corrumpitur*'. Untraced. *forestalled*. Influenced, preoccupied.

9. *reformation*. This religious-political movement, to reform corruptions in Catholicism and return to the purity of the primitive Church, found its first articulation in Martin Luther's three treatises of 1520. *discipline*. The system by which the practices of a church (as opposed to its doctrines) are regulated (*SOED*). *foreward*. Most prominent place. *touched*. Mentioned. *except*. Without. *infer*. Draw the conclusion. *proceed . . . sense*. Derive from purely human concerns, not spiritual. '*Cum . . . estis?*' 1 Cor. 3: 3. *compounding*. Settling a dispute by mutual concession or compromise. *savour*. Have the characteristics of. '*Non . . . pravum*'. Jas. 3: 14–16: 'But if ye have bitter envying and strife in your hearts, glory not, and lie not against the truth. This wisdom descendeth not from above, but is earthly, sensual, devilish. For where envying and strife is, there is confusion and every evil work.' '*Procedere . . . permutationem*'. Untraced. *Arius*. A priest and theologian in Alexandria (*c*.250–*c*.336), responsible for the heresy of Arianism, which denied the true divinity of Jesus Christ, holding that He was not eternal but created by the Father from nothing, as an instrument for the creation of the world. *Gentilism*. Heathen belief or practice. *acknowledgment*. Recognition. *Sabellius*. An early 3rd-century theologian of Roman origin, an exponent of Monarchianism, which attempted to safeguard the unity ('monarchy') of the Godhead but failed to do justice to the independence of the Son, reducing Him to one of a succession of modes or operations of the Deity (*Oxford Dictionary of the Christian Church*). *dissimilitude*. Dissimilarity, diversity. *distinction of persons*. The separately distinguished persons constituting the Trinity: Father, Son, and Holy Ghost. See John 14: 11–17, 26. *several*. Separate. *offices*. Duties, tasks. *dispensations*.

Ordering, management. *scale*. Standard of measurement. *bounds*. Limit, extent. *last*. Most recently. '*posthumi . . . filii*'. Untraced. *amortized*. Deadened, destroyed. *apprehension*. Perceiving, understanding. *touchstone*. Criterion. *try*. Test. *opposite*. Opposed. *removed*. Distant. *participateth . . . appearance*. Shares any similarity or resemblance. *subtle*. Deceitful. *conceit*. Idea. *entertain*. Harbour, receive.

10. *but*. Except. *notorious condemnation*. Unidentified. *rebaptising of children*. Perhaps alluding to the Anabaptists, or 're-baptisers' (literally translated). This radical wing of the Reformation rejected infant baptism in favour of the baptism of adult believers. *much like reason*. Similar import. *re-ordaining of priests*. Some Reformers argued that any priest who had been ordained under the Catholic rule of Queen Mary ought to be re-ordained according to Protestant ceremony. *resolutely maintained*. Seriously upheld. *meet*. Fitting. *abused*. Deceived. *sobriety*. Seriousness. *purged*. Eliminated. *bowels*. Innermost parts. *partial*. Partisan. *affectation*. Earnest pursuit. *conversant in*. Living, passing the time with. '*Consentiamus . . . est*'. Untraced. '*non . . . proximum*'. Untraced. *to plant*. To be founded. *civil*. Secular. *policy*. Political system. *tied*. Forced. '*Qui . . . rebus*'. Untraced. '*Consule . . . Dei*'. Untraced. *admit*. Accept as true. *form*. Organization. *though*. Even if. *parity . . . of ministers*. The Reformers' belief that all priests ('ministers') should have the same status, abandoning any hierarchical system from archbishops and bishops downward (which was the organization of the early Christian Church, preserved in both Roman Catholicism and the Church of England). *confusion*. Disorder, chaos. *ordinary*. Usual, normal. *synods*. Assemblies of the clergy within a particular area convened to discuss church affairs. In the Church of England synods were summoned to deal with specific issues, but in the Presbyterian system the synod (later, General Assembly) was the regular organ of church government. '*numbered . . . weighed*'. Counted on a democratic voting basis, rather than evaluated according to prestige and power.

11. '*Equidem . . . potius*'. Untraced. *abate*. Put an end to. *mischief*. Misfor tune, calamity. *use*. Usefulness. *patriarchs*. High-ranking bishops in one of the ancient sees (e.g. Jerusalem, Rome), in the Eastern Churches, and in Catholicism. *fruits*. Good results, products. *wanton*. Undisciplined, unmanageable. *derogate*. Detract from, disparage. *greatest*. Biggest. *censures . . . churches*. Bancroft's St Paul's Cross sermon of 1589 had given much offence in other countries, especially Scotland. *Platonist*. Probably Philo of Alexandria (de Mas, i. 170). '*Certe . . . occultiora*'. 'Surely the vices of the irascible part of the soul are of a worse kind than those of the concupiscible part, even though more hidden.' *appeared well*. Was exemplified, confirmed. *contentions*. Disputes, quarrels. '*as the waters*'. Prov. 17: 14: 'The beginning of strife is as when one letteth out water: therefore leave off contention, before it be meddled with.' *breach*. Opening, breakthrough. *their part*. Those advocating reform of the Church. *dumb ministers*. Priests who neither preached regularly nor performed other offices. *benefices*. Ecclesiastical livings; properties held by a rector or other church officer. *idle and* *monastical*. One of the Reformers' main claims against Catholicism was that monks, friars, and nuns living in closed religious communities did no work from which society benefited. *continuance ... universities*. The existence of permanent fellowships in Oxford and Cambridge, which included residential rights. *livings ... upon*. Accommodation and stipends to live on. *except*. Object.

12. *delivered*. Rescued. *advanced*. Gone so far as. *define of*. Delineate, specify. *only*. Single, sole. *policy*. Political structure; form of government. *erected*. Established. *magistrate*. A civil officer administering the law (appointed by the state, not by the church). *stay*. Stop, complete their proposals. *footing*. Firm standing place; balance. *descend further*. Proceed with additional proposals. *at their peril*. Taking the risk of punishment for disregarding an injunction or prohibition. *attending of*. Waiting for. *communicate*. Receive Holy Communion. *progression*. Successive positions taken up. *fall to*. Eagerly adopt, take up. *highest strain*. Most extreme stage or place on a scale. *part*. Party. *tenor*. Continuous, consistent course. *opposed*. Confronted. *good times*. The early era. *later superstitious times*. The Middle Ages. *tares*. Weeds. *taught*. Matt. 13: 29–30. *supplant*. Uproot. *After*. Afterwards. *absolute*. Categorical, unconditional. *stiffly*. Obstinately. *needed not*. Was unnecessary. *exasperate*. Irritated. *fallen to*. Reached the point of. *as of*. As if it were. *sect*. Religious faction regarded as heretical. *derogative*. Derogatory. *ministers*. Clergymen. *so as*. So that. *said before*. See above, p. 2. *admonish*. Urge. *alone*. Single, unique form of. *near*. Close (in attitudes and beliefs). *hard*. Obdurate, difficult.

13. *discipline*. System regulating church practices. *want*. Lack. *withal*. Besides, moreover. *staying for*. Waiting for. *demand*. Ask for. *receive*. Accept. *supper*. Holy Communion. *draw*. Withdraw. *draw*. Bring. *mislike*. Dislike. '*Fortasse ... intelligunt*'. 'Perhaps they do not return since they do not realize how much progress they have made'; untraced. *indifferent*. Impartial. *precisely*. Strictly. '*Leges ... acescunt*'. Untraced. '*Qui ... perseverat*'. Untraced. *supplanteth*. Overthrows. *orders*. Coherent structures. '*Morosa ... novitas*'. Untraced. *husbandman*. Farmer. *proyning*. Pruning. *stirring*. Activating thc soil. *lightly*. Easily, readily. '*offers ... parliament*'. Proposals for administrative changes (presumably to check corruptions and abuses). *pertaineth*. Belongs. *acceptation*. Approval. *constitutions*. Administrative organizations. *excommunication*. An ecclesiastical punishment excluding one from receiving Communion or other sacraments. (The ecclesiastical courts that passed this sentence also levied their costs from the person accused.) Bacon repeated this denunciation more forcefully in *Certain Considerations touching the better Pacification and Edification of the Church of England* (1604: see head-note above), describing it as 'the greatest judgment upon the earth, being that which is ratified in heaven, and being a precursory or prelusory judgment to the judgment of Christ in the end of the world'. It was doubly outrageous that bishops should use this severe punishment for trivial offences, and as a way of increasing church income: 'how can it be without derogation to God's honour, and making the power of the keys

contemptible?' (*Works*, x. 121–2). *base process.* Unworthy procedure (legal proceedings bringing a person to court for litigation). *lackey.* Behave in a servile fashion. *duties and fees.* Income for the church. *greatest judgment.* Most severe punishment. *latter day.* The judgement day, at the world's end. *nurse.* Foster, nurture. *yield . . . universities.* Number of qualified theology graduates produced each year. *serve.* Be adequate. *adventureth.* Undertakes, ventures. *prophesying.* Expounding Scripture in meetings outside Church ceremonies. This practice had been forbidden by royal decree in 1562, 'against the advice and opinion of one of the greatest and gravest prelates of this land', as Bacon recorded in *Certain Considerations.* He then recommended that this 'good exercise' be reintroduced, in which 'the ministers within a precinct did meet upon a week-day in some principal town, where there was some ancient grave minister that was president, and an auditory admitted of gentlemen, or other persons of leisure; then every minister successively, beginning with the youngest, did handle one and the same piece of Scripture, spending severally [each] some quarter of an hour or better, and in the whole some two hours; and so the exercise being begun and concluded with a prayer, and the president giving a text for the next meeting, the assembly was dissolved. And this was as I take it a fortnight's exercise; which in mine opinion was the best way to frame and train up preachers to handle the word of God as it ought to be handled, that hath been practised' (*Works*, x. 119–20). Bacon recommends its revival with two additions: first, that the public meeting be followed by a private one in which the ministers involved might 'mutually use such advice, instruction, comfort, or encouragement' as occasion might provide, the elders helping the younger; secondly, that this exercise be 'used in the universities for young divines before they presumed to preach'. If it were feared that it might encourage controversies, that could easily be remedied with a strict prohibition, enforced by the 'grave person president or moderator' (p. 120). *heat of contentions.* Bitterness of controversy. *popular auditory.* An audience of common people. *contained.* Restrained. *urge.* Present earnestly; discuss. *in controversy.* Controversial.

14. *note.* Take note of. *move.* Arouse. *touch.* Affect, concern. '*Injuriae . . . sunt*'. Untraced. *are possessed of.* Hold, maintain. *other.* Other party (the Reformers). *dissembled.* Concealed. *charged.* Accused. *they . . . Caesar.* Matt. 22: 2. *sorted.* Placed in a class. *coupled.* Linked, identified. *Family of love.* A sect founded by Henry Nicholas (*c.*1502–*c.*1580), a native of Münster in Westphalia, which propagated a vague philanthropism, pantheistic and antinomian (holding that the moral law is not binding on Christians). In 1580 Queen Elizabeth issued a proclamation ordering their books to be burned and their members imprisoned. *descry.* Denounce, disparage. *but.* Only. *strait.* Strict. *Swearing men.* Making men swear an oath, *ex officio mero*, which effectively made them incriminate themselves. See head-note above, and for Bacon's repeated denunciation of it in *Certain Considerations*, 114. *blanks.* Documents with blank spaces, to be filled in later. *generalities.* Vague general points (here, incriminating). *compass.* Extent. *certain.* Clearly defined. *captious.* Apt or intended to deceive;

misleading. *strainable*. Burdensome, restrictive. *subscription*. Signing, assenting to. *articles*. Statements of doctrine or belief. '*lacessere ... ecclesiae*'. 'Provoke and irritate the diseases of the church'; untraced. *spend*. Use up. *exercise*. Exert, wear out. The *Resuscitatio* reads 'exercise'; the Bodleian and BL Add. MS have 'crise'; the Harl. MS 'waste'. '*Non ... exigit*'. Untraced. *which*. Who. *exacteth*. Demands and enforces. *inconformity*. Nonconformity. *conscience*. Scruple. *subscribe*. Sign, agree to. *note*. Record, reputation. *defection*. Falling away from faith; or, desertion of one's party. *their*. The Church of England Establishment, with its campaign against dissenting ministers. *easy*. Quick, ready. *silencing*. Removing the right to preach or hold services. *in*. Given. *suppose*. Imagine. *captiously*. Carpingly, raising frequent objections. *watched*. Awaited, expected. *hardly*. With difficulty. *enforced*. Put into practice.

15. *gift*. Ability sent from heaven. *molestations*. Hostile interferences. *style*. Legal title, correct designation. *book ... prayer*. The Book of Common Prayer, the Church of England's official service book (containing the daily offices of Morning and Evening Prayer, the forms for administration of the sacraments and other public and private rites, the psalter and the ordinal), was first published in 1549, issued in a revised form in 1552, repealed by Mary in 1553, but reissued in 1559, with a further revision in 1604. '*tollatur ... certamen*'. 'Let law be removed, and there shall be a contest'; untraced. *either's reasons*. The claims of both parties. *like*. Similar. *have but by rumour*. Only know of through rumours. *comfort*. Encouragement, support. '*restrained ... remiss*'. Untraced. '*Better ... lawful*'. Untraced. *contemned*. Despised, disregarded. *pressed*. Compressed, forced. '*Ira ... Dei*'. Jas. 1: 20. *part*. Party, side. *eager*. Pungent, acid. *fond*. Foolish. *uncivil*. Indecorous, rude. *chargeth*. Burdens with guilt, blame. *persuaded*. Convinced. *mercenary bands*. Groups hired for wages. *to have*. In order to obtain. *spoil*. Plunder; property acquired by confiscation. *hardly*. Harshly. *intelligence*. Agreement, plot. *incendiaries*. Arsonists. *fire*. Set fire to. *rifle*. Rob, ransack. *affect*. Profess. *cognizances*. Distinctive devices, marks. *differences*. Characterizing features. *correspond*. Conform. '*tam ... schismatica*'. Untraced. *impropered*. Appropriated.

16. *zealous*. Among the Reformers 'zeal' (fervent devotion) was a highly prized quality, duly mocked by their opponents: cf. Zeal-of-the-land-Busy, the Puritan parodied in Ben Jonson's *Bartholomew Fair* (1614). *cold*. Lacking in zeal or enthusiasm. *minglers ... profane*. Cf. Ezek. 22: 26, 'and put no differences between holy and profane'. *civil and moral*. One qualified only in secular matters. *denominate*. Name, characterize. *the second table*. The second of two divisions of the Ten Commandments, relating to moral duties (the first, to religious duties). *counterfeited*. Feigned. *St. John ... seen*. 1 John 4: 20. *St. James ... widow*. Jas. 1: 27. *philosophical*. Deriving from pagan philosophy (as opposed to theology). *affection*. Feeling (as opposed to reason). *challenge*. Lay claim to. *knowledge*. Intellectual gifts. *King Edward's*. Edward VI (1537–53), son of Henry VIII, who succeeded to the throne in 1547; his reign strengthened the Calvinist direction of the Church

of England. *beginning... reign.* Queen Elizabeth I (1533–1603), daughter of Henry VIII, conformed outwardly to Catholicism during the reign of Mary (1553–8), but on succeeding to the throne in 1558 supported a moderate form of Protestantism which gave the Church of England increased power and stability. *did... daybreak.* Brought about some enlightenment. *Council of Nicaea.* The first Oecumenical Council, summoned by Constantine in 325, was held at Nicaea (now Iznik) in Bithynia. *censure.* Find fault with. *politiques.* Men governed not by principles but by expediency, purely political considerations. *scholastical.* In the manner of medieval philosophers (schoolmen). *ordering.* Arranging, disposing. *distinctly.* Clearly. *drawing it down.* Applying it. *direction.* Guidance, instruction for life. *authorizing.* Citing appropriate authorities. *warrants.* Conclusive proofs. *becoming.* Suitable to. *St. Paul... wisdom.* 1 Cor. 2: 4. *exhort.* Urge people to praiseworthy conduct. *compunction.* Pricking the conscience; remorse. '*Viri... agemus*'. Acts 2: 37, the Apostles' puzzled reaction to Peter's speech: 'Men and brethren, what shall we do?' In *Promus*, Bacon had noted a passage just before this, Acts 2: 29: 'Viri fratres liceat audenter discere ad vos' ('Men and brethren, let me freely speak unto you'). *resolve.* Settle, answer. '*obiter*'. Incidentally. *as before.* As if addressing. *manners.* Human behaviour. '*bread of life*'. John 6: 35.

17. *toss... break.* Punning on the 'breaking' or expounding of a biblical text by dividing it into its component parts for separate elucidation. *directions.* Instructions. '*ad casus conscientiae*'. 'To cases of conscience'. *warranted.* Justified. *what through.* Either because of. *grounded.* Firmly established. *compendious.* Saving time and labour. *clear.* Elucidate. *asketh.* Requires. *conversation in.* Knowledge of. *carry... equal hand.* Behave impartially. *show of.* Seems to be endorsed by. '*sins... left*'. Untraced in this form, but cf. Deut. 17: 11, 'thou shalt not decline from the things which they shall show thee, neither to the right hand nor to the left'. *bind... bound.* Cf. Matt. 18: 15. '*the... double-edged*'. *Heb.* 4: 12, 'For the word of God is quick, and powerful, and sharper than any two-edged sword...'. *turn back.* Reject. *difference.* Distinction. *midwives... blessed.* Exod. 1: 20, on the Hebrew midwives who disobeyed Pharaoh's command to kill all the male children of Israel, and were therefore blessed by God. *Rahab... spies.* Josh. 2: 1–22, on Rahab, who disobeyed the King of Jericho's order to deliver two spies sent by Joshua to spy out the land. In Heb. 11: 31 St Paul praises her faith, which was justly rewarded. *judgment.* 1 Kgs. 3: 16–28. *Saviour... Emmaus.* The village in which Christ made His resurrection appearance to two of the disciples; see Luke 24: 13–35. *made.* Pretended. *exercise.* Disciplinary suffering. *conceits.* Thoughts, concepts. *Parson's* '*Resolution*'. Robert Parsons (1546–1610), Jesuit missionary, educated at Oxford, who was (with Edmond Campion) leader of the Jesuit Mission to England. His many publications included *The first booke of the christian exercise, appertayning to resolution* (Rouen, 1582), reprinted in various amended forms and commonly known as his *Resolution.* *filial.* Obedient. *entitle.* Qualify, give the right to. *milk... meat.* Heb. 5: 12–14. *confounded.* Confused. '*that... controversies*'. 2 Tim. 2: 23; Titus 3: 9.

18. *inconvenience*. Mischief, disturbance. *express Scripture*. Explicit authority. *embasing*. Discrediting. *conceited*. Fanciful. *mine into*. Undermining. *extremity*. Extreme position. *word... preaching*. The 'word' of God refers to the whole of the Scriptures, both as divine testimony and as texts to be read. Some Reformers, however, took the 'word' as referring primarily to the preacher's authority or inspiration, manifested in sermons. *precedent*. Preceding (to have a sermon preached before Holy Communion). *annihilated*. Cancelled. *liturgies*. Forms of public worship in church. *of the principal*. Principally. *make no question*. Never discuss. *in chair*. Lat. *ex cathedra*: anyone who takes upon himself the authority to address the congregation. *entrench*. Strengthen. '*a tempting of God*'. Exod. 17: 2, 7; Deut. 6: 16; Ps. 78: 18; Mal. 3: 15; Matt. 4: 7; Jas. 1: 13. '*quod bonum... probate*'. 1 Thess. 5: 21: 'Prove all things, hold fast that which is good'. Also noted in *Promus*, no. 253. *assuage*. Moderate, mitigate. *hardness of heart*. In Christian theology the sign of being impervious to grace and charity. Cf. Deut. 15: 7; Mark 16: 14. *rest*. Is vested in, depends on. '*that... love*'. Untraced. *abuse*. Imposture; perversion.

19. *antics*. Clowns, grotesques. *pasquils*. Lampooners. *character*. Role, part. *levels at*. Attacks on. *meet*. Suitable. '*Qui... calumniatur*'. 'He who discusses matters with unqualified people will not be understood but abused'; untraced. *press*. Printing press. *would*. Should. *promotion*. Desire for personal advancement; or, publicizing of a party or issue. *cartels*. Written challenges, defiances. *Cross*. St Paul's Cross, at the NE angle of the old St Paul's Cathedral, was a pulpit at which many public sermons were preached, and government proclamations exhibited. *good temper*. Balanced temperament. *beat upon*. Strive for, insist on. *surseance*. Cessation (of hostilities). *Solon's law*. Plutarch, 'Life' of Solon, 20. In *Promus*, no. 1445, Bacon noted: 'Solon's law that in states every man should declare himself of one faction. Neutralitye'. (No. 1312 reads 'Neutrality'.) *particular*. Individual. *fond*. Foolishly credulous. *calumny*. Malicious misrepresentation. '*neuters ... side*'. Untraced. *simplicity*. Honesty; lacking guile. *insinuation*. Covert manipulation. *grateful*. Welcome. *embarked in*. Committed to, fixed in. *partiality*. Factiousness.

LETTER TO LORD BURGHLEY

William Cecil, Lord Burghley (1520–98) was Queen Elizabeth's Principal Secretary from *1558* to *1571*, from 1572 Lord Treasurer and principal minister. He was Bacon's uncle (on his mother's side), and one of the influential figures (including Sir Francis Walsingham and the second Earl of Bedford, his godfather) from whom Bacon hoped to receive office (see Bacon's letters to Burghley in 1580: *Works*, viii. 12–15). Burghley was instrumental in securing Bacon a seat at Gatton for the 1584 Parliament, although in the event Bacon preferred to sit for Weymouth and Melcombe Regis (a place secured by Bedford). Burghley subsequently secured for Bacon the reversion of the Clerkship of the Star Chamber in 1589 (a post that only materialized in 1608), and was also responsible for his election for Middlesex in the 1593 Parliament. Ironically enough, Bacon's opposition to the Queen's tax demands during this session ruined his public career for the rest of her reign (M. A. Phillips, 'Francis Bacon', in P. W. Hasler (ed.), *The House of Commons* 1558–1603 (London, 1981), i. 374–9).

Bacon was at this point in his career a recognized authority on the law (having been a Bencher of Gray's Inn for four years, a Reader for two, and a member of a legal committee advising the government), but with no intellectual occupation in view. This letter is a manifesto of intent, announcing his commitment to the reformation of natural philosophy from its traditional state as a verbal science to one productive of works, a task which he managed to sustain in the intervals of his public career over the next thirty years. But neither this nor his other requests to noblemen and rulers brought him the hoped-for position from which he could both perform and promote scientific research.

This (undated) letter was first published in the *Resuscitatio*, Supplement (1657, p. 95). My text is from Spedding, *Works*, viii. 108–9, who conjecturally assigns it to 1592.

20. *correspondence*. Relation. *wax*. Grow. *confirmed*. Settled, stable. *painful*. Strenuous. *middle place*. Middling position. *discharge*. Carry out. *Sol*. The sun. *business*. A civic career, in the *vita activa* (public life). *contemplative planet*. Saturn, traditionally linked with the *vita contemplativa*, a life of study (either religious or secular). *self-love*. Selfishness. *Atlas*. According to Greek myth, the god who supported the earth on his shoulders. *estate*. Burghley had helped Bacon's career both at Gray's Inn and in Parliament. *move*. Impel. *my health . . . to get*. I am gifted neither at spending money (by high living) nor earning it (by a profession). *all knowledge for my province*. This apparently grandiose aim corresponds precisely to the goal that Cicero laid down for the ancient orator: 'the art of speaking well, that is to say, of speaking with knowledge, skill and elegance, has no delimited territory, within whose borders it is enclosed and confined. All things whatsoever, that can fall under the discussion of human beings, must be aptly dealt with by him who professes to have this power, or he must abandon the name of eloquent' (*De oratore*, II. i. 5–ii. 6). *purge*. Clear. *rovers*. Distracting, confusing influences. *frivolous*. Trifling, worthless. *disputations*. Controversies in logic: purely verbal exchanges. See Bacon's attack on scholasticism, *Adv.L.*, above, pp. 140–2. *blind*. Not guided by theory or planning. *auricular*. Passed on by hearsay. *impostures*. Pretence, deception. Cf. Bacon's later attacks on alchemy in *Of Tribute*, above, p. 35; *Adv.L.*, above, pp. 143, 201–2 ('auricular traditions . . . impostures'); *Filum Labyrinthi* (*Works*, iii. 497); and *Novum Organum*, i. 85: 'traditiones et auriculares susurros' (*Works*, iv. 84). *spoils*. Damages. *industrious*. Painstaking (based on first-hand experience). *profitable*. Beneficial to mankind in general. *curiosity*. Curiousness. *philanthropia*. Love of mankind (Bacon prefers the Greek word to the less specific Latin equivalent, *humanitas*). *place*. Position in society. *countenance*. Standing, reputation. *wits*. Minds, intellects.

21. *affect*. Desire. *encounter*. Casualness, lack of direction. *any*. Any one. *concurrent*. A competitor. *Anaxagoras*. Presocratic philosopher (*c*.500–428

BC). *voluntary poverty*. According to Diogenes Laertius, ii. 6. *lease*. Land or building. *revenue*. Income. *give over*. Abandon. *care of service*. Concern to serve in the public life. *book-maker*. Compiler of books (disparaging). *pioner*. Miner, one who digs to clear the way for others. *he*. Democritus, to whom Diogenes Laertius (ix. 72) attributes the saying that 'Of a truth we know nothing, for truth is in a well'. For other uses of this quotation see *Promus*, no. 1395 ('Pyonner in the myne of truth'), *Works*, vii. 162, iii. 351, and iv. 343. *rather . . . words*. Ideas, not mere verbal elaboration. *reservation*. Keeping things back, secrecy. *retaining*. Withholding. *Gray's Inn*. Bacon kept rooms at his Inn of Court for many years.

OF TRIBUTE; OR, GIVING THAT WHICH IS DUE

In the first volume (1861) of his *Letters and Life of Francis Bacon* James Spedding included a fragmentary device, consisting of two speeches, 'Mr Bacon in prayse of Knowledge', and 'Mr Bacons Discourse in the praise of his Soueraigne', which he found in the British Museum Library in a composite volume of Bacon's papers copied by ones of Bacon's own secretaries (*Works*, viii. 119–21; Beal BcF 320). A more complete MS, containing the whole device (with a framing plot and two additional speeches), was discovered in the library of Northumberland House in 1867, with the title, 'M^r^ffra: Bacon of tribute or giuing that wch is due' (Beal BcF 319). Spedding published this text as *A Conference of Pleasure, composed for some festive occasion about the year 1592 by Francis Bacon* (London, 1870), although the MS had been badly damaged by fire, with portions of the text missing on every page. Fortunately, a copy of the complete entertainment, with corrections in another hand, headed 'Tribuit or Giving that wch is due' (BcF 320), came into the possession of Peter Beal in the 1970s, and is now in the Kodama Memorial Library of Meisei University, Tokyo. This text forms the basis of my edition.

Spedding recorded that the copy he first printed, in 1861, was undated (the MS volume also contains some of the *Essays* and *Colours of Good and Evil*, printed in 1597), but he assigned it to the year 1592 on the reasonable grounds that the final speech, in praise of Queen Elizabeth, contained many parallels with Bacon's widely circulated tract written to a government commission, *Certain Observations on a Libel published this present year 1592* (*Works*, viii. 143–208; Beal BcF 135–52). But Spedding's suggestions as to the context in which the work was first produced were entirely speculative. Reasoning from the known fact that Bacon was involved in planning and writing a device to be presented by Essex for the celebrations of the Queen's accession day, 17 November 1595 (see below), he suggested that *Of Tribute* was 'drawn up for some similar performance in the year 1592 (*Works*, viii. 119). The only evidence he could cite in support of his suggestion was a letter by N. Fant to Anthony Bacon, on 20 November 1592 (Lambeth Palace MSS 648, 176), recording that the day was 'more solemnized than ever, and *through my Lord of Essex his device*: who, contrary to all the Lords' expectation, came in the morning to the presence, and so to her Majesty's presence, in his collar of Esses, a thing unwonted and unlooked for, and yet hereupon suddenly taken up and used with great liking and contentment of her Highness' (*Works*, viii. 120). Spedding took the phrase he italicized to refer to a 'device' in the sense of a literary composition. However, I think that 'device' here refers, rather, to Essex's costume, his wearing a 'collar of Esses', that is, 'a chain consisting of a series S's' (*SOED*), evidently punning on his own name. (This interpretation is supported by G. B. Harrison, *The Life and Death of Robert Devereux Earl of Essex* (London, 1937), 74). I find no evidence that Bacon's device was presented at court, or anywhere else, on this day or later. The calendar of court performances in E. K. Chambers, *The Elizabethan Stage*, includes for 17 November 1592 only the entry 'Challenge for Shrovetide tilt' (iv. 107), and Chambers does not list *Of Tribute* among the pieces by Bacon known to have been performed in public (iii. 211–14). Perhaps it represents Bacon's first attempt to express the hopes and ambitions he had outlined in his letter to Burghley (above, p. 20), demonstrating his capacities in both the active and contemplative spheres.

In the device four speakers (designated *A*, *B*, *C*, and *D*) meet to argue their case, the first praising 'the worthiest virtue' (Fortitude), the second 'the worthiest affection' (Love), the third pleading for 'the worthiest power' (Knowledge), the fourth celebrating 'the worthiest person' (Queen Elizabeth). The first three speeches are similar in content and style to the speeches Bacon wrote for other devices in the 1590s, but the final speech is much longer and more carefully composed, being in effect a defence of the Queen's policies and government. In annotating it I am indebted to Spedding, and to John Guy's *Tudor England* (Oxford, 1988, 1990).

As for the work's structure, K. A. Hovey has suggested that Bacon modelled several of these early devices 'on the tale of the three young men from the apocryphal book of 1 Esdras, still used "for example of life, and instruction of manners" by the Church of England' ('Bacon's Parabolic Drama: Iconoclastic Philosophy and Elizabethan Politics', in W. A. Sessions (ed.), *Francis Bacon's Legacy of Texts* (New York, 1990), 217–18). In the Geneva Bible we read of the feast of Darius (1 Esd. 3: 1–3) and the competition that 'three yong men of the guard, keepers of the kings bodie', proposed: 'Let every one of us speake a sentence [opinion, verdict], & hee that shall overcome, and whose sentence shall appeare wiser than the others, Darius the King shal give him great gifts.' So each wrote his opinion, and put it under the sleeping King's pillow. 'One wrote, The wine is strongest. The other wrote, The King is strongest. The other wrote, Women are strongest, but trueth overcometh all things' (4–12). The King then summoned his council, and each one spoke in support of his verdict, the first on wine (14–24), the second on the King (4: 1–12), and finally the third speaker—the only one whose name is recorded, Zorobabel (Zerubbabel in the Authorized Version), who first praised women (13–32). But then he rejected all previous arguments by celebrating truth, namely God, who has created the universe: 'Is not he great that maketh these things? therefore the trueth is greater and stronger than all. . . . With her there is no receiving of persons nor difference: but she doeth the things which are just, and abstaineth from unjust and wicked things, and all men favour her works . . . she is the strength and the kingdome, and the power, and Majestie of all ages. Blessed be the God of trueth' (33–40). The people acclaim his speech, the King pronounces him the victor, and offers him whatever he should desire. The young man urges Darius to fulfil his vow to rebuild Solomon's temple in Jerusalem, and the King agrees, giving him the commission to supervise the work (41–63).

Whether or not Bacon was inspired by that happy tale, he was certainly familiar with the forms of disputation and debate so important in Renaissance education, especially at the Inns of Court, and as a lawyer already entered on his career he shows his professional skill in organizing arguments, and refuting them. In this, as in the later devices, he is able to give each speaker a solid argument expressed with appropriate eloquence, but then to make each fresh speaker destroy his predecessor's arguments, until the orator in the favoured final position routs everyone. This accelerating structure is proto-dramatic, and has some interesting parallels with *Love's Labour's Lost*, written at much the same time and (in part, at least) out of a similar rhetorical background. Of the individual speeches, the first, praising Fortitude, deals with a major virtue in classical and Renaissance ethics, and could be profitably read alongside that great model, Cicero's *Tusculan Disputations*. (Bacon repeated some of its arguments in the first letter of his *Advice to the Earl of Rutland* in 1596, above, pp. 69 ff.) The third speech, in praise of Knowledge, displays in embryo many of the ideas which were to inspire Bacon's project to renew natural philosophy, so that it could produce works for the benefit of mankind. Some of its ideas were developed in the speech of the Second Counsellor in the *Gray's Inn Revels* (above, pp. 54–5), but not otherwise mentioned by Bacon until the posthumously published *New Atlantis* (above, pp. 480 ff.). This juxtaposition of the beginning and end of Bacon's life as a writer shows all too clearly the fatal price he paid for having too many interests, lacking the financial independence or institutional support needed to pursue his intellectual bent.

22. *motion*. Suggestion, proposal. *satires*. There was indeed a great vogue for satire in the 1590s, the most scurrilous of which (in the quarrel between Nashe and Harvey) were burned by official command. *virtues*. Several of the virtues discussed here figure in the collection of *Antitheta rerum* which, some thirty years later, Bacon inserted in bk. VI of *De Augmentis Scientiarum* (1623) as an example of a '*Promptuary* or Preparatory Store' of commonplaces that speakers or writers should have in readiness. Bacon arranged these 'acute and concise sentences' on two sides of each topic, giving a *pro* and a *contra* (*Works*, iv. 472). I quote an example of this structure in my head-note to the *Essays* of 1625: below, pp. 713–14. The virtues treated in this *promptuarium* are *Justitia* (no. 20), *Fortitudo* (no. 21), *Temperantia* (no. 22), *Constantia* (no. 23), and *Magnanimitas* (no. 24), followed by *Scientia* (no. 25) (*Works*, iv. 480–2). *good innocent things*. Cf. 'Temperance' (Against): 'I like those virtues which include excellence of action, not dullness of passion' (*Works*, iv. 481). *effect*. Efficacy. *intend*. Look for. impression. Assault, impact. *attach*. Grasp. *grounds*. Fundamental principles. *digested*. Assimilated, comprehended.

confounded. Confused. *prints*. Images, impressions. *Berecynthia's horn*. The horn that was played in honour of Cybele, the oriental mother-goddess, producing disorder and panic: cf. Horace, *Odes* I. xviii. 13 f., and III. ix. 18, and Virgil, *Aeneid*, vi. 785, ix. 82 and 619. *Orpheus's music*. The symbol of harmony and order. *Sibylla's leaves*. Movement of the leaves in Sibylla's cave was the sign that the prophetic *flatus* of Apollo had descended: cf. Virgil, *Aeneid*, vi. 9–10, 74 ff. In the case of the Cumaean Sibyl, according to Varro, the prophetic utterances were inscribed on palm-leaves (*OCD*). *spirits*. The 'vital' or animating forces in the bloodstream, according to Renaissance medicine.

23. *fronteth*. Confronts, faces boldly. *preventing*. Anticipating. *well-collected*. Systematic. *to seek*. Missing. *ravished*. Forcefully seized. *entertaineth*. Deals with. *motions*. Movements. *He looketh . . . remedy*. Cf. 'Justice' (For): 'He that looks steadily at dangers that he may meet them, sees also how he may avoid them' (*Works*, iv. 481). *concluded*. Decided. *dismantled*. Unfortified, unprotected. *appeal*. Accuse, impeach. *balances*. Weighing scales. *every . . . due*. Alluding to the ancient definition of justice in the *Corpus iuris civilis*, 'Iustitia est constans et perpetua voluntas ius suum cuique tribuendi' (*Institutiones*, i. 1). Cf. *Promus*, no. 71: 'Si suum cuique tribuendum est certe et venia humanitati'.

24. *devotion*. Bidding. *divide*. Separate. *expect*. Await. *affect*. Profess; pretend. *"O, but . . . fear to want"*. I have added inverted commas, since this seems to be an interlocutor's imagined response to the speaker's accusation. *vain-glory*. Boasting. *equal tenor of mind*. Cf. *Advice to Rutland*: 'The Stoics were of this opinion that there was no way to attain this even temper of mind but to be senseless; and so they sold their goods to ransom themselves from their evils' (above, p. 70). *I will not fear to want*. Cf. *Adv.L.*, above, p. 251: 'For can it be doubted that there are some who take more pleasure in enjoying pleasures than some other, and yet nevertheless are less troubled with the loss or leaving of them. So as this same "Non uti ut non appetas, Non appetere ut non metuas, sunt animi pusilli et diffendis".' Cf. also 'Temperance' (Against): 'To abstain from the use of a thing that you may not feel the want of it, to shun the want that you may not feel the loss of it, are precautions of pusillanimity and cowardice' (*Works*, iv. 481). *circumstances*. Conditions, states of mind. *afore*. In front, ahead of. *destitution*. Forsaking. *disfortune*. Misfortune. *against the hurt . . . fire*. 'Temperance without Fortitude might teach a man to do without pleasure, but not to encounter pain; and was therefore a provision against heat of sunshine, which warms and comforts, but not against heat of fire, which burns and hurts' (Spedding, *Conference of Pleasure*, 32 n.). *impassible*. Incapable of suffering injury or damage. *with want*. Without fulfilment. *use pleasures . . . spare them*. Cf. *Adv.L.*, above, p. 248: 'So as Diogenes' opinion is to be accepted, who commended not them which abstained, but them which sustained, and could refrain their mind *in praecipitio . . .*'. *contain*. Restrain. *entry*. Access, beginnings. *being limited*. Provided that it is controlled. Cf. *Adv.L.*, above, p. 265: 'As for pleasure, we have likewise determined that the mind ought not to be reduced to stupid,

but to retain pleasure; confined rather in the subject of it than in the strength and vigour of it'. *mar*. My conjectural emendation; the MSS read 'make'. *signal*. Distinguishing mark. *bait*. Allurement. *affection*. Desire.

25. *solid . . . desires*. Cf. 'Fortitude' (For): 'There is nothing either solid in pleasure, or secure in virtue, where fear intrudes' (*Works*, iv. 481). *lightly*. Carelessly, without proper thought. *when*. My conjectural emendation: the MSS read 'Then'. *proverb*. Taken from Theocritus, *Idylls* xii. 2, 'they that yearn grow old in one day' (and cf. Homer, *Odyssey*, xix. 360). Already a proverb in the Middle Ages; found in Erasmus, *Adagia*, III. iii. 86 (no. 2286) as 'uno die consenescere': noted in this form in *Promus*, no. 882. *in a maze*. Bewildered. *surfeit*. Sickness brought on by excessive indulgence. *printeth*. Presses down. *native*. Natural. *rack*. Instrument of torture. *amazement*. Overwhelming fear or apprehension. *season*. Enhance, improve. *taint*. Corrupt, contaminate. *crooked . . . straight*. Cf. Isa. 40: 4, 'Every valley shall be exalted . . . and the crooked shall be made straight . . .'. *comely*. Attractive. *compound*. Settle (by mutual concession or compromise). *civil dissensions*. Disputes within a country. *enableth*. Strengthens. *extern*. External; corresponding to the 'foreign enemies' of the previous sentence.

26. *with*. Confronted by. *conditeth*. Flavours (pleasantly). *stupefaction*. Insensibility, due to extreme pain. *parts*. Bodily parts. *weakness . . . spirits*. In not being able to transmit pain beyond a certain point. *apprehension*. Anticipation (to be in readiness). *custom*. Habitual practice (in enduring). *conscience*. Awareness of having acted rightly. *appetite*. Desire to satisfy natural needs (without extravagance). *baseness of means*. Poverty. *no feeling . . . not being*. This being the 'remedy' for the last of the external ills, 'loss of life', the speaker admits there is no remedy. *not*. My conjectural emendation; the Kodama MS reads 'no'. Cf. *Adv.L.*, above p. 285: 'being without well-being is a curse.' *those . . . nature*. These are the natural remedies, before God's grace intervenes, according to Christian belief. *sergeants*. Officers carrying out legal judgments, arresting offenders, or summoning people to appear in court. *Nothing . . . fear itself*. Cf. *Advice to Rutland*, above, p. 71: 'It teaches us . . . that pain and danger be great only by opinion, and that in truth nothing is fearful but fear itself'; and 'Fortitude' (For): 'Nothing is to be feared except fear itself' (*Works*, iv. 481). *entertained*. Dealt with. *the Stoics*. For Bacon's critique of the Stoic doctrine of rigid emotional self-discipline see *Adv.L.*, above, pp. 239, 259. *hold*. Possess, embody. *meditation of death . . . preparation*. Cf. *Adv.L.*, above, p. 251 on the Stoics' doctrines being 'more fearful and cautionary than the nature of things requireth. So have they increased the fear of death in offering to cure it. . . . it is a terrible enemy against whom there is no end of preparing'; also Essay 2, 'Of Death', above, p. 343–4. *to die . . . died*. Cf. *Advice to Rutland*, above, pp. 71: 'That he which dies nobly doth live for ever, and he that lives in fear doth die continually'. *voluptuous sect*. The Epicureans, whose goals were pleasure, the avoidance of pain, and freedom from anxiety. *conceit and custom*. Thinking and behaving. *Julius Caesar*. For extended accounts of Caesar cf. *Adv.L.*, above, pp. 161 ff. and Bacon's *Imago Civilis Julii Caesaris*

(*Works*, vi. 335–8, 341–5). *effectual.* Effective. *sounding.* Melodious, echoing. *continuate speech.* A formal oration, uninterrupted by urgent business. "*Quirites*". Citizens (in their civil capacity). Cf. Suetonius, 'Life' of Julius Caesar, 70, and *Adv.L.*, above, p. 162. *imprint.* Impress (according to the rhetorical ideal of *movere*).

27. *death.* Bacon draws on Suetonius, 'Life' of Julius Caesar, 82, and Plutarch, 'Life' of Caesar, 66. '*Traitor... thou?*' Cf. Plutarch, 'Life', 66.5: 'the smitten man [cried out] in Latin: "Accursed Casca, what doest thou?".' *decent.* Right, proper. *was nothing astonished.* Retained his presence of mind. *compellation.* Addressing someone by name. *expostulate.* Complain, protest. *want.* Lack. *Descend... thyself.* Reflect, consider. *impoyning.* The reading of the Alnwick MS, which Spedding defended as a version of the French *empoigner*, to grasp. It is supported by the historical sources. In Suetonius' version, after Casca stabbed him, 'Caesar grasped Casca's arm and ran it through with his stylus; he was leaping away when another dagger caught him in the breast' (82). Plutarch describes how Casca gave Caesar the first blow, but not 'a deep one, for which he was too much confused... so that Caesar turned about, grasped the knife, and held it fast' ('Life', 66.4). In Amyot's French translation, the passage reads: 'Casca luy donna par derriere un coup d'espee au long du col, mais le coup ne fut pas grand ny mortel, pource que s'estant troublé, comme il est vray semblable, a l'entree d'une si hardie & si perilleuse entreprise, il n'eut pas la force ny l'asseurance de l'assener au vif. Cesar se retournant aussi tost vers luy, empoigna son espee, qu'il tint bien ferme' (*Les Vies des hommes illustres grecs et romains, comparees l'une avec l'autre par Plutarque de Chaeronee, translatees de Grec en Francois par Messire Iaques Amyot* (Paris, 1583), p. 894, para. 28). *form.* Way of behaving. *causeless.* Spontaneously. *At... infans.* 'But envious fortune and the child of Latona have destroyed me'; untraced, but a regular hexameter. *ill spirit... tent.* Cf. the apparition of Caesar's ghost to Brutus with the ominous message, 'You will see me again at Philippi' (Plutarch, 'Life' of Brutus, 36.3). *covered.* His toga covering his body. Suetonius records: 'Confronted by a ring of drawn daggers, [Caesar] drew the top of his gown over his face, and at the same time ungirded the lower part, letting it fall to his feet so that he would die with both legs decently covered' (82). Plutarch reports the tradition that when Caesar 'saw that Brutus had drawn his dagger, he pulled his toga over his head and sank' (66.6).

28. *compliment.* Mark of courtesy, consideration. *honour of person.* Chaste conduct of the body. *Augustus.* Bacon quotes his last words, and those of Vespasian and Septimius Severus, again in Essay 2, 'Of Death', above, p. 343. *had.* Would have. Cf. Suetonius, 'Life' of Augustus, 79, on his 'serene expression' and 'tranquil face'. *euthanasia.* A gentle and easy death. Suetonius reports that Augustus 'kissed his wife with "Goodbye, Livia: never forget whose husband you have been!" and died almost at once. He must have longed for such an easy exit, for whenever he had heard of anyone having passed away quickly and painlessly, he used to pray: "May Heaven grant the same *euthanasia* to me and mine!"' (ibid. 99). *dainty.* Fastidious.

"Livia . . . farewell". Ibid. *dicacity.* Raillery, banter. *"An . . . god".* Suetonius, 'Life' of Vespasian, 23. *scoffing.* Mocking. *dispatch.* Promptitude, efficiency. *"An' . . . do".* Dio Cassius, lxxvi. 17. *like.* Same. *"It . . . Jupiter".* Plato, *Apology*, 42A. But cf. David Chytraeus, *A Soveraigne Salve for a Sick Soule. A Treatise Teaching the use of patient bearing* . . . tr. W. F. (London, 1590) Sig. D3^{v}: Socrates 'said unto the judges: It is time for me to go away now, that I may die, and for you, that you may live. And whether of the two is better, the immortal gods do know, but I think no man knoweth.' *whether.* Which. *left.* Abandoned. *senator.* According to Bacon's source (Seneca, *de Tranquillitate Animi*, xiv. 4), Julius Canus (not Canius, as in some Renaissance editions), sentenced to death by Caligula after a quarrel, displayed great equanimity. 'Nor up to the very end did he cease to search for truth and to make his own death a subject for debate'. To his philosophy teacher he 'said: "I have determined to watch whether the spirit will be conscious that is leaving the body when that fleetest of moments comes"'. Cf. also Montaigne, *Essais* II. vi, 'De l'exercitation', Pléiade edn., pp. 350–1. *standeth.* Is. *try a conclusion.* Make an experiment. *steel.* Arm, strengthen. *strict.* Severe. *extenuate.* Diminish, disparage. *masterpiece.* The most important feature. *coat of honour.* Mark of distinction. *Other . . . fortune.* Cf. 'Fortitude' (For): 'Other virtues free us from the domination of Vice, Fortitude only from the domination of fortune' (*Works*, iv. 481).

29. *mediocrities.* Middle points between opposites (in the *Nicomachean Ethics* Aristotle defined virtue as the mid-point between excess and deficiency). A contextual sense of 'moderators' is also possible. *mill.* Grind-stone; 'a machine which performs its work by rotary motion, *esp.* a lapidary's machine for cutting, polishing, or engraving gems' (*SOED*). *set.* Applied to. *race.* 'A small vein of spar or ore' (*SOED*); a flaw in a diamond. *a grain.* I prefer this reading (Alnwick) to 'again' (Kodama), the parallel with 'race' suggesting the accepted senses of either 'a small particle' or 'roughness of surface', both being types of imperfection in a diamond. Cf. *Adv.L.*, above, p. 251: 'And therefore men are to imitate the wisdom of jewellers; who, if there be a grain, a cloud, or an ice, which may be ground forth without taking too much of the stone, they help it'; and *New Atlantis*, above, p. 484: 'We also have glasses and means to see small and minute bodies perfectly and distinctly: as . . . grains and flaws in gems, which cannot other wise be seen . . .'. *paragon.* Of surpassing excellence (from the Italian *paragone*, 'touchstone, comparison'. In early 17th-century English a *paragon* was also 'a perfect diamond': *SOED*). *carats.* Units of weight used for precious stones, equivalent to 200 milligrams. *They be the affections.* It is the passions (as distinct from the virtues). *make . . . heroical.* Cf. 'Magnanimity' (For): 'Virtues induced by habit or by precepts are ordinary; those imposed by a virtuous end are heroical' (*Works*, iv. 482). *fascinate and bind.* Terms describing the power attributed to magic, by which one person could dominate another. *sleight.* Manual dexterity; trickery. *practice.* Deception. *fire blown.* Blast furnace. *It is motion, therefore.* This is the reading of the Alnwick MS;

Kodama reads 'It is the motion thereof', which seems to me too limited a sense, given the generalizations about 'motion' in the rest of this passage. *pray in aid of*. Beg the assistance of. *height of mind*. Elevated spirits; magnanimity. *invert*. Alter to the opposite. *[were]*. My conjectural insertion; but the sentence remains difficult. *discovereth*. Perceives. *As... nature*. The Kodama MS reads 'As for the sufferings of nature'. I have inserted from Alnwick the words 'other affections, they be but'. *sufferings*. Passive states, lacking the power of positive action.

30. *thraldom*. Servitude. *supplying*. Relief. *good... true good*. Alluding to the classification of values and goals in classical philosophy: a good absolute, relative, comparative, etc. Cf. *Adv.L.*, above, p. 245. *out of taste*. Disordered, irritable. *want*. Lack. *the season... spring of love*. Cf. Essay 10, 'Of Love', above, p. 358: 'This passion hath his floods in the very time of weakness; which are, great prosperity and great adversity.... Both which times kindle love, and make it more fervent, and therefore show it to be the child of folly.' *steep... plain*. Cf. the allegory of 'Hercules at the crossroads', where Virtue points out to him a stony path leading up a steep hill, but crowned with the temple of honour, while Vice recommends a pleasant path downhill, leading to ruin: Xenophon, *Memorabilia*, II. i. 20–34, and *Adv.L.*, above, p. 147. *hire*. Reward. *commended... fortune*. The closing argument of the previous speech. *ministereth*. Provides. *contemplation*. Matter for thought. *Hercules... two*. Cf. Plato, *Phaedo*, 89C and the Latin proverb, 'Ne Hercules quidem contra duos', 'Not even Hercules was able to fight two opponents'. See e.g. Erasmus' *Adagia*, I. v. 39 (*Collected Works of Erasmus*, xxxi, *Adages*, tr. M. M. Phillips (Toronto, 1982), 419). *ill... beginning*. Virtue must begin by overcoming evil.

31. *exquisite*. Intense, highly susceptible. *simple*. Single. *confluence... assembly*. The text seems incoherent at this point, but no obvious remedy exists. *Pandora*. 'All gifts': the first woman, according to Greek myth, who had a jar containing all evils, which she released: at the bottom remained hope (Hesiod, *Theogony*, 570 ff.; *Works and Days*, 50 ff.). Cf. *Religious Meditations*, above, p. 93. *incident*. Relevant, pertaining to. *Curiosity... delight*. Cf. Aristotle, *Met.* i. 1: 'All men by nature desire to know. An indication of this is the delight we take in our senses' (980a22). *contemplative*. Those who have dedicated their lives to study. *tributary*. Subsidiary. *peregrinations*. Journeys, pilgrimages. *concourses*. Flocking together. *unwonted actions*. Unusual events. *woodbine*. Climbing plant, e.g. hedge bindweed. *grace*. The quality of pleasing. *a fountain*. Spedding's conjectural emendation; the Kodama MS has 'not a field'. *ground*. The background for a design; (*music*) the melody on which a descant is raised. *changes*. Variations in design; (*music*) modulations. *and of the strongest*. The Kodama MS reading; Spedding conjectured 'a journey of strangest', which gives an easier but less distinctive sense. *excitations*. Arousals of excitement; stimuli. *redintegrations*. Renewals, restorations. *attires*. Dress. *quaverings*. Trills, tremolos. *stop*. Note. *Dionysius... pedantius*. Dionysius II (c.397–335 BC), son of Dionysius I, the tyrant of Syracuse, whom Plato tried to transform into the model ruler of a philosophic state.

Deprived of power in 345, he 'became a schoolmaster at Corinth; so complete was his inability to do without the right to rule' (Cicero, *Tusc. D.* III. xii. 27). *would.* Wanted to. *ambitiosa fames.* Hunger for glory. *hardliest.* With greatest difficulty. *aurum potabile.* Potable gold; a drink that alchemists recommended, being supposed to increase longevity. *commandment.* Command. *person ... will.* Body, mind. *popular.* Where the people have power. *voluntary.* Unforced, unconstrained. *heresiarchae.* Leaders of sects, sometimes heretical. *summum bonum.* The highest good: a topic much debated in classical philosophy. *violatores mentium.* Untraced.

32. *compassible.* Able to live or work together; compatible. *disrelish.* Spoil. My conjectural emendation: the Kodama MS reads 'disrelease'; Alnwick reads 'disrelli[ ]', but is then damaged. *that ... his misfortune.* Who is about to suffer. *perpetuating of himself.* Cf. *Adv.L.*, above, pp. 167–8: 'let us conclude with the dignity and excellence of knowledge and learning in that whereunto man's nature doth most aspire; which is immortality or continuance; for to this tendeth generation, and raising of houses and families; to this buildings, foundations, and monuments; to this tendeth the desire of memory, fame, and celebration; and in effect, the strength of all human desires'. *posterity.* The descendants collectively of a person. *carry.* Wield power, influence. *affecting same.* Aiming at the same end (immortality). *[of].* My conjectural insertion: the text seems to lack something here. *high.* Noble. *comfortable.* Comforting, encouraging. *in grace.* In favour. "*archflatterer*". Cf. Essay 10, 'Of Love', above, p. 358: 'whereas it hath been well said that the arch-flatterer, with whom all the petty flatterers have intelligence, is a man's self; certainly the lover is more.' The phrase comes from Plutarch's essay, 'How to tell a flatterer from a friend' (*Moralia*, 48E–F). *concurrence.* Co-operation, agreement. *remission.* A lessening of tension. *languishing and weariness.* Cf. *Adv.L.*, above, p. 249, on man's natural preference for 'Active Good': 'so as it was well said, *Vita sine proposito languida et vaga est.*' *Antaeus.* According to Greek myth, a giant, the son of Poseidon and Earth, who compelled all comers to wrestle with him, and when thrown regained fresh strength from contact with his mother the Earth. *Penelope's web.* The weaving that Penelope, Odysseus' wife, awaiting his return from the Trojan war, carried out during the day, but unpicked at night, on the completion of which she had promised to marry one of her suitors. *entertaineth.* Occupies.

33. *Not ... catcheth.* Not that it does not also affect. *Appius and Virginia.* Appius Claudius was a judge and one of the decemvirs (body of ten men administering Rome and its laws) whose lust for Virginia caused her murder and a people's revolt in 449 BC; cf. Livy, iii. 33. *want ... think.* Have nothing else to think about. *Antony and Cleopatra.* Traditional examples of sensual indulgence. Shakespeare shows them in a much nobler light. *fruit of idleness.* Moralists traditionally associated love with idleness or *otium*, as in Ovid's *Remedia amoris*, 136 ff.: 'Fac monitis fugias otia prima meis ... Otia si tollas, periere Cupidinis arcus' ('No leisure—that's rule | Number one ... Eliminate leisure, and Cupid's bow is broken': tr. P. Green). *the eye ... most affecting*

sense. Cf. Aristotle, *Met*. I. i, and many other sources. *Confer*. Discuss. *meditating*. This word is supplied by the Alnwick MS; Kodama has a gap at this point. *proportionable*. Well-proportioned. *and*. From Alnwick; lacking in Kodama. *decent*. Handsome, comely. *glass*. Mirror. *[the]*. My conjectural insertion. *burning glass*. A lens or concave mirror which concentrates ['gathereth'] the rays of the sun to make a combustible object burn. *Non... error*. 'Amor is not the god, as people claim, but the lover himself, and error'; untraced. 'Amorosus' is post-classical. *patent*. A sign of entitlement or possession; licence. *amare et sapere*. 'Amare et sapere vix Deo conceditur': 'To love and to be wise is hardly granted to the gods' (Publilius Syrus, *Sententiae*, xv): cf. Essay 10, 'Of Love', above, p. 358. *pardoned of*. Excused from making.

34. *accident*. Not essential. *a double... is*. Spedding cites Job 11: 5–6, 'But oh that God would speak, and open his lips against thee; And that he would shew thee the secrets of wisdom, that they are double to that which is!', and comments: 'Knowledge is the reflected image of the thing itself. The mind is only the reflector' (*Conference of Pleasure*, 38). *truth... one*. Cf. *Adv.L.*, above, p. 142. *Are [not] the pleasures of the affections ... satiety?* Cf. *Adv.L.*, above, p. 167: 'Again, for the pleasure and delight of knowledge and learning, it far surpasseth all other in nature: for shall the pleasures of the affections so exceed the senses, as much as the obtaining of desire or victory exceedeth a song or dinner; and must not of consequence the pleasures of the intellect or understanding exceed the pleasures of the affections? We see in all other pleasures there is satiety... But of knowledge there is no satiety, but satisfaction and appetite are perpetually interchangeable.' [*not*]. My conjectural insertion. *perturbations*. Disorders. *imaginations*. Fantasies, illusions (cf. *Adv.L.*, above, p. 138). *estimations*. Judgements. *prospect*. View (cf. *Adv.L.*, below, p. 167). *nature's warehouse*. Cf. *Adv.L.*, above, p. 147. *commodities*. All the goods and inventions that can be produced by properly organized scientific endeavour. *verify*. Assert to be true; support with testimony. *artificers*. Craftsmen. *learned*. Scholars (disputations in Latin were the regular form of university exercises and examinations). *descant upon*. Comment, expatiate. *knit*. Group, relate to. *courses*. Sequences (e.g. of cause and effect). *reduce*. Refer, draw together. *stand up against*. Form an exception or contradiction to their theory. *web*. Cobweb, weaving.

35. *work*. Produce; result in 'works', tangible improvements in knowledge and welfare. *contemplation*. Study. *Grecians*. Bacon thought that reverence for antiquity was a hindrance to new research. *alchemists*. Bacon consistently attacked alchemy as a trial-and-error art, lacking any coherent theory, and subject to both confusion and deception. *in words*. A purely verbal knowledge, unproductive. *auditories*. Assemblies of hearers. "*you Grecians... children*". Said by the Egyptian priest to Solon in Plato's *Timaeus*, 22B: a favourite quotation of Bacon's when polemicizing against too great a reverence for antiquity. Cf. *Promus*, no. 342, and *Adv.L.*, above, p. 150. *imposture*. Deceitful pretence. *auricular*. hearsay. *obscurity*. Alchemy cultivated recondite language to preserve its secrecy. *catching hold of*. Latching

on to, making use of. "*populus vult decipi*" (*ergo decipiatur*). 'The public wishes to be fooled, therefore let it be fooled' (ascribed to Cardinal Caraffa). Noted in *Promus*, no. 344. *vulgar*. Commonplace. *furnace*. Alchemy used heat to effect chemical change. *Aristotle*. Although Bacon acknowledged Aristotle's eminence by the frequency with which he used his work, for polemical purposes he presented Aristotle as symbolizing received and unquestioned knowledge. *eternity . . . heavens*. Cf. *De caelo*, i. 10–12, 279b4–283b22. *be*. Are. *incursions*. Invasions, inroads. *spirit*. In Renaissance natural philosophy *spiritus*, midway between a gas and a liquid, was often conceived as the source of life in both animate and inanimate bodies. *mass*. Solidity. *astronomers*. In Renaissance natural philosophy astronomy was regarded as a highly speculative discipline, dependent on non-verifiable hypotheses, inferior to physics, which could examine matter directly and therefore attain certain knowledge. See Peter Urbach, *Francis Bacon's Philosophy of Science* (La Salle, Ill., 1987), 30–2, 125–34. *car-men*. Charioteers (a mocking reference to the Copernican theory, which made the sun centre of the universe, the earth and other planets revolving around it. Bacon was unconvinced). *double motion*. Bacon seems to be pointing out that apparent angular movement is not the same as true motion in space: because of the earth's daily rotation, the heavens appear to move from east to west; because of its annual rotation, seen against the background of the fixed stars, they appear to have a contrary motion. *abatement*. Reduction. *overgo*. Outrun, move faster than. *interrupted*. Disturbed.

36. *conference of pleasure*. Debate made for diversion, not for academic or business purposes. *seasonable*. Appropriate to the occasion. *this*. Aristotle. *great reputed*. Having a great reputation. *Time . . . discover truth*. Alluding to the old proverb, *veritas filia temporis*, 'truth is the daughter of time'. Cf. *Promus*, no. 341: 'So gyve authors their due as you gyve tyme his due which is to discover truth.' *wits*. Intellects. *universities . . . learn nothing . . . but to believe*. Cf. *Promus*, no. 339: 'In academiis discunt credere' ('In the schools men learn to believe'). *facility*. Easiness (often used by Bacon to describe superficial and uncritical thinking). *temerity to assever*. Rashness, recklessness in making ungrounded assertions. *glory to know*. Seeking knowledge for self-aggrandizement. *end to gain*. Seeking profit, not knowledge. *Seeking things in words*. Confusing the basic distinction between *res* and *verba*. *resting . . . nature*. Prematurely suspending their inquiries in natural philosophy. Cf. *Adv.L.*, above, pp. 139, 198. *match*. Union, as in marriage. For this metaphor, so potent in Bacon's later writings, see *Works*, iv. 19, 27. *vain*. Pointless, futile. *issue*. Progeny, outcome. *gross*. Palpable, striking. *needle*. The compass, used for navigation. *these three*. These inventions were often coupled together as signs of the modern age's vigour. *reserved*. Concealed. *intelligencers*. Informants. *thrall*. Enslaved, captive. *intercepted*. Prevented, hindered. *put*. Forced, made.

37. *magnanimity*. A masculine, heroical virtue, unusual in a woman. *pieces . . . unset*. Gems not fixed in a setting that shows them to advantage. *set*. When set. *meriting*. Deserving merit. *adverse fortune*. The years (1553–8)

in which Elizabeth endured the rule of Mary. *extenuateth.* Lessens, weakens. *alteration of religion.* A 'Device for the Alteration of Religion' was written by an unknown but important court official before Christmas 1558, recommending a return as soon as possible to Protestantism as formulated by King Edward VI (Guy, *Tudor England*, 259). Elizabeth called a parliament in Jan. 1559, which passed two bills, one an 'Act of Supremacy', confirming the Queen as 'Supreme Governor' of the Anglican Church, the other an 'Act of Uniformity' prescribing the Book of Common Prayer as the set form for church worship. *redoubted.* Feared. *adverse.* Opposition. *confirmed.* Firmly established. *reduce.* Bring back, restore. *correspondence.* Relations. *exterminated.* Abolished. *absolute.* Totally, without exception. *wars... neighbours.* At Elizabeth's accession England had not yet negotiated peace treaties with France and Scotland. These were arranged in 1559. *league... one.* Philip II's Spain. *amity... hostility.* Relations between Spain and England deteriorated from 1567 onwards, when Philip II supported the Duke of Alva's campaign in the Netherlands, and open war broke out in 1587 following the execution of Mary Queen of Scots in February, and English intervention in the Netherlands in the summer of that year. *pretence.* Pretension (to rule).

38. *conspiracy.* The Ridolfi plot, 1571–2, in which Robert Ridolfi, a Florentine banker resident in London, acted for the King of Spain and other Catholic enemies, aided by the Duke of Norfolk, in a plot to dethrone Elizabeth and put Mary Queen of Scots in her place. Norfolk was executed in June 1572, shortly after the Queen granted his petition for the care of his children. *conjured.* Conspired with. *art... of words.* Eloquence. *army.* Philip II had begun planning an Armada in Dec. 1585, with the aim of conquering England and thus regaining the Netherlands. Preparation of ships and men went ahead in 1586, delayed only by Drake's raid on Cadiz in Apr. 1587, and by July 1588 Philip had assembled an assault force consisting of 131 vessels, manned by 7,000 seamen and 17,000 troops (Guy, *Tudor England*, 338–41). *council.* The Privy Council, her main advisers. *nobility.* Noblemen. *communicate.* Share. *presence in camp.* The Queen visited her troops in their camp at Tilbury, Essex, 9–11 Aug. 1588. *contempt of profit... offers.* 'That is, as proof of her despising gain, recall her generosity.' *purchases.* Acquisitions. *appointed.* Equipped, fitted out. *progenitors.* Such as Henri VII and Henry VIII. *minority... princes.* James VI of Scotland, later James I of England, b. 1566, proclaimed king in 1567; Henri III of France, b. 1551, acceded to the throne in 1575. *one.* Philip II.

39. *compass.* Circumference. *United... Countries.* Holland, Belgium, Flanders. *traffic.* Trade. *pretences.* 'Assertion of a right' (*SOED*). *drawn in question.* Cast doubt on. *encounter.* Meeting of adversaries in conflict. *policy.* Political expediency. *jailor.* Elizabeth's espousal of Protestantism aroused the suspicion of Queen Mary and her counsellors, leading to Elizabeth being implicated in Wyatt's rebellion (1554). She was for a time imprisoned in the Tower and at Woodstock. *salt.* Pungent. *balm.* Soothing ointment. *breathed.* Lived.

40. *estimation*. Valuation. *metals*. Precious metals (gold, silver). *uncertain... coins*. In 1560 Elizabeth restored the coinage, which had been debased by the issue of coins with a high percentage of alloy, so that the face value of the coinage was well above its intrinsic value. Base silver coins were exchanged for money of a standard weight and fineness. *cozenages*. Frauds. Base money encouraged counterfeiting on a large scale. *overvalue*. Surplus value. *fine*. Fee paid for any privilege. *forfeitures*. Penalties for offences. *set over*. Transferred. *gain*. Profit. *compositions*. Compromises. *promoters*. 'Persons whose business it was to prosecute or denounce offenders against the law; professional informers' (*SOED*). *snaring*. Repressive. *crossed*. Thwarted. *benevolence... rates*. Parliament had two main functions in this period, to raise taxes and to pass laws. *new forms... accepted*. In Mar. 1587 a committee of the House of Commons invited the Lords 'to join "in a contribution or Benevolence for the Charges of the Low Countries' wars" ', but their offer was not taken up by the Queen (Spedding, *Conference of Pleasure*, 42, citing the journals collected by Sir Simonds D'Ewes; cf. *Works*, viii. 65). *in ure*. Into practice. *levied*. Raised. *alienations*. Actions transferring ownership. *purchases and perquisites*. Here I follow the Harley MS; Kodama has only 'perquisites'. *purchases*. Acquisitions of property, especially land. *perquisites*. Casual profits (in addition to revenue). *innocently levied*. Again I prefer the Harley MS; Kodama has only 'innocently'. *tenderness*. Consideration.

41. *dryness*. Parsimony. *donatives*. Official gifts. *pores*. Passages, ducts, especially in an animal body. *pipes*. Tubes. *cavillation*. Chicanery, trickery. *pensions... council*. As the Spanish King did in the time of James I: see below, p. 697. *brand*. Burning log; torch. *Burgundy*. Still a considerable power in the 15th century, it was swallowed up in the 16th. *benignity*. Kindness, generosity. *magnificence*. Sumptuousness in official entertainments. *allowance*. Sum allotted for expenses. *charge*. Cost. *appear*. Can be scrutinized. *cast*. Calculate. *books*. Account books; records. *fee-farms*. Tenure by which land is held with unconditional inheritance, and at a perpetual fixed rent, without other services. *custodies*. Guardianships. *other*. Different. *subtlety*. Cunning; deceitful cleverness. *humorous*. Whimsical, unpredictable. *Tutus... Faustitas*. Horace, *Odes*, IV. v. 17–18, in praise of Augustus: 'For when he is here, the ox in safety roams the pastures; Ceres and benign Prosperity make rich the crops.'

42. *Condit... suis*. Horace, ibid. 29: 'On his own hillside each man spends the day'. *Augustus... marble*. Suetonius, 'Life' of Augustus, 28.3. *traffic*. Trade. *customs*. Revenue from customs duties. *interruptions*. Disturbances. *lady of the sea*. England's dominance as a sea-power, demonstrated in the Armada's defeat, was widely recognized. *abridgement*. Epitome. *universal monarchy*. Global rule. Cf. Essay 29, 'Of the True Greatness of Kingdoms and Estates', above, p. 402. *merit... neighbours*. The credit she has deserved from neighbouring countries. *tapers*. Torches. *Guise*. French ducal house of Lorraine, staunch leaders of the Catholic party during the 16th-century civil wars. The Guise faction gained power in Paris as well as Edinburgh,

provoking English military intervention in 1559–60. *encounter.* Manner, behaviour. *remiss.* Negligent, lenient. *colour.* Pretext. *French . . . Scotland.* French troops had been used to garrison Scottish fortresses, until the Protestant revolution in the summer of 1559 started a reaction which led to Elizabeth's government sending troops and ships in 1560 to drive out the French. *tutele.* Being under the protection of a guardian. *that realm.* Scotland. *reducing.* Bringing under control. *faction.* The Guise. *intelligence.* Information, understanding. *restored . . . peace.* With the help of English troops, in July 1560. *unnestle.* Evict, displace. *ill.* Evil. *afterwards.* In 1570, during the troubles following the assassination of the Regent Murray.

43. *entermise.* Intervention. *again.* At this point Bacon's *Certain Observations on a Libel* (1592–3), which corresponds exactly to this passage, adds the words 'by the King's best and truest servants': cf. *Works*, viii. 133, 187–8. *possessed . . . kingdom.* Edinburgh Castle was surrendered to the English on 28 May 1573. *cunctation.* Delaying. *cavillation.* Quibbling. *rendered.* Returned. *restored . . . hands.* Bacon underlines Elizabeth's care for the young king, then aged 7. *occasions.* Occurrences. *covered.* Disguised, concealed. *queen mother.* Catherine de' Medici (1519–89), Queen of France, wife of Henri II, who assumed political influence during the minority of her sons, Francis II (1559–60) and Charles IX (1560–3), retaining power until 1588. *civil wars.* The series of religious and political conflicts caused by the growth of Calvinism, factionalism among the nobility, and weak royal government. Philip II supported the Catholic Guise faction, making it inevitable that Elizabeth should aid the Huguenots. *shocking.* Causing to collide violently, and so damage. *waste.* Devastate, ruin. *those . . . religion.* Religious leaders (Huguenots). *their aids.* Spedding suggests reading 'her aid'. *giving . . . security.* On 22 Sept. 1562 Elizabeth made a contract with the Prince of Condé for the delivery of Newhaven, with a loan of 100,000 crowns (*Works*, viii. 188 n.). *young king.* Charles IX. *supplanting.* Dispossessing and taking the place of a person, especially by treacherous means. Spedding cites a passage from Stowe's *Chronicle* recording that 'the young King and the Queen mother' were forced by the Guise faction 'to suffer their name and authority to be abused, even to the killing of the King's own unarmed innocent people, the spoiling of his rich towns, the breaking of his best-advised edicts' (*Conference of Pleasure*, 44 n.). *peace.* On 19 Mar. 1563. *would.* Wished. *mortality . . . God.* The *Observations on a Libel* is more explicit: 'but for the great mortality which it pleased God to send amongst our men' (*Works*, viii. 188). Rendered uninhabitable by the pestilence, Newhaven was surrendered on 28 July 1563. *affect.* Desire. *that party.* The French Protestants, who had appealed to her for help in Sept. 1562 (*Works*, viii. 133 n.). *divisions.* Civil dissensions. *edicts.* Proclamations.

44. *cords.* Pieces of rope or leather. *stock.* Handle. *practices.* Treacheries. *play.* Compete, gamble. *title.* Grounds for a claim to the throne; entitlement. *combination.* Association (for an illegal purpose). Cf. *Adv.L.*, above, p. 286. *in substance.* In reality. *furring in.* Spedding took this to be an English form of the French 'fourrer', to stuff, cram; otherwise suggesting

'farcing' or 'forcing'. *Henry the Third.* King of France 1574–89, who gained victories over the Huguenots in 1569 and took an active share in the Massacre of St Bartholomew in 1572. In Dec. 1588 he arranged the assassination of the Duke of Guise, enraging the Catholic League. Joining forces with the Huguenot Henri of Navarre, he was marching on Paris in July 1589 when he was assassinated by a fanatical priest. *pressing.* Urgent. *despair.* Hopelessness. *her.* Elizabeth's. *correspondence.* Concordant or sympathetic response. *Jacobin.* Dominican. *criminous.* Criminal. *worthy . . . crown.* Henri IV (1553–1610), the first Bourbon King of France (1589–1610). Leader of the Huguenot cause, he unified the country in 1593 by the rather desperate measure of turning Catholic. But he thus became a new threat to Protestant Europe, and Bacon could remind the House of Commons in Nov. 1597 'how the French King revolted from his religion' (*Works*, ix. 87). *observing.* Observant. *enlarged.* Generously helped. *Don Antonio.* One of several claimants to the Portuguese throne, left vacant by the death of King Sebastian. Elizabeth approved of help being offered to free Portugal from Spanish rule during the expedition against Spain under Norris and Drake in 1589, which was also supposed to destroy the remains of the Armada. But their attack on Lisbon was unsuccessful, not provoking the hoped-for Portuguese revolt, and instead they sailed for the Azores on a privateering raid which ended disastrously, due to storms and sickness. *mighty invasion.* In Oct. 1590 the Turks were about to send an army of 60,000 men against the King of Poland, but cancelled the attack when the English ambassador intervened, urging that Poland was a vital source of supplies for Elizabeth's navy in the wars against Spain. The Turks expressed in writing their great admiration of Elizabeth (*Works*, viii. 135 n.). *Grand Signior.* The Turkish ruler. *rest.* Dwell on; emphasize. *recommended.* Commended to her care. *confederacy.* Union, alliance. *intercourse.* Dealings, especially mercantile.

45. *offered.* In 1585. *instance.* Urgency, insistence. *partisans.* Zealous supporters. *proverb.* Diogenianus, iii. 33, and Plutarch, 'On Tranquillity of Mind' (*Moralia*, 471D), 'Aratro iacularis' (noted as such in *Promus*, no. 906): 'You shoot with a plough.—To shoot with a plough, is applied to those who do something heedlessly without foreseeing what will happen later, or who attempt to hurt someone else at great inconvenience to themselves' (*Collected Works of Erasmus*, xxxiv (Toronto, 1991), 23). *consumed . . . revenge.* Used up all his resources in seeking revenge. *beginning . . . troubles.* In Aug. 1567 the Duke of Alva marched into the Netherlands with an army of 8,000 Spanish soldiers and 40,000 German, Italian, and Walloon levies, starting a long-drawn conflict. *impeach.* Hinder, prevent. *distract.* Separate, divide. *Duke of Anjou.* François Hercules, Duke of Alençon and Anjou (brother to King Henri III), who courted Queen Elizabeth 1578–82 and then fought in the Low Countries. *pretended.* Alleged, spurious. *apostolic.* Pertaining to the Pope as successor of St Peter. *see.* Office or jurisdiction of the Pope. *donative . . . Spain.* A benefice, within the King's gift. *cell.* A small monastery dependent on some larger house. *parteth.* Divides, shares. *leave . . . rest.* Leave the others free to take part in electing the Pope. *now . . . two.* The

device of naming seven cardinals, to the exclusion of all others, was first used for the election of Gregory XIV in Dec. 1590, and repeated for his two successors in the following years.

46. *incapable.* Unqualified. *incompatible.* Unacceptable; irreconcilable. *conclave.* An assembly of the cardinals for the election of a pope. *colour.* Appearance sake. *quillets.* Small plots of land. *freehold.* A tenure by which an estate can be held by the owner or disposed of, without restriction. *honour.* Seigniory of several manors under one baron. *lordship.* A lord's domain, estate. *Rodamonts and Roytelets.* Generic names for French farmers. *title.* Claim. *absolute.* Having total power. *martial.* Military. *Aragon.* In Sept. 1592 towns in Aragon prepared themselves against Spanish invasion. *Indies.* The West Indies and those parts of South America under Spanish rule. *sconce.* Shelter, defensive fort, or earthwork. *let.* Prevented. *fired.* Inflamed. *usurpeth.* Takes possession of by usurpation. *in contempt of.* Despising. *fly.* Flee. *severally.* Separately. *roses . . . red.* A double allusion, to the two great dynasties of York (the white rose) and Lancaster (the red), and to the ideal colours of a woman's complexion. *between.* By. Elizabeth was the daughter of Henry VIII and Anne Boleyn.

47. *over-curious.* Fastidious. *umbratile.* Spent indoors, in the shadows. *light.* Trivial. *chastest . . . royalest.* Virgil: his supremacy over all other poets had been affirmed by J. C. Scaliger in his *Poetices libri septem* (1561). Bacon assembles a cento of passages from the *Aeneid*, mostly from the scene in bk. 1 where Aeneas admiringly welcomes his mother, Venus, in the attire of the huntress Diana (a common comparison for this virgin queen). *Et . . . Dea.* 'And in her step she was revealed, a very goddess' (i. 405). *Nec . . . sonat.* 'Nor has thy voice a human ring' (i. 328; Aeneas to Venus). *Et . . . honores.* [Venus shed gifts on her son]: 'and on his eyes a joyous lustre' (i. 591). *Indum . . . ebur.* 'As when one stains Indian ivory with crimson dye . . .' (xii. 67 ff.; describing Lavinia blushing at the mention of Turnus' brave deeds). *Et . . . refulsit.* 'Her roseate neck flashed bright' (i. 402). *Veste . . . fluentes.* 'Her flowing robes gathered in a knot' (i. 320). *Ambrosiaeque . . . Spiravere.* 'From her head her ambrosial tresses breathed celestial fragrance' (i. 403 f.). *oweth.* Owns. *compliment.* Courteous behaviour. *character.* Token, indication. *conceit.* Mind. *give . . . to think.* Make someone thoughtful. *adamant.* Magnet. *received.* Spedding suggests reading 'conceived'. *languages.* Elizabeth spoke Latin, French, Spanish, and Italian.

48. *prejudicing herself.* Giving herself prejudices. *appetite.* Readiness. *intentive.* Attentive, assiduous. *covertures.* Dissimulations, pretexts. *censuring.* Judging, assessing. *occasions.* Opportunities. *so.* Such. *reserve.* Set apart, for special use. *no children.* Cf. *Adv.L.*, above, p. 158. "*Brutorum . . . soboles*". 'The only immortality of brutish men is through their offspring' (since they leave no other record of their existence); untraced. *Revolve.* Consider. *line.* Lineage. *spent.* Died out. *slain . . . was.* Cf. Plutarch, 'Life' of Alexander. *impostumes.* Abscesses, cysts. Cf. Suetonius, 'Life' of Augustus, 65, and Bacon's *Apophthegms New and Old*, no. 246: 'Augustus Caesar, out of great indignation against his two daughters, and Posthumus Agrippa, his grandchild; whereof the first two were infamous, and the last otherwise unworthy, would say "That they were not his seed, but some impostumes that had broken from him"' (*Works*, vii. 159–60).

49. "*Generare... divina*". 'Procreation and children are things human; creation and works are things divine'; untraced. *mate*. Partner in marriage. *Of fortune*. Subject to fortune. *passed*. Endured. *quicker*. More intense. *servants... found*. The counsellors who served Elizabeth from the beginning included Sir William Cecil, later Lord Burghley; Sir Nicholas Bacon; Francis Russell (2nd earl of Bedford); Sir Francis Knollys; and Sir Thomas Parry. *accident*. Henri II was killed by accident at a tournament in July 1559. *reduced*. Brought under control. *commodity*. Convenience, utility. *machination*. Intrigue. *ill-compacted*. Badly unified. *Earls... North*. The Earls of Northumberland and Westmorland, Catholic activists, staged a revolt in 1569, linked to the succession of Mary, Queen of Scots, which failed. Northumberland was executed in 1572, as were some 700 of the 5,700 rebels. *Norfolk's plot*. A conspiracy involving the Duke of Norfolk, who was to be married to the Scottish Queen and to restore Catholicism. *prevented*. Anticipated. *Sebastian... Africa*. King Sebastian (1554–78), a grandson of the emperor Charles V, who fell in battle against the Moors at Alcazar. He left no heir. *Portugal*. Under Spanish domination, 1580–1640. *some*. In his *Observations on a Libel* (*Works*, viii. 186) Bacon identifies this person as 'the Genuese', that is, Jeronimo Conestaggio, author of *Dell' Unione del Regno di Portogallo alla corona di Castiglia* (Genoa, 1585). *artificially*. Cunningly, deceitfully. *nourish*. Encourage. *Cardinal's*. The senior churchman who would not nominate a successor to Sebastian, there being 'six or seven competitors to that crown' (*Observations on a Libel*; *Works*, viii. 171). *for*. In. *invasion*. The Armada, preparations for which stretched from 1586 to 1588. *be*. Had been. *League*. The Catholic League, led by the Guise faction. *confirmed*. Established.

50. *animated*. Filled with boldness. *English victory*. Over the Armada, in July 1588. *precedent*. Previously. *dispensed*. Issued dispensations, releasing Catholics from regular church-going. *through*. Thanks to. *eager*. Impetuous. *declare*. Reveal their true nature. *seminaries*. Roman Catholic colleges training priests for mission work in England in the 16th and 17th centuries. *reconcilements*. Reconciliations; readmissions (to the Catholic Church). [*so*]. My conjectural insertion. *conspiracies*. For brief accounts of the many conspiracies against Elizabeth see e.g. William Camden, *The History of the Most Renowned and Victorious Princess Elizabeth, Late Queen of England*, abridged ed. W. T. MacCaffrey (Chicago, 1970) on the Somerville plot in 1583 (p. 164); the Throckmorton plot in 1584 (pp. 171, 175–6); the Parry plot in 1585 (pp. 186–90); the Babington plot in 1586 (pp. 226–37); and the Stafford plot in 1587 (pp. 278–80). To which should be added, at least, the many intrigues of Alvarez de Quadra, the Spanish ambassador (1559–64); the Ridolfi plot in 1571–2; and others. *instruments*. Devices, tools. *Don Juan*. Don John of Austria, outstanding Spanish general, appointed Viceroy of the Netherlands in 1578, but who died of typhoid fever in Oct. of that year. *D'Aubigny*. The Duke of Lennox, d. 1583. *some*. Spedding suggests a reference to the Prince of Parma, nephew to Philip II and Governor-General of the Netherlands, who was hit by a stray bullet, and died on 6 Dec. 1592; but also, perhaps, to the death of Mary, Queen of Scots (*Works*, viii. 142 n.). *seges gloriae*. 'Seed of fame': Cicero, *Pro Milone*, 35. *persuaded*. Convinced. *Leith*. The siege of Leith in 1560, when England assisted Scotland to expel the French, brought in by the Guise faction. *Newhaven*. See above, p. 43. In fact, that was by no means a triumph. *Lammas day*. Spedding (*Conference of Pleasure*, 48) cites Bacon's *Considerations touching a war with Spain* (1624), describing a battle on Lammas Day (1 Aug.) 1578 in which a Spanish force, led by Don John of Austria and the Prince of Parma, attacked the army of the States General, but was repulsed and finally routed by English and Scottish troops under Colonel John Norris (*Works*, xiv. 483–4). *Ghent*. At Ghent in 1582 Norris and a small English body were vastly outnumbered by Parma and the Spanish forces, but carried out their retreat with fewer losses than the enemy (ibid. 485). *Zutphen*. The action of 22 Sept. 1586, in which Sir Philip Sidney was mortally wounded (having dispensed with some of his armour in a gesture of bravado), where Norris and a small English party, thinking to intercept a small Spanish convoy, suddenly found themselves facing a large Spanish force (Spedding, *Conference of Pleasure*, 50–4). *this summer*. In 1592 the campaign in the Low Countries went so well that many English troops could be sent to Brittany. *Portugal*. The abortive expedition of 1589 (see above). *voyages... Indies*. Sir Francis Drake's expedition of 1577–80 (in which he circumnavigated the globe) pillaged Spanish property for the Crown's benefit. *Invincible Navy*. The Armada, outsailed and outgunned by the English fleet, was finally wrecked on its retreat to Spain in Aug. 1588 by Atlantic gales and damaged anchors, with enormous loss of ships and men. *firing*. Setting fire to. *sheep-cot*. Light building for sheltering sheep.

51. *perambulation*. 'The action or ceremony of walking officially round a parish, forest, etc., for the purpose of asserting and recording its boundaries and preserving the rights of possession' (*SOED*): deeply ironic here. *ennobling*. Enriching (with the spoils of shipwreck). *extenuate*. Diminish. *height*. Dignity. *Orbis... tibi*. 'The world's beloved, a song to fame, the eye of heaven, you are the ornament for those who love you, you are your own ornament!' Untraced; but the half-line 'Tu decus omne tuis' is from Virgil, *Eclogues*, v. 34 ('thou alone givest glory to thy people').

A DEVICE FOR THE GRAY'S INN REVELS

In the Elizabethan period, as today, the four London Inns of Court (Lincoln's Inn, Gray's Inn, the Inner Temple, and the Middle Temple) combined the role of Law Faculties at continental universities (training students to be barristers) with that of governing the legal profession. Collegiate organizations, where students resided for several years, often held feasts and other celebrations, especially during the Christmas vacation. In 1594, their revels having been intermitted for several years, the Benchers (or governing body) resolved to hold 'solemn revels' or a 'grand Christmas' on an elaborate scale. These festivities involved hiring professional minstrels and actors, arranging formal dances (part of the students' training in social deportment), banquets, processions through the city, and other diversions. The students elected a Lord of Misrule to preside over the occasion, who was known at Gray's Inn as 'the Prince of Purpoole' (from the name of the manor, *Portpool*, within which the Inn was established). In 1594–5 the festivities lasted from 20 December to Shrovetide (just before Lent); on 28 December 'a comedy of errors (like to Plautus his *Menaechmus*) was played by the players'. In all probability this was Shakespeare's *Comedy of Errors*, performed by the company to which he belonged, the Lord Chamberlain's Men: this was perhaps the closest contact that Bacon and Shakespeare ever had.

Bacon's contribution to the Revels was a device (similar to *Of Tribute*, three years earlier), presented on 3 January 1595, in front of many great persons, including the Lord Keeper (William Cecil, a member of Gray's Inn), the Lord Treasurer, the Vice-Chamberlain, and several other Privy Counsellors. Cecil made a handsome present to the Prince of Purpoole, on this occasion Henry Helmes, 'a Norfolk-Gentleman, who was thought to be accomplished with all good Parts fit for so great a Dignity, and was also a very proper Man of Personage, and very active in Dancing and Revelling'. After a dumb show celebrating the friendship between Gray's Inn and the Temple, and the proclamation of the articles governing the 'Order of the Helmet', a device was presented in which six speakers argued for their chosen way of life, advising the fictional prince in much the same way as the real counsellors present in the audience might have done their sovereign. The ideas contained in these speeches have many parallels with Bacon's writings on civic and political matters, and the speech on behalf of philosophy is a virtual digest of his lifelong programme for the reformation of scientific research, with some details only reappearing in the posthumously published *New Atlantis* (above, pp. 480–9). Bacon's authorship of this section was not recorded when the *Gesta Grayorum* was first published, in mysterious circumstances, in 1688, but in 1861 Spedding's edition (*Works*, viii. 325–43, with selected variant readings) fully authenticated it. A handy (although textually defective) reprint is *Gesta Grayorum, or the History of the High and Mighty Prince Henry of Purpoole, Anno Domini 1594*, ed. Desmond Bland (Liverpool, 1968). Passages within square brackets are Spedding's conjectural additions to the text.

52. *consort-music*. A small group of musicians, often using both string and wind instruments. *not strangers*. Native, not foreign. *running banquet*. A table bearing light refreshments. *good order*. Well-provided. *advise*. Consult. *end*. Goal, policy. *humours*. Whims. *way*. Direction. *set before us*. Propose. *respect . . . affections*. Consideration of our likes or dislikes. *neither guessing . . . our disposition*. Cf. Essay 48, 'Of Followers and Friends', above, pp. 436–7. *Princes' . . . inscrutable*. Prov. 25: 3. *putting . . . yourselves*. Without considering our situation and requirements. *exercise*. Occupation, activity.

53. *difference*. Disagreement. *man-rife*. Abounding in men. *Indian*. The gold-mines of the 'Indies' (a generic name for America, North and South). *Vulcan*. The blacksmith of the gods. *strait*. Strict, scrupulous. *unrevocable*. Binding. *without*. To the outside world. *regards*. Esteem, repute. *try*. Test. *content*. Satisfaction. *invested*. Inaugurated, assumed rule. *lightness*. Frivolity. *nature... proper places*. An Aristotelian principle, still common in Renaissance natural philosophy. *power*. The original reading; Spedding conjectured 'possession'. *particulars*. Individuals. *respect*. A backward survey, retrospect. *warfare*. Cf. 2 Cor. 10: 3. *affect*. Desire.

54. *rovers*. Pirates, marauders. *destruction... preservation*. Cf. Lucan, *Pharsalia*, x. 20, '*felix terrarum praedo*', 'a fortunate robber of territory' (a judgement ascribed to Julius Caesar, standing before the tomb of Alexander the Great in Alexandria); cf. also *Adv.L.*, above, p. 194. *the governing faculties... divine*. Cf. Cicero, *De officiis*, I. xi. 34: 'For since there are two ways of settling a dispute: first, by discussion; second, by physical force; and since the former is characteristic of man, the latter of the brute, we must resort to force only in case we may not avail ourselves of discussion'; and Machiavelli, *The Prince*, ch. 18: 'there are two ways to fight: by using laws and by using force. The former is characteristic of man; the latter of animals. But frequently the former is inadequate and one must resort to the latter. Consequently a prince must perfect his knowledge of how to use the attributes of both animals and men.' *searching out... hid and secret*. Cf. Prov. 25: 2, 'It is the glory of God to conceal a thing; but the honour of kings is to search out a matter.' Noted in *Promus*, no. 234. *lamp... Eye of the World*. Cf. similar metaphors in *Adv.L.*, above, pp. 123–4. *Magi*. Wise men. *Gymnosophists*. Ancient Hindu ascetics. *Ptolemies*. The Macedonian dynasty ruling in Egypt after Alexander's conquest, 323 BC. *Salomon... herbal*. Solomon as a pioneer in natural history was a constant presence in Bacon's work, who idiosyncratically interpreted a passage in the Bible (1 Kgs. 4: 33) as ascribing to him the compilation of a herbal, or book describing plants and their medicinal properties; cf. Introduction, above, p. xxxviii, and *Adv.L.*, above, p. 151. *seen*. Knowledgeable. *Calendar*. Julius Caesar reformed the existing calendar in 46 BC. *Alexander... Aristotle*. Cf. Plutarch, 'Life' of Alexander, 7 (668B). *four... monuments*. The research resources proposed here are described more fully in *New Atlantis*, above, pp. 480 ff. *perfect*. Complete. *divers moulds*. Earths of varying composition.

55. *set*. Planted. *cherished*. Tended. *cabinet*. Museum. *art*. Skill in handicraft. *engine*. Engineering. *shuffle of things*. The succession of time. *want life*. Inanimate objects. *still-house*. Distillery, chemical laboratory. *philosopher's stone*. The alchemists' goal was to discover this substance, supposedly able to transform base matter into gold. *Trismegistus*. Hermes Trismegistus ('thrice greatest') was a legendary figure thought to be a contemporary of Moses, to whom were attributed various mystico-religious texts actually dating from the 1st to the 3rd century AD. *Foundations*. Founding of institutions (usually of an educational or charitable nature). *one*. The same. *fallacy*. A mistake in logic (here, 'begging the question'). *drawn in question*.

Considered. *conclusions of nature*. Arriving at sure results in natural philosophy. *mystical*. Esoteric. *conceipts*. Fanciful notions. *curiosities*. Whims, fancies. *institutions*. Foundations. *orders*. Communities (e.g. for the old or sick). *ordinances*. Regulations governing a society.

56. *cure*. Compensate for. *Tower of Babel*. During the building of which God punished mankind with 'the confusion of tongues': Gen. 11: 1–9. *Alexandria*. Founded by Alexander immediately after his conquest of Egypt in 332 BC. "*I found . . . marble*". Suetonius, 'Life' of Augustus, 28.3. *wallflower*. '*Parietaria*, wallflower, because his name was on so many walls': see *Adv.L.*, above, p. 156, quoting Aurelius Victor. *Constantinople*. The new city built by the emperor Constantine between AD 325 and 330 on the site of old Byzantium. *landflood*. 'Overflowing of land by water from inland sources' (*OED*). *head*. Source. *restrain*. Limit. *Helmet*. Punning on the name of the presiding Prince of Purpoole on this occasion, Henry Helmes. *Absoluteness*. Independence. *policy*. Government.

57. *contemplations*. Studies. *enforced*. Bound, indebted. *that*. Such that. *spend*. Expend. *intend*. Apply yourself to. *security*. Freedom from care; safety. *retired*. Out of public life. *disused*. Unaccustomed. *Your Excellency*. Spedding suggested reading 'Your Excellency's subjects'. *humour*. Disposition, state of mind. *innovation*. Far-reaching alterations; in the Renaissance, often with the connotation of '(political) revolution'. *Constantine . . . Empire*. Constantine conquered the Eastern provinces in AD 324, extending the Roman Empire, a displacement of power which was one cause of its collapse during the following century. *ruinated*. Ruined. *curiosities*. Whims, toys. *in conceipt*. By the mind or judgement. *communicated*. Shared, connected. *transported*. Carried away, enraptured. *put the case by*. Take the example of. *if . . . one show*. If someone were to show. *born . . . year*. Inherit an annual income. *with charge*. At his own expense. *raise*. Increase. "*Milk . . . away*". In *Promus*, no. 553, Bacon noted: 'Milk the standing Cowe. Why follow you the flying.'

58. *overgrowing*. Gaining too much power. *grandees*. Noblemen. *formalities*. Allegiance, conformity. *derogation*. Depreciation, lowering in value. *absolute prerogatives*. The monarch's unrestricted rights and privileges, which override the law. *Joseph*. Who became the adviser of Pharaoh: Gen. 41: 14 ff. *strain up*. Raise to a higher degree of power and prestige. *plats*. Plans. *severally*. Separately. *invention*. Device, project. *in discretion*. Prudently. *artificially*. Skilfully. *formal*. Merely a matter of form. *meriting*. Deserving well of. *memory*. Fame, remembrance. *influence*. According to astrology, the supposed power of the stars to affect human affairs. *standing*. Stagnant: cf. Essay 21, 'Of Delays', above, p. 383. *repairers*. Those who restore or mend. *diversions*. Entertainments. *balm*. Soothing ointment. (In Renaissance Britain the belief survived that monarchs possessed 'the royal touch', able to cure scrofula: cf. *Macbeth*, IV. iii. 140 ff.).

59. *intelligence*. Employees gathering information; spies. *state . . . laws*. The desire to make English law more systematic and functional was for Bacon a

lifelong preoccupation: cf. *The Maxims of the Law* (*Works*, vii. 309–87), and the 'Example of a Treatise on Universal Justice or the Fountains of Equity' in *De Augmentis Scientiarum* (1623), VIII. iii (*Works*, v. 88–110). *snaring*. Obstructive. *press*. Urge on, enforce. *suits*. Litigations. *exactions*. Demanding more money than is due. *alone*. On its own. *confirmed*. Established, strengthened. *decays*. Decrease. *traffic*. Trade. *cankers*. Cancers, ulcerous diseases. *consumption*. Decay, wasting away. *mysteries*. Handicrafts. *with relation*. By narrative. *persuading*. Where the other counsellors 'advise', this one is more informal, suiting his aim. *in hand*. Negotiating. *would*. Wish to.

60. *were*. Would be. *troubled*. At a loss. *triumphs*. Festivities, processions, pageants. *pilgrimages*. Journeys dedicated to some serious end (usually religious). *progresses*. Ceremonial tours of the kingdom made by the ruler, for recreation. *barge*. Pleasure boat. *falleth out*. Happens. *sullen*. Unsociable. *distinction*. Display of excellence. *Conclusion*. Decision. *persuading*. Persuasive. *determine*. Decide. *what time*. However much time. *pensioners*. Attendants, retainers.

OF LOVE AND SELF-LOVE

The last of the proto-dramatic entertainments that Bacon was involved with in the 1590s was produced on behalf of Essex, and presented to the Queen for the festivities celebrating her accession day, 17 November 1595. The entertainment included an outer plot, with the main character, Erophilus (a clear allegory for Essex), declaring his continuing dedication to the Queen by rejecting Philautia, the goddess of self-love, and an inner, self-contained device in which—as in Bacon's *Of Tribute* and his contribution to the *Gesta Grayorum*—a series of speakers argued for his own mode of life. Spedding recorded the existence, in the Gibson collection in the Lambeth Palace library, of two separate MSS in Bacon's own autograph. One (Lambeth MS 936; Beal BcF 309), which Spedding describes as 'a page in Bacon's own most careless hand', outlines the plot of the inner, self-contained device: 'The persons to be three: one dressed like a Heremite or Philosopher, representing Contemplation; the second like a Capitain, representing Fame; and the third like a Counsellor of Estate, representing Experience: the third to begin to the Squire, as being the master of the best behaviour or compliment, though he speak last' (*Works*, viii. 376; facsimile in Beal, p. 25). The other MS which Spedding located (Lambeth MS 933; Beal BcF 308) was Bacon's fair copy of the whole, which Spedding took as his copy-text (*Works*, viii. 378–86), and which I reproduce here.

A letter from Rowland Whyte to Sir Robert Sidney (22 November 1595) gives an eyewitness account of what actually took place: 'My Lord of Essex's device is much commended in these late triumphs. Some pretty while before he came in himself to the tilt, he sent his page with some speech to the Queen, who returned with her Majesty's glove. And when he came himself, he was met with an old Hermit, a Secretary of State, a brave Soldier, and an Esquire. The first presented him with a book of meditations; the second with political discourses; the third with [narrations] of brave-fought battles; the fourth was but his own follower, to whom the other three imparted much of their purpose before he came in. [As these] devised with him, persuading him to this or that course of life, according to their inclinations, comes into the tiltyard unthought upon the ordinary postboy [messenger] of London, a ragged villain all bemired, upon a lean jade, galloping and blowing for life, and delivered the Secretary a packet of letters, which he presently offered my Lord of Essex; and with this dumb show our eyes were fed for that time.

In the after-supper, before the Queen, the [hermit] first delivered a well-penned speech to move this worthy knight to leave his vain following of Love, and to betake him to heavenly meditation; the secretary all tending to have him follow matters of state, the soldier persuading him to the war. But the esquire answered them all, and concluded with an excellent but too plain English, that this knight would never forsake his mistress's love, whose Virtue made all his thoughts divine, whose Wisdom taught him all true policy, whose Beauty and Worth were at all times able to make him fit to command armies. He showed all the defects and imperfections of all their times, and therefore thought his course of life to be best in serving his mistress.

The old man was he that in Cambridge played Giraldy, Morley played the Secretary, and he that played Pedantiq was the soldier, and Toby Matthew acted the Squire's part. The world makes many untrue constructions of these speeches, comparing the Hermit and the Secretary to two of the lords, and the Soldier to Sir Roger Williams; but the Queen said that if she had thought there had been so much said of her, she would not have been there that night, and so went to bed' (*Works*, viii. 374–5).

The other part of the entertainment, the outer plot involving Philautia and Erophilus, has not survived complete, and there is no evidence that it was actually performed. Two fragments of a quite different plot-sequence survive in the Public Record Office, one entitled 'A devise made by the Earle of Essex for the entertainment of the Queene' (S.P. 12, 254/67; see E. K. Chambers, *The Elizabethan Stage*, 4 vols. (Oxford, 1923), iii. 212–13; S. W. May (ed.), *The Poems of Edward De Vere . . . and of Robert Devereux . . . Studies in Philology*, 77 (1980), Texts and Studies no. 5, pp. 88–90). Here Essex's squire conducts a blind Indian Prince and his white Attendant before the Queen, who cures the Prince's blindness, the Attendant then revealing himself as Love, a suitable presence in her court. Spedding reprinted these speeches (*Works*, viii. 387–91), doubting whether this episode belongs to the same device: but E. K. Chambers and S. W. May noted that it mentions Philautia and includes the Latin quotation *amare et sapere*, also used by Bacon. (This sequence may well have been written by Essex, since it includes a sonnet ascribed to him. I agree with Spedding that the prose style seems unlike Bacon's.) The other fragment in the PRO is a speech by Philautia in French, which both Chambers and May recognized as being in the hand of Reynoldes, Essex's secretary. P. E. J. Hammer (*Studies in Philology*, 91 (1994), 167–80) has seized on this fact once again to ascribe authorship of the whole to Essex and his secretariat, and to question Bacon's hand elsewhere in the device, although the manuscript and other historical evidence—which Hammer never considers—is overwhelming. Hammer even fails to mention that the Lambeth Palace drafts in Bacon's handwriting include a letter from Philautia to the Queen in which Bacon has actually added marginal notes to Essex pointing out how he should understand and apply the allegory to his own ends (*Works*, viii. 376–7), further evidence of the central part that Bacon played in drafting this entertainment.

The self-contained device, which I reprint here from Spedding (passages within square brackets are editorial conjectured emendations), is a public debate on the competing claims on a ruler of the academic, military, and political vocations. But now the Squire's retort expresses the criticism that Bacon voices elsewhere in his discussion of ethics and politics, that these careers are sometimes chosen out of self-love, rather than concern for the public good. Particularly impressive is Bacon's ability to think himself into the various personae yet then to attack the weaknesses of their arguments, proof of the thoroughness with which he had mastered both legal procedures and classical rhetoric.

61. *master his.* The old genitive form: master's. *melancholy.* A pathological condition attributed to an excess of one of the four humours ('choler adust'), and characterized by inactivity, together with a pensive or meditative sadness. *brain-sick.* Mad. *busy.* Fussy, meddlesome. *tedious.* Prolix, verbose. *Presence.* The presence-chamber, in which a sovereign receives guests and suitors. *worthy.* Worth. *revolving.* Considering. *table.* Diagram. *to.* Compared to. *conceit.* Thoughts. *mured.* Enclosed. *sweet travelling.* Pleasurable journeys. *universal variety.* The universe in all its diversity. *alms.* Charitable relief, support.

62. *gardens of love . . . fading to-morrow.* In *Promus*, no. 806, Bacon noted: 'Adonis gardens (things of great pleasure, but soon fading)'. Cf. Erasmus, *Adagia*, I. i. 4, 'Adonidis horti', on these pot-plants which bloomed for only a few days, the term being used to characterize 'trivial things, which served no useful purpose' other than 'a brief passing pleasure'. *golden age.* According to classical myth, the first age of the world. *The monuments of wit . . . periods.* Cf. *Adv.L.*, above, p. 167. *Let him . . . now upon a hill . . . former times.* Cf. ibid., and Essay 1, Of Truth', above, p. 342. *for.* Because. "*vitam vitalem*". A 'lively life', a vital existence. *feast . . . throng.* Enjoy enrichment reserved to a few. *light . . . heat.* Have the benefits of knowledge without disadvantages. *in his love.* In the eyes of his beloved. *sense.* Mere perception (lacking judgement). *leave to.* Stop. *exercise.* Occupation. *distasted.* Spoiled.

63. *matches.* Competitions. *looker-on . . . feigned tragedy.* Cf. Aristotle, *Poetics*, xiii (1453a35) and xiv (1453b10–14). *lively.* Living. *straitest.* Strictest. *places.* Occupations. *several.* Individual. *lovers . . . warfare.* A common comparison in Roman love-elegies, and in Renaissance madrigals, cf. Ovid, *Amores*, I. ix. 1: 'Militat omnes amans, et habet sua castra Cupido' ('Every lover is a soldier, and Cupid has a camp of his own'), and

Monteverdi, 'Ogni amante è guerrier' in bk. VIII of the *Madrigals* (1638), text by Rinuccini. *daunted*. Intimidated, overcome. *pantofle*. Slipper. *token*. A thing serving as a sign or symbol. *but*. Only. *unfolded*. Applied, put to use. *sanguine*. Confident. *delightful*. Experiencing delight. *retired*. Withdrawn, solitary.

64. *shifts of humour*. Changes within the humoral system, from one dominant humour (blood, phlegm, choler, melancholy) to another. *cover*. Conceal. *draw*. Attract. *would*. Wish. *alarm*. Signal calling to arms. *discourting*. Discourteous, improper employment. *credulous*. Impressed by. *tender*. Sensitive. *demonstration*. Outward display. *under arrest*. Arrested for non-payment of debts. *service*. Devotion. *cavil*. Frivolous objection. *imputation*. Accusation. *employed men*. Employees. *relations*. Reports. *experimented*. Experienced. *sentences*. Judgements, summings-up. *causes*. Legal cases. *holding of affection*. Maintaining good relations. *confederates*. Associates. *charge*. Expense, damage. *entitling*. Justifying, showing a legal entitlement. *grudging*. General resentment, grumbling. *in appetite*. Ready to work or serve. *most*. Most of all. *charge*. Trouble, inconvenience. *manacles*. Handcuffs. *set for*. Decide, strive towards. *commodity*. Advantage, expediency. *strength*. Power, authority. *Let... affairs*. Some of this 'advice' is ironic, satirizing the weaknesses of statesmen rather than outlining models to be imitated.

65. *drifts*. Aims, purposes. *entertain*. Accept, agree to. *proposition*. Proposal. *circumstances*. Subordinate details. *add... own*. Cf. Essay 52, 'Of Ceremonies and Respects', above, p. 442. *wit*. Intellect. *doth superabound*. Is very abundant. *substance*. Solidity. *obnoxious*. Liable to harm, damage. *oracles... event*. The oracles' replies were often capable of many interpretations. *likeliest... best*. Again, ironic. *true to himself*. Bacon used this phrase in several of the Essays, from the 1612 'Of Faction' (rev. 1625: above, p. 441) to the 1612 'Of Wisdom for a Man's Self' (rev. 1625: above, p. 286) and the 1625 'Of Suspicion' (above, p. 405). It echoes Polonius' advice to Laertes, 'to thine own self be true', and may suggest a common source in the 'Advice literature': cf. Brian Vickers, *Francis Bacon and Renaissance Prose* (Cambridge, 1968), 279 n. 2. *reaches of state*. Politic manœuvres, schemes. *merely*. Absolutely. *particular*. Personal interest. *grateful*. Welcome. *conduit*. Channel. *diligences*. Labours, exertions. *accesses*. Deferential approaches. "*amare et sapere*". Bacon frequently quotes this saying 'Amare et sapere vix Deo conceditur', 'To love and to be wise is hardly granted to the gods', Publilius Syrus, *Sententiae* XV (also in Plutarch's 'Life' of Agesilaus, 415B): cf. *Of Tribute*, above, p. 33, and Essay 10, 'Of Love', above, p. 358. *Wandering*. Rambling, moving purposelessly. *storming*. Vigorous in assault. *hollow*. Insincere, false. *enchanting*. Using magical charms to gain power over people (usually harmful). *statua*. Statue. *vulture... bodies*. On the battlefield. *double heart*. A cloven heart, traditional mark of insincerity. *assurance*. Self-confidence. *forced*. Maintained with effort. *bring in play*. Bring into action. *vouchsafe*. Deign. *severing*. Speaking (or fighting) separately.

66. *would.* Wish to. *chimeras.* Phantasms, fanciful conceptions. *prospect.* Look-out. *dazzling.* Confusion produced by gazing at too bright a light. *inward.* Intimate. *security.* Assurance, guarantee. *cypress.* Symbol of mourning. *bespoken.* Commissioned. *discover.* Reveal. *flights.* Running away (in battle). *engines.* Tricks (from Latin *ingenium*). *motions.* Devices, manœuvres. *be . . . temper.* Run accurately. *practice.* Trickery, deception. *overbuilt.* Over-reached, exceeded the limit. *windfall.* Blown down by the wind. *depend.* Hang downwards. *die-form.* Dice. *lighteth upon.* Lands. *square.* Four-sided figure (in the Renaissance, a metaphor for constancy and solidity). *acknowledgment.* Recognition. *affection.* Love. *merit.* Value. *Antaeus.* In Greek mythology, a giant who compelled all comers to wrestle with him, and was supposedly made stronger when thrown, by contact with his mother the Earth. Hercules defeated him by holding him off the ground and crushing him to death. *of harvest.* Autumn. *wanes.* Decreasing in power (like the moon's visible illuminated area). *Attend.* Pay attention. *beadsman.* One who prays for the soul of another. *shadows.* Alluding to the Platonic doctrine that the visible world is the mere appearance of a deeper reality.

67. *pictures . . . crystals.* Objects kept in a museum. *of the.* Made like. *taketh . . . impressions.* Can adapt to any shape. *water . . . fire.* Earth and water were the two lowest elements, air and fire the highest. *in noise.* In noisy, busy surroundings. *will.* Want to. *advance.* Encourage, promote. *pen.* Be the author of. *accessary to.* Involved in, colluding with. *interruption.* Disturbance, break. *spirits.* Souls. *celestial fire.* Love. *untrue.* Insincere. *Politique.* Politician, schemer. *bondman.* Slave. *bind occasion.* Control fortune. *overwork.* Outmanœuvre. *several.* Separate. *servitudes . . . masters.* Cf. Christ's warning, Matt. 6: 24 (Luke 6: 13): 'no man can serve two masters'. *commissioners.* Representatives. *allowed . . . of.* Acknowledged, accepted the truth of. *set.* Estimated. *fainting.* Weak. *yield.* Concede. *addresses.* Goals, aims in life. *knowledge, fame . . . fortune.* The aims for which the counsellors have argued. *her Majesty.* Queen Elizabeth. For other writings praising the queen see 'Of Tribute', above, pp. 22 ff., and *In felicem memoriam Elizabethae, Angliae Reginae* (*Works*, vi. 291–318). Bacon thought so highly of the Queen that in the first draught of his will, concerning 'that durable part of memory which consisteth in my writings', he asked friends and relatives to take care of his works: 'In particular I wish the elegy which I writ *in felicem memoriam Elizabethae*, may be published' (*Works*, xiv. 540 n.). *liberties.* Privileges. *story.* History. *this . . . day.* 17 Nov., the anniversary of the Queen's accession.

68. *language . . . her.* Queen Elizabeth was skilled in Greek, Latin, and several modern languages. *shop.* Merchandise, store of goods. *her.* Queen Elizabeth.

ADVICE TO THE EARL OF RUTLAND ON HIS TRAVELS

This series of three letters to Roger Manners (1576–1612), the spendthrift fifth Earl of Rutland (who travelled extensively in Europe during 1595–7), was printed by Spedding (*Works*, ix. 6–20) from manuscript copies in Lambeth Palace Library (MS 936, no. 218) and in the British Library (BL Lansdowne MS 238, fos. 158r–159r; BL Add. MS 37232, fo. 97). The first, and longest letter, appeared in print in 1633, in a tiny compilation of travel writings by various authors, *Profitable Instructions; Describing what special observations are to be taken by Travellers in all Nations, States, and Countries* (STC 6789). Spedding firmly ascribed it to Bacon on the grounds of many similarities in thought and style to his published works, but expressed doubts about the authenticity of the third letter. Fortunately the Hulton Papers, recently auctioned by Sotheby's, included a manuscript of this letter (Sale Catalogue, 'Elizabeth and Essex', 14 December 1992, p. 26), a note in Essex's handwriting stating 'This was written yesternight att St Alban's but so ill written as I was fayne to use my mans hand to copy yt out.' Since the Bacon family house was at Gorhambury, near St Albans, and Essex's note implies that the treatise was written for him by someone else, this provides an external argument for Bacon's authorship. As for the second letter, the ascription to Bacon was questioned by N. K. Farmer, Jr. (*Renaissance Quarterly*, 22 (1969), 140–7), who pointed out that other early MS copies exist, in which the authorship is variously attributed, to Fulke Greville (Beal GrF 16–23), in whose *Works* (1633) it was actually printed, and even to Sir Thomas Bodley. However, anyone familiar with the extensive practice of copying manuscripts in this period, for private collections or for exchange, will know that the original authorship—if at all known—was often confused, whether innocently or deliberately (for personal, political, or commercial motives). Farmer's doubts about Bacon's authorship of the second letter were repeated by P. E. J. Hammer (*English Historical Review*, 109 (1994), 26–51; *Studies in Philology*, 91 (1994), 167–80), who has discovered another MS copy of this letter in the hand of Edward Reynoldes, using it to bolster his claim that several of Bacon's early works were in fact written by Essex or by 'one or more of his secretaries'. But the links between the Bacon family and Essex in the mid-1590s were so many—Anthony Bacon collected foreign intelligence for Essex, and lived in Essex House, while it is well known that Francis not only advised Essex but wrote several pieces which were to be circulated as Essex's own compositions—that we can hardly be surprised that Essex's secretariat should have copied their contributions at some point. Rather, this seems to confirm Hammer's analysis of how Essex used the writings of others on topics and occasions where he could 'foster his reputation as a man of gravity and learning'.

Although calling Bacon's authorship in question, neither Farmer nor Hammer bothered to compare the contents of these letters with Bacon's authenticated writings. Anyone who does so will find that the great number of parallels, in thought and style, to the *Essays*, the *Advancement of Learning*, and the early devices (especially *Of Tribute*) leaves no doubt about his authorship. (See Brian Vickers, 'The authenticity of Bacon's earliest writings', forthcoming in *Studies in Philology*.)

Rather than mere 'formula letters', as those critics dismissively call them, these three pieces form a unit (as Spedding saw) which is marked by some distinctive Baconian characteristics. They fall into the genre of 'advice literature', an important form within Renaissance humanism since it fulfilled the humanist ideals that knowledge should be applied to a useful end, and all action dedicated to the benefit of others. The first letter is in fact devoted to ethics, the '*cultus animi*', as Bacon calls it, 'the tilling and manuring of [the] mind', emphasizing the need for the acquisition of knowledge to be intimately linked with the acquisition and practice of virtue. This discussion has many points in common with the speech praising Fortitude in *Of Tribute*, above, pp. 22–9. The second letter addresses the announced topic of travel, and shows again how the Renaissance concept of a *vita activa* regarded travel primarily as a form of practical education, a way of gaining first-hand knowledge about the customs and variety of mankind. The third letter gives more specific instruction about the things to be observed: climate, soil, trade, government, laws, and much else. Readers can note many parallels with Bacon's Essay 18, 'Of Travel', above, pp. 374 ff.

69. *intelligence of state*. Collecting information for the national interest. *mean sufficiency*. Modest abilities. *advertisers*. Informants. *single*. Unsupported by other testimony. *ordinary*. Common. '*sentire . . . meliorem*'. 'To feel himself each day a better man than he was before': Xenophon, *Memorabilia*, I. vi. 9. A favourite quotation, found in *Promus* (no. 95) and in the *Adv.L.*: see above, p. 165. '*cultum animi*'. The cultivation of the mind: cf. *Adv.L.*, above, pp. 255–65. *tilling*. Cultivating. *manuring*. Fertilizing. *forms*. Behaviour, manners. *contrary*. Opposed, refractory.

70. *as well as*. In the same way that. *greatest part of*. Most. *garment*. Outer covering, subject to choice and change. Cf. *Adv.L.*, above, p. 266, and Essay 52, 'Of Ceremonies and Respects', above, pp. 441 f. *counterfeit*. Imposture. *mend*. Correct. *cover*. Conceal. *affect*. Prefer, like. Cf. Essay 11, 'Of Great Place', above, p. 360. *partially*. Biased, prejudiced. *physic*. Medicine. *preservatives*. Medicines protecting against disease or infection. *senseless*. Lacking feeling. The Stoic doctrine of *apatheia*, controlling the passions by repressing them, was often criticized: see e.g. H. W. Sams, 'Anti-stoicism in Seventeenth and Early Eighteenth-Century England', *Studies in Philology*, 41 (1944), 65–78. *grace*. According to Christian belief, all human acquirements are gifts from God. *win . . . upon*. Get the better of. *fear of death*. Some of the major works of classical philosophy, such as Cicero's *Tusculan Disputations*, taught how to overcome this fear. Cf. *Of Tribute*, above, pp. 26 ff., and Essay 2, 'Of Death', above, pp. 343–4. *natural*. Untrained, uneducated. *resolve*. Determine. *virtues*. In the classical tradition the four cardinal ('chief') virtues were Justice, Fortitude, Prudence, and Temperance. Bacon seems to be referring to Aristotle's *Nicomachean Ethics*, which discusses at length such qualities as magnificence ('liberality' or generosity) and magnanimity.

71. *liberal*. Generous. *goods of fortune*. Worldly advantages, such as riches, which depend on fortune rather than virtue. '*beatius . . . accipere*'. 'It is more blessed to give than to receive' (Acts 20: 35). *badge*. Sign, mark. *nothing . . . fear itself*. Cf. *Of Tribute*, above, p. 26. *give with judgment*. Bacon

refers to Seneca's *De beneficiis*, the most extensive classical treatise dealing with the right way to give and receive benefits. *confirm*. Strengthen. *happily*. successfully. *move*. Direct. *imitate the best*. It was one of the main principles of classical ethics that men should select virtuous behaviour to be imitated: in the often-quoted sentence from Plato, 'virtue, once it is seen, is loved' (*Phaedrus*, 250C). *delights*. In Renaissance moral philosophy this term often had a negative connotation, as of 'ephemeral pleasures', more suited to the bodily senses than to the mind. *disease*. Disorder. *discourse of reason*. Rational activity, the property of mankind (not animals), acquired by education and exercise.

72. *caught with*. Infatuated by. *ill graces*. Bad habits. *participate of*. Share, imitate. *affection*. Affectation, mannerisms. *in little*. In small things. *conceived*. Absorbed, comprehended. *wants*. Is lacking. *darings*. Audacious acts. *prize*. Value. *want*. Lack. *for that*. Because. *insufficiency*. Insufficient.

73. *resolved*. Determined. *conference*. Meeting people, social interchange. *nobility*. The landed or aristocratic classes, who still scorned education as only needed by those who have to earn their living. *of all places of*. Throughout. *if it be not*. Apart from. *in despite of*. Against, in contempt of. *clerks*. Scholars. *peace*. The belief that prolonged peace produced idleness and corruption was a commonplace in Roman ethical and political thinking, widely shared in the Renaissance: see Brian Vickers, 'Leisure and Idleness in the Renaissance: The Ambivalence of *otium*', *Renaissance Studies*, 4 (1990), 1–37, 107–54. '*artes luxuriae*'. The arts of luxury, sensual indulgence. *spend their humours*. Discharge the fluids of which the body was thought to be composed, so that they could be replenished. *outrages*. Violent, disorderly actions. *spirits*. Energy, resources. *entered their whelps*. Trained their hunting-dogs. '*Scientia . . . ignorantem*'. 'Science has no enemy except the ignorant person.' For this common saying see H. Walther, *Proverbia Sententiaeque Latinitatis Medii Aevi*, iv (Göttingen, 1966), no. 27590 e, g, h. It is also quoted in e.g. George Puttenham, *The Arte of English Poesie*, ed. G. D. Willock and A. Walker (Cambridge, 1936, 1970), 140. *drawn*. Guided. *course*. Course of study. *liberal arts*. Those practised by a 'free' man: literature, philosophy, history, etc. *but*. Only. *politic*. Political. *lay up*. Store.

74. '*discere . . . recordari*'. 'Learning also consists of recollecting': Cicero, *Tusc. D.*, I. xxiv. 57–8, alluding to Plato's doctrine that all learning consists of recollection: cf. e.g. *Phaedo*, 72–6, *Meno*, 81–5. Cf. *Adv.L.*, above, p. 120 and Essay 58, 'Of Vicissitude of Things': 'Plato had an imagination, that "all knowledge was but remembrance"' (above, p. 451). *draw*. Withdraw (from society). *art*. Subject. *meditation*. Mental exertion. *conference*. Discussion. *in order*. In sequence. *doubt*. Question. *conference . . . study*. Cf. Essay 27, 'Of Friendship', above, p. 394. *whetted*. Sharpened. *sound*. Plumb the depths of. *carping*. Fault-finding. *observation*. Paying careful attention. *coherence*. Interconnection. *counsels*. Advices. *successes*. Results.

75. '*id . . . scimus*'. 'We know that especially whose cause we know'; Aristotle, *Posterior Analytics*, i. 2, 71b9 ff. The early Latin translations read 'tum scimus

cum causam cognoscimus'. *politique*. Politician. *makes*. Amounts to. '*exempla . . . probant*'. 'Examples illustrate [a point], but do not prove it.' A common maxim in logic and rhetoric. *ostentation . . . discourse*. Conversational display. '*turpe . . . laus*'. 'It is disgraceful for a suitor to solicit (his lady's) handmaid, but praise is the handmaid of virtue.' Bacon used this (untraced) quotation in *Promus* (no. 70) and in the 1603 *Apology* (*Works*, x. 141). *praise . . . virtue*. One of the chief principles in epideictic rhetoric: cf. Aristotle, *Rhetoric*, i. 3 (1358b11–1359a6), i. 5 (1362a13), i. 9 (1366a23 ff.). *advertisement*. Missive. *Christmas . . . disorder*. During the Christmas vacation, which lasted until 6 Jan., Twelfth Night, the feast of Epiphany, traditionally a time of licensed misrule (e.g. the crowning of a boy bishop).

76. *warranted*. Given authority. *curious*. Over-precise. *book*. Authority. '*membra . . . confundi*'. 'Parts that have been separated ought not to be confused'; untraced. *mere*. Wholly. *vulgar scandal*. Common and discreditable circumstance. *foundation*. Fundamental beliefs. *country*. France. *two several professions*. Two religious belief-systems (Catholic and Protestant).

77. *contrary humours*. Opposed tendencies. *interest*. Attention, involvement. *practices*. Deceits. *practise upon*. Trick, delude. *deboshed*. Debauched, corrupted. *real*. Solid, genuine. *Of*. About. *traffic*. Trade. *havens*. Ports. *vent*. Commerce, markets. *husbandry*. Domestic economy; agriculture. *traffic*. Movement, dealings. *Intelligencer*. Reporter, spy. *happily*. Perhaps.

78. *proportion*. Relationship. *Many*. Varied, different. *lay impositions*. Levy taxes. *degrees*. Social classes. *inventions*. Provisions, plans. *commodities*. Manufactures. *bring . . . to heads*. Arrange by topics. *spirits*. Sources of energy, life. *captious*. Fault-finding. *collection*. Collecting, catching. *humourists*. Facetious or comical people. *comfortable*. Able to comfort; helpful. *liberal traffic*. Studious activities. *advertisement*. Information, notification. *commodities*. Things of use or value.

79. *posting*. Hasty. *champain*. level, open ground. *straits*. Narrow passage of water connecting two seas or large bodies of water. *passages*. Passes, gorges between mountains. *in the people*. Democracy. *mixed*. State governed by a combination of ruling powers (monarchy, oligarchy, democracy). *subalternate*. Subordinate. *points of state*. Concerning government, statecraft. *get*. Hire.

80. *sprightful*. Vigorous. *go . . . way*. Go astray, get lost. In *Promus*, no. 233 (partly repeated, no. 1240) Bacon noted: 'Melior claudus in via quam cursor extra viam' ('Better is the lame man in the right way, than a swift runner out of the way'). Cf. *Adv.L.*, above, p. 169.

ESSAYS (1597)

The first edition of Bacon's *Essays* (Gibson 1) was a tiny volume, 'a little booke no bigger than a primer', as John Aubrey described it (*Brief Lives*, ed. O. L. Dick (Harmondsworth, 1962), 120). It consisted of seven essays, accompanied by two other short works, a group of 'Religious Meditiations', the *Meditationes Sacrae*, and 'Places of perswasion and disswasion', as the title-page put it, or 'Of the Colours of good and evil, a fragment', as the sub-title calls it. The volume had some success, being reprinted in 1597, 1598, 1606, and 1612 (Gibson 2–5).

This first recension is notable for its extreme bareness of style. Individual sentences are often printed separately, and marked with a paragraph sign, to signal their status as aphorisms, discrete observations embodying wisdom drawn from experience. It is important to realize that the bareness of style is a deliberate, sought-after factor, the corollary of Bacon's choosing the aphoristic form, not a proof that he was somehow lacking in eloquence, or unable to write a coherent argument. Although critics unfamiliar with Bacon's writings as a whole describe his style as 'aphoristic', in fact it constantly varies, as laid down by rhetorical theory, according to the subject-matter treated and the audience addressed. The issue is only secondarily one of style, since the aphorism is properly a distinct form or genre, which Bacon only used when his subject-matter demanded it.

The prime quality for which Bacon valued the aphorism was not the pithiness commonly associated with it but its unsystematic quality, which allowed its user to communicate separate observations without implying any firm connections between them. The advantage of aphorisms was that they would neither foreclose enquiry nor give a premature and spurious coherence to discussions which were not yet definitive and complete. (See Brian Vickers, *Francis Bacon and Renaissance Prose* (Cambridge, 1968), ch. 3, pp. 60–95.) In many ways the best comment on the form of the 1597 *Essays* is a passage in Bacon's Preface to a professional treatise that he composed in that same year, the *Maxims of the Law*, a collection of legal maxims which he had deliberately not arranged in the sequential form of a treatise: 'Thirdly, whereas I could have digested these rules into a certain method or order, which, I know, would have been more admired, as that which would have made every particular rule, through his coherence and relation unto other rules, seem more cunning and more deep; yet I have avoided so to do, because this delivering of knowledge in distinct and disjoined aphorisms doth leave the wit of man more free to turn and toss, and to make use of that which is so delivered to more several purposes and applications. For we see all the ancient wisdom and science was wont to be delivered in that form; as may be seen by the parables of Solomon, and by the aphorisms of Hippocrates, and the moral verses of Theognis and Phocylides: but chiefly the precedent of the civil law, which hath taken the same course with their rules, did conform in my opinion' (*Works*, vii. 321). This combination of pregnant utterance and free form meant that the aphorism could be an appropriate vehicle for quite different areas of knowledge, from Bacon's works in natural philosophy (*Novum Organum*, the *Historia Naturalis et Experimentalis*) to his treatment of politics and ethics, subjects where it was vital to preserve detail of observation without subsuming it into restricting categories. The *Advancement of Learning* contains several important discussions of the aphorism, showing Bacon's high awareness of the need to find the most suitable vehicle for the communication of knowledge (above, pp. 145–6, 234–5, 270–1).

The 1597 *Essays* are Bacon's first attempt to systematize his ideas on politics, especially the individual's proper behaviour in the political arena. They evidently belong to the literature of advice, the conduct-books whose vogue in the Renaissance attracted readers, from courtiers to citizens, bent on self-improvement. The subject-matter concerns the preparation needed by a man entering public life, and although presented separately the topics overlap, to form a miniature treatise. The first essay, on education, also deals with discourse, treated more fully in the second essay. The fourth, on followers and friends, also deals with suitors and expense, which become the subjects of the fifth and sixth essays. A continuity of attitude can be traced, too, in Bacon's constant emphasis on what is 'honourable', a fundamental ethical alignment that many commentators have missed. Bacon's laconic definitions, with the simplest verb-forms of 'to be' ('is' and 'are' recur most frequently), his curt imperatives ('Discern', 'Beware'), and his baldly stated preferences ('It is better') all convey his observations from experience in the plainest and most functional form.

The publishing history of this volume testifies to the celebrity which it enjoyed even before it appeared. As editors and bibliographers have established, it had already been plagiarized before publication, for Edward Monings, in *The Langrave of Hessen his princelie receiving of her Majesties Embassador* (1596), recounting an ambassador's visit to the Hessian court, borrowed several passages from the essay 'Of Studies' to describe the Langrave's 'prince-like' education, 'seasoning his grave and more important studies for ability in judgment, with studies of pastime for retiring . . . and for ornament in discourse . . . wherein he is expert, reading much, conferring and writing much; he is a full man, a ready man, an exact man. . . . For art doth perfect nature and is perfected by experience' (Kiernan, ed. cit., pp. lxv ff.; Wright, ed., *Essays*, 341). Bacon may have been amused to find his words applied to courtly flattery, but the risks involved in the common practice of allowing manuscript copies of your work to be circulated could be less pleasant. When he published these *Essays* in 1597 he dedicated them to his older brother Anthony with an Epistle which referred to another, more serious plagiarism: 'Loving and beloved Brother, I do [behave] now like some that have an orchard ill neighboured, that gather their fruit before it is ripe, to prevent stealing. These fragments of my conceits were going to print; to labour the stay of them had been troublesome, and subject to interpretation; to let them pass had been to adventure [risk] the wrong they might receive by untrue copies, or by some garnishment, which it might please any that should set them forth to bestow upon them. Therefore I held it best discretion to publish them myself as they passed long ago from my pen, without further disgrace than the weakness of the author. And as I did ever hold, there might be as great a vanity in retiring and withdrawing men's conceits (except they be of some nature) from the world, as in obtruding them: So in these particulars I have played myself the inquisitor, and find nothing to my understanding in them contrary or infectious to the state of religion or manners, but rather (as I suppose) medicinable. Only I disliked now to put them out because they will be like the late new half-pence, which though the silver were good, yet the pieces were small. But since they would not stay with their master, but would needs travel abroad, I have preferred them to you that are next myself, dedicating them, such as they are, to our love, in the depth whereof (I assure you) I sometimes wish your infirmities translated upon myself, that her Majesty might have the service of so active and able a mind, and I might be with excuse confined to these contemplations and studies for which I am fittest.' From the records of the Stationers' Company we see that one Richard Serger had obtained a manuscript copy of the *Essays* and entered it in the Stationers' Register—so claiming to be the authorized publisher—as a forthcoming publication, 'a book entitled ESSAYES of M.F.B. *with the prayers of his Sovereigne*'. This additional, wholly spurious piece (whatever it might have been) was evidently the 'garnishment' that Bacon referred to, and which he wanted to disown. Serger had entered his unauthorized title on 24 January, and Bacon took speedy action. His dedicatory epistle was dated 30 January, and on 5 February a fresh entry appeared in the Stationers' Register, assigning to Humfrey Hooper the right to publish 'A booke entitled *Essaies, Religious meditations, Places of Perswasion and Disswasion* by master FRAUNCIS BACON'. Despite the nuisance involved, Bacon had managed to 'labour the stay' or prevent the unauthorized publication by an appeal to the Stationers' Court, who formally cancelled the earlier entry on 7 February, the day on which the *Essays* themselves appeared, so hastily thrust into the world.

My text is from Spedding, *Works*, vi. 521–34 (*Essays*), vii. 229–54 (*Christian Meditations*), and vii. 67–93 (*Colours of Good and Evil*).

Of Studies

81. *for abilities*. To make men able. *retiring*. Retirement (for individual study). *ornament . . . discourse*. Effectiveness in conversation and in society. In classical and Renaissance rhetoric 'ornament' could also refer to a soldier's weapons or kit, that is, to effective communication, not mere display. *expert*. Experienced (in specific fields). *execute*. Do, perform. *censure*. Estimate, evaluate. *sloth*. Laziness (that is, avoiding one's duties to society in the *vita activa*). *humour*. Character, style. *scholar*. One living in a university; remote from the active life. *Crafty*. Cunning. *contemn*. Despise (presumably as irrelevant to cunning practices). *simple*. Innocent (but also 'stupid'). *admire*. Wonder at. *wise*. Prudent, judicious. *teach not their own use*. A course of study can convey the content and methods of a discipline, but not its application. *without . . . above them*. External to, and transcending the scope of any discipline. *Read . . . to weigh and consider*. Cf. the account of 'vanities in studies' in *Adv.L.*, above pp. 140–1. *conference*. Discoursing. *confer*. Discourse. *present wit*. Quick, alert mind. *witty*. Ingenious, full of ideas. *grave*. Serious, weighty.

Of Discourse

81. *Discourse*. See the head-note to this essay in the 1625 version, below, p. 750. *wit*. Mental agility. (In Renaissance psychology 'wit' is the associative faculty, 'judgment' the discriminative one.) *hold*. Maintain. *discerning*.

Discriminating. *should be thought.* That is, the truth. *common-places . . . themes.* Prepared topics. *want variety.* Cf. Plutarch, 'On the education of children' (*Moralia*, 7B–C).

82. *part.* Action. *give the occasion.* Suggest a topic. *present occasion.* Of immediate but ephemeral interest. *argument.* Substantial, more general topic. *reasons.* Statements. *jest.* Joking. *privileged.* Exempt. *matters of state.* Business of national importance. *great persons.* People in power. *present.* Current. *content.* Please others. *apply.* Adapt. *that.* That which. *with good grace.* Decently. *pretendeth.* Aspires. *agreeably.* Fitly, suitably. *continued.* Sustained speech; monologue. *in the turn.* In turning, manœuvring (cf. *Adv.L.*, above, p. 226). *circumstances.* Introductory comments.

Of Ceremonies and Respects

82. *Ceremonies and Respects.* See the head-note to this essay in the 1625 version, below, p. 771. *real.* Sincere, straightforward, lacking social graces. *had need.* Would need to. *parts of virtue.* Merits. *foil.* 'A thin leaf of metal placed under the stone to improve its colour and lustre' (Reynolds). *it is in.* The same holds for. *light gains.* Small profits. *thick.* Frequently. *small matters.* Minor politenesses. *in note.* Noticed. *on holidays.* Rarely. *attain.* Acquire. *forms.* Manners. *so.* Provided that. *and.* Then.

83. *the rest.* Practising them himself. *care.* Take too much trouble. *leese.* Lose. *measured.* Counted, calculated. *comprehend.* Deal with. *breaketh.* Subdues, forces to submit. *again.* In return. *be.* Are. *strange.* Unknown, unfamiliar. *a good title.* A suitable manner. (This is the reading of the 1597 edn.; the 1598 and all subsequent edns. read 'good a little'.) *keep state.* Be a bit formal, dignified. *too much.* Effusive, exaggerated. *apply.* Adapt, accommodate. *upon regard.* Out of personal respect, affection. *facility.* Pliancy. *seconding.* Agreeing with. *motion.* Proposal. *allow his counsel.* Approve of his advice. *alleging.* Adding, producing.

Of Followers and Friends

83. *Costly.* Expensive. *train.* Retinue (punning on 'peacock tail'). *which.* Who. *charge.* Are a charge on. *importune.* Importunate. *suits.* Petitions, requests for favour. *challenge.* Demand. *conditions.* Treatment, favours. *countenance.* Approval, moral support. *Factious.* Partisan. *range.* Ally. *ill intelligence.* Misunderstanding, friction. *states.* Classes of men. *answerable.* Corresponding. *that.* The profession. *him.* The person of high rank; nobleman. *civil.* Decent, orderly. *so.* Provided. *popularity.* Cultivating popular favour.

84. *apprehendeth.* Studies, strives. *advance.* Promote, reward. *desert.* Merit. *no . . . sufficiency.* When neither is clearly superior. *take with.* Have to do with; or perhaps 'employ'. *passable.* Generally acceptable, tolerable. *government.* Normal employment, business. *one rank.* The same capacity. *equally.* In the same way. *claim a due.* Demand favours as if by right. *in favours.* When it comes to distributing favours. *difference and election.*

Choice, discrimination. *officious*. Ready to serve. *all . . . favour*. All depends on the patron's good grace. *make too much*. Take too much notice of; or 'treat favourably'. *hold out*. Sustain. *proportion*. Level of treatment. *one*. Servant or employee. *distracted*. Receive conflicting advice. *gamesters*. Gamblers. *vale . . . hill*. The valley gives the best view of the hill: important people are more transparent to those below them. *wont to be magnified*. Friendship, especially between men, was much celebrated in classical myth and history (Orestes and Pylades, Damon and Pythias, *et al.*) and philosophy. See e.g. Diogenes Laertius, viii. 10, and Aristotle, *Eth. Nic.* ix. 8. *comprehend*. Include, reciprocally depend on.

Of Suitors

84. *Suitors*. This is the title in the 1598 and all subsequent editions; 1597 has 'Of Sutes'. *ill matters*. Evil cases. *which*. Who. *deal*. Behave, act. *life . . . means*. That the matter will succeed in some other way. *second*. Secondary, less important. *cross*. Thwart, obstruct. *make*. Gain, obtain. *sort*. Measure. *equity*. Moral justice. *suit of controversy*. One in a chancery suit or lawcourt, in which a suitor seeks his patron's help to influence the judge's decision (Reynolds). *desert*. Merit. *suit of petition*. One seeking help in obtaining some office for which there were other competitors. *affection*. Liking, partiality. *in justice*. In questions of justice. *countenance*. Influence. *compound*. Settle by agreement, compromise. *carry*. Pursue, prosecute. *depraving*. Speaking ill of. *disabling*. Depreciating. *deal*. Act, take part. *with honour*. Honourably.

85. *distasted*. Disgusted, offended. *abuses*. Deceits. *denying*. Refusing. *at first*. From the outset. *reporting . . . barely*. Giving a frank report of the outcome; or perhaps 'chance of success'. *challenging*. Claiming. *gracious*. Agreeable, deserving thanks. *suits of favour*. Petition (already mentioned). *first coming*. Priority in presenting one's petition. *take little place*. Not give precedence, pre-emptive rights. *his*. The first petitioner. *intelligence*. Knowledge. *note*. Information. *value*. Relevance, import. *simplicity*. Naïvety, stupidity. *right*. Justice. *voicing*. Divulging. *forwardness*. Preparation. *timing*. Choosing the right moment. *which*. That. *easy*. Sure of succeeding. *in*. For.

Of Expense

85. *for honour*. According to one's social status. *limited*. Measured, appropriate. *voluntary undoing*. Freely giving away one's substance. *kingdom of heaven*. Poverty was valued in the ascetic life, as being more likely to earn a place in heaven. *compass*. Limits, control. *abuse*. Larceny. *ordered . . . shew*. Make the best impression. *that*. So that. *less than the estimation abroad*. Smaller than generally imagined. *baseness*. Degradation. *doubting*. Fearing. *in respect*. In case. *broken*. Ruined, bankrupt. *searching*. Examination. *had need*. Had better. *new*. New employees. *subtle*. Deceitful, cunning. *clearing*. Settling the accounts, outstanding debts. *a man's*. One's. *as well*. Equally. *hurt*. Damage. *sudden*. Precipitous. *interest*. Paying interest on a debt.

86. *state*. Estate. *abridge*. Reduce. *petty*. Trivial. *return not*. Are 'extra-ordinary', as discussed at the beginning.

Of Regiment of Health

86. *Regiment*. Control, regimen. *this*. That is, the following observation. *physic*. Medicine. *observation . . . good . . . hurt of*. Plutarch, 'Advice about keeping well' (*Moralia*, 136E f.). *are owing*. Will be paid for. *still*. Always. *free-minded*. Open, carefree. *fly . . . altogether*. Completely avoid medicine. *strange*. Unaccustomed. *accident*. Sudden change, symptom. *respect*. Concentrate on recovering. *action*. Physical exercise. *endure*. Bear exertions. *sharp*. Serious. *tendering*. Careful nursing. *pleasing*. Obliging. *conformable to the humours*. Readily adapting treatment to the bodily state, however imbalanced. *press*. Penetrate to, discover. *regular*. Rigid. *art*. Medical precepts. *faculty*. Professional ability.

Of Honour and Reputation

86. *winning*. Obtaining. *revealing*. Recognition. *disadvantage*. Injustices. *affect*. Pretend to, counterfeit.

87. *inwardly*. In private. *darken*. Obscure, lessen. *given over*. Abandoned. *purchase*. Acquire. *virtue*. Ability. *follower*. Imitator. *temper*. Balance, harmonize. *combination*. Group. *husband*. Manager. *entereth into*. Undertakes. *canker*. An eating sore; corruption. *declaring*. Demonstrating. *a man's*. One's. *felicity*. Good fortune. *policy*. Astuteness. *marshalling*. Hierarchy, disposition. *Conditores*. 'Founders.' *Perpetui principes*. 'Perpetual princes.' *ordinances*. Authoritative decrees. *Liberatores*. 'Saviours.' *compound*. Put an end to. *Propagatores . . . imperii*. 'Enlargers or defenders of empire.' *Patres patriae*. 'Fathers of their country', an honorific title given to many Roman emperors. *which*. Who. *good*. Happy. *Participes curarum*. 'Sharers of care', confidants. *Duces belli*. 'Leaders of war.' *lieutenants*. Second-in-command. *scantling*. Measure. *Negotiis pares*. 'Equals in business.' *places*. Duties. *sufficiency*. Competence.

Of Faction

87. *Faction*. The grouping together of politicians holding a common attitude, often in opposition to the government. It must be remembered that Bacon was writing before the establishment of separate political parties. *new wisdom*. Here ironic: Bacon agrees with Machiavelli (*Discourses*, iii. 27) on the dangers of a ruler continually trying to exploit rival political groups. *fond*. Foolish. *estate*. State. *proceedings*. Actions.

88. *according to the respects of*. Considering the wishes of. *policy*. Both 'astuteness' and 'politics'. *ordering*. Disposing, executing. *general*. Of public advantage. *several*. Distinct, different. *with correspondence to*. According to, suiting the interests of. *Mean*. Of low rank. *adhere*. Associate, stick together. *great*. Distinguished, powerful. *strength in themselves*. Their own resources. *indifferent*. Impartial. *passablest*. Most acceptable. *giveth best way*. Opens the best path (to office). *conjunction*. Cohesion. *extinguished*.

Eliminated. *placed*. Settled in power. *take in*. Join, realign themselves. *enter*. Begin. *lightly . . . it*. Easily comes off the gainer. *balancing*. Equilibrium. *casteth*. Makes one of the scales incline.

Of Negotiating

88. *Negotiating*. A modern equivalent might be 'business dealings'. *third*. Third party. *by . . . self*. In person. *draw*. Obtain, elicit. *breeds regard*. Inspires respect. *instruments*. Go- betweens. *plainer*. Simpler, straightforward. *like*. Likely. *success*. Outcome (good or bad). *somewhat*. Something. *grace*. Ingratiate. *help . . . sake*. Improve upon the facts, so as to please their employer. *sound*. Probe. *afar off*. Indirectly. *short*. Brusque, unexpected. *in appetite*. In a state of desire, expectation. *upon conditions*. Conditionally (expecting a return in service or reward). *performance*. Action performed. *all*. All-important. *demand*. Expect. *which*. That it. *go before*. Be performed first (without guarantees of a return).

89. *party*. Person. *he*. The first agent. *honester*. More creditable. *practice*. Negotiation, diplomacy; sometimes 'trickery'. *discover*. Reveal the true nature of someone. *work*. Control, manipulate. *in trust*. By trusting someone. *passion*. Anger. *at unawares*. Unexpectedly, being surprised. *would*. Want to. *fashions*. Customs. *disadvantages*. Infirmities. *awe*. Influence by awe. *have interest in*. Are involved with. *look for*. Expect.

RELIGIOUS MEDITATIONS [SELECTIONS]

The second work in the 1597 volume was a group of twelve essays on theological topics, in Latin (although announced in English on the title-page). Of these I select eight, in the anonymous translation which was substituted for the Latin text in the 1598 edition (Gibson 3–5) and subsequently. As Spedding argues, the fact that Bacon neither mentions the translation nor includes it in any work published by himself suggests that it was not his own (*Works*, vii. 229).

The significance of these essays within Bacon's development is that they combine discussions of central issues in Christian belief (God's omnipotence, Christ's miracles, heresies) with ethical topics to which Bacon returned in the *Advancement of Learning* and the *Essays*: the need to combine the wisdom of the serpent and the innocence of the dove, the supremacy of charity, the evils of hypocrisy and imposture, and the psychology of atheism. One of the most interesting concerns 'Of earthly hope', for which Bacon had obviously begun to collect ideas when he compiled his *Promus of Formularies and Elegancies* in the Christmas vacation 1594–5, for a whole group of excerpts concern hope (nos. 1280–92—for the manuscript, see Beal, BcF 305; also 561, 773, 1099, 1104, 1117). In the manner of some classical and Christian moralists, this meditation takes a severely critical line against relying on hope, a view repeated in some later writings (*Works*, v. 203, 279–80). Yet hope continued to be a major resource in all Bacon's projects for improving knowledge and benefiting human life: 'I am now therefore to speak touching hope' (ibid. iv. 91; also v. 5). On its subsequent importance in Bacon's career see Michèle

Le Doeuff, 'Hope in Science', in W. A. Sessions (ed.), *Francis Bacon's Legacy of Texts* (New York, 1990), 9–24—who does not, however, cite this text.

89. *OF THE WORKS OF GOD AND MAN* *God . . . good*. Gen. 1: 31. *But . . . spirit*. Eccles. 1: 14. Bacon juxtaposes these two biblical texts elsewhere, notably in the prayer concluding 'The Plan of the Work' prefixed to the *Instauratio Magna* (*Works*, iv. 33). *Wherefore . . . Sabbath of God*. This promise is repeated in the same prayer. *ointment of odours*. Perfumed ointment. *compunction*. Uneasiness of mind after wrong-doing; remorse. *upbraidings*. Reproaches. OF THE MIRACLES OF OUR SAVIOUR *'He . . . well'*. Mark 7: 37. In *Promus*, no. 378, Bacon noted: 'Bene omnia fecit.'

90. *every miracle . . . first creation*. Cf. *Confession of Faith*, above, p. 108–9 and note. *Moses . . . plagues*. Exod. 7–11. *Elias*. Elijah. *shut up . . . earth*. 1 Kgs. 17: 1–7. *brought down . . . captains*. 2 Kgs. 1: 9–12. *bands*. Troops. *Elizeus*. Elisha. *called bears . . . children*. 2 Kgs. 2: 23–4. *Ananias . . . death*. Acts 5: 5. *present*. Instant. *Elymas . . . blindness*. Acts 13: 8–11. *descended . . . dove*. Matt. 3: 16. *'You . . . are'*. Luke 9: 55. *straw*. Chaff. *consummate*. Fully accomplished. *doctrine*. Teaching. *drew . . . nets*. Luke 5: 4–9; John 21: 3–11. *water . . . wine*. At the marriage in Cana, John 2: 1–11. *gladdens the heart of man*. Ps. 104: 15. *fig . . . wither*. Matt. 21: 19–21. *scarcity . . . fishes*. Matt. 14: 15–21; John 6: 5–14. *rebuked the winds*. Matt. 8: 26. *restored*. For collective lists of Christ's miracles, see Matt. 11: 5, 15: 30–1, 21: 14; Luke 7: 21. *motion . . . lame*. Mark 2: 3–12; Matt. 9: 1–8; Luke 5: 17–26. *light . . . blind*. Mark 8: 22–6, 10: 46–52. *speech . . . dumb*. Mark 8: 31–7. *health . . . sick*. Mark 1: 30–1; Luke 4: 38–40. *cleanness . . . leprous*. Matt. 8: 1–4; Mark 1: 40–5; Luke 5: 12–16, 17: 11–19. *right mind . . . possessed*. Mark 1: 23–7, 32–4, 9: 14–29; Luke 4: 31–7, 41. *life . . . dead*. Mark 5 (Jairus' daughter); Luke 7: 11–17 (the widow's son); John 11 (Lazarus). *tribute . . . Caesar*. Matt. 17: 24–7. *OF THE INNOCENCE OF THE DOVE, AND WISDOM OF THE SERPENT* *'The fool . . . heart'*. Prov. 18: 2. Noted in *Promus*, nos. 8, 230. *receiveth*. Understands.

91. *distemper*. Disorder. *complexion*. Constitution, balance. *recured*. Restored to health. *searched*. Probed, examined. *prejudicate*. Prejudiced. *simplicity*. Innocence, naïvety. *throughly*. Thoroughly. *goodness . . . particular*. Selfish, of purely private benefit. *fructifying . . . goodness*. Fertile and productive goodness. For Bacon setting social good above selfish see *Adv.L.*, above, pp. 246–50, 252. *draw on*. Entice, allure. *'deeps of Satan'*. Rev. 2: 24 (his 'deep things', or cunning plots). *insinuation*. Persuasiveness. *'Try . . . good'*. 1 Thess. 5: 21. Noted in *Promus*, no. 253. *election*. Choice. *fountain*. Source. *'Be . . . doves'*. Matt. 10: 16. *wreaths and folds*. The coilings of a serpent's body. *sinks*. Sewers, cesspools. *generality*. Inclusiveness. *'God . . . pure'*. Untraced in this form; but cf. 2 Cor. 3: 5, 'Not that we are sufficient of ourselves to thinke any thing as of ourselves, but our sufficiencie is of God'; and Jude 24: 'Now unto him that is able to keepe you that ye fall not, and to present you faultlesse before the presence of his glorie . . .'. *OF THE EXALTATION OF CHARITY* *Exaltation*. Raising to a high degree.

'*If... him*'. Job 31: 29. *detestation... Job*. Spedding translates 'The protestation of Job'. *publicans... love... enemies*. Matt. 5: 43–8.

92. *satisfactions*. Atoning for an offence by reparation. *bravery*. Ostentation. *greatness of the mind*. Self-importance. *insult*. Exult contemptuously. OF EARTHLY HOPE '*Better... mind*'. Eccles. 6: 9. Noted in *Promus*, no. 1280: 'Melior est oculorum visio quam animi progressio.' *Pure sense... impression*. Pure perception through the senses. In *Promus*, no. 1282, Bacon noted: 'Sufficit praesentibus bonis purus sensus' ('Pure sense suffices for present good'). *government*. Regulation, control. *apprehensions*. Anticipations, especially with fear or dread. In *Promus*, no. 1287, Bacon noted: 'Imaginationes omnia turbant, timores multiplicant, voluptates corrumpunt' ('Imaginations disturb everything, they multiply fears, they corrupt pleasures'). *particular*. Individual detail. *foretelling... present sense*. Cf. Aristotle, *De anima*, iii. 10, 433b 5–10. *beateth upon*. Falls violently and relentlessly. '*The oracles... abuse*'. In the Latin original, 'Fallitur augurio spes bona saepe suo' ('Hope seldom coincides with destiny'): Ovid, *Heroides*, xvii. 234. Noted in *Promus*, no. 1117. '*A... doubts*'. In the original, 'Pessimus in dubiis augur timor' ('In doubtful times Fear still forbodes the worst'): Statius, *Thebaid*, iii. 6). *froward*. Unfavourable, perverse. *fear... some use... hope... unprofitable*. In *Promus*, no. 1288, Bacon noted: 'Anticipatio timor est salubris ob inventionem remedij spes inutilis' ('Fear is a wholesome anticipation on account of its invention of a remedy—hope is useless'). *No shape... prepared be*. In the original, 'Non ulla laborum, | O virgo, nova mi facies inopinave surgit: | Omnia praecepi, atque animo mecum ante peregi': Virgil, *Aeneid*, vi. 103–5 (Aeneas to the Cumaean Sibyl). Noted in *Promus*, no. 380. *sooner... satiety*. In *Promus*, no. 1291, Bacon noted: 'Vitam sua sponte fluxam magis fluxam reddimus per continuationes spei' ('Life, which is fleeting enough of itself, we render more fleeting by a constant succession of hopes').

93. *principal*. Capital, sum invested. *increase*. Interest. *fortitude*. Cf. *Of Tribute*, above, pp. 22 ff. *lightness*. Frivolity, triviality. *poets... hope*. The tradition derives from Hesiod's fable of Pandora's box, which 'scattered pains and evils among men', but retained 'one thing, Hope, only, which did not fly through the door': *Works and Days*, 79–98, tr. Dorothea Wender (Harmondsworth, 1973), who adds the note: 'Hesiod leaves it ambiguous as to whether Hope is the one solace left for men in the now troubled world, or simply one more of the troubles brought by woman' (p. 155). In his *Promus*, no. 1280a, Bacon noted the sentence: 'Spes in dolio remansit sed non ut antidotium sed ut major morbus' ('Hope remained in the jar, but not as an antidote, but as a worse disease'). *counterpoison*. Antidote. Cf. Essay 15, 'Of Seditions and Troubles', above, pp. 369–70, for a more benign view of hope. *imaginations of hope... ungrateful... past*. In *Promus*, no. 1289, Bacon noted: 'Imminens futuro ingratus in praeteritum' ('Springing forward to the future, ungrateful toward the past'). '*I saw... mind*'. Eccles. 4: 15–16: 'I considered all the living which walk under the sun, with the second child that shall stand up in his stead.... Surely this also is vanity and vexation of spirit.' Noted in

Promus, no. 1286, and cf. *Adv.L.*, above, p. 268. *floateth*. Hesitates, wavers. *composition*. Combination, harmonious balance. *rides at anchor*. Rests temporarily (in the Renaissance the anchor was a common emblem for hope). *convenient*. Appropriate, suitable. *so that*. Provided that. *due*. Fitting. *discourse of the mind*. Reasoning process.

94. *unequal*. Inconsistent, unstable. *wandering*. Vague, restless. In *Promus*, no. 1285, Bacon noted: 'Spes facit animos leves tumidos inaequales peregrinantes.' *tincture*. Tinge, tint. *The sum... count*. In the original, 'Vitae summa brevis spem nos vetat inchoare longam' (Horace, *Odes*, I. iv. 15). Noted in *Promus*, no. 1284. *OF IMPOSTORS* *'Whether... youwards'*. 2 Cor. 5: 13. *carriage*. Demeanour. *conversation*. Behaviour. *trances*. Lat. 'extasis'. *ravishment*. Rapture, ecstasy. *appliable*. Compliant. *'I am... all men'*. 1 Cor. 9: 22. *privy*. Sharing the knowledge of. *leaven*. Debasing or corrupting quality. *'Sober... men'*. 2 Cor. 5: 13. *OF THE SEVERAL KINDS OF IMPOSTURE* *'Avoid... called'*. 1 Tim. 6: 20. *'Avoid... fables'*. 1 Tim. 4: 7. *'Let... speech'*. Eph. 5: 6. In *Promus*, no. 252, Bacon noted: 'Hoc autem dico ut nemo vos decipiat in sublimitate sermonis', conflating Eph. 5: 6 with 1 Cor. 2: 1 ('with excellency of speech'). *style and phrase*. Characteristics. *new terms of art*. Bacon disapproved of unnecessary neologizing: cf. above, p. 193. *art... Schoolmen*. Cf. *Adv.L.*, above, pp. 140–2, 224.

95. *clatterings*. Noise and fuss: Lat. 'turbae'. *church poets*. In the Latin original, 'sacri Poetae': presumably poets who wrote didactic works designed to promote a philosophical system, such as Lucretius' exposition of Epicurus' philosophy in *De rerum natura*. *legends*. The translator's laconic version of the original 'vitae patrum', the lives of the Church Fathers. *mystical and profound*. The translator's lame version of the original, 'genus mysticum et gnosticum' or 'mystic and Gnostic style of discourse'. The Gnostics were a heretical Christian sect of the 1st to 3rd centuries AD who claimed to have special redemptive knowledge of God and of the origin and destiny of man (*SOED*). *wit*. Mind. *trained on*. Allured, deceived. *abused*. Deceived. *Of Atheism*. Cf. Essay 16, above, pp. 371 ff. *'The fool... no God'*. Ps. 14: 1. *will*. Wish. *makes not*. Is of no advantage. Cf. Essay 16, above, pp. 371 ff. *resolve*. Assure, satisfy. *position*. Proposition, postulation. *sparkle of our creation light*. Lat. 'igniculus luminis primi', the remnant of the *imago Dei* given to man at the Creation. Cf. *Confession of Faith*, above, p. 109. *'Then... opinion'*. In the original, 'Tunc animus meus accessit ad meam sententiam' (Plautus, *Aulularia*, 383). *fear of government*. In the original, 'metu legis', through fear of the law.

96. *'To deny... enough'*. In the original, 'Negare Deos difficile est in concione populi, sed in consessu familiari expeditum': a digest of Cicero, *De natura deorum*, I. xxii. 60 f. In this dialogue on 'the being and nature of god', Cotta, representing the Academic philosophy, attacks Epicurean theology. 'In an inquiry as to the nature of the gods, the first question we ask is, do the gods exist or do they not? "It is difficult to deny their existence." No doubt it would be if the question were to be asked in a public assembly, but in

private conversation and in a company like the present it is perfectly easy.' *drenched*. Saturated, obsessed. *inculcate*. 'Instil an idea into the mind by forceful admonition or persistent repetition' (*SOED*). *Lucretius... against religion*. In his poem he develops Epicurus' criticism of Greek religious ideas. *enduring*. Suffering. *false of*. Not trusting. '*Who... distrusts it*'. In the original, 'Qui alteri opinionem approbare sedulo cupit, ipse diffidit'; untraced. *politiques*. Politicians. *compass*. Scope. *cunning*. Pretence. *attributed... providence*. Cf. *Adv.L.*, above, p. 272, and Essay 40, 'Of Fortune', above, p. 421. '*have... nets*'. Hab. 1: 16. *knowledge of nature*. In the original, 'physicis'. *a little... to atheism... much... to religion*. Cf. *Adv.L.*, above, pp. 125–6, and Essay 16, above, pp. 371–3.

OF THE COLOURS OF GOOD AND EVIL [SELECTIONS]

In medieval rhetoric the word 'colour' was used for any rhetorical device made by art, but by the end of the 16th century it had acquired the sense 'to disguise, conceal under a fair appearance'. Bacon placed this fragmentary treatise within the genre of rhetoric known as deliberative, that practised in assemblies, according to Aristotle's canonic definition of the three rhetorical genera (*Rhetoric*, i. 3, 1358b6 ff.). Deliberative oratory, whether used by 'private counsellors' or public speakers, 'urges us either to do or not to do something', arguing for the acceptance or rejection 'of a proposed course of action': hence Bacon's alternative title, 'Places [that is, *loci* or 'seats' of argument] of perswasion and disswasion'. In his *Promus of Formularies and Elegancies* (1594–5) Bacon had already assembled the raw material for a substantial treatise, including two sets of what he calls 'Semblances or popularities of good and evill with their regulations for deliberacions' (no. 1440; cf. nos. 1246–364 *passim*, 1441–60). This first published version, modestly entitled 'a fragment', includes only ten short sections from the hundred or so in the *Promus*. Although he mentions the collection briefly in the *Advancement of Learning* as filling a gap in the 'attendances' or auxiliary helps in rhetoric, Bacon quotes only one sophism and its redargution (above, p. 240). When he revised and enlarged that book for its Latin translation, the *De Augmentis Scientiarum Libri IX* (1623), he reprinted the original 'fragment' and added two further colours (*Works*, iv. 458–72), still only a fraction of the material he had collected.

In each section Bacon sets out a sophism, or apparently sound proposition, and then refutes it. He accurately described the corrective function of such analyses: 'to make a true and safe judgment, nothing can be of greater use and defence to the mind, than the discovering and reprehension [refutation] of these colours, shewing in what cases they hold, and in what they deceive... being performed, it so cleareth man's judgment and election, as it is the less apt to slide into any error' (above, p. 97). While belonging mostly to the first stage of rhetoric, the invention of sound arguments, Bacon also linked this fragmentary treatise with one of the major goals of all rhetoric, to achieve persuasion by *movere* (moving the hearers' passions to endorse whatever the orator is urging). In 1597 he writes that such refutations 'quicken and strengthen the opinions and persuasions which are true', for if reasons be delivered not 'heavily and dully' but are 'varied and have more life and vigour put into them by these forms and insinuations, they cause a stronger apprehension, and many times suddenly win the mind to a resolution' (above, p. 97). In 1623 Bacon was more laconic: 'their use is not more for probation than for affecting and moving. . . . Therefore these points and stings of words are by no means to be neglected' (*Works*, iv. 458).

Bacon's unacknowledged inspiration for the *Colours of Good and Evil* may have been Aristotle's *De Sophisticis Elenchis*, which, as one of the logical treatises in the *Organon*, formed part of many university curricula. However, the only Aristotelian debt Bacon acknowledged is to *Rhetoric*, i. 6–7, a discussion of the nature and degrees of good, which Bacon treats rather unfairly, pronouncing Aristotle's labours 'in three points defective': in giving only a few examples, not adding 'answers', nor seeing their use for moving the feelings (above, p. 240; *Works*, iv. 458). In fact, Aristotle's discussion is quite extensive, but focused to a different end. Bacon had invoked Aristotle in an unpublished letter (*c.*1597) addressed to Lord Mountjoy, apparently intended to accompany a presentation copy of this treatise. Bacon announces that he is sending Mountjoy 'the last part of the best book [? the *Rhetoric*] of Aristotle of Stagira, who (as your Lordship knoweth) goeth for the best author. . . . [I] do freely acknowledge that I had my light from him, for where he gave me not matter to perfect, at the least he gave me occasion to invent.' After some remarks in the polemical mode that he elsewhere recommended as 'discoursing scornfully of the philosophy of the Grecians' (*Works*, xi. 64), Bacon concedes that although he has 'brought in a new manner of handling this argument to make it pleasant and lightsome', he does not 'pretend so to have overcome the nature of the subject, but that the full understanding and use of it will be somewhat dark, and best pleasing tastes of such wits as are patient to stay the digesting and soluting unto themselves of that which is sharp and subtile' (*Works*, vii. 69–71).

97. *deliberatives.* In *Promus*, no. 1278, Bacon noted: 'In deliberatives and electives'. *popularities.* Arguments appealing to popular attitudes. *circumstances.* All the external conditions affecting an act or event. *weak.* Untrained in logic and argument. *quicken.* Give life to. *fine and fastidious.* Over-scrupulous. *insinuations.* Arguments designed to win over hearers by using an indirect or oblique approach. Bacon subsequently defines rhetoric as 'imaginative or insinuative reason' (above, pp. 218, 239). *apprehension.* Understanding, conception. *resolution.* The solving of a doubt or problem. *reprehension.* Refutation. *hold.* Are valid. *fallaxes.* Fallacies, flawed or sophistic arguments. *elenches.* Refutations. '*Quod . . . esset*'. 'That which has relation to truth is greater than that which has relation to opinion. And the proof that a thing has relation to opinion is this: it is what a man would not do, if he thought it would not be known' (tr. Spedding). A rather garbled version of this Colour is in *Promus*, no. 1270. *Epicures.* Epicurean philosophers, great rivals of the Stoics. *felicity.* Good fortune, happiness. *player.* Actor. *left.* Deprived.

98. *out of heart and countenance*. Disheartened, disconcerted. '*bonum theatrale*'. Public good. Cf. *Adv.L.*, above, p. 247. *Populus... plaudo*. Horace, *Satires*, I. i. 66, 'The people hiss me, but at home I clap my hands for myself.' *Grata... pudorem*. Theocritus, *Idylls*, xxvii. 69, in the Latin translation of Eobanus Hessus: 'her eyes downcast, her heart glad within'. *subtile*. Not easily understood, abstruse. '*propter... popularem*'. For the sake of public appearance. '*maxime... reverere*'. Pythagoras, *Golden Verses*, 12: 'a man should above all reverence himself.' *in solitudine*. When alone. *in theatro*. When in company. *percase*. Perhaps. *supposition*. 'A thing held to be true and taken as the basis of an argument' (*SOED*). *reprehend*. Refute, find fault with. *opinion*. Reputation. *election*. Choice. *causa impulsiva*. Impelling, motivating cause. *causa constituens... efficiens*. Constitutive, establishing cause. *doing*. Performance. *form*. Manner, formula. *of which*. That one. *delicacy*. Luxury, self-indulgence. *attend*. Wait on. '*nota... esset*'. 'It is known that whatever is chosen to be done for the sake of opinion, not for the sake of truth, would not be done secretly if it could not be done publicly.' '*Quod... malum*'. 'That which is next to a good thing is good; that which is far off, is evil' (tr. Spedding). *severed*. Separated.

99. *consenting*. Agreeing. *quartered*. Lodged (as in military 'quarters'). *nature... in similitude*. Tendency of similar things. *middle region*. Cf. Aristotle, *Meteorology*, i. 12, 348a15 ff. Wright (p. 348) quotes from Thomas Blundevile's *Exercises* (1594) on the Aristotelian division of the air into three regions, the 'highest region being turned about by the fire', the 'lowest region' being 'hotte by the reflexe of the sunne, whose beames first striking the earth, doe rebound backe againe to that region'. Hence 'the middle Region is extreame cold by contra opposition, by reason that it is placed in the midst betwixt two hotte Regions, and therefore in this Region are bred all cold watry impressions, as frost, snow, ice, haile, and such like' (fo. 179b). *per antiperistasin*. By resistance or reaction; 'opposition of circumstances'. In *Promus*, no. 3, Bacon noted: 'In circuitu ambulant impii—honest by antiperistasis'. The quotation is from Ps. 12: 8, 'The wicked walk on every side, when the vilest men are exalted.' *environing*. Surrounding, enclosing. *pleasantly*. Wittily. *taken hold of*. Caught up, responded. *propter*. Because of. *amplitude*. Bulk, especially width, breadth. *in their kind*. By nature. *engross*. Amass, collect from all quarters. *next*. Nearest. *spread*. Spreading. *pined*. Starved, deprived. '*divitis... servi*'. 'The servants of a rich man are most servants.' In *Promus*, no. 886, Bacon noted: 'Spartae servi maxime servi' ('The slaves of Sparta were the greatest of slaves'), from Erasmus, *Adagia*, IV. ix. 15. *next*. Preceding. *extenuate*. Diminish in size. *astronomers*. This word was used interchangeably for 'astrologers'. *conjunction*. 'The apparent coincidence or proximity of two celestial objects as viewed from the earth' (*SOED*): thought by astrologers to have ominous significance. *aspect*. 'The position of one planet with respect to the others, as they appear from the earth' (*SOED*). '*saepe... boni*'. Ovid, *Ars amatoria*, ii. 662: 'Vice often hides itself in the neighbourhood of virtue'; cf. *Adv.L.*, above, p. 278. *sanctuary-*

men. Those who sought sanctuary in a sacred place, where they were traditionally immune to arrest by the civil authorities. *inordinate*. Disorderly, unrestrained.

100. *confines*. Frontiers. '*The ... whole*'. Matt. 9: 12. *nearness of*. Associating with. *rioters*. Dissolute people. '*Quod ... malum*'. 'The ill that a man brings on himself by his own fault is greater; that which is brought on him from without is less' (tr. Spedding). *imputation*. Accusation. *attemper*. Moderate, reduce. *gemination*. Doubling. *Seque ... malorum*. Virgil, *Aeneid*, xii. 600: 'She calls herself the cause of all this ill' (tr. Dryden). *extremities*. Extreme measures; excessive behaviour in crises. *without*. Outside. *evaporation*. Gradual dispersion. *Nemesis*. The Greek goddess of retribution. *expostulation*. Protest, reproof. *Atque ... mater*. Virgil, *Eclogues*, v. 23: [Daphnis' mother, 'clasping her son's piteous corpse'] 'cried out on the cruelty of both gods and stars'. *suffocateth*. Smothers, overwhelms. *in nostra potestate*. In our own power. '*That ... errors*'. From *Phil.* i and iii. *declination*. State of decline.

101. *degrees*. Perhaps 'stages towards knowledge or virtue': Epictetus' main work, the *Enchiridion*, means 'hand-book or manual'. *extern*. External. '*The worst ... neither*'. *Enchiridion*, 5. *Leve ... onus*. Ovid, *Amores*, I. ii. 10: 'Balance makes a burden light.' *true ... cleaving to themselves*. Pursuing their own self-interest. *ill*. Badly. *shift*. Pretext, evasion. *ado*. Fuss. *against ... consents*. Cf. Essay 8, 'Of Marriage and Single Life', above, p. 354. *if they be never so ill used*. However badly they are treated.

ADVICE TO FULKE GREVILLE ON HIS STUDIES

This little treatise was printed by Spedding (*Works*, ix. 21–6) from a manuscript in the Bodleian Library (Tanner MSS, vol. lxxvi, fo. 82). Although originally ascribed to the Earl of Essex, Spedding had no hesitation in pronouncing it to be by Bacon, and indicated several parallels in thought and style to the discussion of epitomes in the *Advancement of Learning* and elsewhere. His ascription to Bacon was strengthened by Vernon F. Snow, who discovered an earlier manuscript in the Public Record Office (PRO, SP.14/59, fo. 4^r–5^v) and published it with a helpful commentary ('Francis Bacon's Advice to Fulke Greville on Research Techniques', *Huntington Library Quarterly*, 23 (1960), 369–78). I reproduce his text here, modernized and with a few corrections, by kind permission of the *Huntington Library Quarterly*.

Paul E. J. Hammer, however (*Studies in Philology*, 91 (1994), 167–80), has challenged this ascription to Bacon, alleging it to be the work of Essex and his secretariat. Hammer has identified the MS published by Snow as being in the hand of Reynoldes, Essex's personal secretary, but on its own, of course, this proves nothing; and no other evidence is cited. Once again this fact seems rather to strengthen Hammer's analysis of how Essex used (or commissioned) compositions by others to rehabilitate himself in the eyes of the Queen as a serious scholar and reliable counsellor. Indeed, there is a well-known letter from Bacon to Essex (not cited by Hammer), dated 4 October 1596, in which Bacon counselled Essex to change his behaviour at court in order to modify the dangerous impression he gives of being '*opiniastre*; not rulable', threateningly affecting 'a martial greatness'. Bacon's politic advice was that Essex should 'pretend to be as bookish and contemplative as ever you were' (*Works*, ix. 40–5). Bacon not only gave this advice but was prepared to implement it by writing this *Advice*, as he had done with the *Advice to Rutland on his Travels*. While disputing Bacon's authorship, Hammer again did not bother to examine the contents or style of this letter: a more detailed account (Brian Vickers, 'The authenticity of Bacon's earliest writings', forthcoming in *Studies in Philology*) will present stronger advice to confirm the ascription.

The addressee is the poet Sir Fulke Greville (1554–1628), a friend and political ally of Bacon. 'Both resided together in Essex House on the Strand at various times during the last decade of Elizabeth's reign', both shared the political patronage of the Earl of Essex before his fall in 1599, and corresponded when apart (Snow, art. cit., 377). Greville may have asked Bacon for advice on his studies when preparing his philosophical poem 'A Treatise of Human Learning': see *Poems and Dramas of Fulke Greville, First Lord Brooke*, ed. Geoffrey Bullough (London, 1932), especially i. 52–62, indeed Bullough argued that Greville's poem is in part a counterstatement to Bacon's *Advancement of Learning*. It is significant, then, that Bacon, who criticized elsewhere the narrowing effects of university studies being limited to theology, law, and medicine, should warn Greville that orthodox graduates would be 'strangers to the books your abridgers should read' (above, p. 104), suggesting that Greville had planned research on a wider scope. One interesting point that Hammer has established (art. cit., 174–5) is that in about 1590 Greville had offered a yearly stipend to John Coke, a young fellow of Trinity, Cambridge, if he would 'commit [his] work in the arts to him alone', and thus 'live by means of letters'. It is not clear what the offer actually implied, or whether Coke ever took it up, but it does show that Greville was attracted by the idea of having some university men devil for him.

102. *get*. Hire. *gather*. Collect information. Bacon himself employed several research assistants to collect and transcribe notes on topics he was interested in: cf. *Works*, xi. 61–2. *heads . . . common places*. Arrangement of material by headings or topics, which Bacon praised elsewhere, e.g. in *Adv.L.*, above, p. 224, 229. *Epitomes*. For Bacon's disapproval of epitomes see *Adv.L.*, above, pp. 179, 236 and Essay 'Of Studies', above, p. 439. *Justinian*. The *Corpus iuris civilis*, a digest of Roman law, was commissioned by the emperor Justinian in the 6th century AD. *Littleton*. The *Tenures* (*c*.1481) of Sir Thomas Littleton, a treatise on land and property, regarded in the 16th century as a fundamental authority on English law. *logic*. The major logical treatise of Peter Ramus was the *Dialectica* (Paris, 1543; 21 different versions by 1577): see p. 638 below. The simplification of the traditional logic course that Ramus proposed was often regarded with suspicion, as encouraging superficiality: cf. *Adv.L.*, above, p. 235–6 and Richard Hooker, *Of the Lawes of Ecclesiasticall Politie* (1593), I. vi. 4. *physics*. The *Physicae, seu de Naturae*

Philosophia Institutio (1574) of Cornelius Valerius. *politics*. The *Politica sive Civiles Doctrinae* (1589) of Justus Lipsius (1547–1606), distinguished Belgian humanist; an English translation appeared in 1594. *Art of War*. Machiavelli's *L'arte della guerra* (1521) appeared in English in 1560 (repr. 1573, 1588). *history*. I have modernized the spelling here and elsewhere (Bacon writes 'story'). *Carion*. John Carion, French historian, whose *Chronicorum libellus* (1543) was reissued with further commentary by Philip Melanchthon in 1544, and Johannes Funccius in 1554. *Bibliotheque Historien*. Snow argues that this refers to Lancelot Voisin's *L'histoire des histoires* (1599): this is the main evidence for dating this essay *c.*1599. *calendars*. Lists, registers. *Civilian*. Practitioner of civil law. *naturalist*. Natural philosopher. *Mercator's... Map*. Gerardus Mercator (1512–94), Flemish mathematician and geographer, famous for his projection technique, more accurate than any previous system, and for many detailed maps, e.g. *Orbis terrarum compendiosa descriptio* (1587). In later versions of this argument Bacon substituted the name of Ortelius: cf. *Adv.L.*, above, p. 237. *'Multum non multa'*. '[Study] much, not many things' (Seneca, *Epist.* I. xxxiii—an attack on the over-use of extracts and epitomes). *tedious*. Prolix. *reason*. Reasonable. *would*. Want to.

103. *post*. Haste. *matter of art*. Treatises on practical subjects. Bacon's point is that to reproduce the conclusions without the justifying argument would be useless. *positions*. Principles, conclusions. *arguments*. Topics for discussion. *Florus's epitome*. The *Epitome bellorum omnium annorum DCC* (*c.* AD 140) of Lucius Annaeus Florus, sometimes described as an abridgement of Livy's history of Rome, *ab urbe condita libri*, written between 29 BC and AD 11, but often at variance with it. *conceive*. Grasp mentally, comprehend. *conceit*. Thought, intellect. *'Nusquam... conjuncta'*. 'Nowhere, as a rule, is service given without recompense, nor recompense except for service; toil and pleasure, most unlike in nature, have been linked together in a sort of natural bond' (Livy, V. iv. 4). *'id scimus'*. Cf. *Advice to Rutland*, above, p. 75. *Curtius*. Quintus Curtius Rufus, rhetorician and historian, published (*c.* AD 69) a history of Alexander, of which only parts survive. *Plutarch*. The 'Parallel Lives of the Greeks and Romans', written *c.* AD 105–15, one of the most celebrated historical sources of antiquity. Bacon made much use of Plutarch for his own account of Alexander, in *Adv.L.*, above, pp.159–61. *humours*. Personal characteristics. *accidents*. Events, happenings. *title*. Heading, topic.

104. *popular*. Endorsed by the people. *grandees*. Noblemen of the highest rank. *'Quae... sua'*. 'And that proceeding from slender beginnings it has so increased as now to be burdened by its own magnitude' (Livy, I, Pref. 4). *'Tempora... conficiunt'*. 'There are times in which the very might of a powerful people is destroying itself' (ibid.). *reduced*. Recorded, assigned. *book-makers*. Compilers of anthologies or books of excerpts, often arranged alphabetically. *idle heads*. Unnecessary, irrelevant headings. *paper-book*. Notebook. *physic*. Medicine. Universities, from the Middle Ages to the 19th century, only recognized three subjects for higher study: theology, law, and medicine.

105. *matter . . . words.* Alluding to the traditional terms for subject-matter (*res*) and its expression (*verba*), Bacon's point is that inexperienced readers prefer verbally pleasing writing, however superficial, to really profound argument. *mean.* Intend, purpose. *humanity.* The *studia humanitatis*, characteristic of Renaissance humanism, comprised history, poetry, rhetoric, philosophy, and literature. *Demosthenes . . . Cicero's.* For the same ranking of these orators see *Adv.L.*, above, p. 131. *for that.* Because. *Tacitus.* See further praise in *Adv.L.*, above, pp. 182–3, and for that historian's popularity at this time, K. C. Schellhase, *Tacitus in Renaissance Political Thought* (Chicago, 1976). *matters.* Events, the 'stuff' of history. For other praise of Thucydides see *Adv.L.*, above, p. 180. Bacon never mentions Herodotus by name. *natural, moral.* The two main branches of philosophy. *waded.* Studied, proceeded. *eyes.* Eyesight. *lapidaries.* Artificers who cut or engrave precious stones. *proverb.* See *Advice to Rutland*, above, p. 73 and *Adv.L.*, above, p. 252. *Phormio . . . Hannibal.* See *Adv.L.*, above, p. 252. *conceit.* Thought-processes. *suggestion.* Prompting, stimulus. *judicious.* Able to judge. *dehort.* Dissuade, advise against. *end.* Aim, goal. *distasted.* Displeased.

106. *Make.* Keep.

A CONFESSION OF FAITH

Bacon's *Confession of Faith* was first printed in 1641, in three different editions, two of which have the subtitle 'Penned by an Orthodox man of the reformed Religion' (Gibson 200–2). It was included in *The Remaines of the Right Honorable Francis Lord Verulam* (1648: Gibson 218), and reprinted by William Rawley in the *Resuscitatio* (1657: Gibson 226). Rawley's source was a manuscript (BL Harl. MS 1893: Beal BcF 154) which Spedding judged to be 'a copy in the hand of one of Bacon's servants, and the oldest I have met with' (*Works*, vii. 216), and took as his copy-text, collating it against two other MSS in the British Museum Library (Beal BcF 156, 158). Peter Beal listed nine further copies in his *Index* (BcF 155, 157, 159, 160–5), and has discovered an additional one since (private communication). The *Remaines* claimed that Bacon had written the piece 'about the time he was solicitor general', but Rawley merely says that he composed it many years before his death. Spedding, observing that his copy-text describes it as being 'by Mr Bacon', dated it some time before Bacon was knighted, in the summer of 1603. Otherwise, we know nothing about the text, nor what might have occasioned it.

The Confession of Faith was an important document in the sixteenth-century Reformation, embodying the principles of faith of a Protestant Church, as it became an urgent necessity to define Protestantism against Rome. Such a declaration also had a legal force, for whoever publicly opposed it in a territorial Church where it was accepted had either to leave that territory, or be penalized. Professor Basil Hall (private communication) writes: 'I know of no instance of a private person (even a cleric of standing) writing a Confession of faith to clarify his own mind or to organize his thinking. Someone regarded as a heretic or a dangerous thinker might present in writing an account of his beliefs to an authorized court (Anabaptists did so sometimes) but it would not have the force of a Confession of Faith for the state or the Church where he presented it. (Such statements were in any case called *Articles*, not Confessions.) I cannot think of anyone writing a Confession as Bacon did. Men left papers with an account or even summary of their religious ideas but did not call them a Confession that I know of. Bacon here is to my mind unique.'

As for the form and content of this work, they also seem unique. Bacon neither copied the pattern of the Creeds accepted by the English Church as being based on Scripture, nor did he model his work on the major confessional statements in existence. This series began with the Lutheran *Augsburg Confession*, mainly written by Melanchthon, which was presented at Augsburg to the Emperor, Charles V, and published in 1531 (substantially revised in 1540). It is divided into two parts, the first epitomizing in 21 articles Lutherans' beliefs on the doctrines of the Creeds (e.g. on the Trinity, and the person of Christ), so declaring that they in fact maintained the proper Catholic faith. To this element Melanchthon added a statement of those scripturally founded beliefs (e.g. on Justification and merit), with which Luther opposed what he considered to be additions and perversions by the medieval Church. The second part of the Augsburg Confession reviewed clerical abuses and called for remedies. Reformed Church confessions were subsequently published in Switzerland (the *First Helvetic Confession* of 1536, compiled by Bullinger, which is basically Zwinglian; the *Second Helvetic Confession* of 1566, drafted by Bullinger, but now under Calvinist influence), and in France (Calvin's *Gallican Confession* of 1559, also called the 'Confession of La Rochelle', 1571). All these activities were closely followed by the Protestant community in England, needless to say, and translations were produced, such as *A confession of fayth, made by common consent of divers reformed churches beyonde the seas* (1568 and 1571), which contains the second *Helvetic Confession* and Calvin's *Gallican Confession*. In England the series of Articles authorized by the sovereign in the course of the sixteenth century, culminating in the Thirty-nine Articles (1563, rev. 1571), can be seen as comparable institutional affirmations of belief, but in that typically English compromise, the 'via media'. (See e.g. Philip Schaff, *The Creeds of Christendom*, 1877 (6th edn. 1931); B. J. Kidd (ed.) *Documents Illustrative of the Continental Reformation*, 1911.)

The proposal to adopt a confession of faith by a state declaring its Protestant beliefs (such as Luther's Saxony, Zwingli's canton of Zurich, the city-state of Geneva) was a theological and political act, produced at an important turning-point in the life of that community, and capable of provoking further crises. What events in Bacon's life, external or internal, provoked him to set down his personal *Confession of Faith*, there is no way of knowing. For its structure, Bacon borrows almost nothing from previous models, and rearranges doctrinal themes in a very personal way. As Spedding rightly noted, it is 'digested and reproduced in a form of [Bacon's] own; in which the several parts of the scheme are exhibited in logical coherency' (*Works*, vii. 216). The general sequence discusses God, the Creation, and Christ (like the Apostles' Creed), but in a fluid, not mechanical way.

Although giving special emphasis to the Creation, Bacon did not follow what might seem to be the obvious model, the so-called 'hexaemeral literature', inspired by the *hexameron* or six days' work by God from which the world and its inhabitants emerged (see e.g. Arnold Williams, *The Common Expositor: An Account of the Commentaries on Genesis 1527–1633*, Chapel Hill, NC, 1948). This tradition largely comments on the biblical account a verse at a time: Bacon picks out only a few passages.

For its content, Bacon's *Confession* shows many points of contact with Calvinist theology. Like most English Protestants, Bacon accepted much of Calvin's doctrine while remaining a loyal member of the Church of England (as we see from his 1589 *Advertisement touching [its] Controversies*, above, p. 1), being neither a Puritan nor a separatist. John Calvin (1509–64), the most influential writer among the Reformers, published the first version of his *Christianae Religionis Institutio* in 1536, and enlarged it several times until its final edition in Latin (1559) and French (1560): an English translation was published by Thomas Norton in 1561. (I quote throughout from *Institutes of the Christian Religion*, tr. F. L. Battles, ed. J. T. McNeill, London, 1961.) Bacon's family background brought him close to Calvinism. His mother, Lady Ann Bacon (1528–1610) was a life-long supporter of the Protestant cause, and gave financial and other help to numerous militant Reformers. Her elder son Anthony (1558–1601), who served abroad on diplomatic business between 1580 and 1591, had contacts with many Protestants in France and Switzerland. In 1581–2 he lodged with Beza in Geneva, who, 'out of respect to him', dedicated his *Meditations* to Lady Ann. Both Francis and Anthony Bacon studied with the Protestant theologian Lambert Daneau, who dedicated his commentary on the minor prophets (Geneva, 1586) to Anthony.

The Calvinist nature of Bacon's *Confession* is recognizable from the way he presents some of its major doctrines: the elect and the reprobate, the manner of Christ's mediation, the corruption of man and nature by the Fall, with the consequence that sinners are subject to the wrath of God. Bacon followed Calvin on a number of central issues. He describes predestination, God's eternal counsel, as 'the perfect centre of God's ways', and emphasizes that Adam fell by his own choice, contrasting the 'liberty of the creature' with the 'liberty of God's grace' in the work of Christ saving the elect. As Basil Hall has pointed out (private communication), 'Bacon is trying to avoid the denial of free-will to Adam, and of our responsibility for our sins: Calvin simply affirmed the paradox'—unlike Beza, whose determinist views made even the Fall, by which free will was completely destroyed, belong to God's eternal plan. Yet, as my commentary will show, on many issues Bacon went his own way, not wanting to tie himself too closely to doctrinal positions, and being equally reluctant to get involved in current controversies, such as those concerning the covenants of works and grace (the so-called 'two covenant theology' of the Federal Theology), and the sacraments. Of his several debts to Calvin, Professor Hall draws special attention to Bacon's interpretation of Jacob's dream of a ladder ascending to heaven (Gen. 28: 12) as one 'whereby God might descend to his creatures, and his creatures might ascend to God'. This plainly derives from the idiosyncratic annotation given by Calvin in his

French Bible (repeated in the English Geneva Bible, 1560 and later), and in his *Commentary on Genesis*.

One point of special interest to students of Bacon is the way his underlying concern with natural philosophy keeps coming into this confessional statement, as in his quite personal emphasis on the laws of nature. Bacon affirms his belief that God 'created heaven and earth, and all their armies and generations, and gave unto them constant and everlasting laws, which we call Nature, which is nothing but the laws of creation' (above, p. 108). Bacon is obviously aware of the concept of natural law, derived from the Stoics, according to which 'one eternal and unchangeable law', given by God, 'summons [all people] to duty by its commands, and averts from wrong-doing by its prohibitions' (Cicero, *De re publica*, III. xxii. 33). This 'true law is right reason in agreement with Nature', applicable to all men equally, as Cicero writes in *De legibus* (I. x. 29; xii. 33), for all 'creatures who have received the gift of reason from Nature have also ... received the gift of Law, which is right reason applied to command and prohibition'. Bacon referred to natural law, in terms very close to Cicero's, in the concluding section of the *Advancement of Learning* (above, p. 290), and also invoked it, appropriately enough, in his legal writings (*Works*, vii. 529, 644–7, 663–5; xiv. 237). This legal and moral conception of a *ius naturale*, a universal justice prior to any formal legal code, was developed in various ways by both patristic and medieval philosophers, in works well known to the Renaissance. But Bacon is using the term 'laws of nature' in a different way, to describe the regular recurrence of natural and physical phenomena, what we would call scientific laws. The *Oxford English Dictionary* (2nd edn., 1989), under 'Law, *sb*. III.17.a', defines law in the scientific sense as 'a theoretical principle deduced from particular facts, applicable to a defined group or class of phenomena, and expressible by the statement that a particular phenomenon always occurs if certain conditions be present'. The first occurrences it can cite in English for 'laws of nature' in this sense both date from 1665, the *Philosophical Transactions of the Royal Society*, and Robert Boyle's *Occasional Reflections*. Bacon's *Confession of Faith* antedates these by over 60 years, and credit must be given him for this pioneer usage. (See also Paolo Rossi, 'Francis Bacon, Richard Hooker e le leggi della natura', *Rivista Critica di Storia della Filosofia*, 32 (1977), 72–7.)

As has recently been shown (J. E. Ruby, 'The Origins of Scientific "Law"', *Journal of the History of Ideas* 47 (1986), 341–59), the ancient use of 'law' for natural phenomena was derived from classical and biblical notions of a legislation laid down by God or Nature. But the Renaissance knew of other traditions in which divine legislation played no part, deriving from the Roman use of *regula* or rule in the sense of guideline or standard. In this context *lex* was originally used to describe principles laid down by authorities, principles which were subsequently seen as inherent in the nature of things. Both medieval optics, in its study of the laws of reflection and refraction, and Renaissance mathematics used the term 'law' to describe nature as a set of intelligible, measurable, predictable regularities. Renaissance astronomers (Regiomontanus, Copernicus, Peucer, Melanchthon) also used 'law' to describe the observable motions of the heavenly bodies, without reference to God. Some astronomers continued to praise God as the creator of the universe, the wisest and most orderly of craftsmen, but the concept of physical laws had been freed from the notion of divine legislation.

Seen against this background Bacon's concept of law is somewhat old-fashioned. It belongs more with the biblical notion of a divinely appointed cosmic order, as expressed in the Psalms or in several Old Testament books. Bacon follows the old tradition of regarding the Bible, the record of God's words and works, as a contribution to scientific knowledge, praising 'that excellent book of Job' in the *Advancement of Learning* (above, p. 151) as 'pregnant and swelling with natural philosophy', including cosmography and astronomy. Bacon's allegiance to older habits of thought may derive from his lack of initiative concerning the relevance of mathematics to natural philosophy. He was not unusual among philosophers at the end of the sixteenth century in failing to make this transition, but we can now judge that it seriously limited his scientific achievement. As A. C. Crombie has recently written, Bacon 'did not grasp the profound change being brought about in natural science by mathematical thinking. So when he specified the object of science as natures, and synonymously forms or laws, he held in view not laws expressing functional dependence such as were being made the object of quantitative experimental inquiries, but the configurations and processes of matter' (*Styles of Scientific Thinking in the European Tradition*, 3 vols., 1994, i. 629). Another reason for Bacon's neglect of the mathematical sciences was that he took as his model disciplines in natural philosophy not astronomy nor optics but the sciences of matter, inorganic and organic, what we would today call chemistry and biology.

Where Bacon is unusual (or so it seems on the basis of our current knowledge) is not only in accepting the divine creation of the universe (the orthodox belief, shared by Kepler, Galileo, Descartes and many other philosophers), but in taking biblical accounts of the creation literally, and in developing, in an equally literal way, the implications for natural philosophy of the Fall, in particular its effects on the laws of nature. These laws 'began to be in force when God first rested from his work and ceased to create', on the sixth day of creation. But, Bacon reasons, since God's curse on Adam included nature ('Because thou . . . hast eaten of the tree . . . cursed is the earth for thy sake', Gen. 3: 17), the laws of nature, like the rest of creation, must have 'received a revocation in part'—that is, a 'calling back' or reduction in energy and efficacy. The only exception to this general falling off is God's power to work a miracle, which Bacon sees as a 'new creation', but only partly and indirectly, 'not violating Nature, which is his own law upon the creature'. (The same definition of miracles recurs in the *New Atlantis*, above, p. 464.) Otherwise, the consequences of 'the fall of man' have been uniformly disastrous, for 'the image of God in man was defaced, and heaven and earth which were made for man's use were subdued to corruption by his fall'. Bacon does not go into detail here about the nature of this corruption as it affects the natural world, but in his later work he frequently reverts to the consequences for human cognition, deprived of the *Imago Dei* which had constituted the intellectual light first bestowed on Adam. In the later works, the Fall is seen as having brought about a divorce of the human mind from nature, and destroyed man's original dominion over the creation. The importance of natural philosophy, once properly renewed in Bacon's *Instauratio Magna*, is that it will restore the 'commerce between the mind of man and the nature of things ... to its perfect and original condition', before the Fall. The resulting remarriage 'between the mind and the universe, the divine goodness assisting', will undo the evil of the Fall, and Bacon prays that from it 'there may spring helps to man, and a line and race of inventions that may in some degree subdue and overcome the necessities and miseries of humanity' (*Works*, iv. 7, 27; also 247–8, 320, 405). This inspiring vision was yet to be formulated, but it is a direct response to the profession of belief set down here.

But the major significance of this work is as a record of Bacon's religious beliefs. Those modern critics who have labelled him an atheist should note the terms in which his chaplain, William Rawley, described this text in the 'Epistle to the Reader' prefixed to *Resuscitatio*, that important collection of minor works published in 1657: 'for that *Treatise* of his *Lordships*, inscribed *A Confession of the Faith*; I have ranked that in the Close of this whole *Volume*: Thereby to demonstrate to the world, that he was a *Master* in *Divinity* as well as in *Philosophy* or *Politicks*, and that he was versed no less in the *saving knowledge* than in the *universal* and *adorning knowledges*. For though he composed the same many years before his death, yet I thought that to be the fittest place, as the most acceptable Incense unto *God* of the *Faith* wherein he resigned his *Breath*; the Crowning of all his other Perfections and Abilities; and the best Perfume of his Name to the *World*, after his death.' (Sig. B3^{r}.)

My text is derived from Spedding, *Works*, vii. 215–26 (based on the British Library MS, collated with that in the *Resuscitatio*). In annotating it I have been much helped by the suggestions of the late Professor Basil Hall, whose comments are recorded as (B.H.). I have added at the end two of Bacon's prayers (from *Works*, vii. 257–60), which cannot be dated precisely.

107. *no nature ... matter ... spirit.* The three realms of living creatures and the natural universe; inanimate nature; angels. *Father ... persons.* The doctrine of the Trinity, the three persons in God, was a major issue in Christian theology from its beginnings. Calvin devotes an important chapter to this topic, *Inst.* I. 13 ('In Scripture, from the Creation onward, we are taught one essence of God, which contains three persons'). He subsequently discusses Father (I. 14 ff.), Son (II. 6 ff.), and Holy Spirit (III. 1 ff.). *jealous.* 'Of God: demanding absolute faithfulness and exclusive worship' (*SOED*). *Mediator.* One who mediates between God and man: applied especially to Jesus Christ (cf. 1 Tim. 2: 5). A central concern in Calvin's *Institutio*: cf. I. 2.1, 6.1; II. 6.1–2, 8.1, 10.1–2, 11.4, 12.1–13.1, etc. *Lamb of God ... worlds.* Cf. Rev. 13: 8 (Geneva Bible): 'Therefore all that dwell upon the earth, whose names are not written in the booke of life of the Lambe, which was slaine from the beginning of the world', with the gloss: 'As God ordained before all beginning, and all the sacrifices were as signes and sacraments of Christ's death'. Bacon 'wants it to be clear that man and creation cannot continue effectively without the energizing power of the Lamb, Christ as Mediator, and this must be seen as from before creation' (B.H.). *counsel.* Purpose. *descended.* Proceeded. *enjoyed . . . only for ever.* That is, 'he would have enjoyed forever . . . Godhead only'. (Such variations from what seems to us normal word order are common in Bacon's early writings, as in other Elizabethan prose.) *communicate.* Share, have a living relationship with. *one person of the Godhead.* Christ. *particular.* One individual, out of all creatures: Christ partakes of the natures of man and God. *ladder . . . descend to his creatures . . . ascend to God.* The Geneva Bible (1560, 1576) at Gen. 28: 12 includes a marginal note: 'Christ is the ladder whereby God & man are joined together, and by whome the Angels minister unto us: al graces by him are given unto us, and we by him ascend into heaven.' For a fuller statement of this interpretation, as against the traditional Jewish view of the ladder as symbolizing divine providence, see Calvin, *Commentaries on the first Book of Moses called Genesis*, tr. John King (Edinburgh, 1847), ii. 112–14. In *Inst.* II. 9.2 Calvin links this passage with John 1: 51, 'Afterward you will see heaven opened, and the angels of God ascending and descending upon the Son of Man', commenting: 'how excellent [Christ's] advent is, he has marked through opening by it the gate of heaven, that each one of us may enter there'. *in the same light and degree.* Not all creatures (men) receive the same illumination, nor to the same extent, leading them to election. *some . . . stand . . . fall . . . restored in their state.* Cf. *Inst.* III. 21 ('Eternal election, by which God has predestined some to salvation, others to destruction') and the following three chapters. But whereas Calvin recognizes only two states, the elect and the reprobate, Bacon, wrestling with the problem of people in the world, creates three categories: the elect, who keep their innocence although the Fall has occurred; the elect who fall but can be restored by God's grace; and the reprobate who fall and are left, enjoying success in this world, perhaps, but suffering eternal damnation (B.H.). *under wrath and corruption.* Calvin frequently describes God's wrath against sinners as the consequences of Adam's fall and our resulting corruption: cf. *Inst.* II. 1.8, 8.4, 10.18, 16.2, 16.3. *in the virtue of.* By the power or efficacy of. *perfite.* Perfect. *centre . . . God's ways.* A metaphor from geometry (the circle), or astronomy (planets revolving around a central point). A Christocentric theology was common to Reformers of many kinds, e.g. Luther, Calvin, and Hooker. *chose . . . should be united.* Cf. *Inst.* II. 12, that 'Christ had to become man in order to fulfill the office of mediator'; II. 13, on how 'Christ assumed the true substance of human flesh', with proofs of 'Christ's true manhood'; II. 14, on 'How the two natures of the Mediator make one person'; etc. *elected.* Chose. *a small flock.* Cf. *Inst.* III. 21.7. *express.* Manifest by external signs. *riches . . . glory.* That God's designs toward mankind display his glory is a constant theme in Calvin: cf. *Inst.* I. 5.1. *ministration of angels.* On God's providing angels for the help of mankind cf. *Inst.* I. 14.5–7. *reprobate.* Those rejected by God; hardened in sin. In Calvin's discussion of predestination mankind is divided into the elect, or chosen of God (cf. Rom. 8: 29–30), and the reprobate, who are damned: cf. *Inst.* II. 7.8 .ff.; III. 21–3. Bacon 'prefers to emphasize that all nature, all creatures, angels and devils, all temporal evolving, is under the counsel of God; all this redounds to the majesty of his being, and to his merciful bringing order and justice to reign' (B.H.). *administration.* Management, government. On God as perpetually active in governing the world, especially by contrast with Epicurus' notion of an idle and indifferent God, cf. *Inst.* I. 2.1–2, 16.1–17.

108. *dispensation ... times.* Dispensing the various epochs of creation, until the Day of Judgement, when time will be suspended. *ambages.* Indirect or roundabout paths. *one.* Identical with; the same as. *times and seasons ... known.* Cf. Acts 1: 7, 'It is not for you to know the times or the seasons, which the Father hath put in his own power.' *Word.* The creating *logos* of John 1: 1–14. *estate.* State, condition. *evil ... creature.* Calvin repeatedly emphasizes that 'Man proceeded spotless from God's hand' (*Inst.* I. 15.1), having been created with the 'pre-eminent endowments [of] reason, understanding, prudence, and judgment ... In this integrity man by free will had the power, if he so willed, to attain eternal life ... Therefore Adam could have stood if he wished, seeing that he fell solely by his own will', and 'in destroying himself he corrupted his own blessings' (*Inst.* I. 15.8). 'Although he would agree with this concept, Bacon is making a special point, viz. God is not the author of evil, since evil and vanity derive from man in his liberty, while God offsets human evil by the liberty of his grace. Bacon is trying to meet in this way the Catholic Church's criticism that Calvinism made God the author of evil' (B.H.). *restitution.* For God's mercy see *Inst.* I. 2.1, 2.8.3–5; III. 12.4 8, etc., and for God the redeemer see the whole of bk. II. But Bacon implies by restitution 'more than mercy, namely the restoring of that order of Nature that makes things go reasonably and smoothly, over against man's foolish, evil and vain choices, causing corruption. Again Bacon is concerned about something that the God oriented Calvin by-passed, and again he shifts the focus to his own concern for the laws of nature and how they order the universe' (B.H.). *prescience.* Foreknowledge. Where Calvin believes that God not only foreknew the Fall but arranged it (*Inst.* III. 23.7), Bacon 'merely argues that God foreknew the Fall and therefore turned it to the working of his counsel' (B.H.). *created Spirits.* Calvin, discussing the angels as ministers of God, writes: 'one may easily infer that he, to whom they devote their effort and functions, is their Creator', and goes on to argue that 'there should be no question that they are also his creatures', dismissing as only fruitless, speculations about 'the time and order in which they were created' (*Inst.* I. 14.3–4). *armies.* Multitudes. *generations.* That which is generated. Cf. Gen. 2: 4, 'These are the generations of the heavens and of the earth when they were created.' The Geneva Bible glosses: 'Or, the originals and beginnings'. *everlasting laws ... Nature.* The concept of nature having laws governing all its workings seems to be an original emphasis within Bacon's natural philosophy: see head-note above. Calvin, by contrast, takes natural law in its traditional social and ethical senses, linking it with

conscience and with the Christian's duties to society (as in Rom. 2: 14–15): cf. *Inst.* II. 2.22, 7.3–4, 8.1–2. *times.* Eras, epochs. *matter... without forms.* Gen. 1: 2. *interim.* Interval, intervening time. The six days of Creation, separated by night (Gen. 1: 5, 8, 13, 19, 23, 31). *curse.* Gen. 3: 14 ff., esp. 17 ('cursed is the ground for thy sake'), and 19 ('In the sweat of thy face shalt thou eat bread, till thou return into the ground; for out of it wert thou taken: for dust thou art, and unto dust shalt thou return'). *privation.* Act of depriving. *virtue.* Inherent quality, efficacy. *end of the world.* Where Calvin demonstrates how the Day of Judgement will be revealed (*Inst.* II. 16.17; III. 25.6, 9–12), Bacon's belief that 'the manner whereof is not yet revealed' implies that a new creation will take place at the end of the world (B.H.). Cf. Rev. 21: 1: 'And I saw a new heaven and a new earth: for the first heaven and the first earth were passed away.' *revocation.* Partial hindrance: the Fall 'calls back', in part, the ongoing energies of the Creation, affecting the laws of nature. *Sabbath.* Gen. 2: 2–3: 'God rested on the seventh day from all his work which he had made.' The Geneva Bible adds the gloss: 'For he had now finished his creation, but his providence still watcheth over his creatures, & governeth them.' *providence.* Cf. *Inst.* I. 16.1, rejecting the notion of God as 'a momentary Creator', since 'he is also everlasting Governor and Preserver—not only in that he drives the celestial frame as well as its several parts by a universal motion, but also in that he sustains, nourishes, and cares for everything he has made, even to the least sparrow'; also I. 16.2–17.14. *by compass.* In a roundabout way. *upon.* For, applying to. *produced.* Gen. 1: 11 ff., 'Let the earth bring forth grass', etc., where the Latin text repeats the word *producat.* Cf. *Adv.L.*, above, p. 215, and *Sylva Sylvarum*, v. 401 (*Works*, II. 475). *breathed.* Gen. 2: 7: 'God formed man... and breathed into his nostrils the breath of life.' *spirits.* Angels. Cf. *Inst.* I. 14.3–12.

109. *miracles... new creations.* Cf. *Religious Meditations*, p. 89–90 and headnote above. Calvin defends the authenticity of miracles, and their function in God's plan as aids to faith (*Inst.* I. 8.5–6, 13.13, 14.2, 16.2, 7, etc.). 'Bacon, however, is worried by these breaches in the ordered laws of Nature, and calls them new creations to overcome this, creations which then help redemption, greater than Nature. Bacon is again independent, finding a theological explanation for this problem in natural philosophy' (B.H.). *pass.* 'Situation or point in the course of any affair; *esp.* a critical position, a predicament' (*SOED*). *but.* Only; except. *in his own image.* Cf. Gen. 1: 27, and *Inst.* I. 15.3. The Geneva Bible glosses: 'man was created after God in righteousness and true holinesse, meaning by these two wordes all perfection, as wisedome, truth, innocencie, power, etc.' *reasonable soul... sovereignty.* Cf. *Inst.* I. 15.6–8. The sovereignty is that of Adam's over all the creatures, when he named them (Gen. 2: 19–20). For Bacon man can regain that dominion over nature in part through a reformed natural philosophy. *defection.* Desertion. *Good and Evil... beginnings.* Calvin argues that 'Adam was denied the tree of the knowledge of good and evil to test his obedience and prove that he was willingly under God's command', his pride being the cause of his sin (*Inst.* II. 1.4; and *Commentaries on Genesis*, ed. cit., i. 125–8). 'Bacon agrees, but adds that man's proud self-assertion led not only to disobedience but to making himself the master of good and evil by deciding for himself how they are to be understood. Calvin goes on at great length here to challenge the arrogance of philosophers and others who wanted to decide how to run their own lives and by-pass God's demands. Bacon agrees' (B.H.). *imagined.* Imaginary, supposed. *light.* Source of knowledge, illumination, which properly comes from God; Adam wished for himself to be as God. Cf. *Adv.L.*, above, pp. 123, 150. *moved.* Proposed, instigated. *suggestion.* Temptation, incitement to evil. Cf. Gen. 3 and *Inst.* III. 1.4.1. Calvin sees the Fall primarily as man's fault; Bacon attributes more responsibility to Satan, seeing the devil as an instrument to incite man (who is willing to listen) to the denial of God. Cf. the Geneva Bible's marginal notes to Gen. 3: 1 ('Now the serpent was more subtill [deceitful] than any beast of the field, which the Lord God hath made...'): 'As Satan can change himselfe into an Angel of light, so did he abuse the wisedome of the serpent to deceive man', 'God suffered Satan to make the serpent his instrument, and to speake in him'; and 3: 6 (Eve gave the fruit 'to her husband ... and he did eat'): 'Not so much to please his wife, as moved by ambition at her perswasion'. Bacon could have got his theme of the devil's agency from section VIII.i of the Heidelberg Catechism (still regarded as having confessional status today, as in the Netherlands Reformed Church). (B.H.). *defected.* Fallen, deserted. *of.* As a result of. *death and vanity.* Cf. *Inst.* III. 9.1–2 on 'the vanity of this life'. *defaced.* Destroyed. *subdued to corruption.* One of the most powerful emphases in Calvin's *Institutio* is on the miseries of human 'ignorance, vanity, poverty, infirmity, and—what is more—depravity and corruption' resulting from Adam's original sin (I. 1.1), which affects the whole of nature: *Inst.* II. 1.8–11, 2.12–25, 3.4–5, 5.15, 18–19, etc. *frustrate.* Made ineffectual, counteracted. *promise.* On God's promise to mankind cf. *Inst.* III. 13.4, 17.6, 18.4, 20.13, etc. *wrought.* Brought about, fulfilled. *several.* Different, distinct. *light of nature.* The original light given to man at creation, the *imago Dei* (Gen. 1: 26 f.) which was not entirely lost by the Fall. Cf. Bacon's *Adv.L.*, above, p. 290. Some authorities identified this 'image of God' with human reason. *accuse.* Call to account; charge with a crime: Cf. Gen. 4: 8 ff., Cain's murder of Abel, with God's rebuke: 'What hast thou done? the voice of thy brother's blood crieth unto me from the ground. And now art thou cursed from the earth...' *written law.* Cf. *Inst.* II. 7.1, on the moral law revealed to the chosen people through Moses, as foreshadowing Christ. Bacon seems to be referring literally to the Torah, the first five books of Moses, or Pentateuch (Gen. to Deut.), which still form the basis of Jewish religious law. *opened.* Explained. *prophet... law.* On Christ as teacher and prophet cf. *Inst.* II. 8.51–9, 9.4 15.1–2; IV. 19.30. *figures.* Signs, or 'types', foreshadowing some future state or event. *rites... law.* Cf. *Inst.* II. 7.1, 7.16, 8.28, 9.3–4, on the ceremonies enjoined in the law as 'types' foreshadowing the full and clear revelation of the gospel, at which the ceremonies cease. *continual.* Continuous. *Church of the Jews.* Jewish beliefs and practices, as recorded in the Old Testament. *pregnant.* Suggestive, prophetic.

110. *shadow.* On the Old Testament as foreshadowing the New cf. *Inst.* II. 8.15, 33, 9.1, 3–4, 10, 11.1–4, 14, etc. *evangile.* The message of the redemption of the world through Christ. *overshadowing.* 'Sheltering or protecting with some superior power or influence' (*SOED*). Cf. Luke 1: 35. *Word . . . made flesh.* Cf. John 1: 14, and *Inst.* II. 1–4.1, arguing that we should not understand John's words as implying 'that the Word was turned into flesh or confusedly mingled with flesh. Rather, it means that . . . he who was the Son of God became the Son of man—not by confusion of substance, but by unity of person'; also II. 14.6–7. *Substance . . . nature.* Cf. *Inst.* II. 14.1: 'For we affirm his divinity so joined and united with his humanity that each retains its distinctive nature unimpaired, and yet these two natures constitute one Christ.' *so.* To that extent. *catholicly.* Universally, without exception. *Deipara.* Mother of God (as a title of the Virgin Mary). This term (unusual in a work of Calvinist theology) occurs, e.g., in the *Canons and Decrees of the Council of Trent* (1564), and in the *Profession of the Tridentine Faith* (1564); cf. Schaff, *op.cit.* ii. 201, 209. *worker.* Creator, producer. *quickened.* Given life. *sacrificer and sacrifice.* Cf. *Inst.* III. 4.30. *satisfaction.* Cf. *Inst.* III. 4.26, arguing that 'Christ has provided full satisfaction', that is, 'since he alone is the Lamb of God, he also is the sole offering for sins, the sole expiation, the sole satisfaction'. *price.* Ransom. Cf. Matt. 20: 28: 'Even as the Son of man came not to be ministered unto, but to minister, and to give his life a ransom for many.' *finisher of the ceremony.* Christ has by his own sacrifice abolished the sacrifices of the Jewish law; the ceremonial law is now finished. *corner-stone.* Cf. Eph. 2: 20, 'Jesus Christ himself being the chief corner-stone'. *Lord of Nature.* An unusual epithet for Christ: Christ's miracles are as new creations (B.H.). *superior to the Angels.* Cf. *Inst.* I. 15.10, where Calvin, opposing the excessive importance given to angels in the Catholic Church, quotes Col. 1: 16 ff. on Christ being superior to the whole of creation: 'For by him were all things created, that are in heaven, and that are in earth, visible and invisible . . . And he is before all things, and by him all things consist . . . And, having made peace through the blood of his cross, by him to reconcile all things unto himself; by him, I say, whether they be things in earth, or things in heaven.' 'Bacon is saying that because we are in Christ (the redeemed) we are transfigured for God's new heaven and earth. He may have had in mind Rev. 22: 8, 9, where the Angel says that Christ is supreme and only to be worshipped' (B.H.). *Pilate . . . Romans.* Cf. Luke 3: 1. This passage is unusually explicit, as if Bacon were emphasizing the historical reality of these events, which took place in this town, at this time, with these actors. *raised himself.* If Bacon means this literally, he is attributing to Christ what the Scripture attributes to God. Calvin sees the Resurrection as God's work, but occasionally writes of Christ 'rising again': cf. *Inst.* II. 16.13. *by the space of.* To the extent of. *divers.* Several.

111. *merits.* 'The essential Protestant doctrine, that God gives us salvation through Christ, and whose merits alone save us gratuitously, out of God's mercy, thanks to the faith given by Christ through the Holy Spirit inspiring us to perform good works through his grace. Bacon agrees here with the

common Protestant doctrine, as found in the Augsburg, the Second Helvetic, and the Gallican Confessions' (B.H.). *do away*. Cancel. *effectual*. Efficacious. *regenerate*. On the Holy Spirit as initiating regeneration (that is, restoring the image of God in its original integrity), cf. *Inst.* III. 3.9–10, etc. *free grace*. Calvin emphasizes God's 'free grace' in *Inst.* II. 17.1; III. 13.5, 22.9, etc. *quickeneth*. On God's vivification or quickening of man, imparting 'the desire to live in a holy and devoted manner, a desire arising from rebirth', see *Inst.* III. 3.3; IV. 15.5, 17.5, 19.8. *conceiveth*. Forms, gives physical life to. *member*. Participant in, constituent part of. *imputation*. 'Attribution to believers of the righteousness of Christ and to Christ of human sin' (*SOED*). *figureth*. Represents as an image or type (of Christ). *lively*. 'Of or pertaining to life; necessary to life; vital' (*SOED*). *God's elect*. Cf. *Inst.* III. 21–4. *in regard of*. In comparison with. *more or less*. Greater or smaller. *sacraments*. Cf. *Inst.* IV. 14–17. A sacrament was a church service or rite held to have been instituted by Christ himself. Calvin (like other Reformers) recognized as sacraments, or 'testimonies of divine grace toward us, confirmed by an outward sign, with mutual attestation of our piety towards him' (IV. 14.1), only two, Baptism and the Eucharist, believing that they alone were attested in Scripture. 'Bacon is affirming their importance: we are required to be Church members, where alone we can receive these and other gifts of grace' (B.H.). *covenant ... children*. 'The Church receives our children, as God received their fathers into the realm of grace. Puritans had already begun to introduce the "Federal theology" (*foedus* = covenant) into England, but Bacon does not endorse it, confining 'covenant' to either an explanation of the two sacraments (which were both instruments of the covenant of grace), or to saying that as Israel of old was in covenant with God, so are we. Cf. the passage two paragraphs later, on "the outward works of God's covenant", that is, the covenant of grace shown in God's saving works' (B.H.). *reading*. Study of Scripture, either privately or (sometimes) collectively under the guidance of a minister. *censures*. Judicial (especially ecclesiastical) sentences. Cf. *Inst.* IV. vii. 6–7, xii. 1–10. *society of the godly*. Community of Christian believers. Cf. *Inst.* IV. 1.7. *vocation*. The calling of the elect to salvation. *derogating*. Detracting from, impairing. *tradition*. 'A doctrine usually regarded as having divine authority without written evidence, especially the laws held by the Pharisees to have been delivered by God to Moses' (*SOED*). *were*. Existed. *book ... closed*. As a body of record and doctrine completed and entire. *as to*. As far as, concerning. *Ark ... preserved*. Cf. the description of the chest containing the two tables of the law in Exod. 25: 18–21, 37: 1 ff. and Heb. 9: 4 ff., etc.

112. *catholic*. Universal. *spouse*. The metaphor of the Church as the bride of Christ is ancient. Calvin referred to Christ as the *sponsa ecclesiae*, and the long-established (since St Hippolytus and Origen in the 3rd century) allegorization of the Song of Solomon survived in the Geneva Bible, which prefaces that work with this note: 'In this song Salomon by most sweet and comfortable allegories and parables describeth the perfect love of Jesus Christ, the true Salomon & King of peace, and the faithful soule of his Church, which he hath sanctified and appointed to be his spouse, holy, chaste, and without reprehension.' *gathered*. Collected from, constituted by. *faithful dissolved*. Christians who have died. *faithful militant*. Christians on earth warring against evil. *mysteries*. The Greek word μυστεριον was translated in the Vulgate as 'sacrament' (*Inst*. IV. 14.2, 19.36), and was taken as referring to the two mysteries of the Baptism and Eucharist. As a highly controversial issue in contemporary Europe, affecting the English Church, it may be a sign of prudence that Bacon refrains from making any detailed statement about the sacraments (B.H.). *succession*. Line of descent; uninterrupted transmission of spiritual authority. *from*. By. *gift*. Divine grace. *inward anointing*. Receiving a directly, inwardly transmitted office, without external consecration. 'As an Anglican Bacon presumably accepts the Ordinal, which declares that consecration is given to clergy. He is emphasizing that men must have a true vocation, and wants an inward anointing for clergy. After that the Church can lay on hands by the Bishop, which, in the Church of England Ordinal, replaces the anointing of Catholics' (B.H.). *rest . . . labours*. Cf. Matt. 11: 28–30, Heb. 4: 3–11. *changed*. Transfigured (at the Resurrection). Cf. 1 Cor. 15: 51–2, and *Inst*. III. 25.4. *when . . . only*. When only the Godhead, that is, the Trinity, existed. *mystery*. The time between the Creation and the Day of Judgement, during which God's purposes are hidden. *revelation of the sons of God*. At the last day, the manifestation of God's glory and judgements: cf. Rom. 2: 5, 8: 19, and *Inst*. III. 25.3–12.

TWO PRAYERS

These two prayers were first published by Tenison in his *Baconiana* (1679; Gibson 237). The first prayer is very similar to one found (in Latin) in the 'Preface' to the *Instauratio Magna* of 1620 (*Works*, i. 131; iv. 20); the second prayer is virtually identical with the concluding paragraph of 'The Plan of the Work' that follows (*Works*, i. 145; iv. 33). My text comes from Spedding, *Works*, vii. 257–60.

113. *pilgrimage*. Cf. Ps. 119: 54, Heb. 11: 13, and S. C. Chew, *The Pilgrimage of Life* (New Haven, 1962). *days few and evil*. Cf. Job 10: 20, 14: 1. *prejudice*. Injure materially. *gates of sense*. The five senses. *intellectual night*. Deprived of revelation; cf. 'Intellectual Light' below. *given . . . Faith's*. Cf. Matt. 22: 21; Mark 12: 17. *Light . . . Creatures*. Gen. 1: 3. *Intellectual Light*. The *imago Dei*, the illumination that Adam received from God: cf. *Confession of Faith*, above, p. 109 and *Adv.L.*, above, p. 290. '*everything . . . good*'. Gen. 1: 10, 12, etc. *complacency*. Tranquil pleasure. '*all . . . Spirit*'. Eccl. 1: 14, 2: 11, 17. *acquiesce*. Rest satisfied. *sweat . . . brows*. Cf. Gen. 3: 19. *Sabbath*. Resting-place. *largeness*. Liberality, donation. *family of Mankind*. Cf. Eph. 3: 14–16, in the Geneva Bible: 'I bow my knee unto the Father of our Lord Jesus Christ (Of whom is named the whole family in heaven and in earth), That he might graunt you according to the riches of his glory, that yee may bee strengthened by his spirit in the inner man.' The marginal gloss on the word 'family' reads: 'The

faithfull which died before Christ came, were adopted by him, and make one family with the Saints which yet remain alive.'

A LETTER AND DISCOURSE TO SIR HENRY SAVILE, TOUCHING HELPS FOR THE INTELLECTUAL POWERS

Sir Henry Savile (1549–1622) was one of the most distinguished classical scholars of his generation. He published a translation of four books of Tacitus' *Histories* in 1591, and was one of the translators who prepared the Authorized Version of the Bible. His sumptuous edition of the Greek works of John Chrysostom (8 vols., Eton, 1610–13) was outstanding; he also produced editions of Xenophon's *Cyropaedia* and Bradwardine's *De Causa Dei contra Pelagium*. Savile became Provost of Eton College in 1596, and was knighted in 1604; Bacon's treatise dates from somewhere between those points. It was first printed in Rawley's *Resuscitatio* (1657); Spedding's text (*Works*, vii. 97–103) came from a manuscript copied by one of Bacon's secretaries, with alterations in Bacon's hand (BL Sloane MS 629, fos. 246–50: Beal BcF 198). It is another example of Bacon using the genre of advice literature to communicate his thinking on specific topics. Many of the ideas on mankind's learning capacity sketched here are worked out more fully in the *Essays* and the *Advancement of Learning*.

114. *policy*. Matters of importance to the state. *laws*. Official decrees (here, concerning education). *large*. Copious, abundant. *conceit*. The faculty of forming ideas, concepts. *intended*. Directed attention to. *confirmed*. Strengthened. *nearer the mark*. More accurately. *stronger bow*. To shoot longer distances. *office*. Duty, task. *quench . . . temper*. Technique in making steel, cooling the heated iron in water to make it stronger. *broken*. Broached, begun. *tamquam . . . agens*. Something different acting the same way. *apt*. Appropriate. 'Αριστον μεν ὕδωρ'. 'The best of things is water': Pindar, *Olympian Odes*, i. 1, probably from Aristotle, *Rhetoric*, i. 7, 1364a26, illustrating the argument that 'what is often useful surpasses what is seldom useful'.

115. *gratulation*. Celebration, compliment. *unlucky*. Unfortunate. *Faber . . . fortunae*. 'Every man is the maker of his own fortune': a saying attributed to Appius Claudius (*fl.* 300 BC). See *Adv.L.*, above, pp. 271–86, and Essays 40, 'Of Fortune' and 23, 'Of Wisdom for a Man's Self', above, pp. 420 and 386–7. *hortative*. Exhortation, encouragement. *high*. Excessive. *compass and fathom*. Understand, control. *accidents*. Whatever happens. *drifts and reaches*. Subterfuges, schemes. *as of*. Compared to. *felicity*. Good fortune. *providence*. God's protective care and foreknowledge. *Faber . . . sui*. 'The maker of his own intellect'. Noted in *Promus*, no. 357; cf. *Adv.L.*, above, p. 271. *bend*. Apply. *cover*. Conceal. *shew and demonstration*. Outward appearance. *notwithstanding*. All the same. *bind*. Commit. *amendeth*. Improves, corrects. *stonds*. Obstructions. *clear*. Open, free. *perfectest*. The idea of man as the crown of creation derives from a Renaissance philosophical tradition in opposition to medieval religious

treatises emphasizing human corruption. See e.g. Pico della Mirandola's oration *De dignitate hominis*, in Ernst Cassirer *et al.*, *The Renaissance Philosophy of Man* (Chicago, 1948; repr. 1963), 215–54. *impression.* Influence. Cf. *Adv.L.*, above, pp. 206–7.

116. *tumblers.* Acrobats. *funambulos.* Rope-walkers. *penances.* Self-inflicted mortifications. *strict . . . superstition.* Ascetic monastic orders. *report.* By Cicero, *Tusc. D.* II. xiv. 34, in a long discussion of how pain can be overcome. *scourged.* Whipped. *made.* Habituated. *wreath and cast.* Contort. *propriety.* Property. *put to.* Challenged, made to do something. *gregarius funambulo.* 'A rope-dancer belonging to a group', hence of no striking individuality. *diffidence.* Distrust, doubt. *prejudging.* Dismissing. *holdeth.* Applies. *Possunt . . . videntur.* 'They do it because they think they can do it' (Virgil, *Aeneid*, V. 231). Noted in *Promus*, nos. 425 and 1243; cf. *Adv.L.*, above, p. 175. *sequestered.* Excluded.

117. *painful.* Difficult. *travel.* Travail, hard work. *maniable.* Manageable. *opinion.* Shared beliefs. *apprehension.* The action of learning, mode of apprehending. *infused.* Introduced, instilled. *tradition.* Handing down of information. *institution.* Instruction, teaching. *wrought.* Formed. *disputation.* Debates (important exercises in school and university education). *obversant.* Well known. *affection.* Passion. *corroborate and confirm.* Strengthen, support. *though.* That. *election.* Choice. *inclineth.* Directs. *inceptions.* Commencements. *rudiments.* Imperfect beginnings, preliminary stages. *known.* Familiar. *palliation.* Alleviation, temporary relief (but no cure). *pretend.* Simulate, feign. *institutions.* Forms of instruction. *politic traffic.* Political affairs. *suffocating.* Suppressing. *that.* So that. *risus jussus.* 'Laughter by command'. *lachrymae coactae.* 'Constrained tears': politicians emulate the skills of actors. *rest.* In the *Resuscitatio* William Rawley adds a marginal note at this point: 'These that follow are but indigested Notes' (part 1, p. 178).

118. *Si . . . scibili.* 'If only people would debate all things knowable from both sides'; untraced in this form (probably post-classical). The exercise of disputing a topic *in utramque partem*, taking each side in turn, was especially practised by the New Academy (followers of Aristotle, 1st century BC), and had important applications in rhetoric. See Cicero, *De finibus*, V. iv. 10; *De oratore*, III. xxi. 80; and Quintilian, XII. i. 33–6. *ex tempore.* Without preparation; improvised. '*Stans . . . uno*'. 'Standing on one foot' (Horace, *Satires*, I. iv. 10). *memory narrative.* Training the memory to recall sequences of events, used in legal disputes. *sophists.* Students of logic. *Jo. ad oppositum.* 'To play the opposite part'. *Artificial memory.* Aids to memorizing, originally used by orators memorizing their speeches: cf. *Rhetorica ad Herennium*, III. xvi. 28 ff., and *De Aug.* V. v (*Works*, iv. 435–7). *holpen.* Helped. *buffons.* Clowns, crude jesters. *draw.* Lead, reduce. *conversation.* Social intercourse. *obiter, si videbitur.* 'In passing, should anything turn up'. *Tully's opinion . . . imitate.* For Cicero's advice to budding orators to choose a model for imitation see *De oratore*, II. xxii. 90–xxiii. 98. There is a fuller

discussion in Quintilian, x. ii. *Plato*. Bacon lists four Platonic dialogues; the '*Litigiosus*' is known in English as the *Sophist*. *Schola sua*. 'His school': Aristotelian or peripatetic philosophers who commented on his works. *Elenchs*. The *Sophistici Elenchi*, one of Aristotle's logical works, dealing with fallacious arguments. *Organon*. The canon of Aristotle's logic. *Hermogenes*. Rhetorician of the 2nd century AD, whose *Progymnasmata* (preparatory exercises) and *Peri ideon* were well known in the Renaissance. *Neoterics*. Modern authors. *Ramus*. Pierre de la Ramée (1515–72), French author of books on philosophy, logic, and rhetoric. *Agricola*. Rudolph Agricola (1443/4–85), Dutch humanist, whose *De inventione dialectica* (1479) was popular in the 16th century. *Nil sacri*. '[Thou art] no divinity.' The story occurs in Erasmus, *Adagia*, I. vii. 37; see *Adv.L.*, above, p. 140. *Lullius*. The Catalan philosopher Raimund Lull (*c*.1235–1315): see *Adv.L.*, above, p. 237 and note. *Typocosmia*. A treatise by the philosopher Alessandro Citalini (a religious exile who lived in England): see *Adv.L.*, above, p. 237. *Cooper's Dictionary*. The *Thesaurus linguae Romanae & Britannicae* (1565) of Thomas Cooper, one of the first major Latin–English dictionaries. *Mattheus*. Untraced; perhaps Bacon was referring to Hadrian Junius, *Nomenclator omnium rerum propria nomina variis linguis explicata indicans* (Augsburg, 1555), Englished by John Higins as *The Nomenclator, or Remembrancer . . . conteining proper names and apt termes for all thinges under their convenient Titles* (London, 1585). *Agrippa*. Heinrich Cornelius Agrippa (1486–1535), German philosopher, whose *De incertudine et vanitate scientiarum atque artium declamatio* (1526) was a semi- ironic rejection of all human knowledge in favour of divine revelation. *imitation*. An important process in Renaissance theories of literary composition. See e.g. Juan Luis Vives, *De tradendis disciplinis* (1531), IV. iv, in *Vives on Education*, tr. F. Watson (Cambridge, 1913), 189–200. *similitude*. In *Sophistici Elenchi*, ii. 9, 184: see *Adv.L.*, above, p. 223. *Exordia Concionum*. 'The beginnings of speeches': see *Adv.L.*, above, p. 223 and *De Aug*. v. iii (*Works*, iv. 423). *precept*. Cf. *De oratore*, II. xxxiv. 145–7, and *Adv.L.*, above, p. 223. *difference*. Distinction. *tempering*. Tuning, adapting. *framed to the life*. Adapted to ordinary use. *work*. Bring about. *per partes*. Using each part at a time. *per consequentiam*. Connecting them. *per se*. In itself. *per accidens*. By circumstance. *lightness*. Superficiality. *stayed*. Concentrated. *mathematics*. Cf. Essay 'Of Studies' (1597), above, p. 81. *roam*. Wander.

119. *framing*. Devising. *veluti . . . prima*. Horace, *Satires*, I. i. 25 f.: 'even as teachers sometimes give biscuits to children to coax them into learning their ABC.' *inveterate*. Settled, incorrigible. *Slubbering*. Performing quickly and carelessly. *clean*. Correctly, without faults. *would*. Should. *breathing*. *Pause*. *heads*. Topics, appropriate headings. *tables*. Notes. *discourse*. 'Speak or write at length on a topic' (*SOED*).

THE ADVANCEMENT OF LEARNING

In the autumn of 1605 (probably late October) appeared *The Twoo Bookes of Francis Bacon. Of the proficience and aduancement of Learning, diuine and humane*. The Stationers' Company Register, the official organ which safeguarded a publisher's right to print a text, shows separate entries for the two books on 19 August and 19 September, with the note that they were 'to be printed bothe in Englishe and Lattine'. Bacon had evidently planned a version in the 'international language', but this did not materialize until the vastly expanded *De Dignitate & Augmentis Scientiarum Libri IX* appeared in 1623. The printing history of the 1605 edition reveals signs of haste. As Spedding observed, the two books 'were not printed at the same time. The first ends with a half-sheet, and the second begins upon a fresh one with a new signature', which suggests that the first was printed before the second was completed (*Works*, iii. 256), while all the printed copies are extremely erratic in page-numbering and proof-correcting (Gibson 81). Furthermore, as every reader notices, the style of Book One is more finished and elaborate, while that of Book Two shows less polish and balance, reproducing a more direct, indeed urgent form of utterance. Bacon's ideas were always extending themselves unpredictably, their interrelation at any one time being provisional, so that any large-scale work that he undertook to complete by a specific point always risked seeming hasty or incomplete.

The letters that Bacon wrote accompanying the copies presented to friends and patrons reveal both the demands made on his time by his public career, and the role he set himself in the reform of knowledge. Between the death of Queen Elizabeth on 24 March 1603 and the meeting of James's first Parliament on 19 March 1604, Bacon had little to do in the way of public duties; by contrast, from March until December of that year he was extremely busy with legal and parliamentary affairs. The Parliament which had been fixed for February 1605 was postponed until October, and Bacon probably used these two periods of enforced 'leisure' to compose the *Advancement of Learning*. Sending a copy to the Earl of Salisbury, Chancellor of Cambridge University, Bacon wrote: 'I present your Lordship with a work of my vacant time, which if it had been more, the work had been better. . . . But sure I am the argument is good, if it had lighted upon a good author. But I shall content myself to awake better spirits, like a bell-ringer, which is first up to call others to church' (*Works*, x. 253–4). (That metaphor recalls the marvellously eloquent sentence ending his discussion of philosophy, above, p. 288.) Writing to Sir Thomas Bodley, who had refounded the library of Oxford University two years earlier, Bacon quoted one of his favourite texts from the Psalms, 'My soul hath long dwelt with him that hateth peace. I am for peace' (Ps. 120: 6–7), to convey his sense of having been occupied on tasks inimical to his true gifts, which, with a punning allusion to the theatre (to 'hold a book', like the prompter, rather than to 'play a part', like an actor), he here defined as lying in the secular *vita contemplativa*, of intellectual activity: 'I think no man may more truly say with the Psalm *Multum incola fuit anima mea*, than myself. For I do confess, since I was of any understanding, my mind hath in effect been absent from that I have done; and in absence are many errors which I do willingly acknowledge; and amongst the rest this great one that led the rest; that knowing myself by inward calling to be fitter to hold a book than to play a part, I have led my life in civil causes; for which

I was not very fit by nature, and more unfit by the preoccupation of my mind. Therefore calling myself home, I have now for a time enjoyed myself; whereof likewise I desire to make the world partaker. My labours (if I may so term that which was the comfort of my other labours) I have dedicated to the King; desirous, if there be any good in them, it may be as the fat of a sacrifice, incensed to his honour: and the second copy I have sent unto you, not only in good affection, but as a kind of congruity, in regard of your great and rare desert of learning. For books are the shrines where the Saint is, or is believed to be: and you having built an Ark to save learning from deluge, deserve propriety in any new instrument or engine, whereby learning should be improved or advanced' (*Works*, x. 253).

A more personal tone appears in the letter to his old friend Toby Matthew, then in Italy: 'I have now at last taught that child to go, at the swadling whereof you were. My work touching the *Proficiency* and *Advancement of Learning*, I have put into two books; whereof the former, which you saw, I count but as a Page[-boy] to the latter. I have now published both; whereof I thought it a small adventure to send you a copy, who have more right to it than any man, except Bishop Andrewes, who was my inquisitor' (*Works*, x. 256). Lancelot Andrewes (1555–1626), Dean of Westminster from 1601 to 1605, was a friend of Bacon's student days, with whom he enjoyed a long and warm relationship. From Bacon's allusion he seems to have commented on the *Advancement of Learning* in manuscript, for in 1609 Bacon sent him a copy of his *Cogitata et Visa* with the request that he 'be so good now as when you were the Dean of Westminster' to 'mark unto me whatsoever shall seem unto you either not current in the style, or harsh to credit or opinion, or inconvenient for the person of the writer; for no man can be judge and party; and when our minds judge by reflexion on ourselves they are more subject to error' (*Works*, xi. 141).

The *Advancement of Learning* played an important part in Bacon's plans for the reformation of knowledge, as announced in his 1592 letter to Burghley (above, p. 20). There he recorded his hope that, if he could 'purge' knowledge 'of two sorts of rovers [misleading trends], whereof the one with frivolous disputations, confutations, and verbosities, the other with blind experiments and auricular traditions and impostures', he hoped to be able to promote 'industrious observations, grounded conclusions, and profitable inventions and discoveries; the best state of that province'. Both targets are addressed in the *Advancement*, the first in the brilliant satire on scholasticism (above, pp. 140. ff.), the second in the outline of some fundamental criteria for scientific research (above, pp. 193 ff.). But the scope of the *Advancement* extends far beyond that described in the letter to Burghley.

Book One has a dual task, first to defend knowledge in general from all its enemies, ecclesiastical and secular, then to argue for its dignity and value. Bacon must have been keeping notes for some years under the heading 'attacks on and defences of learning', for the amount and range of historical material he is able to cite is remarkable. Negative attitudes to human learning were long-standing, stretching from Plato (in some modes) and the sceptic Sextus Empiricus to St Paul and the Church Fathers, but they were also contemporary, indeed the issue was contemporary and urgent. As Geoffrey Bullough showed ('Bacon and the Defence of Learning', in *Essential Articles*, 93–113), during the Renaissance learning came under attack from several camps. The Christian Church had always had a divided attitude to classical—that is, pagan—learning. On the one hand, the very tools of knowledge, from the *trivium* of grammar, rhetoric, and logic, which formed the basis of both grammar-school and university education, to the more advanced study of the *quadrivium* (arithmetic, music, geometry, astronomy)—all these fundamental disciplines derived both their form and substance from Greece and Rome. The Church Fathers had solved that problem by a combination of assimilation and redefinition, using moral allegory to turn the pagan deities and heroes into emblems of virtue and vice, but under the pressures of the Reformation and Counter-Reformation the dislike of 'pagan learning' reappeared, in old and new forms. One sign of the break-up of the earlier settlement was the Protestant iconoclasm which destroyed images and denounced the heathen gods; another was the objection of theologians (both Catholic and Protestant) to physics and astronomy as tending to pry into God's secrets, 'curiosity' becoming a pejorative word. (See e.g. Howard Schultz, *Milton and Forbidden Knowledge*, New York, 1955.) Another sign of the hostility of religious groups towards human learning in general was the tendency of the extremely pious (e.g. the French Protestant Du Plessis Mornay, in *De la Vérité de la Chrétienne Religion*, 1581) to identify any commitment to classical learning with atheism. In fact, of course, Renaissance humanism willingly embraced the study of both God and man, and all intellectuals were practising Christians. Such accusations, although lacking substance, became increasingly hostile as fears of religious and civil wars spread across Europe. For all these reasons Bacon insisted on keeping Scripture and science separate, a tactically wise move, as Galileo's unhappy experience with the Church was to show. The pursuit of learning, Bacon is at pains to emphasize, in no way damages Christian belief or Christian observances.

The second enemy of learning that Bacon addressed was also external, deriving from the civic sphere, the objection by 'politiques'—men involved in public and political life—that a dedication to acquiring knowledge unfits men for practical life. One recurring debate in this period was whether or not the occupations of the scholar and the soldier were compatible. This debate runs from Castiglione's *Il Cortegiano* (1528; bk. 1, chs. 42–6) through the work of Gascoigne (whose motto, *Tam Martio quam Mercurio*, proclaimed his competence in both fields), Sidney, and on to Montaigne. (See e.g. James Supple, *Arms versus Letters: The Military and Literary Ideals in the 'Essais' of Montaigne*, Oxford, 1984.) Bacon dismisses that argument by instancing the number of military leaders who were also outstanding scholars, and for the related charge that learning disables men from becoming counsellors (government advisers) he inverts the argument: bad leaders have been precisely those who practised politics while lacking a theoretical over-view. Men of learning, Bacon adds, being 'perfect in those plain grounds of religion, justice, honour, and moral virtue', are likely to give better advice than the corrupt hangers-on in court. Equally effective is his argument that the scholar's virtues of sifting evidence, industrious research, and laborious pursuit of the truth are qualities that can be easily translated from the *vita contemplativa* (which he understood in the secular sense of study, removed from social duties) to the *vita activa* of public life. Indeed, Bacon himself is one example of many outstandingly gifted men in this period, from More to Milton, who moved from the study to the council chamber, showing that in both worlds they were fulfilling the ideal of Cicero and the Florentine humanists that all educated men should act *pro bono publico*.

The final 'discredit' to learning that Bacon deals with is not external but internal, deriving from 'learned men themselves', and affecting their fortune, their manners, or the nature of their studies. Here Bacon redefines and deflects some common criticisms: scholars are poor—yes, but also virtuous; the occupation of teaching the young lacks prestige—but it is at the same time crucial to educating both rulers and citizens; scholars are unworldly—but are also ready to sacrifice themselves for the good of their countries or master; they are not ambitious—but are therefore not corrupt. While not yielding to common prejudices for this third internal discredit to learning, Bacon goes on to deliver a critique of his own, denouncing the 'errors and vanities... in studies, whereby learning hath been most traduced'. This section (above, pp. 138–44) is one of the most carefully composed parts of the book, a brilliant union of argument and style (see Brian Vickers, *Francis Bacon and Renaissance Prose* (Cambridge, 1968), 111–12, 122–30, 160–4), mounted to attack those vices which prevent learning from fulfilling its proper role in the *vita activa*. They include (according to Renaissance humanists, at least, wishing to distance themselves from their predecessors) medieval scholasticism, which supposedly limited itself to elaborate and servile commentaries on the text of Aristotle, or indulged in endless and fruitless disputations; Ciceronianism, an unhealthy offshoot in Renaissance humanism which encouraged a narcissistic imitation of the purely superficial features of Cicero's style; all forms of credulity, whether involving narratives of miracles by Christian martyrs, fabulous stories of the natural world (as in the writings of Pliny), or the gross impostures of the occult arts (astrology, magic, alchemy). According to Bacon, the common failing in all these perversions lay in their not recognizing the true goal of knowledge, which is to be applied 'to the benefit and use of men... for the glory of the Creator and the relief of man's estate'. Thanks to this critique, and to his reorientation of science towards the goal of 'works', inventions and processes that would transform nature, Bacon influentially united the Christian notion of charity with the philanthropic motives espoused by classical and Renaissance philosophers of the *vita activa*.

To move from subject-matter to form, we can see that the first book of the *Advancement of Learning* is constructed according to the principles of that branch of classical rhetoric known as epideictic oratory, which used the complementary techniques of praise and blame, *laus* and *vituperatio*, on matters involving human ethics. Having defended learning from blame, and made some discriminating attacks himself, Bacon ends Book One with a stirring account of the dignity and excellence of learning. This complementary account also matches structurally the opening criticism. To the objections of churchmen with which he began he now opposes divine testimonies to 'the dignity of knowledge', from the acts of God in creating the universe, through the role of angels as instruments 'of knowledge and illumination', down through biblical times and into Church history up to the present (above, pp. 148 ff.). To the objection of 'politiques' he now opposes a series of arguments, copiously illustrated from his reading in ancient and modern history: that inventors have always been worshipped (above, pp. 153–4); that learning restrains human evil (above, p. 154); that countries ruled by learned princes have enjoyed happy times (above, pp. 154 ff.); that learning positively helps martial virtue, as seen in the twin careers of Alexander the Great and Julius Caesar (above, pp. 158 ff.); and, finally, that learning helps moral and private virtue (above, pp. 164–5 ff.). Although some of the earlier controversies (happily) no longer affect our pursuit of knowledge, this final section, on the pleasures and profit of learning, still speaks directly to us, especially Bacon's account of how prolonged study 'disposeth the constitution of the mind not to be fixed or settled in the defects thereof, but still to be capable of growth and reformation' (above, p. 165).

If Book One corresponds to an epideictic oration, Book Two shifts from *res* to *verba*, from words to things, as a Renaissance writer might have put it, for Bacon now presents himself as constructing a map or globe of the whole intellectual world, paying particular attention to the undeveloped areas: 'Wherefore I will now attempt to make a general and faithful perambulation of learning with an inquiry what parts thereof lie fresh and waste, and not improved and converted by the industry of man; to the end that such a plot made and recorded to memory may both minister light to any public designation, and also serve to excite voluntary endeavours' (above, p. 174; also p. 299). To appreciate the significance of Bacon's metaphor we need to recall the precise meaning of 'perambulation', namely 'The action or ceremony of walking officially round a parish, forest, etc., for the purpose of asserting and recording its boundaries and preserving the rights of possession' (*SOED*). Through Bacon, mankind is reminded of the knowledge that it possesses; his 'written account of a survey or tour of inspection' (ibid.) is emblematic also of the mind's domination of the natural world that he so often celebrated and wished to advance. For this reason his map or 'plot' will not be a passive replica of what has already been taught or said, but an inventory of what needs to be done. Bacon's ability to render abstract thought-processes in concrete imagery allows him to create a map of learning which appeals simultaneously to the intellectual explorer, interested in a *terra incognita* as something to be discovered, as well as to the settler or farmer who intends to cultivate it. The survey of human knowledge he provides in Book Two has been compared to medieval and Renaissance encyclopedias. The comparison is justified in that Bacon uses a division or partition of the sciences having some similarities with the encyclopedic tradition (although actually, and more originally, derived from Galen's classification of the human faculties). Yet the crucial difference is that Bacon is concerned not with positive achievements, the cataloguing of existing knowledge, but with identifying deficiencies. This is an encyclopedia of lacunae, so to speak, of gaps in our knowledge that need to be filled, a list of desiderata that will demand much labour over many years. The Latin version of 1623, the *De Augmentis Scientiarum*, expanded to nine books, concluded with a list summing up all the areas that Bacon had marked as deficient, including all his concrete proposals for new areas of science ('Novus Orbis Scientiarum, sive Desiderata': *Works*, i. 838–40).

The range of Book Two is remarkable by any criteria, a truly Renaissance *tour de force* which few of Bacon's contemporaries would have had the courage to attempt. Two general points need to be made before we can estimate his achievement, one intellectual, one practical. The first concerns the key terms of the title, *Of the proficience and advancement of Learning*, Bacon's dynamic notion of the body of knowledge as something that can be continually extended. As was shown many years ago (see John Bury, *The Idea of Progress*, 1920), Bacon was among the first to formulate the concept of scientific progress at the level of discoveries and inventions (always adding the corollary that improved practical skills would help alleviate human want and suffering). The idea was new, so that here, as elsewhere, Bacon had to use older words while extending their meaning. 'Proficience' at that time had the connotations of skill in performing an activity; 'advancement' could refer to a pupil's development at school, or a man's rising within society; while 'progress' was used to describe the tours that a sovereign might make of her kingdom, journeys that would start from and return to base. Bacon's achievement was to apply all these concepts to knowledge itself, establishing a novel concept of knowledge as being infinitely expandable, so that the identification of obstacles was merely the first step to overcoming them. As Michèle Le Doeuff puts it in her recent French edition, Bacon's task was to devise a concept of progress which would determine how that concept itself could become possible (*Du progrès et de la promotion des savoirs* (Paris, 1991), p. xx).

Secondly, given Bacon's lifelong commitment to transferring theory to practice, changes would need to be made in society, both to make scientific progress possible and to respond to its results. As we can see from the speech he wrote for the Gray's Inn Revels in 1594 (above, pp. 54 f.), Bacon realized that science could only develop if energies were co-ordinated within an institutional framework, as in the four centres of learning he proposed there (a library, a garden, a museum, a laboratory). The extant institutions in England, the universities of Oxford and Cambridge, dedicated to teaching a humanist-inspired curriculum, were essentially geared to producing graduates in the three recognized professions (law, medicine, theology). The notion that they should also devote themselves to research in order to generate new knowledge was unheard of. (It was quite typical that the Bacon-inspired research group which emerged at Oxford in the 1640s and 1650s should have ultimately resulted in the founding of the Royal Society in 1660, an institution outside the university frame.) The only person capable of initiating such fundamental changes at that time was the sovereign, which accounts for Bacon's frequent appeals to King James to undertake the necessary *opera basilica*, works for a king, which could permit the renovation of learning. In particular, James is called on at the opening of Book Two to approve a reform of universities which would provide regular, adequate positions and salaries for university teachers, and to promote the founding of colleges for the 'arts and sciences at large', including the provision of scientific research institutes. The *Advancement of Learning* is simultaneously a defence, a celebration, and a programme for action.

The survey of knowledge which fills most of Book Two is an account of extant achievements, many sections judging a discipline 'well handled', but more importantly of deficiencies and omissions, identified so that they can be corrected. 'This I speak', as Bacon puts it in one place, 'not to disable the mind of man, but to stir it up to seek help' (above, p. 222). It is arranged partly according to the traditional divisions of the sciences (natural philosophy; mathematics; logic; rhetoric; ethics; politics), but partly according to Bacon's own idea of how subjects should really be linked up, a concept which caused him to propose new sciences and new groupings of extant sciences. Having given the term 'Metaphysic' a completely new sense (above, pp. 193 ff.), and subordinated physics to it, he now places mathematics within this new framework, justifying this new arrangement on the grounds that 'we have endeavoured in these our partitions to observe a kind of perspective, that one part may cast light upon another' (above, p. 200). This awareness of the benefits to be derived from lateral movement among the sciences, transferring categories and methods from one discipline to another (which has accounted for so many fundamental developments in our time), is one of Bacon's most impressive qualities. We see it again in his justification for a history of technology (using 'history' in the original classical sense of 'enquiry', systematic study, not necessarily chronological), which will help fulfil his lifelong concern that 'natural philosophy . . . shall be operative to the endowment and benefit of man's life'. Such a study, Bacon believes, 'will not only minister and suggest for the present many ingenious practices in all trades [crafts], by a connexion and transferring of the observations of one art to the use of another, when the experiences of several mysteries [crafts] shall fall under the consideration of one man's mind; but further it will give a more true and real illumination concerning causes and axioms than is hitherto attained' (above, p. 177).

This concern for the possible connections between branches of learning otherwise detached from each other informs Bacon's bold aim to establish 'one universal science, by the name of *Philosophia Prima*', or 'the fountains of knowledge', which will consist of 'profitable observations and axioms' common to several sciences, to be formulated and investigated at a higher level of abstraction (above, pp. 189 ff.). The substantive content of Bacon's argument, noting deficiencies and proposing specific research programmes, was an inspiration to his successors. Even more inspiring was his new conception of the progress of research itself as a self-propelling activity, which can always be improved by reflection on the problems at issue. As he formulated the principle, 'every degree of proceeding in a science giveth light to that which followeth; which light if we strengthen, by drawing it forth into questions or places of inquiry, we do greatly advance our pursuit' (above,

p. 224). Bacon summed up this fundamental idea in the memorable axiom that *Ars inveniendi adolescit cum inventis*, 'every act of discovery advances the art of discovery' (ibid.).

Any writer attempting such a wide-ranging survey will inevitably reveal both strengths and weaknesses. There are topics on which Bacon provides no more than a standard account. In the case of theology he does so in distinctly guarded terms, ostentatiously refusing to meddle in the Church's affairs. In the case of mathematics, a subject which had yet to demonstrate its centrality to the experimental sciences, his account is perfunctory, passing over the already substantial achievements made in the sixteenth century. Another weak spot in the work is astronomy, for, like many of his contemporaries, Bacon was not convinced of the superiority of Copernicus' theory of a heliocentric universe, even though he was undoubtedly familiar with some of Galileo's work, and perhaps with Kepler's. On the other side, we can already see in this early survey of knowledge original ideas that will be brought to partial fruition in the *Novum Organum* or 'New Instrument' of the sciences (1620). The discussion here of the 'forms' underlying matter (above, pp. 196 ff.) can be seen as a sketch for the extended discussion in Book Two of that work, where we also find a more detailed account of induction than that given here (above, pp. 221 ff.). The most famous of these topics for Bacon's later work is his account of the 'idols' (illusions, fallacies) governing human understanding (above, pp. 226 ff.), outlined here, but like so much else in his work, continually elaborated over the years, although probably never fully thought through.

The most impressive parts of this 'perambulation of learning', it seems to me, are those directly concerning mankind, in body, mind, and society. Bacon's account of medicine, considering the state of knowledge then, is perceptive and critical, opening up the topic in a constructive fashion (above, pp. 205 ff.). On the verbal arts, grammar, logic, and rhetoric, he writes with the authority of a theoretician and practitioner, already much experienced, making several original proposals (above, pp. 219 ff.). Of particular importance to him was to establish a proper study of men and women as individuals acting, and being acted on, in society, topics which embrace the modern study of psychology and social anthropology. His discussion of the human 'appetite and will' begins with the complaint that students of psychology 'have despised to be conversant in ordinary and common matters', although life chiefly consists in them (above, p. 244). To supply this deficiency Bacon offers an account of the 'Platform or Nature of Good' (above, pp. 245–55), the basic role of ethics in human life, followed by a discussion of the 'Regimen or Culture of the Mind' (above, pp. 255–65). These are both long, varied, and original discussions, not overlooking the minutiae of behaviour. Another substantial and original discussion concerns 'Negotiation or Business' (above, pp. 266–86), where 'business' does not have the commercial connotations that it has today, but refers rather to human negotiations in general, what we might call 'the presentation of the self in everyday life'. The originality of these discussions may excuse, or at least explain their rather

ANALYSIS

BOOK I. OF THE EXCELLENCY OF LEARNING AND KNOWLEDGE.

Dedication to the King (120–2). The treatise in two parts

1. The excellency of learning and knowledge.
2. What has been done for the advancement of learning, with the defects of the same.

(A) Discredits of Learning (122–48), arising from

- I.
  1. The zeal and jealousy of divines (122–6).
  2. The severity and arrogancy of politicians (126–31)
  3. The errors and imperfections of learned men, from their fortune, manners, or the nature of their studies (131–7)
- II. Errors and vanities in the studies themselves (137–44)
  1. Delicate learning (138–40)
  2. Contentious learning (140–2)
  3. Fantastical learning (142–4)
- III. Peccant humours (144–8) which are
  1. Affectation of antiquity and novelty (144–5)
  2. Distrust of new discoveries (145)
  3. Conceit that the best opinions prevail (145)
  4. Premature reduction of knowledge to arts and methods (145–6)
  5. Neglect of universality or *philosophia prima* (146)
  6. Too great reverence for the intellect (146)
  7. Mixture of knowledge with men's inclinations (146)
  8. Impatience of doubt (146–7)
  9. Dogmatic delivery of knowledge (147)
  10. Aim which men propound to themselves (147)
  11. Mistaking of the end of knowledge (147–8)

- (B) Dignity of Learning (146–8) shown by
  - I. Divine testimony (148–53) in
    - The work of creation (149)
    - Spirits (149)
    - Light (149)
    - The Sabbath (149)
    - The contemplative life of Adam and Abel (149–50)
    - The inventors of arts (150)
    - The learning of Moses (150–1), Job (151), Solomon (151–2), Christ (152), the Apostles (152), the Fathers (152), the Jesuits (152–3)
  - II. Human proofs (153–68) from the
    - Supreme honours paid to inventors of arts (153–4)
    - Influence of learning upon
      - The intercourse of man and man (154)
      - States under learned princes (154–64)
        - In peace; e.g. the Roman Emperors (155–8), Q. Elizabeth (158). In war (158–64); e.g. Alexander (159–61), Caesar (161–3), Xenophon (163–4)
      - Private virtue (164–5)
      - Superiority of man to beasts (165–6)
      - Fortune and advancement (166)
      - Pleasure (167)
      - Immortality (167–8)

diffuse structure. It is as if Bacon was working in a new area with a quantity of specific observations that he was reluctant not to communicate entire, but not yet having elaborated a theoretical framework which could adequately differentiate them. Hence his liking for the aphorism, that unsystematic form.

Although some sections of the *Advancement* are disproportionately long, and we often find Bacon himself apologizing for having exceeded the scope of an illustration, much of the detail remains inherently interesting, and in my experience readers cannot predict in advance which passage might illuminate their own private concerns. Some of these sections have an added significance in Bacon's work as a whole. His discussion of psychology and the influence of social groups was part of a long-standing project for a scientific description of the interplay between internal and external forces, and finds a complementary articulation in the *Essays* of 1625 (see below, pp. 710 ff.). Even after frequent rereading, these *Two Books of the proficience and advancement of Learning* do not lose their ability to stimulate the reader to intense thinking about the nature of knowledge, and to evoke admiration of the mind that made them.

My text is based on Spedding's (*Works*, iii. 255–491), checked against that of W. A. Wright, who differs over several readings corrected in the early editions. I have not reproduced the typographical emphases of the 1605 printing, with its notoriously inconsistent use of italics and capitals. In annotating it I have benefited above all from the Victorian edition by William Aldis Wright (Clarendon Press, 2nd edn., 1873), which formed the basis for the useful modern edition by Arthur Johnston (Oxford, 1974). I have also consulted two recent commented translations by continental scholars expert in Bacon's philosophy: the Italian version by Paolo Rossi in his admirable selection (nearly 900 pages) of *Scritti Filosofici di Francesco Bacone* (Turin, 1975, 1986), and the French version by Michèle Le Doeuff, *Du progrès et de la promotion des savoirs* (Paris, 1991). Collating these four editions, however, revealed many inconsistencies in their documentation of Bacon's quotations and allusions, so I have checked them all against original texts (in the Loeb editions, where available).

I have added the useful analyses of the two books prepared by Wright for his edition.

Book One

120. *the Law ... offerings.* According to Mosaic law burnt offerings were made to Jehovah both daily and at other times. *proceeding upon.* Resulting from. *wanting.* Deficient. *respective.* Appropriate. *oblation.* Offering. *propriety.* Individual character. *inquisitive.* Prying. *presumption.* Taking upon oneself more than one's position warrants. *Scripture ... inscrutable.* 'The heart of Kings is inscrutable' (Prov. 25: 3). *capacity.* Mental power. *Plato's opinion ... remembrance.* See *Phaedo*, 72E–76A and *Meno*, 81B ff. *native.* Natural, innate. *tabernacle.* Temporary dwelling. *sequestered.* Separated, cut off. *Scripture ... sea.* 1 Kgs. 4: 29, on Solomon. *portions.* Parts. *composition ... admirable.* A wonderful combination of intellectual powers. *should.* Would.

121. '*Augusto . . . fuit*'. 'Augustus had a fluent way of speaking, such as became a sovereign' (Tacitus, *Annals*, xiii. 3). *savoureth*. Is marked by. *affectation*. Earnest pursuit; or 'studied display'. *framed*. Formed. *pattern*. Model. *holding . . . subject*. Appropriate to a subject; not prince- like. *felicity*. A happy grace. *civil estate*. Civic situation. *virtue . . . fortune*. Two key (and often opposed) terms in Renaissance thought, as defining internal and external influences on human life. *regiment*. Training. *time was*. Before James acceded to the throne in 1603. *fruit*. James had a daughter, Elizabeth, and two sons, the Princes Henry (1593–1612) and Charles (1600–49). *universality*. All-round knowledge. Bacon identifies James with the Renaissance ideal of *l'uomo universale*. *amplification*. Rhetorical exaggeration. *literature . . . human*. Alluding to the traditional metaphor of two books, one containing God's words (the Bible), the other containing His works (the world). Bacon quotes below Cicero's definition of wisdom as 'rerum divinarum et humanarum scientia' (*Tusc. D.* IV. xxvi. 57). *revolve*. Reflect upon. *peruse*. Review. *Caesar the dictator*. Julius Caesar (*c.*100–44 BC), who achieved absolute power in the last five years of his life. *Marcus* [Aurelius] *Antoninus*. Emperor AD 161–80, best known for his *Meditations*, a digest of Stoic philosophy. *Graecia*. Greece. *lines*. Lineage. *seemeth much*. Appears a great thing. *extractions*. Digests, summaries. *shews*. Appearances. *countenance*. Favour, encourage. *prefer*. Promote. *so as*. So that. *triplicity*. Threefold combination.

122. *ancient Hermes*. Hermes Trismegistus ('thrice greatest'), an apocryphal figure to whom a corpus of philosophical and theological writings (written between the 1st and 3rd centuries AD) was attributed. *universality*. Being learned in many subjects. *propriety inherent*. Innate property. *difference*. Distinguishing mark. *undervalues*. Deprecations. *positively*. Explicitly laid down. *affirmatively*. Assertively. *framed particulars*. A detailed plan. *excite*. Stimulate. *cogitations*. Thoughts. *deliver*. Rescue, free. *severally*. In several ways. *zeal*. 'Ardent feeling manifested in love, jealousy, or righteous indignation' (*SOED*). *arrogancy*. Arrogance. (The English language had at this time yet to develop a consistent spelling for such word forms.) *politiques*. Statesmen, politicians. *of*. One of. *accepted of*. Accepted. *serpent*. In the Garden of Eden: cf. Gen. 3. *Scientia inflat*. 'Knowledge puffeth up' (1 Cor. 8: 1). *censure*. Judgement. *no end . . . flesh*. Eccles. 12: 12. *spacious knowledge . . . anxiety*. Eccles. 1: 18. *contristation*. Sadness. *that . . . philosophy*. Col. 2: 8. *spoiled*. Corrupted.

123. *arch-heretics*. Leaders of heresy. *learned times . . . atheism*. Cf. Essay 16, 'Of Atheism', above, pp. 371–3. *second causes*. God, creator of the world, is the first cause; second causes are the modes by which he operates in the world through the agency of physical laws. Cf. *Confession of Faith*, above, p. 108. *derogate from*. Take away, detract from. *discover*. Reveal. *grounds*. Basic arguments or assumptions. *universality*. The universe and all it contains. *give names*. Adam's ability to name the creatures (Gen. 2: 19–20), which Bacon elsewhere interpreted as showing the 'pure light of natural knowledge': *Works*, iii. 219. *proprieties*. Particular natures. *proud knowledge* *of good and evil.* Gen. 2: 16–17; 3: 1–19; cf. *Confession of Faith*, above, p. 109. *intent.* Purpose. *form.* Essential determinant principle. *Salomon.* The old form of the name, as in the Geneva Bible (1560); cf. Eccles. 1: 8. *inquisition.* Enquiry, investigation. *continent.* Container. *content.* Contained. *calendar or ephemerides.* An almanac; cf. Eccles. 3: 1–11. '*God . . . to the end*'. Eccles. 3: 11. *decent.* Appropriate. *capable of.* Able to receive. *joyeth.* Rejoice. *vicissitude.* Alternation, regular change. *raised.* Elevated. *ordinances and decrees.* Laws of nature. *summary.* Most important. *referred.* Attributed. *as of.* Such as. *ill conjunction.* Poor coordination. *tradition.* Transfer, communication. *nothing parcel.* No part.

124. *place.* Biblical text, locus. *rule over.* Proclaim, determine. '*The spirit . . . secrets*'. Prov. 20: 27. *receit.* Receptive power. *out-compass.* Grow beyond its proper bounds (which include not meddling into God's mysteries, the vice of 'curiosity'). *taken.* Absorbed. *corrective.* Antidote. *nature.* Kind. *ventosity.* Windiness. *sovereign.* Superior, potent (often applied to medical remedies). '*knowledge . . . buildeth up*'. 1 Cor. 8: 1. *bloweth up.* Inflates. Cf. St Bernard, *Cantica Canticorum*, Sermon 36, *Life and Works*, tr. S. J. Eales, vol. iv (London, 1896), 233–8. *buildeth up.* Strengthens. *delivereth.* Pronounces. '*If I spake . . . cymbal*'. 1 Cor. 13: 1. *referred.* Assigned, applied. *sounding.* Hollow. *glory.* Ostentation. *meriting.* Deserved. *censure.* Judgement. *Salomon.* Eccles. 12: 12. *redoundeth.* Proceeds or arises from. *that . . . philosophy.* Col. 2: 8. *places.* Texts. *coarctation.* Restriction. *distaste.* Annoyance, discomfort. *repining.* Complaining. *as touching.* With respect to. *expound.* Express. *place.* Passage. '*I saw well . . . both*'. Eccles. 2: 13–14. *recedeth.* Differs from. *roundeth.* Roams, wanders. *accident.* Something present but not necessarily so; non-essential.

125. *wonder . . . knowledge.* Cf. Plato, *Theaetetus*, 155D; Aristotle, *Met.* i. 2, 982b 12. *framing.* Drawing, deriving. *particular.* Individual case. *ministering.* Imparting. *carefulness.* Anxiety. *Lumen siccum.* A dry light. '*Lumen . . . anima*'. 'The dry light is the best soul'. Heraclitus, Frg. 118, as in Stobaeus, *Florilegium*, III. 5.7–8. *Lumen . . . maceratum.* A blurred light (the metaphor is from substances being damaged by steeping in liquid). Cf. Bacon's letter to Sir Toby Matthew of February 1609: '. . . there is almost no *lumen siccum* in the world, but all *madidum* and *maceratum*, infused in affections and bloods or humours . . .' (*Works*, xi. 132), and *Apophthegms*, no. 268: 'Heraclitus the Obscure said: *The dry light was the best soul.* Meaning, when the faculties intellectual are in vigour, not wet [*Resuscitatio* reads 'drenched'] nor, as it were, blooded by the affections' (*Works*, vii. 163). *humours of the affections.* Whims of the passions. *stood upon.* Insisted upon. *sensible.* Perceivable by the senses. *broken knowledge.* Incomplete, fragmentary knowledge: 'contemplation broken off, or losing itself' (*Works*, iii. 218). *one of Plato's school.* Philo Judaeus, *De somniis*, I. xiv. 84. *terrestrial globe.* The earth. *celestial globe.* The constellations of the heavens. *proceeded.* Happened. *divers.* Various. *great.* Greatly. *waxen wings.* Like Icarus, who put on wings made of wax and feathers, but flew too near the sun; when his wings melted, he fell into the sea and was drowned. *conceit.* Idea.

which. Who. *Will you . . . gratify him?* Job 13: 7, 9. *by*. Through. *in favour towards God*. For God's sake. *unclean sacrifice*. Not made according to religious observances. *atheism . . . religion*. See Essay 16, 'Of Atheism', p. 371. *entrance*. Beginning stages. *next*. Nearest. *induce*. Bring about.

126. *dependence*. Subordination. *causes*. Second causes. *Providence*. God's protective government. *nature's chain*. The 'great chain of being', a hierarchical principle believed to pervade the universe, from the chair of Zeus down to earth and through all its realms, in the sequence human, animal, vegetable, mineral. Cf. Homer, *Iliad* viii. 18 ff., and Plato, *Theaetetus*, 153C–D. *weak conceit*. Feeble, ungrounded notion. *sobriety*. Sober-mindedness, seriousness. *the book of God's works*. The created world, which can be understood by 'philosophy' (natural and moral), as 'God's word' is understood by 'divinity' or theology. *endeavour*. Strive after. *proficience*. Improvement. *charity . . . swelling*. For the good of others, not for personal pride. *politiques*. Politicians, involved in the *vita activa*, and tending to scorn the *vita contemplativa* of research and study. *be*. Are. *soften*. Make weak or effeminate. Cf. Montaigne, *Essais*, i. 25, 'Du Pedantisme': 'Les exemples nous apprennent, et en cette martiale police [martial government] et en toutes ses semblables, que l'estude des sciences amollit et effemine les courages, plus qu'il ne les fermit et aguerrit' (Pléiade edn., p. 143). *mar*. Damage. *pervert*. Lead astray. *policy*. Public affairs, politics. *curious*. Fussy, over-scrupulous. *irresolute*. Lacking in decisiveness. *positive*. Categorical. *strictness*. Dogmatic assertion. *incompatible*. Incongruous. *differing from*. Unsuited to. *dissimilitude of examples*. Discrepancy between their studies and contemporary life. *travails*. Labours. *privateness*. Privacy (opting out of the public life or *vita activa*, where every virtuous person's duty lay). *Cato*. Marcus Porcius 'Censorius' (234–149 BC), appointed censor of Rome in 184 BC, a defender of traditional Roman values against 'foreign' innovations. Cf. Plutarch, 'Life' of Marcus Cato, xxii; Cicero, *Tusc. D.* IV. iii. 5 ff., *De oratore*, II. xxxvii. 155; Pliny, *Nat. Hist.* vii. 31. *in embassage*. On an embassy. *give . . . despatch*. Dismiss, discharge. *at unawares*. Unexpectedly. *challenging*. Laying claim to. *Grecians*. Greeks. *'Tu . . . artes'*. 'Remember thou, O Roman, to rule the nations with thy sway—these shall be thy arts' (*Aeneid*, vi. 851–3; *Promus*, no. 1013). *Anytus*. Recorded as one of the accusers in Bacon's main source for Socrates' trial (Plato, *Apology*, 19–23).

127. *science*. Knowledge, learning. *countenance of gravity*. Appearance of importance. *warrant*. Attest. *that pair*. Alexander and Caesar were coupled by Plutarch in his *Parallel Lives of the Greeks and Romans*. Bacon returns to this pairing later (above, pp. 159 ff.), with Plutarch his main source. *Aristotle's scholar*. Aristotle tutored Alexander between the ages of 13 and 16. See Plutarch's 'Life' of Alexander, 7.1 ff. *Epaminondas*. Outstanding Theban general (*c*.418–362 BC), a pupil of Lysis the Pythagorean, who led a retired life until his fortieth year. He joined the patriots in 379, after his fellow-citizens expelled the Spartans, and led the Theban army to victory over the Spartans and their allies at Leuctra (371). *Xenophon*. Greek historian, essayist, and military commander (*c*.428–354 BC), who fought a famous campaign against the Persians. *abated*. Beat down. *by how much*. In the same proportion as. *a*. This word was added in the 1640 edn. *much ... age*. At about the same age. According to Aristotle, the body is strongest from 30 to 35, the mind at 49 (*Rhetoric*, ii. 14, 1390b 8 ff.). *policy*. Political Science. *enable*. Qualify. *empiric physicians*. Doctors who work by trial and error, having no theoretical knowledge other than observation of phenomena. *pleasing*. Plausible, superficially effective. *receits*. Recipes. *adventurous*. Rashly bold. *complexions*. Temperaments: the balance of the four qualities, hot, wet, cold, and dry, resulting from the mixture of elements in the human body. *peril of accidents*. Symptoms of disease. *practice*. Practical experience, lacking theoretical knowledge. *grounded*. Thoroughly instructed. *besides*. Outside. *of doubtful consequence*. Of uncertain outcome, dangerous. *contrariwise*. On the contrary.

128. *instance contradictory*. Exception. (Such inversions of normal word-order, by modern conventions, are common in Bacon's prose, and sometimes deliberate, for stylistic emphasis.) *ordinary*. Customary. *politic men*. Politicians. *extenuate and disable*. Underrate and disparage. *Pedantes*. Schoolmasters. The Pedant (such as Holofernes in *Love's Labour's Lost*) was one of the stock comic figures in the Italian *commedia dell'arte* in the 16th century. Montaigne also objected to such mockery in *Essais*, i. 25, 'Du Pedantisme' (Pléiade edn., p. 132). *in minority*. Not having attained full age; needing a guardian. *traduce*. Mock, expose to contempt and ridicule. *magnified*. Celebrated, extolled. *Nero ... Seneca*. Nero became Emperor in AD 54, aged 17 (Suetonius, *Nero*, 7; Tacitus, *Annals*, xiii). Seneca, having been his tutor since AD 49, was able to exercise some control over his violent vices. *Gordianus ... Misitheus*. Gordianus III became Emperor of Rome in AD 238, aged about 13. In 241 he appointed his father-in-law Misitheus prefect of the praetorian guard, in effect as adviser and administrator; but Gordianus was assassinated in 244. *contentation*. Contentment. *Alexander Severus*. Marcus Aurelius Severus Alexander (AD 208–35), who became Emperor of Rome in 222, owed much to the counsel of his mother and grandmother. *Pius Quintus*. A Dominican friar whose rule as Pope from 1565 to 1572 encompassed the Christian defeat of the Turks at Lepanto. *Sextus Quintus*. A Franciscan friar, Pope from 1585 to 1590, who restored the papal finances and built extensively in Rome. *estate*. Statesmanship. *to seek*. Deficient. *points of convenience*. Matters of expediency. *accommodating*. Adaptability. '*ragioni di stato*'. 'Reasons of state', policies (often expedient) designed to preserve the existence and power of a state. *terming them*. Wright (p. 272) cites G. Gatena's *Vita di Pio V* (Rome, 1587), 31, for the Pope's disapproving comment on the maxim, 'Chi non sà simulare non sà regnare' ('Who does not know how to simulate does not know how to rule'). *recompense*. Compensate. *physic*. Medicine. *one man's life*. Wright (p. 272) suggests as the correct reading 'even of one's man's life'. *sort*. Agree, fit with. *immediate*. Proximate. *countervail*. Outweigh. *hold way*. Keep pace. *common purse*. Civic or national treasury. *seducements*. Seductions from. *pretended*. Alleged.

129. *insinuate*. Introduce indirectly, by winding courses. *ministereth*. Furnishes, supplies. *every*. Each. *resolve*. Decide. *positive*. Holding strict positions. *regular*. Rigidly adhering to rules. *demonstrative*. Capable of demonstration. *cautions of application*. Warnings against wrongly applying (e.g.) general principles to specific instances. *rectify*. Correct. *Clement*. Clement VII, Pope from 1523 to 1534, whose alliance with Francis I of France against Charles V (Holy Roman Emperor) led to the latter's sacking of Rome, and whose excommunication of Henry VIII provoked England's schism with Rome. As V. Luciani notes ('Bacon and Guicciardini', *PMLA* 62 (1947), 96–113), 'Guicciardini's analysis of Clement's errors is found in Books xvi–xx' of his *Storia d'Italia* (p. 101 n.): but see esp. xvi. 12. *lively*. Vividly. *Cicero*. See *Epist. ad Att.* xvi. 7. *painted out*. Depicted. *apace*. Swiftly. *Phocion*. See Plutarch, 'Life' of Phocion, 2.4, 3. *Ixion*. A mythical character who tried to seduce Hera, wife of Zeus; but Zeus made a cloud in her image, on which he begot centaurs. See Pindar, *Pythian Odes*, ii. 21 ff., and below, p. 201. *hold*. Restrain. *vaporous or imaginative*. Taking as real those things that are insubstantial or imaginary. *Cato the second*. Marcus Porcius Cato (*c*.95–46 BC), great-grandson of Cato the censor, a rigid exponent of republicanism who committed suicide when defeated by Julius Caesar at the battle of Thapsus. See Cicero, *Epist. ad Att.* ii. 1. *Antipodes*. People living on the opposite side of the world, popularly supposed to walk upside down. *tread opposite*. Oppose. *hireling*. Mercenary. *beareth . . . up*. Gives prestige. *wear*. Suffer from wear and tear. *occasion*. Opportunity. *valours*. Reputation for prowess in battle. *designments*. Enterprises. *according to*. In harmony with.

130. *purchase*. Thing acquired. *softness*. Effeminacy. '*Quidam . . . est*'. 'Some men live so much in the shade, that whenever they are in the light they seem to be in trouble' (Seneca, *Epist.* iii. 6, loosely quoted). *leisure*. The ethos of the *vita activa* tended to stigmatize any activity not directly connected with work as idleness or *otium* (in a negative sense). Bacon defends the concept of leisure as 'free or unoccupied time', but characteristically prefers it to be employed on studies, not pleasure. *expecteth*. Awaits. *tedious*. Slow. *dispatch*. Promptness. *Demosthenes*. See Plutarch, 'Life' of Demosthenes, 8.2 f. (a story in fact told of Pytheas). *smell of the lamp*. Show signs of laboured preparation. *So as*. So that. *doubt*. Fear. *expulse*. Drive out, displace. *prejudice*. Harm. *both*. That is, learning and business. *of*. For. *depravation*. Defamation. *all*. Any. *tread*. Walk. *by*. With the help of. *gentle*. Courteous, agreeable. *maniable*. Manageable. *churlish*. Rough, intractable. *thwart*. Perverse. *clear*. Demonstrate the truth of. *judgment of Cato*. Quoted above, p. 126, and contradicted by him subsequently studying Greek: Plutarch, 'Cato', ii. 6; Cicero, *Acad. Quest.*, ii. 2, *De senectute*, i. 3. *blasphemy*. Slander.

131. *tongue*. Language. *to the end*. In order. *affected gravity*. Pretended seriousness. *Virgil's verses*. Quoted above, p. 126. *brave*. Challenge. *taking*. Appropriating. *two first Caesars*. Julius and Octavius (Augustus). *best poet*. Bacon's high estimate of Virgil was shared by many Renaissance scholars, notably J. C. Scaliger in his *Poetices libri septem* (1561). *historiographer*. Historian. *antiquary*. Student or collector of antiquities. *second*. To Demosthenes. *accusation of Socrates*. Quoted above, pp. 126–7. Bacon erroneously places Socrates' trial during the rule of the Committee of Thirty, 404–403 BC: it took place after democracy had been restored. Rossi (p. 145) suggests that Bacon may have been following Seneca, *De tranquillitate animi*, v. 2–3; Wolff (i. 39) suggests Montaigne, *Essais*, i. 19. *envious*. Malicious. *revolution*. In the sense 'complete rotation'. *state*. Form of rule. *heroical*. Heroic; as recorded by Diogenes Laertius, ii. 43. *accumulate*. Heaped up. *for*. As being. *humorous*. Capricious. *redargution*. Refutation. *Castor and Pollux*. Twin sons of Zeus and Leda, identified with the constellation Gemini. *lucida sidera*. 'Shining stars' (Horace, *Odes*, I. iii. 2). *influence*. A term from astrology. *groweth unto*. Attaches to. *cleaveth*. Adheres. *fastest*. Most closely. *accidental*. Not essential. *handled*. Discussed, treated. *in hand with*. To treat of. *derogations*. Low esteem.

132. *by reason*. Because. *lucre*. Gain. *common place*. Theme for composition, often implying unoriginality. *Machiavel. Discourses*, III. i, referring to St Francis and St Dominic. *borne out*. Compensated for. *delicacy*. Addiction to pleasure; luxuriousness. *had*. Would have. *rudeness*. Lack of culture. *civility*. Civilization, refinement. *paradoxes*. Contradictions; 'things contrary to common opinion', as Cicero defined the genre in *Paradoxa Stoicorum*. Such topics as 'That poverty is preferable to riches' were argued as serious moral positions in the tradition of the philosophical paradox, and were common themes in school and university disputations. '*Caeterum . . . fuerit*'. 'If I am not led away by the task which I have undertaken, there never was a republic greater, more religious, and more abounding in good examples than the Roman, nor one that so long withstood avarice and luxury, or so much honoured poverty and thrift' (Livy, Preface). *that person . . . Caesar*. Sallust, the historian, reputedly the author of the 'Speech on the State Addressed to Caesar in his Later Years', from which (i. 8) the following quotation comes. *summary*. Important. *estimation*. Valuing, esteem. '*Verum . . . erunt*': 'But these, and all evils, will disappear when wealth is no longer honoured, and when the magistracies and other objects of general ambition are not procurable by money'. '*rubor . . . color*'. 'A blush is virtue's colour': attributed to Diogenes the Cynic in Diogenes Laertius, vi. 54. '*paupertas . . . fortuna*'. 'Poverty is virtue's fortune.' '*Qui . . . insons*'. 'He that maketh haste to be rich shall not be innocent': Prov. 28: 20; *Promus*, no. 10. '*Buy the truth . . . knowledge*'. Prov. 23: 23; *Promus*, no. 9. *means*. Wealth. *obscureness*. Obscurity. *contemplative*. Those devoting their lives to study. (Bacon uses the concept in a secular, not a religious sense.) *taxed with*. Accused of. *civil life*. Public life, the *vita activa*.

133. *consonancy*. Harmony. *conceits*. Conceptions. *allowing*. Approving. *Cassius and Brutus*. As the murderers of Julius Caesar, their images were not carried at the funeral of Junia Tertia (wife of Cassius and half-sister to Brutus). '*Eo . . . visebantur*'. 'They outshone them all, from the very fact that they were not to be seen' (*Annals*, iii. 76). *meanness*. Lowness. *traduced to*

contempt. Mockingly paraded. *transferred*. Applied. *disesteeming*. Depreciating. *reduce*. Convert, bring back. *popularity of opinion*. Common belief. *curious*. Careful. *seasoned*. Made fit for use. *corroborate*. Well-established. *terms*. Definite portions of time. *use*. Are accustomed. *applications*. Appliances. *Rabbins*. Teachers of theology; here, Isaac Abrabanel in his commentary on Joel. See also Essay 42, 'Of Youth and Age', above, p. 425. '*Your young men . . . dreams*'. Joel 2: 28. *apparitions*. Appearances, manifestations. *ape*. Imitator. *tutors*. Guardians. '*quo . . . deteriores*'. 'The better they are, the worse they are' (Diogenes Laertius, vi. 46), that is, the stronger their superstition (Catholic beliefs), the worse. In the *De Augmentis*, designed for an international audience, Bacon deletes these criticisms of the Jesuits, and speaks warmly of their pedagogy, especially for its encouragement of school (Latin) drama: *Works*, i. 709–11; iv. 494–6. '*talis . . . esses*'. 'You are so good that I wish you were on our side': Plutarch, 'Life' of Agesilaus, xii. 5, perhaps in Bacon's own translation (Wolff, ii. 86). *temperatures*. Temperaments. '*abeunt . . . mores*'. 'One's studies pass into one's character': Ovid, *Heroides*, xv. 83; *Promus*, no. 1121.

134. *indifferent*. Impartial. *not*. Not anything. *to*. In. *contend*. Endeavour. *reduce*. Lead back. *honesty*. 'High and honourable character' (Wright). *too great height*. Too high a standard. *caveats*. Cautions. *walks*. Occupations. *Solon*. See Plutarch's 'Life', 15. *Plato*. Cf. *Crito*, 50D ff., and the Seventh Epistle, 331C ff.; Cicero, *Epist. ad Fam*. I. ix. 18. *contestations*. Controversies. *Caesar's counsellor*. Sallust, in the speech referred to above, i. 5. '*Non . . . sunt*'. 'Do not attempt to restore things to their original institutions which, because of the long corruption of morals, are held in contempt.' *Cicero, ad Att*. II. i. 8. '*Cato . . . Romuli*'. 'Cato's opinions are excellent, but occasionally he does harm to the state: for he speaks as though he were living in Plato's Republic, and not in the filth of Rome.' *Cicero. Pro Murena*, xxxi. 65. *expound*. Interpret, explain. *prescripts*. Prescriptions. '*Isti . . . consisteremus*': 'Those very teachers of virtue themselves seem to have fixed the standard of duty somewhat higher than nature can bear, in order that when we strive our utmost to attain it, we may nevertheless reach the level we ought to.' '*Monitis . . . meis*'. 'I do not act up to my own precepts' (Ovid, *Ars amatoria*, ii. 548). *fault*. Bacon is being ironic, since such dedication to the well-being of others is for him a fundamental ethical value. See bk. II, above, pp. 245 ff., and the *Essays*. *Demosthenes. On the Chersonese*, 71. *as*. As that. *Quinquennium Neronis*. In the first five years of Nero's reign Seneca was both tutor and counsellor. *held on*. Persisted with.

135. *casualty*. Uncertainty. *vocation*. Calling; a mode of life regarded as requiring dedication. *end*. Goal. *ordainment*. Ordination. '*Ecce . . . lucrefeci*'. 'Lo, I have made profit for you', and not 'Lo, I have made profit for myself': cf. Matt. 25: 20 (parable of the talents). *universality*. The general good. *lines*. As in geometry. *ship of estates*. The whole community. *cockboat*. Small boat holding very few people. *make good*. Support, defend. *stand in*. Stand firm during. *parts*. Parties. *versatile*. Changeable, variable. *of*. Belonging to. *carriage*. Conduct. *tender sense*. Sensitivity. *fast*. Firm,

settled. *tax*. Censure. *allowance*. Approval. *excusation*. Excuse. *probably*. Plausibly. *applying*. Adapting. *application*. Compliance. *exquisite*. Elaborate, minute. '*Satis . . . sumus*'. 'Each of us is enough of an audience for the other': ascribed to Epicurus in Seneca, *Epist.* vii. 11. Cf. Essay 10, 'Of Love', above, p. 358. *yield*. Concede. *disperse*. Spread. *wanteth*. Lacks. *speculative*. Prying into. *wind*. Manipulate. *cloven*. Split in two, insincere.

136. *ingenuous*. Noble, high-minded. *custom*. See Herodotus, *Hist.* i. 99. *Levant*. East. *bent*. Crooked. *Scripture*. Prov. 25: 30. *decency*. Decorum. *discretion*. Propriety. *consequence*. Inference. *Themistocles*. See Plutarch's 'Life', 2.2 f. *uncivilly*. Rudely. *seen*. Versed. *passages*. Transactions. *are to seek*. Are at a loss. *punctual*. Small. *Plato*. *Symposium*, 215B. As Spedding noted, Socrates is actually compared to the images of Silenus, and Bacon probably fused that passage with one in the prologue to *Gargantua* where Rabelais describes these images as 'little boxes, like those we may now see in the shops of Apothecaries, painted on the outside with wanton toyish figures . . . but within those capricious caskets were carefully preserved and kept many rich jewels, and fine drugs' (tr. Urquhart, 1653). *gallypots*. Glazed earthenware jars. *antiques*. Grotesque figures. *confections*. Preparations. *levities*. Improprieties. *courses*. Forms of behaviour. *divers*. Several. *professors of*. Pretenders to. *trencher philosophers*. Scholars who flattered the great so as to be invited to eat with them. *Lucian*. See the satirical dialogue 'On Salaried Posts in Greek Houses', xxxiii f. *uncomely*. Awkwardly. *doubted*. Feared. *Cynic*. The word literally means 'doglike'. *Du Bartas*. See *La Création du monde* (1578), 'Le second jour'. Hecuba was Helen's grandmother. *Faustina*. Younger daughter of Antoninus Pius, b. *c*. AD 125. 'Ancient authority groundlessly interpreted her lively temperament as a sign of faithless and disloyal character' (*OCD*). *price*. Value.

137. *as*. For instance. *for that*. Because. *intitle*. Designate. *morigeration*. Obsequiousness. *application*. Deference. *men in fortune*. Scholars who had no institutional employment depended on private patrons for their livelihood. *Diogenes*. In fact the answer was given to Dionysius by Aristippus, a disciple of Socrates, according to Diogenes Laertius, ii. 69. *Aristippus*. Ibid. ii. 79. *tender*. Sensitive. *him*. The philosopher Favorinus, according to Spartianus, 'Life' of Hadrian, 15. *his best*. In his best manner. *reason*. Reasonable. *disallowed*. Disapproved. *errors*. In the original sense of *errare*, to go astray. *vanities*. Futile or worthless activities or attitudes. *intervened*. 'Come in as something extraneous' (*SOED*). *censure*. Critical assessment. *separation*. Division, discrimination. *aspersion*. Damaging report. *scandalize*. Slander, malign. *deprave*. Defame. *the state*. Its original pure condition. *virtue*. Inherent strength, purity. *upon*. Of. *blemish*. Stigmatize. *taint*. Tarnish. *exact animadversion*. Specific criticism.

138. *unto*. With reference to. *popular*. Ordinary, general. (Bacon's detailed critique follows in bk. II.) *vain*. Useless, fruitless. *frivolous*. Paltry, valueless. *curious*. Fussy, over-elaborate. *in reason*. In theory. *in experience*. In

practice. *fall out*. Happen. *distempers*. Disorders, diseases. *fantastical*. Fanciful, capricious. *contentious*. Quarrelsome. *delicate*. Affected, obsessed with niceties of style. *altercations*. Controversies. *Martin Luther*. Bacon associates (rather loosely) the Reformation with the Renaissance recovery of Greek, which really took place in 15th-century Florence. The Reformation constituted a 'second wave' of Greek scholarship, for specifically theological purposes (Melanchthon was appointed to the chair of Greek at Wittenberg in 1518). Bacon then, equally loosely, identifies the Reformation hostility to medieval (Catholic) theology with Renaissance humanism, which frequently mocked medieval scholastic philosophy for its 'barbarous', non-classical Latin and its schematic use of logical categories. The whole sequence is a brilliant polemic, but unites a number of quite distinct traditions. *discourse of reason*. Process of reasoning. *province*. Sphere of action. *solitude*. Intellectual isolation. *no ways*. In no way. *make a party*. Take sides. *divinity... humanity*. Sacred and secular studies. *revolved*. Considered. *exquisite travail*. Painstaking study. *pressing*. Emphasizing. *kind of writing*. Bacon now shifts to a satiric account of Ciceronianism, a short-lived fashion in the early 16th century by which some fanatical admirers would only use Latin words that Cicero had used. For a detailed account of this movement, substantiating Bacon's critical evaluation, see J. W. Binns, *Intellectual Culture in Elizabethan and Jacobean England: The Latin Writings of the Age* (Leeds, 1990), 202–4, 270–90. *precipitated*. Brought about. *primitive*. Ancient. *schoolmen*. Scholastic philosophers (9th to 14th century). *terms of art*. Technical terminology. *circuit of speech*. The need to avoid circumlocution was one of the reasons sometimes given for inventing neologisms, a practice of which Bacon disapproved. *pureness*. Purity. *Pharisees*. Strict observers of the laws and rituals laid down by Moses. '*Execrabilis... legem*'. 'This multitude which knoweth not the law are accursed': John 7: 49. The quotation fits the present argument, but not if we recall its biblical context, the Pharisees' disapproval of Christ. *price*. Value. *vulgar sort*. Common people.

139. *affectionate*. Zealous, devoted; also 'affected'. *copie*. Latin 'copia', fullness of expression (as in 'cornucopia'). *words... matter*. That is, expending more care on the 'verba' or style than on the 'res' (or subject-matter). *choiceness*. Affected expression. *round and clean*. Polished and smooth. *sweet falling*. Syntactical parallelism made with words having the same endings, the rhetorical figures *homoioptoton* and *homoioteleuton* (*similiter cadens* and *similiter desinens*). *varying*. Repeating in artfully varied forms. *watery*. Bacon puns on the other connotations of having a 'flowing' or fluent style. *Osorius*. Jeronimo Osorio (1506–80), a theologian known as 'the Portuguese Cicero', whose publications included a treatise *De gloria*, supposedly a lost work by Cicero. *Sturmius*. Johann Sturm (1507–89), head of the Strasbourg Gymnasium, a prolific writer on rhetoric and style, known as 'the German Cicero'. *periods*. Sentences, prose-style. Sturm published *De periodis* (Strasbourg, 1550). *imitation*. Sturm published *De imitatione oratoria* (Strasbourg, 1574). *Carr*. Nicholas Carr (1524–68) succeeded Sir John Cheke as Regius Professor of Greek at Cambridge in 1547. *Ascham*. Roger Ascham (1515–

68), Reader in Greek at Cambridge and tutor to Queen Elizabeth I, praised Cicero repeatedly in his *Scholemaster* (1570), and was a friend of both Osorius and Sturm. *Erasmus*. In 'Juvenis and *Echo*' (*Colloquies*), the last three letters of 'Cicerone' echoing the Greek word for 'ass' (ὄνε). '*Decem* ... *Cicerone*'. 'I have spent ten years reading Cicero.' *is*. The 1605 text omitted this word; later editions restored it. '*secundum* ... *minus*'. 'To a greater or lesser degree.' *but* ... *have*. That this should not have. *operation*. Effect. *first letter* ... *limned book*. The initial capital of a 'patent' (official document) was usually 'limned' (illuminated), ornamented with a drawing or design, a common practice also with manuscripts. *flourishes*. Ornamental forms. *Pygmalion's frenzy*. The deluded condition of falling in love with the statue he had made (Ovid, *Met*. x. 243). *all one*. The same. *sensible*. Full of sense. *plausible*. Worthy of applause. *for*. Notwithstanding. *just period*. Appropriate conclusion or stopping-point. *be to*. Needs.

140. *civil occasions*. In public life, where the ability to express oneself is important. *conference*. Conversation. *counsel*. Advice. *persuasion*. Oration. *discourse*. Formal discussion of a topic in speech or writing. *minion*. Darling. '*Nil sacri es*'. 'Thou art no divinity': from Erasmus' *Adagia*; *Promus*, no. 724. *respective*. Referring. *extensive*. Capable of being extended, far-reaching. '*Devita* ... *scientiae*'. 'Turn a deaf ear to empty and worldly chatter and the contradictions of so-called knowledge' (1 Tim. 6: 20). *badges*. Signs. *suspected*. Imagined to be faulty; dubious. *falsified*. Shown to be erroneous. *strictness of positions*. Dogmatic, rigid assertions. *solid*. Substantial, wholesome. *subtile*. Abstruse, elusive. *vermiculate*. 'Intricate, winding like the moving of a worm' (Wright, ed. cit.). *quickness*. Animation. *spirit*. The source of motion. *degenerate*. Debased. *agitation*. Disturbance. *creatures*. Creations. *spider* ... *web*. Cf. *Promus*, no. 797a: 'Ex se finxit velut araneus' ('he fabricated out of himself like a spider'), from Erasmus, *Adagia*. *cobwebs*. Le Doeuff (ed. cit. 302) cites an ancient tradition (going back to Diogenes Laertius, vii. 161) of describing dialectic as producing arguments like spiders' webs. See also *Novum Organum*, i. 95 (*Works*, i. 201; iv. 92–3). *curiosity*. Niceness.

141. *knowledge*. Body of knowledge. *objections* ... *solutions*. Scholastic philosophy did indeed use a formal sequence of *quaestiones, objectiones*, etc. See above, p. 235 and note. *faggot*. Broom made of twigs. Bacon alluded to Aesop's fable (no. 52) of the three quarrelling sons whose father showed them that a strong broom can be broken if its twigs are snapped one at a time: an image of the destructiveness of quarrels. '*Verborum* ... *pondera*'. 'That he broke up the weight of the matter by his verbal subtleties': Quintilian, x. i. 130. '*Quaestionum* ... *soliditatem*'. 'They broke up the solidity of the sciences by hair-splitting questions.' *fair*. Large, well-proportioned. *great*. Strong, large. *watch candle*. Small candle, night-light. *proved*. Tested. *cavillation*. Quibbling objection. *Scylla*. A beautiful maiden whom Circe, jealous of Glaucus' love for her, transformed into a composite being, the upper body that of a woman, the lower parts the tail of a fish surrounded by dogs. '*Candida* ... *monstris*'. 'With howling monsters girt about her white waist'

(Virgil, *Eclogues*, vi. 75.) *proportionable*. Suitable, appropriate. *decisions*. Conclusions. *out of their way*. Erring. *meet*. Agree. *digladiation*. Wrangling. '*Verba . . . otiosorum*'. 'It is old men's idle talk': the rebuke of Plato by Dionysius the Sicilian tyrant, in Diogenes Laertius, iii. 18. Cf. *Novum Organum*, i. 71 (*Works*, iv. 72). *travail of wit*. Mental work. *undertakers*. Those who take on themselves a task or business (ironically meant).

142. *fierce with dark keeping*. Angry, like animals (or madmen) kept in the dark. *unequal*. Uneven, inaccurate. *foulest*. Most wicked, vile. *representation*. Image, likeness. *one*. The same. Cf. *Of Tribute*, above, p. 34. *direct beam*. Ray of light. *simplicity*. Freedom from guile; naïvety. *concur*. Coincide. '*Percontatorem . . . est*'. 'Shun the inquisitive man, for he is a babbler' (Horace, *Epist*. I. xviii. 69. *fame*. Rumour, hearsay. '*Fingunt . . . creduntque*'. 'Those who are prone to invent are also prone to believe': *Annals*, v. 10, adapted. *facility*. Readiness. *of credit*. In believing. *fact*. Something done. *art*. A discipline, such as logic or mechanics. *received*. Accepted. *registered*. Written down, as if historical fact. *had a passage*. Were current, generally accepted. *politic*. Cunning. *poesies*. Poems. *badges*. Marks, signs. *antichrist*. Enemy of Christianity. *scandal*. Discredit. *Plinius*. Pliny the Elder (AD 24–79) whose *Naturalis Historia* in 37 books was a compilation of all kinds of fact and fiction. *Cardanus*. Girolamo Cardano (1501–76), a physician and mathematician whose two encyclopedic works, *De Subtilitate Rerum* (1550) and *De Varietate Rerum* (1557), similarly mingled reliable information and credulous report. *Albertus*. Albertus Magnus (*c*.1200–80), a scholastic author of several uncritical compilations.

143. *fraught with*. Full of. *Aristotle*. The *Historia animalium* is an authentic work, whereas the book of 'prodigious narrations', or *De Mirabilibus Auscultationibus*, is spurious. *history*. At this time implying not a sequence of events over time but rather, in the original Greek sense of *historein*, any form of 'enquiry': the systematic study of a discipline (natural history) or physical phenomenon (as in Bacon's own *History of Life and Death*). *had . . . intelligence*. Corresponded. *imagination*. In the Middle Ages and Renaissance the imagination was often regarded with distrust as a source of delusion. *Natural Magic*. An eclectic mixture of magic and physics, developed in the Renaissance, which studied physical phenomena deriving from 'occult' or unobservable causes (e.g. magnetism). *pretences*. Intentions, aims. *pretendeth*. Proposes as end or object. *concatenation*. Connection. *superior globe . . . inferior*. Of the heavens (the constellations) and the earth. *reduce*. Restore. *mixtures. Compound substances*. *derivations . . . these ends*. 'The subsidiary channels leading to these ends and the modes in which they have been followed' (Wright, p. 277). *veil over . . . writings*. Bacon's consistent attacks on the occult tradition for its protective devices of secrecy were historically important in helping to formulate a concept of scientific knowledge as demanding open discussion. See e.g. Paolo Rossi, *Francis Bacon: From Magic to Science* (London, 1968). *auricular*. Hearsay. *Aesop*. *Fables*, no. 33. *mould*. Earth. *dictators*. Absolute authorities. *stand*. Stand firm, unchallenged.

144. *consuls*. The 1605 text reads 'counsels', corrected in the Errata to 'consuls'. Spedding suggested 'counsellors'. *stay*. Standstill. *arts mechanical*. Involving machines. *comes shortest*. Achieves least. *leeseth*. Loses, destroys. *artillery*. Constructing machines for throwing missiles. *grossly*. Clumsily. *imbased*. Debased. *in one*. To the same purpose. *one*. Individual. *depraved*. Misrepresented. *illustrated*. Illuminated; also, 'made illustrious'. '*Oportet ... credere*'. 'The learner should take things on trust' (Aristotle, *Sophistici Elenchi*, ii, 165b3). '*Oportet ... iudicare*'. 'The learned man should exercise his judgment.' *as time*. Provided that time. *peccant humours*. Unhealthy tendencies. *intrinsic*. Internal, hidden. *traducement*. Misrepresentation. *affecting of*. Liking for. *time*. In the ancient myth, Chronos (the Greek for 'time'), father of Zeus, devoured his children at birth until foiled by Earth. *malice*. Evil disposition. *envieth*. Begrudges. *deface*. Destroy. '*State ... ea*'. 'Stand in the ancient paths and look about, and see which is the straight and right way, and so walk in it' (Jer. 6: 16). *reverence*. Respect. *make a stand*. Pause. *taken*. Understood.

145. '*Antiquitas ... mundi*'. 'What we call antiquity is the youth of the world.' Ellis (*Works*, i. 459) cites 2 Esdras 14: 10, 'For the world hath lost his youth, and the times begin to waxe olde', and Giordano Bruno, *La Cena de le Ceneri*, 'noi siamo più vecchi ed abbiamo più lunga età, che i nostri predecessori', *Dialoghi Italiani*, rev. ed. G. Aquilecchia (Florence, 1985), i. 39. *Lucian*. In fact Seneca (in Lactantius, *De falsa religione*, i. 16). *other*. Other of. *old time*. Former ages. *septuagenary*. 70 years old. *Pappia*. A law passed during the reign of Augustus Caesar, forbidding a person of 60 or over to marry anyone under 50. *generation*. Able to procreate. *prejudged*. Prematurely deemed. '*Nil ... contemnere*'. 'It was nothing but daring to despise groundless fears' (Livy, ix. 17). *Columbus*. In his discovery of America in 1492, a project long dismissed as absurd. *relation*. In law, the treatment of a decision as applying to a date earlier than that on which it was made; retrospective validity. *after ... examination*. After various opinions have been thoroughly examined. *still*. Constantly. *like*. Likely. *give passage*. Allow through; accept. *popular*. In the corresponding passage in *Valerius Terminus* (an early draught of the *Advancement*), where Bacon also uses this metaphor of time as a river on which only light things float, he writes that 'howsoever governments have several forms, sometimes one governing, sometimes few, sometimes the multitude; yet the state of knowledge is ever a *Democratie*, and that prevaileth which is most agreeable to the senses and conceits of people' (*Works*, iii. 227). Le Doeuff (p. 305) cites a passage from *De Principiis* in which Bacon laments that Presocratic philosophy, particularly that of Democritus, should have sunk, while the lighter systems of Aristotle and Plato survived (*Works*, iii. 84). *peremptory*. Admitting no refusal; in law, an argument that, if upheld, ends an action. *reduction*. Organization. *arts and methods*. Systematic and complete accounts of a topic. *knit and shape*. Complete their physical development. *aphorisms*. Self-contained observations (not necessarily pithy) of physical or other phenomena.

146. *comprehended*. Summed up. *succeed*. Result from. *universality*. Universal principles. '*philosophia prima*'. The basic principles common to all philosophy, discussed further in bk. II, above, pp. 189 ff. *cease*. Cause to cease. *discovery*. Perception, insight. *tumbled up and down*. Gone round and round. *intellectualists*. (A word invented by Bacon): those who think that knowledge comes from the intellect alone. *Heraclitus*. According to Sextus Empiricus, *Against the Logicians* i. 133. Wolff (i. 243–4), believing that Bacon was otherwise unfamiliar with Sextus, suggests Plutarch, 'Of Superstition', 3 (*Moralia*, 166C) as a more likely source, but notes that Bacon could have derived the saying from Henricus Stephanus's collection of Presocratic fragments, *Poiésis Philosophiké* (Paris, 1573), 130. *invocate*. Invoke, as if calling up a spirit. *divine*. Make out or interpret by supernatural or magical insight. *used*. Been accustomed to. *applied*. Practised. *tincture*. Colouring. *Proclus*. Greek Neoplatonist philosopher (*c*. AD 412–85), who synthesized many strands of late classical philosophy into one comprehensive theological metaphysic. He wrote a commentary on Euclid's *Elements*. *primogeniture*. Having the rights of the first-born child (to inherit property). *severally*. Separately, individually. *Gilbertus*. William Gilbert (1540–1603), whose *De Magnete* (1600) was the greatest contribution to knowledge yet made by an English scientist. Bacon is not objecting to the observational and experimental parts of the book, which he admired, but to its speculative conclusion ascribing a universal role to magnetism. See Peter Urbach, *Francis Bacon's Philosophy of Science* (La Salle, Ill., 1987), 109–21, and Suzanne Kelly, *The De Mundo of William Gilbert* (Amsterdam, 1965), 75–96. '*Hic . . . recessit*'. 'This man has not gone outside the limits of his own art': describing Aristoxenus; Cicero, *Tusc. D.* I. x. 20. *conceits*. Conceptions; but also 'fanciful ideas'. '*Qui . . . pronunciant*'. 'Those who take into account only a few considerations easily make dogmatic pronouncements' (*De generatione et corruptione*, i. 2, 316a5 ff.).

147. *two ways of action*. Bacon alludes to two famous classical moralizing analogies, Hesiod's *Works and Days*, 287 ff.: 'Inferiority can be got in droves, easily: the road is smooth, and she lives very near. But in front of Superiority the immortal gods set sweat; it is a long and steep path to her, and rough at first. But when one reaches the top, then it is easy, for all the difficulty' (tr. M. L. West, *Hesiod*, Oxford, 1988); and the related fable of Hercules at the crossroads, faced with the choice between a broad and pleasant path, which descended to vice, and a steep and narrow one, ascending to virtue. See Xenophon, *Memorabilia*, II. i. 21–34 and Erwin Panofsky, *Herkules am Scheidewege* (Leipzig and Berlin, 1930). *tradition*. Communication. *magistral*. Dogmatic. *ingenuous and faithful*. Candid and accurate. *for practice*. Practical, 'how to do it' treatises. *disallowed*. Disapproved. *vein*. Style, manner. '*Nil . . . videretur*'. 'Fearing nothing so much as to appear doubtful about anything' (Cicero, *De natura deorum*, I. viii. 18). *asseveration*. Emphatic declaration. *scope*. Aim, object. *professors*. Those who offer or profess competence in a subject. *certain*. Safe, secure. *commenter*. Commentator. *patrimony*. Heritage. *upon*. Out of. *enable*. Give power. *profession*. Livelihood. *the benefit and use of men*. This is one of the most important emphases in Bacon's redefinition of the goal and function of science, that it be not theoretical but operative, designed to relieve the miseries of human life. *wandering*. Whimsical. *estate*. Condition. Cf. St Bernard, *Cantica Canticorum*, Sermon 36, *ed.cit*., p. 235, and John of Salisbury, *Policraticus*, Bk. 7, ch. 15.

148. *straitly*. Tightly. *Saturn ... Jupiter*. These two planets were assigned to the contemplative and active lives by Macrobius in his *Commentary* on Cicero's *Dream of Scipio*, I. xii. 14. *Atalanta*. According to the myth, she would only marry the man who beat her in a race, and defeated all her suitors until Milanion threw down golden apples as they were running: she stopped to pick them up and lost. '*Declinat ... tollit*'. 'She leaves the path and picks up the rolling gold' (Ovid, *Met*. x. 667). *Socrates*. As recorded by Cicero, *Tusc. D*. v. iv. 10. *converse*. Dwell. *manners and policy*. Moral and political philosophy. *bond-woman*. Female slave. *opened*. Explained, exposed. *peccant humours*. Unhealthy tendencies. *proficience*. Progress. '*Fidelia ... malignantis*'. 'Faithful are the wounds of a friend, but the kisses of an enemy are deceitful' (Prov. 27: 6). *freely*. Openly, generously. *laudative*. Eulogy. *amplification*. Exaggeration. *arch-type*. Original pattern from which copies are made. Ellis (*Works*, i. 464) cites Philo Judaeus, *De Opificio Mundi*, who expounds the first 5 verses of Genesis as referring to 'the formation in God's mind of the archetype or exemplar of the visible universe'. *platform*. Model. *sobriety*. Sound judgement. *is original*. Is generated by God.

149. *wisdom*. One of the books in the Vulgate Bible was called Sapientia; Protestants relegated it to the Apocrypha. Bacon alludes to ch. 7, vv. 24–6. *as the Scriptures call it*. Prov. 8: 22–31. *virtue*. In Bacon's time this word had the connotations of 'innate strength, excellence'. *subsistence*. Substance. *disposing*. Arranging. *for any thing which appeareth*. Despite anything which seems to the contrary. *disposition*. Arrangement. *note of difference*. Distinction. *made*. Gen. 1: 1. *supposed*. Fictitious. *Dionysius*. Known as the 'Areopagite', to whom were ascribed *The Celestial Hierarchies* (a synthesis of Christianity and Neoplatonism, dating from the 5th century). Bacon seems to draw here on Hooker, *Of the Lawes of Ecclesiasticall Politie* (1593), I. iv. 1–2. *ministry*. Serving God by executing his commands. *spirits and intellectual forms*. Angels. *light*. For Bacon's continuing use of this metaphor for knowledge see *New Atlantis*, above, pp. 472, 486. *incorporal*. Incorporeal. *distribution of days*. The six days into which God divided Creation. *set down*. Gen. 2: 15: God put Adam 'into the garden of Eden to dress it and keep it'. *contemplation*. Study, sustained thought. *exercise*. Employing one's faculties. *experiment*. Testing something; gaining knowledge. *reluctation*. Struggle, violent effort. The Geneva Bible adds the gloss to Gen. 2:15: 'God would not have man idle, though as yet there was no need to labor.' *sweat of the brow*. This was the later punishment, after the Fall: cf. Gen. 3: 19. *first acts*. Gen. 2: 19–20. *summary*. Most important. *imposition*. Bestowing ('whatsoever Adam called every living creature, that was the name thereof').

150. *touched.* Mentioned (above). *supposition.* Assumption, (mistaken) belief. *the end.* In order. *depend . . . himself.* Ellis (*Works*, i. 465) cites a close parallel in Aquinas, *Summa Theologiae*, Sec. Secund. q. 163a. 2. See also *Confession of Faith*, above, p. 109. *the story or letter.* The literal, or narrative level (as opposed to the other three levels of interpretation: moral, allegorical, and anagogical. See above, pp. 295, 296). figured. Represented. *persons.* Gen. 4: 20. *primitive.* Original. *contemplative life.* Ellis (*Works*, i. 466), observing that the standard biblical types of the active and contemplative lives are Rachel and Leah in the Old Testament, Mary and Martha in the New, notes that Philo Judaeus takes Cain as representing energies and thoughts devoted to ourselves, Abel as referring all things to God. *husbandman.* Farmer. *election.* Choice. *within.* Among. *inventors . . . metal.* Jubal and Tubalcain (Gen. 4: 21–2). *confusion of tongues.* Men's attempts to build the tower of Babel to reach to heaven were punished by God with the proliferation of languages (Gen. 11). *imbarred.* Prevented. *first pen.* First writer of sacred law. *Scriptures.* Acts 7: 22. *seen in.* Knowledgeable. *Plato. Timaeus*, 22B; a rebuke often cited by Bacon in his polemical mode. Cf. *Promus*, no. 342, and *Of Tribute*, above, p. 35. *prefiguration.* A 'representation beforehand by a figure or type' (*SOED*). In Christian biblical interpretation the Old Testament was seen as prefiguring the New. *difference.* Distinguishing mark. *impression.* Stamp, lasting character. *observe.* Take note of; point out. *moral . . . reduction.* Moral implication. *If . . . unclean.* Lev. 13: 13–14. *noteth.* Indicates. *nature.* Natural philosophy. Ellis (*Works*, i. 467) states that Giordano Bruno has 'a similar view of the book of Job': cf. *La Cena de le Ceneri*, in *Dialoghi Italiani*, ed. cit. i. 124–5.

151. *aspersion* Sprinkling, intermixture. '*Qui . . . nihilum*'. 'He stretcheth out the north over the empty space, and hangeth the earth upon nothing' (Job 26: 7). *pensileness.* Suspended condition. *finiteness.* (Assuming the earth to be hung in space.) '*Spiritus . . . tortuosus*'. 'By his spirit he hath garnished the heavens; his hand hath formed the crooked serpent' (Job 26: 13). '*Nunquid . . . dissipare?*' 'Canst thou bind the glittering stars of the Pleiades, or scatter the array of Arcturus?' (Job 38: 31) '*Qui . . . Austri*'. 'Which maketh Arcturus, Orion, and Pleiades, and the secrets of the south' (Job 9: 9). *takes knowledge of.* Recognizes. *depression.* Astronomical term for the angular distance of the pole beneath the horizon. *climate.* Region of the earth. *generation.* Procreation. '*Annon . . . me?*' 'Hast thou not poured me out as milk and curdled me like cheese?' (Job 10: 10). '*Habet . . . vertitur*'. 'Surely there is a vein for silver, and a place for gold, where they refine it. Iron is taken out of the earth, and brass is molten out of the stone' (Job 28: 1–2). *so forwards.* So forth. *Salomon's petition.* For 'an understanding heart to judge thy people, that I may distinguish between good and bad' (1 Kgs. 3: 5–9). *before.* To. *terrene.* Earthly. *donative.* Gift. *verdure.* Vegetation. *moss.* The biblical text to which Bacon alludes is usually translated (in both the Geneva and the Authorized versions): 'And he spake of trees, from the cedar tree that is in Lebanon even unto the *hyssop* that springeth out of the wall' (1 Kgs. 4: 33). Bacon's idiosyncratic use of 'moss' follows 'the rendering of Junius and Tremellius' (Wright, ed. cit. 279). *rudiment.* Half-way form or stage between two species. *inquisition*. Investigation. '*The glory . . . out*'. Prov. 25: 2; *Promus*, no. 234.

152. *commandment*. Command. *conference*. Cf. Luke 2: 41–50. *gift of tongues*. Cf. Acts 2: 1–4. '*vehicula scientiae*'. 'Carriers of knowledge' (Acts 2: 1). *election*. Choice. *the faith*. Christianity. *immediate*. 'Direct, without any intervening medium or agency' (*SOED*). *abase . . . knowledge*. St Augustine, *The City of God*, xiv. 38. *vicissitude*. Alternation, revolution of time. *waited on*. Attended. *with*. By. *learnings*. Branches of knowledge. *only learned*. The only educated man. *had his pen most used in*. Wrote the greater part of. *edict*. By Julian the Apostate in 363, forbidding Christians to set up schools. See Ammianus Marcellinus, *Rer. gest*. 22.10, 25.4; Augustine, *Confessions*, viii. 5. *engine*. Contrivance. *machination*. Plot, scheme. *sanguinary*. Bloodthirsty. *emulation*. Ambitious rivalry. *jealousy*. Vehement feeling. *Gregory the first*. Pope 590–604, once thought to have destroyed pagan temples and books. *opinion*. Reputation. *censure*. Judgement, verdict. *humour*. Whim. *pusillanimity*. Littleness of mind. *contrariwise*. In the Byzantine Church. *Scythians*. The Tartars invaded the Gothic empire AD 375. *Saracens*. The Arabs under Abubekr conquered Syria AD 633–9. *obnoxious*. Blameworthy.

153. *renovation*. Again Bacon gives credit to the Reformation (16th century) for the much wider revival of learning that had been initiated by Petrarch and the Italian Renaissance (late 14th century). *knowledges*. Sciences. *Jesuits*. Members of the Society of Jesus, a Catholic order founded by Ignatius Loyola and others in 1534 to propagate the faith among unbelievers, and which developed a highly efficient educational system. *provocation*. Incitement. *reparation*. Restoration. *the Roman see*. The papacy. *illustration*. Illumination; also 'making illustrious or famous'. *magnify*. Pss. 19, 104. *construe*. Interpret. '*You err . . . God*'. Matt. 22: 29. *two books*. Of God's words, and of His works: see above, p. 126. *secured*. Saved. *signed*. Stamped. *obtain to*. Attain, achieve. *separately*. By itself. *apotheosis*. 'Deification': Bacon recalls Herodian, *Hist*. iv. 2, probably in Politian's translation (Wolff, ii. 146). But cf. also Dio Cassius, lxxiv. 4. '*relatio . . . divos*'. 'The raising of a mortal to the rank of a deity'. *inward assent*. Tacitly.

154. *middle term*. Intermediate stage. *extirpers*. Those who root out and destroy. *worthies*. Heroes. *circle*. Limits. *for a latitude of*. Within the limits of. '*aura leni*'. 'With a gentle breeze', as God did when he visited Elijah: 1 Kgs. 19: 12 (Vulgate). *agitation*. Disturbance. *which grow from man to man*. Which men impose on each other. *lively*. Vividly. *relation*. Narrative, story. *Orpheus' theatre*. The theatre (or 'beholding place', as Puttenham terms it) was the assemblage of animals so charmed by the music of Orpheus' lyre that they forgot their natural enmities. See Philostratus, *Imagines*, vii, and Bacon's later exposition of this fable in *The Wisdom of the Ancients* (*Works*, vi. 720–2). *game*. Hunting. *accords*. Harmonies. *commonwealths and popular estates*. Republics and democracies. *then . . . kings*. Plato, *Republic*, 5. 473C–D.

155. *customs.* Habitual behaviour. *illuminate.* Enlightened. *refrain.* Restrain. *peremptory.* Dictatorial, destructive. *near hand.* Near. *ward.* Defend. *Domitianus . . . Commodus.* Two vicious tyrants, between whose reigns (96 to 180) six enlightened emperors ruled Rome. *for.* In. *respects.* Considerations. *model.* Compendium. *Domitian.* Emperor 81–96; his dream is recorded in Suetonius, *Lives of the Caesars*, 'Domitian', xxiii. *matter.* Subject-matter. *vulgar.* Familiar. *declamation.* A rhetorical exercise or address. *infolded.* Intricately structured. '*neque . . . Apollo*'. 'Apollo does not always bend the bow' (Horace, *Odes*, II. x. 19 f.). The metaphor implies a relaxation of vigilance. *only.* Merely. *Nerva.* Emperor 96–98. *glance.* Passing reference. '*Postquam . . . libertatem*'. 'The divine Nerva reconciled things which did not go together before, namely, sovereignty and freedom' (*Agricola*, iii). *missive.* Message. *proceeding upon.* Deriving from. *Telis . . . nostras*: 'Phoebus, with thy darts revenge our tears': *Iliad*, i. 42, in the Latin translation of Xylander (1592 and earlier). The quotation was applied to Nerva by Dio Cassius in his *Roman History*, lxviii. 3.4 (Wolff, ii. 8, 53). Trajan invoked retribution on the Danubians. *Trajan.* Emperor 98–117, trained as a soldier, not a scholar, but a great benefactor of libraries and learned men. His reign saw the literary production of Juvenal, Tacitus, Pliny the Younger, and Plutarch. *not learned.* See Dio Cassius, lxviii. 7.4.

156. '*He . . . reward*'. Matt. 10: 41. converser. Associate. *noted.* Specially noticed. *legend.* Legendary. *Gregorius Magnus.* St Gregory, Pope 590–604. Legend had it that his prayers delivered Trajan from hell and secured his salvation: cf. Dante, *Purg.* x. 73–6; *Parad.* xx. 43–5. *envy.* Hostility. *excellency.* Excellence. *caveat.* Warning, proviso. *certificate.* Evidence. *Plinius Secundus.* Pliny the Younger, a grateful recipient of Trajan's patronage: see his *Epist.* x. 96–7. *Hadrian.* Emperor 117–38, a man of wide culture. See Dio Cassius, lxix. 3. 11. *curious.* Eager to learn. '*God . . . than I*'. Plutarch records this anecdote of Philip, father of Alexander, in 'Table-Talk', ii. 1.17, *Moralia*, 634D. *curiosity.* Desire for knowledge. *inducement.* Incentive. *matched.* Paired, placed alongside. *Appollonius.* A Pythagorean philosopher of the first century AD, reputed to have worked miracles. Hadrian respected both him and Christ, but the person who set up their pictures in a shrine was in fact Alexander Severus. *weal.* Welfare. *subject.* People. *as.* That. *wall flower.* See Aurelius Victor, *Epitome de Caesaribus*, xli. 13. *of.* For. *survey.* Hadrian visited all the provinces of Rome, including Britain, where in AD 119 he extended the wall from Solway to the mouth of the Tyne. See Dio Cassius, lxix. 9, and Spartianus, 'Life' of Hadrian, 12–13. *assignation.* Instructions. *re-edifying.* Rebuilding.

157. *passages.* Fords, passes. *policing.* The regulating and governing of a town. *commonalties.* Corporations. *franchises and incorporations.* Guarantees of freedom and municipal independence. *lapses.* Falling into ruin. *Antoninus Pius.* Emperor 138–61, whose reign was a period of remarkable peace and prosperity. *subtile.* Intricate, abstruse. *schoolman.* Scholastic philosopher, versed in logic. *untaxed.* Uncensured. '*cymini sector*'. See Dio Cassius, lxx. 3.3; *Promus*, no. 891. *least.* Smallest. *settled.* Purposeful. *scruples.* Doubts.

all. Any. *fiction*. Pretence. '*half a Christian*'. Acts 26: 28, and Dio Cassius, lxx. 3.1. *giving way to*. Promoting. '*Divi fratres*'. Divine brothers. The Romans attributed divinity to emperors. Marcus Aurelius, Emperor 161–80, appointed his brother Lucius Commodus as co-ruler during his lifetime. *adoptive*. Adopted. *softer*. Less demanding. *Martial*. Writer of satiric epigrams, a genre much lower than Virgilian epic. *long*. Lucius Verus died of apoplexy AD 169; Marcus Aurelius survived until AD 180. *Julianus*. Julian the Apostate, Emperor 361–3, whose satire *Caesars, or the Banquet* (312A ff., 328C) uses the old satyr Silenus as a *persona* to criticize the faults of all the Roman emperors save Julian himself. Bacon probably read it in the Latin translation by Cantoclarus (Paris, 1577): Wolff, ii. 159. *pasquil*. Lampoon. *nether*. Lower. *Marcus Philosophus*. Marcus Aurelius, whose *Meditations* was one of the most widely read works of Stoic ethics. *gravelled*. Lost for words. *carp at*. Find fault with. *glance*. Satirical allusion. *Commodus, Caracalla, and Heliogabalus*. Emperors during the periods 180–92, 211–17, and 218–22: all three were depraved tyrants. '*Quomodo . . . Antoninus*'. 'Let the name of Antoninus be as the name of Augustus' (Lampridius, *Life of Severus*, 5–10). 'Augustus' was adopted as an official title by successors of that emperor.

158. *style*. Title, designation. *painted forth*. Depicted. *table*. Picture. *queen Elizabeth*. Bacon frequently praised the Queen, unhappy though their personal relations had been, in the 1592 device, *Of Tribute* (above, pp. 37 ff.), in his Letter to the Lord Chancellor touching the history of Britain (*Works*, x. 249–50), and in the posthumously published *In felicem memoriam Elizabethae, Angliae Reginae* (*Works*, vi. 291–318). She was outstandingly learned, with a fluent command of French, Italian, and Latin, a sound knowledge of Greek and Spanish. For her studies see e.g. Roger Ascham, *The Schoolmaster* (1570), and C. Pemberton (ed.), *Queen Elizabeth's Englishings . . .* (London, 1899); G. B. Harrison (ed.), *The Letters of Queen Elizabeth* (London, 1935); G. P. Rice, Jr. (ed.), *The Public Speaking of Queen Elizabeth* (New York, 1951); and L. Bradner (ed.), *The Poems of Queen Elizabeth I* (Providence, RI, 1964). *Plutarch*. Best known for his *Parallel Lives of the Greeks and Romans*, a series of biographies of famous men, most of which pair a Greek and a Roman. *accustomed*. Was used. *duly*. Punctually. *exceed*. Exaggerate. *regiment*. Government. *truth . . . established*. By the 1559 Acts of Uniformity and Supremacy, establishing the Church of England. *prerogative*. The monarch's absolute power. *sortable*. Appropriate. *convenient*. Proper. *discontents*. Grievances. *of herself*. Unmarried. Cf. *Of Tribute*, above, p. 48. *to*. As regards. *temperature*. Disposition. *enablement towards*. Promoting. *affections*. Devotion.

159. *with*. By. *Callisthenes*. A cousin and pupil of Aristotle's. *notably*. Conspicuously. *bare*. Bore. *solution*. Answer. *touching*. Concerning. *cabinet*. Pliny, *Nat. Hist.* vii. 29.107–8. *Darius*. Darius III, surnamed Codomannus (d. 330 BC), King of Persia, defeated by Alexander. *set forth*. Published. *nature*. Natural philosophy. *science*. Knowledge. *use*. Application. *scholastical*. Pedantic. *handle*. Treat. *humour*. Whim. *Diogenes*. The Cynic philosopher who lived frugally in a tub to prove that true

happiness consists in freedom from desires and appetites. One day 'when he was sunning himself in the Craneum, Alexander came and stood over him and said "Ask of me any boon you like". To which he replied, "Stand out of my light"' (Diogenes Laertius, vi. 38). *state*. Determination. *said*. Plutarch, 'Life' of Alexander, 14.3. *inverteth*. Reverses. '*Plus . . . dare*'. Seneca, *De beneficiis*, v. iv. 4. *that . . . lust*. Cf. Plutarch, 'Life' of Alexander, 22.3; 'How to tell a Flatterer from a Friend', *Moralia*, 65F; Seneca, *Epist*. VI. vii. 12. Seneca adds that deficiency (weariness) and intemperance (lust) are 'indications of mortality'. *liker*. More likely. *Democritus*. Another believer in the moderation of desire and the happy mean.

160. *that speech*. See Plutarch, *Apophthegmata* 180E; Seneca, *Epist*. lix. 12. *liquor*. The ichor which flowed in the veins of gods. *Homer*. *Iliad*, v. 335 ff. *reprehension of logic*. Detecting a fallacious argument. *happed*. Happened. *say*. Plutarch, 'Life' of Alexander, 74.2 f. *matter*. Point (i.e. being far from home they believed that no one could disprove their accusations). *adoration*. Alexander had himself worshipped like a god, the common practice of oriental rulers. *moved*. Proposed. *purpose*. Topic. *Turn your style*. Speak to the contrary (from the Latin *vertere stylum*, 'turn your pen', i.e. make use of the flat end of your pen for erasure): Plutarch, 'Life' of Alexander, 53.3 ff. *despite*. Spitefulness. *translation*. Transferring a word from its literal to a figurative sense (the Latin term for metaphor was *translatio*). *taxed*. Criticized. *tyrannous*. Tyrannical. *habit*. Dress. *all purple within*. Full of ambition (purple was worn by kings): Plutarch, *Apophthegmata*, 180E 17. *Arbella*. A city near the Tigris, where Alexander defeated Darius in 331 BC. *lights*. Of camp-fires. *as*. As if. *answered*. Plutarch, 'Life' of Alexander, 31.5.

161. *weigh*. Consider. *embraced*. Endorsed. *said*. Plutarch, 'Life' of Alexander, 47.5. *taxation*. Reprehension. *model*. Scale. *said*. Plutarch, 'Life' of Alexander, 29.4. *Hope*. That is, hopes of reward: Plutarch, 'Life' of Alexander, 15.2. *cast up*. Calculated. *portion*. Lot, destiny. *estate . . . largesses*. Caesar had huge debts at this time. *Henry*. 3rd Duke of Guise (1550–88), head of the Catholic League, a faction of extremists. *obligations*. By making people indebted to him he could claim their service when needed. *Julius Caesar*. For his account Bacon draws on the lives of Caesar by Plutarch and Suetonius. *company*. Companions. *permanent*. Surviving. *Commentary*. Caesar described his *De bello Gallico* (7 books, one for each year from 58 to 52 BC) and *De bello civili* (3 books) as *commentarii*, a term originally referring to notebooks of recent or contemporary events, usually kept by an actor in them. As published they constituted a new literary genre, autobiographical in character, written in a plain style as a groundwork for a fuller history, yet directed to the reading public. *real passages*. Describing actual things and events. *lively images*. Vivid pictures. *propriety*. Appropriateness. *De Analogia*. Now lost, 'a careful treatise on the principles of correct Latinity', as Cicero described it (*Brutus*, lxxii ff.), presumably intended 'to determine uncertain points of language by the analogy of cases which were free from doubt' (Ellis, *Works*, i. 476).

162. *'vox ad placitum'*. Current usage. *'vox ad licitum'*. Correct speech. *reduce*. Reform, establish rules for a standard Latin. *picture ... reason*. Making words, which are the images of things, correspond with what reason suggests to be real. *computation*. The corrected, so-called Julian calendar, decreed in 46 BC, which fixed the year at 365.25 days, with a leap year every fourth year. It remained the standard calendar until modified by Pope Gregory XIII in 1582 (whose reform was not adopted in England until 1752). *Anti-Cato*. A lost refutation of Cicero's panegyric to Marcus Porcius Cato (leader of the republicans and Caesar's adversary). Cf. Plutarch, 'Life' of Julius Caesar, 54; Aulus Gellius, *Noctes Atticae*, xiii. 9. *victory of wit*. Intellectual supremacy. *Apophthegms*. Witty sayings of famous men; Caesar's collection (mentioned by Cicero, *Epist. ad Fam.* ix. 16) has not survived. Bacon published a collection of his own, *Apophthegms New and Old* (1625), and had gathered many more, which were posthumously published: see *Works*, vii. 111–84. *make ... tables*. To be a mere recorder. The Romans wrote on tablets of wood coated with wax. *take*. Record. *oracle*. Wise utterance. '*Verba ... defixi*'. 'The words of the wise are as goads, and as nails driven deep in' (Eccles. 12: 11). *delectable*. Delightful. *first*. Suetonius, 'Life' of Julius Caesar, 70; Appian, *Bell. Civ.* ii. 93. *Milites*. Soldiers. *Quirites*. Citizens (in their civil capacity). *cashiered*. Dismissed from service. *by expostulation thereof*. By making this protest. *crossed*. Put out. *suffer*. Allow. *second*. Suetonius, 'Life' of Caesar, 79; Appian, *Bell. Civ.* ii. 108. *affect*. Aspire to. *set on*. Provoked. *put it off*. Dismissed. '*Non ... Caesar*'. 'I am not King, but Caesar.' *searched*. Closely examined. *signify*. Indicate.

163. *of great allurement towards*. Powerfully promoting. *last*. From Plutarch, 'Life' of Caesar, 35. *war declared*. After declaring war against Pompey and the senatorial party. *taking himself up*. Restraining himself. *took it upon him*. Arrogated it to himself. *appeared*. Suetonius, 'Life' of Caesar, 77. *resolution*. Decision. *dictature*. Lucius Cornelius Sulla (138–78 BC), dictator of Rome 82–79 BC. *skill of*. Understand. *Xenophon*. Famous general whom Bacon referred to earlier (above, p. 127), a pupil of Socrates, who also wrote works of history, philosophy, and economics. *Cyrus ... Artaxerxes*. The expedition of Cyrus against his brother took place in 401 BC. Cyrus was killed, and Xenophon led the Greeks home through enemy territory, an exploit recorded in his *Anabasis*. ('The up-country march'). *voluntary*. Volunteer. *love ... of Proxenus*. Xenophon, *Anabasis*, iii. 1.4. *conversation*. Consorting with. *in message*. As a messenger. *the great king*. Special title given to the King of Persia. *divers*. Various members. *conferred*. Consulted. *Xenophon. Anabasis*, ii. 1.10–15, as stated in the Stephanus edn. of 1561; modern texts give the saying to Theopompus. *virtue*. Courage, innate resourcefulness.

164. *that*. That which. *abused*. Deceived. *parley*. Informal conference between combatants, usually protected by a truce. *high countries*. Mountainous regions inland. *after*. Afterwards. *Thessalian*. Jason of Pherae planned to invade Persia but was murdered in 370 BC before he could do so.

Agesilaus. Invaded the western provinces of Persia in 396–394 BC, but was recalled. Bacon contrasts Agesilaus and Jason more fully in 'Of the True Greatness of Britain' (*Works*, vii. 50). *Macedonian.* Alexander the Great. *ground.* Basis, example. *private.* Personal, individual. *Scilicet... feros.* 'Certainly a faithful study of the liberal arts refines manners and decreases barbarism' (Ovid, *Epistulae ex Ponto*, II. ix. 47–8). *accent.* Emphasis. *fideliter.* 'Faithful'. *levity.* Making light of serious matters; also 'instability, fickleness' (*SOED*). *acquainting.* Accustoming. *turn back.* Reject. *offers and conceits.* Notions and fancies. *accept of.* Accept. *wadeth in.* Gets to know. *contemplation.* Study. *throughly.* Thoroughly. '*Nil... terram*'. 'There is no new thing under the sun' (Eccles. 1: 9). *play of puppets.* Puppet show. *adviseth.* Considers. *motion.* Manipulation. *spacious.* Widely extended. *services.* Military engagements. *for a passage.* To gain possession of a pass or ford. '*It seemed... went of*'. Plutarch, 'Life' of Agesilaus, 15.4. *advertised.* Informed. *battles... frogs... mice.* In the mock-Homeric epic, *Batrachomyomachia.* *went of.* Were current about. *frame.* Structure. *earth... dust.* Seneca, *Nat. Quaest.* I, Praef. 10. *except.* Excepted. *whereas.* Where.

165. *fro.* From. *mitigateth.* Lessens, reduces. *seasoned.* Imbued. *Epictetus.* A Stoic philosopher (1st century AD). *pitcher of earth.* Earthenware jar. '*Heri... mori*'. 'Yesterday I saw a brittle thing broken and today a mortal die': Epictetus, *Enchiridion*, chs. 3 and 26, and Simplicius' commentary. *concomitantia.* Going together. *Felix... avari.* 'Blessed is he who has been able to win knowledge of the causes of things, and has cast beneath his feet all fear and unyielding Fate, and the howls of hungry Acheron!': *Georg.* ii. 490–2; *Promus*, no. 348. *opening.* Freeing. *exulcerations.* Ulcers. '*rationem totius*'. 'The essence of the whole'. Cf. Eccles. 12: 13. *descend into himself.* Observe, criticize. Cf. Plato, *Alcibiades*, I. 133B. '*suavissima... meliorem*'. 'That most happy state, to feel one's self becoming a better man day by day' (Xenophon, *Memorabilia*, I. vi. 9). See also *Promus*, no. 95, and *Advice to Rutland*, above, p. 69. *colour.* Conceal, or show in a favourable light. *ill.* Bad. *still.* Constantly. *amendment.* Improvement. *veritas... perturbations.* Ellis (*Works*, i. 481) cites Aquinas, *Summa Theol.* I. 16. 4. See *Of Tribute*, above, p. 34. *veritas and bonitas.* Truth and goodness. *print.* Impression (of a seal on wax). *prints.* Stamps. *descend.* Are precipitated by. *commandment.* Command. *according... commanded.* Wolff (ii. 251) cites Aristotle, *Politics* i. 3, 1254a25: 'that rule is the better which is exercised over better subjects'.

166. *disparagement.* Disgrace. *put off.* Lost, discard. *generosity.* Nobility. *holden.* Held. *putteth himself forth.* Endeavours. *victorque... Olympo.* 'The conqueror gives laws to willing peoples and takes the road to heaven' (*Georgics*, iv. 561–2). *chair of estate.* Chair of government. *Satan.* Rev. 2: 24. *similitude.* Likeness. *as.* That. *Homer.* By Plutarch in the *Moralia* ('Sayings of Kings and Commanders', 175D; 'Hiero', iv), an argument much favoured by Renaissance humanists. *largesses.* Gifts of money. *advanced.* Promoted. *in case.* In the case. *carried away.* Got the dominance over.

ANALYSIS

BOOK II. OF WHAT HAS BEEN DONE FOR THE ADVANCEMENT OF LEARNING HUMAN AND DIVINE, WITH THE DEFECTS OF THE SAME.

- Dedication to the King (169–75) defining
  - Acts of merit towards learning (169–71) as regards
    - 1. Places of learning (170)
    - 2. Books of learning (170)
    - 3. The person of the learned (170)
  - Defects of places of learning (171–5)
    - 1. All dedicated to professions and none to sciences at large (171)
    - 2. Smallness of reward for lecturers (171–2)
    - 3. Want of apparatus for experiments (172)
    - 4. Neglect of consultation in governors and of visitation in princes (172–3)
    - 5. Want of mutual intercourse between the Universities of Europe (173–4)
    - 6. Want of public appointment of writers or inquirers into the less known branches of knowledge (174)

- (A) Human learning divided into
  - I. History (175–86)
    - Natural (175–8)
      - Of creatures (175–6)
      - Of marvels (176–7)
      - Of arts (177–8)
    - Civil (178–86)
      - Memorials (178–9)
      - Antiquities (179)
      - Perfect histories (179–80); their deficiencies (180–6)
        - Chronicles
        - Narrations.
        - Lives.
    - Ecclesiastical
      - History of the Church (184)
      - Of prophecy (184–5)
      - Of Providence (185)
    - Literary (defective)
      - Appendices to history, orations, letters, sayings (185–6)
  - II. Poesy (186–8)
    - Narrative (187)
    - Representative (187)
    - Allusive (187–8)

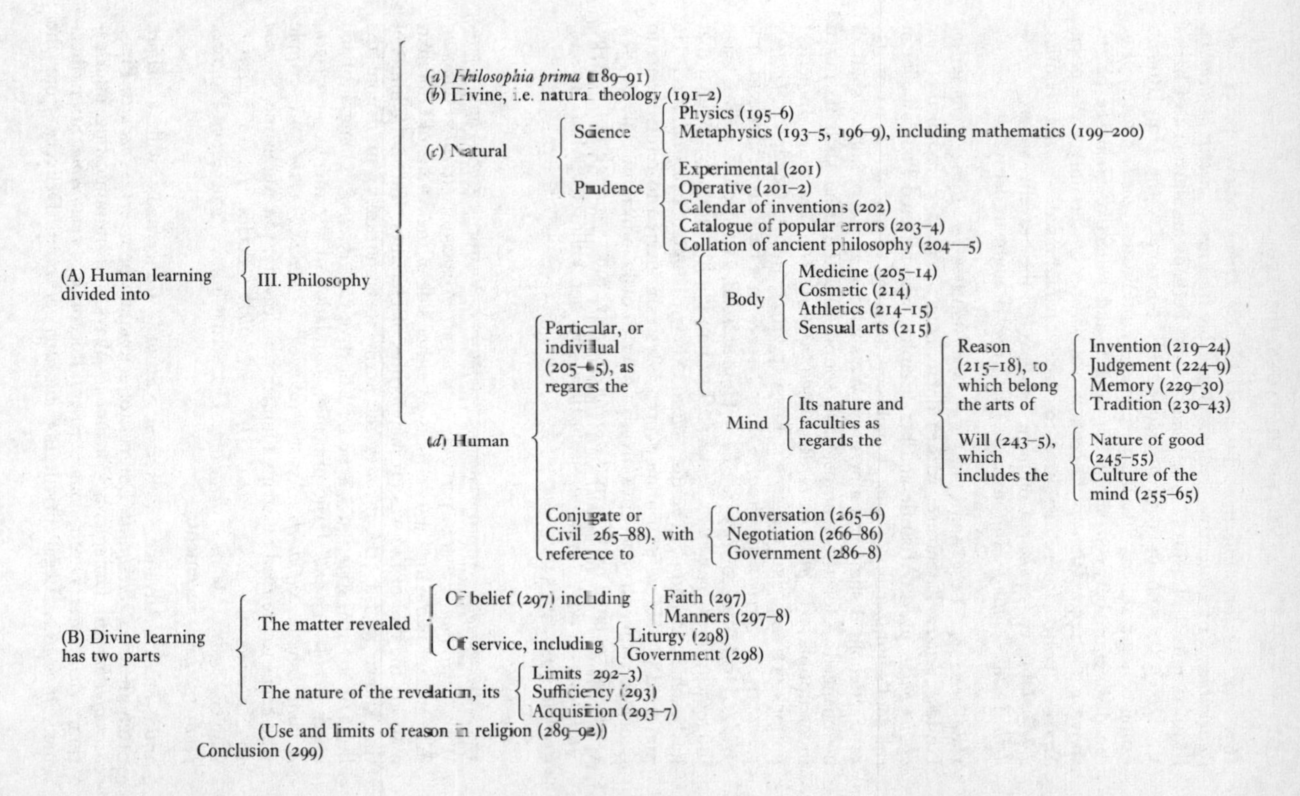
(A) Human learning divided into
III. Philosophy
(a) *Philosophia prima* (189–91)
(b) Divine, i.e. natural theology (191–2)
(c) Natural
Science
Physics (195–6)
Metaphysics (193–5, 196–9), including mathematics (199–200)
Prudence
Experimental (201)
Operative (201–2)
Calendar of inventions (202)
Catalogue of popular errors (203–4)
Collation of ancient philosophy (204–5)
(d) Human
Particular, or individual (205–65), as regards the
Body
Medicine (205–14)
Cosmetic (214)
Athletics (214–15)
Sensual arts (215)
Mind
Its nature and faculties as regards the
Reason (215–18), to which belong the arts of
Invention (219–24)
Judgement (224–9)
Memory (229–30)
Tradition (230–43)
Will (243–5), which includes the
Nature of good (245–55)
Culture of the mind (255–65)
Conjugate or Civil (265–88), with reference to
Conversation (265–6)
Negotiation (266–86)
Government (286–8)
(B) Divine learning has two parts
The matter revealed
Of belief (297) including
Faith (297)
Manners (297–8)
Of service, including
Liturgy (298)
Government (298)
The nature of the revelation, its
Limits (292–3)
Sufficiency (293)
Acquisition (293–7)
(Use and limits of reason in religion (289–92))
Conclusion (299)

167. *for*. As for. *affections*. Passions. Cf. *Of Tribute*, above, p. 34. *of consequence*. Consequently. *verdure*. Agreeable freshness. *deceits of pleasure*. Illusory, unreal pleasures. *ambitious princes*. Perhaps an allusion to Charles V (1500–58), who in 1555 abandoned his position as the Holy Roman Emperor for a solitary life in the monastery of Yuste. *good... simply*. For the distinction between simple (or single) and collective good see bk. II, above, p. 245. *fallacy*. Deception. *accident*. Uncertainty, exception. *Suave... ventis*. Lucretius, ii. 1–10; freely paraphrased. *battles*. Armies. *descry*. Observe. *vulgar*. Common. *motions*. Of the heavenly spheres. *tendeth*. Leads. *infinite*. Innumerable. *decayed*. Brought to decay.

168. *leese of*. Lose some of. *still*. Constantly. *in*. Into. *consociateth*. Unites. *letters*. Literature, written documents. *magnified*. Extolled. *some of the philosophers*. Aristotle and his disciples, who rejected Plato's belief in the soul's immortality. *point*. Conclusion, opinion. *changed*. Cf. 1 Cor. 15: 51–2, 'Behold, I shew you a mystery. We shall not all sleep, but we shall all be changed... for the trumpet shall sound, and the dead shall be raised incorruptible...' *disclaim in*. Renounce. *rudiments of the senses*. Imperfect notions derived from the senses. Rossi (p. 192) cites A. Cesalpino, *Quaest. perip*. II. 8. *probation*. Proof. *Aesop's cock*. In Phaedrus, *Fables*, iii. 12. *Midas*. Ovid, *Met*. xi. 153–4. *Paris*. Homer, *Iliad*, xxiv. 25–9. *judged*. Gave judgement for. '*occidat... imperet*'. 'Let him kill his mother, provided that he becomes emperor' (Tacitus, *Annals*, xiv. 9). '*qui... immortalitati*'. 'He preferred his old wife to immortality.' Ulysses preferred to return home to Penelope rather than accept Calypso's offer to make him immortal: *Odyssey*, v. 214–20, as cited by Cicero, *De oratore*, I. xliv. 196. *figure*. Emblem, type. '*Justificata... suis*'. 'Wisdom is justified by her children': Matt. 11: 19; *Promus*, nos. 249, 346.

Book Two

169. *convenience*. Appropriateness. *sojourner*. Visitor. *issue*. Family. *in*. With. *affection*. Passion, partiality. *transport*. Mislead. *Hercules' Columns*. The ancient name for the two rocks on either side of the Straits of Gibraltar, marking the end of the western wanderings of Hercules, and so the limits of the known world. Bacon uses the term metaphorically, to denote the limit of any investigation. *actively*. Effectively. *dilating*. Going into too much detail. *ground*. Basic principle, foundation of argument. *overcomen*. Accomplished. *supplieth*. Assists, relieves. '*claudus... viam*'. 'A cripple who keeps to the road outstrips a runner who leaves it': St Augustine, *Sermon* 169; *Promus*, nos. 233, 1240. '*If... prevaileth*'. Eccles. 10: 10. *election*. Choice. *mean*. Means. *derogating*. Detracting from. *progression*. Progress. *proficience*. Advancement.

170. *raise*. Advance. *conversant about*. Concerned with. *comfort*. Strengthen. *spring-heads*. Sources (of a fountain or river). *conduits*. Pipes. *accomplishments*. Ornaments. *traditions*. Ways of communicating information. *conferences*. Discussions. *receipt*. Reception. *comforting*. Strengthening. *franchises*. Legal privileges. *discharge of cares*. Delivery from the

burden of worries. *Principio . . . aditus*. 'First seek a settled home for your bees, whither the winds may find no access' (*Georgics*, iv. 8–9). *imposture*. Deception. *reposed*. Laid up as in store. *impressions*. Printings. *countenancing*. Encouragement. *designation*. Appointment. *readers*. Lecturers. *laboured*. Elaborated. *prosecuted*. Investigated.

171. '*Difficile . . . praeterire*'. 'It is difficult to mention everybody; it would be ungracious to omit anyone' (Cicero, *Oratio post reditum in Senatu*, xii. 30). *Scriptures*. Phil. 3: 13. *find it strange*. Wonder. *professions*. The traditional triad of divinity, medicine, and law, to which university teaching at this time was confined. Bacon pleads for a new course of studies 'at large', freed from such restrictions, to act as a source and stimulus for intellectual development. *the ancient fable*. See Livy, ii. 32, and Shakespeare's *Coriolanus*, I. i. 96–163. *universality*. The study of general principles. *progression*. Progress. *in passage*. In passing. *dotations*. Endowments. *professory*. The teaching of one special branch of study (as in a single professorship). *malign . . . influence*. Malignant power over (terms from astrology). *it proceedeth*. The result is. *solitude*. Complete absence. *causes of estate*. National affairs. *enablements . . . estate*. Qualifications for serving the state. *in order*. In sequence. *lectures*. Lectureships. *readers*. Lecturers. *sufficient*. Competent. *appropriate . . . labour*. Devote his whole energy. *continue . . . age*. Spend his whole life. *function and attendance*. Discharging and attending to the duties of his office. *mediocrity*. Moderate fortune or status in society. *that . . . action*. 1 Sam. 30: 22–4.

172. *carriage*. Baggage. *ill*. Badly. *attended*. Looked after. *entertainment*. Support. '*Et . . . nati*'. 'Feeble offspring may repeat the leanness of the sires' (Virgil, *Georgics*, iii. 128). *Minerva*. The Roman goddess of wisdom, hence any intellectual pursuit. *Vulcan*. The Roman god of fire and furnaces, hence alchemy. *operative*. Effective, productive. Bacon's whole aim was to encourage a science 'of works', having concrete results, for human benefit. *physic*. Medicine. *only*. Alone. *instrumentals*. Tools. *spheres*. Either 'planispheres' (flat projections of the earth's globe) or 'armillary spheres' (skeleton celestial globes). *instituted*. Established. *commodity*. Convenience. *simples*. Plants used for medical remedies. *anatomies*. Dissections. *respect*. Involve. *main*. Major. *disclosing*. Discovery. *Daedalus*. In Greek myth, an architect and engineer. *engine*. Machine. *spials*. Spies. *bills*. Lists of expenses. *intelligence*. Information. *intelligencers*. Enquirers. *advertised*. Informed. *Alexander*. Cf. Pliny, *Nat. Hist.* viii. 17. *History*. Enquiry; any systematic study of a topic. For Bacon the 'history of nature' involves collecting observations of natural phenomena; the 'Arts of nature' are the experimental sciences, which work upon and alter nature by art. *travail*. Labour in. *visitation*. Inspection. *inconvenient*. Unsuitable. *your Majesty's*. In King James's *A Counterblaste to Tobacco* (1604).

173. *for suspect*. Suspected. *scholars*. Students, undergraduates. In this period boys entered university between the ages of 13 and 18. *unripe*. This criticism was made by other humanists: cf. J. L. Vives, *De ratione dicendi*

(1553), *Praefatio*, in his *Opera omnia*, ed. G. M. y Siscar (Valencia, 1782–90; repr. London, 1964), ii. 91–2; and L. D. Green (ed.), *John Rainolds's Oxford Lectures on Aristotle's 'Rhetoric'* (Newark, NJ, 1968), 98–9. *logic and rhetoric*. Forming, with grammar, the *trivium*, the first stage in classical, medieval, and Renaissance education. *gravest*. Important, fundamental. *ornament*. In rhetoric the word implied 'effective persuasive power', not just 'decoration'. *dispose*. Arrange (the process of *dispositio*). *matter*. The rhetorical term for subject-matter. *unfraught*. Unequipped, bare. *Cicero*. See *De oratore*, III. xxvi. 103 ('one has to begin by accumulating a supply of matter'), and *Orator*, xxiv. 80 ('For we [orators] do have after a fashion a stock-in-trade in the stylistic embellishments'). *sylva*, *supellex*. The 'forest' or collection of material basic to all intellectual activity; and the 'furniture' or 'equipment' needed for an experiment or operation. *fitteth*. Is befitting. *invention and memory*. The five stages of composition in rhetoric were invention (finding subject-matter), disposition (arranging it into an effective structure of argument), elocution (putting it into persuasive language), memory (memorizing the speech), and delivery (reciting it with appropriate gestures). '*in verbis conceptis*'. 'In prepared language.' *extemporal*. Extemporized. *be*. Should be. *set into*. Apply themselves to. '*Hoc . . . suscipiatis*'. 'I have some ideas as to the way in which this may be brought about, and many others may be thought of; I ask you to reflect on the matter' (Cicero, *Epist. ad Att.* ix. 7). *precedent*. Preceding.

174. *intelligence mutual*. Exchange of information. *orders and foundations*. Religious organizations, such as the Jesuits, with their regional governors. *contract*. Establish. *communalties*. Trade guilds. *anointment*. Anointing. *superinduceth*. Brings about. *God . . . Father of . . . lights*. Jas. 1: 17. *designation*. Appointment. *inducement*. Introduction. *prosecuted*. Followed, carried out. *surcharge*. Surfeit. *serpent of Moses*. In fact Aaron: see Exod. 4 and 7. *enumerate*. Mentioned separately. '*opera basilica*'. 'Works for a king' (Plautus, *Trinummus*, IV. iii. 23). *image in a crossway*. Signpost at a crossroads. *go*. Travel. *inducing*. Introductory. *set forward*. Promoted. *perambulation*. Tour. *fresh and waste*. Uncultivated, unused. *converted*. Transformed. *plot*. Plan, outline. *voluntary endeavours*. The generosity of private persons subsidising research. *redargution*. Refutation. *incomplete prosecutions*. Unfinished investigations. *unmanured*. Uncultivated.

175. *affection*. Loving concern. '*it is not . . . wise*'. From Publilius Syrus, *Sententiae*, xv. *indifferently*. Impartially. '*Nam . . . viam*'. 'He who kindly shows the bewildered wanderer the right road' (Cicero, *De officiis*, i. 16). *censure*. Judge. *curiosities*. Trifling matters. '*Dicit . . . via*'. 'The slothful man says there is a lion in the path' (Prov. 22: 13, 26: 13). '*Possunt . . . videntur*'. 'Strong they are, for strong they deem themselves' (*Aeneid*, v. 231). *asketh*. Requires. *demand*. Ask (not implying right or authority). *impertinent*. Irrelevant. *parts*. This division of the sciences uses a traditional concept of the mental faculties (deriving ultimately from Galen, but much developed in medieval and Renaissance treatises on the soul, or what we would call psychology), according to which separate ventricles in the brain

contained the memory, imagination, and reason (see e.g. E. Ruth Harvey, *The Inward Wits: Psychological Theory in the Middle Ages and the Renaissance* (London, 1975)). This system was widely disseminated, but Bacon's use of it to classify knowledge seems to be original: see G. T. Olivieri, 'Galen and Francis Bacon: faculties of the soul and the classification of knowledge', in D. R. Kelley and R. H. Popkin (eds.), *The Shapes of Knowledge from the Renaissance to the Enlightenment* (Dordrecht, 1991), 61–81. *oracle*. Divine revelation. *diverse*. Different. *supernumerary*. Additional. *Historia Literarum*. 'History of Letters', that is, of all written records.

176. *Polyphemus*. The Cyclop had only one eye, which was put out by Ulysses: *Odyssey*, ix. *divers*. Several. *jurisconsults*. Lawyers, teachers of jurisprudence. *schools*. University faculties. *relations*. Narratives. *just story of learning*. An accurate history of knowledge. *removes*. Changes of place. *curiosity*. Concern with strange or abstruse things. *Historia... Errantis*. 'History of Erring Nature.' *in course*. In its due order. *nature... varying*. Unusual and anomalous states. *altered or wrought*. Reshaped by human action. *Arts*. Trades, human skills. *deflexion*. Turning aside. *generations*. Giving birth. *motions*. Movements. *proprieties*. Properties, natures. *severe*. Searching, accurate. *Heteroclites*. Things deviating from the ordinary rules. Cf. *Novum Organum*, i. 45; ii. 28: *Works*, iv. 55–6, 68–9. *popular errors*. Bacon probably knew of a recent attempt to supply this lack in one area: Laurent Joubert's *Erreurs populaires au fait de la médecine et régime de santé* (1578), since in his 1608 notebook, the *Commentarius Solutus*, one memorandum is 'To procure an History of Marvailes, Historia naturae errantis or variantis, to be compiled with judgment and without credulity, and all the popular errors detected; Viscentius, Jubart, Plyny, Hystorie of all sorts for matters strange in nature' (*Works*, xi. 65). Sir Thomas Browne was inspired by Bacon's comment to compile his *Pseudodoxia Epidemica, or Enquiries into Vulgar and Common Errors* (1646). *on foot*. In circulation. *countenance*. Approving, confirming (without enquiry). *opinion*. Erroneous belief. *similitudes*. As in Lyly's *Euphues* (1578, 1580), where great use is made of comparisons drawn from Pliny and other fabulous compilations. *called down*. Decried, exposed.

177. *Aristotle*. The treatise *De Mirabilibus Auscultationibus* is spurious. *nothing less than*. Anything but; by no means. *curious*. Inquisitive in unimportant matters. *Mirabilaries*. Collections of marvels: see above, p. 143 note. *axioms*. Propositions, general principles. *intelligence*. Information. *hounding*. Pursuing. *sorceries*. Fortune-telling by casting lots. *divinations*. Foretelling of future events. *speculation*. Enquiry. *inquisition*. Investigation. *example*. King James was the author of *Demonologie* (1597). *merely*. Completely. *sincerely*. Without falsification, accurately. *impertinent*. Irrelevant. *Historia Mechanica*. 'History of Mechanical Arts', in this period associated with manual labour and held by many to be beneath the dignity of philosophical enquiry. Bacon's project to found an operative science, producing works, made him understand the value of technology when properly applied. *Wrought*. Worked upon. *collections*. Summaries. *familiar and*

vulgar. Common, everyday. *subtleties*. Unusual phenomena. *Plato. Hippias major*, 288A–291A. For other arguments that no object is unworthy scientific study see *Parmenides*, 130E, *Sophist*, 227, *Statesman*, 266D (Wolff, i. 24–5). *inquisitor*. Searcher. *wandering ... inductions*. Bacon saw the Socratic technique of argument as prefiguring his own concept of scientific induction (*Novum Organum*, i. 105; *Works*, i. 205, iv. 97–8), but criticized it as 'wandering', undirected.

178. *think much*. Grudge, take ill. *allege*. Quote. *irony*. Ironical speech. *the philosopher*. Thales, as reported by Plato, *Theaetetus*, 174A. *aloft*. Upwards. *Aristotle. Politics* i. 1, 1252a18 ff., i. 3, 1253b4 ff.; *Physics*, i. 1, 184a 10 ff. *his*. Its. *inquireth*. Investigates. *conjugations*. Relationships. *policy*. Government. *concordances*. Agreement. *loadstone*. Magnet. *fume*. Vapour, smoke. *subtile*. Minute, over-precise. *sublime*. Vapourized (a term from alchemy). *delectable*. Directed to pleasure, not use. *mysteries*. Crafts, trades. *axioms*. Scientific principles. *crossed*. Obstructed. *Proteus*. A mythical being who could be made to prophesy only when prevented from changing shape by being held firm. See Virgil, *Georgics*, iv. 386 ff. *straitened*. Confined. *passages*. Courses. *vexations of art*. Scientific examinations that disturb nature's order. *perfect*. Complete.

179. *casually*. By chance. *Caesar*. His 'Commentaries' are in fact histories of the Gallic and Civil Wars: see above, p. 603. *of the world*. In the world. *Antiquities*. Antiquarian collections, such as William Camden's *Remaines, concerning Britaine* (1604, 1614). '*tanquam ... naufragii*'. 'Like the planks of a shipwreck.' *story*. Historical narrative. '*tanquam ... mista*'. 'As it were imperfectly compounded', an unstable mixture bound to disintegrate. Cf. *Works*, x. 94. *Epitomes*. Abridgements of histories, such as those by the Roman historians L. Annaeus Florus and Aurelius Victor. Cf. Bacon's remarks on Ramus, above, p. 236, and his *Advice to Greville*, above, pp. 103 ff. *fretted*. Eaten away. *pretendeth*. Intends. *verity*. Truth. *passages*. Proceedings. *as*. That. '*maxima ... suspendens*'. 'The greatest things depending on the smallest.' *resorts*. Sources, springs.

180. *Narrations and Relations*. Bacon refers to three of the most celebrated works of classical historiography written by observers and participants, Thucydides' *History of the Peloponnesian War*, Xenophon's *Anabasis* (on the Persian expedition), and Sallust's *Catilina* (on the Catiline conspiracy). *story ... length*. An extensive period of history. *spaces*. Gaps. *conjecture*. Powers of invention. *distribution*. Classification (the rhetorical process of *partitio*). *illustrate*. Make illustrious. *exemplar*. Conspicuous. *to*. Attached to. *Antiquities*. Studies of archaic Greece and older cultures. *holpen*. Helped. '*caput ... condit*'. 'She hides her head among the clouds': Virgil, *Aeneid*, iv. 177; *Promus*, no. 1081. *perfect course of history for Graecia*. The whole course of Greek history from Theseus to Philopoemen, head of the Achaean League, who died in 182 BC and was nicknamed 'last of the Greeks'. *what time*. When. *drowned*. Were drowned. *for Rome*. The period of Roman history from Romulus to Justinian (d. AD 565). '*ultimus Romanorum*'.

'The last of the Romans' (a phrase originally used of Brutus and Cassius). *Herodianus.* Le Doeuff (p. 317) thinks this must be an error for 'Claudianus' (370–404); but Bacon is probably referring to Herodian of Syria, a subordinate official in Rome early in the 3rd century AD, whose *Tes meta Markon basileias historiai*, in 8 books, covered the period from M. Aurelius to Gordian III, AD 180–238. First published by Aldus Manutius in 1503, it had at least 14 editions in the 16th century, including a major one by Stephanus (Paris, 1581): see C. R. Whittaker's Loeb Library edn., 2 vols. (London, 1969, 1971). Bacon cited it earlier (p. 153), probably from Politian's Latin translation (Wolff, ii. 148). *supplied.* Helped. *supererogation.* Doing more than is required. '*curiosus . . . republica*'. 'A meddler in the affairs of other states' (Cicero, *De officiis*, I. xxxiv. 125).

181. *obliquity.* Unfairness. *largest author.* George Buchanan, in his *Rerum Scoticarum Historia* (1582), much criticized by King James in his *Basilicon Doron* (1599, 1603). *one history.* John Speed fulfilled this desideratum in his *History of Great Britain* (1611), from the beginnings up to the reign of James I. *passed.* Past. *Ten . . . Two Tribes.* The kingdoms of Israel and Judea, respectively. See *Works*, x. 98 for an elaboration of this point. *Uniting of the Roses . . . Kingdoms.* From 1485 to 1603. *adeption.* Obtaining. *answerable.* Corresponding. *pilot.* Henry VII. *sufficient.* Able. *king.* Henry VIII. *alteration . . . ecclesiastical.* The shift from Catholicism to Protestantism. *minor.* Edward VI. *offer.* Attempt. *usurpation.* The attempt of the Duke of Northumberland to place Lady Jane Grey on the throne. '*febris ephemera*'. 'A day-long fever.' *queen.* Mary, married to Philip of Spain. *queen.* Elizabeth I. *impression.* Effect. *operation.* Influence. '*Antiquam . . . matrem*'. 'Seek out your ancient mother' (Virgil, *Aeneid*, iii. 96). *ancient . . . Britain.* '*Britain*, after the OE period, was for long used only as a historical term, but in 1604 James I & VI was proclaimed "King of Great Britain"' (*SOED*). *period.* Conclusion. *peregrinations.* Wanderings. *trepidations.* Vibrations. *waverings.* Oscillations. *prelusive.* Preliminary. *Lives.* Biographies. *find strange.* Wonder at.

182. *most.* Mostly. *elogies.* Panegyrics. *invention.* By Ariosto, *Orlando Furioso*, xxxiv. 89–xxxv. 22. *ancient fiction.* Of the three fates, Clotho, Lachesis, and Atropos. *waited upon.* Attended. *shears.* Of Atropos, who cut the thread of life at death. *Lethe.* One of the four rivers of Hades. *consecrate.* Consecrated, dedicated to a god. *affections.* Feelings. *ventosity.* Pompous conceit. *Animi . . . egentes.* 'Souls that care not for great renown' (Virgil, *Aeneid*, v. 751). *root.* Principle, basis. '*non . . . desivimus*'. 'Men do not despise praise until they have ceased to do anything that deserves it' (Pliny, *Epist.* iii. 21). '*Memoria . . . putrescet*'. 'The memory of the just is blessed, but the name of the wicked shall rot' (Prov. 10: 7). *consumeth.* Perishes. *style.* Title or formula. '*felicis . . . memoriae*'. 'Of happy, of pious, of good memory'. '*bona . . . defunctorum*'. 'Good fame is all that a dead man can possess' (Cicero, *Philippics*, ix. 5). *waste.* Untreated. *Relations.* Accounts. *pen.* Writer. *particularity.* Details. *nursery garden.* Where plants are grown for transplantation.

183. *accoupleth*. Joins. *Annals*. Narrative of events year by year. *Journals*. Daily records. *estate*. State. *giving... touch*. Briefly alluding to. '*Cum... mandare*'. 'It suits the dignity of the Roman people to reserve history for great achievements, and to leave such details to the city's daily register' (*Annals*, xiii. 31). *contemplative heraldry*. The ranking of books (like noble persons) on a scale of importance. *degrees*. Ranks in society. *embase*. Debase. *novelty*. Fashion. *upon point of*. Taking into account. *Ahasuerus*. See Esther 6: 1. *take rest*. Sleep. *Journal*. See Plutarch's 'Life' of Alexander, 23.2, 76, etc. *navigations*. Sea voyages. *politic discourse*. Concerning policy or government, such as Machiavelli's *Discourses*, which Bacon later singles out for praise: above, pp. 270–1. *Ruminated*. Meditated. *manifoldly*. In many ways. *Cosmography*. Study of the earth, combining geography and history. *regiments*. Forms of government. *mathematics*. Often associated with astronomy. *configurations*. Positions relative to the heavenly bodies. *through-lights*. Windows on both sides of a room.

184. *Nosque... Vesper*. 'And when on us the rising Sun first breathes with panting steeds, there glowing Vesper is kindling his evening rays' (Virgil, *Georgics*, i. 250–1). *demonstration*. Reasoning. *enterprised*. Undertaken. *word*. Motto. *plus ultra*. 'Further yet': the motto of Emperor Charles V. *non ultra*. 'No further.' *imitabile fulmen*. 'Imitable thunder' (i.e. the invention of gunpowder). *Demens... fulmen*. 'Madman! to mimic the storm-clouds and inimitable thunder' (Virgil, *Aeneid*, vi. 590). *imitabile coelum*. 'Imitable sky'. *voyages... earth*. Magellan was the first navigator to sail round the world, 1519–22; Drake followed in 1577–9. *coevals*. Contemporaries. '*Plurimi... scientia*'. 'Many shall pass to and fro, and knowledge shall be increased' (Dan. 12: 4). *through passage*. Traversing. *giving place*. Yielding. *periods or returns*. A metaphor from astronomy: the completion of a planetary cycle. *with*. As. *militant church*. The church on earth as constantly warring against the powers of evil. *fluctuant*. Floating on the waves. *Noah*. Gen. 6: 13 ff. *wilderness*. Moses, Exod. 31: 7. *temple*. Of David, 2 Sam. 7: 2. *remove*. Departure to another place. *sincerity*. Freedom from falsification; correctness. *according*. Corresponding. *in hand*. Concerned. *Historia Prophetica*. 'History of Prophecy.' *relatives*. Related points.

185. *sorted*. Classified. *latitude*. Freedom of application. *thousand... day*. Ps. 90: 14; 2 Pet. 3: 8. *punctually*. Exactly. *springing*. Welling up, like water. *germinant*. Sprouting, germinating. *height*. High degree, culmination. *correspondence*. Agreement. *natural man*. Man lacking the benefit of divine revelation. Cf. 1 Cor. 2: 14. *behold... tabernacle*. Those who have received divine illumination. *establishment*. Strengthening, confirmation. *without God in the world*. Eph. 2: 12. '*he... it*'. Hab. 2: 2, but not understood in the sense given by the Authorized Version, where 'run' means 'hasten to carry on the tidings'. *sensual*. Lacking reason. *hasten by*. Rush past, ignore. *bend... cogitations*. Take notice of, concentrate on. *passage and race*. Course of life. *present*. Represent. *Appendices*. The *De Augmentis* (*Works*, iv. 314–15) expands the remarks on letters, but (unusually) omits (despite the chapter title) the division of orations. Perhaps something has

been lost. *inducements*. Preambles. *passages*. Transitions. *receit*. Reception. *pleadings*. Speeches in a lawcourt. *counsel*. Political advice. *laudatives, invectives*. The two forms of epideictic rhetoric, praise (of virtue) and blame (of vice). *reprehensions*. Reproofs. *advertisements*. Informations. *directions*. Instructions. *passages*. Courses.

186. *advised*. Well-considered. *present*. Impromptu. *Apophthegms*. Memorable speeches by famous people (see above, p. 162). In the *De Augmentis* Bacon adds: 'Neither are Apophthegms themselves only for pleasure and ornament, but also for use and action. For they are (as was said) "words which are as goads" [*secures aut mucrones verborum*: Cicero, *Epist. ad Fam.* ix], words with an edge or point, that cut and penetrate the knots of business and affairs. Now occasions are constantly returning, and what served once will serve again; whether produced as a man's own or cited as an old saying. Nor can there be any question of the utility in civil matters of that which Caesar himself thought worthy of his labour' (*Works*, iv. 314). *insist*. Dwell on. *cells*, *domiciles*. The ventricles of the brain, supposedly the location of specific faculties (reason, imagination, memory). See above, p. 175 and note. *offices*. Functions; but also 'places of business'. *measure*. Metre. *restrained*. Controlled. *licensed*. Free. '*Pictoribus . . . poetis*'. 'Painters and poets [have always been allowed to take what liberties they would]' (Horace, *Ars poetica*, 9–10). *styled*. Called ('Feigned History'). Wright suggests as the correct reading: 'which may be *so* styled'. *verse*. That 'feigning', or fiction, is the real defining mark of poetry, not whether it uses metre, had recently been affirmed by Sir Philip Sidney: see *Apology for Poetry* (1595), ed. G. K. Shepherd (London, 1965), 103. *in proportion*. In terms of symmetry, harmony, balance. *true history*. Bacon's juxtaposition of history with poetry (i.e. epic, tragedy) owes something to Aristotle's *Poetics* (ix, 1451a30 ff.), but more to Renaissance theories of literature's superior scale and power. Cf. Sidney, ed. cit. 99–100. *successes*. Results of an action. *alternative*. Alternating. *conferreth*. Contributes.

187. *magnanimity*. Nobility of feeling. Cf. Sidney: 'Heroical' (epic) poetry 'maketh magnanimity and justice shine throughout all misty fearfulness and foggy desires' (ed. cit. 119). *delectation*. Pleasure. *erect*. Inspire. *shews*. Appearances. *buckle*. Bend. *insinuations*. Intertwinings, connections. *consort*. Fellowship. *access*. Acceptance, admission. *estimation*. Esteem. *Representative*. Drama, to which Bacon devotes more space in the *De Augmentis*: *Works*, iv. 316. *Allusive*. Symbolical, figurative. *remembered*. Mentioned. *state*. National affairs; statesmanship. *Parabolical*. Based on comparison or similitude; allegorical. *conceit*. Conception. The *De Augmentis* expands this point: 'Parabolical Poesy is typical History, by which ideas that are objects of the intellect are represented in forms that are objects of the sense.' *the Seven*. The 'Seven Wise Men of Greece', whose teaching was recorded in brief sayings. *hieroglyphics*. Wright (p. 292) notes that Bacon is probably drawing on Plutarch, 'Of Isis and Osiris' 10 (*Moralia*, 354D ff.), which links hieroglyphs with the instance of parabolical wisdom added in the Latin text, the 'tesserae Pythagorae'. *sharp*. Acute. *subtile*. Intricate, abstruse.

wanted. Lacked. *sensible*. Easily understood; perceptible. *examples*. Drawn from historical events. *fit*. Fitting. *illustrate*. Exemplify. *retire*. Put into obscurity, conceal. *involved*. Enfolded. *divine poesy*. The traditional interpretation of biblical texts in four senses: literal, allegorical, moral, anagogical. *heathen*. Pagan poetry dealing with the gods. *fall out*. Happen. *Fame*. Rumour.

188. *Illam . . . progenuit*. 'Her, 'tis said, Mother Earth, provoked to anger against the gods, brought forth last, as sister to Coeus and Enceladus' (Virgil, *Aeneid*, iv. 178–80). *Expounded*. Interpreted. *taxations*. Censures. *Pallas*. But according to Homer (*Iliad*, i. 396–404) it was Thetis who called Briareus to free Zeus after the other Olympian gods had shackled him. *Briareus*. A mythical giant with a hundred arms and fifty heads. *come in*. Side with, join. *Chiron*. The teacher and doctor among the Centaurs, who, according to Greek myth, educated divine children and heroes, Achilles, Asclepius, and Jason. *Machiavel. The Prince*, ch. 18. *discipline*. Training. *encounters*. Instances. *was*. Came. *devised*. Invented. *framed*. Constructed. *Chrysippus*. Stoic philosopher (b. 280 BC) mocked by Cicero (*De natura deorum*, I. xv. 38–41) for finding Stoic doctrines in the stories of the gods. *contention*. Effort. *pleasure*. Pleasing fictions. *figure*. Figurative. *no opinion*. For Bacon's later views on myths see *The Wisdom of the Ancients* (1609): *Works*, vi. 607–764, and the *De Augmentis*: *Works*, iv. 317–35. *Scripture*. The discovery of moral lessons in Homer, and other allegorical approaches, begun in the 4th century BC, was especially practised by the Neoplatonists. In the prologue to *Gargantua* Rabelais makes the same sceptical judgement on the allegorizers. *inwardness*. Hidden sense. *upon*. From. *third*. It should be 'second'. *beholding*. Indebted. *judicial*. Depending on judgement, the third term in faculty psychology (after memory and imagination).

189. *informed*. Animated. *cumulative*. Formed by accumulation, not by organic growth (nor by revelation). *water*. Piece of water. *Philosophia . . . scientiarum*. 'Primary philosophy, or the spring-head of the sciences.' In *Met*. vi. 1 Aristotle refers briefly to 'natural science' (which 'deals with things which are inseparable from matter') as constituting the 'first philosophy', since it investigates 'the principles and the causes of the things that are' (1025b1–1026b32), and in *Physics*, ii. 2 he writes that its task is 'to efine the mode of existence and essence of the separable' (194b14). Bacon uses the idea in a much wider sense. *circumferred*. Carried round. *reverted*. Turned back. Ellis (*Works*, i. 540) cites Plutarch, '*De curiositate*', *Moralia*, 515D ff. *Humanity*. Not just the humanities in our sense, but any study involving mankind. *difference of*. Differences within. *distributions*. Classifications, categories. *several lines*. Distinct lines, forming geometrical figures, abstract and two-dimensional. *Primitive*. Primary. *Summary*. Most important. *common*. Shared. *rhapsody*. Confused mixture. *commixed*. Mixed up. *depredation*. Robbing, plundering. *height of terms*. Exaggerated terminology. *substantive*. Substantial. *several*. Separate. *notion*. Conception; theoretically. *pursued*. Followed out. *Extern*. External. *of force*. Necessarily.

190. *virtue*. Efficacy. *participles*. Sharing the nature of two different classes (such as moss, midway between putrefaction and a plant; or fish that can fly: see *Works*, i. 283, 543–4; iv. 169–70, 339–40). *mere*. Complete. *Adjuncts*. Non-essential attributes; ancillaries. *resuming*. Repetition. *speech or argument*. That is, as topoi used for rhetorical ornament. The *De Augmentis* adds: 'With regard to these and similar things in the discussion of Transcendentals there is a deep silence; for men have aimed rather at height of speech than at the subtleties of things. Wherefore I wish the real and solid inquiry, according to the laws of nature and not of language, to have a place in Primitive or Summary Philosophy' (*Works*, iv. 340). *subtlety*. Intricate or abstruse discussion. *profitable ... axioms*. On axioms common to several sciences see Aristotle, *Posterior Analytics*, i. 10, 76a36 ff. This topic was discussed in the 16th century by several writers within the Aristotelian tradition, such as Melanchthon (in his *Loci communes rerum theologicarum*, 1521) and Bartholomeus Viottus (in *De demonstratione libri quinque*, 1560): see N. W. Gilbert, *Renaissance Concepts of Method* (New York, 1960), 109, 119, 155–7. '*Si ... inaequalia*'. 'If equals be added to unequals, the result will be unequal' (Euclid, *Elements*, i. 4). *coincidence*. Exact agreement. *commutative*. Relating to exchange; corrective justice, which judges the transaction and not the person (punished according to the crime). See Aristotle, *Eth. Nic.* v. 2–3, 1130b30 ff. *distributive*. Giving to each according to his merits. See ibid. v. 3–5, 1131b32 ff. '*Quae ... conveniunt*'. 'Things that are equal to the same are equal to each other' (Euclid, *Elements*, i. 1). *syllogisms*. Logical forms which unite propositions agreeing in a middle term. '*Omnia ... interit*'. 'All things change but nothing is lost' (Ovid, *Met.* xv. 165). *quantum*. Quantity, totality. Bacon repeats elsewhere this principle, that 'there is nothing more certain in nature than that it is impossible for any body to be utterly annihilated; but that as it was the work of the omnipotency of God to make somewhat of nothing, so it requireth the like omnipotency to turn something into nothing': *Sylva Sylvarum*, i. 100 (*Works*, ii. 383). '*Didici ... auferre*'. 'I know that whatever God does lasts for ever; to add to it or subtract from it is impossible' (Eccles. 3: 14). *ground*. Fundamental principle. *Machiavel. Discourses*, III. 1. *reduce*. Bring back. *ad principia*. To their origins or first state. *Persian Magic*. According to Plato, *Alcibiades*, 121E, the tutors to the princes of Persia were versed in both politics and magic. *affection*. Emotion. *trope*. A short cadence at the end of a melody. *slide*. Move away. *close*. Cadence in music. *trope ... expectation*. For this 'trope' (here in the normal rhetorical sense, 'turning' a word or phrase from its normal to a transferred sense), namely *aprosdoketon*, or 'cheating expectations by taking words in a different sense from what was intended', see Cicero, *De oratore*, II. lxiii. 255, 284, 289, and Quintilian, VI. iii. 24 (Wright, p. 294).

191. *quavering*. A trill or tremolando. *stop*. A held note (cf. 'double-stopping' on the violin). *Splendet ... pontus*. 'The sea glitters beneath [the moon's] dancing beams' (Virgil, *Aeneid*, vii. 8–9). *one*. The same. *glass*. Mirror. *strait*. Confined space. *similitudes*. Rhetorical comparisons, sometimes collected for use as verbal ornament, as in Erasmus' *De parabolis*. Bacon's interest is not in words alone, but in the reality of the natural phenomena involved. *disclosing*. Discovery. *abridgment of art*. Shortening of the process of understanding and controlling nature. *Berecynthia*. Cybele, the *Magna Mater* of the gods, who appears in the prophetic vision that Anchises communicates to his son Aeneas in the underworld, 'clasping a hundred of her children's children'. *omnes... tenentes*. 'All denizens of heaven, all tenants of the heights above' (Virgil, *Aeneid*, vi. 783–7). *light*. Divine illumination. *convince*. Refute. *inform*. Teach. *confess*. Acknowledge. *world... image of God*. See Macrobius, *Commentary on the Dream of Scipio*, ii. 12, and St Augustine, *The City of God*, vii. 23. *extract*. Summary. Bacon frequently rejected the related idea, the human body as microcosm corresponding to the universe as macrocosm. *the... hands*. Gen. 1; Ps. 8: 3–6. *induce*. Argue inductively, from observation. *inforce*. Urge, press home.

192. *divers*. Various writers. '*Da... sunt*'. 'Give unto faith the things that are faith's': Matt. 22: 21. *golden chain*. The Great Chain of Being: see above, p. 126. *touching*. Concerning. *prejudice*. Injury. *commixed*. Mixed. *Otherwise it is of*. The situation is different concerning. *angels and spirits*. A large body of doctrine about angels and other spirits was taken over and elaborated by the medieval Church from late classical, especially Neoplatonist sources. '*Let... not*'. Col. 2: 4, 18; *Promus*, no. 252. *sublime*. Inflated, specious, deceiving. *that*. That which. *fantastical*. Fabulous, unreal. *creature*. Anything created. *ground*. Basis. *grounded*. Well founded. *inquiry*. Investigation. *gradations*. Traditionally, the cosmos was divided into categories of ascending intelligence and power: mineral, vegetable, animal, human, angelic, and divine. *revolted*. Lucifer and the other fallen angels who rebelled against God and became devils in Hell. *much more*. All the more. *science*. Study. '*We... stratagems*'. 2 Cor. 2: 11, referring to the devil's schemes and tricks to hurt mankind. *challenge*. Accuse, reprehend. *fabulous*. Based on fables, untrue.

193. *sabbath*. Resting place. *that... caves*. In Diogenes Laertius, ix. 72. *inculcate*. Insist, instil by frequent repetitions. See Paracelsus, *Liber Meteororum* (1566), ch. 4. *Vulcan*. The furnace, basic to alchemy. *compendiously*. Briefly. *ambages*. Circuitous ways or methods. *pioners*. Miners. Cf. *New Atlantis*, above, p. 486. *scholastical*. Academic. *Speculative, and Operative*. Theoretical (enquiring into causes), and practical, yielding works. As the *De Augmentis* adds, 'The one searching into the bowels of nature, the other shaping nature as on an anvil' (*Works*, iv. 343). *Prudence*. Practical wisdom, as distinct from theoretical knowledge. *direction*. Enquiry. *reintegrate*. Restore. *Natural Magic*. See above, p. 143 n. This point is omitted in the *De Augmentis* (*Works*, iv. 343). *acception*. Accepted meaning. *invention*. Finding. *Theory*. Speculation (into causes). *Physic*. Here physics, knowledge of the phenomenal universe. *Metaphysic*. Traditionally the study of the first principles of things, such concepts as being, substance, essence, time, space, cause, identity. Bacon retains the word, since he disapproves of neologisms,

but uses it in an idiosyncratic sense. *mistaking*. Being misunderstood. *that*. That which. *affectionate*. Eagerly desirous. *stand with*. Be consistent with. *Aristotle*. In his surveys of extant knowledge, such as the *Metaphysics*, which attacks the Presocratics and Pythagoreans.

194. *confound*. Overthrow. *reprove*. Refute. *glory*. Ostentation. '*Veni ... recipietis*'. 'I am come in my Father's name, and you do not receive me; if another comes in his own name, you will receive him' (John 5: 43). '*Eum recipietis*'. 'You will receive him.' *Aristotle*. The *De Augmentis* adds: 'so wonderful for the acuteness of his mind' (*Works*, iv. 345). *his scholar*. Alexander the Great. *emulate*. Vie with. *to conquer all opinions*. The *De Augmentis* adds: 'and to establish for himself a kind of despotism in thought' (ibid.). *bitter*. Satirical. *like*. Similar. '*Felix ... praedo*'. 'A fortunate robber of territory, who was a bad example to the world'. So 'A fortunate robber of learning': adapted from Lucan, *Pharsalia*, x. 20 f. *ground*. Lay the foundation of. *antiquity*. Tradition. *proficience*. Progress. *keep way with*. Conform to. '*usque ad aras*'. 'So far as is consistent with religious obligations.' '*eadem ... vocabula*'. 'The names of the magistracies have remained the same' (Tacitus, *Annals*, i. 3). *acception*. Acceptation. *Summary*. General, universal. *confounded*. Confused *promiscuous*. Mixed indiscriminately. *indifferent*. Belonging to all alike. *adventive*. Coming from without, inessential. *essences*. Things in themselves.

195. *in nature ... logically*. As they function in the real world, not in theoretical discussions. *conceit*. Conception. *abstracted*. Abstract. *fixed*. Permanent. *supposeth*. Assumes, presupposes. *transitory*. Ephemeral, subject to change, like the physical world. *platform*. Design (the principle of teleology). *sensible*. Easily perceived, understood. *Formal and Final Causes*. According to Aristotle (*Physics*, ii. 3, 194b16 ff.; *Met*. i. 3, 983a 26 ff.; v. 2, 1031a24 ff.), the efficient cause is that which acts, the material cause is that which is acted on (as when fire melts wax, the former is the efficient, the latter the material cause of the effect produced). The formal cause is that which determines an object (including its properties) to be as it is; while the final cause is that for the sake of which any effect takes place. *derivation*. The new sense that Bacon ascribes. *situate*. Placed. *respective*. Special, relative. *Limus ... igni*. 'As this clay hardens, and as this wax melts in one and the same flame' (Virgil, *Eclogues*, viii. 80–1). *induration*. Hardening. *colliquation*. Melting. *matter*. Material. *collected*. As a coherent whole. *diffused*. Spread out. *distributed*. Divided into parts. *seeds*. Atoms, component parts. *Configuration*. Spatial arrangement. '*de ... rerum*'. 'Of the world, of the universe of things.' *gloss*. A marginal explanation.

196. *deficient*. Here the *De Augmentis* (*Works*, iv. 347–55) adds a long discussion of astronomy and astrology. *deserted*. Left uncultivated. *Metaphysica ... rerum*. 'Metaphysics, or, of the forms and final causes of things.' *nugatory*. Futile, useless. *inveterate*. Firmly established. *inquisition ... differences*. The discovery of the laws governing the ultimate constituents of matter and their transformation was for Bacon a major goal of scientific enquiry. Although they only reached maturity in the *Novum Organum* (1620), many of his key ideas (such as rejecting the scholastics' doctrine that Forms are unknowable because beyond the senses' perception) were expressed here and in other works of the early 1600s, such as *Valerius Terminus* (*Works*, iii. 215–52) and *Temporis Partus Masculus* (iii. 527–39). *invention*. Discovery. *Forms*. 'In Aristotelian and scholastic philosophy, the essential determinant principle of a thing; that which makes something (*matter*) a determinate species or kind of being' (*SOED*). *Plato*. Bacon consistently criticizes Plato's concept of Ideas (as expounded e.g. in *Republic*, 6. 507 ff., 7. 523 ff., and *Philebus*, 15 ff.) as being too abstract, remote from the material universe. Aristotle had made similar criticisms: see e.g. *Met*. vii. 15 (1040a8–64), xiii. 4 (1078b31), xiii. 9 f. (1068a18–1087a25). *wit of elevation*. Perceptive intelligence. *descry*. Discern. *confined*. Kept in place. *determined*. Bounded. *advise*. Consider. *except*. Excepted. '*Formavit . . . vitae*'. 'God formed man from the dust of the ground and breathed into his nostrils the breath of life' (Gen. 2: 7). '*Producant . . . terra*'. 'Let the waters bring forth; let the earth bring forth' (Gen. 1: 20, 24). *compounding*. Combination. *transplanting*. Grafting on to another organism. *perplexed*. Confused. *enquired*. Investigated. *in gross*. As a whole. *composition*. Combining. *induceth*. Brings into being. *consist*. Are composed of. *levity*. Lightness. *tenuity*. Thinness. *upheld*. Sustained. *that part*. And not concerning final causes. *define of*. Define.

197. *the cause of Whiteness*. An early example of Bacon's scientific method, much expanded in the *Novum Organum*: *Works*, iv. 156–8. *subtile*. Fine, intricate. '*vehiculum formae*'. 'The carrier of the form.' As Le Doeuff (pp. 323–4) notes, the carrier also 'involves movement or modification, for the form is the fixed law of a transformation'. *laboured*. Worked on, developed. *invention*. Discovery. Bacon argues that natural philosophers have never made a sustained enquiry into detailed instances. *recess*. Withdrawal, distance. '*vita . . . longa*'. 'Life is short, art [that is, the acquisition of knowledge or the mastery of a technique] is long' (Hippocrates, *Aphorisms*, 1. i.) *uniting*. That is, the *De Augmentis* adds, 'by collecting and uniting the axioms of the sciences into more general ones, and such as may comprehend all individual cases' (*Works*, iv. 361). *pyramides*. Here the arrangement of knowledge from particular to general, from the lowest or broadest to the highest and most abstract forms. *history*. Enquiry. *next*. Next to. *vertical point*. Peak, summit. '*Opus . . . finem*'. 'God hath made every thing beautiful in his time: also he hath set the world in their heart, so that no man can find out the work that God maketh from the beginning to the end' (Eccles. 3: 11). *the*. Wright (p. 297) suggests reading 'three': the Latin version has 'tres moles giganteae'. *depraved*. Corrupted (by original sin); not knowing God. *Ter . . . Olympum*. 'Thrice did they attempt to pile Ossa on Pelion, and over Ossa to roll leafy Olympus' (Virgil, *Georgics*, i. 281–2). *sancte*. 'Holy, holy, holy!' (Rev. 4: 8). *dilatation*. Dilation, expanded description. *concatenation*. Joining together. This provides the subject-matter of physics. *uniform law*. To be studied by metaphysics, as Bacon defines it. *Plato*. See *Parmenides*, 165–6; *Philebus*, 16.

As Wright (p. 297) and Wolff (i. 20 f.) observe, Bacon is reproducing Plato's idea via Neoplatonic intermediaries, such as Proclus and Ficino. *by scale*. By degrees. *charged*. Burdened, complicated. *valueth*. Makes more valuable. *enfranchise*. Liberate.

198. *flexuous*. Winding. *'latae ... viae'*. 'To the wise there are broad paths everywhere'. Untraced; Le Doeuff (p. 324) suggests a conflation of Prov. 3: 17 and 4: 26. *sapience*. Wisdom. *'rerum ... scientia'*. 'Knowledge of things divine and human' (Cicero, *De officiis*, I. xliii. 154, and *Tusc. D.* IV. xxvi. 57). *in simili materia*. In similar matter. *superinducing*. Superadding, imposing on. *Matter*. Material cause. *'Non ... offendiculum'*. 'As you walk you will not slip, and, if you run, nothing will bring you down' (Prov. 4: 12). That is, the *De Augmentis* adds, the ways of knowledge are neither confined to particular methods, nor liable to be defeated by accidental obstructions (*Works*, iv. 363). *liable*. Subject to. *misplaced*. In the department of Physics instead of Metaphysics. *order*. Arrangement. *improficience*. Lack of progress. *intercepted*. Put a stop to. *stay upon*. Stop at. *satisfactory*. Superficially plausible. *specious*. Deceptively attractive. *Plato*. For his teleological arguments see e.g. *Timaeus*, 45–8, 68E; *Phaedo*, 97 ff. *fall*. Run aground on. *flats*. Shallows. *discoursing*. Produced by discourse, purely verbal, not referring to anything in reality. *to say*. Bacon criticizes the following positions as tautologous explanations of phenomena in terms of function. The examples come from Plato (*Timaeus*, 44D ff.), Aristotle (*Physics*, ii. 8, 199a25), and Galen (*De Usu Partium*). *quickset*. Hedge (of living plants). *solidness*. Solidity. *station*. Standing-place. *mansion*. Dwelling-place. *collected*. Discussed. *impertinent*. Irrelevant. *remoras*. Sucking-fish, thought to obstruct a ship in its course by clinging to its side. *slug*. Retard. *Democritus*. Valued by Bacon as an early exponent of atomism (the doctrine that nature was created not by a divine decree but by the chance movement of atoms). See Cicero, *Tusc. D.* I. xi. 22; Diogenes Laertius, ix. 44–5. *others*. These early atomists included Epicurus, whose system was versified in Lucretius' *De rerum natura*.

199. *frame of things*. Structure of the universe. *able*. Being able. *proofs*. Experiments, tests. *recital*. Enumeration of a series of facts or statements. *excursions*. Extensions. *vastness*. Emptiness, desolation. *track*. Tract, expanse of land. *rendered*. Given; by Cicero in *De natura deorum*, II. lvii. 142. *impugn*. Call in question, fault. *pilosity*. Hairiness. *incident*. Naturally attaching. *Muscosi fontes*. 'Ye mossy fountains' (Virgil, *Eclogues*, vii. 45). *incident*. Liable to happen. *outwardest*. Outmost. *adjacence*. Nearness. *intention*. Purpose, aim. *politique*. Politician. *characters and impressions*. Forms and marks (a metaphor from printing). *special*. In the special sense used here. *said*. Above, p. 190. *determined*. Definite, bounded.

200. *Democritus and Pythagoras*. See Aristotle, *De anima*, i. 2; *Met.* i. 4, 5; Iamblichus, *Vit. Pythag.* xii. 59. *figure*. Shape. *champain*. Open, unenclosed. *particularity*. Concern with minute particulars. *material*. Significant. *partitions*. Divisions of the topic into its constituent parts. In the *De Augmentis* mathematics is firmly assigned as auxiliary to natural philosophy (or physics), where it properly belongs: *Works*, iv. 369–70. *perspective*. Correct relation between more and less important (as in optical perspective). *merely*. Absolutely. *continued*. Continuous. *dissevered*. Broken up into parts. *auxiliary*. Subsidiary. *incident*. Incidental, subordinate to. *invented*. Discovered. *subtlety*. Precise detail. *sort*. Class. *Enginery*. The art of constructing engines, machines. *wandering*. Vague, unconcentrated. *inherent*. Fixed. *intervenient*. Incidental. *Mixed Mathematics*. Applied mathematics, involving matter, a topic discussed in more detail in the *De Augmentis*: *Works*, iv. 370–2. *disclosed*. Opened up, unfolded.

201. *Operative*. Applied to works or action. *incidence*. Coincidence. *purposed experiment*. These types of experiment point on to the concept of '*experientia literata*' described below, and further developed in the *De Augmentis*: *Works*, iv. 413–21. *divers*. Various, different. *empiric*. Someone having only practical, not theoretical knowledge: able to apply 'ingenuity and sagacity', as the *De Augmentis* puts it, but not natural philosophy (*Works*, iv. 366). *coastings*. Sailing near the shore. '*premendo . . . iniquum*'. 'Hugging the hostile shore' (Horace, *Odes*, II. x. 3). *direction*. Guidance. *Naturalis . . . major*. 'Natural Magic or the major Operative Physics.' *in books*. Such as *Magia Naturalis Libri XX* (1589) of Giovanni Battista della Porta. *Sympathies and Antipathies*. Traditional concepts, uncritically extended by the occult tradition, by which all things in the universe are said to have an inherent likeness or unlikeness, defined by such superficial criteria as appearance, taste, texture, or colour. *hidden proprieties*. Concealed properties, such as the supposed 'signatures' of things, revealing their essential nature. *disguisement*. Disguising. *king Arthur of Britain*. Sir Thomas Malory's *Le Morte D'Arthur* (*c*.1470), printed by Caxton in 1485. *Hugh of Bordeaux*. A celebrated medieval romance. *de vero*. 'In reality.' *fable*. See Pindar, *Pythian*, ii. 21. *figure*. Type, emblem. *chimeras*. Fabulous creatures. *entertain*. Harbour, cherish. *vaporous*. Vain, boastful. *hold so much of*. Pertain to. *pretence*. Expressed aim.

202. *Weight . . . Colour . . . Pliant . . . Fragile*. Tests to distinguish metals from each other. *mechanique*. Mechanism. *medicine*. Or 'tincture', the substance supposedly able to produce transmutation. *projected*. In the final stage of alchemy, the throwing (projecting) of the powdered 'philosopher's stone' into the heated vessel in order to transmute base metal into gold. *arefaction*. Drying out. *assimilation*. Conversion into a similar substance. *clearing*. Evaporation, purifying. *spirits*. Subtle and invisible fluid components of matter. *depredations*. Corrosive effects, wearing away. *humours*. Bodily fluids. *ambages*. Circuitous methods. *motions*. Bodily exercises. *scruples*. Small amounts. *receit*. Recipe, concoction. *Inventarium . . . humanarum*. 'Inventory of man's inventions.' *latitude*. Scope, freedom. *relative thereof*. Metaphysics. *plausible*. Specious. *note*. Observation worth noting. *artificial*. Skilfully made. *optatives*. Things to be desired. *potentials*. Possibilities. *Speculation*. Investigation. *mariner's needle*. Navigational compass.

203. *move*. Arouse. *Non . . . sylvae*. 'We sing to no deaf ears; the woods echo each note' (Virgil, *Eclogues*, x. 8). *Borgia*. Pope Alexander VI, describing the French expedition under Charles VIII in 1494. Bacon's source is the *Mémoires* of Philip de Commines (ed. J. Calmette and G. Durville (Paris, 1924–5), iii. 81). *Continuatio . . . naturae*. 'Continuation of the Problems in nature. Catalogue of the falsehoods lurking in natural history.' *Considerative*. Requiring consideration or reflection. *non liquets*. In Roman law, the judge's declaration of being unable to decide on the accused's guilt or innocence. The 'skilful proposing of doubts', the *De Augmentis* adds, 'is no despicable part of science', since 'new doubts are daily arising' (*Works*, iv. 357). *Aristotle's Problems*. The *Problemata naturalia*, a collection of problems in various sciences attributed to Aristotle. *collected into assertion*. Presented as an established fact. *draw*. Draw out, produce. *use*. Increase. *should*. Would. *advised*. Considered. *solicitation*. Entreaty. *countervail*. Outweigh. *received*. 'Allowed as just', the *De Augmentis* explains, or 'authorised', adding that at this stage a doubt 'immediately raises up champions on either side, by whom this same liberty of doubting is transmitted to posterity' (*Works*, iv. 358). *accordingly*. In accordance with. *scholars*. Academics. *brought to resolution*. Completed. *decarded*. Set aside.

204. *material*. Relevant. *calendar of popular errors*. See above, pp. 176–7. *pass*. Are accepted. *apparently*. Openly. *imbased*. Spoiled. *De . . . philosophiis*. 'Concerning the ancient philosophies.' *diversity of sects*. The Presocratic philosophers (much criticized by Aristotle), whose surviving works Bacon would like to see collected. *Ottomans*. In 1574 Amurath IV of Turkey had his brothers strangled on his accession, as did his successor Mahomet in 1596; cf. Shakespeare, 2 Henry IV, v. ii. 48. *magistrality*. Dogmatism. *phaenomena in astronomy*. Like many classical and Renaissance thinkers, Bacon judged astronomy inferior to natural philosophy (physics), in that it only attempted to explain visible phenomena, and so could be satisfied with various models, or adaptations of them. (Thus it lacked one of the main features of Bacon's scientific theory, the use of 'negative instances' to eliminate irrelevant data.) As he adds in the *De Augmentis*, 'the same phenomena, the same calculations, are compatible with the astronomical principles both of Ptolemy and Copernicus' (*Works*, iv. 359). Physics, by contrast, investigating the laws of matter and motion, aspired to certainty. *received astronomy*. The Ptolemaic system, with the earth at the centre of the universe, and the planets supposedly revolving in spheres around it. *eccentrics and epicycles*. Modifications to the Ptolemaic system introduced to explain visible discrepancies from circular motion. *Copernicus*. His heliostatic theory, expounded in *De Revolutionibus Orbium Coelestium* (1543), was not an unequivocal improvement on the geocentric theory, and was only validated by the revised theories developed by Kepler, Galileo, and Newton, who evolved a fully heliocentric model. *indifferently*. Equally. *experience*. The Renaissance did not distinguish clearly between 'experience' and 'experiment'. *Aristotle saith*. In *Physics*, i. 1, 184b12. *understandingly*. Intelligently. *de . . . philosophiis*. 'Concerning the ancient philosophies', the *De Augmentis* adds, such as the 'Lives' of the philosophers (Diogenes Laertius), Aristotle's confutations (of the Presocratics, in the *Metaphysics*), and the scattered notices in Lactantius, Philo, Philostratus, etc. (*Works*, iv. 359). *severedly*. Separately. (Emended in the Errata to the 1605 edn. from 'severely', in the text.) *faggoted*. Bundled up together, like sticks. *Plutarch*. In the compilation attributed to him, *De placitis Philosophorum*. *foreign*. Dissimilar, irrelevant. *Tacitus*. In his *Annals*. *inducements*. Causes, circumstances leading to an event or action. *Suetonius*. In his *Lives of the Caesars*, which is organized by topics, not chronologically. *titles*. Headings, topics.

205. *dismembered by articles*. Cut up into separate points. *Paracelsus*. Theophrastus Paracelsus Bombastus of Hohenheim (1493–1541), Swiss alchemist whose semi-mystical medical theories aroused much controversy. *Severinus*. Peter Severinus (1542–1602), a Danish physician who made a digest of Paracelsian theory in *Idea Medicinae Philosophicae* (Basle, 1571). *Telesius*. Bernardino Telesio (1508–88), author of *De Rerum Natura* (Naples, 1565, 1570, much enlarged 1586), who rejected Aristotelian doctrines for a system based on sense experience and nature. *Donius*. Anton Francesco Doni (1513–74), a materialist philosopher, author of *De natura hominis* (Lyons, 1581). *pastoral*. A philosophy which contemplates the world placidly and at its ease: cf. *De principiis*, *Works*, v. 491. *Fracastorius*. Girolamo Fracastoro (1470–1553), author of a Latin poem *Syphilis* (Verona, 1530) and works on astronomy, contagious diseases, sympathy, and antipathy. The *De Augmentis* omits Donius and Fracastorius, substituting the name of 'Patricius the Venetian, who sublimed the fumes of the Platonists' (*Works*, iv. 359). *absoluteness*. Independence, categorical assertiveness. *Gilbertus*. William Gilbert (1544–1603): see above, p. 146. *demonstrations*. Logical arguments, proofs. *Xenophanes*. Greek natural philosopher, 6th century BC. In the *De Augmentis* Bacon substitutes the more appropriate name of Philolaus, a Pythagorean philosopher contemporary with Socrates, presumably referring to his belief that the earth moved in a circle (Diogenes Laertius, viii. 85). *Radius Directus... Refractus... Reflexus*. The direct, refracted, and reflected beam. *inequality*. Unevenness (i.e. the inability of the human mind to know God). *resteth*. Remains. *ancient oracle*. Over Apollo's temple at Delphi was written the admonition 'know thyself'. See Plato, *Alcibiades*, 124, *Protagoras*, 343B. *by how much*. In the same proportion as. *term*. Limit. *intention*. Mind, understanding. *lines and veins*. Divisions that still show the original connections between sciences, unlike 'sections', which are cut away. Cf. Seneca, *Epist*. lxxxix. 1–17. *Cicero*. In *De oratore*, III. xix. 72. *repugnant to*. Inconsistent with. The *De Augmentis* (1623) adds that Copernicus' theory 'has now become prevalent' (*Works*, iv. 373). *destituted*. Abandoned. *Humanity*. The knowledge of man. *segregate*. Separately. *congregate*. Collectively.

206. *Conjugate*. United. *Civil*. Social. *constitute*. Establish. *emancipate*. Set free. *dignity... miseries*. These were favourite topics among Renaissance humanists, as in Pico della Mirandola's *De hominis Dignitate*, and Garzoni's *De miseria humana*. See Charles Trinkaus, *'In Our Image and Likeness': Humanity and Divinity in Italian Humanist Thought* (London, 1970). The *De*

Augmentis adds that while the miseries of humanity have been abundantly discussed, the triumphs have not, adding a brief specimen (*Works*, iv. 374–5). *adjuncts*. Subordinate aspects. *concordances*. Agreements. *Pars... corporis*. 'The Parts of Physiognomy, concerning the gesture or motion of the body.' *Intelligence*. Interchange of information. *Offices*. Obligations, duties. *Impression*. Effect on the mind. *Prenotion*. Foreknowledge. *Aristotle*. The *Physiognomica* falsely ascribed to Aristotle briefly discusses connections between bodily and mental characteristics. *Hippocrates*. See his *Regimen IV* on dreams and health. *superstitious... arts*. See e.g. the *De Humana Physiognomonia* (1586) of G. B. della Porta. *Physiognomy*. A science that attempted to classify and explain human character in the light of facial expression and bodily proportion. The *De Augmentis* adds more detail both on physiognomy and dreams (*Works*, iv. 376–7). *Dreams*. The science of dream interpretation, of which the most famous example in classical antiquity was the *Oneirocritica* of Artemidorus. *factures*. Shapes, structures. *art*. Human enquiry. '*As... eye*'. James I, *Basilicon Doron* (Edinburgh, 1599), 135. But cf. Horace, *Ars poetica*, 180–1. *subtile*. Crafty, cunning. *most*. The greatest part. *dissimulations*. Feignings, pretence. *direction*. Useful guide. *business*. Affairs involving negotiation.

207. *antistrophe*. Correspondence. *affects*. Dispositions; in the Latin, *temperamentum*. *Superstition*. Le Doeuff (ed. cit. 327) notes that this term often has for Bacon the Ciceronian sense of 'an excess of religious zeal'. *phrensies*. Frenzies, states of excitement. *pretendeth*. Intends, claims. *exhibit*. Administer a remedy. *exhilarate*. Cheer up. *corroborate*. Strengthen. *scruples*. Minute considerations. *regiment*. Course of treatment. *Pythagoreans*. Followers of Pythagoras (6th century BC) had various dietary prohibitions, particularly beans. *Manicheans*. Followers of Mani (AD 216–76) were forbidden to eat meat. *Mahomet*. Muslim law prohibits the consumption of pork or wine, and orders a month's fast in Ramadan. *exceed*. Go too far. *Ceremonial Law*. See Lev. 3: 17, 7: 22–7, 17: 11. *faith*. Christianity. *serene*. Untroubled. *macerations*. Mortifications. *figurative*. Metaphorical. *prescripts*. Rules. *conceive*. Imagine, think. *question*. Call in question. *compatible with*. Participating in. *accidentia animi*. (Unfavourable) symptoms of the mind. *Imagination*. Bacon shared the Renaissance belief in the danger of the imagination as a quasi-autonomous power, able to create and respond to images without rational control, although he also gave it a nobler, indeed divine status: see above, pp. 217–18. *imaginant*. One possessed by the imagination. See also *Works*, ii. 642 ff.; iii. 167. *pestilent airs*. Some Renaissance thinkers still held the ancient belief that contagious diseases were caused by *miasma*, bad air. Others held a germ-theory, in which diseases were transmitted by infected individuals. *sovereign*. Extremely effective.

208. *a Delian diver*. Socrates judged a work by Heraclitus so obscure that 'it needs a diver from Delos to get to the bottom of it': Diogenes Laertius, ii. 22; *Promus*, no. 851. *de communi vinculo*. Of the common tie. *domiciles*. In classical and Renaissance psychology the human faculties were supposedly located in specific parts of the brain or body. *occupate*. Occupy. *controverted*. Controversial. *Plato*. Inaccurate reference to *Timaeus*, 69–71. Plato actually lodged the concupiscent part of the soul not in the liver (the seat of the imagination, according to him) but between the diaphragm and the navel. *allowed*. Accepted. The *De Augmentis* adds: 'Neither again is that arrangement of the intellectual faculties (imagination, reason, and memory) according to the respective ventricles of the brain, destitute of error' (*Works*, iv. 378). *which*. That which. *four kinds*. Cf. Cicero, *Tusc. D.*, IV. xiii. 30, and *Advice to Rutland*, above, p. 69. *Cure*. Bacon means 'care' rather than 'cure' in the modern sense: cf. J. Boss, 'The Medical Philosophy of Francis Bacon (1561–1626)', *Medical Hypotheses*, 4 (1978), 208–20. *Cosmetic*. The art of beautifying the body. *Voluptuary*. Belonging to pleasure. *eruditus luxus*. 'Refined luxury' (*Annals*, xvi. 18). *subtility*. Intricacy, complexity. *higher*. To more general issues. *Microcosmus*. The idea that the body of man reproduced in miniature the contents and structure of the macrocosm, or universe, goes back to Democritus and gave rise to many speculative analogies in the Middle Ages and Renaissance. See e.g. Plato, *Timaeus*, 43 ff.; *Philebus*, 29; Diogenes Laertius, vii. 142–3; and Rudolph Allers, 'Microcosmus: From Anaximander to Paracelsus', *Traditio*, 2 (1944), 319–407. *Paracelsus*. He indeed tended to apply such analogies literally: see Brian Vickers, 'Analogy versus Identity: The Rejection of Occult Symbolism, 1580–1680', in Vickers (ed.), *Occult and Scientific Mentalities in the Renaissance* (Cambridge, 1984), 95–163. The *De Augmentis* extends this criticism: *Works*, iv. 379–80. *the alchemists*. In particular, followers of Paracelsus, such as Severinus and Croll. *extremely compounded*. Made up of the widest range of components. *dressings*. Seasonings. *mansion*. Dwelling-place.

209. *compounded*. Complex. *Purumque . . . ignem*. 'Pure and unmixed aetherial sense is left, mere air and fire' (Virgil, *Aeneid*, vi. 746–7). '*Motus . . . loco*'. 'The motion of things is rapid out of their place, quiet in their place.' See also Essay 11, 'Of Great Place', above, p. 361. *distemper*. Disturb, disorder. *Apollo*. As described by Ovid, *Met*. i. 518 ff. *curious*. Made with care. *by consequent*. In consequence. *masterpieces*. Actions of masterly ability. *successes*. Outcomes. *cause*. Case. *taxed*. Criticized. *mountebank*. Quack doctor. *Aesculapius*. Roman god of medicine. *Circe*. Daughter of Helios and Perse, a goddess powerful in magic. *Ipse . . . undas*. 'Jupiter himself with thunder hurled down to the Stygian floods Apollo's offspring [Aesculapius], the inventor of medicine and art' (Virgil, *Aeneid*, vii. 772–3). *Dives . . . lucos*. '[Circe's land] where with incessant song the wealthy daughter of the sun makes her inaccessible groves resound' (*Aeneid*, vii. 11). '*If . . . wise?*' Eccles. 2: 15.

210. *intend*. Occupy themselves with. *of*. Some of. *humanists*. Students of the humanities. *every*. Each. *seen*. Skilled; more learned in their other occupation than in medicine. The *De Augmentis* adds: 'Nor does this happen, in my opinion, because . . . [physicians] have so many sad and disgusting objects to deal with that they must needs withdraw their minds to other things for relief (for "he that is a man should not think anything that is human alien to him"),' but rather that the difference between competence and excellence in a doctor is not matched by their financial reward or reputation, and patients still 'depend on physicians with all their defects'. But these considerations 'tend rather to inculpate physicians than to excuse them. For instead of throwing away hope, they ought to put on more strength' (*Works*, iv. 382). *mediocrity*. Competence. *courses*. Customs, practices. *occasion*. Circumstance, motive. *observation*. 'The careful watching and noting of an object or phenomenon for the purpose of scientific investigation' (*SOED*). *faculty*. Power, influence. *spirit*. 'A subtle highly refined substance or fluid formerly supposed to permeate and animate the blood and chief organs of the body' (*SOED*). *shells*. 'Mussel-shells containing the artist's pigment, to be mixed with a medium before use' (*SOED*). *discern*. Distinguish between. *buffon*. Clown. *pantomimus*. A dancer who represented traditional themes in dumb show. *incomprehensions*. Failures to understand. The Latin reads 'acatalepsias'. *sense*. Sight. *Et . . . erunt*. 'And since diseases vary, we will vary our arts. There are a thousand forms of disease; there shall be a thousand methods of healing' (Ovid, *Remedia amoris* 525–6, adapted). *shadowed*. Portrayed. *son*. Aesculapius was traditionally the son of Apollo and Coronis. *miracles*. For Christ's miracles involving curing and resurrection see e.g. Matt. 8: 1–17, 28–32; 9: 2–8, 18–33; and *Religious Meditations*, above, p. 90. *Caesar*. Matt. 17: 24–7. (In fact, this miracle was wrought for the tax which was due to the Temple.)

211. *professed*. Taught (as an accepted system, not freshly researched). *iteration*. Repetition. *impulsions*. Impelling causes or occasions. *accidents*. Unfavourable symptoms. *preservations*. Directions for maintaining health. *enumerate . . . place*. List, but not arrange in order of importance. *Narrationes medicinales*. 'Medical histories.' *Hippocrates*. Greek physician (469–399 BC), around whose name a large corpus collected, including many inauthentic works. The *De Epidemiis* attributed to him contains case histories. *judged*. Decided. *foreign*. From another subject (law). *reserved*. Restricted. *intend*. Take pains. *Anatomia comparata*. 'Comparative anatomy.' *Anatomy*. The study of anatomy had recently been revolutionized by Andreas Vesalius (1514–64), whose *De humani corporis fabrica* (Basle, 1543) gave the first accurate account based on the dissection of human cadavers (Galen's anatomy had used monkeys). But Vesalius had not pursued research into internal medicine, a step Bacon advocates. *figures*. Shapes. *collocations*. Arrangements. *passages*. Bodily channels not visible to the naked eye. *nestling*. Locations, breeding-places of the humours, traditionally: choler in the gall-bladder, melancholy in the spleen, phlegm in the kidneys. *footsteps*. Traces, progress. *impressions*. Influences. *anatomies*. Dissections. *casual*. Depending on special cases. *facture*. Structure. *cause continent*. A technical term in Celsus (see below) for the immediate cause of a disease, persisting as long as the disease remains. *mechanic*. Mechanism. *alterative*. Designed to produce change. *accommodate*. Adjusted, brought into harmony. *palliate*. Alleviated. *familiar*. Common, widely used. *subtile*. Finely textured, minute.

212. *anatomia vivorum*. The dissection of living creatures. *Celsus*. Roman physician, *fl.* AD 14–27, whose *De re medica* was rediscovered by Guarino in the 15th century. *reproved*. Ibid. i, *Praef.* *slightly*. Dismissively. *purgaments*. Excretions. *devastations*. Ravages. *imposthumations*. Abscesses. *exulcerations*. Ulcers. *discontinuations*. Want of cohesion of parts. *consumptions*. Wasting of the body by disease. *dislocations*. Displacements of an organ. *obstructions*. Blockages (e.g. constipation). *repletions*. Eating or drinking to excess. *preternatural*. Outside the ordinary course of nature. *stones*. Gall- or bladder-stones. *carnosities*. Morbid fleshy growths. *historically*. Systematically, as in the 'natural histories' Bacon recommended. *contribution . . . experiences*. Collecting individual case-histories. *artificially*. According to the art of medicine. *Inquisitio . . . insanabilibus*. 'Further investigation into incurable diseases.' *Sulla*. Roman dictator who had thousands of his enemies executed in 81 BC: Plutarch, 'Life' of Sulla, 31. *triumvirs*. Octavius, Antony, and Lepidus, who listed 2,300 proscribed persons after the murder of Julius Caesar. *proscriptions*. Death sentences. *perfect*. Completed. *law of neglect*. Passing over negligently. *De . . . exteriore*. 'Of physical Euthanasia'; a gentle and easy death. *Euthanasia*. Bacon seems to have introduced this word in English (the *OED*'s first citation is from Joseph Hall in 1646). In the *De Augmentis* (*Works*, iv. 387) he adds: 'This part I call the inquiry concerning *outward Euthanasia*, or the easy dying of the body (to distinguish it from that Euthanasia which regards the preparation of the soul)'—i.e. the spiritual *ars moriendi*. *dolors*. Physical sufferings. *passage*. Transition, from life to death. *Augustus*. Suetonius, 'Life' of Augustus, 99. *Antoninus*. See Capitolinus' Life, 12. *kindly*. Natural. *Epicurus*. Greek philosopher (342–270 BC), whose philosophies of atomism and hedonism were extremely controversial. *ingurgitation*. Consumption. *epigram*. Diogenes Laertius, x. 15 f., in the Latin translation of Sambucus (Antwerp, 1566).

213. *Stygian*. Of the river Styx, one of the nine rivers of the underworld. *deplored*. Abandoned as hopeless. *attendances*. Ministrations. *Medicinae experimentales*. 'Experimental medicines': derived from experience and directed to specific ends. *receipts of propriety*. Treatments (prescriptions) appropriate to a particular disease. *magistralities*. Remedies 'devised by a physician for a particular case; not in the pharmacopoeia' (*SOED*). *quid pro quo*. This for that; an expression from pharmacy. *commanding*. Controlling. *treacle*. Theriaca, a remedy invented by Andromachus (Nero's physician), which compounded many antidotes to poisons. See Celsus, *De re medica*, v. 23. *mithridatum*. A similar compound, named after Mithridates VII, King of Pontus, who supposedly made himself immune to poison by drinking antidotes. *diascordium*. A medicine invented by the Italian scholar and physician Girolamo Fracastro or Fracastorius (1483–1533), from various herbs. See his *De contagione*, iii. 7. *severely*. Strictly. *religiously*. Scrupulously. *confections of sale*. Commercial, ready-made medicines. *for readiness*. For immediate, general use, not prepared for particular complaints. *purging*. Discharging waste matter from the body. *opening*. Freeing obstructions. *empirics*. Physicians who work on a trial-and-error basis, lacking theoretical knowledge. *happy*. Successful. *learned*. University-trained in

medical theory, often without practical experience. *holding*. Persisting with. *probations*. Tests; in old collections of receipts the words '*probatum est*' are often written against those found to be effective. *magistral*. Specific (*ad hoc*) remedy. *composition*. Make-up. *consuls*. Representatives of the ruling class or senate. *tribunes*. Representatives of the people. *best*. Here, as in so many areas, Bacon is calling for a union of practical and theoretical knowledge. *Imitatio ... medicinalibus*. 'Imitation of nature in baths and medicinal waters.' *find strange*. Wonder at. *mineral medicines*. Paracelsus, opposing the Galenic tradition of simples (vegetable-based remedies), advocated ('extolled') the use of mineral derivatives, and wrote a treatise on the benefit of mineral waters. *Medicinable*. Medicinal. *virtues*. Qualities. *tincture*. An imparted physical quality (an alchemical term). *vitriol*. Metallic sulphates, used in the arts or medicinally. *may*. Can. *reduced*. Concentrated, distilled. *compositions of art*. Man-made preparations. *temper*. Proportionate mixture. *commanded*. Controlled, regulated.

214. *Filum ... medicinarum*. 'The physician's clue, or the sequence of remedies.' *particular*. Detailed. *prescripts*. Prescriptions. *compendious*. Summary, quick-working. *order*. A coherent conception or plan. *pursuit*. Steady application. *pursuance*. Following out. *every day's devices*. Improvised, unplanned. *providence*. Foresight, prevision. *project*. Plan, design. *superstitious*. Zealously observed. *straight way*. Matt. 7: 13–14. *civil*. Relevant to the whole community. *effeminate*. Unmanly, over-refined. *reverence ... ourselves*. Cf. in *New Atlantis* the people of Bensalem's belief 'that the reverence of a man's self is, next to religion, the chiefest bridle of all vices' (above, p. 477). *decoration*. Cosmetics. *nor handsome ... to please*. Spedding suggested that the proper wording should be 'nor wholesome to use, nor handsome to please'. *largely*. Broadly. *patience*. Endurance. *hardness*. Hardiness. *practices*. Habitual actions. *tumblers*. Acrobats. *savages*. Popularly supposed to be able to endure great pain. The *De Augmentis* adds 'the stupendous strength of maniacs' (*Works*, iv. 394). *dive ... containing respiration*. The ability of divers to hold their breath for a long time. *are down*. Have been discontinued.

215. *mediocrity ... use*. Competence in such things is sufficient. *Arts*. The *De Augmentis* specifies these, 'according to the senses themselves': the pleasures of the eye (painting, architecture, gardens, clothes, jewellery) and the pleasure of the ears (music) constitute the 'two purest' senses. Arts gratifying the other senses are 'more allied to luxury' (*Works*, iv. 395). *in state*. Established, settled. *liberal*. The liberal arts are those 'worthy of a free man' (opposed to a slave); general intellectual culture. *declination*. Decline. *doubt*. Fear. *practices joculary*. Juggling and conjuring. *tabernacle*. Dwelling-place. *Mind*. Bacon uses the terms 'mind' and 'soul' interchangeably. *native or adventive*. Present at birth, or subsequently introduced 'from outside'. *in a maze ... in a way*. In a confused, not in a clear-cut or direct fashion. *really*. In actuality. *producat*. 'Let the earth bring forth' (Gen. 1: 11 ff.), the divine act of creation. *inspired*. As with 'inspiration' below, in the literal sense of 'breathe into', as the spirit or divine afflatus entered the human body. *Divination*. The mantic arts were widely developed in antiquity, and accepted as part of state religion, especially in Rome. *Fascination*. Sorcery, the casting of a spell. *argument*. Deduction. *concluding upon*. Inferring from. *tokens*. Presages, omens. *presention*. Presentiment.

216. *inducement*. Introduction. *experimental*. Derived from experience. *Chaldean*. The Persians were traditionally credited with having invented astrology. *Astronomer*. In the Renaissance 'astronomy' and 'astrology' were often used interchangeably. *conjunctions*. The apparent proximity of two planets. *aspects*. The ways in which the planets 'look upon' each other from their relative positions, as viewed from the earth, to which astrology attached varying significances, good or bad. *accidents*. Symptoms. '*O... invenerit!*' 'A city for sale, and doomed to speedy destruction, if it finds a purchaser' (Sallust, *Bell. Jug*. XXXV. 10). *impertinent*. Irrelevant. *referred over*. Deferred, to the 'particular knowledges' in which artificial divination is found (astronomy, medicine, politics, etc.). *primitive*. Original, primordial. *influxion*. Inflowing, influence. *prenotion*. Foreknowledge. *extasies*. Trances, states of nervous excitement in which the soul was thought to be liberated from the body. *observances*. Keeping a way of life (e.g. meditation, an ascetic diet). *consist in*. Retain its essential features. *regiment*. Mode of life. *retiring... within itself*. As in meditation and mystical experiences. *save*. Except. *elevation*. Excitement. *noted*. Denoted. *fury*. Inspired frenzy; see Ovid, *Met*. ii. 640. *intensive*. Strenuously directed upon. See also *Sylva Sylvarum*, Century X (*Works*, ii. 640–72). *imaginant*. One possessed by the imagination. *spake*. See above, p. 207. *pretended*. Spurious. *much one with*. Equal to. Cf. Paracelsus, *De vi imaginativa*. *miracle-working faith*. Cf. Matt. 17: 20. *secret*. Unobservable. *contagion*. Communication of disease by direct or indirect contact.

217. *civil*. Popular, widely accepted. *fortify*. Render more effective. *material*. Relevant. *palliation*. Excuse. *Ceremonial*. Magic performed in a ritual, often by summoning demons or spirits. See e.g. H. C. Agrippa, *De occulta philosophia* (1533), bk. III, or Marlowe, *Dr Faustus*. *Characters*. Magical signs (both letters and numbers). *sacramental*. Bound by an oath or solemn obligation. *fix*. Concentrate. *cogitations*. Thoughts. *raise*. Intensify. *opposing*. Repugnant. '*In... tuum*'. 'In the sweat of thy brow shalt thou eat thy bread' (Gen. 3: 19): God's curse on Adam and Eve when expelling them from the Garden of Eden, where all things flourished without the need for labour. *propound*. Hold up as an aim. *Faculties*. In the *De Augmentis* (v. 1) Bacon adds a prefatory paragraph: 'The doctrine concerning the Intellect... and the doctrine concerning the Will of man, are as it were twins by birth. For purity of illumination and freedom of will began and fell together; and nowhere in the universal nature of things is there so intimate a sympathy as between truth and goodness. The more should learned men be ashamed, if in knowledge they be as the winged angels, but in their desires as crawling serpents; carrying about with them minds like a mirror indeed, but a mirror polluted and false' (*Works*, iv. 405). In the succeeding discussion Bacon links the distinction between understanding and appetite more clearly to that between logic and ethics: 'Logic discourses of the Understanding and

Reason; Ethic of the Will, Appetite, and Affections: the one produces determinations, the other actions', while 'imagination is as a common instrument to both,—both reason and will' (*Works*, iv. 405–6). *Appetite.* Desire or inclination. *Affection.* Passion. *Position.* Laying down a law. *nuncius.* Messenger, ambassador. *sendeth over.* Reports, transmits. *Voluntary Motion.* Actions governed by the will. *Janus.* Roman god with a double face. '*Quales . . . sororum*'. 'Such as sisters' faces should be' (Ovid, *Met.* ii. 14). *usurpeth.* Appropriates. *Aristotle.* In *Politics*, i. 5, 1254b4 ff., freely adapted by Bacon.

218. *raise.* Elevate. The *De Augmentis* adds the important rider that while 'in matters of faith and religion our imagination raises itself above our reason', this is not because 'divine illumination resides in the imagination; its seat being rather in the very citadel of the mind and understanding; but that the divine grace uses the motions of the imagination as an instrument of illumination, just as it uses the motions of the will as an instrument of virtue' (*Works*, iv. 406). *similitudes.* Comparisons. *types.* Symbols, especially 'a foreshadowing in the Old Testament of a person or event of the Christian dispensation' (*OED*). *impression.* Pressure, influence. *paint.* By using the rhetorical 'colours', expressive devices. *recommendation.* Communication, report. *Reason . . . Imagination.* The Latin version makes the point more clearly: 'for when by arts of speech men's minds are soothed, inflamed, and carried hither and thither, it is all done by stimulating the imagination till it becomes ungovernable, and not only sets reason at nought, but offers violence to it, partly by blinding, partly by incensing it' (*Works*, iv. 406). *former division.* Of the mental faculties, into Memory, Imagination, and Reason. *his.* Its. *mentioned.* See above, pp. 207–8. *Insinuative.* Winding itself into. *rational.* 'Which regards logic', the *De Augmentis* substitutes (*Works*, iv. 407). *subtility.* Confusing minuteness. *spinosity.* Having thorns. '*pabulum animi*'. 'Food for the mind' (Cicero, *Acad. Quaest.* ii. 41.128). '*ad . . . carnium*'. 'To the flesh-pots [of Egypt]' (Num. 11: 4–6). *manna.* Food from heaven miraculously provided for the Israelites in the wilderness. *comfortable.* Strengthening. *taste well.* Enjoy. '*lumen siccum*'. 'Dry light.' *offend.* Wound or hurt physically. *Aristotle.* In *De anima*, iii. 8, 432a1 3. *a nearer shoot.* An arrow nearer the target.

219. *Intellectual.* 'Logical', the *De Augmentis* calls them. *invent.* Find out. *Custody.* Storing. *Tradition.* The communication of information. *Invention is of two kinds.* Bacon contrasts rhetorical with scientific *inventio*: the former merely 'finds' what is already known, recorded in notebooks or other written sources; the latter discovers new knowledge. *defunct.* Deceased person. *fetch.* Purchase. *pretend.* Claim; intend. '*cuique . . . credendum*'. 'To believe every artist in his own art.' For a loosely related discussion see Aristotle, *Prior Analytics* i. 30, 46a17, 24. *Celsus.* Preface to *De re medica*, but describing the attitudes of the Empirics, to whom (as a Rationalist) Celsus was opposed (Spedding, *Works*, i. 617). *after.* Afterwards. *discoursed.* Written down; invented. *Theaetetus.* An incorrect reference, revised in the *De Augmentis* to read 'Plato more than once intimates' (*Works*, iv. 408). Ellis (ibid. i. 617) and

Wright (p. 304) proposed *Philebus*, 17. Le Doeuff (pp. 338–9) suggests a debt to Averroës, who distinguishes the most general principles (furthest from action) from the least general (closest to action), and interposes a third category, cogitation, the empirical power which adapts generalities so that they can be realized in matter. Averroës takes his examples from politics and medicine. See R. Lerner (ed.), *Averroës on Plato's Republic* (Ithaca, NY, 1974), 21–2, 46. *arts-man*. One skilled in the liberal arts. *inventions and originals*. Most myths concerning the discovery of inventions useful to man attribute them to accident rather than to human design ('Reason') or purposeful experiment ('Art'). *Dictamnum ... sagittae*. 'With a mother's care [Venus] plucks from Cretan Ida a dittany stalk, clothed with downy leaves and purple flower; not unknown is that herb to wild goats, when winged arrows have lodged in their flank' (Virgil, *Aeneid*, xii. 412–15).

220. *Omnigenumque ... Minervam*. 'Her [Cleopatra's] monstrous gods of every form, and barking Anubis, opposed to Neptune, Venus and Minerva, are wielding their weapons': ibid. viii. 698–9, contrasting the Egyptians' theriomorphic with the Romans' anthropomorphic gods. *Prometheus*. Traditionally the inventor of fire and of all arts useful to mankind. *intelligence*. Communication. Bacon's knowledge of how natives in the West Indies make fire comes from Joseph de Acosta, *Naturall and Moral Historie of the East and West Indies* (1590; Engl. tr. 1604), iii. 2. As Wright (p. 305) notes, the English edition mistranslated the Spanish *palos* as 'stones' instead of 'sticks'. *European*. One of the earliest uses of this word; the *OED*'s first citation is from Richard Knowles, *History of the Turks* (1603). *occasion*. Incidental cause. *beholden*. Indebted. Bacon's example of the wild goat for surgery derives from the Virgil passage just quoted. The example of the ibis, or black stork, which can give itself enemas with sea water, comes from Plutarch, 'De sollertia animalium,' *Moralia*, 967A–974A, or from Pliny's *Nat. Hist.* viii. 27 (both instances being mentioned by Montaigne in *Essais*, ii. 12, 'Apologie de Raimond Sebond' (Pléiade edn. 440). The accidental discovery of gunpowder by a German monk (Bertholdus Schwartz, according to one tradition, an alchemist who kept powder of sulphur, for medicinal use, in a pot, into which one day a spark fell, causing an explosion which lifted the lid) comes from Polydore Virgil, *De Rerum Inventoribus*, 2nd edn. (Basle, 1521). *chance*. Cf. Aristotle, *Eth. Nic.* vi. 4, 1140a10 ff., a discussion of art or applied science as a rational activity 'contriving and considering how something ... may come into being', which nevertheless recognizes the element of chance: 'in a sense chance and art are concerned with the same objects; as Agathon says, "art loves chance and chance loves art"'. *other*. Different. *Ut ... Paulatim*. 'That practice, with meditation, might by degrees hammer out the arts' (Virgil, *Georgics*, i. 133). *ure*. Use. *intending*. Striving. '*Usus ... vincit*'. 'Practice applied to one thing often accomplishes more than nature and art' (Cicero, *Pro Balbo*, 20, 45). *Labor ... egestas*. 'Incessant labour and want, in hardships urgent, surmounted every obstacle' (Virgil, *Georgics*, i. 145–6). '*Quis ... χαῖρε?*' 'Who taught the parrot to say "Hello"?' (Persius, *Satires*, Prol. 8). The following instances of the raven and the ant come ultimately from Plutarch ('De Sollertia') or from Pliny, *Nat. Hist.* x. 43 and xi. 30; but both also figure in the same essay by Montaigne. *drowth.* Drought. *extundere.* Hammer out. *importeth.* Denotes. *paulatim.* By degrees.

221. *induction.* The art of propounding general laws from the observation of particular instances. *Plato.* Le Doeuff (p. 339) cites Cicero's remark (*Topics*, x. 42) that the Greek term *epagoge* should be rendered by Latin 'inductio', and that 'this induction is used several times by Socrates in his dialogues'. *pretended.* Professed, claimed. *vicious.* Faulty, unsound. *traduced.* Slandered, misrepresented. '*Aërei ... dona*'. 'Heaven's gift, the honey from the skies' (Virgil, *Georgics*, iv. 1). *act.* Enact, bring about. *instance contradictory.* A 'negative' instance, one that disproves a provisional conclusion: see the more detailed discussion in the *Novum Organum*, *Works*, iv. 56, 97–8, 147–9. Le Doeuff (p. 340) cites Sextus Empiricus, *Outlines of Pyrrhonism*, II. xv. 204, for the argument that an incomplete enumeration renders induction invalid, since 'some of the particular [instances] omitted in the induction may contravene the universal'. *assure.* Guarantee. *of a side.* On one side. *rested upon.* Stopped at; chosen. *Issay.* Jesse, in 1 Sam. 16: 1–13. *failed of.* Overlooked, neglected. *dogmaticals.* Systems of dogma. *lictores.* 'Whifflers', servants attending Roman magistrates whose duty it was to push the crowd aside. *viatores.* 'Sergeants', who summoned persons before the magistrates. *ad ... turbam.* 'For the summoning of the crowd.' *seducement.* Seduction. *become ... child.* Matt. 18: 3; Luke 18: 17; Mark 10: 5. *attending.* Applying oneself to. *cannot be deduced ... of nature.* In studying the natural world syllogisms are of no use, since they deal with words, not—as the *De Augmentis* adds—with 'things natural, which participate of matter' (*Works*, iv. 411). *touch.* Testing. *reduction.* A simplification, as in the form of a syllogism (a premiss, a middle term, and a conclusion arising out of them). *moralities.* Moral philosophy. *apply.* Adapt. *natural philosophy.* The *De Augmentis* deletes this exception, taking the more severe line that 'in Physics, where the point is not to master an adversary in argument, but to command nature in operation, truth slips wholly out of our hands, because the subtlety of nature is so much greater than the subtlety of words; so that, syllogism failing, the aid of induction (I mean the true and reformed induction) is wanted everywhere, as well for the more general principles as for intermediate propositions' (*Works*, iv. 411). *argument.* A purely verbal construct, having no necessary consequences for reality. *satisfactory.* Merely plausible, superficially convincing. *quae ... est.* 'Which procures assent but can do no work.' The whole thrust of Bacon's proposed reforms was to encourage a natural philosophy that would be 'operativa', productive of works, to the benefit of mankind.

222. *Words ... things.* Bacon alludes to Aristotle's *De interpretatione*, i. 1, which defines words as 'symbols of affections in the soul' (or 'mental impressions', in most modern translations), these 'affections in the soul' being in turn 'likenesses of actual things' (16a3–8). See also above, p. 230 for another (abbreviated) reference to this text. *grossly.* Clumsily. *variably.* Unsystematically. *first digestion.* Initial operation. *Sceptics.* Philosophers

who believed it impossible to know whether we know anything or not. The *De Augmentis* is more explicit in condemning scepticism for doubting the evidence of the senses: *Works*, iv. 412. *Academics*. The New Academy, under Arcesilaus and Carneades (3rd and 2nd centuries BC), taught the doctrine of *acatalepsia*, that nothing can be known with certainty. *Socrates*. His 'irony' took the form of disclaiming knowledge of a topic in order to make his interlocutors expose their knowledge or lack of it. '*Scientiam ... simulavit*'. 'An affectation of knowledge under pretence of ignorance': cf. Cicero, *Acad. Quaest.* ii. 5.15. *disable*. Disparage. *Tiberius*. Tacitus, *Annals*, i. 7.11. *acatalepsia*. A term used by Sceptic philosophers (attacking the Stoics) to argue that reality is 'non-apprehensible'. See Sextus Empiricus, *Outlines of Pyrrhonism*, I. xxv. 200, and Cicero, *Acad. Quaest.* ii. 6.8. *doubt*. Suspect. *copie*. Copiousness, fertility. *variable*. Inconstant; or, 'varied in expression, using rhetorical devices'. *progresses*. Long journeys for pleasure, without a specific goal. *both*. The *De Augmentis* adds 'but much more among the Sceptics' (*Works*, iv. 412). *subtilty*. Wright (p. 306) suggests that the true reading should be 'in *simplicity* and integrity', since the Latin reads 'simpliciter et integre'. *charged ... upon*. Attributed to. *cavillations*. Quibbling objections. *certify*. Give information of. *immediately*. Directly, without anything intervening. *instruments*. Bacon was aware of the growing importance of scientific instruments for observing and measuring natural phenomena. In the *De Augmentis* he adds that the whole aim of his existence is 'to make the mind of man by help of art a match for the nature of things' (*Works*, iv. 412). *compass*. Pair of compasses. *Experientia ... naturae*. 'Learned experience and interpretation of nature.' *digested*. Divided, organized. *Experientia literata*. 'Educated experiment', i.e. which can read and write, and so be directed to specific ends, as opposed to random enquiry. *Interpretatio Naturae*. Whereas the previous approach moves from one experiment to another, Bacon's 'interpretation of nature' constitutes a new scientific method, using induction to generalize axioms (scientific laws) from observation and experiment, which axioms will in their turn generate new experiments. In the *De Augmentis* Bacon explicitly identifies this approach with the one he expounds in the *Novum Organum*. *rudiment*. An initial stage. *that*. That which.

223. *resummon*. Recollect. Cf. Bacon's critique of 'invention' as finding words, not things, above, p. 219. *Suggestion*. The process by which an idea brings to the mind another idea by association or natural connection. *schools*. Universities. *Chase*. Hunt. *so as*. Provided that. *discerned*. Distinguished. *artificial erudition*. Systematic learning. *Aristotle*. In *Sophistici Elenchi* ('Sophistical Refutations') xxxiv. 185a2 ff. *should*. Would. *bespoken*. Has orders for. *customed*. Patronized. *that ... store*. Matt. 13: 52. *Places*. Commonplaces or topics that frequently recur, and which should be collected, rhetoricians recommended, as forming 'the secret places where arguments reside, and from which they must be drawn forth' (Quintilian, V. x. 20). *equity*. General principles of justice (as against specific provisions of the law). See Aristotle, *Rhetoric*, I. xv, 1375a26 ff. *testimony*. Evidence given in a court of law. *broken unto*. Trained. *premeditate*. Thought out in advance. '*in*

thesi'. 'In propositions' (Cicero, *De oratore*, II. xxxii. 137–xxxiv. 147; *Orator*, xiv. 44–6). *particular*. Individual case. *put to*. Attach. *Demosthenes*. A collection of 56 *Prooemia* (stock openings to forensic and political speeches) perhaps written by Demosthenes for his own use between 349 and 346 BC, and first published in 1572. Rhetoric taught that the prooemium should be used by the orator to capture the audience's goodwill (*captatio benevolentiae*) by presenting himself in a good light: e.g. Cicero, *De inventione*, I. xvi. 20 ff. *shears*. Scissors.

224. *handling*. Discussion: see the section on rhetoric, above, pp. 240–1. *which I term*. Bacon seems to be drawing attention to his neologism. *Places*. The *loci* (headings, categories) in which knowledge has been stored. *excite*. Move, stimulate. *probably*. Plausibly. *minister*. Suggest, prompt. *within ourselves*. In the internal debate common in philosophical thinking. *ap-prompt*. Prompt, stimulate. *Plato*. *Meno*, 80D; also Aristotle, *Posterior Analytics*, i. 1, 71a27 ff. *before*. In front of. *revolve*. Consider, think over. *Topics*. The *topoi*, or *loci communes*, were general heads of argument, which could be generated by using the categories laid down in the *Topica* of Aristotle or Cicero. *to*. Of. *inartificial*. Not suited to the methodical exposition used in an 'ars' or practical treatise. *variable*. Varying; hard to systematize. *subtile*. Over-ingenious, too precise. *receive*. Admit. '*Ars . . . inventis*'. 'Every act of discovery advances the art of discovery.' Bacon reiterates this important principle in his mature philosophical works, e.g. in the last sentence of the *Novum Organum*, bk. I (*Works*, iv. 115). *going of a way*. Moving along a path or road. *arts of Judgment*. Logical proof. *which*. Referring to 'arts'. *which*. Referring to 'Judgment'. *hath a coincidence*. Resembles, overlaps.

225. *the same . . . judgeth*. The same mental process is involved in both discovery and evaluation. *all one*. The same. *the sense*. Sense-experience (which is also direct and coherent). *otherwise*. It is different. *mean*. The syllogism moves from an initial premiss (assumed to be universally valid) through a middle term to a conclusion. Bacon's objection is that the process of abstraction needed to generalize these terms breaks up the mental process. *exciting*. Initiating, setting the process in motion. *spoken*. See above, p. 222. *agreeable*. Suitable—an ironic judgement on the weaknesses of the human mind. *Aristotle*. In *De motu animalium*, ii, 669b7 ff.; iii, 69a12 ff. *quiescent*. Unmoving. *Atlas*. Homer, *Odyssey*, i. 52–4; Virgil, *Aeneid*, iv. 481–2. *bare up*. Supported. *axle-tree*. The fixed bar on the ends of which the wheels of a carriage revolve. *conversion*. Rotation. *within*. In their minds. *peril of falling*. Vertigo. *reduction*. 'The process of establishing the validity of a syllogism by showing that the contradictory of its conclusion is inconsistent with its premisses' (*SOED*). *agreed*. Admitted. *elected*. Chosen. '*Probation ostensive*'. An argument that proves either the same conclusion or one which implies it. *per incommodum*. 'By absurdity', or *reductio ad impossibile*: an argument proving indirectly not that the original conclusion is true, but that it cannot be false. *pressing*. Forcing. *number . . . to be*. That is, greater or less, depending on the intermediate stages. *degrees*. Stages of an

argument. *direction*. Instruction. *consequence*. Logical sequence. *incident*. Relevant, pertinent. *Analytics*. The logical treatises of Aristotle, constituting the third part of the *Organon*, which discuss the process of demonstration or proof. *expedite*. Speedy, unencumbered. *sophisms*. Ingenious but false arguments, sometimes deliberately deceptive. *illaqueations*. Entangling in arguments. *redargutions*. Refutations.

226. *Elenches*. Cautions (in logic: the refutation of a proposition that has been syllogistically defended). *Seneca*. *Epist*. xlv. 8. *besides his answer*. Unable to reply. *abuse*. Deceive. *Aristotle*. In *Sophistici Elenchi*. *persons*. Also implying *personae*, 'masks' or characters in a fictitious work, especially a dialogue or drama. Plato's treatment of the Sophists often makes them comic characters, butts for Socrates' arguments. *Sophists*. The *De Augmentis* specifies Gorgias, Hippias, Protagoras, Euthydemus, and the rest (*Works*, iv. 430). *Socrates*. See Plato, *Theaetetus*, 150B ff. *infirm*. Render questionable. *fallace*. Fallacy. *the*. That the. *caption*. Deception, specious objections. *difference*. Distinction. *be*. Is. *sophisters*. Sophists. *in the turn*. In tricks and evasions. *latitude*. Scope. *are laboured*. Have been worked on. *adjuncts of essences*. Substances (essences) and their attributes: common topics in Aristotelian and scholastic philosophy. Bacon, like many Renaissance humanists, rejected such discussions as sterile, arguing here that the endless dispute over defining the key terms in fact should serve as a warning against the dangers of ambiguity. *intervene*. Interfere. *and the like*. The *De Augmentis* adds: 'Much, Little, . . . Active, Passive, Motion, Rest, Entity, Non-entity, and the like' (*Works*, iv. 430). *tribes*. Groups. *predicaments*. The Latin term (*predicamenta*) for the Greek 'categories', Aristotle's enumeration of 'existences', all things capable of being named, according to the *summa genera* or most extensive classes in which things could be distributed. They were ten in number: substance, quantity, quality, relation, action, passion, time, place, position, and habit. *impression*. Impact, pressure. *Rhetoric*. In a later section, above, pp. 238–9. *fallacies*. These are the inherent weaknesses in the human mind which Bacon subsequently labelled the 'Idols' (from the Greek *eidolon*, in the sense of a mental image, a false and misleading appearance). This section is expanded in the *De Augmentis* (*Works*, iv. 431–4), but for the full development of Bacon's analysis see *Novum Organum*, i. 39–68: *Works*, iv. 53–69.

227. *appertaineth*. Pertains, is suited. *rectify*. Correct. *clear . . . glass*. A smooth-surfaced, accurately reflecting mirror. *reflect*. Be reflected. *enchanted*. Subjected to a magic spell (usually thought to be harmful). *imposture*. Fraudulent deception, illusion. *delivered*. Freed, rescued. *reduced*. Corrected, restored to its proper condition. *false appearances . . . general nature of the mind*. These are the 'Idols of the Tribe': see *Novum Organum*, i. 41, 45–52. *consonant*. In agreement, accordant. *privative*. Absent (instance). *hitting*. Agreeing, coinciding. *countervails*. Outweighs. *Diagoras*. In Cicero, *De natura deorum*, III. xxxv. 89; Diogenes Laertius, vi. 59. *scaped*. Escaped. *Advise*. Consider. *invocate*. Pray to. *perfect circles*. The truth of Bacon's comment is borne out by Kepler's admission (in his *Astronomia nova*,

1609) of the great difficulty he had in correlating Tycho Brahe's observations of the movements of Mars (traditionally, all planets revolved on circular orbits) with the only geometrical figure that would account for them, an ellipse with two centres. Bacon, however, remained ignorant of Kepler's work. *eccentrics*. Some astronomers, disturbed by the many adjustments of the planetary model needed to account for observed phenomena (such as an 'eccentric', a circle or orb not having the earth precisely in its centre), wished to abolish them altogether. Fracastaro, in *Homocentricorum sive de stellis liber* (1538), proposed homocentric spheres, with the earth as centre. *monodica*. (Properly, *monadica*) singular, unique. See *Novum Organum*, i. 45 and ii. 28 (*Works*, i. 165, 281–2; iv. 55–6, 168). '*sui juris*'. Like to nothing but themselves. *conjugates*. Things related to, and so resembling each other. *Fire*. Empedocles (and after him Heraclitus) designated fire not only one of the four elements, but the most important. *keep square with*. Correspond to. *opened*. Revealed. *similitude*. The supposed resemblance between the arts and actions of man and the operations of nature. *communis mensura*. '[Man] the measure of all things': the doctrine of the Sophist Protagoras, according to Plato, *Theaetetus*, 152A, and Aristotle, *Met*. x. 1, 1053a35 ff. Bacon uses the phrase to mean man as the 'model' for all things. *Anthropomorphites*. A monastic sect, 4th century AD onwards, which took the biblical statement 'God created man in his own image and likeness' (Gen. 1: 26–7) literally, assigning human bodily parts to God. *gross*. Ignorant. *Epicurus*. See Cicero, *De natura deorum*, II. xvii. 45–6. *answerable*. Corresponding. *Velleius*. Ibid. I. ix. 22. *should have*. Did. *Aedilis*. Roman magistrate in charge of buildings, who adorned them on feast days with lights and statues. *that should have*. Whose job it was to. *work master*. God as maker of the world (the *demiourgos* in Platonic philosophy).

228. *works*. Ornamental structures. *orders*. Patterns. *frets*. Architectural ornaments, carved on roofs or ceilings, consisting of intersecting lines in relief. *posture*. Shape. *again*. Further. *nature and custom*. The 'Idols of the Cave', distorted perceptions deriving from the individual's make-up and habits. See *Novum Organum*, i. 42, 53–8. *Plato*. *Republic*, 7. 514A ff. *continued*. Kept continually. *grot*. Grotto. *absurd*. Incongruous, irrational. *persons*. Bodies. *included*. Shut up. *complexions*. Constitutions, dispositions. *recalled to examination*. Corrected by the continual contemplation of nature at large. *ran briefly over*. Discussed: see above, p. 146. *Elenchi . . . adventitiis*. 'Important refutations, or concerning the idols of the human spirit, innate and imposed'. *by words*. The 'Idols of the Market-place', distortions inherent in language: see *Novum Organum*, i. 43, 59–60, which adds a fourth category, 'Idols of the Theatre', false appearances derived 'from the various dogmas of philosophies' (i. 43, 61–5). *sort*. Class of persons. '*Loquendum . . . sapientes*'. 'To speak as the common people do, to think as the wise men do' (Aristotle, *Topics*, II. ii, 110a14–22). *Tartar's bow*. Tartar horsemen could shoot arrows behind them, while riding away from pursuers. See Plato, *Laches*, 191A ff. *pervert*. Turn upside down. *almost*. For the most part. *caution*. Precaution. *import*. Affect, have a bearing on. *De . . . demonstrationum*. 'Concerning the analogy of proofs.' *touched*. Discussed.

229. *by congruity*. Analogical reasoning, involving proportional relations. *a notoribus*. 'From things better known': Aristotle, *Prior Analytics*, ii. 5, 57b18 ff.; *Posterior Analytics*, ii. 13, 97a5 ff. *every*. Each. *facility*. Easiness, laziness. *remiss*. Loose, not strict. *character*. Script, alphabet. *For*. As for (as often in this section). *conjugation*. Connection, relation. *place*. See above, pp. 230–1. *disposition*. Distribution. *collocation*. Arrangement. *digest of commonplaces*. Observations or quotations arranged by topics. *retardation*. Delaying the progress or development. *relaxation*. Diminution, weakening. *forward*. Keen, eager. *pregnant*. Teeming with ideas. *entry*. Keeping a notebook. *essence*. Essential importance. *copie*. Abundance. *contracteth*. Focuses, concentrates. *art*. The *ars memoriae* was developed in classical rhetoric to help the orator learn his speech by heart. See *Rhetorica ad Herennium*, III. xvi. 28–xxiv. 40, and *Novum Organum*, ii. 26 (*Works*, iv. 161–3). In the Renaissance it was expanded indiscriminately, and even had magical properties ascribed to it. *points*. Degrees, heights. *dexterous*. Handy, suitable.

230. *cavil*. Frivolous objection. *funambuloes*. Rope-walkers. *baladines*. Ballet-dancers. *worthiness*. Worth. *intentions*. Concepts, directions. *Prenotion*. 'A notion of something prior to actual knowledge of it' (*OED*). *Emblem*. A picture expressing a moral or allegory, sometimes accompanied by a poem or motto. *dischargeth*. Removes, frees. *practique*. Practice. *distinguish*. Assert distinctly, decide (as announced at the beginning of bk. II, above, p. 174). *transitive*. Concerned with transmitting. *Tradition*. Communication of knowledge. *Aristotle*. In *De interpretatione*, i. 1, 16a3–8. Bacon makes another abbreviated reference (as above, p. 222) to Aristotle's four-part scheme, in which 'words are symbols of affections in the soul, and written marks symbols of spoken sounds'. The 'affections of the soul' (or 'mental impressions'), in turn, are 'likenesses of actual things'. *differences*. Forms of differentiation, sufficient to explain the variety of notions. *express*. To convey the cogitations of one man to another. *commerce*. Dealings, conversation. *divers*. Some people. *use*. Practice. *the high Levant*. The Far East. Bacon's probable source was Acosta's *Naturall Historie of the Indies*, which (mistakenly) reported that the characters in Chinese 'signifie no partes of distinctions, as ours do, but are figures and representations of things, as of the Sunne, of fire, of a man, of the sea, and of other things', a shared script 'understood of them all, though the languages' spoken 'are many and very different.... So as things being of themselves innumerable, the letters likewise or figures which the Chinese use to signifie them by, are in a maner infinite.' *Characters Real*. Non-iconic signs (such as the components of a nautical flag code) that convey the meaning of a word or concept, but have no phonetic or analogical relation with the thing represented. *radical*. Having an etymological stem or root.

231. *De... rerum*. 'Concerning the notes of things.' *note*. Mark, sign. *ad placitum*. Arbitrary. *contract*. Convention, agreement. In *De interpretatione*, i. 2, 16a19 ff., Aristotle said that 'A name [or word] is a spoken sign significant by convention', as in the accepted usage within a speech community.

Hieroglyphics. Sacred characters used in ancient Egyptian picture-writing. *Gestures.* The fourth stage in rhetorical composition, *actio* or *pronuntiatio*, taught a repertoire of gestures designed to make the oration more effective. In the increasing systematization of rhetoric during the Renaissance, ever-larger manuals appeared, and some writers claimed that gesture constituted a language in itself. *continued.* Sustained, uninterrupted. *impresses.* Crests or heraldic devices, uniting a visual image and motto. *emblems.* Pictorial devices conveying a moral meaning. *transitory.* Ephemeral, not recorded. *abide not.* Have no permanent record. *Periander.* In Aristotle, *Politics*, iii. 8, 1284a26 ff.; Herodotus, *Hist.* v. 92; Livy, i. 54. *feigning.* Fiction, fantasy. *names from reason.* The idea that words represent concepts by their innate form (rather than by social convention), which Bacon rejects, is discussed and refuted in Plato's *Cratylus*, but was influential in the magical arts. *intendment.* Intention. *by reason.* Because. *reverent.* Deserving respect. *mint.* Shaping place. *words . . . tokens . . . conceits.* Alluding again to Aristotle's semantic theory in *De interpretatione*: above, p. 222, 230. *moneys.* Coins. *fit.* Proper, fitting. *another kind.* That the Chinese used paper money had been known since the travels of Marco Polo. *reintegrate.* Restore. *in.* To. *his fault.* Adam, breaking God's prohibition on eating from the tree of knowledge. *first general curse.* 'In the sweat of thy brow shalt thou eat thy bread' (Gen. 3: 19). *second general curse.* The confusion of tongues: Gen. 11: 6–8. *vulgar.* Vernacular. *turned.* Changed. *learned tongues.* Languages learned from books; Greek, Latin. *natures.* Kinds. *popular.* Used by the general public. *as well.* Both.

232. *sparsim.* Dispersedly. *reduced.* Organized, collected. *Accidents.* Accessory qualities. *measure.* Metre. *elevation.* Intonation. *accent.* 'Prominence given to syllable by stress or (in some languages) by pitch' (*SOED*). *curious.* Careful to excess, over-precise. *ancient measures.* Classical metres, based on quantity rather than accent, and therefore unsuitable to modern languages, as the *De Augmentis* observes more pointedly: *Works*, iv. 443. *Coenae . . . cocis.* 'The dinner is for eating, and my wish is | That guests and not that cooks should like the dishes' (Martial, *Epigrams*, ix. 83). *expressing.* Representing, imitating. *servile expressing antiquity.* Slavish imitation of classical models; likely to violate the Renaissance concept of decorum, the appropriateness of style and genre to subject-matter. '*Quod . . . novum*'. 'There is nothing more new than an old thing that has ceased to fit.' *Ciphers.* The use of ciphered language in secret military or political communication was occasionally practised in the Renaissance. Bacon's main source was Blaise de Vigenère, *Traicté des Chiffres, ou secrètes manières d'escrire* (1587). *nulls . . . non-significants.* Dummy figures occurring in a cipher language but having no significance. (Bacon introduced both words into English.) *infolding.* Encoding. *preferred.* Recommended. '*omnia per omnia*'. Anything under cover of anything. *quintuple.* Fivefold. *at most.* At the most. *Ciphering.* Writing in cipher. *relative.* Complement. *by supposition.* Commonly assumed. *muster-roll.* List of soldiers belonging to a troop or regiment, drawn up for parades.

233. *marks.* Notes, observations. *Seat of the Estate.* National capital. *studies.* Some copies of the 1605 edition read 'labours studies': Spedding argued that since the words duplicate each other 'studies' alone was the true reading. Wright (p. 170) reads 'labours and studies'. *Method... controversy.* Several controversies in Bacon's time surrounded the French scholar Pierre de la Ramée, or Peter Ramus (1515–72), a vociferous critic of Aristotle (to whom he was still much indebted). Ramus evolved a method of composition or instruction which handled all topics by a strict progress from general to particular, dividing each heading into two parts, each part into a further dichotomy. Here Bacon is probably referring to an acrimonious dispute in the 1580s between two Cambridge dons, Everard Digby (Aristotelian) and William Temple (Ramist): see N. W. Gilbert, *Renaissance Concepts of Method* (New York, 1960), 202 ff., and L. Jardine, *Francis Bacon: Discovery and the Art of Discourse* (Cambridge, 1974), 59–65. Bacon often criticized the rigidity of Ramistic method: see e.g. *Works*, iii. 530, and *The Philosophy of Francis Bacon*, ed. and tr. B. Farrington (Liverpool, 1964), 17, 19, 63–4. For more detailed studies see Walter Ong, SJ, *Ramus, Method and the Decay of Dialogue* (Cambridge, Mass., 1958), and Nelly Bruyère, *Méthode et dialectique dans l'œuvre de La Ramée* (Paris, 1984). *fall at words.* Disagree, dispute. *proceeding.* Progress. *placed.* Ramus advocated a radical reform and simplification of traditional disciplines and their teaching. He broke up the five parts of rhetoric, assigning *inventio* and *dispositio* to dialectic, *elocutio* and *pronuntiatio* to rhetoric, and omitting *memoria*. Then, following Cicero and Quintilian, he identified *dispositio* with *iudicium*, subsuming under it the study of the axiom or proposition, the syllogism, and method. See e.g. his *Dialectica*, where bk. I handles 'De Inventione', bk. II 'De Judicio', devoting the last four chapters to method. *delivered.* Transmitted. *material.* Pertinent. *progression.* Advancement. *perfection.* Completeness. *Magistral.* A dogmatic or doctrinal exposition of information to be believed, and not questioned. *of Probation.* To be tested, examined. In the *De Augmentis* this second type is called 'initiative', as in ceremonies admitting a newcomer to a group: see *Works*, iv. 449. '*via... interclusa*'. 'A way deserted and shut off' (Cicero, *Pro Caelio*, xviii. 42). *present satisfaction.* An instant but superficial settling of the question. *glory.* Ostentation. *De... scientiarum.* 'Of the true method of the sciences, or the method of delivery to posterity.' *spun on.* Spun continuously, without a break, and so capable of extension. *intimated.* Made known.

234. *induced.* Arrived at by induction. *prevented.* Knowledge which we already possess. '*secundum... minus*'. To a greater or lesser extent. *consent.* Agreement. *assured.* Secure. *slips.* Cuttings. *is as of.* Resembles. *method... mathematiques.* Extant methods of teaching mathematics. *shadow.* Inkling. *ure.* Use. *inquisition.* Testing, enquiry. *discretion.* Discernment. *Enigmatical.* Esoteric, deliberately obscure—a way of restricting communication to the initiated, as in the occult arts. *Disclosed.* Exoteric, openly communicated. *whereof.* In the former 'pretence'. *remove.* Dissuade. *in... in.* Either in... or in. *Aphorisms.* Self-contained observations, sometimes universally relevant but not systematically developed, allowing for

expansion and correction as knowledge develops. See Brian Vickers, *Francis Bacon and Renaissance Prose* (Cambridge, 1968), ch. 3. *formal.* Regular. *discourses.* Discussions. *sensible.* Easily understood. *writing in Method.* Using systematic exposition. *trieth.* Tests. *except ... should be.* Unless they would be. *suffice.* Is competent.

235. *Tantum ... honoris.* 'Such is the power of order and connexion, such the beauty that may crown the commonplace' (Horace, *Ars poetica*, 242–3). *shew.* Pretence, display. *demonstration ... circle.* A system which seems complete, but is only a circular argument, not progressing. *satisfy.* Are superficially convincing. *particulars, being dispersed.* Natural events are discrete, not subordinated to an overarching pattern or system. As the *De Augmentis* puts it, 'as actions in common life are dispersed, and not arranged in order, dispersed directions do best for them' (*Works*, iv. 451). *broken.* Broken off, incomplete. *invite.* Encourage. The *De Augmentis* adds: 'aphorisms ... invite others to contribute and add something in their turn' (*Works*, iv. 451). *secure.* Give a sense of security. *at furthest.* Having reached the highest stage of development. *Questions ... Determinations.* The scholastic method, developed in the early Middle Ages, which proceeded from the *lectio* of a text to locate the *quaestiones* (questions or points needing clarification), giving rise to a *disputatio* or debate and to 'determinations', solutions to the questions at issue. Bacon had attacked the rigidity of scholastic methodology, and its fault of dissipating energy on trivial issues, in Book 1, above, pp. 140 ff. *proceeding.* Progress. *go about.* Endeavour. *hold.* Stronghold. *sum.* Main goal. *come in.* Agree. *would.* Should. *piece enemy.* Wright (p. 311) suggests here a metaphor from chess; other editors understand 'a fortified place, stronghold'. *delivery.* Transmission. *preoccupations and prejudgments.* Prepossessions, prejudices. *Policy.* Politics, 'immersed' in matter, the contingent phenomena of human behaviour, hard to generalize. *uniformity of method.* The 'one and only' method of Ramus, as Bacon mockingly describes it in the *De Augmentis*, 'was a kind of cloud that overshadowed knowledge for awhile and blew over: a thing no doubt both very weak in itself and very injurious to the sciences. For ... these men press matters by the laws of their method, and when a thing does not aptly fall into these dichotomies, either pass it by or force it out of its natural shape. ... And therefore this kind of method produces empty abridgments, and destroys the solid substance of knowledge' (*Works*, iv. 448–9). *taketh the way.* Takes steps. *expulsed.* Expelled. *press.* Pressure. *allow well.* Approve of: above, p. 224. *particular.* Specific, individual. *tradition.* Communication. *judgment.* That is, method. *another.* Different. *agreeable.* In accordance with received opinions. *Aristotle.* Perhaps *Eth. Nic.* vi. 3, 1139b18: 'Now what knowledge is, if we are to speak exactly and not follow mere similarities' (presumably a reference to Plato, *Theaetetus* 197A ff.), but with no mention of Democritus. Ellis believed that Bacon had inadvertently put the name of Democritus for Plato (*Works*, i. 667), but it recurs in *De Augmentis*. Wolff (op. cit. i. 193–4) suggests *De anima*, i. 2, 405b12 ff., where Aristotle reviews various theories of the soul, including Democritus', and comments: 'The language they all use is similar;

like, they say, is known by like; as the soul knows everything, they construct it out of all the principles.'

236. *tax.* Criticize. *conceits.* Conceptions. *conceived.* Understood. *translations.* Metaphors (the Latin term for metaphor is *translatio*). *trivial.* Elementary (the first stage of education comprised a *trivium* of three subjects: grammar, logic, rhetoric). *mark.* Note. *consonant.* Suitable. *pray in aid of.* Call in to one's assistance. Cf. Plato, *Statesman*, 277D. *De... traditionis.* 'The wisdom of tradition.' *vulgar.* Familiar. *Analysis.* In logic, resolving a sentence or proposition into its constituent elements and considering their interrelation. *Systasis.* The opposite process, combining elements (synthesis). *Cryptic.* Concealment, using ciphers (discussed earlier, p. 232). *stood upon.* Considered. *remembered.* Mentioned. *quantity and figure.* As architecture must consider the size and shape of the constituent materials, so scientific methodology must consider its propositions in the same terms. *limitation and manner.* The scope and nature of propositions. *rules of Propositions.* The principles of demonstration, as laid down by Aristotle in *Posterior Analytics*, i. 4, 73a–74a, and adapted by Ramus in his *Dialectica*, as providing general rules to be fulfilled by the first principles of every science ('axioma artium'), that they should be true in all cases (καθ όλου), homogeneous or true *per se* (καθ αὑτό) and universal (κατα παντος). Ramus then named these rules 'the law of truth', 'the law of justice', and 'the law of wisdom'. *Epitomes.* The Ramist method, which treated all topics in the same dichotomous manner, was used by some writers to produce digests or summaries, a practice that Bacon, like other Renaissance scholars, frequently criticized as a 'canker' or cancer eating up knowledge. See N. W. Gilbert, *Renaissance Concepts of Method* (New York, 1960), 73, 114. *fables.* As of the dragons which protected the garden of the Hesperides and the Golden Fleece. *fall upon.* Resort to. *well conducted.* Clearly organized, acute. *convertible.* Transferrable from one area of knowledge to another. *non-promovent.* Failing to progress. *incurring.* Running into themselves, in a circular argument, as in Ramus's 'law of wisdom', which demanded that the converse of a proposition must be true as well as the proposition itself. *De... axiomatum.* 'The production of axioms.' For Bacon 'axiom' is synonymous with 'proposition'. *utmost.* Outer. *limit.* Govern, define.

237. *latitude... sciences.* The way one science or body of knowledge impinges on others. *longitude... action.* The extent to which a science can be applied in the world. καθ αυτό. Ramus's 'law of justice', that the parts of a proposition must be essentially connected together. *latter.* Bacon criticizes Ramus for not considering what level of detail would be appropriate to various types of logical or scientific method. *material.* Pertinent. *to scorn.* To be scorned by. *map.* Of Abraham Tortelius (1527–98), to whose *Theatrum orbis terrarum* (1570), the first great world atlas, was prefixed a map of the world called *Typus Orbis Terrarum.* *glasses.* Mirrors. *filed.* Polished. *chrystalline.* Clear, accurately reflecting. *how far forth.* To what extent. *Lullius.* The Catalan polymath Raimond Llull (1235–1316), whose 'art' was a method of superimposing on each other a limited number of categories

(written on wheels divided into segments), this mere juxtaposition supposedly generating new knowledge. Bacon's dismissal of it as an 'imposture' which delivers only 'a shew of learning' echoes that of H. Cornelius Agrippa in his *De Vanitate* (1530), who described it as 'a monstrous Arte', suited rather to produce an 'outwarde shewe of the witte, and to the ostentation of Learninge, than to gette knowledge': *Of the Vanitie and Uncertaintie of Artes and Sciences*, tr. James Sanford (1569), ed. C. M. Dunn (Northridge, Calif., 1974), 56. *Typocosmy*. A system intended as an aid to learning, in which words or terms are grouped according to types or classes (as in Roget's *Thesaurus*). Alessandro Citolini's *Tipocosmia* was published at Venice in 1561. For a similar humanist's complaint against 'short-cut' guides to learning, which delivered a superficial acquaintance with words, but no understanding of the subject-matter, see Erasmus on the *Ars notoria*: *The Colloquies of Erasmus*, tr. C. R. Thompson (Chicago, 1965), 458–61. *give . . . countenance*. Give the appearance. *fripper's*. An old-clothes shop. *broker's*. Second-hand dealer, pawnbroker. *ends*. Remnants. *Illustration of Tradition*. The elucidation of discourse; but also 'making it more illustrious, enhancing it'. *God to Moses*: Exod. 4: 16; 7: 1. *disabled*. Disparaged. '*Sapiens . . . reperiet*'. 'The wise in heart shall be called prudent but he that is eloquent shall attain greater things' (Prov. 16: 21).

238. *emulation of Aristotle*. Cf. Cicero, *De oratore*, III. XXXV. 141, *Tusc. D.* I. iv. 7; Quintilian, III. i. 14. *Reason to Imagination*. The *De Augmentis* adds that 'Rhetoric is subservient to the Imagination, as Logic is to the Understanding' (*Works*, iv. 455). *Illaqueation*. Sophism, an entangling in argument. *Impression*. Pressure, influence. The *De Augmentis* reads 'by juggleries of words' ('per Praestigias Verborum'): *Works*, i. 671; iv. 455. *Morality*. Moral philosophy. *negotiation*. Dealings. *wrought*. Influenced. *importunity*. Persistency; pressurizing. *undermined*. Confused, deceived. *Inconsequences*. Illogical or fallacious arguments. *importuned . . . Observations*. The *De Augmentis* adds, 'by assiduity of impressions and observations' (*Works*, iv. 455). *fill the imagination*. The *De Augmentis* adds: 'with observations and images' (*Works*, iv. 456). *second*. Support. *ex obliquo*. Obliquely. *Plato*. *Gorgias*, 462E ff. *esteem of*. Reckon, judge. *resembling*. Comparing. *speech . . . more conversant . . . good than . . . evil*. Cf. Aristotle, *Rhetoric*, i. 1, 1355a22: 'Rhetoric is useful because things that are true and things that are just have a natural tendency to prevail over their opposites . . .' *colouring*. Presenting in a favourable light. *Thucydides*. See *History of the Peloponnesian War*, iii. 40, 42.2. *hold on*. Support, side with. *causes of estate*. National affairs. *Plato*. *Phaedrus*, 250D, and Cicero, *De officiis*, I. v. 14.

239. *Chrysippus*. Cicero, *De finibus*, iv. 18–19, *De oratore*, II. XXXVIII. 159, *Tusc. D.* II. xviii. 42. *insinuations*. Indirect arguments. In rhetoric, *insinuatio* is a process used in the beginning of a speech to win the audience's goodwill. *Video . . . sequor*. 'I see the better course and approve it, but I follow the worse' (Ovid, *Met.* vii. 20). *practise*. Conspire. *confederacy*. Union. *the affection . . . sum of time*. Cf. Aristotle, *De anima*, iii. 10, 433b5 ff. *the present*. Short-term benefits. *De . . . privati*. 'The wisdom of private discourse': that

is, a speech delivered to a small group, as opposed to a public assembly. *the worse part*. Bacon echoes Aristotle's defence of rhetoric against Plato's charges (*Rhetoric*, i. 1; 1355a22–1355b8), but goes considerably further in giving rheotoric an ethically positive role. See Brian Vickers, 'Bacon and Rhetoric' in *Cambridge Companion*. *doctrines of contraries*. *Rhetoric* i. 1, 1355a30 ff.: 'we must be able to employ persuasion, just as deduction can be employed, on opposite sides of a question, not in order that we may in practice employ it both ways (for we must not make people believe what is wrong), but in order that we may see clearly what the facts are.... No other of the arts draws opposite conclusions: dialectic and rhetoric alone do this. Both these arts draw opposite conclusions impartially.' *as... palm*. The traditional equation of logic with the fist (combative), rhetoric with the open palm (persuasive), goes back to the Greek philosopher Zeno, as recorded by Cicero, *De finibus*, II. vi. 17, *Orator*, xxxii. 113; and Quintilian, II. xx. 7. *at large*. Extended. *Aristotle*. In *Rhetoric*, i. 2, 1356a25 ff.: 'rhetoric is an offshoot of dialectic and also of ethical studies. Ethical studies may fairly be called political.... Neither rhetoric nor dialectic is the scientific study of any one separate subject: both are faculties for proving arguments.' As Wolff (i. 215) notes, Bacon's distinction between logic and rhetoric is not Aristotle's, who would classify both as *doxa* (opinion), inferior to *episteme* (knowledge). *indifferent*. Impartial. *proofs... auditors*. Aristotle, *Rhetoric*, i. 2, 1356a14 ff. *Orpheus... Arion*. 'An Orpheus in the woods, an Arion among the dolphins' (Virgil, *Eclogues*, viii. 56). *respectively*. In terms adapted to the persons addressed. *want*. Lack.

240. *leese*. Lose. *volubility*. Quickness in turning from one subject to another; versatility. *curious*. Fussy. *Colores... comparati*. 'The colours of good and evil, simple and comparative', i.e. apparent, proofs. Bacon had appended a collection of the 'Colours of Good and Evil' to his first publication, the 1597 *Essays* (above, pp. 97 ff.), further expanding it in *De Augmentis* (*Works*, iv. 459–72). *attendances*. Subordinate items, appendices. *Aristotle*. In the *Rhetoric*, ii. 24, 1400b34 ff. and elsewhere (*Topics*, *Sophistici Elenchi*), Aristotle recommended the study of sophisms, specious but fallacious arguments. Bacon advises professional speakers, whether in rhetoric or politics, to prepare in advance refutations of such statements. *Sophisma... malum*. *Sophism*: 'What men praise is good; what men condemn is bad.' *Redargutio... merces*. *Refutation*: 'The merchant praises what he wants to sell' (Horace, *Epist*. II. ii. 11). *Malum... gloriabitur*. 'A bad buy, a bad buy, says the purchaser, but he goes off to brag about it' (Prov. 20: 14). *Elenches*. Logical refutations. *annexed*. Attached. *probation*. Proving. *impression*. Affecting (the rhetorical doctrine of *movere*, moving the listener's passions). *percussion*. Blow. *raised*. Affected, upset. *Hoc... Atridae*. 'This would the prince of Ithaca wish, and the sons of Atreus give large sums to purchase' (Virgil, *Aeneid*, ii. 104). As Wolff (ii. 239) notes, Bacon may be recollecting Aristotle, *Rhetoric*, i. 6, 1362b30 ff., on arguments of the type of 'that is good the contrary of which is to the advantage of our enemies', illustrated with a very similar quotation from Homer, 'Surely would Priam exult' (*Iliad*, i. 256).

resume. Repeat. *before*. See above, p. 223. *furniture*. Ornamenting, decking out. *readiness*. Preparedness, speed. *shop*. Store. *Antitheta*. Opposed positions. In the *De Augmentis* Bacon expanded his treatment of *antitheta rerum* to 47 examples (*Works*, iv. 472–92). For one example see the Notes to the *Essays* (1625), pp. 713–14 below. *Formulae*. Sentences or phrases appropriate as openings, conclusions, or linking passages, which Bacon advised the orator to have ready at hand. He followed his own advice in the Christmas vacation of 1594, setting down from memory some 1,650 proverbs, sayings, and formulae in several languages which he had been rehearsing. Spedding printed a selection of these (*Works*, vii. 187–210); the full text was published (unfortunately along with some deluded ideas about Bacon's authorship of Shakespeare) by Mrs H. Pott, as *Promus of Formularies and Elegances* (London, 1883). *Antitheta rerum*. 'The antitheses of things.' *pro et contra*. Arguing for or against an issue was a fundamental exercise in all humanist education, but especially in the more detailed study of law and rhetoric. *large and laborious*. Expansive, painstaking. *prolixity of entry*. Too verbose a beginning. *wish*. Like. *cast up*. Summed up. *acute*. Penetrating. *bottoms*. Skeins or balls on which thread is woven. *unwinded*. Unwound.

241. *Pro . . . legislatorem*. '*For the words of the law*': 'Interpretation which departs from the letter is not interpretation but divination'. 'When the letter is departed from, the judge becomes a law-giver.' *Pro . . . singula*. '*For the intention of the law*' (in opposition to the previous category). 'The sense according to which each word is to be interpreted must be collected from all the works together.' *decent*. Appropriate. *apt*. Suitable. *indifferently*. Equally well. *excusation*. Excuse. *well*. Proper. *casting*. Planning. *Deliberative*. The branch of rhetoric involving persuasion and dissuasion to action, as in politics. Distinguished from judicial rhetoric (establishing guilt or innocence), and from demonstrative or epideictic rhetoric (dispensing praise or blame). *So . . . future*. Le Doeuff (p. 350) cites Demosthenes, *Phil.* i. 2. *Pedantical*. Pedagogical, involving teachers. *proper*. Own. *concerneth chiefly writing of books*. The editions of 1605, 1629, and 1633 read 'concerneth chiefly *in* writing of books'. Wright (p. 314) suggests as the true reading, '*consisteth* chiefly *in* writing', corresponding to the Latin version, 'in scriptione librorum consistit'. *relative*. Related. *incidently*. Incidentally. *critics*. In the original sense of 'correctors of a text'. *false*. Falsely. '*Demissus . . . sportam*'. 'He was let down in a basket' (Acts 9: 25; 2 Cor. 11: 33). *mended his book*. Emended his copy. *per portam*. 'By the gate.' *out of*. Outside, unknown to. *corrected copies*. The *De Augmentis* adds: 'Moreover, to speak truly, unless critics are learned in the sciences which the books they edit treat of, their diligence is not without its danger' (*Works*, iv. 494). *resteth*. Consists of. *over usual*. Too customary. *blanch*. Leave unnoticed. *plain*. Clear, obvious. *times*. The period in which the text was written. *censure*. Critical evaluation. The famous example of such guides to reading was bk. XII of Quintilian's *Institutio Oratoriae*, evaluating the Roman authors to be studied in schools. The *De Augmentis* adds: 'This last office is indeed, so to speak, the critic's chair; which has certainly in our age been ennobled

by some great men,—men in my judgment above the stature of critics' (*Works*, iv. 494). *election*. Choice. *syntax*. Systematic arrangement. *pursuit*. Sequence.

242. *Pedantical*. Pedagogical. The *De Augmentis* adds: 'As for the Pedagogical part, the shortest rule would be, "Consult the schools of the Jesuits", for nothing better has been put in practice. Nevertheless I will as usual give a few hints, gleaning an ear here and there. I am clearly in favour of a collegiate education for boys and young men; not in private houses, nor merely under schoolmasters. For in colleges there is a greater emulation of the youths amongst themselves; there is also the sight and countenance of grave men, which tends to modesty, and forms their young minds from the very first after that model; and in short there are very many advantages in a collegiate education' (*Works*, iv. 495). *difference*. Division, part. *refrain*. Restrain, hold in check. The *De Augmentis*, having repeated the injunction to 'avoid abridgments and a certain precocity of learning, which makes the mind over bold, and causes great proficiency rather in show than in fact', adds: 'Also let some encouragement be given to the free exercise of the pupils' minds and tastes; I mean, if any of them, besides performing the prescribed exercises, shall steal time for other pursuits to which he is more inclined, let him not be checked' (ibid.). *courses*. Sequence. Spedding conjectured 'cases'. For the following discussion cf. Essay 38, 'Of Nature in Men', above, pp. 417–18. *press*. Push forward. *propriety*. Property, nature. *bird-witted*. Having a short concentration-span. *one . . . begin*. One must begin afresh. *apt*. The consideration of a pupil's aptitude for one kind of study rather than another was a new feature of humanist educational theory, given best expression by Juan Luis Vives in *De tradendis disciplinis* (1531). *Cicero*. In *De oratore*, I. xxxiii. 150. *intermission*. Temporarily ceasing. *wronging or cherishing*. Either damaging or protecting. *noted*. By Machiavelli, *Discourses*, i. 19 (who refers to four kings). *tutors*. Guardians. *manurance*. Cultivation, training. *contention*. Exertion. *Tacitus*. In *Annals*, i. 16–22. *playing*. Acting. *combustion*. Feverish excitement.

243. *committed*. Arrested. *suddenly*. Quickly. *mortalest*. Most deadly. *stage*. In the *De Augmentis* Bacon adds a passage recommending the use of drama in education, praising the Jesuit schools for this practice: 'It is a thing indeed, if practised professionally, of low repute; but if it be made a part of discipline [education], it is of excellent use. I mean stage-playing: an art which strengthens the memory, regulates the tone and effect of the voice and pronunciation, teaches a decent carriage of the countenance and gesture, gives not a little assurance, and accustoms young men to bear being looked at' (*Works*, iv. 496). *period*. Conclusion. *divisions*. Classification of topics. *next*. Nearest. *like*. Likely. *cabinet*. Small chamber. *of a*. Of the same. *particular*. Private. *were like*. Would be likely. *several*. Different. *had been*. Were. *parts*. Factors. *without*. Beyond. *Appetite and Will*. The passions and moral choices of man, the subject-matter of ethics. In the *De Augmentis* (vii. 1; *Works*, v. 3) the word 'appetite' is omitted, and a sentence added: 'The will is governed by right reason, seduced by apparent good, having for its spurs the passions, for its ministers the [bodily] organs and voluntary motions.'

244. '*Ante ... vitae*'. 'Above all, my son, keep thy heart, for out of it proceed the actions of life' (Prov. 4: 23). *carriage*. Holding, moving. *framing*. Forming. *exemplars*. Patterns. *marks*. Goals. *frame*. Prepare (timber) for use in building; hew out, shape. *true*. Accurately prepared. *conformable*. Disposed to conform. *moral virtues ... reward and punishment*. Topics discussed by Aristotle, *Eth. Nic.* ii. 1; x. 9. *glances and touches*. Hints, brief references. Bacon is being unfair to Aristotle who explicitly announced in *Eth. Nic.* that 'the present inquiry does not aim at theoretical knowledge like the others (for we are inquiring not to know what excellence is, but in order to become good, since otherwise our inquiry would have been of no use), we must examine the nature of actions, namely how we ought to do them; for these determine also the nature of the states that are produced' (ii. 1, 1103b26 ff.). *barks*. Ships. *ordinary ... matters*. In the *De Augmentis* Bacon extends his criticism of how, out of 'innate pride and vainglory, men have chosen those subjects of discourse, and those methods of handling them, which rather display their own genius than benefit the reader. For myself', Bacon adds, 'I may truly say that both in this present work, and in those I intend to publish hereafter, I often advisedly and deliberately throw aside the dignity of my name and wit (if such thing be) in my endeavour to advance human interests; and being one that should properly perhaps be an architect in philosophy and the sciences, I turn common labourer, hodman, anything that is wanted; taking upon myself the burden and execution of many things which must needs be done, and which others through an inborn pride shrink from and decline.' Moral philosophers, he concludes, in preferring 'rhetorical ornaments ... have for the most part passed over' those things 'which are of most use for practice' (*Works*, v. 4). *subtilities*. Minute, abstruse points. *compounded*. Joined together, built up. *check*. Rebuke. '*Nocet ... sui*'. [Let 'young men ... be roused to the matter, and not to the style; otherwise] eloquence does them harm, making them enamoured of itself, and not of the subject' (Seneca, *Epist.* lii. 14). '*Quae ... meliore*'. 'If you do what I advise, you will not only praise the orator at the time, but soon afterwards yourselves also, by reason of the improved state of affairs': Demosthenes, *Olynthiacs*, ii. 31 (in the Latin tr. by Wolf, Basle 1572). *so excellent parts*. Such great ability. *husbandry*. The business of a farmer, managing an estate. *Nec ... honorem*. 'And well I know how hard it is to win with words a triumph herein, and thus to crown with glory a lowly theme' (*Georgics*, iii. 289–90).

245. *suborn*. Furnish, equip (Lat. *subornare*). *Georgics*. Virgil's four *Georgics* deal with several types of farming or cultivation, including plants, animals, and bees. *tillage*. Ploughing, tending. *heroical*. Exemplified by heroes. *primitive*. Original. *Platform*. Model. *accommodate*. Adapt. *Simple*. Composed of a single element. *disputations*. Discussions. *as*. Like, equivalent to. *heathen divinity*. Bacon alleges that discussions of the highest good, or happiness as the goal of existence, occupied the same place in pagan philosophy as theology does in Christianity (a rather unkind criticism). He may be referring to Aristotle, *Eth. Nic.*, e.g. i. 2, 1095a15 ff., i. 7, 1097a27 ff., and i. 9, 1099a21 ff. *discharged.* Dismissed. *Aristotle.* In *Eth. Nic.* i. 9, 1100a2 ff., and *Rhetoric*, ii. 12, 1389a30 ff. *we.* The *De Augmentis* adds: 'instructed by the Christian faith' (*Works*, v. 5). *minority.* Subordinate state. *height of style.* Intense eloquence. (Rhetoric classified style into three levels, high, medium, and low.) '*Vere . . . Dei*'. 'It is true greatness to have in one the frailty of a man, and the security of a god' (Seneca, *Epist.* liii. 12). *sobriety.* Soberness, moderation. *receive.* Accept. *labours.* Bacon refers mostly to Aristotle's ethics and to Cicero, *De officiis*. *Positive.* Practical; concerning matters of fact and experience. *postures.* Relationships. *intrenched.* Defended. *triplicity of Good.* A threefold good, relating to mind, body, and estate, as discussed in Aristotle's *Eth. Nic.* i. 8. *contemplative . . . active life.* See *Eth. Nic.* x. 6–8; the opposed claims of the *vita activa* and *contemplativa* formed a major topic in Renaissance moral philosophy: see e.g. Brian Vickers (ed.), *Arbeit, Musse, Meditation: Studies in the 'vita activa' and 'vita contemplativa'* (Zurich and Stuttgart, 1991). *reluctation.* Struggle, violent effort. *secured.* Achieved. *honesty and profit.* Aristotle, *Rhetoric*, i. 6. Bk. I of Cicero's *De officiis* discusses *honestas*, bk. II *utilitas*, bk. III the conflict between them. *balancing of virtue with virtue.* Aristotle, *Eth. Nic.* iii and iv; Cicero, *De officiis*, I. xiv. 160. *laboured.* The *De Augmentis* adds: 'and that the ancients have done their work admirably therein, yet so as the pious and earnest diligence of divines, which has been employed in weighing and determining duties, moral virtues, cases of conscience, the bounds of sin, and the like, has left the philosophers far behind' (*Works*, v. 6). This judgement, of the superiority of Christian to pagan ethics, was common throughout the Renaissance, both Catholic and Protestant.

246. *strings.* Smaller roots. *resume.* Recall. *open.* Expound. The *De Augmentis* is more explicit: 'endeavour to open and cleanse the fountains of morality, before I come to the culture of the mind' (*Works*, v. 6). *total.* Entity. *substantive.* Having a separate and independent existence. *latter.* The *De Augmentis* adds: 'The former of these may be termed "Individual or Self-good", the latter the "Good of Communion"' (*Works*, v. 7). *particular.* 'Belonging to or affecting a part' (*SOED*). *sympathy.* Affinity, correspondence. *loadstone.* Magnet. *affection.* Attraction. *patriot.* Devotee, supporter. *massy.* Dense, compact. *go forward.* Extend the argument. *divulsion.* Tearing asunder. *continuance.* Continuity. *engraven.* Impressed on. *in commission . . . famine.* Being commissioned to make provision for a famine. *instance.* Urgent entreaty. '*Necesse . . . vivam*'. 'It is necessary that I go, not that I live' (Plutarch, 'Life' of Pompey, 50). *discipline.* Form of teaching (here, ethics). *depress.* Disvalue. *saints of God.* Moses in Exod. 32: 32, and St Paul in Rom. 9: 3. *anathematized.* An 'anathema' is an intensified curse or ecclesiastical denunciation. *razed out.* Erased absolutely: that is, not even surviving in an after-life. The *De Augmentis* adds 'rather than that their brethren should not obtain salvation' (*Works*, v. 7). *against Aristotle.* In the final book of the *Eth. Nic.* (x. 7–8) Aristotle argues that the contemplative life is preferable for the philosopher.

247. *Pythagoras.* As reported by Cicero, *Tusc. D.* v. iii. 8–9. But where Cicero uses the story to justify the philosopher's life of 'contemplation', Bacon, appropriating it for the *vita activa*, rebukes inactivity within society. ('Hiero' is a mistake for Leon, tyrant of Phlius). See also *Apophthegms*, no. 251, which adds a gloss: '*I am one of them that come to look on.* Meaning it of philosophy and the contemplative life' (*Works*, vii. 160), and Montaigne, *Essais*, i. 26, 'De l'institution des enfans' (Pléiade edn. 157–8). *utter.* Sell. *theatre.* Literally, a 'beholding-place', as Puttenham defined it. Spedding suggests as a source St Augustine's description of the sufferings of St Paul: 'in the theatre of this world where he was made a spectacle to God, angels, and men, fighting a lawful and great fight' (*The City of God*, xiv. 9). But cf. e.g. Felice Figliucci, *Della filosofia morale libri dieci sopra i dieci libri dell'Etica d'Aristotele* (Venice, 1552), who finds that the contemplative life, being superior, 'is reserved for angels and other pure spirits, while moral and virtuous action and works are appropriate for men'; cit. E. Garin, *Italian Humanism: Philosophy and Civil Life in the Renaissance* (Oxford, 1965), 172. *God.* As in Gen. 1, where after each of the six days' work 'God saw that it was good'. Cf. also Essay 11, 'Of Great Place', above, p. 360: 'For if a man can be partaker of God's theatre, he shall likewise be partaker of God's rest.' '*Pretiosa . . . ejus*'. 'Precious in the sight of the Lord is the death of his saints': Ps. 116: 15; *Promus*, no. 347. *place.* Text, biblical authority. *civil death.* Being in effect dead to society. *regular.* Subject to a religious rule; belonging to a monastic order. *professions.* Religious orders (entered after a declaration or vow of obedience to a shared belief-system). *simple.* Purely. *office.* Duty, activity. *Moses . . . mount.* Exod. 19–23. *Henoch.* Enoch, father of Methuselah, in Gen. 5: 24, and Jude 14. *Contemplative.* One devoted to contemplation. For some Renaissance writers the 'contemplative life' implied a life of piety and Christian meditation; Bacon usually applies it to studying, in a secular sense. *finished.* Complete, self-contained. *casting beams.* Communicating with. Cf. Cicero, *De officiis*, I. xliv. 158: 'Every duty, therefore, that tends effectively to maintain and safeguard human society should be given the preference over that duty which arises from speculation and science alone.' *controversies . . . society.* Zeno (d. 263 BC), a leading Stoic philosopher who, like Plato's Socrates, believed that happiness resulted from virtue exercised in society. *simply or attended.* Individual or communal virtue. The *De Augmentis* adds: 'which is ever concerned with the duties of life' (*Works*, v. 8). *exercises.* Use, practice. *other side.* All the opinions that follow are opposed to Zeno and Socrates. *Cyrenaics.* A sect founded by Aristippus of Cyrene (*fl.* 366 BC) which, like the Epicureans, defined happiness as the enjoyment of pleasure and the avoidance of pain. *pleasure.* This term (*hēdonē*; 'voluptas') carried mostly negative connotations in classical moral philosophy, being usually seen as opposed to reason and virtue. *comedies of errors.* In Roman comedy, Plautus or Terence. *habits.* Clothes. *the first age.* The Golden Age, with its perpetual spring, as described by Ovid, *Met.* i. 89 ff. *after.* In the manner of. *Herillus.* A heterodox Stoic (3rd century BC), a disciple of Zeno, who made free choice the major factor in human happiness: cf. Cicero, *De finibus*, IV. xv. 40 f.; Diogenes Laertius, vii. 165.

The *De Augmentis* adds 'that exploded [discredited] school of Pyrrho and Herillus.' Pyrrho, the founder of scepticism, aimed at imperturbability of mind: since certainty in knowledge is not possible, we should not worry about things. *clearness*. Composure, balance. *reluctation*. Reluctance. *Anabaptists*. A German Protestant sect, founded in 1521, which rejected infant baptism and practised Christian communism.

248. *motions*. Unpredictable movements. *to point of*. In the direction of. *censureth*. Criticizes, qualifies. *Epictetus*. See his *Enchiridion*, 1–7. *liable*. Subject, subservient. *Consalvo*. Gonzalo Fernandez Consalvo Cordoba (1453–1515), as reported by Guicciardini, *Storia d'Italia*, vi. 7: 'But that pusillanimitie of the captaynes was resolutely rejected of Consalvo with this answere worthy the greatnes of his heart.... So for my parte I desire rather to be presently buryed ten foote deepe in the grounde whereon we stand, than by giving back one foote to prolong my life an hundred yeres' (tr. G. Fenton (London, 1579), 224C). *signed*. Attested. *a good ... feast*. Prov. 15: 15. *Epictetus*. See Aulus Gellius, *Noctes Atticae*, ix. 2. *Aristotle*. *Rhetoric*, i. 5, 1361b4 ff. *intend*. Occupy himself with. The *De Augmentis* adds: 'and accordingly abstained from an infinite variety of things, depriving himself as it were of the use of his body in the meantime' (*Works*, v. 10). *as*. Just as. *proper*. Appropriate. *Diogenes*. Referring either to Diogenes Laertius, ii. 75, a saying of Aristippus, or vi. 29, a similar one of Diogenes. *sustained*. Persevered. *refrain*. Bridle, rein in. *in praecipitio*. In a steep place, on a precipice. *tenderness*. Sensitiveness. *application*. Compliance, adaptability. *for avoiding*. In order to avoid. '*e tela crassiore*'. 'Of a stouter web.' Cf. *Promus*, no. 392; *Charge touching Duels*, above, p. 311; and Essay 57, 'Of Anger', above, p. 450. *Promus and Condus*. An officer who dispenses stores (a butler), and one who collects or keeps them (a steward): cf. Plautus, *Pseudolus*, 2.2.14, and *Promus*, no. 819: 'Promus magis quam condus'. The *De Augmentis* adds: 'whereof the latter [good active], which is active and as it were the Promus, seems to be the stronger and more worthy; and the former [good passive], which is passive and as it were the Condus, seems to be inferior' (*Works*, v. 11).

249. *dilate*. Expand, propagate. *generation*. Procreation. '*Beatius ... accipere*'. 'It is more blessed to give than to receive' (Acts 20: 35). *soft*. Effeminate, unmanly. *but esteemeth*. But he esteemeth. *effecting*. Achieving. *state*. Stability. '*Magni ... tardius*'. 'We think it a great matter to be a little longer in dying' (Seneca, *Nat. quaest.* II. lix. 7). '*Ne ... diei*'. 'Do not flatter yourself about tomorrow, for you never know what a day will bring forth' (Prov. 27: 1). '*Opera ... eos*'. 'Their works follow them' (Rev. 14: 13). *proceeding*. Progress. *latitude*. Extent. '*Cogita ... potest*'. 'Think how often you do the same thing over and over. Food, sleep, play come round in a perpetual circle; a man might wish to die, not only from fortitude or misery or wisdom, but merely from weariness of life' (Seneca, *Epist.* lxxvii. 6). Modern texts read *libido*, not *ludus*. *sensible*. Aware of. *inceptions*. Beginnings. *recoils*. Setbacks. *reintegrations*. Recoveries, restorations. '*Vita ... est*'. 'Life without a purpose is unsettled and languid' (Seneca, *Epist.* xcv.

46). The *De Augmentis* adds: 'And this befalls as well the wise as the foolish. . . . [For] we see that the greatest kings who might have at command everything which can gratify the sense, have yet sometimes affected mean and frivolous pursuits (as was the passion of Nero for the harp, of Commodus for gladiatorial combat, of Antonnius for chariot-driving, and the like); which nevertheless they esteemed more of than of the whole abundance of sensual pleasures; so much pleasanter it is to be doing than to be enjoying' (*Works*, v. 11–12). *incidence into*. Coincidence with. *respect*. Consideration. *private*. Peculiar to a particular person. *contrary*. Opposed. *gigantine*. Giant-like, rebellious, like the giants in Greek mythology, who warred against the gods. *Lucius Sulla*. Dictator of Rome, whose epitaph for himself was that no man ever surpassed him in doing good to his friends or harm to his enemies. *infinite*. Innumerable. *model*. Scale. *as*. According as. *Theomachy*. Striving against God. *pretendeth*. Tends. *Conservative*. Conserving, retaining. *Perfective*. Capable of being improved or completed.

250. *intention*. Purpose, scope. *supposed*. Conceived. *signature*. The act of impressing. *conserving*. Preserving. *in state*. In the original condition. *greater*. The *De Augmentis* adds: 'For in all things there are some nobler natures to the dignity and excellence whereof inferior natures aspire as to their sources and origins' (*Works*, v. 12). *Igneus . . . origo*. 'They have a fiery vigour and a heavenly origin' (Virgil, *Aeneid*, vi. 730). *assumption*. Taking on. *tumble*. Move restlessly. *remove local*. Change of place. *mean*. Means. *estuation*. Mental agitation. *pleasures*. Cf. Aristotle, *Eth. Nic.* vii. 4–12, x. 1–7. *softest*. Most effeminate. *lowest*. Most debased. *difference*. Distinction. *sincereness*. Correctness, purity. *superinduced*. Brought about. *equality*. Evenness of mind or temper; equability. *vicissitude*. Variety, frequent changes. *impression*. Effect, power. *Whether of*. Which of. *Socrates*. The debate with Callicles in Plato, *Gorgias*, 491–5.

251. *ill words*. Quarrels. *supports*. Supporters. *Epicures*. Epicureans. *clearing*. Dispelling. *compassing*. Achieving. *shew of advancement*. Mere appearance of progress (towards the perfection of nature). *in a circle*. Circular movement. *second question*. The *De Augmentis* specifies: 'as to whether a man's nature may not be capable of tranquillity of mind and vigour of fruition both' (*Works*, v. 14). '*Non . . . diffidentis*'. 'Enjoy not, that you may not desire; desire not, that you may not fear. This is the precaution of cowardice and pusillanimity': Plutarch, 'Life' of Solon, 7; probably in the translation by Cruserius (Wolff, ii. 82). *cautionary*. Full of cautions. *discipline*. Training, education. *preparation to die*. Plato, *Phaedo*, 64, 67D; Cicero, *Tusc. D.* I. xxx. 74 ff.; and Montaigne, *Essais*, i. 20: 'Que philosopher c'est apprendre à mourir'. *terrible enemy . . . no end of preparing*. Le Doeuff (p. 355) notes Montaigne's later opinion that 'la préparation à mort donne plus de tourment que n'a faict la souffrance': *Essais*, iii. 12, 'De la phisionomie' (Pléiade edn. 1028). Cf. also Erasmus' colloquy *Funus* for the argument that the way to death is commonly more painful than death itself: *The Colloquies of Erasmus*, tr. C. R. Thompson (Chicago, 1965), 357–73. *the poet*. The *De Augmentis* adds: 'for a heathen'. *Qui . . . Naturae*. 'He who would reckon the close of life among nature's gifts' (Juvenal, x. 358–9). Cf. Essay 2, 'Of Death', above, p. 344. *harmonical.* Harmonious. *breaking.* Training, especially of horses. *unapplied.* Independent; not adapted to other people. *ground.* A plain-song or melody. *changes.* Modulations. *breaketh.* Trains. *stops and passages.* Technical terms for lute-playing: chords and runs. *set song.* Written-out piece of music, fully composed. *voluntary.* A musical piece performed spontaneously or of one's free choice, especially as a prelude to some more elaborate piece. *civil.* Civic, the *vita activa* in society. *grain . . . cloud . . . ice.* All types of flaw in a jewel: 'ice' is defined by *OED* (citing this passage in Bacon as its first occurrence) as 'a congelation or crystalline appearance resembling ice'. Cf. *Of Tribute*, above, p. 29: the virtues 'are as the mill when it is set upon a rich diamond. Here it grindeth out a race and there a grain, to make it more neat and paragon.' *abate.* Reduce in size, value. *magnanimity.* Nobility of feeling, generosity.

252. *inward disposition.* Frame of mind given to self-examination. *science.* Knowledge. *framing.* Preparing timber for use in building. *mechanicals.* Mechanics, the science of constructing machines. *on work.* Working. *conjugation.* Combination. *conformity.* The conformation of men to the business of society. *respective.* Particular. *dispersed.* Fragmentary, uncoordinated. *challenge.* Claim. *place.* Station in life. The *De Augmentis* adds: 'But treatises on matters of this kind which do not savour of experience, but are only drawn from a general scholastic knowledge of the subject, are for the most part empty and unprofitable' (*Works*, v. 15). *looker-on.* Spectator. gamester. Games-player (not necessarily a gambler). *vale . . . hill.* The valley gives the best view of the hill. Why Bacon should call this proverb 'more arrogant than sound' is explained by the qualification added in the *De Augmentis*: 'concerning the censure of the people on the actions of their superiors' (*Works*, v. 15). Cf. also the dedication to Machiavelli's *Prince*, justifying this attempt of 'a man of humble and obscure condition' to 'discuss and direct the government of princes', just as 'landscape painters station themselves in the valleys in order to draw mountains or high ground'. *materially.* Soundly, to the point. *speculative.* Knowledgeable in theory, lacking practical experience. *of.* About. *Phormio's argument.* In Cicero, *De oratore*, II. xviii. 75–6. Phormio was a Peripatetic philosopher, who one day 'held forth for several hours upon the functions of a commander-in-chief and military matters in general', without having any experience of them. Hannibal's candid comment was that 'he had seen many old madmen but never one madder than Phormio'. *dotage.* Senility, imbecility. *honoris causa.* 'As a token of respect.' *book.* James's *Basilicon Doron*, an instruction book addressed to his eldest son Prince Henry, was first published in 1599, when he was King James VI of Scotland, and reissued after his accession to the English throne in 1603. *aspersion.* Intermixture.

253. *distempered.* Disordered, disturbed. *dizziness.* Confusion. *leese . . . order.* Get lost in their own plan. *convulsions.* Contortions. *cramp . . . impertinent.* Cram in irrelevancies. *savouring.* Smelling. *beareth.* Endures. *spirits.* Innate faculties of movement. *extern.* External. *pastors.* Shepherds. *leese.*

Lose, erase. *cause of judicature*. Probably the case of Sir Francis Goodwin in 1604. *free monarchy*. James's *The True Law of Free Monarchies* (1598, 1603). *circle*. Extent. *allege*. Cite. *prime*. Chief. *decencies*. Proprieties. *pro Marcello*. A speech delivered in 46 BC praising Caesar for his lenience in pardoning Marcellus. *table*. Picture. *to his face*. With him in the audience. *doubt*. Hesitate. *De... artibus*. 'Of fraudulent and evil arts.' *Relative*. Corresponding. *cautels*. Deceits. *cynically*. Satirically, bitterly. The *De Augmentis* adds: 'after the manner of Lucian' (*Works*, v. 7). Lucian (2nd century AD) was a brilliant and irreverent satirist. *sever*. 'Divide by cutting, slicing, esp. suddenly and forcibly' (*SOED*).

254. *matter*. Subject-matter. '*Quaerenti... obviam*'. 'A scorner seeks wisdom and does not find it; to one of understanding, knowledge comes easily' (Prov. 14: 6). *Basilisk*. A fabulous creature, with fiery death-dealing eyes and breath. *prevent*. Act before or more quickly than someone else. *Machiavel*. Bacon approved of Machiavelli's *Prince* (1513) and *Discourses* (1531) for their realism in the observation of human affairs, achieved by using a descriptive attitude rather than the prescriptive one normally found in treatises on politics and ethics. *serpentine... innocency*. See Matt. 10: 16, and Bacon's *Religious Meditations* (1597), 'Of the Innocency of the Dove and the Wisdom of the Serpent' (above, pp. 90 f.). Ellis and Spedding (*Works*, v. 17 note) suggest the probable source in Charron, *De la sagesse* (1601): 'Il faut tempérer et marier l'innocence colombine en n'offensant personne avec la prudence et astuce serpentine en se tenant sur ses gardes et se préservant des finesses, trahisons, et ambuches d'autrui' (bk. II ch. 10, para. 5; English tr. by S. Lennard [London, n.d.; before 1612], p. 325). *baseness*. Low position in the natural scale; supineness. *volubility*. Rolling or twisting motion. *lubricity*. Slipperiness. *envy*. Hostility. *unfenced*. Unprotected. *upon*. To. *simplicity*. Innocence, purity. *reaches*. Tricks, deceptions. '*Non... ejus*'. 'A fool will not listen to the words of the wise, unless you first tell him what is in his own heart' (Prov. 18: 2). The *De Augmentis* adds: 'But this part, touching respective cautions and vices, we set down as deficient, and will call it by the name of "*Serious Satire*", or the *Treatise of the Inner Nature of Things*' (*Works*, v. 18). The Latin version, *Tractatus de Interioribus Rerum*, recalls the subtitle Bacon gave to the Latin translation of the 1625 *Essays*, *Sermones Fideles, Ethici, Politici, Œconomici: sive Interiora Rerum*. See headnote to the 1625 *Essays*, below, p. 710. *companies*. Groups of people, having a common goal (not just commercial). *politic bodies*. Political groups. *society*. The *De Augmentis* adds: 'for that is referred to policy', or politics (v. 18). *framing*. Forming. *persons*. The *De Augmentis* adds: 'towards the preservation of those bonds of society' (v. 18). *Lucius Brutus*. Who executed his sons for conspiring to restore the Tarquins (Livy, ii. 5). *Infelix... minores*. 'It was an unhappy deed, whatever posterity might say of it' (Virgil, *Aeneid*, vi. 822). *doubtful*. Dubious. *M. Brutus and Cassius*. In Plutarch's 'Life' of Brutus, xii. 2 f. *feel*. Sound out.

255. *cast forth*. Threw out (for discussion). *tyranny*. Absolute power. *civil war*. The *De Augmentis* adds: 'While a third set affirmed, according to the

doctrine of Epicurus, that it was unfit for wise men to endanger themselves in the cause of fools' (*Works*, v. 18). *ensue of.* Result from. '*Aliqua ... possint*'. 'Some things must be done unjustly, that many may be done justly' (Plutarch, *Precepts of Statecraft*, 817F–818A). '*Authorem ... habes*'. 'Present justice is in your power, for that which is to come you have no security' (Plutarch, 'Advice about keeping well', *Moralia*, 135). *in present.* At the present time. *De ... animi.* 'The culture of the mind.' *husbandry.* Cultivation. *statua.* Statue. '*Necesse ... quomodo*'. 'It is necessary then to speak of virtue, both what it is and whence it proceeds. It would be almost useless to know what virtue is, but be ignorant of the ways and means of acquiring it. Therefore we must inquire not only to what kind virtue belongs, but also how it may be obtained. For we wish to be acquainted with the thing itself, and to gain possession of it; and we shall not fully succeed in this, unless we know both the whence and the how': *Magna Moralia*, i. 1, 1182a1 ff., in the Latin version of George Valla. See also *Eth. Nic.* ii. 1, 1103b26 ff. (quoted above, p. 649). *iteration.* Repetition, rhetorical emphasis. *inculcate.* Urge, emphasize. The *De Augmentis* adds: 'although he does not himself pursue it' (*Works*, v. 19). *Cato the second.* Marcus Porcius Cato (95–46 BC), military tribune and enemy of Caesar, who brought the Stoic philosopher Athenodorus back with him from Greece, and spent the last night of his life reading Plato's *Phaedo*. '*non ... vivendi*'. 'Not for the sake of arguing like a philosopher, but that he might live like one' (Cicero, *Pro Murena*, xxx. 62). '*De ... nemo*'. 'Everyone thinks about the parts of his life, no man thinks about the whole': Seneca, *Epist.* lxxi. 2. '*Qui ... aegrotat*'. 'Those who are sick without feeling pain, are sick in their minds' (Hippocrates, *Aphorisms*, ii. 6). *awake the sense.* Revive their powers of sensation. *cure ... minds.* This section on the 'Culture of the Mind' yielded many topics for the 1612 version of the *Essays*, enlarged further in 1625: see head-note to the *Essays*, below, pp. 713–14. *preferred unto.* Recommended to. As Le Doeuff notes (p. 356), the Middle Ages had made philosophy as a whole subordinate to theology: Bacon exempted natural philosophy from that position, but allowed moral philosophy to remain there—evidently because he found the other-centred Christian ethics superior to anything in the classical tradition. *the Psalm.* Ps. 123: 2. *discern of.* Interpret.

256. *written inquiry.* Many ethical treatises in the Renaissance were commentaries on the works of Aristotle. Bacon wants this area of knowledge, also, to be based on the direct observation of reality. *propound.* Put forward for consideration. *particularity.* Detail. *acquit ourselves.* Be excused for. *application.* Adaptation. *husbandman.* Farmer. *accidents.* Unfavourable symptoms. *cure.* Care. *without.* Outside. *Vincenda ... ferendo.* 'All fortune may be overcome by suffering' (Virgil, *Aeneid*, v. 710); hence: 'All nature may be overcome by suffering.' *dull.* Indistinctly felt. *Accommodating or Applying.* Adapting oneself. *article.* Rubric or heading.

257. *radical.* Fundamental. *commixture.* Mixture. *in passage.* In passing. *mediocrities.* Mid-points. (Aristotle defined virtue as the mid-point between excess and deficiency.) *Aristotle.* This subject is in fact discussed in *Eth.*

Nic. iv. 2, 1122a19 ff. For Aristotle magnanimity (*megalopsychia*) is 'the crown of the virtues', including 'goodness and nobility', a state of excellence rewarded with honour. *intend.* Aim at, take on. *to.* Spedding conjectured 'to intend' as the correct reading. *divide themselves.* Deal with several different matters simultaneously. *exactly well.* With careful attention to detail; meticulously. *pusillanimity.* Mean-spiritedness. *return of time.* Implying a cyclic notion of time, as if 'coming round again'. '*Jam...fovetque*'. 'She [Juno] begins to attend and nurse his project while it is yet in the cradle' (Virgil, *Aeneid*, i. 18). *longanimity.* Patience, long-sufferingness. *Aristotle.* Who does discuss this in *Eth. Nic.* iv. 6. *conversation.* Behaviour. *soothe.* Flatter. *cross.* Oppose. *indifferent.* Not involving one personally. *ministery.* Service. *suppeditation.* Assistance. *planets.* The human temperaments listed are those that astrology claimed were influenced, respectively, by the planets Saturn, Jupiter, Mars, the Sun, Venus, Mercury, and the Moon. *Relations.* Contemporary narratives of the conclaves at which Popes were elected. *handsomely.* Aptly, skilfully. *lively.* Vividly. *conference.* Business. *dry.* Impassive. *formal.* Precise, stiff. *real.* Sincere, straightforward.

258. *humorous.* Whimsical, moody. *certain.* Trustworthy, confident. *huomo di prima impressione.* The impulsive man. *huomo di ultima impressione.* The cautious man. *wandereth.* Loses its way. *fixed.* Treated in a steady, systematic manner. *found.* Established. *conclude.* Lay down as a conclusion. *posies.* Bunches of flowers. *confectionary.* Place where confections or sweetmeats are made, sometimes medicinal. *receits.* Recipes. *those... nature.* Many of these topics are treated in the *Essays*. *region.* Climate. *magistracy.* Holding the office of magistrate. *per...gradus.* 'By leaps, by degrees'. '*benignitas...est*'. 'His generosity is like that of a young man' (Plautus, *Miles gloriosus*, III. i. 40). '*Increpa...dure*'. 'Rebuke them sharply.' '*Cretenses...pigri*'. 'The Cretans are always liars, vicious brutes and lazy gluttons': Titus 1: 12–13, where St Paul quotes from Epimenides. '*Sed... adversae*'. 'For the most part the desires of kings are as changeable as they are strong, and often contradictory' (Sallust, *Bell. Jug.* cxiii. 1). *mendeth.* Improves, corrects. '*Solus... melius*'. 'Vespasian alone was changed for the better' (*Histories*, i. 50). *defeateth.* Ruins. '*Qui...possunt*'. 'Who are not able to digest great prosperity': *Olympian Odes*, i. 55, in the translation by Henricus Stephanus (Paris, 1560, 1566), referring to Tantalus (Wolff, ii. 16–17). '*Divitiae... opponere*'. 'Though riches increase, set not your heart upon them' (Ps. 62: 10). *Aristotle. Rhetoric*, ii. 12–17: a more extensive discussion than Bacon's 'touched a little...in passage' (as if in passing) would suggest. *grounds and moulds.* Soils and fertilizers. *complexions and constitutions.* Psychological and physical types. *indiscretion.* Lack of discrimination. Just like medicine, Bacon argues, moral philosophy must take account of individual differences. *empirics.* Medical practitioners lacking scientific knowledge, who work by trial and error. *affections.* Passions. *medicining.* Administering medicine to. *it...first.* The correct order of proceeding is, first....

259. *in order.* In sequence. *perturbations . . . affections.* Emotional disturbance, agitation. *ancient politiques.* Solon, in a fragment attributed to him. See also Cicero, *Pro Cluentio*, 49. *popular estates.* Democracies. *divers volumes of Ethics.* In addition to the best-known *Nicomachean Ethics* modern scholarship confirms the attribution to Aristotle of both the *Eudemian Ethics* and the so-called *Magna Moralia.* *the affections.* In *Rhetoric*, ii. 1–11. But the passions are also discussed in *Eth. Nic.* i. 13, ii. 5–6. *pretermitteth.* Does not mention. *pleasure and pain.* In *Eth. Nic.* ii. 3, 1104b4 ff., and ii. 5, 1105b22 ff. *travails.* Pains. *second hand.* Of Zeno, the Greek founder of Stoicism, no written works survive, only fragmentary reports. *subtilty.* Minute discriminations. *curiosities.* Trifles. *active.* Applied to, or stimulating action. *particular writings.* Plutarch and Seneca both wrote treatises on Anger and on Consolation, while Plutarch wrote on 'tenderness of countenance' or shyness. *doctors.* Teachers. *painted forth.* Depicted. *contained . . . degree.* Restrained from leading to actions, and so becoming more intense. *fortify.* Gain strength. *inwrapped.* Interconnected. *fly.* Chase flying, hunt. *percase.* Perhaps. *recover.* Regain health, composure. *praemium . . . poena.* Rewards and punishments. Cf. Machiavelli, *Discourses*, i. 24, and (more closely) Guicciardini, *Ricordi*, no. 134: 'All men are by nature more inclined to good than evil, and there are none who would not more gladly do right than wrong, other things being equal. Yet man's nature is so weak, and so frequent in the world are the occasions which invite one to evil doing, that men easily allow themselves to be driven away from what is good. For this reason wise legislators invented rewards and punishments simply to hold men firm in their natural inclinations, through hope and fear' (Guicciardini, p. 35). *consist.* Subsist.

260. *those points.* Another list of topics needing investigation, to which Bacon devoted a number of *Essays* in 1612 and 1625. *as.* According as. *determinate.* Definite, dominant. *moralities.* Moral principles, ethics. *receipts and regiments.* Prescriptions and regimens for living. *compounded.* Made up. *described.* Wright (p. 320) suggests 'prescribed' as the correct reading. *Aristotle.* In *Eth. Nic.* ii. 1, 1103a19 ff. *consist.* Remain settled. *peremptory.* Positively fixed. *stand.* Stop. *latitude.* Freedom. The *De Augmentis* substitutes: 'intension and remission'. *strait.* Narrow. *wand.* Stick. *use.* Application, repetition. *allegeth.* Asserts. *allowing.* Accepting. *superinducing.* Inducing, causing. *too high a strain.* Too demanding, with too great expectations. *facility.* Easiness. *insatisfaction.* Dissatisfaction. *look.* Expect.

261. *several.* Different. *gain . . . step.* Make progress. *work out.* Get rid of. *knots.* Obstacles. *stonds.* Impediments. *inclined.* *Eth. Nic.* ii. 9, 1109b4 ff. (but Aristotle's discussion is not 'by the way'). The *De Augmentis* adds: 'so it be without vice' (*Works*, v. 25). *pretend.* Aim, aspire. '*tamquam . . . agendo*'. 'While you are attending to something else.' *conducted.* Regulated. *another nature.* Second nature. *ape.* Imitator. *counterfeit.* As the *De Augmentis* glosses, 'imitating nothing to the life' (*Works*, v. 26). *operation.* Effect. *caution.* Prudence, precaution. *direction.* Instruction. '*vinum daemonum*'.

'The wine of devils.' A fusion of remarks by St Augustine on the 'vinum erroris' in Terence (*Confessions*, I. xvi. 26), and St Jerome, that 'Daemonum cibus est carmina poetarum' ('the song of the poets is the food of demons': *Epist.* 146). Bacon's probable source is Henry Cornelius Agrippa, *De incertitudine et vanitate scientiarum et artium declamatio* (1530), ch. 4, who writes: 'Hinc Augustinus Poësim vocat vinum erroris, ab ebriis Doctoribus propinatum. Hieronymus eam daemonum cibum appellat' (*H. Cornelii Agrippae... Operum* (Leiden, 1550), 16). In James Sanford's Engl. tr. (1569, 1575): 'then Poetes write marveilous thinges, when they are mad or droncke. For this cause Augustine calleth Poetrie, the Wine of errour, ministred by drunken Doctours. Hierome nameth it the meate of Divels' (*Of the Vanitie and Uncertaintie of Artes and Sciences*, ed. Catherine M. Dunn (Northridge, Calif., 1974), 34). See now S. Chaudhuri, *Notes and Queries*, NS 34 (1987), 226–7. *Aristotle.* In *Eth. Nic.* i. 3, 1095a2 f.; the text actually reads 'political science', under which Aristotle subsumed ethics. Cf. Shakespeare, *Troilus and Cressida*, II. ii. 165 ff. *settled from.* Calmed down, quietened. *attempered.* Well-balanced. *popular.* Vulgar. *in their parasites' coats.* Shown in their true habit as the hangers-on of vice. *revolved.* Reflected on. *seasoned.* Matured. *differences.* Distinctions; criteria for evaluating. *utility.* Personal advantage, profit. '*Prosperum... vocatur*'. 'Successful crime is called a virtue' (Seneca, *Hercules Furens*, 251). '*Ille... diadema*'. 'For the same crime one man is hanged, another is crowned' (Juvenal, xiii. 105).

262. *positively.* Practically, as behaviour to be imitated. *Machiavel.* In *Discourses*, i. 10, freely paraphrased. Machiavelli's point is that Caesar's power was so great that none dared criticize him while he lived. *odious.* Deserving of hatred or repugnance. *reserved.* Excepted. *the world.* The *De Augmentis* reads 'the unconverted world' (i.e. non-Christian): *Works*, v. 27. *precise.* Strict in observing rules. *incompatible.* Unsociable. '*In... magistro*'. 'The divine and excellent qualities we see in Marcus Cato are his own; but the defects we sometimes find are derived not from his nature but from his teachers' (Cicero, *Pro Murena*, xxix. 61). *proprieties.* Properties, forms of behaviour. *accurate.* Worked out with care. *ground.* Basis. *state.* Condition. *perfect.* Free from defect. *evil.* The *De Augmentis* adds: 'out of the calendar' (*Works*, v. 27). *redemption.* Atonement for an offence. *inception.* Beginning. *de novo.* Anew. *compendious.* 'Containing the essential facts in a small compass' (*SOED*). *summary.* Concise. *reducing.* Bringing back, restoring. *true.* Loyal. *at once.* Simultaneously. *carver.* Sculptor.

263. *rude.* Unworked (from Lat. *rudus*, broken stone). *at one time.* All at once. *passage.* Movement, transit. *invested of.* Endowed with. *precedent.* Already existing. *which.* With regard to which. '*Immanitati... virtutem*'. 'To brutal vice we may naturally oppose that divine or heroical virtue which is above humanity' (Aristotle, *Eth. Nic.* vii. 1, 1145a18 ff. Here 'oppose' implies 'make correspond to', or 'juxtapose'). '*Nam... vitio*'. 'For as beasts are incapable of virtue or vice, so likewise is the Deity, for this latter state is something higher than virtue, as the former is something other than vice' (ibid. 1145a24 ff.). *celsitude.* Eminence. *funeral oration.* In fact, Pliny's

Panegyric to Trajan (i. 74), delivered in AD 100, when the Emperor was living. *pattern*. Model. *passages*. Observations. *Charity*. In the Christian sense of *agape*, love of the other. *bond of Perfection*. Col. 3: 14. *comprehendeth*. Includes. *vain*. Devoid of real value. '*Amor ... vitam*'. 'Love is a better teacher for human life than a left-handed Sophist': Stobaeus, *Florilegium*, IV. 20a. 10 (not Menander, in fact, but a fragment of Anaxandrides). *carry*. Behave. *preceptions*. Precepts. *dexterously*. Adroitly, skilfully. *prize*. Value highly. *sophist*. Pretender. *Xenophon*. In his *Symposium*, i. 10. *advance*. Improve, enlarge. *transgressed*. Exceeded the limits prescribed by God. '*Ascendam ... Altissimo*'. 'I will ascend and be like the highest' (Isa. 14: 14).

264. '*Eritis ... malum*'. 'You shall be as gods, knowing good and evil' (Gen. 3: 5). *similitude of*. Likeness to. '*Diligite ... injustos*'. 'Love your enemies, do good to them that hate you, and pray for them that despitefully use you and persecute you; that you may be the children of your father which is in Heaven, who makes his sun to rise on the evil and on the good, and sendeth rain on the just and on the unjust' (Matt. 5: 44; Luke 6: 27–8). *platform*. Model. '*Optimus Maximus*'. 'Best and Greatest.' '*Misericordia ... ejus*'. 'For his mercy is over all his works' (Ps. 145: 9). *Wherefore*. For which reason. *pretermitted*. Omitted. *judgeth well*. The *De Augmentis* substitutes: 'he is welcome to his opinion; but in the mean while let him remember that I am in pursuit, as I said at first, not of beauty but of utility and truth' (*Works*, v. 29). *Demosthenes. De Falsa Legatione*, 46. *Sunt ... manes*. 'Two gates of sleep there are, whereof the one is said to be of horn, and thereby an easy outlet is given to true shades; the other gleaming with the sheen of polished ivory, but false are the dreams sent by the spirits to the world above' (Virgil, *Aeneid*, vi. 893–6). Wolff (ii. 231) describes Bacon's interpretation of these lines as idiosyncratic, and suggests the influence of Servius' commentary. *sobriety*. Sober-mindedness, seriousness. *vaporous*. Intoxicating, deluding the brain. *braver*. Finer, more beautiful. *segregate*. Separate. *good of. ... body ... mind*. Cf. *Tusc. D.*, IV. xiii. 30, and *Advice to Rutland*, above, p. 69. *rational ... knowledges*. That is, with reference to what is known in reason and morals. *sound*. Strong, steadfast. *decency*. Comeliness, balance. *sever*. Are separated. *from*. Because of.

265. *carriage*. Deportment, bearing. *sufficiency*. Ability, capacity. *reformed*. Morally improved. *become themselves*. Act in a graceful or becoming manner. *sometimes*. The *De Augmentis* adds: 'while others, though perhaps endowed with all these three, yet from a stoical severity and insensibility have no pleasure in the virtuous actions which they practise' (*Works*, v. 30). *reduced to stupid*. Rendered insensible. *retain*. Be able to experience. *Civil*. Civic, involving man in society. *immersed in matter*. Densely circumstantial; existing as contingent phenomena. *hardliest*. With most difficulty. *reduced to axiom*. Abstracted or synthesized, so as to reveal general truths. *Cato*. In Plutarch's 'Life' of Cato, 8. *go*. To go. *difficile*. Hard to construct. *framing*. Fashioning. '*Sed ... suorum*'. 'Howbeit the people had not yet set their hearts aright to the Lord God of their fathers' (2 Chron. 20: 33). *as*.

Like. *engines*. Machines. *out of frame*. Out of order. *seven good years*. Gen. 41. *well grounded*. Having good foundations. *bear out*. Endure. *subverted*. Upset. *respects*. Considerations. *qualify*. Modify, moderate. *Conversation*. Social intercourse, involving both speech and behaviour. *Negotiation*. Dealings with other people. *comfort, use, and protection*. As the *De Augmentis* explains, 'comfort against solitude, assistance in business, and protection against injuries' (*Works*, v. 32). *do . . . sever*. Are found separately. *affected*. Aimed at. *into*. Upon. *government*. The *De Augmentis* (viii. 1) adds: 'For as gesture in an orator, though an external quality, is held of such account as even to be preferred to those other parts which appear more important and internal; so in a man of business conversation and the management thereof, though employed on external objects, finds, if not the highest, yet at all events an eminent place. For look what an effect is produced by the countenance and the carriage of it' (*Works*, v. 33). *Nec . . . tuo*. 'Let not your looks your word betray': Ovid, *Ars amatoria*, ii. 312; *Promus*, nos. 985, 1026.

266. '*Nil . . . clausum*'. 'It is of no use to have an open door and a shut countenance': from Quintus Tullius Cicero (attrib.), *De petitione consulatus*, xi. 44. *Caesar and Cicero*. In Cicero, *Epist. ad Att.* ix. 12, 18. *depending*. Impending. *carriage*. Behaviour. '*Ne . . . suae*'. 'That I may not appear to be either arrogant or servile; the arrogant man is unmindful of others' (Livy, xxiii. 12). *intended*. Studied, aimed at. '*quid . . . transferre*'. 'What is more unseemly than to carry the manners of the stage into real life?': a sentence Bacon quoted again in the *De Augmentis*, vi. 3. *from*. Against. '*Amici . . . temporis*'. 'Friends are the thieves of time.' *intending of*. Attention to. *discretion*. Concern with what is right or fitting in one's personal conduct. *honour*. Source of distinction. *urbanity*. Social polish. *becometh*. Is fitting. *puntos*. Polite formalities. *decency*. Decorum. '*Qui . . . metet*'. 'He that considereth the wind shall not sow, and he that looketh to the clouds shall not reap' (Eccles. 11: 4). Cf. Bacon's Essay 52, 'Of Ceremonies and Respects', above, pp. 441 f., for this whole passage. *curious*. Fastidious. *making*. Conformation. *strait*. Narrow, close-fitting. *elegantly handled*. In many Renaissance 'conduct books', such as Castiglione, *Il Cortegiano* (1528), Della Casa, *Galateo* (1558), and Guazzo, *La Civile Conversatione* (1574). *De . . . gerendis*. 'Of the conduct of affairs.' From the Latin *negotium*, the transaction of business, including the tasks and responsibilities of public life. *Negotiation*. Cf. the 1597 and 1625 Essays, 'Of Negotiation', above, pp. 88–9 and 435–6. *derogation*. Depreciation. *note*. Observation. *adage*. Cf. Advice to Rutland (above, p. 73) for the 'English proverb . . . that the greatest clerks are not the wisest men'. If really a proverb, it has not been traced. Wright suggests Diogenes Laertius, ix. 1: Heraclitus' saying, 'Much learning does not teach understanding'.

267. *for*. As for. *virtue . . . meditation*. Representing, respectively, the active and contemplative lives. *advertisements*. Informations. *as the other*. As of the others. *mean*. Moderate. *outshoot*. Surpass. *is laboured*. Has been studied. *reduced*. Systematized. *saddest*. Most serious. *professors*.

Exponents. *Cicero*. In *De oratore*, III. xxxiii. 133–5. *then*. In the times before his own. *opinion*. Reputation. *Place*. Market-place (here, the Roman Forum). *would use*. Wanted to have. *particular*. Individual. *cases*. Instances; but with some legal connotations, as with 'causes' (actions), used twice in this sentence. *Q. Cicero*. This book on the art of canvassing votes, ascribed to Cicero's younger brother, Quintus Tullius, was written in 63 BC to assist Marcus' election to the consulship. *on foot*. Under way. *popular elections*. Those where the people vote. *Scriptures*. 1 Kgs. 4: 29. *positions*. Maxims. '*Sed . . . tibi*'. 'Do not pay attention to everything men say, or you may hear your servant disparage you' (Eccles. 7: 21).

268. *concluded*. Demonstrated. *stay*. Stop, cessation. *judged*. See Plutarch, 'Life' of Pompey, 20, and 'Life' of Sertorius, 27.3. '*Vir . . . requiem*'. 'If a wise man goes to law with a fool, he will meet abuse or derision, but get no remedy' (Prov. 29: 9). *undertaking*. Contending with. *lighter*. Less serious. *engagement*. Enterprise. *heat*. Anger, quarrel. *change copy*. Change style or tactics. *no ways*. In no way. '*Qui . . . contumacem*'. 'Pamper a slave from boyhood, and in the end he will prove ungrateful' (Prov. 29: 21). *pitch*. Level. *unkindness*. Ingratitude. '*Vidisti . . . ignobiles*'. 'If you see a man that is quick in his business, he will stand before kings, not among common men' (Prov. 22: 29). *dispatch*. Performance, execution. *sufficient*. Competent, capable. '*Vidi . . . eo*'. 'I have studied all life here under the sun, and I saw his place taken by yet another young man' (Eccles. 4: 15). Cf. *Promus*, no. 1286, and *Religious Meditations*, above, p. 93. '*Plures . . . meridianum*'. 'There be more that worship the rising sun than the sun setting or at midday' (Tacitus, *Annals*, vi. 46; Plutarch, 'Life' of Pompey, 14 f.). '*Si . . . maxima*'. 'If your ruler breaks out in anger against you, do not resign your post; submission will make amends for great fortunes' (Eccles. 10: 4). *upon*. Out of, in reaction to. *retiring*. Withdrawing, opting out. '*Erat . . . pauperis*'. 'There was a small town with few inhabitants, and a great king came against it, and besieged it and raised great siege works against it. There was in it a poor, wise man, and he by his wisdom saved the city. But no one remembered that poor wise man' (Eccles. 9: 14–15). '*Mollis . . . iram*'. 'A soft answer turns away anger' (Prov. 15: 1). *present*. Immediate. '*Iter . . . spinarum*'. 'The path of the sluggard is a tangle of weeds' (Prov. 15: 19). '*Melior . . . principium*'. 'Better is the end of a speech than the beginning' (Eccles. 7: 8; *Promus*, no. 235). *taxed*. Attacked. *inducements*. Preambles, introductions.

269. '*Qui . . . veritatem*'. 'He who shows favour in judgement does not do well; even for a piece of bread will that man depart from the truth' (Prov. 28: 21). Cf. Essay 11, 'Of Great Place', above, p. 361. *were*. Had. *briber*. Taker of bribes. *lightly*. Easily. *facile*. Pliant, easily swayed. '*Vir . . . fames*'. 'A poor man that oppresses the poor is like driving rain, which produces a famine' (Prov. 28: 3). *extremity*. Extreme hardship. *necessitous extortions*. Forcefully extracting money, reducing people to the bare necessities of life. *fable*. See Aristotle's *Rhetoric*, ii. 20, 1393b23 ff. (a story resembling Aesop's fable of the fox and the fleas). '*Fons . . . impio*'. 'Like a muddied spring or a tainted well is a righteous man who gives way to a wicked one' (Prov. 25: 26).

judicial. Belonging to judges. *exemplar*. Conspicuous. *in ... world*. Publicly performed. '*Qui ... homicidii*'. 'He who robs his father and mother and says he has done no wrong, is a murderer' (Prov. 28: 24). *use to*. Usually. *extenuate*. Minimize. *upon*. Towards. '*Noli ... furioso*'. 'Never make friends with an angry man, nor keep company with a bad-tempered one' (Prov. 22: 24). *election*. Choice. *espouse*. Involve, contract. '*Qui ... ventum*'. 'He who brings trouble on his own family inherits the wind' (Prov. 11: 29). *domestical*. Domestic. *deceived*. Disappointed. '*Filius ... suae*'. 'A wise son brings joy to his father, a foolish son is his mother's bane' (Prov. 10: 1). *good ... ill proof*. Turning out well, badly. '*Qui ... foederatos*'. 'He who conceals another's offence seeks his goodwill, but he who harps on something breaks up friendship' (Prov. 17: 9). *reconcilement*. Reconciliation. '*In ... egestas*'. 'In every good work there is abundance; but where there are many words there is commonly penury' (Prov. 14: 23). '*Primus ... eum*'. 'In a lawsuit the first speaker seems right, until another steps forward and cross-questions him' (Prov. 18: 17). *causes*. Legal actions. *possesseth*. Prepossesses, influences opinion. *hardly*. With difficulty. *abuse*. Imposture, deceit.

270. '*Verba ... ventris*'. 'The words of the double-tongued man seem artless, and they go down into the innermost parts of the belly' (Prov. 18: 8). *set*. Carefully prearranged, elaborate. *artificial*. Full of art. *liberty*. Freedom, candour. *simplicity*. Innocence. '*Qui ... generat*'. 'He that instructs a scorner gets to himself shame, and he that rebukes the wicked gets himself a blot' (Prov. 9: 7; *Promus*, no. 226). *tender*. Offer. *contumely*. Insolent reproach. '*Da ... sapientia*'. 'Give opportunity to a wise man and he will increase his wisdom' (Prov. 9: 9; *Promus*, no. 350). *brought into habit*. Made customary, fully absorbed. *swimming ... conceit*. Superficial. *quickened*. Revived, animated. *redoubled*. Increased. *amazed*. Bewildered. '*Quomodo ... prudentibus*'. 'As the face is reflected in the water, so is the heart of man manifest to the wise' (Prov. 27: 19). *glass*. Mirror. '*Qui ... erit*'. 'The wise man will know how to adapt himself to all sorts of characters' (Ovid, *Ars amatoria*, i. 760). *authority*. 'Power to influence action, opinion, belief, etc.' (*SOED*). *attended*. Accompanied. *the Eagle*. Symbolizing insight and power. *discourse*. Discussion. *broken*. Divided up into constituent parts. *deducements*. Deductions. *wisdom ... fable*. See Bacon's *De Sapientia Veterum* (1609), the 'Wisdom of the Ancients', a collection of 31 classical fables reinterpreted (often idiosyncratically) to show their relevance to Bacon's ideas on natural philosophy, politics, and other topics: *Works*, vi. 607–764. *for*. As for. *vicegerents*. Substitutes. *supplies*. Stores. *mark*. Target. *variable*. Of different kinds or degrees; hard to systematize. *negotiation*. Transaction of business. *Machiavel*. In the *Discourses*.

271. *view*. Sight. *attendeth upon*. Accompanies. *point of order*. Question of arrangement. *ground*. Basis. *history*. Inquiry. *control*. Regulate, dominate. *supply*. Assist, support. *servile*. Subordinate. *histories*. Wright (p. 325) suggests 'history' as the correct reading. *as*. All the early editions read 'is'. Spedding reads 'as'; Wright (p. 325) emends to 'because it is'. *Cicero*. The *Epist. ad Att.*, together with the three other collections by Cicero, rediscovered by Petrarch in the late 14th century, were key documents for the Renaissance understanding of Roman history and civilization. *great.* Wright (p. 325) suggests 'greater', the Latin text at this point reading *magis. particular.* Detailed. *sapere . . . sibi sapere.* 'To be wise' and 'to be wise for oneself'. Cf. *Promus*, no. 1001: 'Necquiquam sapit qui sibi non sapit' ('He is wise to no purpose who is not wise for himself'), from Erasmus' *Adagia*, and Essay 23, 'Of Wisdom for a Man's Self', above, pp. 386–7. *counsel.* Advising others. *pressing.* Pursuing, promoting. *sever.* Are separated. *take . . . knowledge of.* Recognize. '*Nam . . . sibi*'. 'For a wise man fashions his fortune for himself' (Plautus, *Trinummus*, II. ii. 82). *comical.* Writer of comedies. '*Faber . . . propriae*'. 'Every man is the architect of his own fortune': a saying attributed to Appius Claudius (*fl.* 300 BC), which became a key concept in Bacon's thought, as can be seen from *Promus*, no. 357; the *Discourse to Sir Henry Savile*, above, pp. 115–16; and the essays 'Of Fortune', and 'Of Wisdom for a Man's Self'. '*In . . . videretur*'. 'Such was his force of mind and genius that wherever he had been born he would have made his fortune' (Livy, XXXIX. 40). *position.* Proposition, maxim. *estate.* State. *particular.* Item; the anecdote comes from Plutarch's 'Life' of Sulla, vi. 3 ff.

272. *high.* Conceited. '*Dicis . . . ipsum*'. 'My river is mine own, it was I who made it' (Ezek. 29: 3). *prophet.* Hab. 1: 16. '*Dextra . . . adsint.*' 'Let this right hand, my god, and the hurtling weapons I poise, now be my aid' (Virgil, *Aeneid*, X. 773). *confidences.* Excessive assurance, presumption. *felicity.* Good fortune. *Felix, not Magnus.* In Plutarch's 'Life' of Sulla he is called 'the Fortunate', not 'the Great'. *Caesarem . . . ejus.* 'You carry Caesar and his fortune' (Plutarch, 'Life' of Caesar, 38). *Faber . . . vitae.* 'The architect of fortune, or of the conduct of life.' '*Sapiens . . . astris*'. 'A wise man shall rule the stars': a saying ascribed to Ptolemy by Cognatus, and repeated in innumerable handbooks to astrology. '*Invia . . . via*'. 'There is no path that virtue cannot tread' (Ovid, *Met.* XIV. 113). *spurs.* Encouragements. *stirrups.* Supports. *resolution.* Steadfastness of purpose. *presumption.* Arrogance. *sensible of.* Sensitive to. *died.* Suetonius, 'Octavius', 99. *Plaudite.* 'Give your applause', traditionally the closing words of Roman comedy. *conscient.* Conscious. *heads.* Headings, topics. *yield.* Acknowledge. *impositions.* Burdens. *politique.* A shrewd and expedient man. *because.* In order that. *pragmatical.* Skilled in business. *holdeth . . . of.* Behaves like.

273. *should.* Could. *globe of crystal.* Perhaps alluding to a mystical passage in Rev. 4: 6, which Bacon later quotes (p. 286) and interprets as describing 'the government of God, that this globe, which seemeth to us a dark and shady body, is in the view of God as crystal'. Cf. also Bacon's metaphor of this book as constituting 'a small Globe of the Intellectual World' (above, p. 299). But Wright (p. 376) suggests that Bacon's distinction between a 'globe of matter' and a 'globe of crystal or form' derives from Empedocles' notion of a *sphairos aisthetos* and a *sphairos noetos*, as interpreted by Proclus in his *Commentary on the Timaeus*, 160D; cf. also Simplicius' commentary on Aristotle's *Physics*, 7b. *esteem of.* Reckon. *end.* Goal—that is, there are higher things in life than worldly success. *organ.* Instrument, means of

action. *respects.* Considerations. *summary.* Essential, effective. *Momus.* The god of raillery, who complained that Vulcan, in making man, had not provided a window in the breast: Lucian, *Hermotimus*, 20. *obnoxious.* Exposed. *dependances.* Dependents. *opposites.* Opponents. '*Sola ... noras*'. 'Thou alone knowest the hour for easy access to him' (Virgil, *Aeneid*, iv. 423). *what.* Which. *on foot.* Under way. *import.* Mean. *another.* Different. *surety.* Guarantor. '*Consilium ... illud*'. 'Counsel in the heart of man is like deep water; but a man of understanding will draw it out' (Prov. 20: 5). *knowledge ... individuals.* Such information is hard to generalize, since it derives from many discrete particulars. *sinews of wisdom.* A saying of Epicharmus, quoted by Cicero, *Epist. ad Att.* i. 19, and by Q. Cicero in *De petitione consulatus*, x. 39. Cf. also Guicciardini, *Ricordi*, no. 201: 'But this is the main thing: do not trust anyone without good reason' (ed. cit. 51). *surprised.* Unprepared. *set.* Carefully composed. *purposed.* Deliberated. '*fronti nulla fides*'. 'Trust not to a man's face' (Juvenal, ii. 8).

274. *Q. Cicero.* In *De petitione consulatus*, xi. 44. *close.* Secret. '*Etenim ... conjectaverat*'. 'He guessed that he was offended by his looks' (Tacitus, *Annals*, i. 12). *carried.* Delivered. '*Magis ... oratione*'. 'It was in words too laboured and specious to be taken for what he really felt. But of Drusus, he said less, but more earnestly, in a style of sincerity': ibid. i. 52. '*velut ... subveniret*'. 'Of a kind of struggling speech', but then again 'he spoke with more freedom when he was speaking in a man's favour': ibid. iv. 31. *commanded.* Controlled. '*vultus jussus*'. 'Prescribed face.' *sever.* Remove, detach. *fashions.* Modes of behaviour. *wandering.* Digressing. *drily and hardly.* With stiffness and difficulty. *pledges.* Tokens of something to come. '*Fraus ... fallat*'. 'The treacherous man begins by being faithful in small things, that he may afterwards deceive with greater profit' (Livy, xxviii. 42). *Italian.* Referring to the proverb 'Chi mi fa più caresse chè non suole | O m'a ingannato, o ingannar mi vuole'. *For.* As for. *favours.* Kindnesses. '*Alimenta socordiae*'. 'Sops to feed sloth': Demosthenes, *Olynthiacs*, iii. 33, in Wolf's Latin tr. (Basle, 1572). *hollow.* Insincere. *reconcilement.* Reconciliation. *advanced.* Promoted. '*simul ... largitur*'. 'At the same time he bestows tribuneships and governments on his friends' (Tacitus, *Histories*, iv. 39). *desolate.* Render desolate, forlorn. *dependances.* Dependents. *waters.* Urine (the main diagnostic source in early medicine). *flattery.* Insincerity, deception. *advantage.* Heightened effect. *incensing.* Exasperating. *forth of.* Out of. '*Audita ... regnaret*'. 'These words drew from Tiberius the voice, so rarely heard, of his secret heart; he retorted upon her, in a Greek verse, that she was thus offended because she did not reign' (*Annals*, iv. 52). *Vino ... ira.* 'Tortured with wine and wrath' (Horace, *Epist.* I. xviii. 37 f.).

275. *true to themselves.* Self-reliant; pursuing their own ends. *settled.* Composed. *upon.* Out of. *heat.* Anger. *bravery.* Daring, bravado. '*Di ... truth*'. *Promus*, no. 267. *fame.* Rumour, report. *light.* Unreliable. '*Verior ... emanat*'. 'The truest character comes from a man's own household' (Q. Cicero, *De petitione consulatus*, v. 17). *weakest ... wisest.* The *De Augmentis* specifies: 'the weaker and the more simple', and 'the wiser and the more

close' (secretive): *Works*, v. 62. *nuncio*. Papal ambassador to a foreign government. *lieger*. Permanent representative. *did*. Would. *shoot over*. Overshoot the mark. *compass reaches*. Roundabout schemes or tricks. *Di... credi*. *Promus*, no. 266. *account*. Count upon. *other*. Different. *take measure and scale*. Form a just estimate. *inscrutable*. The *De Augmentis* identifies '(as the Scriptures declare) inscrutable' (cf. Prov. 25: 3), and adds: 'But every private person is like a traveller striving earnestly to arrive at the end of his journey where he may rest; whereby it is not difficult to conjecture what he will do, and what he will not do. For if it be a means to his end he will probably do it; but if opposed to his end, he will probably not do it' (*Works*, v. 63). *inform ourselves in*. Instruct ourselves with regard to. *humour*. When one of the four 'humours' gained 'predominancy', according to Renaissance medical theory, it resulted in imbalanced behaviour. '*metus... rimatur*'. 'He wrought on [Nero's] fears' (Tacitus, *Annals*, xiv. 57). Bacon has conflated this episode, where the 'mischievous arts' of Tigellinus Ofonius caused the death of Rubellius Plautus, with a later one (ibid. xvi. 18), where his jealousy of Petronius (not Petronius Turpilianus but Caius Petronius, the *elegantiae arbiter*) provoked another malicious plot.

276. *inwardness*. Close friendship, intimacy. *look*. Insight, knowledge. *according to the diversity... trust and inwardness*. Cf. Guicciardini, *Ricordi*, no. 186: 'One cannot in fact always proceed according to one settled rule. If it is often a mistake to talk too much even with friends—I mean of matters which ought to be kept secret—on the other hand, if you show your friends that you are reserved with them, it is the way to make them reserved with you. For nothing inspires people to confide in you more than the belief that you are confiding in them. Thus if you do not tell anything, it removes the possibility of your learning anything from others. Hence in this and many other matters you must proceed by distinguishing the qualities of persons, circumstances, and times, and for that discretion is needed' (ed. cit. 47). *which*. Who. *intelligenced*. Informed. *liberty of speech*. Openness, frankness. *importeth*. Is important. *reducing*. Recalling, turning back. *serene*. Tranquil, unperturbed. *Et... servare*. 'I both wish to do this and also to keep to any rule' (Epictetus, *Enchiridion*, 4). '*Et... addiscere*'. 'I both wish to do this and also to learn something from it.' *stayed*. Dwelt on. *answereth*. Is equivalent. *stay*. Control. *draw on*. Induce. *but only*. Only. *St. James*. Jas. 1: 23–4. *suddenly*. Quickly. *sorteth with*. Agrees with. *fit*. Well adapted. *dissonant*. Conflicting. *close*. Concealed.

277. '*Alia... via*'. 'The habits of Tiberius were different' (*Annals*, i. 54). *make election*. Choose. *engaged*. Already involved. *Valentine*. Caesar Borgia (1476–1507), son of Pope Alexander VI, who was trained up for an ecclesiastical career, becoming a cardinal and archbishop of Valencia. But in Aug. 1498 he abandoned his church career for a military one, being created Duke of the Valentinois: Guicciardini, *Storia d'Italia*, iv. 5: '... the Cardinall (joyning with the intencion of his father) entred one morning into the consistorie, and with an action in speeche and gesture farre from the office and modestie of his profession, besought his father and the whole colleadge of Cardinalls, that seeing he entred not into the priesthood of his proper will and disposicion, that they woulde make it lawfull to him, to leave the dignitie and the habitte to follow the exercise [i.e. of arms] whereunto his destinie and inclinacion drew him' (tr. Fenton, 1579, p. 203). *to*. For. *parts*. Qualities. *concurrents*. Rivals. *solitude*. Absence of people. *like*. Likely. *Caesar*. As reported in Plutarch's 'Life' of Caesar, 3. *civil*. Civic, in rhetoric and politics. *popular*. Appealing to public audiences. *dependances*. Dependents. *Caesar*. Caesar's friends (Antony, Dolabella, Balbus, Pollio) were not outstanding, but loyal: as the *De Augmentis* adds, 'displaying an infinite affection for Caesar, but arrogance and contempt towards everyone else' (*Works*, v. 66). *effectual*. Effective, energetic. *solemn*. Grave, decorous. *carriages*. Forms of behaviour. *Cicero. Epist. ad Att.* ix. 10. '*Sulla . . . potero?*' 'Sulla could do it, why not I?' *abused*. Deceived. *unlikest*. Most unlike. *pressing the fact*. Urgently pursuing the business in hand. *insist*. Dwell on. *well opening*. That is, the *De Augmentis* adds, 'the best way to set himself forth to advantage' (*Works*, v. 66). *artificial*. Skilful. *staying upon*. Emphasizing. *sliding from*. Evading. *cherishing*. Protecting. *circumstances*. External conditions affecting (e.g. extenuating) action. *gracing*. Enhancing.

278. *exposition*. Explanation. '*Omnium . . . ostentator*'. 'He had a certain art of setting forth to advantage everything he said or did' (*Histories*, ii. 80). *though*. Even though. *manners*. Polite behaviour. *policy*. Prudence or skill in the conduct of public affairs. '*Audacter . . . haeret*'. 'Slander boldly, for some of it will stick' . . . 'boldly sound your own praise, for some of it will stick': cf. Plutarch, 'How to tell a Flatterer from a Friend', *Moralia*, 65D, and Guicciardini, *Ricordi*, no. 37: 'Always deny what you do not wish to be known, and affirm what you wish to be believed, for . . . a bold statement or denial often sows doubt in the mind of those who hear it' (ed. cit. 15). *with*. In the eyes of. *government*. Self-possession, control. *ingenious*. Ingenuous. *unsafety*. Insecurity. *careless*. Unstudied (the quality of easy mastery, scorning effort, that Castiglione described as *sprezzatura*). *taxing*. Censuring. *gracing*. Praising, complimenting. *putting down*. Crushing. *injury*. Offensive speech. *solid*. Serious, substantial. *want*. Lack. *ventosity*. Vanity. *height of the winds*. When strongest. *flourishes*. Ostentatious embellishments. *disvalued*. Depreciated. Le Doeuff (p. 364) cites Montaigne's Essay 'De l'exercitation' (ii. 6; Pléiade edn. 359 f.) for a similar discussion of under- and over-valuing oneself. *obtruding*. Thrusting forward, without invitation. '*Cave . . . delectat*'. 'If he take so much delight in a little thing, he will be thought unused to greater things' (*Rhetorica ad Herennium*, IV. iv. 7). *Colour*. By a specious appearance, pretext. *Confidence*. Self-assurance. *difference*. Differentiation, careful choice. *wants*. Defects. *construction*. Interpretation. '*Saepe . . . boni*'. 'Vice often hides itself in the neighbourhood of virtue' (Ovid, *Ars amatoria*, ii. 662).

279. *pretend*. Falsely assume. *shadoweth*. Lies closest to. *affect gravity*. Profess seriousness. *use*. Practice. *industries*. Devices, crafty expedients. *depress*. Disparage. *passeth*. Surpasses (in impudence). *face out*. Brazen out. *failing*. Weak. *except*. Take exception to. *number*. The *De Augmentis* adds:

'and the least open to objection, and seeming to suspect it themselves they will ask your opinion of it' (*Works*, v. 69). *carriage*. Behaviour, conduct. *dismantled*. Stripped of disguise. *injury*. Insolence, contumely. *dulceness*. Sweetness. *facility*. Pliancy. *sparkles*. Sparks. *rescussing*. Rescuing. *person*. Body. *fortune*. The *De Augmentis* adds: 'as in the case of persons deformed, illegitimate, or disgraced. Whence men of this nature, if ability be not wanting, commonly turn out fortunate' (*Works*, v. 69). '*Idem . . . decebat*'. 'He continued the same, when it no longer became him' (Cicero, *Brutus*, xcv). *men . . . turn*. Men are left behind in their previous place when circumstances change. '*versatile ingenium*'. 'A versatile nature'; *Livy*, xxxix. 40. *make departures*. Depart from their normal behaviour. *felicity*. Good fortune. *nature*. Natural. *viscous*. Adhesive, sticky. *inwrapped*. Engrossed in themselves. *conceit*. Idea. *Machiavel*. *Discourses*, iii. 9. Fabius Maximus (d. 203 BC) was nicknamed 'Cunctator', the Delayer, for his successful tactics in the war against Hannibal. *bias*. Inclination.

280. *period*. Conclusion. *Demosthenes*. *Phil*. i. 46. *have*. Receive. *ward*. Defensive gesture. *lothness*. Unwillingness. *leese*. Lose. *passed*. Past. *ply*. Inclination. *Tarquinius*. As recorded in Aulus Gellius, *Noctes Atticae*, i. 19. See Bacon's Essay 21, 'Of Delays', above, p. 383. *simple*. Single. *politic*. Expedient, skilfully contrived. *wheels of our mind . . . wheels of fortune*. Cf. Guicciardini, *Ricordi*, no. 31: 'Even those who would ascribe everything to prudence and virtue and exclude as far as possible the power of fortune must at least agree that it is very important to come upon, or be born into, times when your own best qualities are of value. One may quote the example of Fabius Maximus, who acquired such great reputation from his dilatory nature. For he found himself in the kind of war in which eagerness would be pernicious, and delay useful. . . . Any man who could adapt his nature according to the conditions of the times—most difficult and maybe impossible—would be so much the less ruled by fortune' (ed. cit. 13). *concentric*. Moving around the same centre. *voluble*. Capable of revolving. '*Fatis . . . Deisque*'. 'Yield to destiny and the gods' (Lucan, *Pharsalia*, viii. 485). *passable*. Manageable. *foil*. Defeat. *most*. Majority. *repugnancy*. Contradiction, inconsistency. *Demosthenes*. *Phil*. i. 39, in Wolf's Latin translation (Basle, 1572). *high*. Eloquent, impressive. '*Et . . . cogantur*'. 'As the captain leads the army, so should wise men lead affairs; they should get that done which they think good to be done, and not be forced to follow at the heel of events.' *plot*. Conceive, initiate new projects. *accommodate*. Adapt to circumstances. *take in*. Absorb. The *De Augmentis* rephrases: 'cannot take advantage of accidental opportunities' (*Works*, v. 74). *mediocrity*. Middle way. *declaring*. Disclosing. '*qualis . . . mari*'. Prov. 30: 19, 'the way of a ship out at sea'. *sourdes menées*. Underhand practices. *set . . . work*. Put in motion. *opening*. Revealing. '*Dissimulatio . . . illaqueant*'. 'Dissimulation breeds mistakes in which the dissembler himself is caught.'

281. *free*. Open. *professed*. Avowed. *Sulla*. In Plutarch's 'Life' of Sulla, 38.4 (Wright); Wolff (ii. 89) suggests ibid. 3.3. *Caesar*. In Plutarch's 'Life' of Caesar, 11. 2. '*Alter . . . tyrannus*'. *Epist. ad Att*. x. 4: 'He does not refuse, but in a manner demands, to be called what he is—tyrant.' '*Ita ... liceat*'. Ibid. xvi. 15: 'By the hope of attaining his father's honours.' *tyranny*. Absolute power. *help*. Emphasize. *whereat many men*. The reading of the 1629 and 1633 edns.; 1605 has 'and men'. *did it*. Seemed to say what he felt. *Pompey*. Gnaeus Pompeius (106–48 BC), Roman general and statesman, appointed sole consul in 52, defeated by Caesar at Pharsalus in 48. '*Occultior ... melior*'. 'A more reserved but not a better character' (*Histories*, ii. 38). '*ore ... inverecundo*'. 'Of honest tongue and shameless mind' (Sallust, as quoted by Suetonius, *On Grammarians*, xv). *engines*. Plots, contrivances 'requiring *ingenium* or skill' (Wright). *fain*. Forced. *by colour*. Under the pretext. *doubt*. Uncertainty. *tedious*. Troublesome. *casual*. Uncertain, subject to accident. '*Et ... composita*'. 'She was a match for the diplomacy of her husband and the dissimulation of her son' (*Annals*, v. 1). *substantially*. In a sound or solid manner.

282. *proportions*. Relative value. *shew*. Appearance. *sense*. Appealing to the senses. *access*. Having contact with. *purchase*. Value. *ever*. Always. '*Haec ... agebat*'. 'He did all these things with great energy' (*De bello civili*, i. 30). *greatest*. The *De Augmentis* clarifies: 'men often deceive themselves, in thinking that if they procure the assistance of any man of worth and reputation, they are certain to succeed; whereas it is not the greatest but the fittest instruments that finish the work both quickest and best' (*Works*, v. 72). *marshalling*. Arranging in methodical order. *amendment*. Correction, improvement. *remove*. Removal. *Machiavel*. In *Discourses*, ii. 10. Cf. Cicero, *Philippics*, v. 2.5, 'Nervos belli pecuniam infinitam'. *populous*. Well-peopled. *voucheth*. Cites. *Solon*. Lucian, *Charon*, 10–12. *peremptory*. Fixed, unchangeable. *extreme*. Extremely. *after-game*. Second game, return match. The *De Augmentis* has: 'to restore a falling reputation' (*Works*, v. 73). *matter*. Subject-matter, affairs. *preposterous*. Inverted order, placing the first last. *intend*. Direct attention to. *instance*. Priority, urgency. '*Quod ... agamus*'. 'Let us to the task in hand' (Virgil, *Eclogues*, ix. 66).

283. '*Sed ... tempus*'. 'But time meanwhile is flying, flying beyond recall' (Virgil, *Georgics*, iii. 284). *professions of burden*. Involving hard work and responsibility. *painful*. Taking pains, diligent. *divines*. Preachers, theologians. *politic*. Skilfully contrived. *plots*. The *De Augmentis* continues: 'plots to advance their fortunes. Moreover, in the courts of princes and in commonwealths you will find that the ablest persons both to improve their own fortunes and to assail the fortunes of others are those who have no duty to perform, but are ever occupied in this study of advancement in life' (*Works*, v. 74). *interlace*. Vary by mixing. *under*. Subordinated to. *somewhat else*. That is, to turn his labour taken therein to some other use. *exact an account*. Take stock of his doings. *reap*. Benefit. *mind*. Concern oneself with. '*Haec ... omittere*'. 'These things ought you to do without leaving the others undone' (Matt. 23: 23; Luke 11: 42). *way*. Road. *retire*. Withdraw. *plash*. Shallow pool. *moved*. Proposed. *dry*. Dry out. *construed*. Interpreted. '*Et ... amaturus*'. 'Love as if you were sometime to hate, and hate as if you were sometime to love'. In *Rhetoric*, ii. 13, 1389b23 f., Aristotle

attributes this saying to Bias (*fl.* 550 BC); also Cicero, *De Amicitia*, 16. *betrayeth.* Disappoints, frustrates. *utility.* Usefulness. *spleens.* Fits of anger. *humorous.* Fanciful, capricious.

284. *would not.* Do not want to. *imaginative.* Imaginary. *in the air.* Unfounded. *hardlier.* With more difficulty. *tractates.* Treatises. *patterns.* Examples. *plain.* Straightforward. *Cicero.* In *De oratore* and *Orator.* *pleader.* Lawyer. *a Prince... a Courtier.* The treatises by Machiavelli and Castiglione. *mould.* Model. '*bonae artes*'. 'Good, honest arts.' *Machiavel.* In *The Prince*, ch. 17. *cumber.* An encumbrance. *other. Prince*, ch. 18. *obnoxious.* In dread of punishment, submissive. *low.* Cast down. *in strait.* In need, dependent. '*Cadant... intercidant*'. 'Let friends fall, provided our enemies perish with them' (Cicero, *Pro Rege Dejotarus*, ix. 25. *Triumvirs.* See Shakespeare's *Julius Caesar*, IV. i. *every.* Each. *protestation.* Solemn affirmation, principle. *L. Catilina.* Lucius Sergius Catilina (*c.*108–62 BC), Roman conspirator whose plots were revealed and routed by Cicero. *trouble.* Disorder. *to the end.* In order. *droumy.* Turbid, muddy (where fish are more easily caught). *unwrap.* Liberate; promote. '*Ego... restinguam*'. 'If fire breaks out among my possessions I will put it out not with water but with destruction' (Cicero, *Pro Murena*, xxv. 51; Sallust, *Catiline*, xxxi. 9). *Lysander.* In Plutarch's 'Life' of Lysander, 8.4. *comfits.* Sweets. *positions.* Principles, maxims. *pressing.* Pursuit, urging. *compendious.* Quickly brought about. *ways.* Paths, roads. *about.* Longer.

285. *map.* Description. '*all things... spirit*': Eccles. 2: 11. *cards.* Maps. *directions.* Instructions (as, how to find one's way). *well-being.* Living a life of virtue ('well-doing'). *Quae... vestri.* 'What recompense, O friends, can I hold out worthy such deeds? The best is that ye have—God's blessing and your proper nobleness' (Virgil, *Aeneid*, ix. 252–4). '*He hath... thing*'. Job 15: 35. *sabbathless.* Unresting. *tribute.* Paying what is due. *a tenth.* A 'tithe' or a tenth of the produce of land and stock, originally allotted for church purposes. *seventh.* The sabbath day, devoted to God: cf. Exod. 20: 10. *grovelling... earth.* Worshipping earthly things. '*Atque... aurae*'. 'The particle of the divine spirit cleaves to the ground' (Horace, *Satires*, II. ii. 79). *flatter.* Deceive. *ill.* Evilly. *that either... died.* Aurelius Victor, *Epitome de Caesaribus*, i. 28 and Spartianus, 'Life' of Septimius Severus, 18. *mischief.* Wickedness. *purposed.* Intended. *Charles the fifth.* King of Spain and Emperor of Germany, who abdicated in 1556 in favour of his son, Philip II. The saying is ascribed to him after raising the siege of Metz. *tastes.* Judgements (perhaps alluding to the Fall). '*Primum... vobis*'. 'Seek ye first the kingdom of God, and all those things shall be added unto you' (Matt. 6: 33).

286. '*Primum... oberunt*'. 'Seek ye first the good things of the mind, and the rest will be either supplied, or their loss will not be felt.' Untraced; perhaps Bacon's coinage, modelled on the biblical quote preceding. Le Doeuff (p. 366) suggests an imperfect recollection of the opening of Epictetus' *Enchiridion.* *sand.* Matt. 7: 24–7. *Te... es.* 'I took thee, Virtue, for a reality, but I

find thee an empty name': Brutus' dying words, according to Dio Cassius, xlvii. 49. Wolff (ii. 59) notes that Bacon took this quotation not from Xylander's Latin translation of Dio Cassius, but from Plutarch, 'Of Superstition', *Moralia* 165A. *Government*. Political rule and administration. *retired*. Hidden. *Totamque ... miscet*. 'The mind works in each member of the frame, and stirs the mighty whole' (Virgil, *Aeneid*, vi. 726–7). *participate of*. Share. *inward*. Secret. *profound*. Deep-seated. *passages*. Transactions, courses. *hardly*. With difficulty. *reduced to demonstration*. Explained; made evident. *wisdom of antiquity*. Ancient myth. *shadows*. Relics, images. *Giants'*. Their rebellion against Zeus is recorded in Hesiod's *Theogony*. *futility*. Idle talkativeness, inability to keep silent. Cf. Homer, *Odyssey*, xi. 582–600; Cicero, *Tusc. D.* I. v. 10, IV. xvi. 35. *Sisyphus*. Punished for betraying the designs of the gods. *Tantalus*. Punished for divulging the secrets of Zeus. *particulars*. Detailed instances. *discourses*. Discussions. *reserved*. Reticent. *toward*. In relation to. *shady*. Obscure. '*Et ... crystallo*'. 'And before the throne there was a sea of glass, like unto crystal' (Rev. 4: 6). *necessities*. Needs. *combinations*. Associations of people for an illegal purpose. *intelligences*. Information. *assisted*. By the Privy Council.

287. *certificate*. Declaration, qualification. *contended*. Competed. *that ... peace*. Told of Zeno in Plutarch, 'Of Garrulity', *Moralia*, 504A; Diogenes Laertius, vii. 24. *De ... juris*. 'The wisdom of a legislator, or the fountains of equity.' *imaginary laws*. Plato, in the *Republic* and *Laws*; Sir Thomas More, in *Utopia*. *received*. Accepted. *fountains of justice*. Principles of equity. *tinctures*. Colourings. *platform*. Model. *doubtfulness*. Ambiguity. *meum and tuum*. Mine and thine (referring to belongings). *into*. Upon. *apt*. Appropriate. *penned*. Written, draughted. *Texts ... or Acts*. Either promulgated in writing (like the Ten Commandments, or the Roman Ten Tables), or established by the decree of the authority responsible. *vast*. Extensive. *crossness*. Intricacy; or perhaps 'inconsistency, contradiction'. *causes*. Legal proceedings. *emergent*. Arising, in progress. *judicially*. In a lawcourt. *responses*. Judgments given after specific enquiries. *conferences*. Legal discussions. *pressed*. Enforced. *mitigated*. Moderated, made less severe. *discretion*. 'A court's degree of freedom to decide a sentence, costs, procedures, etc.' (*SOED*). *strict*. Rigorously enforced. *censured*. Assessed. *animation*. Imparting life to. *a work*. In the *De Augmentis* Bacon substitutes for this brief discussion an extended 'Example of a Treatise on Universal Justice or the Fountains of Equity, by Aphorisms' (*Works*, v. 88–109).

288. *laws*. Comprising common law, 'the unwritten law of England, administered by the King's courts, based on ancient and universal usage, and embodied in commentaries and recorded cases' (*SOED*), and statute law, laid down in statutes passed by Parliament. *civil laws*. The *Corpus iuris civilis* or digest of Roman law made for the Emperor Justinian in the 6th century AD. '*non ... usus*'. 'A gift meant for different uses' (Virgil, *Aeneid*, iv. 647). Cf. *Charge touching Duels*, above, p. 313. *action*. Involving social practices and policies. '*si ... imago*'. 'If the image never lies' (Virgil, *Eclogues*, ii. 27). *nothing*. In no respect. *third visitation*. After the Greeks and Romans, the

modern age. *wits.* Intellects. *lights.* Illumination. *by.* From. *experiments.* Experience. *popularity.* Democratic character, involving constant citizen participation. *consumption.* Wasting away. *diverted.* Distracted. *vollies.* Flights (as of birds). Cf. Tacitus, *Annals*, vi. 28 on the phoenix, which, whenever it appears, attracts 'a great escort of common birds'. *propriety.* Nature. *time... truth.* Alluding to the old saying, 'Truth is the daughter of time'. *persuasion.* Conviction. *only if.* If only. *one... other.* From each other. *enterprise.* A hazardous or arduous undertaking. *quality.* Accomplishment. *magnificence.* Liberality of expenditure. *to.* For, upon.

289. *reprehension.* Censure, rebuke. *they.* The labours. '*Verbera... audi*'. 'Strike, but hear': Plutarch's 'Life' of Themistocles, 11; *Apophthegmata*, 185B; *Promus*, no. 1134. *so.* Provided that. *Sabbath.* Harbour, resting-place. *prerogative.* Exclusive right, privilege. *reluctation.* Reluctance, struggle. *accounted... for righteousness.* Rom. 4: 22. *Sarah.* She laughed when God told Abraham that she would bear a son in her old age, but Abraham believed: Gen. 18: 12. *natural reason.* Human understanding, lacking divine enlightenment. *suffereth from sense.* Is at the mercy of the senses. The *De Augmentis* adds: 'sense, which is the reflection of things material' (*Works*, v. 112). That is, knowledge derives from the *anima sensibilis* (that part of the soul or mind receiving the impressions of material things through the senses), faith from the *anima rationalis*, the rational soul, which derives its impressions from things spiritual. *such one.* Such a one. *authorised.* Gifted with authority. *Otherwise it is.* It is different. *faith shall cease.* 1 Cor. 13: 12. *grounded.* Based. '*Coeli... gloriam Dei*'. 'The heavens declare the glory of God': Ps. 19: 1. '*Coeli... voluntatem Dei*'. 'The heavens declare the will of God.' '*Ad... istud*'. 'To the law and to the testimony: if men do not according to this word [it is because there is no light in them]' (Isa. 8: 20). '*Love... unjust*'. Matt. 5: 44–5. *it... applauded.* This applause ought to be given. '*Nec... sonat*'. 'This voice has not a human ring' (Virgil, *Aeneid*, i. 328). *light of nature.* Unaided human reason. *So.* Inversely (the *De Augmentis* reads *Quinetiam*: *Works*, i. 831). *fall upon.* Come across. *libertine.* Indulging the pleasure of the senses. *expostulate with.* Protest against. *opposite.* Contrary. '*Et... negant*'. 'What nature grants the envious laws deny' (Ovid, *Met.* x. 330–1). *Dendamis.* In Plutarch's 'Life' of Alexander, 65.2.

290. *several.* Different. *inward.* Innate. *conscience.* An innate knowledge or consciousness. *sparkle.* Spark. Le Doeuff (p. 370) cites St Jerome's notion of 'a spark of conscience inextinguished in the soul of Adam', which patristic writers described as *synderesis*, the innate moral sense. *first estate.* Original state, before the Fall. *participant.* Partaking. *discerning.* Understanding. *inform.* Teach, instruct. *latitude.* Scope. *our... God.* St Paul, in Rom. 12: 1. *old law.* The covenant of the Old Testament. *signification.* Meaning. *non-significants... surd characters.* Lacking sound or meaning. *golden mediocrity.* *Aurea mediocritas*, the golden middle point. *law of Mahumet.* Islamic law. *interdicteth.* Forbids. *imposture.* Le Doeuff (p. 370) suggests a sense derived from the Ciceronian *imponere*, as in 'imposing laws by force on a people'. But *OED* would only recognize this sense in the form 'imposure', an imposition. Perhaps the text is faulty. *admit . . . difference*. Both admits and rejects the use of reason and disputation, with just limitations. *direction*. Instruction, guidance in life. *illustration*. Exemplification. *probation*. Proof. *vouchsafeth*. Deigns, condescends. *sensible*. Comprehensible, perceivable.

291. *grift*. Graft. *secondary*. In a secondary degree. *respective*. With respect to. *placed*. Set forth, laid down. *analogy*. Proportional relationship, similarity. *nature*. Natural philosophy. *examinable*. Capable of being examined. *medium or syllogism*. Form of argument that generates conclusions through a middle stage within itself, and has no necessary correlation with natural phenomena (which, Bacon believed, can only be comprehended by induction, reasoning from observed particulars towards general laws). *inferior*. Subordinate. *posita*. Positive. *placita*. Arbitrary. *wit*. Skill (as opposed to chance). *positive*. Explicitly laid down as fixed positions. *ad placitum*. As it pleases the authority giving the rules. *artificial*. Depending on art, skill. *placita juris*. Decrees, decisions of law. *field*. Scope, opportunity. *placets*. Assent. *De . . . divinis*. 'The legitimate use of reason in divine subjects.' Cf. Hooker, *Of the Laws of Ecclesiasticall Politie*, i. 8–9, iii. 8–9. *dialectic*. 'The art of critically investigating the truth of opinions' (*SOED*). Le Doeuff (p. 371) cites Aristotle's definition (*Topics*, i. 1), according to which dialectic implies a syllogism based on probable premisses. *for that*. Because. *mine*. Dig, explore (here, in the realm of forbidden knowledge). *enucleating*. Extracting as a kernel. *inferences*. Drawing conclusions from premisses. *contradictories*. Propositions that contradict each other. *sensible*. Comprehensible. '*Quomodo . . . senex?*' 'How can a man be born when he is old?' (John 3: 4). This was Nicodemus' objection to Christ's statement that 'except a man be born again, he cannot see the kingdom of God'. *scandalized*. Horrified. *show*. Appearance. '*Quid . . . me*'. 'What is this that he saith unto us, A little while, and ye shall not see me; and again, a little while, and ye shall see me' (John 16: 17). *defined of*. Defined. *opiate*. Narcotic, to assuage pain or induce sleep.

292. *curious*. Minute; over-inquisitive. *positive*. Categorical, beyond question or dispute. *upon*. Out of. *doctor*. Teacher: St Paul. '*Ego . . . Dominus*'. 'I, not the Lord' (1 Cor. 7: 12). '*Secundum . . . meum*'. 'This is my opinion' (1 Cor. 7: 40). *positions and oppositions*. Categorical statements, provoking refutations. *style*. Manner of speaking. '*Non . . . Dominus*'. 'Not I, but the Lord.' *Salomon*. Prov. 26: 2. *Divinity*. The science of divine things, theology. *informed*. Taught, by human agency. *revealed*. Disclosed by God. *sufficiency*. Adequacy. *perfective*. Capable of being perfected. *light*. Illumination. *dispensation*. The ordering of events by divine Providence. *times*. Ages, periods. *material*. Pertinent. *De gradibus . . . Dei*. 'On the degrees of unity in the kingdom of God.' *advice*. Opinion. '*Why strive you?*' Exod. 2: 11–13. '*You . . . you?*' Acts 7: 24–8. *penneth*. Sets down. *league*. Connection.

293. '*He . . . against us*'. Matt. 12: 30 (which actually reads 'with me . . . against me'). '*He . . . with us*'. Luke 9: 50. *without seam*. John 19: 23. *chaff*. Ps. 1: 4; 35: 5. *tares*. Weeds; Matt. 13: 29. *latitude*. Extent. *merely*. Absolutely. *disincorporate*. Dissevered, excommunicated. *methodical*. According to method or order. *solute*. Loose, discursive. *at large*. Not in a sequence or system. *Jacob's well*. John 4: 13–14. *cistern*. Reservoir, water-tank. *derived*. Drawn off, as in channels. *subject*. Liable. *to corrupt*. To become corrupt. *scholastical divinity*. Medieval theology, with its attempts to sum up the whole of philosophy, divine and human, into one system. *reduced*. Systematized. *summary*. Containing the chief points; compendious. *compacted*. Compact. *purpose . . . dilate*. Attempts to summarize a topic result in the expansion of it. *sum*. Gist, summary. *deduced*. Set down, conveyed. *titles*. Treatment by topics. *fathers*. Church fathers: the patristic commentators. *Master of the Sentences*. The 12th-century theologian, Peter Lombard, Bishop of Paris, wrote a 'Sum' of theology in 4 books called *The Sentences*, drawn mainly from St Augustine. From the 13th to the 16th centuries this work was an obligatory text for all students of theology (Rossi, p. 353). *civil law*. Roman law. *jurisconsults*. Lawyers. *Tribonian*. Head of the group of 16 lawyers commissioned by Justinian (*c*. AD 530) to compile a complete body of Roman law, the *Digest*. *methods*. Systems.

294. *satisfactory*. Superficially convincing. *compaction*. Solidity, firmness of construction. *several*. Separate. *grounds*. Premisses, basic authority. *the weaker do you conclude*. The weaker are your conclusions. *dilute*. Enfeebled. *artificial*. Systematic, highly organized. *suspect*. Suspicious. *round*. Brought to a perfect finish. *abrupt*. Broken off. '*O altitudo . . . ejus!*' 'O the depth of the wisdom and knowledge of God! How incomprehensible are his judgments, and his ways past finding out' (Rom. 11: 33). '*Ex . . . scimus*'. 'We know in part' (1 Cor. 13: 9, 12). *supplies*. Additional help from. *supposition*. Assumption. *presumption*. Arrogance. *institutions*. Digest of the elements of a subject. *deducement*. Deduction. *prejudicial*. Harmful. *curious*. Ingenious, excessively detailed. (Some copies of the 1605 edn. read 'ruinous'.) *warranted*. Authenticated. *in the author*. As concerns their author. *draw on*. Involve as a consequence. *inditer*. Composer, author. '*He . . . glory*'. Prov. 25: 27. '*No man . . . live*'. Exod. 33: 20. '*When . . . deep*'. Prov. 8: 27. '*Neither . . . man*'. John 2: 25. '*From . . . works*'. Acts 15: 18. *former two*. First two. *contained*. Restrained.

295. *sobriety*. Sober-mindedness. *anagogical*. One of the four levels of biblical interpretation in the Middle Ages, concerning the spiritual sense applying to the life to come. *prevent*. Anticipate. '*Videmus . . . faciem*'. 'For now we see through a glass darkly, but then face to face' (1 Cor. 13: 12). *as . . . as*. To the extent of. *glass*. Mirror. *aenigma*. Riddle. *dissolution*. Disintegration. *Aliment*. Food, nourishment. *overcome*. Digest, transform. *worketh . . . upon*. Acts on, effects. *intoxication*. Poisoning, drunkenness, madness. *latter*. The philosophical interpretation. *Paracelsus*. Alchemical followers of Paracelsus sometimes interpreted the Bible to support their own theories, e.g. claiming 'what generates is an element' (referring to Gen. 1: 11,

21, 24), or seeing their 'triad' of elements—sulphur, salt, mercury—as being symbolized by the Trinity. The *De Augmentis* adds that 'the beginnings' of this form of literal interpretation 'came from the Rabbis and Cabalists' (*Works*, v. 117). *scandalizing*. Slandering. *traducing*. Maligning. '*Heaven ... pass*'. Mark 13: 31. *seek the dead ... living*. Luke 24: 5. *lavers*. Wash-basins (as in 1 Kgs. 7: 38). *ark of the testimony*. 'The wooden chest which contained the tables of Jewish law and was kept in the holiest place of the Tabernacle and of the First Temple in Jerusalem' (*SOED*). Cf. Exod. 25 ff.; Heb. 9: 4. *in passage*. In passing. '*Authoris ... authoritas*'. 'What a man says incidentally about matters which are not in question has little authority.' *cabalists*. Interpreters of the Jewish mystical doctrine, the Cabbala. '*Noli ... time*'. Matt. 24: 35; Rom. 11: 20: 'Be not overwise, but fear.'

296. *answers*. As in Luke 9: 47–8. *impertinent*. Irrelevant. *elect*. Those who (according to Calvinist theology) have been chosen by God for eternal life. *latitude*. Extent. *respectively*. Relatively. *sense*. In the fourfold allegorical interpretation the four senses or levels of meaning were literal, allegorical, moral, and anagogical. So 'Jerusalem' could signify at the literal or historical level, the holy city; allegorically or figuratively, the Church militant; tropologically or morally, the just soul; and anagogically or analogically, the Church triumphant. *typical*. Based on a 'type', a 'figure' or symbolic fore-shadowing. *indulgent*. Apt to indulge. *light*. Too free with. *Emanationes ... positivas*. 'The emanations of the Scriptures into positive doctrines.' *common places*. Books arranged by topics. *art*. Systematic subject for study. *harmonies*. Collations of passages on the same subject from different writings, arranged so as to exhibit their consistency, as in a 'harmony of the Gospels'. *vanish*. Sermons were not usually printed, except those given by distinguished preachers on important state occasions. '*Absit ... verbo*'. 'Without meaning to offend' (Livy, ix. 19).

297. *forty years*. It is significant that Bacon should date the revival of preaching in England from the accession of Queen Elizabeth, 1558. *largeness*. Public dissemination. *exhortations*. The *Certain Sermons or Homilies Appointed to be read in Church* (1548, 1563) were sermons published by the government and regularly read aloud in church, covering essentials of religious, moral, and political behaviour. *continuance*. Continuous form. *informed*. Taught. *jealous*. Demanding absolute faithfulness and exclusive worship. Cf. Exod. 20: 5. *admitted into part*. Given partial recognition, in a polytheistic system. *respect*. Be concerned with; esteem. *so*. As long as. *Liturgy*. Forms of public worship. *Government*. The internal administration of the church. *respective*. Relative. *election*. The exercise of God's sovereign will in the predetermination of certain persons to salvation. *consummation*. Cf. John 19: 30, 'consummatum est'. *privatively*. Absent, lacking. (Spedding's emendation of the 1605 reading, 'privately'.) *reprobate*. One rejected by God. *discloseth*. Exposes to view. *edition*. Promulgation, publication.

298. *Infirmity*. Moral weakness or frailty. *it*. Sin. *libertine*. Acknowledging no religious or moral law. *thought, word, or act*. Some editors cite Plato, *Protagoras*, 348D, but cf. also the general confession in the service for Holy Communion: 'we acknowledge and bewayle oure manifold synnes and wyckednesse, whiche we from tyme to tyme most grevously have committed, by thoughte worde and deede, against thy divine Majestie . . .': Book of Common Prayer (1588 and later), and Jas. 3: 2. *deducing*. Deriving. *conscience*. Thought, consciousness. *breaking*. Dividing. *consent*. Agreement in feeling and belief. *sacraments*. 'A religious ceremony or act regarded as imparting spiritual grace to the participants or having spiritual benefits . . . in Protestant Churches either of the rites of baptism and the Eucharist' (*SOED*). *seals . . . covenant*. Confirmations of the mutual engagement between God and the believer. *visible word*. Material manifestations of divine purpose. *visible*. Material. '*in . . . veritate*'. 'In spirit and in truth' (John 4: 23–4). '*vituli labiorum*'. 'Offering of the lips' (Hos. 14: 2). *retribution*. 'Recompense or return for a service, merit, etc.' (*SOED*). *sealed petitions*. Private requests, delivered individually to the power concerned. *patrimony*. 'The estate or property belonging by ancient right to an institution, etc., *spec*. the ancient endowment of a church' (*SOED*). *franchises*. 'Legal immunities or exemptions from a particular burden or jurisdiction, granted to an individual, a corporation, etc.' (*SOED*). *declinations*. Falling away. *privative*. Depriving a person of some benefit. *supposing*. Imagining. *your Majesty*. In King James's *Daemonologie* (1597), iii.6.

299. *a*. The same. '*Quasi . . . acquiescere*'. 'Rebellion is as the sin of witchcraft, and stubbornness as the sin of idolatry' (1 Sam. 15: 23). *tares*. Weeds. *occupate*. Occupied. *converted*. Reclaimed. *in melius*. 'For the better.' *in aliud*. 'To be different.' *amendment*. Improvement. *proficience*. Progress. *difference*. Dispute. *preoccupate*. Preoccupy, prejudice. *objection*. Scruple, disagreement. *prejudiced*. Harmed. *litigious*. Contentious. *invented*. Founded. *question*. Calling in question, raising doubts. *preferment*. Promotion. *repulse*. Rebuff. '*tanquam . . . sacrificii*'. 'As the fat of the sacrifice' (Lev. 3: 16). *incensed*. Burnt. *your Majesty*. The *De Augmentis* (1623) omits the reference to King James, adding a different conclusion: 'Meanwhile I am reminded of the sarcastic reply of Themistocles to the ambassador, who coming from a small town used great words, "Friend, (said he), your words require a city". And certainly it may be objected to me with truth, that my words require an age; a whole age perhaps to prove [test] them, and many ages to perfect them. But yet as even the greatest things are owing to their beginnings, it will be enough for me to have sown a seed for posterity and the Immortal God; whose Majesty I humbly implore through his Son and our Saviour; that He will vouchsafe favourably to accept these and the like offerings of the human intellect, seasoned with religion as with salt, and sacrificed to His Glory' (*Works*, v. 119). *bounden*. Indebted.

ESSAYS, 1612

The success of Bacon's *Essays* in their first version (see above, p. 544) was one motive for preparing the new and enlarged edition that appeared in 1612.

Another motive was undoubtedly to fill the lacunae he had noted in the *Advancement of Learning* for enlarging our knowledge of humanity by detailed studies of men and women from a psychological and sociological viewpoint (see above, pp. 255 f., 260 f., with notes). This goal reached its complete fulfilment in the final version of the *Essays* in 1625 (see below, pp. 713–14), with its treatment of 58 topics; the 38 handled in 1612 form an intermediate stage.

That the *Essays* formed an important part of Bacon's overall programme for the reform of knowledge, despite his occasionally deprecatory references to them as a mere diversion from more serious work (so displaying the courtly virtue of *sprezzatura*, disdaining effort and seriousness), can be seen from a manuscript preserved in the British Library, MS Harleian 5106 (Beal BcF 203), containing 34 essays (including all but six of the essays printed in 1612, and two not included in that collection: 'Of Seditions and Troubles' (cf. *Works*, vi. 589–90) and an enlarged version of the 1597 'Of Honour and Reputation'). This manuscript, transcribed some time between 1607, when Bacon became Solicitor-General, and 1612, when the new edition appeared, is in the handwriting of one of Bacon's scribes, and contains interlineations in Bacon's own hand (full description in Kiernan, pp. lxxi ff.; *Works*, vi. 535–6). Of particular significance for the serious role of the *Essays* in organizing information about mankind is the title that this manuscript bears: *The writings of Sr Francis Bacon Knt, the King's Sollicitor Generall: in Moralitie, Policie, and Historie.*

This triple study of the ethical, political, and historical constraints and influences on human behaviour give the *Essays* their unique quality. In content, they are contributions to knowledge, drawn from Bacon's own experience and reading; but in form they eschew system, preferring either aphoristic utterance or a loose, discursive structure. Valuable evidence of Bacon's own conception of the *Essays*' form, and their literary models, is provided by the dedication to Prince Henry, which Bacon had originally planned to prefix to the printed volume. His sudden and tragic death on 6 November 1612 destroyed this plan, but a copy of Bacon's dedication has also survived in the British Library (Birch MS 4259, fo. 155: Beal BcF 204), and is of great interest:

> To the most high and excellent Prince Henry, Prince of Wales, Duke of Cornwall and Earl of Chester.
>
> It may please your Highness:
>
> Having divided my life into the contemplative and active part, I am desirous to give his Majesty and your Highness of the fruits of both, simple though they be.
>
> To write just treatises requireth leisure in the writer and leisure in the reader, and therefore are not so fit, neither in regard of your Highness' princely affairs, nor in regard of my continual services; which is the cause that hath made me choose to write certain brief notes, set down rather significantly than curiously, which I have called *Essays*. The word is late, but the thing is ancient, for Seneca's *Epistles to Lucilius*, if one mark them well, are but *Essays*, that is, dispersed meditations, though conveyed in

> the form of Epistles. These labours of mine I know cannot be worthy of your Highness, for what can be worthy of you? But my hope is, they may be as grains of salt, that will rather give you an appetite than offend you with satiety. And although they handle those things wherein both men's lives and their pens are most conversant, yet (what I have attained, I know not), but I have endeavoured to make them not vulgar, but of a nature whereof a man shall find much in experience, and little in books; so as they are neither repetitions nor fancies. (*Works*, xi. 340)

Bacon's conception of the *Essays* as 'brief notes, set down rather significantly than curiously'—that is, concerned with pregnancy of meaning rather than nicety of form—and as 'dispersed meditations' well describes their unsystematic nature ('dispersed' implying not being organized around a single argument). But he also describes them as 'meditations', a word which (like 'contemplation' or 'contemplative') has for him secular connotations, as of prolonged study, rather than the religious associations found in other writers of this period. Their substance, further, is of a kind found 'much in experience, little in books', which makes them neither 'repetitions', reworkings of topoi within a literary tradition, nor 'fancies', purely imaginary constructs unanchored in social reality. This double claim, for a density of observation derived from lived experience, was recognized by a kindred spirit, Samuel Johnson. His judgement of the 1625 *Essays*, as recorded by Reynolds, was that 'their excellence and their value consisted in being the observations of a strong mind operating upon life; and in consequence you find there what you seldom find in other books' (*Johnsonian Miscellanies*, ed. G. B. Hill (Oxford, 1897), ii. 229).

My text is from Spedding, who reprinted the 1612 collection complete (*Works*, vi. 541–91). I have selected three essays. One of them, 'Of Friendship', was totally rewritten for the 1625 volume (cf. above, pp. 390 ff.). The two others, 'Of Religion', and 'Of the True Greatness of Kingdoms', were subject to massive expansions (and some deletion) in 1625: the differences between the two stages is valuable evidence for the development in Bacon's thinking on these major issues in contemporary society.

Of Religion

300. *jealous God.* Exod. 34: 14. *good fellows.* Jovial companions, who enjoy food and drink. *so.* In such a manner. *Lucretius. De rerum natura*, i. 101–2. *Tantum... malorum.* 'To such ill actions religion could persuade a man.' *massacre.* The St Bartholomew's Eve massacre of 1572, in which some 13,000 Protestants were murdered. *powder treason.* The conspiracy of Guy Fawkes (a Catholic), to blow up the King, his ministers, and the members of both Houses of Parliament, 5 Nov. 1605. *epicure and atheist.* Lucretius versified Epicurus' philosophy, which emphasized the importance of pleasure, and denied that a deity had created the world. *madmen of Münster.* The Anabaptists, an extreme sect of Protestants, who refused to recognize infant baptism, and were ready to die for their beliefs. They achieved supremacy in Münster under Jan Matthys (d. 1534), and spread across Europe. *engine.* Instrument. *'I will... the highest'.* Isa. 14: 12–14. These words are in fact spoken by the evil King of Babylon, but the early Church Fathers argued that the devil spoke them (Reynolds, pp. 31–2). *firing*. Setting fire to. *dove*. Cf. Matt. 3: 16, the form in which the Holy Spirit appeared at the baptism of Christ: an emblem of peace and divine protection. *bark*. Boat, ship (traditionally associated with Peter). *assassins*. Muslim fanatics, sent out (during the Crusades) to murder Christian leaders. *learning*. Sciences. *Mercury rod*. The staff with which Mercury guides souls to the underworld (Homer, *Odyssey*, xxiv. 1–5; Virgil, *Aeneid*, iv. 242–4): an emblem of protection and authority. *facts*. Deeds (Lat. *facta*). *supports*. Supporters. *would*. Should. '*Ira . . . Dei*'. 'The wrath of man worketh not the righteousness of God' (Jas. 1: 20).

Of Friendship

301. *meeting*. Mere encounters, lacking any true exchange. *union*. The joining or fusion of two or more solid bodies. *natural*. Such as gravity. *violent*. Created by a percussion or explosion. *wanteth*. Lacks. *fortitude*. The virtue which enables misfortune to be borne with dignity and control. *painful*. Troublesome, tedious. *exquisite*. Consummate, highly cultivated. *hireling*. One who serves for hire; a mercenary. *smother*. Concealment, withdrawal from society. *communicate*. Share. *evaporate . . . affections*. Drive out bodily humours and passions in the form of vapour. *and . . . but*. If only. *facility*. Ease, indolence; or 'convenience'. *imposition*. A thing imposed or inflicted. *sincerity*. Straightforwardness, honesty. *reckoning*. Debit account. *Perfection . . . speculation*. The notion of perfect friendship is purely imaginary.

Of the True Greatness of Kingdoms

301. *Greatness*. Expansive power, to acquire new territories. See the head-note to this essay in the 1625 edn., below, pp. 748–9. *speech*. Saying, recorded by Plutarch, 'Life' of Themistocles, ii. 4. *in challenge*. As a reproach, objection. *in censure*. As a critical assessment. *touch*. Play. *fiddle*. Meaning: (*a*) 'play the instrument', (*b*) cheat. *solace*. Recreation. *pretty*. Apt, pleasing. *estate*. State. *procedure*. Manner of proceeding, conduct. *distressed*. Afflicted with trouble.

302. *vulgar*. Common people. *amplitude*. Physical extent. *doth . . . measure*. Can be measured. *doth . . . computation*. Can be computed. *appear*. Be displayed, computed. *musters*. Registers. *charts*. Plans. *Certainly . . . earth*. This sentence is omitted in 1625: above, p. 398. *compared*. In Matt. 13: 31; Mark 4: 30–2. *least*. Smallest. *get up*. Sprout, grow. *apt*. Capable, suitable. *stem*. Stalk, bearing leaves; a race or family. *ordnance*. Military store or supplies. *sheep . . . skin*. Cf. Matt. 7: 15. *mercenary aids*. Soldiers of other nationalities, who serve for pay (and are thus always likely to quit for higher pay elsewhere). *waged*. Paid wages (their only loyalty). *spread*. Display. *mew*. Moult, shed. On the dangers of relying on mercenaries see Machiavelli, *Discourses*, ii. 10; *Prince*, ch. 12. *blessing*. 'Grace given by God': cf. Gen. 49: 9, 14. *meet*. Coincide. *between burdens*. Loaded down. *tributes*. Taxes, levies. *swain*. Country labourer. *driven . . . heart*. Disheartened, demoralized. *coppices*. Woods of small growth for periodic cutting. *staddles*. Young trees left standing in a plantation after the removal of the underwood.*infantry . . . nerve of an army*. Cf. Machiavelli, *Discourses*, ii. 18. *not . . . helmet*. Only one man in a hundred fit to be a soldier. *poll*. Head. *Terra . . . glebae*. *Aeneid*, i. 531, 'A land powerful in arms and richness of soil'. *owners . . . labourers*. Farmers should own the land they work, and not be mere hirelings.

303. *within-doors arts*. Crafts practised in a building or factory. *nice manufactures*. Skilled occupations. *require*. Need the use of. *pain*. Bodily labour, exertion. *broken of*. Worn out by. *shall*. Are to. *succours*. Coming to the aid of. *slothful peace*. The idea that peace is dangerous to a country, bringing idleness and disease, goes back to the ancient Romans, especially Cato and Livy. See e.g. Brian Vickers, 'Leisure and Idleness in the Renaissance: The Ambivalence of *otium*', *Renaissance Studies*, 4 (1990), 1–37, 107–54. *liberal of naturalization*. Generous in allowing resident foreigners to become citizens. *capable of greatness*. Able to grow in extent and power. *rest upon*. Stop at; will not grow beyond. *tribe*. Founding family. *stirp*. The stock of a family; offspring. *want*. Lack. *body . . . branches*. The trunk of the tree is not strong enough to support the branches. *receipt of*. Recipe for. *greatness*. Greater extent, area. *No man . . . stature*. Matt. 6: 27; Luke 12: 25. *model*. Plan or frame on a small scale: the microcosm. *ordinances*. Laws, decrees. *manners*. Customs. *to*. For the benefit of. *left to chance*. That is, not coherently planned over a long period of time.

THE CHARGE TOUCHING DUELS

Duelling, as a way of settling private quarrels, was a serious social problem in the late sixteenth and early seventeenth centuries. In an age when every gentleman was entitled to wear a sword, resort to personal violence was the quickest way of asserting one's honour. Honour was indeed the crucial issue, the notion of good repute in the eyes of society—usually, among gentlemen of a similar rank—which continued to express that individualistic sense of identity found already in Homer's heroes, totally absorbed in their *timē* or honour. English society in this period, as Mervyn James has admirably shown (*English Politics and the Concept of Honour 1485–1642*, *Past and Present*, suppl. 3 (Oxford, 1978); also in his *Society, Politics and Culture: Studies in Early Modern England* (Cambridge, 1986), 308–415) was in a delayed transition between two incompatible concepts of honour, each having opposed philosophical and political implications. The older idea, prevalent up to the sixteenth century, regarded honour and virtue as deriving from blood and lineage, affecting not only individuals but their wider social connections through families and kinship groups. Although having positive effects in times of war, where loyalty to one's leader encouraged feats of chivalric bravery, in peacetime it was more of a nuisance, hypersensitive to anything which might be perceived by society as an insult to one's status. 'Blot, discredit, shame—these are all keywords in the vocabulary of honour' (James, *Concept of Honour*, 30). As Bacon wrote in the *Advancement of Learning*, men who succeed in that 'honour of urbanity' and become socially accepted find fulfilment, 'whereas those that have defect in it do seek comeliness by reputation: for where reputation is, almost every thing becometh; but where that is not, it must be supplied by *puntos* and compliments' (above, p. 266)—and by physical violence if offended.

This essentially individualistic notion persisted in the age of Elizabeth, where duelling increased in popularity. One cause of its increase was the arrival of the new Italian sword, the lighter and more flexible rapier, useless on the battlefield but an ideal weapon for peacetime combats. Italian treatises on the new fencing also encouraged both the fashion for the *duello* and the elaboration of a very touchy etiquette of honour, laying down the precise degrees by which two men predisposed to quarrel could engage, or disengage themselves (a sequence described by Bacon as 'being before-hand or behind-hand'). In *Romeo and Juliet* Shakespeare makes Tybalt report on the new fencing style ('*alla stoccata* carries it away'), itemizing the stages by which a quarrel could be 'incepted' (as Bacon puts it), in the brawls between the serving-men of the houses of Capulet and Montague (I. i. 1–115; III. i. 1–196). Mercutio mockingly describes Tybalt as 'a duellist, a duellist; a gentleman of the very first house, of the first and second cause' (II. iv. 23 ff.), referring to the stages within the duelling-code at which a gentleman had to seek satisfaction for his honour. The fussy rituals involved could provide materials for satire, as in *As You Like It*, where Touchstone defines 'the quarrel on the seventh cause' as constituting one 'upon a lie seven times remov'd'. Obligingly, Touchstone fulfils Jacques' request to 'nominate in order now the degrees of the lie': 'O sir, we quarrel in print, by the book, as you have books for good manners. I will name you the degrees. The first, the Retort Courteous; the second, the Quip Modest; the third, the Reply Churlish; the fourth, the Reproof Valiant; the fifth, the Counter-Check Quarrelsome; the sixth, the Lie with Circumstances; the seventh, the Lie Direct. All these you may avoid but the Lie Direct; and you may avoid that too, with an If' (V. iv. 46 ff.). Yet not enough gentlemen in Jacobean London made use of 'the only peacemaker', that mutually given 'if' by which one could escape from a quarrel without losing face. Contemporary letters refer frequently to duels fought to the death, the quarrelling parties sometimes taking ship for the Continent to escape English law.

Opposed to this ego-based chivalric survival was a newer code of honour, emphasized in sixteenth-century England by the Protestant Church and by exponents of classical humanism (the system which dominated both grammar-school and university education). In this new code honour was seen as the reward for virtue, a notion that united the altruistic or other-centred emphasis in both classical and Christian ethics with the civic value of obedience. As Mervyn James has shown, early seventeenth- century England formulated a new concept of civil society in which the State and its sovereign claimed a monopoly over both honour and violence (*Concept of Honour*, 2, 18, 22). Only the king could award honours; only the State could employ violence. Thanks to the union of classical and Christian ethics initiated by such humanists as Sir Thomas Elyot and Roger Ascham, a 'moralization of politics' took place in which both Church and State supported a single, cohesive source of power and authority against 'the traditional dissidence of honour, with its turbulent emphasis on autonomy and will' (ibid. 27, 45 ff.). Bacon's acceptance of this Christian humanist tradition naturally made him oppose duelling. In his essay 'Of Vain-Glory' (boasting) written originally for the 1612 collection, he commented on that 'competitive assertiveness' which Mervyn James has singled out as basic to the mentality of honour: 'They that are glorious must needs be factious; for all bravery stands upon comparisons. They must needs be violent, to make good their own vaunts. Neither can they be secret, and therefore not effectual' (*Works*, vi. 585). All such ostentatious display must take place in public. Like other opponents of duelling, Bacon conceded that factious emulation could have its use, for 'in cases of great enterprise, upon charge and adventure, such composition of glorious natures doth put life into business' (ibid. 586). In his 1625 revision of the essay he added the corollary, that 'In military commanders and soldiers, vain-glory is an essential point; for as iron sharpens iron, so by glory one courage sharpeneth another' (above, p. 444).

In peacetime, however, as a lawyer and Crown servant (appointed Solicitor-General in 1607, Attorney-General in 1613), Bacon could see no virtue in individuals (or 'particulars', as he calls them) settling their differences violently. After a series of notorious duels in 1613, the King issued another proclamation forbidding them, and about this time Bacon addressed to him a brief 'Proposition for the Repressing of singular Combats or Duels' (reprinted in *Works*, xi. 397). Some of these proposals were incorporated in the more formal *Charge* (or 'injunction') which Bacon addressed to the Star Chamber at the first sitting of that court in Hilary Term, 26 January 1614, published later in the year as *The Charge of Sir Francis Bacon, Knight, his Majesties Attourney-generall, touching Duels, upon an information in the Star-chamber against Priest and Wright. With the decree of the Star-chamber in the same cause*. It may seem remarkable that such a case should appear before the Star Chamber, but it shows the importance that the Crown attached to this issue. As a historian of English law has pointed out, it was particularly appropriate that 'an extraordinary court, without right to jury trial, [should] try such cases in the interest, not of private rights or vengeance, but the collective rights of the state' (Daniel R. Coquillette, *Francis Bacon*, Jurists: Profiles in Legal Theory (Edinburgh, 1992), 155). The Star Chamber 'was, in essence, the Privy Council sitting in a judicial capacity', and despite its later bad reputation was still respected as a check on lawlessness 'in situations where ordinary courts could not be expected to function properly' (ibid. 167). This could happen where powerful nobles or political figures might intimidate a local court, or where the jury might not punish the violation sufficiently severely. For what was involved here was the category of 'victimless crimes', those in which there was no clear distinction between aggressor and victim, but where both were 'consenting adults' (ibid. 167–8). Such crimes were not against individuals but against the State, and this was rightly 'a Star-chamber matter'.

The case had been transmitted to the Star Chamber by the Justices of Peace in Surrey, against '*William Priest*, gentleman', and '*Richard Wright*, esquire' (*Works*, xi. 409), concerning 'the confession of the said defendant *Priest* himself, that he having received some wrong and disgrace at the hands of one *Hutchest*, did thereupon in revenge thereof write a letter to the said *Hutchest* containing a challenge to fight with him at single rapier, which letter the said *Priest* did deliver to the said defendant *Wright*, together with a stick containing the length of the rapier, wherewith the said *Priest* meant to perform the fight: whereupon the said *Wright* did deliver the said letter to the said *Hutchest*, and did read the same to him; and after the reading thereof, did also deliver to the said *Hutchest* the said stick, saying that the same was the length of the weapon mentioned in the said letter. But the said *Hutchest* (dutifully respecting the preservation of his Majesty's peace) did refuse the said challenge, whereby no farther mischief did ensue thereupon' (ibid. 413). The court found both men guilty, fining them 500 marks each, committing them to the Fleet prison, and ordering them to appear before the next assizes in Surrey to 'publicly . . . acknowledge their high contempt and offence against God, his Majesty, and his laws, and show themselves penitent for the same' (p. 415). Bacon was charged 'to have special care to the penning of this decree', which was then to be 'published and made known in all shires of this kingdom' (p. 416).

In fulfilling his public duties in this exemplary case Bacon took his stand firmly on natural law as applying to both Church and State, invoking St Paul's famous formulation, 'Vengeance is mine, I will repay, saith the Lord' (Rom. 12: 19), and reiterating the King's authority over all civil affairs. It is typical of the moralizing of politics made by Protestant humanism that Bacon should assert the superiority of fortitude, a major Christian and secular virtue, over valour, mere bravery in combat, and that he should assert his fundamental endorsement of the classical-Renaissance concept of the *vita activa*, declaring that 'A man's life is not to be trifled away, it is to be offered up and sacrificed to honourable services, public merits, good causes, and noble adventures'—civic ideals to which his whole life was dedicated. Also in line with the growing rejection of honour as a justification for duelling is his attack on it as 'a kind of satirical illusion and apparition of honour; against religion, against law, against moral virtue', and his emphasis on the new image of society, by which 'the King's power and authority' becomes the arbiter of honour. This new order dispenses a private citizen of the need to 'acquit his reputation' by involving himself in 'a vain and unnecessary hazard', making personal quarrels yield to the authority of 'the State' and the monarch: 'The fountain of honour is the King'.

As for the legal issues involved, duels were not only against natural law but 'were also a particularly dangerous challenge to the law because those participating in them affirmed them to be *good*' (Coquillette, *Francis Bacon*, 168). This is the point of Bacon's argument that 'both in divinity and in policy, offences of presumption are the greatest. Other offences yield and consent to the law that it is good, not daring to make defence, or to justify themselves; but this offence expressly gives the law an affront, as if there were two laws, one a kind of gown-law, and the other a law of reputation, as they term it' (above, p. 305). Bacon also demolished the claims on behalf of duelling that placed it on the same level as self- defence or crimes of passion: duelling obviously involved forethought. Invoking classical and biblical history, Bacon accurately defines the duel as 'a monstrous child of this later age', finding 'no shadow of it in any law divine or human'. Then, demonstrating, as has been said, 'an exhaustive knowledge of civil law precedents' (Coquillette, ibid. 169), Bacon shows that the duel was unknown to the Romans, 'and sure they would have had it, if there had been any virtue in it' (above, p. 309). And as for the claim that 'the law hath not provided sufficient punishment... for contumely of words', Bacon easily disposes of it by itemizing all the 'degrees of injury to the person' recognized by English law and coming within the jurisdiction of the courts (above, pp. 310–11). And he mockingly dismisses the scale of values invoked by such a phrase as 'a mortal wound to the reputation', citing one of his favourite quotations, General Consalvo's remark that a gentleman's honour should be made 'of a good strong warp or web, that every little thing should not catch in it'.

Bacon also expresses a more personal concern with the human waste produced by duelling, deploring the 'miserable effect, when young men full of towardness and hope, such as the poets call *Aurorae filii*, sons of the morning, in whom the expectation and comfort of their friends consisteth, shall be cast away and destroyed in such a vain manner' (above, p. 305). (One of his own man-servants had been convicted of manslaughter for killing a man in a duel in November 1613: Coquillette, ibid. 167.) A more personal judgement can also be seen in his bitter comments on contemporary history in the encouragement given to the duelling mentality by the aggressiveness of the French king, Francis I. The whole tract is an excellent example of how Bacon, in the duties of public life, could bring to bear the qualities of intellect and humanity that mark so many of his writings.

I reprint Spedding's edition of this text (*Works*, xi. 399–409), omitting the court's decree (ibid. 409–16), with its verbose summary of Bacon's speech.

304. *Charge*. Injunction; official instruction. *cause*. Legal case. *touching*. Concerning. *reclaim*. Remedy. *greater persons*. Of higher rank. *presence*. Assembly, often with the connotation of 'assembled dignities'. The Star Chamber Decree of 26 Jan. 1614 actually records 'The Presence' (the names are printed on a separate page in the *Resuscitatio*, 1657, to underline the awesomeness of the occasion, and the importance attached to this test case); namely 'Thomas, Lord Ellesmere, Lord Chancellor of England. Henry, Earl of Northampton, Lord Privy Seal. Charles, Earl of Nottingham, Lord High Admiral of England. Thomas, Earl of Suffolk, Lord Chamberlain. Edward, Lord Zouche. William, Lord Knolles, Treasurer of the Household. Edward, Lord Wotton, Controller. John, Lord Stanhop, Vicechamberlain. Sir Julius Caesar, Knight, Chancellor of the Exchequer. George, Lord Archbishop of Canterbury. John, Lord Bishop of London. Sir Edward Coke, Knight, Lord Chief Justice of England. Sir Henry Hobart, Knight, Lord Chief Justice of the Common-pleas'. *on foot*. Under way, already prosecuted. *published*. Announced. *mischief*. Wickedness, cause of evil. *dog... lion*. A proverb, which *The Oxford Dictionary of English Proverbs* (D 443) traces back to Jacobus de Voragine and Chaucer, on to Surrey and Shakespeare (*Othello*, II. iii. 266).

quality. High status. *leave*. Stop. *vilified*. Debased. *barbers' surgeons*. Hairdressers who also acted as surgeons and dentists (without medical qualifications). *mechanical*. Involved in manual labour. *supply*. Make up for. *doubt*. Fear. *particular*. Individual case. *stick not*. Do not scruple.

305. *resolution*. positive intention. *extorted*. Torn away from. '*Mihi* ... *retribuam*'. 'Vengeance is mine; I will repay, saith the Lord' (Rom. 12: 19). *bear the sword*. Wear a sword: a right enjoyed by gentlemen. *assail*. Attack. *banding*. Forming armed bands. *trooping*. Forming private armies. *alliances*. Associations of families and larger groups (e.g. York, Lancaster). *accidents*. Unfortunate events. *distempered*. Disordered. *unperfect*. Abnormal. *inflammations*. Violent social or political disturbances. *policy*. Politics. *presumption*. Unwarranted seizure or usurpation. Bacon emphasizes this crucial point, that to defend duelling is in effect to claim the right to set up a rival law. *it*. The law. *gown-law*. Official legal system. *reputation*. Honour, good name: a concept much invoked by duellists. *Paul's*. St Paul's Cathedral: symbolizing the Church authorities. *Westminster*. The Houses of Parliament, which enact English laws. *Ordinary tables*. Dining-rooms in a public house: a frequent source of quarrels. *year-books*. The books of reports of cases in the English lawcourts between the 13th and 16th centuries. *statute-books*. Containing the laws or decrees made by a monarch or legislative authority. *pamphlets*. Such as *Vincentio Saviolo his Practise. In two bookes. The first intreating of the use of the rapier and dagger. The second, of honor and honorable quarrels* (London, 1595). Bk. II is largely a translation of G. Muzio, *Il Duello*. *receive*. Accept, endorse. *towardness*. Promise. '*Aurorae filii*'. 'Sons of the dawn' (Virgil, *Aeneid*, i. 751). *vain*. Futile. *gentle*. Well-born; entitled to bear a coat of arms. *adventured*. Risked. *make* ... *day*. Make a day's fighting victorious. *disfurnisheth*. Deprives, disables.

306. *credit*. Reputation. *proclamation*. Royal decree. King James issued one renewing the ban on duelling in Oct. 1613, following the notorious quarrel between the Earl of Essex and Henry Howard: cf. *Works*, ix. 396. *bewitching*. Casting a malefic spell over someone. *bear* ... *minds*. Have a high opinion of their own distinction. *species*. Appearance. *apparition*. Illusion, sham. *vain discourses*. Speeches or writings encouraging vanity. *green*. Immature. *staid*. Dignified, responsible. *value*. High rank, worth. *dominations*. Higher powers (in Christian theology, the fourth order of the ninefold celestial hierarchy, between 'thrones' and 'virtues'). *Scripture*. Cf. e.g. Eph. 6: 12, 'For we wrestle not against flesh and blood, but against principalities, against powers ...' (Geneva Bible, adding a gloss identifying 'Satan the spiritual enemy, who is most dangerous: for he is over our heads'). The 'powers' are aerial spirits, often conceived as malignant: Rom. 8: 38; Eph. 2: 2, 3: 10; Col. 2: 15. *Fortitude*. For Bacon's emphasis on this virtue cf. the first speech in *Of Tribute*, above, pp. 22 ff., and the first letter in his *Advice to Rutland*, above, pp. 69 ff. *grounds*. Motives, occasions. *ledgier*. Light, frivolous. *It is in* ... *in*. The same rule holds in ... as in. *liberality*. Generosity. *wish*. Hope. *effectual*. Effective. *particular*. Individual. *must* ... *down*. That must be put down.

307. *acquitted*. Cleared. *it*. Duelling. *Charles the ninth*. King of France, 1560–74. *interessed*. Injured. *none*. No one. *sober*. Serious. *disinterest*. Rid of concern with; release. *cockered*. Pampered, indulged. *humour*. Disposition, appetite. *Commissioners Marshall*. Not in *OED*; presumably Crown officials having jurisdiction over martial encounters (? a survival from the medieval court). *compounding*. Making, creating. *otherwise*. On the other hand. *punctual*. Punctilious, accurate. *conceits*. Opinions. *what's ... behind-hand*. In advance, in arrears with regard to payments or other obligations. (Here, questions of precedence; or the various stages—as in the 'middle acts' mentioned below—by which offence is given, and taken.) *would*. Should. '*In ... peccat*'. 'So that everyone should be punished in the same way that he has offended'. Untraced; but cf. Cicero, *De legibus*, III. xx. 46, 'ut in suo vitio quisque plectatur'. *aspect*. Way of looking, referring to the astrological doctrine of planetary 'aspects' (the angles subtended by planets supposedly 'looking' at each other, which were thought to be favourable or unfavourable). *continueth*. Maintains. *banished ... presence*. Expelled from the court, or proximity to the monarch. Bacon had made this proposal to King James earlier (*Works*, ix. 397), following the King's proclamation in Oct. 1613. *discipline*. Authority. *Queen and Prince*. Queen Anna of Denmark and Prince Charles.

308. *severity*. Strictness; harsh punishment. *capital*. Involving loss of life. *presently*. Instantly. *had*. Fetched. *prevent*. Anticipate. *middle*. Intermediate. *vex*. Disturb physically. *branches*. Subordinate manifestations. *excepted*. Objected. *insidious*. Treacherous. *contumely*. Contemptuous insult. *Lie*. Accusation of falsehood. *difference*. Distinction. *misadventure*. Accidental homicide. *cities of refuge*. Sanctuaries. *put ... flight*. Forced to flee. *accident*. Chance. *revenger of blood*. One trying to avenge a death. *gotten*. Reached. *inflamed*. Angered. *advised*. Deliberate, considered. *prepensed malice*. Premeditated; with malice aforethought. *indulgence*. Licence, privilege. '*ira furor brevis*'[*est*]. 'Anger is a short madness' (Horace, *Epist*. I. ii. 63). *restrained*. Restricted, forbade. *but to a case*. Apart from a single case. *took*. Caught. *in the manner*. In the act; *in flagrante delicto*. *single*. Simple, mere. *vied*. Resulting from rivalry.

309. *Cain*. Gen. 4: 2–15. *Lamech*. A descendant of Cain. Cf. Gen. 4: 16–24: 'And Lamech said unto his wives ... I have slain a man to my wounding, and a young man to my hurt. If Cain shall be revenged sevenfold, truly Lamech seventy and sevenfold' (23–4). *braving*. Showing bravado, ostentatious defiance. *presumptuous*. Arrogant. *generous*. Noble, magnanimous. *righted*. Avenged. *bare show*. Resembled. '*Fas ... doceri*'. 'It is lawful to learn even from an enemy'. (Ovid, *Met*. iv. 428; *Promus*, no. 1068.) *that*. That which. *Emperor's*. Presumably, ruler of the Holy Roman Empire. *Bassaes*. Rulers (pashas). *whether*. Whichever. *Seigneur's*. Chief ruler's. *barbarous*. Uncivilized. *later*. Latter. '*pugna ... provocationem*'. 'Single combat by challenge', a legendary practice in the Roman army, especially against barbarians. *absolute*. Independent; all-powerful. *particulars*. Individuals. *licence*. Permission. *David ... Goliath*. 1 Sam. 17: 4–54.

'*Let*... *us*'. 2 Sam. 2: 14. *example*... *Naples*. Unidentified. *singles*. Individual combat. *judicial*. According to law.

310. *entertained*. Maintained. *lot*. Portion, share assigned by chance or divine agency. Mervyn James writes: 'perhaps the most fundamental tenet of honour belief is that Fate, irrational, incomprehensible and uncontrollable, rules over human history' (*Concept of Honour*, 7). *controverted*. Disputed. '*Taliter*... *accidit*'. Untraced. 'Likewise one sees fighting men implore God in the hope that he will reveal himself and work a miracle, so that the one with a just cause will win—which often fails, however.' *warrant*. Authorization. *whence*. From which source. *but only*. Except. *practice and toleration*. A custom that was tolerated. 'Duels were occasionally permitted by the states of Western Europe but only under conditions of legal safeguard, and with the consent of public authority which was increasingly rarely given' (James, *Concept of Honour*, 13, citing John Selden, *The Duello, or Single Combat*, 1610). *pretended*. Claimed. *fillips*. Slight blows or injuries. *lie given*. Accusation of lying. *Solon's answer*. 'On being asked why he had not framed any law against parricide, he replied that he hoped it was unnecessary' (Diogenes Laertius, i. 59). *fantastical*. Fanciful, irrational. *civilians*. Practitioners of civil law (not criminal; concerning private relations between members of a community). *lie*. Be applicable. *Francis the first*. King of France, 1515–47. *set on*. Instigated, promoted. 'In England the duel was always considered a French custom', for 'the vogue for duelling grew alarmingly in Henri IV's time [1553–1610] . . .': John Stoye, *English Travellers Abroad 1604–1667*, rev. edn. (London, 1989), p. 39. *deep*. Deeply. *taxed*. Criticized. *defy*. Declaration of defiance; challenge to fight. *current*. Accepted. *bear*. Put up with. *fountain*... *learning*. A highly ironic phrase. *but*. Except. *exquisite*. Carefully chosen; ingenious. *fouled*. Dishonoured. *breath*. Unimportant utterances. *touch*. Reproach. *light*. Trivial. *person*. Individual; body.

311. *battery*. Physical beating or menace. *maim*. Bodily injury. *bastinadoes*. Blows with a stick, especially on the soles of the feet. *apprehension*. Perceiving. *Consalvo*. Gonzalo Fernandez Consalvo Cordoba (1453–1515), distinguished Spanish general. Cf. *Adv.L.*, above, p. 248, and Essay 57, 'Of Anger', above, p. 450. *cobweb-lawn*. Finely spun cotton. *greatness of mind*. Magnanimity, a virtue in classical ethics. *tender*. Hypersensitive. *capacity*. Legal competency. *ground*. Fundamental principle. *felony*. Grave crime, usually involving violence. *combination*. Criminal association. *practice*. Conspiracy, intrigue. *heinous*. Highly criminal, atrocious. *inceptions*. Beginnings. *suborning*. Bribing; procuring. *deposed*. Made a written deposition. *material*. Relevant, pertinent. *major*... *minor*. The first two parts of a syllogism. *appointment*... *field*. Arranging the time and place for a duel. *gild*. Adorn, disguise. *terms*. Conditions. *of*. From. *conclusion*. To the syllogism. *censure*. Judgement, assessment. *are to*. Tend towards. *would* . . . *not*. Do not wish. *of*. Out of.

312. *part*. Role (as Attorney-General). *Counsel learned*. Qualified legal adviser. *art*. Practice. *advertised*. Informed. *any*. Any one. *have granted*.

Receive permission to use. *'Ne exeat regnum'*. 'He may not leave the Kingdom.' *giant...sea.* Duels were often fought abroad, to escape English law. *motion.* Formal application. *'fur de se'...'felo de se'.* 'A thief of himself...a felon of himself': someone who commits suicide (at this time a criminal act). *satisfying.* Complying with. *contra...dignitatem.* 'Against the [King's] crown and dignity.' *suit.* Request. *fond.* Foolish. *puppetry.* Masquerade. *stir coals.* Arouse aggressive feelings. *countenanced.* Approved, encouraged.

313. *noblesse.* Nobility. *would.* should. *'Non...usus'.* 'A gift besought for no such end!' (the sword of Aeneas, with which Dido kills herself) (Virgil, *Aeneid*, iv. 647). *'ipsi viderint'.* 'Let them consider.'

THE CHARGE AGAINST SOMERSET FOR POISONING OVERBURY

Sir Thomas Overbury (b. 1581), best known for his *Characters* (1614), died in the Tower of London on 15 September 1613. His death was put down to the normal hazards of incarceration in London's prisons, but two years later rumours began to circulate that he had been poisoned. Sir Edward Coke, Lord Chief Justice, was entrusted with the first enquiries, from which it emerged that a conspiracy had taken place, involving Overbury's under-keeper, Richard Weston, one Mrs Turner, who had received poisons from the apothecary Franklin, and the apothecary's boy who had administered a poisoned clyster, all this having been done with the knowledge of the Lieutenant of the Tower, Sir Gervase Hellwysse. It subsequently transpired that all these parties had been installed just before Overbury's arrest, which had also been manipulated. Overbury's enemies had arranged for him to be offered a diplomatic post abroad, to get him out of the way, and when he declined they had him arrested and imprisoned on a charge of contempt. Overbury was friend and personal adviser to the Scottish courtier, Robert Carr (*c.*1588–1645), one of James's protégés, on whom the King had showered favours, making him Viscount Rochester in 1611, Privy Counsellor in 1612, unofficial secretary in that year (after Salisbury's death), and finally in November 1613 creating him Earl of Somerset. At this time Rochester was in love with Frances Howard, Countess of Essex, daughter of the Earl of Suffolk and great-niece of the still influential Earl of Northampton, who was planning to have her marriage to the 3rd Earl of Essex annulled on the grounds of impotence. Overbury opposed the new marriage, apparently reluctant to lose his influence over Rochester, with his easy access to the King's favours, and the Countess retaliated with this poison plot. Her marriage to Essex having been annulled in September, in December she married Somerset (the occasion on which the *Masque of Flowers* was presented, at Bacon's expense: see Introduction, p. xxiii above). Somerset received further favours from the King, becoming Treasurer of Scotland, and in 1614 acting Lord Keeper of the Privy Seal (in succession to Northampton). When enquiries into Overbury's murder established conclusive evidence against the Countess, she confessed her guilt. Somerset, however, denied complicity, and the Crown case against him as an accessory to murder before the fact rested on the evidence of his self-incriminating behaviour after enquiries started. He burned a number of his own letters written to Northampton at the time of the murder, and directed Sir Robert Cotton to affix false dates to the letters which he had received then from Northampton and Overbury. He further abused his official position by having officers of the law sent to Weston's house, to seize correspondence between Weston and Mrs Turner, and even had a message conveyed to Turner while under arrest. Somerset was arrested in turn, imprisoned until his wife's trial, at which she pleaded guilty (24 May 1616), to which his own followed the next day.

Bacon's involvement in these proceedings, much of it behind the scenes, was set out in detail by Spedding (*Works*, xii. 208–23, 228–32, 262–72, 275–346). His first public appearance was on 10 November 1615, in connection with some disorder at Weston's execution on 25 October, when several unruly gentlemen (Sir Thomas Wentworth, Sir John Hollys, Thomas Lumsden, and others) had questioned Weston on the scaffold as to whether Overbury had really been poisoned, so calling in question the Crown's proceedings. Bacon delivered a formal *Charge* against those involved (ibid. 213–23), and as Attorney-general he subsequently led the prosecution against the Countess (ibid. 297–304) and Somerset himself (ibid. 307–20, 330–1, 334–5). During the preparations for the trial he also wrote a number of important letters to James, outlining the prosecution's strategy for the King's approval, whose marginal comments or 'postils' survive (ibid. 231–2, 263–5, 265–8, 270–1, 275–80, 281–3, 285–9, 290–2, 292–4, 295–6). The King's approval was necessary, since he was involved in several ways. First, the murder had taken place while Overbury was in official custody, so the Crown's innocence had to be publicly demonstrated. Secondly, although Somerset had been the King's principal favourite, he was being gradually displaced in James's affections by George Villiers (subsequently Duke of Buckingham), and his relationship to James had deteriorated so rapidly that in July 1615, just before the murder was disclosed, Somerset had petitioned the King for a general and extraordinary pardon. James agreed to it, but both the Solicitor-General (Yelverton) and the Lord Chancellor (Ellesmere) had refused to seal such an unprecedented document (unless, Ellesmere said, he might have a pardon himself for doing so). Given this special relationship it was vital that justice should be seen to be done, without personal considerations. Thirdly, Somerset had been involved in secret transactions with Spain, concerning Prince Charles's proposed marriage with the Infanta Maria, and he threatened to reveal some secrets (political or personal) which could damage James's government. The whole situation was so delicate that James kept a close eye on the proceedings. To avoid provoking Somerset in an open court, Bacon was explicitly instructed not to indulge in invective or abuse (as Coke had already done in the preliminary trials, pronouncing the accused guilty of heinous crimes before the evidence had even been heard), nor to digress to more general issues.

Considered in the light of these pre-imposed constraints, Bacon's *Charge against Somerset* is a masterpiece of conciseness and self-discipline, expounding the material accusations without emotional excess. In his *Charge* against Wentworth and Hollys the previous November, before the delicacy of the King's position had become evident, Bacon was able to speak more freely, expressing righteous indignation at the crime of poisoning: 'The offence of impoisonment is most truly figured in that device or inscription which was made of the nature of Caius Caligula, that he was *lutum a sanguine maceratum* [Suetonius, 'Life' of Tiberius, 57.1; *Promus*, no. 732, as 'Dust trampled with bloode'], mire mingled or cemented with blood. For as it is one of the highest offences in guiltiness, so it is the basest of all others in the mind of the offenders. Treasons *magnum aliquid spectant*: they aim at great things; but this is vile and base. I tell your Lordships what I have noted, that in all God's books both of the old and new testament, I find examples of all other offences and offenders in the world in their kinds, but not any one of impoisonment or an impoisoner. I find mention of some fear of casual impoisonment: when the waters were corrupted and bitter, they came complaining in a fearful manner; Master, *mors in olla* [2 Kgs. 4: 40; *Promus*, no. 92]. And I find mention of beasts and serpents; *the poison of asps is under their lips* says the Psalm [Rom. 3: 13]; but I find no example in a human creature of a malicious and a murderous impoisonment' (xii. 214–15). The King's prosecution, Bacon claimed, displayed the royal virtue of justice, for which 'God hath of late raised an occasion, and erected as it were a stage or theatre, much to his honour, for him to show it and act it, in the pursuit of the violent and untimely death of Sir Thomas Overbury, and therein cleansing the land from blood. For, my Lords, if blood spilt pure cry to heaven, much more blood defiled with poison is not only a loud cry to God's ears, but (to use the Scripture phrase without niceness [coyness]) it is also a stench in his nostrils [Amos 4: 10], I say in the nostrils of God and man' (xii. 214).

Metaphors from the theatre seem to have been much in Bacon's mind during these proceedings (see Brian Vickers, 'Bacon's Use of Theatrical Imagery', in W. A. Sessions (ed.), *Francis Bacon's Legacy of Texts* (New York, 1990), 21–33). In a letter to Villiers (5 May 1616) he described the prosecution as literally staging its case: 'I have distributed parts to the two serjeants' (Crown attorneys), as if they were actors being assigned their roles, having instructed them not to strike too high a note, 'that the matter itself being tragical enough, bitterness and insulting be forborne' (xii. 285). Should the Countess 'confess the indictment', he wrote a week later, 'no evidence ought to be given. But because it shall not be a dumb show, and for his Majesty's honour in so solemn an assembly, I purpose to make a declaration of the proceedings to this great work of justice from the beginning to the end' (xii. 291). When she came to trial and pleaded guilty, Bacon had to pray judgment against her, and did so, according to one observer, with great respect, 'nothing being aggravated against her by any circumstance, nor any invective used'. In his prosecution speech Bacon drew on the theatre again to create an extended metaphor for the conspiracy and its discovery: 'The great frame of justice (my Lords) in this present action, hath a Vault, and it hath a Stage; a Vault where these works of darkness were contrived; and a Stage, with steps, by which they were brought to light. . . . I will reserve [the case against Somerset] until tomorrow, and hold myself to that which I called the stage or theatre, whereunto indeed it may be fitly compared: for that things were first contained within the invisible judgments of God, as within a curtain, and after came forth, and were acted most worthily by the King, and right well by his ministers' (xii. 299). Bacon extends this theatrical imagery to take in Weston, Turner, Hellwysse, and Franklin, 'being but the organs and instruments of this fact [deed], the actors and not the authors' (p. 302), who had all been executed. He develops it one stage further, before wittily dismissing it: 'But (my Lords) where I speak of a stage, I doubt I hold you upon the stage too long' (p. 304). The following day, delivering his speech against Somerset, he kept to the careful path James had set him. (Both accused were found guilty, but James had no intention of executing them and they were released after six years in the Tower.)

Apart from one emendation I reprint Spedding's text, which he based on the version given in Tenison's *Baconiana* (1679), collated against two manuscripts in the British Library, Harl. MS 2194 and Add. MS 1002. Spedding gave occasional readings from a further copy in Cambridge University Library (Ee. IV. 12).

314. *accessary*. A person who helps in, or is privy to any act, especially a crime. *before the fact*. Involved before the deed takes place. *favour... King*. Somerset had been one of James's intimate favourites. *place*. Somerset was a Privy Counsellor, entrusted to conduct James's correspondence (though without the title of secretary), Treasurer of Scotland, and acting Lord Keeper of the privy seal. *sensible*. Aware. *justice*. In his *Charge* against Wentworth, Hollys, and Lumsden (10 Nov. 1615) for having questioned Weston on the scaffold, Bacon expounded at greater length: 'The King among many his princely virtues, is known to excel in that proper virtue of the imperial throne, which is Justice. It is a master virtue, unto which the other three are ministrant and do service. Wisdom serveth to discover, and discern of innocency and guiltiness. Fortitude to prosecute and execute. And temperance, so to carry justice as it be not passionate in the pursuit, nor confused in involving persons, nor precipitate in time' (*Works*, xii. 213–14). *expatiate*. Speak or write at great length. On 5 May 1616, Bacon addressed to Villiers, James's new confidant, the outline of the charge he intended to make against Somerset, on which James made 'postils', or marginal notes. Where Bacon proposed that Somerset should not 'be charged with anything by way of aggravation otherwise than as conduceth to the proof of the impoisonment', the King approved: 'I have commandit you not to expatiate nor digresse upon any other points, that maye not serve clearlie for probation or inducement of that pointe quhairof he is accused' (*Works*, xii. 286). *invectives*. Vituperative, abusive speeches. *materially*. Pertinently. *Strains*. Exertions. *play prizes*. Compete for rewards. *blazon*. Proclaim, announce boastfully. *lanthorn*. Lantern. *to be able*. This is the reading of the MSS; the *Baconiana* reads 'and be able', an anacoluthon. Spedding conjectured 'upright and steady'. *vantages*. Additions. *order*. Bacon refers to the rhetorical technique of *partitio*, a formal division of the topics of a speech or other work, which he

frequently used: see Vickers, *Francis Bacon and Renaissance Prose* (Cambridge, 1968), 30–59.

315. *Scripture-phrase.* 'As I live, saith the Lord, though Coniah the son of Jehoiakim king of Judah were the signet upon my right hand, yet would I pluck thee hence' (Jer. 23: 24). *put . . . off.* Remove, reject. *colour.* Pretence. *moral.* Involving ethical, not criminal matters. *Cain.* He murdered his brother Abel: Gen. 4: 8. *primogeniture.* Since Cain was the first-born. *population.* According to Genesis, mankind is descended from Cain. *Abner . . . Joab.* See 2 Sam. 3: 6–39. *respited.* Reprieved from death. *discourse.* Speak at length. *tables.* Commandments: 'Thou shalt not kill' being the second. *first table.* The first five commandments, concerning moral matters. *lieutenancy.* Delegated authority or command. '*nostri . . . sanguinis*'. '[Neither] of our kind nor of our blood.' *Italian.* Renaissance Italy was notorious for crimes of poisoning, particularly by the Borgia family. *court of Rome.* The Pope's dwelling: Alexander VI, Rodrigo Borgia, Pope 1492–1503. *grievous.* Atrocious (of a crime). *refection.* Refreshment with food and drink. '*his . . . snare*'. Ps. 69: 22. In his *Charge* against Wentworth and the others Bacon was fuller: 'I have sometime thought of the words in the psalm, *let their table be made of a snare*: which certainly is most true of impoisonment; for the table, the daily bread for which we pray, is turned to a deadly snare. But I think sure that that was meant of the treachery of friends that were participant of the same table' (*Works*, xii. 215). *hardly.* With difficulty.

316. '*cum sonitu*'. By making a noise. *discover.* Reveal. *say-cup.* One used to test ('assay') drink before it is presented to a king or noble. *discolour . . . distaste.* Cause a change of colour or taste, so revealing the presence of poison. '*Quis . . . erit?*' 'Who will ever be safe?': from Anselm, *Carmen de Contemptu Mundi* (Migne, *Patrologia Latina*, vol. 158, col. 697A). '*concidit . . . vulnere*'. *Aeneid*, x. 781 (reading 'sternitur infelix . . .'): 'he falls, alas! by a wound meant for another'. *sagitta . . . night.* Ps. 91: 5. *all one.* All the same. *pledge.* Toast (an assurance of fidelity and goodwill). *21st . . . eight.* 1529–30. *barm.* The froth on the top of fermenting malt liquors, used to leaven bread. *pottage.* Oatmeal porridge. *alms basket.* Containing food remnants for the poor. *accident.* Event. *out of.* Deprived of. *respondent.* Responsible. *young princes.* Edward, Prince of Wales, and the Duke of York, sons of the dowager Queen Elizabeth, murdered by the order of Richard III. *impunity.* Exemption from punishment. *book-examples.* Instances from literature. *Livia.* 58 BC–AD 29, wife of the Emperor Augustus (who awarded her the honorific name 'Julia Augusta') and mother of Tiberius and Drusus. Historians formerly believed rumours that she poisoned her husband, but this is no longer accepted. Cf. Tacitus, *Annals*, i. 3–14, v. 1.

317. *Parysatis.* Daughter of King Artaxerxes of Persia; her children included Cyrus (her favourite) and Artaxerxes Mnemon. When Cyrus was executed for plotting against his brother, in revenge Parysatis poisoned Stateira, Artaxerxes' wife. Because both queens mistrusted each other, they would only eat from the same dishes prepared by the same cook, but Parysatis took a bird that had been served them, and 'cut it in the middest with a little

knife, the which was poysoned onely on one of its sides, and gave that half which was poysoned unto Stateira', who died 'with grievous panges and gripings in her bowels': Plutarch, 'Life' of Artaxerxes, 19; tr. North. *triers.* Judges. *Salomon's spirit.* Revealed in adjudicating between the two mothers: 1 Kgs. 3: 16–28. *collected.* Deduced (the true relationship). *affection.* Strong feelings. *concluded.* Convinced. *procurement.* Instigation. *abetting.* Assisting in a crime. *confecting.* Preparing, from various ingredients. *commixing.* Blending. *familiar.* Everyday. *ways.* Public roads. *by colour.* Under pretence. *that.* So that. *lets slip.* Releases. *lodges.* Discovers the lair of a buck or stag. *raises.* Rouses (from its lair). *puts . . . out.* Drives into the open. *toil.* Net or nets, forming an enclosed area into which a hunted animal is driven. *grossness.* Magnitude.

318. *break.* Divide up. *open.* Outline. *take . . . aim.* Prepare your defence. *without confusion.* Observers of the trial noted that—apparently at Bacon's urging—Somerset was provided with pen and ink to take note of the charges against him: *Works*, xii. 312 n., 331. *interest.* Involvement for his own advantage. *oracle of direction.* Infallible guide, adviser. *Thrasonical.* Bragging (after Thraso, a boastful soldier in Terence's *Eunuchus*). *better teacher.* King James. *rounding.* Whispering. *secrets of estate.* National secrets. *Secretary.* Administrative post in the King's household. *settled.* Systematic. *tables.* Lists of contents. *Council-table.* The King's closest advisers. *made a play.* Duped. *ciphers.* Codes. *jargons.* Coded, otherwise meaningless symbols. One of the letters written by the Countess to the Lieutenant of the Tower concerned some tarts and jelly sent to Overbury as containing 'letters', jargon for 'poison' (*Works*, xii. 268–9). In his long brief to the King (5 May 1616) Bacon outlined his intended course in the prosecution: 'I mean to show likewise what jargons there were and cyphers between them, which are great badges of secrets of estate, and used either by princes and their ministers of state, or by such as practise against princes. That your Majesty was called *Julius* in respect of your empire, the queen *Agrippina* (though Somerset now saith it was *Livia* and that my lady of Suffolk was *Agrippina*); the bishop of Canterbury *Unctius*; Northampton, *Dominic*; Suffolk, first *Lerma*, after *Wolsey*; and many others; so as it appears they made a play both of your court and kingdom, and that their imaginations wrought upon the greatest men and matters' (ibid. 288). Further evidence cited at the trial explained that Agrippina referred to Nero's mother, following Northampton's remark 'that Prince Henry if he reigned would prove a Tyrant'. Northampton was known as the Dominican because of his Catholic sympathies. *work.* Commit a crime. *practise.* Plot an evil act.

319. *best . . . vinegar.* Common principles in Aristotelian natural philosophy; cf. also the proverbs, *Corruptio optimi pessima est*, 'there is nothing so ill as the corruption of the best', and 'it is the best wine that makes the sharpest vinegar' (*Oxford Dictionary of English Proverbs*, 145, 860). In *Promus*, no. 571, Bacon noted: 'Beware of the vinegar of sweet wine' (repeated, no. 910). *his.* Overbury's. *projected.* Planned. *Lord Privy Seal.* Henry Howard, Earl of Northampton, sometimes alleged to have been involved in the murder plot. He was certainly no friend of Overbury. *oppugn.* Oppose. *counted.* Judged.

Howards. Northampton's family. *bargains*. Struggles. *civility*. Courtesy. *naught*. Wicked. *ballads*. Contemporary popular songs gave their version of the poisoning. *under his girdle*. In his power.

320. *menaces . . . secrets*. Threats to disclose damaging information. *crossed*. Thwarted. *furies*. Avenging spirits. *mineral*. Fundamental. *mine into*. Undermine. *crossed*. Cancelled. *preventions*. Obstructions. *lay . . . up*. Imprison, confine. *under-hand*. Covertly. *contempt*. Disrespect of the monarch's lawful commands. *close*. Strictly confined. *exhibiting*. Administering. *note*. Notice.

321. *fifteen days*. These events took place between 21 Apr. 1613, when Overbury was committed to the Tower, and 7 May, when the first poison was administered (*Works*, xii. 316). He did not die until 15 Sept. *trunks* Channels. *several*. Different. *hit his complexion*. Be suitably toxic for his constitution. *say-mistress*. Tester. *present*. Immediately effective. *chase*. Pursue. In his *Charge* against Wentworth and the others Bacon commented that 'to have a man chased to death in such manner . . . by poison after poison, first roseaker, then arsenick, then mercury sublimate, then sublimate again; it is a thing would astonish man's nature to hear it' (*Works*, xii. 216). *Mithridates*. King of Pontus (*c*.120–65 BC), whose diet of prophylactics made him immune to poison. *treacle*. Medicinal salve used as an antidote. *preservatives*. Medicines giving protection from disease or infection; prophylactics. *chid*. Blamed. *persons*. According to reports of the court proceedings, Bacon specified 'that kind of persons who keep in a tale of fortune-telling when they have a felonious intent to pick their pockets or purses' (*Works*, xii. 316 n.). *lights*. Rays of light, direct and reflected.

322. *dispatch*. Speed. *detect*. Display. *clip*. Cut, crop. *misdate*. Somerset asked Sir Robert Cotton either to cut off the original dates on the letters or to redate them to disguise their coherence. *pardon*. Somerset had persuaded James to issue a general pardon for himself (*Works*, xii. 329–30). *convert*. Turn, direct. *heads*. Topics. *two Serjeants*. Serjeant Montague, who had to prove the most difficult part of the case, that Somerset's hand was in the poisoning itself (*Works*, xii. 325–6), and Serjeant Crew, who had to detail Somerset's attempts to destroy the evidence (ibid. 326–9). *light*. Trivial. *elevation . . . pole*. Exact details of their seriousness.

323. *toy*. Trifle. *doubt*. Fear. *coadjutor*. Assistant. '*periculum . . . vincitur*'. 'Danger is overcome by dangerous trial.' *flashes*. Sudden, transitory bursts of flame or light. *middle acts*. Intermediate stages of a conspiracy, stopping short of a crime. *train*. Trick, stratagem.

324. *wrought*. Worked. *extremities*. Acute pains. *denied*. Refused permission. *advertisements*. Reports. *in progress*. On a ceremonial tour. *on work*. Into action. *recognition*. Acknowledgement. *entertainment*. Diversion. *dark*. Obscure. *bottom*. Full extent.

325. *urged*. Charged, accused. *implicative*. Which implicates someone in a crime. *none*. No one. *formal*. Meticulous.

LETTER TO KING JAMES

Bacon's public career, blocked by Elizabeth as a punishment for having opposed her tax demands in the 1593 Parliament, finally made headway under James. Appointed Solicitor-General in 1607, Attorney-General in 1613, he was made Privy Councillor in 1616, Lord Keeper in 1617, and Lord Chancellor in 1618. In that year he was elevated to the House of Lords as Baron Verulam, and on 27 January 1621 was created Viscount St Alban. As he gratefully wrote to the King, marking the eighth promotion he had enjoyed under James, 'this is the eighth rise or reach, a diapason in music, even a good number and accord for a close' (*Works*, xiv. 168). The first major scientific work of his to be published, the *Instauratio Magna* (including the *Novum Organum*) appeared in a sumptuous edition in October 1620, which also announced his planned *Historia Naturalis et Experimentalis* (*Works*, xiv. 119–20, 122, 129–31). From this absolute pinnacle of his career he was plunged, within three months, into 'a close' of quite a different kind. In March 1621 he was accused of corruption, found guilty by the House of Lords, dismissed from Parliament and all offices, temporarily imprisoned in the Tower, fined £40,000, and banned from coming within the verge (the radius of twelve miles around the court). Few public figures in Britain have experienced so rapid and so complete a reversal.

Bacon's fall has been subject to more misunderstanding, ill-informed indignation, and malicious misrepresentation than any other event in his life. It has been taken out of its historical context and evaluated according to anachronistic criteria, those governing modern ideals of justice and the desired freedom of public servants from any form of corruption. It has been treated as a purely legal matter, as if it only involved the Chancery Court over which he presided as Lord Chancellor; but in fact it was part of a much wider political dispute, involving both personal and institutional conflicts, dating back many years. Bacon's impeachment was the culmination of several struggles, which converged on his person and his office, but which were originally directed against other targets, who could not be openly attacked.

To begin with his actual performance as a judge in the four years from taking his seat in the High Court of Chancery on 7 May 1617. A recent detailed study by a historian of English law, Daniel R. Coquillette (*Francis Bacon*, Jurists: Profiles in Legal Theory (Edinburgh, 1992), 196–211), has confirmed the estimate of other authorities that Bacon performed his duties with exemplary efficiency and conscientiousness. By working 'immensely hard', making the court sit mornings and afternoons, he had cleared the backlog of cases by 8 June 1617, and continued to maintain a high work-load, the court dealing with up to 1,700 cases a year. His *Ordinances in Chancery* (*c.*1618) succeeded in fixing the practice of the court for over two centuries, and laid great emphasis on 'expedition and fairness'. He introduced legislation to block vexatious litigation, and took care to replace judges who were personally implicated in the cases they were trying. His decrees show that he took very seriously the Chancellor's statutory responsibility as 'the guardian of those subject to abuse, including the aged, the handicapped, and minors'. He defended widows and women mistreated by their husbands and fathers, in one case with a damaging effect on his political career and standing at court.

In 1616, as Attorney-General, he had supported King James's assertion of the royal prerogative on the issue of 'commendams', ecclesiastical positions in the King's gift which represented valuable patronage to the Crown. The King's opponent was Edward Coke, Bacon's lifelong rival, and now Chief Justice of the King's Bench, who wanted to abolish the practice. But James exerted his authority, forced Coke to capitulate, and dismissed him from office in November 1616, with Bacon as the legal officer responsible for carrying out the dismissal. The following year Coke, wanting to ingratiate himself with George Villiers, now Marquis of Buckingham and the King's favourite, offered the hand of his daughter in marriage, together with a huge dowry and annual income, to Buckingham's brother John. Coke's wife, Lady Hatton, opposed the marriage and escaped with her daughter, only for Coke to pursue them and seize the girl by force. Appealed to by Lady Hatton for redress, Bacon, chairing the Council in the absence of the King and Buckingham, ordered Coke to deliver his daughter up. James was furious, reversed the Council's orders, reprimanded Bacon, and himself gave away Coke's daughter in marriage to Sir John Villiers. It took Bacon over a year to regain the King's favour; meanwhile Coke had greatly strengthened his position in Parliament (Coquillette, *Francis Bacon*, 218–22)—a position which he was able to exploit in 1621 as joint leader of the attack on Bacon. This incident shows the degree to which personal and institutional conflicts intertwined.

Bacon's conduct of the Chancery Court was, with two exceptions, exemplary. He 'issued over four thousand orders and decrees' in his time presiding over the court, many of them 'being marked by sensitivity, restraint and principle. Almost none were reversed after his impeachment' (ibid. 210). Of the two exceptions, one was institutional, involving the so-called 'bills of conformity', issued on application by the Court of Chancery, which granted the King's protection to bankrupts and thus prevented creditors from collecting their debts. This humane provision to protect people whose businesses or personal finances had collapsed was deeply unpopular with financiers, and provided, as Robert C. Johnson has shown ('Francis Bacon and Lionel Cranfield', *Huntington Library Quarterly*, 23 (1960), 301–20), the occasion for Cranfield to lead a violent attack on Bacon in the quarrelsome Parliament of 1621 (see below). The other exception to Bacon's exemplary conduct as Lord Chancellor was personal, and the formal cause of his impeachment, namely that he was found guilty of having accepted gifts, two years earlier, from suitors whose trials were still pending. Although he regularly took pains to intervene in cases where his subordinate judges might be guilty of a conflict of interest or improper influence, Bacon was remarkably careless as regards his own behaviour. In his audience with the King on 16 April, he was ready to 'confess . . . that at my new-yeares tides and likewise at my first coming in (which was as it were my wedding), I did not so precisely as perhaps I ought examine whether those that had presented me had causes before me or no' (*Works*, xiv. 236). Bacon never denied that he had accepted these gifts, but could justly declare that he had not allowed them to affect his decision, which actually went against the suitors. Although no one can condone Bacon's behaviour, and although he certainly sought no excuse, as the harrowing letters and personal memoranda, deeply humiliated and self-recriminating, show (ibid. 213–16, 225–301), judged by contemporary standards Bacon was not corrupt. He was certainly at fault in not supervising his servants more closely, as Spedding showed (ibid. 563–4), and by the standards of his age he was naïve in not concealing the gifts, as others did. But his accusers were also hypocritical in making him the scapegoat for a generally accepted practice. Bacon could claim with some justice that 'I . . . partake of the abuse of the times' (above, p. 326).

Bacon's offence needs to be seen in the context of a growing dislike of abuses at the Jacobean court involving money and office, which had started with the King's notorious sale of titles. James created nearly 1,000 knighthoods in the first year of his reign, accepting money from those honoured, and even allowing favoured courtiers to sell titles openly. The rank of baronet was created in order to raise money for the Crown (the going rate was £1,095 but it soon became cheapened by overuse, the price dropping to £200), while peerages could be had for £10,000. The key figure controlling these transactions was Buckingham, who had become all-powerful by 1620, making enormous profits for himself, his family, and hangers-on. Another form of enrichment for many of the most high-ranking courtiers was to accept gratuities and pensions from England's recent enemy, Spain, with whom James had negotiated a peace in 1604. It was estimated that such pensions were worth £9,125 a year, and the English ambassador to Spain was outraged when he discovered this practice by decoding secret documents; but the King turned a deaf ear to his complaints. Public offices were also bought and sold, and courtiers regularly competed with each other to obtain posts guaranteed to be lucrative, for those who had bought office might extort payment from those who needed their help. As recent historical and sociological studies have shown, practices at James's court which seem corrupt to us—bribery, misappropriation, nepotism—were actually common in all early modern governments. (In our own time European governments used to observe complacently that it was only in 'third world countries' that one needed to bribe officials: few countries can now throw that stone.) Since officers of the Crown in this period were not paid a regular salary out of the public purse, they depended on the sale of favours and the exploitation of patron–client relationships which, although sometimes ritualized in the form of a lavish New Year's gift, were often frankly commercial. Crown servants were in effect paid by those who needed to use government services. (See e.g. Joel Hurstfield, *Freedom, Corruption and Government in Elizabethan England*, London, 1973; Linda L. Peck, *Court Patronage and Corruption in Early Stuart England*, London, 1990.)

The growing sense of outrage at corruption in King James's court is the sign of a definite change in moral standards between 1610 and 1630. But, as Linda Peck has shown, this new morality was also put to political ends by the opposition groups within Parliament, who proceeded ever more vigorously against the King's officials, using accusations of bribery as their chief weapon. The 1610 Parliament had imprisoned Sir Stephen Proctor for misuse of office and for taking bribes, formally requesting the King not to protect his servants from arrest and attainder. The King's expectation that a docile House of Commons would fulfil his wishes were rudely disappointed at every session, and a series of conflicts caused him to suspend the 1614 Parliament, expressing surprise that 'my ancestors should ever have permitted such an institution to come into existence' (cit. D. M. Loades, *Politics and the Nation 1450–1660: Obedience, Resistance and Public Order* (London, 1974), 331). James summoned no new parliament between 1614 and 1621, raising money by other resources, so that when it was finally reconvened in 1621 much resentment had built up, particularly against the corrupt favourite Buckingham.

One of Parliament's major grievances when it reassembled was the detested system of monopolies and patents, which gave their possessors (or purchasers) the exclusive rights to trade in a particular commodity imported from abroad, or to enjoy the profits of an invention or technical innovation. Power to issue patents and monopolies rested with the Crown, and both Elizabeth and James were guilty of using their power to reward not only deserving servants but favoured courtiers. Since such gifts were under royal prerogative, monopolies were administered by court officials who could seize goods, and imprison and fine offenders in the monarch's name. Citizens enjoyed no right of appeal against these actions, and monopolies could easily be abused for blackmail and extortion. Sir Nicholas Bacon had objected against the practice in Elizabeth's reign, and his son spoke on the issue in the parliaments of 1597–8 and 1601, when he loyally defended the royal prerogative to grant or refuse patents and monopolies, but expressed his gratitude at the Queen's (never fulfilled) promise to abolish the most offensive ones (*Works*, x. 27–9; M. A. Phillips, 'Francis Bacon', in P. W. Hasler (ed.), *The House of Commons 1558–1603* (London, 1981), i. 378). Three previous parliaments of James (1606, 1610, and 1614) had attacked monopolies, but to no avail. As a prominent legal officer of the Crown from 1613 to 1621, despite his sincere dislike of monopolies (*Works*, xiii. 49; xiv. 145–9, 151–2, 182–3), Bacon had no choice but to co-operate with a corrupt system endorsed by the Crown. He actually advised James to rescind some of the most obnoxious cases, but the King, whose own protégés were profiting most from the system, ignored the advice. In any case, James was too dependent on the sale of monopolies as a source of Crown revenue ever to abolish them, although he made occasional token gestures to suppress some of the less lucrative ones. Since Bacon was loyal to both James and Buckingham, the parliamentary opposition group, led by Coke and Cranfield, made him their main target.

The 1621 Parliament brought many grievances into the open. Bacon was soon caught up in the campaign against monopolies, especially two in which Buckingham was heavily involved, one for the manufacture of gold and silver thread (on which was blamed the shortage of bullion that provoked an economic crisis in 1620–1), and one for the patent to license inns. Buckingham's brothers, Sir Edward and Sir Christopher Villiers, had profited from both, and in the inn-licensing they were joined by the notoriously corrupt Sir Giles Mompesson. The reformers' extraordinarily violent campaign against monopolies descended at times to a witch-hunt, all legal procedures suspended: Mompesson's deputy, Sir Francis Michell, was sent to the Tower for misuse of office without having been charged, or allowed to defend himself (*Works*, xiv. 182–208; R. Zaller, *The Parliament of 1621: A Study in Constitutional Conflict*, Berkeley, 1971). Coke and Cranfield—both of whom had earlier supported, and profited from monopolies—then decided to attack the legal referees who had certified the patent awarded to Mompesson. The referees or legal advisers, of whom Bacon was the most prominent, had the task of investigating and licensing patents of monopoly as having been properly drawn up, and approved by the King or Buckingham. Their role was hardly to question the whole system (which would in any case have merely aroused the King's anger), so there is something hypocritical in Cranfield's demands, in February 1621, that the referees be brought to book in order to save the King's honour (Johnson, 'Bacon and Cranfield', 309–10). Cranfield reiterated this demand in parliamentary debates until the King placed pressure on him not to proceed against the referees, openly warning the Commons on 10 March not to attack his ministers. Cranfield's campaign was 'in effect a direct attack on Bacon' (ibid. 310), and being blocked in this matter he shifted tack, to the bills of conformity approved by the Chancery Court. On 14 March he denounced the protection they offered to bankrupts as 'an intolerable grievance', knowing full well that Bacon was directly involved, and he appealed to the King to dissociate himself from the injustices of his courts. James, his position weakening, conceded the point within a week, and by the end of that month Cranfield had successfully elicited a royal proclamation to abolish the bills of conformity.

While this intense investigation into Bacon's court was being carried on, the fact that he had accepted gifts came to light, and two aggrieved suitors were found whose gifts had not produced verdicts in their favour, and who were willing to publicly testify against him (*Works*, xiv. 212–14). A few years earlier, or in a less heated context, these gifts might have been seen as having been carelessly accepted by a public figure who, 'unlike many of the other royal officials, . . . had not used his position to amass a large fortune in either money or in lands' (Johnson, 'Bacon and Cranfield', 313). In April 1621 Sir John Bennett, Judge of the Prerogative Court of Canterbury and the Court of High Commission, was charged with bribery and corruption. As a contemporary diarist noted, 'there was found in his custody two hundred thousand pounds in coin. He was as corrupt a judge as any in England, for he would not only take bribes of both parties, plaintiff and defendant, but many times shamefully begged them' (Peck, *Court Patronage and Corruption*, 187). Unable to attack Buckingham, but determined to prosecute Bacon, the reformers revived the long obsolete procedure of impeachment, whereby the Commons acted as accusers and the Lords as judges, an unfamiliar proceeding in which the rights of the accused to defend himself were not clearly defined. Bacon tried to discover what form the actual proceedings would take (*Works*, xiv. 215–16), but once he realized that three parliamentary committees, after several weeks' investigation, had produced witnesses to some twenty-eight cases where he or his servants were accused of improperly accepting gifts or loans, he decided not to defend himself on individual points but to plead guilty (ibid. 245–67). As recent researches have shown, nineteen of these charges were without substance, witnesses were named but not examined, and in some cases false testimony was given. An air of desperate haste attended the proceedings.

All the same, Bacon's 'surrender made it impossible for James to defend him' (Loades, *Politics and the Nation*, 357). Bacon may have already realized that the King was not going to stand by his loyal officials if this meant a direct and possibly awkward confrontation with Parliament. Some commentators have speculated that Bacon was ordered by the King to sacrifice himself to save Villiers. One pathetic detail showing his dependance on these masters is provided in the rough draught of a letter to Buckingham on 14 March 1621, where Bacon had written, 'I hope the K. and your Lp. will keep me from oppression': on second thoughts he crossed it out (*Works*, xiv. 213 n.). Whatever the explanation, James and Buckingham abandoned Bacon to his fate, in effect making him the scapegoat for their own unpopular policies. As Bacon bitterly commented on his role as the focus for Parliament's hostility to the administration, 'whosoever was the justest judge, by such hunting for matters against him as hath been used against me, may for a time seem foul, specially in a time when greatness is the mark [target] and accusation is the game. And if this be to be a Chancellor, I think if the great seal lay upon Hounslow Heath, nobody would take it up' (ibid. xiv. 213).

In many ways the most remarkable feature of the proceedings is Bacon's lack of concern before the charges came out into the open, proof of his sense of his own innocence: as he wrote to Buckingham on 14 March, 'I know I have clean hands and a clean heart' (ibid.). In some notes he made for an audience with the King on 16 April Bacon wrote: 'There be three degrees or cases (as I conceive) of gifts and rewards given to a judge: 'The first is, Of bargain, contract, or promise of reward, *pendente lite* [while a case is in progress]. And this is properly called *venalis sententia*, or *baratria*, or *corruptelae munerum*. And of this my heart tells me I am innocent; that I had no bribe or reward in my eye or thought, when I pronounced any sentence or order. The second is, A neglect in the Judge to inform himself whether the cause be fully at an end or no, what time he receives the gift; but takes it upon the credit of the party, that all is done; or otherwise omits to enquire. And the third is, when it is received *sine fraude*, after the cause ended; which it seems by the opinion of the civilians [experts on civil law], is no offence' (*Works*, xiv. 237–8). Pronouncing himself 'to be as innocent as any born upon St. Innocents day' of the first charge, for the second he wrote: 'I doubt [fear] in some particulars I may be faulty'. Yet, while defending himself as having practised justice, and as being uncorrupt in the normal sense of the word, he admitted that on an absolute, not a relative scale, he had indeed shared 'the abuse of the times'. So he could affirm, paradoxically, that 'I was the justest judge that was in England these fifty years. But it was the justest censure in parliament that was these two hundred years' (ibid. 560). The final paradox is this coexistence within his personality of the two opposed states of innocence and experience. As Spedding commented, the naïvety by which he could accept gratuities without recalling 'the peril of meddling with any thing that could be construed into corruption' (ibid. 211–12, 562), reveals a fundamental 'simplicity' or unworldliness. For all his extended exposition of the arts of negotiation and the need, in human affairs, to blend the innocence of the dove and the cunning of the serpent, Bacon did not always apply the fruits of such knowledge to his own career.

The aftermath of Bacon's impeachment includes several ironies. On 26 March the King addressed the House of Lords, 'recognising their authority as the Supreme Court of Justice', as Spedding summarizes him, 'confirming their privileges, reminding them of their duties, advising them to be guided by precedents in the times of good government; leaving judgment [on Bacon] wholly to them, and declaring his readiness to carry their sentence into execution; and moreover, in return for their kind dealing with himself, announcing his intention to "strike dead" the three Patents principally complained of—Inns, Ale-houses, Gold and Silver Threads;—a temperate, judicious, honest speech; with which the Lords were much gratified' (ibid. 226–7). The parliamentary reformers had eliminated two major grievances (the bills of conformity and the monopolies controlled by Buckingham), had embarrassed the King and his favourite, and reasserted their power—but only temporarily: Villiers and Mompesson were still profiting from corruption in 1623. Bacon's career was ruined, but (as if in recognition that he had been a scapegoat) moves were taken to soften the other blows. His imprisonment in the Tower only lasted three days, and the huge fine was never collected. Yet his apparently generous pension of £1,200 a year lay in the hands of Cranfield, now promoted to Lord Treasurer, who withheld payment until Bacon was forced to beg. Buckingham and Cranfield between them then used their delaying-tactics with the pension, together with the continuing ban on Bacon's freedom of movement, as a lever to force Bacon to sell York House—where his father had lived before him—to Buckingham, who had long coveted it. Bacon resisted as long as he could, but was finally compelled to sell it to Cranfield as the middleman, who passed it on to Buckingham (Johnson, 'Bacon and Cranfield', 313–19).

The shifting and unstable power relationships at James's court that expelled Bacon and elevated Coke and Cranfield to power, coupled with the King's total lack of a reciprocal sense of obligation to loyal servants, made anyone's position fragile. Coke had ingratiated himself by marrying his daughter to Buckingham's brother, and Cranfield similarly helped secure his career in 1620 by promising to marry Buckingham's cousin: after which he was admitted to the Privy Council and created Baron Cranfield the following year (ibid. 305). But Coke's increasingly stubborn defence of parliamentary liberties against the Crown infuriated James, who dissolved Parliament, arrested Coke and other leaders of the 'turbulent party', and confined him to the Tower for seven months in 1621–2. As for Cranfield, his financial zeal as Lord Treasurer in imposing spending cuts balanced the budget but antagonized many courtiers, and 'he made the fatal error of offending both

Buckingham and Prince Charles by withholding payment of money to them and opposing their plan of dissolving the Spanish marriage treaties'. Cranfield's enemies, including Coke, now fully supported by the favourite and the prince, attacked him in the 1624 Parliament, and after several weeks' pressure, the House of Commons charged him on 15 April 'with several acts of corruption, malfeasance and neglect of office'. He was impeached before the House of Lords on 12 May, where he vigorously defended himself but was still found guilty, confined to the Tower, deprived of all offices, heavily fined, and forced into exile (ibid. 319–20). Historians have observed the bitter irony that the methods he invented to destroy Bacon should have been used to oust him too, a sequence which inevitably recalls those aphorisms from the Essay 'Of Great Place': 'The rising unto place is laborious; and by pains men come to greater pains; and it is sometimes base; and by indignities men come to dignities. The standing is slippery, and the regress is either a downfall, or at least an eclipse, which is a melancholy thing' (above, p. 359). The consequences for Britain were more serious, for the loss of such capable public servants as Bacon and Cranfield meant that Buckingham could continue his corrupting influence over the new ruler. For Bacon, at least, there was the consolation that he could resume work on his other career, so long neglected.

326. *gemitum columbae*. 'The moaning of doves.' *higher flight*. As a result of the sustained enquiries between 15 and 19 Mar., Bacon fell seriously ill (*Works*, xiv. 215–16). *enter . . . myself*. Examine my conscience. *materials*. Material causes, grounds. *immoderate*. Extreme. '*suavibus modis*'. In the smoothest ways. *conversation*. Behaviour. *carriage*. Demeanour. *use to*. Usually. *credit*. Reputation; public career. *sepulture*. Burial. *ingenuity*. Nobility of character, ingenuousness. *trick up*. Disguise. *cavillations*. Legal quibbles. *voidances*. Evasions. *hardness of heart*. State of being deprived of God's grace. *thirst . . . streams*. Ps. 42: 1.

327. *abyssus*. Bottomless, unfathomable cavity. *usufructuary*. 'A person who has the right of enjoying the use of and income from another's property without destroying, damaging, or diminishing the property' (*SOED*). *oblation*. Offering.

LETTER TO LANCELOT ANDREWES

As the shattering effect of his fall from office diminished, Bacon took up the threads of his various careers. In publishing *The Historie of the Raigne of King Henry the Seuenth* (1622) he displayed his abilities both as a historian and as a counsellor. Wanting to regain his profession as a lawyer, he offered the King 'A Digest to be made of the Laws of England' (*Works*, xiv. 358–64). He still hoped for public office, either as Provost of Eton (ibid. 406–10), or in Parliament (453–5), but without success. To show his continuing grasp of current political issues, he drew up notes on usury and its uses (413–20), notes and letters of advice to Buckingham (440–51), notes of a speech concerning war with Spain (455–69), and (also in 1624) a fully worked-out treatise, *Considerations touching a War with Spain* (469–505).

A related (and unfinished) project, which took the literary form of a dialogue, was the *Advertisement Touching a Holy War*, written in 1622 and published posthumously by William Rawley in *Certaine Miscellany Works*, 1629 (*Works*, vii. 9–36; xiv. 367–71). Bacon prefixed to this dialogue (a brilliant and tantalizing fragment) a dedicatory epistle to his lifelong friend, Lancelot Andrewes, which is significant in two respects. First, it shows Bacon coming to terms with his disgrace by drawing comparisons between himself and three distinguished figures of antiquity who had also suffered banishment and humiliation, Demosthenes, Cicero, and Seneca. The comparison serves as a consolation, and as an admonition to bear misfortune with fortitude. But it also functions as a self-justification for Bacon's continuing activity as would-be Counsellor of State, hoping to be useful to his countrymen, an involvement with politics exemplified by Demosthenes alone of his three role-models. Secondly, it shows which priorities he attached to his career as a writer, and what he hoped to achieve in his remaining years.

My text is from Spedding, *Works*, xiv. 371–4.

328. *consolations... calamity in others.* The *consolatio* in classical antiquity indeed advised the sufferer to find solace in the fact that others shared similar misfortunes: cf. e.g. Cicero, *Tusc. D.* III. xxiv. 58–9; Seneca, *Ad Marciam*, xii. 4–5; Plutarch, 'Consolatio ad Apollonium', *Moralia*, 106B–C. *quicker.* More lively. *tendereth... satisfaction.* Offers as consolation, compensation. *that... us.* Cf. Eccles. 1: 9–10. *savoureth.* Suggests the presence of. *particulars.* Individual cases. *sentence.* As pronounced by a judge or tribunal. *night-work.* Work done during the night; of lesser value or significance. *tables.* Pictures, records. *Demosthenes.* Demosthenes (384–322 BC) was banished from Athens to Troezen in 324 for allegedly accepting a bribe from Harpalus, Alexander's treasurer, who had fled to Athens with much of his master's money. The evidence is dubious, however, and in his exile Demosthenes continued to be active in the campaigns against Alexander. In 323 he was recalled from exile, his fine having been paid from public funds, and he was honourably reinstated. Bacon's main source is Plutarch's 'Life' of Demosthenes (which is paired with that of Cicero), 25–7. *Cicero.* Cicero (106–43 BC) was driven out of public life in 58 by the ascendancy of his enemies at Rome, led by the tribune Clodius. On Clodius' eclipse in 57 Cicero was recalled by law and compensated, resuming his practice at the Bar. See Plutarch's 'Life' of Cicero, 31–3. *Seneca.* Seneca (*c.*5 BC–AD 65) had achieved eminence as an orator and writer when Claudius became Emperor in 41, in which year the emperor's wife Messalina secured his banishment to Corsica on the improbable charge of adultery with the notorious Julia Livilla, sister of Caligula. He remained in Corsica 'broken and unstoically despondent' (*OCD*) until 49, when he was recalled and appointed tutor to Nero. For his banishment see his *Ad Helviam*, iii–x. *durst.* Dared. *contracted.* Brought together.

329. *wrought upon.* Affected. *softened.* Made weaker or effeminate. *nothing but... epistles.* According to Plutarch, Cicero 'passed his time [in exile] for the most part in dejection and great grief, looking off towards Italy like a

disconsolate lover, while in his spirit he became very petty and mean by reason of his misfortune, and was more humbled than one would have expected in a man who had enjoyed so lofty a discipline as his' (32.4). For his sadness see e.g. the *Epist. ad Att.* iii. 8–21. In his concluding 'Comparison' (*synkrisis*) of the two orators Plutarch describes Cicero's banishment as a 'most honourable result, since he had rid his country of baleful men. . . . However, Cicero spent his exile idly, remaining quietly in Macedonia' (4.1–2). *penal.* Liable to punishment. *tempest of popularity.* Tumult of the populace. *bribery . . . disloyalty.* According to Plutarch (25.3–5) Demosthenes knowingly accepted a bribe from Harpalus (treasurer to the Athenians' greatest enemy). In the 'Comparison' Plutarch describes his banishment as 'disgraceful, since he had been convicted of theft' (4.11). Modern authorities are not so positive. *epistles . . . extant.* Six letters of Demosthenes have come down to us under his name, of which four may be genuine (see J. A. Goldstein, *The Letters of Demosthenes*, New York, 1968). The second describes his banishment. *solitary island.* Corsica. *mean.* Middling position, between two extremes. *business.* Public affairs. *books . . . all ages.* Seneca's works written in exile include the two *consolationes, Ad Polybium* and *Ad Helviam Matrem*, and the philosophical treatises *De providentia, De constantia*, and (perhaps) *De ira*. *dedications.* Bacon may be referring to the dedication of *De clementia* to the Emperor Nero in laudatory terms, hopefully intending to guide him towards the ideal of a merciful ruler: cf. I. i. 1–9. *talent . . . given me.* The gifts of God bestowed on men: cf. Matt. 25: 14–30. *mounts.* Financial institutions, especially banks. *perpetuity.* Immortality. *break.* Go bankrupt. *Instauration.* In 1620 Bacon had published his *Instauratio Magna*, comprising the second part of that project, the *Novum Organum* (2 books only; more were projected) and the *Parasceve ad Historiam Naturalem et Experimentalem* ('Preparative Towards a Natural and Experimental History'). The 'Plan of the Work' included shows how far Bacon was from realizing his grand design. '*si . . . imago*'. 'If the image never lies' (Virgil, *Eclogues*, ii. 27). *testimonies.* See e.g. the letters from Bacon to Isaac Casaubon (*Works*, xi. 146–7), to Father Baranzano, 30 June 1622 (ibid. xiv. 374–8), and to Father Fulgentio, *c*.Oct. 1625 (ibid. 531–4). *break the order of time.* The 6 parts into which the *Great Instauration* was divided were listed in 'Plan of the Work' as '1. The Division of the Sciences. 2. The New Organon; or Directions concerning the Interpretation of Nature. 3. The Phenomena of the Universe; or a Natural and Experimental History for the foundation of Philosophy. 4. The Ladder of the Intellect. 5. The Forerunners; or Anticipations of the New Philosophy. 6. The New Philosophy; or Active Science' (*Works*, iv. 22–33). *draw . . . sense.* Make easier to grasp (by citing concrete instances from the natural world). *Natural Story.* For a brief outline of what Bacon meant by a Natural History see e.g. *Works*, iv. 27–31. His *Historia Naturalis et Experimentalis* was intended to be published in monthly instalments (ibid. xiv. 360–1, 395, 398–9), but Bacon only ever issued the *Historia Ventorum* ('History of the Winds') in Nov. 1622, and the *Historia Vitae et Mortis* ('History of Life and Death') in Jan. 1623. For these and other remaining texts see *Works*, ii. 7–305 (Latin), iv. 125–400 (English).

Inquisition. Probably the fourth part of the *Instauratio*, in which he proposed to 'set forth examples of inquiry and invention according to my method, exhibited by anticipation in some particular subjects' (ibid. iv. 31).

330. *aspersion*. Sprinkling in of an ingredient; admixture. *second book... Partition of Sciences*. For the *De Augmentis Scientiarum Libri IX* (1623) Bacon reused bk. I of the *Advancement of Learning* virtually intact, but divided bk. II, much expanded, into 8 books. He finally conceived it as comprising the first part of the *Instauratio*. *hold*. Believe, intend. *civil person*. Role within civil law (barrister, judge, Lord Chancellor). *work touching Laws*. Probably the 'Treatise on Universal Justice or the Fountains of Equity, by Aphorisms; one Title of it', included as a specimen in bk. VIII of the *De Augmentis* (*Works*, v. 88–109). *middle term*. Mean. *reverend*. Deserving respect. *obnoxious*. Subject, submissive. *digest... nation*. Bacon's project to resystematize English law dates back to the 1590s, in his interventions in the Parliaments of 1593 and 1601 (see M. A. Phillips, 'Francis Bacon', in P. W. Hasler (ed.), *The House of Commons 1558–1603* (London, 1981), i. 373, 378–9), in his 1597 treatise on the *Maxims of the Law* (*Works*, vii. 313–87), and was echoed many times subsequently (e.g. vii. 302–3; xiv. 353–4, 357–64). *of assistance*. Needing assistants. *in contemplation*. In mind; under serious consideration. *dowries*. Gifts, benefits. *place*. Political office. *King Henry the Seventh*. This *History* had been published in Mar. 1622 (*Works*, xiv. 352). *embracement*. Undertaking, exertion. *revolving*. Considering. *City... Temple*. Secular and sacred realms. *oblation*. Gift offering.

331. *exoriare aliquis*. Virgil, *Aeneid*, iv. 625: 'Exoriare, aliquis nostris ex ossibus ultor', Dido's words before committing suicide: 'Arise from my ashes, unknown avenger!' *platform*. Architectural plan. *draw on*. Lead to, involve as a consequence. *work*. The dialogue, *Advertisement Touching a Holy War* (*Works*, vii. 9–36).

POEMS AND PSALMS

Bacon's compositions in verse have received little attention. Spedding printed in his notes (but not in the text) two poems ascribed to Bacon on external evidence. One of them ('The man of life upright, whose guiltless heart is free') is ascribed to Bacon in a manuscript collection of the British Library (Bibl. Regia, 17.B.L.), with no other authority beyond the copyist's attribution (*Works*, vii. 268–9). We now know that it first appeared in *A Book of Ayres*, 1601, and has been attributed to Thomas Campion: cf. Beal CmT 89–95. A much stronger case can be made for 'The world's a bubble', a translation of an epigram from the *Greek Anthology* (ix. 359) ascribed to Poseidippos (*c*.310–260 BC). This translation was stated to be Bacon's by Thomas Farnaby, a contemporary who had known him, in his collection of Greek epigrams, *Florilegium epigrammatum Graecorum, eorumque Latino versu a varijs redditorum* (London, 1629). It was also included as an appendix ('an Excellent Elegy by... the Prince of English Oratory') in Joshua Sylvester,

Parthea; or Divine Wishes and Meditations (London, 1630), and in an anthology by Henry Birkhead and Henry Stubbe, *Otium literatum, sive miscellanea quaedam premata* (Oxford, 1658). Farnaby printed Poseidippos' poem together with a Latin translation, and then added Bacon's English version with the note 'Huc elegantem V.C.H. et Domini Verulamii παρωδίαν adjicere adlubuit'. I give Spedding's blank verse translation:

> What life shall a man choose? In court and mart
> Are quarrels and hard dealing; care at home;
> Labours by land; terrors at sea; abroad,
> Either the fear of losing what thou hast,
> Or worse, nought left to lose; if wedded, much
> Discomfort; comfortless unwed: a life
> With children troubled, incomplete without:
> Youth foolish, age outworn. Of these two choose then;
> Or never to be born, or straight to die. (*Works*, vii. 271)

Poseidippos' poem, 'on the miseries of human life', is an exercise in paradox belonging to the negative pole (*vituperatio*) within the genre of epideictic rhetoric, and drawing on a number of stock ideas. As A. S. F. Gow and D. L. Page record, several adaptations or responses to it survive in the florilegia (*The Greek Anthology: Hellenistic Epigrams* (Cambridge, 1965), ii. 501–2). In this anthology, indeed, it is followed by a poem ascribed to Metrodorus (ix. 360) which is an optimistic refutation of it, a matching poem praising human life.

These paired poems were well known during the Renaissance, and often imitated. Erasmus translated both of them (*Adagia*, II. iii. 49), as did the Scottish humanist George Buchanan in his first book of *Epigrams* (i. 37 and 38, in *Poemata* (Leiden, 1621), 129–30: see the classic studies by James Hutton, *The Greek Anthology in Italy* (Ithaca, NY, 1935), 530; *The Greek Anthology in France* (Ithaca, NY, 1946), 701 ff.). As H. J. C. Grierson showed ('Bacon's Poem "The World": Its Date and Relation to certain other Poems', *Modern Language Review*, 6 (1911), 145–56; repr. in his *Essays and Addresses* (London, 1940), 221–37), both epigrams were also translated by Nicolas Grimald, and appeared as poems 151 and 152 in *Tottel's Miscellany* of 1557 (see *Tottel's Miscellany (1557–1587)* ed. H. E. Rollins (Cambridge, Mass., 1928, 1929; rev. edn. 1956), i. 105, ii. 236–8). In *The Arte of English Poesie* (1589; but written 10–15 years earlier) George Puttenham, evidently familiar with Grimald's, presented his own version of the paired poems to illustrate the figure '*Antipophora* or Figure of responce' (*Arte*, ed. G. D. Willocks and A. Walker (Cambridge, 1936, 1970), 205–6). Puttenham's versions are referred to by other Elizabethan writers, and Sir John Beaumont, in turn, made a verse-translation of both epigrams (in his *Bosworth-field* (1629) 35–6), confirming the vogue that they enjoyed in the 1590s. (Beaumont's version was also attributed to William Drummond of Hawthornden, which shows how easily mistakes of attribution could occur in an age where poetry circulated freely in manuscript copies.)

Some contextual details link Bacon's poem with John Donne and Henry Wotton, who were all connected with Essex in the 1590s. Bacon's poem was printed, evidently from a copy in Wotton's possession, in the *Reliquiae Wottonianae* (1651) and attributed to 'Ignoto' (subsequent editions give it to Bacon). As Grierson showed, Donne echoed it in his verse-letter to Sir Henry Wotton, 'Sir, more than kisses, letters mingle soules' (see note below), which is usually dated 1597. This suggests that Bacon's translation was made around 1597–8.

In his version Bacon expands and varies the original text, in particular achieving some striking metrical effects with his combination of one long line (a pentameter), rhyming couplet-wise with one short (four-syllable) line. Paul Fussell, Jr., in *Poetic Meter and Poetic Form* (New York, 1965), 182–3, has given a sensitive analysis of the poem's structure, observing that the word 'span' in the second line refers to the distance between the tip of the thumb and the tip of the little finger, when the hand is fully extended, so that 'line 2 measures this distance in its size as well as in its statement.... The principle of formal parallelism beautifully operates in the short line here: once the shortness of the experience contained in the short lines has been emphasized by line 2, we associate the experiences implied by lines 4 and 6 likewise with brevity. All these short, light lines are preparing us for the contrast of the mighty weight of the concluding couplet, the heaviest part of the stanza structurally, and exploited by Bacon so that it becomes the weightiest part—that is, the conclusion, introduced by *then*—of the argument as well. We notice too that it is only the concluding couplet that is wholly independent and complete grammatically: it is the only predication in the poem which elides no verbs. The student of the semantics and logic of rhymes can profit from considering the kinds of literal and ironic meaning that are generated from such sound resemblances as *man–span*; *womb–tomb*; *years–fears*; and *trust–dust*. Bacon's use of the shape of his stanza has resulted in an impressive density, the sort of density that we can never really exhaust.'

My text is from Spedding, with some emendations from Grierson. 'The world's a bubble' was a popular poem in manuscript collections, for Beal lists over 50 copies (BcF 1–52), including one set to music by Cesare Morelli for the use of Samuel Pepys, and he has subsequently located another 2 (personal communication). It was sometimes given the titles 'Upon the miserie of Man', or 'Ode against man's life' (Gibson 248*).

No problems of attribution apply to Bacon's version of the Psalms, for he himself published *The Translation of Certaine Psalmes into English verse*, dated 1625 (Gibson 163), but already circulating in December 1624 (*Works*, vii. 265). Bacon had made these versions during an illness earlier in 1624, and dedicated them 'To his very good friend Mr. George Herbert', with the following epistle: 'The pains that it pleased you to take about some of my writings I cannot forget; which did put me in mind to dedicate to you this poor exercise of my sickness. Besides, it being my manner for dedications, to choose those that I hold most fit for the argument, I thought that in respect of divinity and poesy met, (whereof the one is the matter, the other the stile of this little writing), I could not make better choice. So, with signification of my love and acknowledgment, I ever rest Your affectionate Friend.' (Herbert wrote three Latin poems to Bacon in the period 1620–1, as Public Orator at Cambridge (an office he held from 1620 to 1628), and contributed another to the memorial anthology published in 1626 by William Rawley: see *Manes Verulamiani*, ed. W. G. C. Gundry (London, 1950) for a facsimile edition and translation; and *The Latin Poetry of George Herbert: A Bilingual Edition*, tr. M. McCloskey and P. R. Murphy (Athens, Oh., 1965), 166–73.) This translation of the Psalms evidently had some special value to Bacon, for it figures in his last will and testament as the first of his 'Legacies to my friends: I give unto the right honourable my worthy friend the marquis Fiath, late lord ambassador of France, my books of orisons [prayers] or psalms curiously [carefully] rhymed' (*Works*, xiv. 541–2).

The standard metrical version of the Psalms during Bacon's lifetime was that by Thomas Sternhold and John Hopkins, first issued (incomplete) in 1549, expanded until complete, and then published in innumerable versions up to 1640 and beyond. Their versions, issued as the Anglo-Genevan Psalter, printed both separately and in complete editions of the Old Testament, were included in both the English Bible and the Geneva Bible throughout this period, with differing prose glosses and commentary. Other writers issued their own versions, partial or complete, including Sir Thomas Wyatt, Archbishop Matthew Parker, William Alexander, George Wither. The most distinguished translation, by Sir Philip Sidney and his sister Mary, Countess of Pembroke, remained in manuscript at Penshurst Place until first published in 1823: see *The Psalms of Sir Philip Sidney and the Countess of Pembroke*, ed. J. C. A. Rathnell (Garden City, NY, 1963). Theirs are perhaps the only English versions to have a claim to be poetry in their own right.

Bacon translated seven psalms altogether: nos. 1, 12, 90, 104, 126, 137, and 149, of which I have selected five. Bacon's treatment is not a word-for-word translation but a form of expanded paraphrase, 'in which', as Spedding put it, 'the inevitable loss of lyric fire and force is in some degree compensated by the development of meanings which are implied or suggested by the original, but not so as to strike the imagination of the modern reader; so that the translation serves for a kind of poetical commentary; and, though far from representing the effect of the original in itself, holds up a light to read it by' (*Works*, vii. 266). Bacon's expansive treatment indeed brings out implications that are not obvious to the casual reader, recreating circumstance and effect. But Bacon does more than add glosses: comparison with the original will show that he has practised a technique known in the Renaissance as 'varying', rephrasing a text or argument in a different form, setting and preserving his own conventions of tone and style, and adding variety to the sometimes repetitive originals. I have quoted the Geneva Bible's version of Psalm 90 in the notes, to provide a specimen reference-point for Bacon's elaboration of the given text.

As concerns versification, like other Renaissance translators, Bacon employed different verse-forms according to the structure and mood of the psalm. He uses iambics throughout, octosyllabics for Psalm 1, alternating 8 and 6 syllables for no. 126, and pentameter for the rest. For nos. 1 and 12 he uses an 8-line stanza, rhyming abab, cdcd; for nos. 90 and 137 he uses a 6-line stanza, rhyming ababcc; for no. 126 a 4- line stanza, rhyming abab; and for no. 104, the longest, he uses couplets, an appropriately relaxed form for this long narrative account of the Creation. While no great claims can be made for these translations, Bacon used the freedom given by his technique of expansion and variation to add his own emotional response to that of the psalmist, so they can be judged a not altogether negligible instance of 'divinity and poesy met'.

The world's a bubble

332. *limns*. Paints. *on*. Grierson's reading: Spedding has 'the'. *a*. Grierson; Spedding reads 'the'. *These*. Grierson; Spedding reads 'Some'.

Psalm 90

333. The Geneva Bible describes this as 'A Prayer of Moyses, the man of God'. The first two verses (corresponding to Bacon's st. 1) are: 'Lord, thou hast bin our habitation from generation to generation. | Before the mountains were made, and before thou hadst fourmed the earth, and the world, even from everlasting to everlasting thou art our God.'

intercept. 'To stop the natural course of light; to cut off light from anything' (*SOED*). *frame . . . stage*. A variation on the 'theatrum mundi' metaphor. The Geneva Bible adds the marginal gloss: 'before the foundations of the world were laid'. *Both . . . sleep*. Verses 3 and 4 correspond to st. 2: 'Thou turnest men to destruction: again thou sayest, Returne, yee sonnes of Adam. | For a thousand yeres in thy sight are as yesterday when it is past, and as a watch in the night'. *turns*. Cf. 'turnest' in the biblical text, and 'Returne, yee sonnes of Adam' (that is, 'to the dust from which you are made'). *Thou . . . again*. St. 3 corresponds to verse 5: 'Thou hast overflowed them: they are as a sleepe: in the morning hee groweth like the grasse.' A gloss adds: 'Thou takest them away suddenly as with a flood.' *term*. A portion of time. *At . . . sins*. St. 4 corresponds to verses 6 and 7: 'In the mourning it flourisheth & groweth, but in the evening it is cut down and withereth. For we are consumed by thine anger, and by thy wrath are we troubled.' *pins*. 'Small pieces of wood or metal . . . used for fastening or holding together parts of a structure' (*SOED*). *attend*. Pay attention *Thou . . . end*. St. 5 corresponds to verses 8 and 9: 'Thou hast set our iniquities before thee, and our secret sinnes in the light of thy countenance. For all our dayes are past in thine anger: we have spent our yeares as a thought.'

334. *The life . . . life?* St. 6 corresponds to verse 10: 'The time of our life is threescore yeeres and tenne, and if they be of strength, fourescore yeeres: yet their strength is but labour and sorrow: for it is cut off quickly, and wee flee away.' *But . . . again*. St. 7 corresponds to verse 11: 'Who knoweth the power of thy wrath? for according to thy feare is thine anger.' A gloss adds: 'If mans life for the brevitie be miserable, much more if thy wrath lie upon it, as they which feare thee, onely know.' *Teach . . . death*. St. 8 corresponds to verse 12: 'Teach us to number our dayes, that wee may apply our hearts

unto wisdome.' A gloss on 'wisdome' adds: 'Which is by considering the shortnesse of our life, & by meditating the heavenly joyes'. *Return . . . rejoice.* St. 9 corresponds to verses 13 and 14: 'Returne, O Lord, how long? and be pacified toward thy servants. Fill us with thy mercy in the morning; so shall we rejoyce and be glad all our dayes.' *Begin . . . be.* St. 10 corresponds to verses 15–17: 'Comfort us according to the dayes that thou hast afflicted us, and according to the yeeres that we have seene evill. Let thy worke bee sene toward thy servants, and thy glory upon their children. And let the beauty of the Lord our God be upon us, and direct thou the worke of our hands upon us, even direct the workes of our hands.'

Psalm 104

334. The Geneva Bible describes this as 'An excellent Psalme to praise God for the creation of the world, and the governance of the same by his marveilous providence'. It has obvious echoes of the account of creation in Genesis. Bacon's 120 lines correspond to 35 biblical verses. The opening three verses (which Bacon expands to 18 lines) read: 'My soule, praise thou the Lord: O Lord my God, thou art exceeding great: thou art clothed with glory and honour. Which covereth himselfe with light as with a garment, and spreadeth the heavens like a curtaine. Which layeth the beames of his chambers in the waters, and maketh the cloudes his charet, and walketh upon the wings of the winde.'

sounding. Resounding. *thy works and wondrous ways.* The Geneva marginal gloss on verse 1 reads: 'The Prophet sheweth that we need not enter into the heavens to seek God, forasmuch as all the order of nature, with the proprietie and placing of the elements, are most lively mirrors to see his majestie in.' *blaze.* Emblazon, adorn. *turn . . . sight.* The rays emitted from human eyes are deflected or defeated.

335. *compass.* Circumscribed; curved. *unpent.* Unrestrained. *removing tent.* One that can be removed (for the biblical text: 'and spreadeth the heavens like a curtain'). *counterpoise.* 'A weight which balances another weight or establishes equilibrium against a force' (*SOED*). The biblical text reads: 'He set the earth upon her foundations, so that it shall never move.'

336. *Plaining.* Complaining. *exhilarate.* Make cheerful. *travellers.* 'Animals that can go quickly or for long distances' (*SOED*). *conies.* Rabbits. Cf. Prov. 30: 26. *ask . . . God.* The biblical text is: 'The lyons roare after their prey, and seeke their meat at God', the gloss adding: 'That is, they onely find meate according to Gods providence who careth even for the brute beastes.' *Then man goes forth.* The gloss adds: 'To wit, when the day springeth; for the light is as it were a shield to defend man against the tyrannie and fierceness of beasts.' *unto . . . fall.* Become the responsibility (domain) of.

337. There . . . . pan. These lines correspond to verse 26: 'There goe the ships, yea that Leviathan, whom thou hast made to play therein.' For Leviathan, a mythical sea-monster, a cross between the crocodile and the whale, see Job 40: 25–41. *All life . . . dust and vanity.* These lines correspond

to the second part of verse 29: 'if thou take away their breath, they die, and returne to their dust', on which the gloss is: 'As by thy presence all things have life: so, if thou withdraw thy blessings, they all perish.' *breath.* The Geneva Bible reads 'spirit'. *amain.* Exceedingly, with full force. *The earth . . . smoke.* These lines correspond closely to verse 32: 'Hee looketh on the earth and it trembleth: hee toucheth the mountaines, and they smoke.'

Psalm 126

337. *We seem'd . . . stand.* In the Geneva Bible verse 1 reads: 'When the Lorde brought againe the captivitie of Zion, we were like them that dreame.'

338. *The heathen . . . day.* Verse 2, 2nd part, reads: 'then said they among the heathen, The Lord hath done great things for them.' *O Lord . . . mouth.* Verse 4 reads: 'O Lord, bring againe our captivitie, as the rivers in the South', the gloss adding: 'It is no more impossible to God to deliver his people, than to cause the rivers to runne in the wildernesse and barren places.' *So.* Provided.

Psalm 137

337. *When as . . . stream of tears.* Bacon's first two stanzas are an expansion of verse 1: 'By the rivers of Babel we sate, and there we wept, when we remembred Zion.' The gloss adds: 'That is, we abode a long time and albeit the country was pleasant, yet could it not stay our teares. . . .' *cruel masters . . . our misery.* The marginal gloss to verse 3 is: 'The Babilonians spake thus in mocking us, as though by our silence we should signifie that we hoped no more in God.'

339. *'Down . . . ground'.* The biblical text, verse 7, is: 'Rase it, rase it to the foundation thereof.' Cf. Ezek. 25: 12 ff. *happy he . . . stones.* The biblical text, verse 9, is: 'Blessed shall hee be that taketh and dasheth thy children against the stones.'

Psalm 149

339. *on your secret beds your spirits raise.* Verse 5 reads: 'Let the Saints bee joyfull with glory: let them sing loud upon their beds', the gloss adding: 'He alludeth to that continuall rest and quietnesse, which they should have, if they would suffer God to rule them.'

340. *revenge.* The gloss to verse 7 is: 'This is chiefly accomplished in the kingdom of Christ, when Gods people for just causes execute Gods judgements against his enemies: and it giveth no libertie to any to revenge their private injuries.' *expect.* Await.

ESSAYS OR COUNSELS, CIVIL AND MORAL (1625)

In April 1625 Bacon published the third and final version of the work by which he had achieved most fame in his own day, and indeed since. In dedicating it to the Duke of Buckingham, chief favourite and confidant to King James, Bacon wrote: 'I do now publish my Essays; which, of all my other works, have been most current; for that, as it seems, they come home to men's business and bosoms. I have enlarged them both in number and weight; so that they are indeed a new work' (*Works*, vi. 373). By 'most current' he meant 'most in circulation', for to the earlier editions of 1597, reprinted four times, and 1612, reprinted seven times, he could list the Italian translation, *Saggi Morali* (including also the *De Sapientia Veterum*), which was published in London in 1617 and reprinted seven times in both England and Italy during his lifetime (Gibson 33–43: for which version Bacon provided the essay 'Of Seditions and Troubles', not published in English until 1625); and two French translations published in 1619, one by Arthur Gorges, printed in London, the other by his faithful French translator, I. Baudoin, issued in Paris (Gibson 44–7). As for the English version of the *Essays*, Gibson records some 30 editions by 1706, and there have been several hundred since then.

He had obviously enlarged them 'in number', for while the 1597 volume had contained 10 essays, that of 1612 contained 38, of which 29 were new, and the rest corrected or enlarged in varying degrees. The 1625 volume, finally, contains 58 essays, of which 20 are new, the remainder being further enlarged and corrected. (The best way to follow this double enlargement is provided by Edward Arber, *A Harmony of Bacon's Essays*, 1871, which prints in three parallel columns the three published texts, together with the manuscript version preceding the 1612 volume. See also Brian Vickers, *Francis Bacon and Renaissance Prose* (Cambridge, 1968), 217–31, 298–9.) The *Essays* of 1597, being essentially aphoristic, rather resisted expansion: their average length in 1597 is 325 words; as revised in 1612, 400 words; as further revised in 1625, 550 words. Those of 1612, by contrast, being already more discursive, and making more use of the opening division of topics (*partitio*), averaged 490 words originally, rising to 980 in 1625. The new essays in 1625, by contrast, average 950 words: like other writers, Bacon found the alteration of an already extant text a greater invitation to expansiveness.

Rewriting a book, although unusual today, was quite common in the Renaissance. Sir Philip Sidney rewrote his *Arcadia* on a wholly new plan, unfortunately left unfinished at his death. Montaigne's *Essais* appeared in 1580 (divided into two books), and were enlarged in 1588 (now in three books, with over 600 additions to Books I and II). At his death Montaigne left behind a copy of the 1588 edition which included profuse marginal additions, amounting to about a quarter of the work's total length. Erasmus' *Praise of Folly* (*Moriae Encomium*) went through no less than seven revisions of one kind or another, not all of them beneficial to its overall coherence (or so it seems to me). But Bacon's revisions of his book were unusual in that he claimed to have also enlarged them 'in weight', that is, seriousness. Writing to the Italian scholar Fulgentio in the autumn of 1625, and describing his current publishing projects, Bacon announced that he was having not only his scientific but also his 'moral and political writings' translated into Latin, including 'the little book which in your language you have called *Saggi Morali*. But I give it a weightier name, entitling it "Faithful Discourses—or the Inwards of things"' (*Sermones Fideles, sive Interiora rerum*) (*Works*, xiv.

531–3). Archbishop Tenison, in the collection of minor works that he edited in 1679, says that Bacon gave to the Latin edition 'the Title of *Sermones Fideles* after the manner of the Jews, who call'd the words Adagies, or Observations of the Wise, Faithfull Sayings; that is, credible Propositions worthy of Firm Assent and ready Acceptance. And (as I think) he alluded more particularly, in this title, to a passage in Ecclesiastes [12: 10, 11], where the Preacher saith that he sought to find out *Verba Delectabilia* (as Tremellius rendreth the Hebrew), 'pleasant Words' (that is, perhaps, his Book of Canticles); and *Verba Fidelia* (as the same Tremellius), 'Faithful sayings'; meaning it may be, his Collection of Proverbs. In the next Verse, he calls them 'Words of the Wise', and so many Goads and Nails given *ab eodem Pastore*, from the same Shepherd [of the flock of Israel]' (*Baconiana*, 60–1).

This concern to address serious issues is announced in the new title, *Essayes or Counsels, Civill and Morall*, and is explicitly commented on in the dismissive opening of Essay 37, 'Of Masques and Triumphs': 'These things are but toys, to come amongst such serious observations.' By describing them as 'Counsels', writings giving advice, Bacon aligned himself with an important Renaissance genre, the 'Advice to Princes', what we would today call political science. The projects described in 'Of Building' and 'Of Gardens' are certainly conceived on a grand scale, some being (as Bacon says of Essay 29, 'Of the True Greatness of Kingdoms and Estates') truly 'an argument fit for great and mighty princes to have in their hand' (above, p. 397). The dedication to Buckingham was appropriate in that, as the courtier closest to the monarch, he was in the best position to influence James towards responsible government, and it is no accident that several precepts from the *Essays* are echoed in the 'Letter of Advice' that Bacon had addressed to Villiers in 1616, when he had just been made Duke of Buckingham and publicly declared the King's favourite (*Works*, xiii. 27–56). Other essays, such as 'Of Great Place' and 'Of Nobility', similarly address issues that largely concern those living at the upper levels of society. Yet at the same time, the reason Bacon gave for the *Essays* being his most popular work, that 'they come home to men's business and bosoms', describes their dual appeal, explicitly recognized in the new subtitle, 'Civil and Moral', that is involving both the *civis* or public life and the *mores* or behaviour of private individuals.

These were not empty claims; indeed the *Essays* could hardly have achieved the eminence they have enjoyed for several centuries had they been merely entertaining chatter. The true significance of the *Essays*, as R. S. Crane was the first to show ('The Relation of Bacon's *Essays* to His Program for the Advancement of Learning', in *Essential Articles*, 272–92), was to provide treatments of the lacunae in knowledge identified in Book Two of *The Advancement of Learning*. In particular, these involved lacunae in 'moral' and 'civil' knowledge, that is 'The Regiment or Culture of the Mind', a discipline which should prescribe 'rules how to subdue, apply, and accommodate the will of man' to 'the nature of good'—uniting psychology and ethics. Bacon noted the lack of any systematic 'descriptions of the several characters and tempers of man's natures and dispositions', including 'those impressions of nature, which are imposed upon the mind by the sex, by the age, by the region, by health and sickness, by beauty and deformity, and the like, which are inherent and not extern; as sovereignty, nobility, obscure birth, riches, want, magistracy, privateness, prosperity, adversity, constant fortune, variable fortune, rising per saltum, per gradus and the like' (above, p. 258). As Crane argued, it was to fulfil these desiderata that (already in 1612) Bacon wrote the essays 'Of Youth and Age', 'Of Beauty', 'Of Deformity', 'Of Nobility', 'Of Great Place', 'Of Riches', and 'Of Fortune', expanding them all in 1625 and adding a new essay for this group, 'Of Adversity'. Bacon's diagnosis of the absence of 'active and ample observations' concerning 'the diseases and infirmities of the mind . . . the perturbations and distempers of the passions' (above, p. 259), gave rise to the essays 'Of Love', 'Of Envy', and 'Of Anger', as well as to many incidental comments in other essays on the effect that the passions have in human affairs.

Turning from the formative influences of nature and external fortune to 'those points which are within our own command, and have force and operation upon the mind to affect the will and to alter manners', the topics which Bacon listed as deficient included 'custom, exercise, habit, education, example, imitation, emulation, company, friends, praise, reproof, exhortation, fame, laws, books, studies'. Correspondingly, he provided treatments 'Of Custom and Education', 'Of Praise', 'Of Nature in Men', 'Of Friendship', and 'Of Fame' (unfinished). These links reveal that, at the time when he was writing the *Advancement of Learning* (*c.*1603–5), Bacon was already planning an expanded edition of the *Essays* to supply a contribution to cultivating those 'parts' of learning which 'lie fresh and waste, and not improved by the industry of man'. (It seems probable that the MS version of the 1612 set was begun in 1607.) Under the heading 'civil knowledge' Bacon included three topics: 'wisdom of behaviour', or conversation, in the wider sense of social intercourse; 'wisdom of business', or negotiation; and 'wisdom of state', or government. It is significant that five of the essays first published in 1597—'Of Discourse', 'Of Ceremonies and Respect', 'Of Negotiating', 'Of Followers and Friends', and 'Of Faction'—already dealt with topics that Bacon subsequently noted in the *Advancement* as deserving to be discussed, suggesting that his idea for a coherent and systematic treatment of this area of human affairs already existed in embryonic form. All these essays were enlarged in 1612, while the new ones written for that volume and for 1625 fulfil the desiderata Bacon had noted for treatises on the 'doctrine of advancement in life' ('Of Vain-Glory', 'Of Dispatch', 'Of Boldness', 'Of Delays', and 'Of Simulation and Dissimulation'), and for discussions of the art of government ('Of Seditions and Troubles', 'Of the True Greatness of Kingdoms and Estates', 'Of Empire', and 'Of Plantations'). The fact that, as Crane also showed (pp. 276–8), many of these essays absorb material first published either in the *Advancement of Learning* or as occasional papers in Bacon's political career, helps us to see the *Essays* as forming an integral part of Bacon's overall scheme for the reform and improvement of knowledge concerning individual and social life.

An interesting insight into the intellectual world of these essays can be gained from a recent computation of the most frequently used words (D. W.

Davies and E. Wrigley (eds.), *A Concordance to the Essays of Francis Bacon* (Detroit, 1973), 373–92), which shows some of his mental preoccupations. The two most frequently used words are 'man' (331 times) and 'men' (260). Bacon's sustained focus on human behaviour in its ethical, psychological, and political contexts is revealed by their dominance and by the high frequency of such related words as 'nature' (82), 'persons' (79), 'times' (79), 'business' (75), 'state' (75), 'time' (71), 'use' (68), 'counsel' (67), 'people' (59), 'envy' (57), 'place' (45), 'princes' (45), 'fortune' (36), 'honour' (36), 'estate' (32), 'greatness' (30), 'power' (27), 'custom' (26), and 'states' (23). The movement towards ethical discrimination, in particular the singling out of behaviour to be praised and imitated, can be seen in the dominance of those words giving a positive evaluation: 'great' (223), 'good' (170), 'well' (115), 'best' (71), 'better' (62), 'true' (54), 'God' (50), 'wise' (47), 'kind' (46), 'virtue' (44), 'religion' (31), 'fair' (29), 'fit' (29), 'greatest' (29), 'love' (29). Negative evaluation, by contrast, is less frequent: 'ill' (25), 'evil' (14), 'bad' (10), 'worse' (14), 'worst' (7), 'foul' (3), 'hate' (1), 'vice' (5), 'devil' (7), and 'atheism' (11). Clearly such rough-and-ready listing is only the first step towards a proper analysis, yet it does show the density of Bacon's concern with psychology and ethics.

However, our awareness of their functional role in the 'small Globe of the Intellectual World' which Bacon made, in order to promote 'amendment and proficience' (above, p. 299), does not mean that the *Essays* have a unitary form or style. They fascinate, indeed, by their very diversity. As Anne Righter put it, 'the 1625 edition ... is an accumulation of disparate pieces as difficult to generalize about, or to connect internally, as Donne's *Songs and Sonets*, and it is to be read in a not dissimilar fashion' ('Francis Bacon', in *Essential Articles*, 300–21, at 317–18). Some essays are organized systematically, with an opening *partitio* that sets out the heads of the argument, a technique that Bacon used more frequently in the 1625 volume. Such are 'Of Counsel', 'Of Great Place', 'Of Seditions and Troubles', and 'Of Judicature', among others. But announcing the topics does not tie Bacon down to a tidily sequential treatment of them, and even in these systematic essays the reader is in for several surprises. Other essays are constructed rather by the association or juxtaposition of ideas and metaphors, such as 'Of Delays', or demand to be read as a 'metaphysical' poem is read. 'Of Truth' begins with one of Bacon's most striking quotations. "What is Truth?", said jesting Pilate; and would not stay for an Answer'. As Anne Righter comments: 'The rifle-shot of this opening, the little imaginative explosion, is a familiar Bacon technique and frequently imitated. Less imitable, however, is the curious configuration of the space which separates this first sentence from the one which follows. "Certainly there be, that delight in Giddinesse; And count it a Bondage, to fix a Beleefe; Affecting Free-Will in Thinking, as well as in Acting". It is not merely that these two sentences are of a markedly different kind: the second simply does not move forward from the first in any fashion which we normally associate with the logic of prose. The movement performed is deliberately oblique in a way that forces the reader in part to create the link himself. A passive attitude here, or even a very rapid perusal of the page, is fatal to the essay' (p. 319). That observation well describes the unpredictable, thought-provoking movement of Bacon's prose.

The non-linear nature of the *Essays* will be evident on any careful reading. But fascinating evidence that this was a deliberate effect created by the author is provided by the *Antitheta rerum*, one of several aids in acquiring argumentative skills that Bacon had originally developed for his own use. These 'antitheses of things', adumbrated in the 1594–5 *Promus* (nos. 1234–47), are briefly mentioned and illustrated in the *Advancement of Learning* (above, pp. 240–1), and were finally published in the *De Augmentis* (vi. 3). Commenting on Cicero's recommendation that the orator should 'have commonplaces ready at hand, in which the question is argued and handled on either side', Bacon urges that all topics which will be frequently discussed should be 'studied and prepared beforehand; and not only so, but the case exaggerated both ways with the utmost force of the wit, and urged unfairly, as it were, and quite beyond the truth. And the best way of making such a collection, with a view to use as well as brevity, would be to contract those commonplaces into certain acute and concise sentences; to be as skeins or bottoms of thread which may be unwinded at large when they are wanted' (*Works*, iv. 472). Bacon modestly offers 'a few instances of the thing, having a great many by me'. In fact, his inclusion of some 47 topics amounts to quite a substantial treatise (iv. 473–92), and he concludes, rather self-ironically, that 'these Antitheses (which I have here set down) are perhaps of no great value; but as I had long ago prepared and collected them, I was loth to let the fruit of my youthful industry perish—the rather because (if they be carefully examined) they are *seeds* only, not *flowers*' (p. 492). Of the 47 titles, no less than 26 correspond to titles of the *Essays*, the first six being: 'Nobility; Beauty; Youth; Health; Riches; Honours'. Under each topic brief sentences are arranged for and against, as here:

Riches

For

They despise riches who despair of them.

It is envy of riches that has made virtue a goddess.

While philosophers are disputing whether virtue or pleasure be the proper aim of life, do you provide yourself with the instruments of both.

Virtue is turned by riches into a common good.

Other goods have but a provincial command; only riches have a general one.

Against

Of great riches you may have either the keeping, or the giving away, or the fame, but no use.

Do you not see what feigned prices are set upon little stones and such rarities, only that there may be some use of great riches?

Many men while they have thought to buy everything with their riches, have been first sold themselves.

I cannot call riches better than the baggage of virtue; for they are both necessary to virtue and cumbersome.

Riches are a good handmaid but the worst mistress.

That collection of 'acute and concise sentences' looks like the raw material for Essay 34, 'Of Riches', and comparison will show that Bacon indeed used some of them for that essay, but that he did not arrange them into a continuous flow of argument. The apparent continuity given by the essay as a whole is in fact deceptive, for in reading it the reader's mind needs to be able to shift ground, to see the topic from different sides in rapid succession. To think of these constituent arguments as being arranged into opposed poles is actually too simple, for the finished essays deal with considerations that are complementary rather than opposed. But seeing them set out in this antithetical form helps make us aware of the unpredictable, endlessly suggestive effect that Bacon's *Essays* have, existing as 'skeins or bottoms of thread' to be unwound by the reader's mind in a variety of directions and lengths, according to each person's age and experience of life. In this respect the *Essays* fulfil Bacon's own ideal for the communication of knowledge, which should avoid the faults committed in that usual 'contract of error between the deliverer and the receiver; for he who delivers knowledge desires to deliver it in such form as may be best believed, and not as may be most conveniently examined; and he who receives knowledge desires present satisfaction, without waiting for due inquiry' (*Works*, iv. 449). The proper method, by contrast, is one in which 'knowledge . . . is delivered to others as a thread to be spun on', stimulating the receiver to examination and enquiry, a dynamic process intended to provoke thought but not exhaust it. For that reason their endings are often abrupt: the last word is not, cannot be spoken.

In annotating the *Essays* I have been primarily concerned to clarify Bacon's meaning, especially in the cases of words still in use which had radically different connotations in the early seventeenth century. In this task I have been able to benefit from the rich accumulation of knowledge brought together in previous editions (some of which also go into more historical detail concerning Bacon's quotations and allusions), in particular those by W. Aldis Wright (London, 1862; 3rd rev. edn., 1865); Edwin Abbott (London, 1876); S. H. Reynolds (Oxford, 1890); and Michael Kiernan (Oxford, 1985). The most detailed edition of the *Essays* yet published, still hardly known in the English-speaking world (Kiernan does not refer to it), is by Mario Melchionda: *Gli 'Essayes' di Francis Bacon: Studio introduttivo, testo critico e commento* (Florence, 1979). Conceived on the most expansive lines (the introduction runs to 204 pages; the text occupies 196 pages; while the notes run to 255 pages, large-format), this work is a mine of information and deserves to be widely known as the most detailed edition yet produced of any work by Bacon.

1. *Of Truth*

341. '*What is Truth?*' John 18: 38, after Pilate has delivered up Christ for execution. 'Jesus answered . . . for this cause came I into the world, that I should bear witness unto the truth. . . . Pilate saith unto him, What is truth?'; *Promus*, no. 241. *jesting*. Scoffing. *there be that*. There are some who. *giddiness*. Constantly changing opinions. *count*. Consider. *fix a belief*. Hold fast to any belief. *affecting*. Liking. *that kind*. The Sceptics. *discoursing*

wits. Rambling minds. Reynolds (p. 9) suggests that Bacon may be referring to the sceptical treatise by Franciscus Sanchez, *Quod nihil scitur* (1576). *veins*. Inclination. *imposeth*. Lays restraints on. *One . . . Grecians*. Lucian of Samosata (*c*. AD 120–80), in his dialogue *Philopseudes* ('The lover of lies'), 1. *at a stand*. At a loss. *make for*. Conduce to. *day-light*. Window. *masks and mummeries*. Splendour, shows. *triumphs*. Triumphant processions. *daintily*. Elegantly. *sheweth*. Looks. *carbuncle*. Precious stone of a red, fiery colour, supposed to shine in the dark. *imaginations . . . would*. Wishful fantasies. *indisposition*. Languor, inability to do anything. *One . . . Fathers*. See above, p. 261 note. Bacon probably follows Henry Cornelius Agrippa, who in *De incertitudine et vanitate scientiarum et artium declamatio* (1530) quotes in successive sentences the two patristic authorities whom Bacon conflates. *vinum daemonum*. 'Wine of demons'. *filleth*. Takes hold of, overpowers. *howsoever*. Although. *depraved*. Corrupted. *judgments*. Perceptions.

342. *affections*. Passions. *truth*. Here in the sense of right reason. *belief*. Trust in. *creature*. Thing created. *days*. The six days of creation. *light of the sense*. Eyesight (Gen. 1: 2–5). *light of reason*. Gen. 1: 26–7. *sabbath*. Time of rest, cessation of labour. *illumination . . . Spirit*. A genitive: 'the divine spirit illuminates the works of creation'. *still*. Always, ever since. *The poet . . . sect*. Lucretius (*c*.94–55 BC) described the Epicurean philosophy favourably in *De rerum natura*, from which Bacon quotes the opening of bk. II. *beautified*. Adorned. *yet*. Despite belonging to that inferior sect. (The Epicureans, who believed that the world was created by chance, not by a divine act, and that the goal of philosophy was to avoid pain, were traditionally dismissed as atheistical hedonists. See *Adv.L.*, above, pp. 227, 247.) *adventures*. Alternating fortunes. *commanded*. Overgone. *serene*. Calm. *prospect*. Observing the spectacle. *pity*. Also meant 'piety'. *so*. Provided. *swelling*. Arrogance. *move*. Have as its motive force. *rest*. Enjoy peace. *turn . . . poles*. As a planet moves upon its axis, with constancy. *civil business*. Politics, public affairs. *clear and round*. Straightforward, honest. *allay*. Alloy (mixture of base metal in coinage, allowed by law). *work . . . embaseth*. The metal is easier to work with, but loses value. *winding*. Devious. *goings*. Movements. *Montaigne*. *Essais*, ii. 18, 'Du démentir' (Pléiade edn. 649–50), himself quoting Plutarch, 'Life' of Lysander, 4. *weighed*. Considered. *highly*. Solemnly. *peal*. Of the bell; here, the summons announcing the Day of Judgement. *foretold*. Luke 18: 8.

2. *Of Death*

343. *Death*. See M. Walters, 'The Literary Background of Francis Bacon's Essay "Of Death" ', *Modern Language Review*, 35 (1940), 1–7. *fear . . . dark*. Cf. Seneca, *Epist*. lxxxii. 15. *go*. Walk. *contemplation*. Consideration. *wages of sin*. Rom. 6: 23. *tribute*. Something owing. For the topos of life as being a loan only, see e.g. Seneca, *Nat. Quaest*. vi. 32.6, and *Consolatio ad Polybium*, x. 2 ff.; Lucretius, *De rerum natura*, iii. 971, etc. *weak*. Inconsistent (since it is natural). *vanity*. Futility. *books of mortification*. Catholic treatises on mortifying the flesh in this world to avoid punishment in the next, which often included discussions of the 'Four Last Things' (death, judgement,

heaven, and hell). *should think*. Consider. *corrupted*. Infected with decay. *dissolved*. Disintegrating. *parts*. Organs. *quickest of sense*. Most sensitive. *him*. Seneca. *natural*. Guided only by nature, not by the grace of Christian revelation. '*Pompa . . . ipsa*'. 'It is the trappings of death that scare us more than death itself': freely quoted from Seneca, *Epist*. xxiv. 14. The idea is developed at length by Montaigne, *Essais* i. 20, 'Que philosopher, c'est apprendre à mourir' (Pléiade edn. 79–95). *blacks*. Mourning clothes. *obsequies*. Funeral rites. *shew*. Make it seem. *mates*. Overpowers. *win . . . him*. Defeat death. *Honour aspireth to it*. The 1612 edn. and the Latin translation of 1638 add the phrase 'delivery from *Ignominy* chuseth it' (Kiernan, p. 181). But perhaps Bacon omitted it here since the next phrase duplicates it (Abbott, ed. cit. ii. 114). *pre-occupateth*. Anticipates death (by committing suicide): cf. Seneca, *Epist*. xxiv. 23 and lxx. 5–8. *Otho*. Marcus Salvius Otho, emperor AD 69: Tacitus, *Histories*, ii. 49. *tenderest*. Weakest, most delicate. *niceness*. Fastidiousness. '*Cogita . . . potest*'. 'Reflect how long you have been doing the same thing. The desire to die may be felt, not only by the sensible man or the brave or unhappy man, but even by the man who is merely surfeited' (Seneca, *Epist*. lxxvii. 6, adapted). *would*. Wishes to. *upon*. From. *to do*. At doing. *good spirits*. Men of courage. *Augustus Caesar*. Emperor 31 BC–AD 14. '*Livia . . . vale*'. 'Farewell, Livia, and forget not our married life' (Suetonius, 'Life' of Augustus, 99). *Tiberius*. Emperor AD 14–37. '*Jam . . . deserebant*'. 'Tiberius was now losing his vigour and vitality, but not his dissimulation' (Tacitus, *Annals*, vi. 50). *Vespasian*. Emperor AD 69–79. *stool*. Toilet. '*Ut . . . fio*'. 'As I suppose, I am on the point of becoming a god' (Suetonius, 'Life' of Vespasian, 23). *Galba*. Servius Sulpicius Galba, emperor AD 68–9. *sentence*. Sententia, pithy saying. '*Feri . . . Romani*'. 'Strike, if it be for the benefit of the Roman people' (Plutarch, 'Life' of Galba, 20). *Septimius Severus*. Emperor AD 193–211. *despatch*. Speed in settling business. '*Adeste . . . agendum*'. 'Be ready, in case anything remains for me to do' (Dio Cassius, lxxvi. 17). *bestowed . . . cost*. Made too much fuss about. This is the burden of Montaigne's complaint about Seneca in *Essais* iii. 12, 'De la phisionomie' (Pléiade edn. 1027–9).

344. *he*. Juvenal, *Satires*, x. 358 (inaccurately quoted). '*qui . . . naturae*'. 'Who considers the end of life among the gifts of nature.' *earnest pursuit*. Pursuing some great cause. *in hot blood*. In the heat of combat. *somewhat*. Something. *dolours*. Pains. *canticle*. Song. '*Nunc dimittis*'. 'Lord, now lettest thou [thy servant] depart [in peace]' (Luke 2: 29). *obtained*. Attained. *hath*. Is marked by. '*Extinctus . . . idem*'. 'He, too, [disliked in life] will win affection when his light is quenched' (Horace, *Epist*. II. i. 14).

3. *Of Unity in Religion*

344. In this essay Bacon is especially concerned with unity within the Church of England, a topic on which he had written two earlier treatises, the *Advertisement touching the Controversies*, *c*.1589–91 (above, pp. 1–19 ff.) and *Certain Considerations touching the better Pacification and Edification of the Church of England*, 1604 (*Works*, viii. 74–95; x. 103–27). In the 1612 edn. this essay is simply called 'Of Religion'; in the Latin translation, 'De Unitate

Ecclesiae'. In revising the 1612 version for 1625, Bacon (unusually) deleted several passages: see pp. 301–3 above and Kiernan, pp. 11–16. *band.* Bond. *contained.* Held together. *constant.* Coherent, rigorous. *belief.* System of beliefs. *doctors.* Teachers. '*jealous God*'. Exod. 20: 5, 34: 14. *mixture.* Dilution. *Bounds.* Limits. *Means.* Ways of attaining. *all in all.* The highest good. *without.* Outside. *schisms.* Divisions within the church over religious doctrine. *scandals.* Discredits to religion caused by its adherents. *manners.* Behaviours. *solution of continuity.* 'The separation from each other of normally continuous parts of the body by external or internal causes' (*SOED*). *corrupt humour.* Diseased component (the four constituent humours, whose balance was thought essential to bodily health, were: red bile, yellow bile, choler, melancholy). *pass.* Point, stage. '*Ecce in deserto ... in penetralibus*'. Christ's warning to his disciples against the 'false Christs and false prophets' who will declare a second coming: 'Wherefore if they shall say unto you, "Behold, he is in the desert", go not forth; "behold, he is in the secret chambers", believe it not': Matt. 24: 26 (Vulgate). *conventicles.* Religious meetings (corresponding to the *desertum*). *voice.* Saying. *Doctor.* Teacher. *Doctor ... Gentiles.* St Paul. *propriety.* Special nature.

345. *If ... mad.* 1 Cor. 14: 23. *with ... tongues.* In several languages. *avert.* Alienate. '*to sit ... scorners*'. Ps. 1: 1. *light.* Superficial, trivial. *vouched.* Appealed to as an authority. *master of scoffing.* François Rabelais (*c.*1490–1553), in the library catalogue of the Abbey St Victor: *Pantagruel*, ch. 7. *sets down.* Lists. *morris-dance.* A grotesque dance performed on feast-days (derived from the Moriscos or Moors). *them.* The heretics. *posture.* Attitude. *cringe.* Servile bow, distortion (Lat. tr. 'gestus deformitatem'). *by themselves.* Of their own. *move.* Arouse. *worldlings.* Non-believers. *politiques.* Corrupt politicians, schemers. *contemn.* Despise. *establisheth.* Fortifies. *distilleth.* Yields. *importeth exceedingly.* Is extremely important. *zelants.* Zealots. '*Is it ... me*'. 2 Kgs. 9: 18–19. *Peace ... party.* Such zealots are not seeking peace but more partisan followers. *contrariwise.* On the contrary. *Laodiceans.* Inhabitants of Laodicea, reproved in Rev. 3: 16 for being 'lukewarm' in their religious faith. *accommodate.* Compromise, settle differences. *points.* Disputed issues. *taking part.* Siding with. *witty.* Ingenious. *reconcilements.* Reconciliations. *arbitrement.* Arbitration. *league of Christians.* Covenant, alliance. *penned.* Set out. *cross.* contradictory. '*He ... against us*'. Matt. 12: 30 (which actually reads 'with me ... against me'). '*He ... with us*'. Mark 9: 40. *discerned.* Discriminated. *merely.* Wholly, altogether. *order.* Discipline. *good intention.* Goodwill. *done already.* Performed. *less partially.* In a partisan manner. *model.* Limited outline, scheme. *one of the fathers.* St Augustine. '*Christ's ... seam*'. John 19: 23–4. '*the church's ... colours*'. Ps. 45: 14, a saying frequently applied to the mystic marriage between Christ and the Church. *vesture.* Garments. '*In ... sit*'. 'Let there be variety in the garment, but not division': *Enarratio in Psal. xliv*, xxiv. But, as Reynolds (pp. 26–7) notes, neither in the passage quoted, nor in those of other patristic authors, 'are there any excuses made for differences of opinion on the most minute points of doctrine or church

polity'. Wright (p. 293) gives St Bernard as the source, and notes eight other occasions when Bacon uses this quotation. *two*. Two different things.

346. *subtilty*. Excessive refinement in argument. *ingenious*. Clever, insubstantial. *differ*. Disagree. *one thing*. The same thing. *not*. This term is not a true negation here. *frail men*. Humankind being weak. *intend*. Mean. *accepteth*. Approves, receives with favour. '*Devita ... scientiae*'. 'Avoid profane novelties of terms, and oppositions of science falsely so called': 1 Tim. 6: 20 (Vulgate). See *Adv.L.*, above, p. 140. *are not*. Do not exist. *terms*. Definitions. (The subject or predicate in a logical proposition.) *as*. That. *grounded*. Based. *implicit*. Entangled. *for all colours ... dark*. Cf. the expression 'in the dark all cats are grey', and Ruth Tarselius, '"All colours will agree in the dark": A Note on a feature in the style of Francis Bacon', in *Essential Articles*, 293–9. *pieced up*. Put together. *Nebuchadnezzar's image*. Dan. 2: 31–43. *cleave*. Adhere, cling to. *incorporate*. Unite into one body. *muniting*. Fortifying. *dissolve*. Annul—used of laws (Lat. *dissolvere*). *deface*. Destroy. *two swords*. Luke 22: 38. *third sword ... Mahomet's*. Of a holy war, within Christendom. *sanguinary*. Bloody. *practice*. Scheming, conspiracy. *authorise*. Legitimize. *ordinance*. Command. *dash ... second*. Oppose the first commandment, enjoining duty to God, against the second, enjoining duty to men. (See Exod. 32: 19 for Moses shattering the Decalogue's stone tablets.) *beheld*. Considered, contemplated. *Tantum ... malorum*. 'To such ill actions religion could persuade a man': Lucretius' comment on the myth telling of Agamemnon's sacrificing Iphigenia (*De rerum natura*, i. 101–2). *massacre in France*. The Catholics' massacre of *c*.13,000 Protestants on St Bartholomew's Eve and after, Aug. 1572. *powder treason of England*. On 5 Nov. 1605 (the Catholic conspiracy of Guy Fawkes).

347. *Epicure and atheist*. Epicurean and unbeliever (according to Christian attitudes). Lucretius followed Epicurus, who denied both that the gods concerned themselves on man's behalf, and that the human soul was immortal. *Anabaptists*. A sect of radical Protestants who refused to recognize civil authority, would not allow their children to be baptised, asserted the equality of all men under divine illumination, and used violence to obtain its ends (unsuccessfully). *furies*. 'Avenging or tormenting evil spirits' (*SOED*). '*I will ... Highest*'. Isa. 14: 12–14. These words are in fact spoken by the evil king of Babylon, but early patristic writers argued that the devil was the speaker (Reynolds, pp. 31–2). *personate*. Impersonate, act the part of. *bring him in*. Bring on to the stage. *dove*. Matt. 3: 16. The form in which the Holy Spirit appeared at the birth of Christ: an emblem of peace and divine protection. *bark*. Ship (as in the traditional image of the 'bark of Peter'). *Assassins*. Muslim fanatics, sent out (during the Crusades) to murder Christian leaders. *doctrine*. Teaching. *learnings*. Sciences. *as*. As though. *Mercury rod*. The staff with which Mercury guides souls to the underworld, an emblem of protection and authority. Cf. Virgil, *Aeneid*, iv. 242–4 and Homer, *Odyssey*, xxiv. 1–5. *facts*. From the Latin *facta*, deeds. *the same*. Those actions and opinions producing disaster to society. *counsels*. Considerations, advice. *would*. Should, ought. *Ira ... Dei*. 'The wrath of man

worketh not the righteousness of God': Jas. 1: 20. *a wise father*. Not traced. *ingenuously*. Candidly. *persuaded*. Urged, recommended. *interested*. Personally involved, for their own advantage.

4. *Of Revenge*

347. *Revenge*. For the destructive effects of duelling in Jacobean England see Bacon's *Charge touching duels*, above, pp. 304 ff. Despite Bacon's condemnation of the practice, 'the essay's tone' is one of 'detached exploration' (Kiernan, p. 186). *wild*. Uncultivated (as against civilized laws). *but*. Merely. *office*. Use, function. *part*. Role. '*It . . . offence*'. Prov. 19: 11. *purchase*. Obtain. *merely*. Altogether. *thorn*. Prickly bush.

348. *wrongs*. Such as lies and insults. *take heed*. Beware, lest that. *before hand*. In the advantage. *two for one*. Two punishments against one: that inflicted by his enemy, and that which the law inflicts as a punishment for his illegal act of revenge. *Some . . . cometh*. Cf. Aristotle, *Rhetoric*, ii. 3, 1380b20 ff., also Guicciardini, *Ricordi*, no. 202: 'Those who take their revenge in such a way that the victim does not realize whence it comes, can be said to do so merely to satisfy hatred and spite. It is more generous to do so openly so that everyone knows whence it comes, and it may then be interpreted as being done less from hatred and desire for revenge than for honour, that is, so as to be known as a man who will not suffer offence' (Guicciardini, p. 51). *the party*. The adversary. *arrow . . . dark*. Ps. 90: 5–6, 'the arrow that flies by night'. *Cosmus*. Cosimo I de' Medici, Duke of Florence 1537–69; source untraced. *neglecting*. Negligent. *desperate*. Disconsolate. *better tune*. More trustful. *Shall we . . . also?* Job 2: 10. *in a proportion*. Proportionately. *studieth*. Plots, schemes. *green*. Fresh, unhealed. *fortunate*. Bringing good fortune. *Caesar*. The assassination of Julius Caesar ultimately brought the prosperous rule of Augustus. *Pertinax*. The assassination of Publius Helvius Pertinax in AD 193 led to the efficient rule of Septimius Severus. *Henry the Third*. His murder in 1589 led to the reign of the heroic Henri IV and the Edict of Nantes (1598), guaranteeing tolerance to the Huguenots. *mischievous*. Malicious, destructive.

5. *Of Adversity*

348. *high*. Proud. *Seneca*. *Epist*. lxvi. 29. Cf. also Boethius, *De consolatione philosophiae*, II. viii: 'Fortune, when she is opposite, is more profitable to men than when she is favourable. For in prosperity, by a show of happiness and seeming to caress, she is ever false, but in adversity when she sheweth herself inconstant by changing, she is ever true. . . .prosperity, with her flatterings withdraweth men from true goodness, adversity recalleth and reclaimeth them many times by force to true happiness.' *after*. According to. *wished*. Desired. *admired*. As extraordinary. *be*. Mean, imply. *command over nature*. Self-command, in restraint of natural impulse. *higher*. More exalted. *too high*. Too lofty, moral ('heathens' not having access to Christian revelation). *in one*. At the same time. *security*. Freedom from care. '*It is . . . Dei*'. Paraphrased from *Epist*. liii. 12. *transcendences*. Exaggerations, hyperboles.

figured. Represented. *strange fiction*. A myth recorded in Apollodorus, *Bibliotheca*, II. v. 10. *mystery*. Hidden meaning, known only to the initiate.

349. *pitcher*. Vessel. (In the classical sources the vessel was made of gold, not clay.) *lively*. Vividly. *thorough*. Through. *in a mean*. In moderate language. *morals*. Ethics. *benediction*. Blessedness. *David's harp*. The Book of Psalms. *hearse-like airs*. Funeral hymns (perhaps the Penitential Psalms: nos. 6, 32, 38, 51, 102, 130, 143). *carols*. Joyous dance-songs. *felicities*. Good fortune. *distastes*. Discomforts, annoyances. *lively work*. Vivid pattern or picture. *sad*. Sober, dark-coloured. *ground*. Background. *lightsome*. Bright. *incensed*. Set on fire (as incense is). *crushed*. Compressed, squeezed. *discover*. Reveal. Cf. Guicciardini, *Ricordi*, no. 164: 'Men's good fortune is often their greatest enemy, for it causes them to become wicked, frivolous, insolent. Thus it is a better test of a man to withstand good fortune than adversity' (ed. cit. 42).

6. *Of Simulation and Dissimulation*

349. *Simulation*. Pretending to be that which one is not. *Dissimulation*. Concealment (under a feigned semblance). Bacon's treatment of this topic overlaps that of Francesco Guicciardini, in his *Ricordi*, a collection of observations and precepts assembled during the 1520s, and available in numerous editions in various languages during the 16th century. Cf. e.g. no. 104: 'dissimulation is condemned and hated yet it is much more profitable [than openness], and frankness is helpful to others rather than oneself. But because it is undeniably a fine quality I would praise the man who in his daily life lives freely and openly and uses dissimulation only rarely and for some very important reason. Thus you will acquire the reputation of being free and frank and will enjoy the popularity of those so regarded; yet in matters of the greatest importance you will profit most by dissimulation, the more so since having the reputation of sincerity, your duplicity will more easily be believed' (ed. cit. 28). *faint*. Weak. Cf. Bacon's similar judgement in *Adv.L.*, above, pp. 280–1. *policy*. Sagacity in public affairs. *asketh*. Requires. *wit*. Understanding. *politiques*. Men in public life. *Tacitus*. *Annals*, v. 1. *sorted*. Adapted herself to. *arts*. Devices. *Mucianus*. Commander of Syrian troops; cf. Tacitus, *Histories*, ii. 76. *closeness*. Secrecy. *properties*. Characteristics. *several*. Different. *as*. That. *secreted*. Kept concealed. *half lights*. In part; or 'by twilight' (the Latin translation has 'tanquam in crepusculo'). *state*. Statesmanship, government. *Tacitus*. *Annals*, iii. 70; *Agricola*, xxxix. *poorness*. Disadvantage. *obtain to*. Attain.

350. *close*. Secretive. *particulars*. Individual cases. *going softly*. Treading warily. *well*. Properly. *openness*. Sincerity. *name of certainty*. Reputation for reliability. *managed*. Trained (in formal manœuvres). *passing*. Exceedingly. *opinion*. Good reputation. *Closeness*. Secrecy. *Reservation*. Reservedness. *without hold . . . is*. Without revealing himself (a 'hold' is a grip in wrestling). *arguments*. Tokens, proofs. *that*. That which. *Simulation*. Pretending to be something one is not. *industriously and expressly*. On purpose. *first*. Cf.

Adv.L., above, p. 276. *secret man.* One who keeps secrets. *blab.* Teller of secrets. *babbler.* Foolish, idle talker. *discovery.* Disclosure. *close ... open.* As the confined air in a room gives way to the colder air from outside. *in that kind.* In the same way that a confessor does. *while.* For. *discharge.* Relieve. *impart.* Share with, communicate. *mysteries are due to secrecy.* People who can keep a secret earn others' confidences. *uncomely.* Improper, unseemly. *reverence.* Respect. *talkers.* Talkative persons. *futile.* Unable to hold their tongue (Lat. *futilis*, 'easily pouring out'). *vain.* Lightminded. *withal.* In addition. *set it down.* Take note, as of an important truth. *politic.* Shrewd, statesmanly. *part.* Point. *face give ... leave.* That his facial expression does not reveal beforehand what the tongue has to say, or does not contradict what he has said. Cf. *Adv.L.*, above, pp. 265–6. *discovery.* Revealing. *tracts.* Traits, features. *by how much.* Inasmuch as. *it.* The facial expression. *marked.* Noticed. Cf. *Adv.L.*, above, p. 206. *followeth ... upon.* Derives from. *suffer.* Allow. *keep an indifferent carriage.* Maintain an impartial bearing. *beset.* Assail.

351. *absurd.* 'The Lat. *absurdus* is applied to the answer given by a deaf man (*surdus*) which has nothing to do with the question; hence it signifies, deaf to reason, unreasonable' (Wright, p. 354). *equivocations.* Using a word in more than one sense, intending to deceive. *oraculous speeches.* Obscure hints, often deceptive. *scope.* Space for free movement. *skirts or train.* Subordinate, attending circumstances. *false profession.* Declaration, pretence. *rising.* Arising. *of.* From. *fearfulness.* Timidity. *main.* Major. *ure.* Use, practice. *surprise.* Make use of surprise. *published.* Disclosed. *alarum.* Signal calling men to arms. *fair.* Honourable. *engage.* Commit. *manifest.* Open. *go through.* Follow it up. *take a fall.* Be thrown (another metaphor from wrestling). *fair.* Just, probably. *freedom of thought.* Reserving judgement. *proverb.* Cf. *Adv.L.*, above, p. 275: 'Di mentira, y sacras verdad.' *to set it even.* To balance the account. *shew.* Token, sign. *spoil ... mark.* Prevent the arrow's direct ('round') flight to its target. *perplexeth.* Leaves in doubt. *conceits.* Thoughts. *belief.* Credit. *composition.* Constitution. *temperature.* Temperament (ideally, a balance between the humours). *to ... opinion.* Have a reputation for frankness. *in habit.* In practice. *seasonable use.* At the opportune moment.

7. *Of Parents and Children*

351. *utter.* Reveal.

352. *mitigate.* Reduce, render less painful. *remembrance.* Thought. *perpetuity.* Perpetuating oneself. *generation.* Procreation. *memory.* 'Exemption from oblivion' (Dr Johnson). *proper.* Peculiar. *noblest works ... images of minds ... bodies have failed.* Plato, *Symposium*, 208B–209A. *foundations.* Charitable institutions. *which.* Who. *express.* Give shape to. *care of.* Concern for. *most.* Greatest. *raisers of their houses.* Founders of a family or dynasty. *kind.* Kin, family. *work.* Serving to perpetuate the family. *children and creatures.* That is, both 'kin' and 'created beings', corresponding to 'kind' and 'work' earlier. *affection.* Regard. *unequal.* Unfair. '*A wise ... mother*'.

Prov. 10: 1. Bacon explains this obscure sentence in the *De Augmentis* as meaning that 'a wise and prudent son is of most comfort to the father, who knows the value of virtue better than the mother.... But the mother has more sorrow at her son's ill fortune... both because the affection of a mother is more gentle' and perhaps because she feels she may have spoiled him (*Works*, v. 40). *respected*. Treated with respect. *wantons*. Spoiled, dissolute. *allowance*. Financial support. *harmful*. Pernicious. *base*. Bad. *shifts*. Subterfuges (to get money). *sort with*. Associate with. *surfeit*. Gorge themselves. *proof*. Result of experience. *keep... not their purse*. Withhold money. *emulation*. Strife, contention. *sorteth to*. Results in. *kinsfolks*. Relatives. *of the lump*. Lot (here lineage). *pass not*. Derive from. *a like matter*. The same thing. *the blood happens*. As the family traits come out. *betimes*. In good time. *courses*. Ways of life. *apply*. Bend, conform. *affection or aptness*. Attitude or inclination. *cross*. Thwart. *precept*. Proverb, maxim. '*optimum... consuetudo*'. 'Choose what is best: custom will make it pleasant and easy': Pythagoras, according to Plutarch, 'De Exilio', 8 (*Moralia*, 602C).

8. *Of Marriage and Single Life*

353. *hostages*. Pledge or security (which Fortune can damage). *impediments*. Obstacles. *which*. Who. The 1612 text (*Works*, vi. 547) adds: 'which have sought eternity in memory, and not in posterity'; omitted 1625. *affection*. Love. *means*. Generosity. *reason*. Reasonable. *thoughts*. Considerations for the future. *end with*. Reach no further than. *impertinences*. Irrelevances. *bills of charges*. Expenses. *because*. In order that. *except*. Object. *charge*. Burden, responsibility. *abatement*. Reduction. *humorous*. Whimsical. *sensible of*. Sensitive to. *as*. That. *near to* Almost. *girdles*. Belts. *shackles*. Chains. *light to*. Quick, prone to. *fugitives*. Outlaws. *doth well with*. Is proper for. *churchmen*. Ecclesiastics. *ground*. Here, society. *pool*. Here, the family. *indifferent*. Neutral. *facile*. Easily swayed, fickle. *hortatives*. Exhortations. *vulgar*. Common. *discipline*. Training. *exhaust*. Used up. *inquisitors*. Legal officers charged with cross-examinations. *Grave natures*. Serious men. '*vetulam... immortalitati*'. 'He preferred his old wife to immortality': *Odyssey*, v. 135, perhaps via Cicero, *De oratore*, i. 44, or Plutarch's 'Gryllus' (*Moralia*, 985F, 989A), in a Latin translation. *froward*. Bad-tempered, contrary. *bonds*. Guarantees. *wise*. Of sound judgement.

354. *mistresses*. Lovers. *So as*. So that. *quarrel*. Motive (here: reason for marrying at any age). *he... answer*. Thales, according to Diogenes Laertius, i. 26, and Plutarch, 'Table-Talk', iii. 6.3 (*Moralia*, 654C). *price*. Value. *make good*. Make up for.

9. *Of Envy*

354. On a remarkable stylistic feature of the Essay, the fact that no less than 21 of its 76 head clauses begin with the explanatory 'For', see Ruth Tarselius, '"All colours will agree in the dark": A Note on a feature in the style of Francis Bacon' (*Essential Articles*, 293–9). Reynolds (p. 62) observes that Bacon uses 'envy' in two senses: in the private sphere it has its ordinary meaning, and sometimes 'malevolence'; in the public sphere it means

'discontent, disaffection'. For a partial source see Aristotle, *Rhetoric*, ii. 10 (on envy), and Cicero, *Tusc. D.*, III. ix. 19–21. *affections*. Passions. *fascinate*. To 'cast a spell over with a look'. On the tradition in Renaissance occultism for 'fascination' as beginning through the eye and overpowering the imagination, see Melchionda, p. 433. Reynolds (p. 62) cites Plutarch's 'Table-Talk' (v. 7; *Moralia*, 680C ff.) on 'people who are said to cast a spell and to have an evil eye'. This power derives from 'effluences', the 'most active stream of such emanations' deriving from the eye. 'For vision, being of an enormous swiftness and carried by an essence [*pneuma*] that gives off a flame-like brilliance, diffuses a wondrous influence' (681A), especially in conjunction with love, 'that most powerful and violent experience of the soul', which makes the lover 'melt and be dissolved when he looks at those who are beautiful, as if he were pouring forth his whole being towards them' (681A–B). The power of vision is active, as well as passive, for 'the body is sympathetically affected when the mind is subject to any influence'. So 'envy, which naturally roots itself more deeply in the mind than any other passion, contaminates the body too with evil.... When those possessed by envy to this degree let their glance fall upon a person, their eyes, which are close to the mind and draw from it the evil influence of the passion, then assail that person as if with poisoned arrows' (681D–E). Another classical source is Cicero, *Tusc. D.*, III. ix. 20. Cf. also *Sylva Sylvarum*, X. 944 (*Works*, ii. 653), and Wright (p. 298) for a similar passage in G. B. della Porta. *vehement*. Intense. *imaginations*. Fantasies. *suggestions*. Temptations. *into the eye*. Thought to be the strongest of the senses, hence most vulnerable to external influences. *points*. Characteristics. *scripture*. Mark 7: 22. *influences*. 'Streams' from the stars. *evil aspects*. Angular relationships between the planets or zodiacal signs and the earth; some of which were supposed to be inauspicious. *still*. Always. *ejaculation ... eye*. Renaissance optics and psychology believed that the eye perceived by emitting beams of light; so here envy is thought of as being physically transmitted. *curious*. Careful, meticulous. *percussion*. Blow. *glory*. Display (of success or potency). *sets an edge*. Sharpens. *spirits*. Vital powers (within the body, 'spirit' was thought to be an extremely fine liquid, easily vaporized). *outward parts*. Bodily surface. *curiosities*. Fruitless questions. *handle*. Discuss. *apt to*. Prone to. *virtue*. In Bacon this word often has the sense of Latin *virtus* and Italian *virtù*, 'innate strength'. *good ... evil*. Prosperity ... disgrace. *who*. He who. *wanteth*. Lacks. *whoso*. Whosoever. *come ... hand*. Get even with. *depressing*. Bringing low. *busy*. Meddlesome, prying. *inquisitive*. Unduly curious. *ado*. Fuss, activity. *estate*. Interests, affairs. *needs*. Of necessity. *play-pleasure*. Vicarious pleasure, as when watching a play.

355. *but*. Only. *gadding*. Restless, rushing hither and thither. '*Non ... malevolus*'. 'No one is inquisitive without wishing for the worst' (Plautus, *Stichus*, i. 3.54). *new men*. Those who have newly acquired rank, were not born to it (Lat. *novi homines*). *deceit of the eye*. Optical illusion. *come on*. Advance. *old men*. Cf. Aristotle, *Rhetoric*, ii. 10: 'We also envy those who have what we ought to have, or have got what we did have once.

Hence old men envy younger men' (1388a19 ff.). *mend . . . case*. Improve his own condition. *light upon*. Happen to. *thinketh to*. Expects to. *wants*. Disadvantages. *affecting*. Aiming at. *Narses*. (*c*.478–*c*.578), an outstanding Byzantine general, who reclaimed Italy from the Ostrogoths. *Agesilaus*. 444–360 BC, King of Sparta, who had one leg shorter than the other (Plutarch, 'Life' of Agesilaus, ii. 2). *Tamberlanes*. Mongol conqueror, *c*.1336–1405, lame according to some sources. *rise*. Regain power, fortune. *levity*. Instability. *vain-glory*. Vanity, ambition. Cf. Aristotle, ibid., 'Ambitious men are more envious than those who are not . . . Indeed, generally, those who aim at a reputation for anything are envious on this particular point' (1387b31 ff.). *cannot want work*. Always have occasion to feel envy. *Hadrian*. Publius Aelius Hadrianus, Emperor AD 117 38. *artificers*. Skilled workmen. *vein*. Inclination. *near kinsfolks*. Cf. Aristotle, ibid.: 'We feel [envy] towards equals . . . in birth, relationship, age, disposition, distinction, or wealth' (1387b23 ff.); 'we envy those who are near us in time, place, age, or reputation. Hence the line "Ay, kin can even be jealous of their kin"' (1388a5 ff.). *fellows in office*. Companions sharing the same rank. *raised*. Promoted. *upbraid unto*. Reproach with. *pointeth at*. Directs attention to (contemptuously). Cf. Aristotle, ibid.: 'We also envy those whose possession of or success in a thing is a reproach to us: these are our neighbours and equals; for it is clear that it is our own fault we have missed the good thing in question; this annoys us, and excites envy in us' (1388a17 ff.). *incurreth*. Obtrudes itself. *note*. Notice. *redoubleth*. Echoes, increases. *sacrifice*. Abel's offering (Gen. 4: 3–5). *no body to look on*. No spectators to see his disgrace, and so justify his envy. *advanced*. Promoted, awarded distinctions. *comparing . . . kings*. Cf. Aristotle, ibid., on the envy we feel towards 'our fellow-competitors': 'We compete with those who follow the same ends as ourselves . . . who are after the same things; and it is therefore these whom we are bound to envy beyond all others. Hence the saying, "Potter against potter"' (1388a9 ff.). *unworthy*. Undeserving. *coming in*. Prominence. *contrariwise*. On the contrary.

356. *per saltum*. 'By a leap'. *travels*. Travails, labours. Reynolds (p. 66) cites Plutarch, 'On inoffensive self-praise', *Moralia*, 544D: men 'seldom or never envy such as have bought [glory] very dear, with many travails and great dangers'. *hardly*. With difficulty. *healeth*. Covers up. *deep*. Profound. *sober*. Serious-seeming. *politique persons*. Politicians. *bemoaning*. Lamenting. *quanta patimur*. 'How much we suffer!' *abate*. Beat down, blunt. *engrossing*. Monopolizing, acquiring. *pre-eminences*. Ranks. *well*. Satisfied. *do sacrifice*. Make concessions. *of purpose*. Purposely. *crossed*. Thwarted. *overborne*. Put down, overpowered. *carriage*. Bearing, reaction to. *that*. i.e., by crafty disclaimers. *disavow fortune*. Disown it, as having favoured him more than he deserved. *want in*. Lack of. *lot*. Spell (Lat. *sors*, which 'gave its name to the practisers of witchcraft, *Sorcerers*, Latin *sortiarii*': Wright, p. 299). According to Renaissance witchcraft theory, a maleficent spell could only be removed by being transferred to some other person or animal. *derive*. Transfer.

357. *turn*. Trick, purpose. *violent and undertaking*. Impetuous, enterprising. *so*. Provided. *yet*. However, all the same. *ostracism*. Temporary banishment. *eclipseth*. Makes disappear. *This*. (Public envy.) *discontentment*. Discontent; 'ressentiment'. *Sedition*. See Essay 15, below. *sound*. Healthy. *tainteth*. Corrupts. *gotten*. Begotten. *traduceth*. Slanders. *ill odour*. Bad reputation. *won*. Gained. *plausible*. Praiseworthy, deserving applause. *beat . . . upon*. Affect. *estates*. States, republics. *general*. Widespread. *estate*. State. *handled*. Discussed earlier. *touching*. Concerning. *importune*. Pressing, insistent. '*Invidia . . . agit*'. 'Envy keeps no holidays.' *some*. Some one. *pine*. Long for. *devil*. 'The word *devil* means *slanderer*, i.e. one that slanders God to men, and . . . slanders men to God' (Abbott, ed. cit. ii. 139). *tares*. Weeds in a cornfield. '*The envious . . . night*'. Matt. 13: 25 (although the biblical text refers not to an envious man but to 'his enemy'). *subtilly*. 'Subtly', often in the pejorative sense of 'craftily, cunningly'. *prejudice*. Harm.

10. *Of Love*

357. *beholding*. Indebted, obliged. *matter*. Plot.

358. *syren*. A mythical creature, half-woman, half-bird, whose song lured sailors to destruction (Homer, *Odyssey*, xii. 39). Cf. Bacon's *De Sapientia Veterum*, 31: 'The Sirens, or Pleasure' (*Works*, vi. 762–4). *fury*. In Greek myth, winged woman with snakes for hair, goddess of vengeance. *great spirits*. Outstanding rulers and statesmen. *business*. Employments, tasks. *weak*. Inconstant. *Marcus Antonius*. Infatuated by Cleopatra (as in Plutarch and Shakespeare). *half-partner*. Joint ruler (with Octavius Caesar). *Appius Claudius*. Judge and one of the decemvirs (body of ten men administering Rome and its laws), whose lust for Virginia caused her murder and a people's revolt in 449 BC; cf. Livy, iii. 33. *inordinate*. Ungovernable. *poor*. Squalid, feeble. '*Satis . . . sumus*'. 'Each of us is enough of an audience for the other' (Seneca, *Epist*. vii. 11, who quotes the saying to prove that the opinion of the multitude is of no account). *idol*. Image or puppet. *higher purposes*. Nobler ends. *strange*. Remarkable. *braves*. Distorts by exaggeration. *hyperbole*. Verbal exaggeration. *comely*. Becoming, suitable. *phrase*. Diction (as well as in the thought). *said*. By Plutarch, 'How to tell a Flatterer from a Friend', 11 (*Moralia*, 48E–F): the flatterer can appeal to 'Self-love, whereby every man being the first and greatest flatterer of himself, he can be very well content to admit a stranger to . . . confirm that good self-conceit and opinion of his own'. *have intelligence*. Have an understanding with; conspire. '*That . . . wise*'. A sentence first found in Publilius Syrius and in Plutarch's 'Life' of Agesilaus, but given great currency by Erasmus' *Adagia*. *reciproque*. Mutual. *reciproque*. Return, reciprocity. *poet's relation*. Poetic narrative of the choice of Paris, who preferred beauty (Venus) to wisdom (Pallas) and power (Juno), as in Ovid, *Heroides*, xvi. 163–6. *figure*. Describe. *esteemeth*. Values. *floods*. Overwhelming power. *fervent*. Intense. *admit*. Allow entrance. *keep quarter*. Stay in its proper place. *sever*. Keep apart.

359. *check*. Interfere. *no ways*. In no way. *true . . . ends*. Constant in pursuing their own interests. *martial men*. Soldiers. Cf. Aristotle, *Politics*, ii. 9, 1269b27 ff. *given*. Prone, liable to. *ask*. Require. *motion*. Impulse. *maketh*. Creates. *embaseth*. Degrades.

11. *Of Great Place*

359. *great place*. Public eminence (the Latin translation renders it 'De Magistratibus & Dignitatibus'). Cf. *Adv.L.*, above, p. 271. *state*. Republic. *fame*. Reputation. *persons*. Body, self. *seek power . . . lose liberty*. Reynolds (p. 77) cites Seneca, *Consolatio ad Polybium*, vii. 2. *place*. High position. *base*. Despicable. *indignities*. Dishonourable acts. '*Cum . . . vivere*'. 'When you are no longer what you have been, you see no reason why you should wish to live' (Cicero, *Epist. ad Fam.* vii. 3). *reason*. Appropriate. *impatient of*. Irritated by. *privateness*. Private or retired life. *the shadow*. The obscurity of home (as against the openness of public life). The Latin translation reads 'umbram et otium'. *townsmen*. Town-dwellers. *had need to*. Would have to. *fain*. Gladly. *the puzzle of business*. The confused bustle of public life. *tend*. Attend to. '*Illi . . . sibi*'. 'Death comes heavily on him who dies well-known to the world, a stranger to himself' (Seneca, *Thyest.* ii. 401–3). *In place*. In a position of power. *licence*. Opportunity. *will*. Desire.

360. *can*. Be able (an unusual usage). *aspiring*. Ambition. *good thoughts*. Good intentions. *in act*. Into action. *vantage and commanding ground*. Superior position, able to do good. *end*. Aim. *motion*. Activity. *conscience*. Consciousness. *theatre*. Spectacle (a life of good works). Cf. *Adv.L.*, above, p. 247. *rest*. In heaven (as God rested on the seventh day of Creation). '*Et . . . nimis*'. 'And God turned to look upon the works which his hands had made, and saw that all were very good' (Gen. 1: 31). Cf. also Augustine, *Confessions*, xiii. 36.51. *discharge*. Carrying out. *globe*. 'A complete or perfect body' (*SOED*). *carried themselves ill*. Behaved badly. *set off thyself*. Elevate, distinguish yourself. *taxing*. Censuring. *direct*. Instruct. *bravery*. Ostentation. *set it down*. Propose. *Reduce*. Bring back. *first institution*. Beginning. Cf. *Adv.L.*, above, p. 190, and Machiavelli, *Discourses*, iii. 1. *degenerate*. Degenerated. *positive*. Self-assured. *peremptory*. Autocratic. *express*. Explain. *right*. Prerogative. *stir*. Bring up. *de facto*. In fact. *voice*. Proclaim. *direct in chief*. Rule through intermediaries. Cf. Plutarch, 'Precepts of Statecraft', *Moralia*, 812C (Reynolds, p. 78). *Embrace and invite*. Accept and solicit. *helps*. Aid. *touching*. Concerning. *execution*. Carrying out of duties. *accept of*. Receive with favour. *facility*. Pliancy, being over-ready to yield. *go through*. Finish, complete. *interlace*. Intersperse, mix up together. *taking . . . offering*. Gifts and bribes. *used*. Practised. *professed*. Displayed, declared. *variable*. Inconsistent.

361. *steal*. Perform secretly. *inward*. Familiar, intimate. *by-way*. Indirect path. *close*. Secret. *roughness*. Coarse or brutal treatment. *needless*. Superfluous. *taunting*. Mocking. *importunity*. Allowing oneself to be pressurized or manipulated by others. *respects*. Considerations, or 'favours' (to the wrong people). '*To respect . . . bread*'. Prov. 28: 21. Cf. *Adv.L.*, above, p. 269.

anciently. By various classical writers, including Plutarch (*Lives*), and in the later proverb found (e.g.) in Erasmus' *Adagia*, 'Magistratus virum indicat'. '*Omnium . . . imperasset*'. 'A unanimous verdict would have judged him fit to rule, if he had never ruled' (*Histories*, i. 49). Proclaimed emperor in AD 68, Galba was so unpopular that he was murdered within a year. '*Solus . . . melius*'. 'Vespasianus was the only emperor that was improved by empire' (ruling power) (ibid. 50). Emperor AD 69–79, Vespasianus' rule brought peace and prosperity. *sufficiency*. Ability, administrative competence. *affection*. Benevolence. *assured*. Sure. *honour*. Celebrity, fame (in Renaissance ethics, properly the reward for or recognition of virtuous behaviour for the public good). *amends*. Improves. *in nature . . . in their place*. According to Renaissance astronomy and physics, linear motion (often caused by attraction) was distinguished from circular motion (as in a planet's orbit). *winding stair*. Indirect or devious path. *to side . . . self*. To take sides. *placed*. Put in place or position. *tenderly*. Carefully. *debt*. One owed to your predecessor's reputation. *called*. Consulted. *sensible*. Sensitive. *remembering . . . place*. Do not stand on your dignity.

12. *Of Boldness*

361. *trivial grammar-school text*. A trite or elementary text, as taught in the *trivium* or three beginning subjects of education (grammar, logic, rhetoric) between the ages of 10 and 16. *Demosthenes*. The greatest Athenian orator (?384–322 BC); the anecdote is recorded in Pseudo Plutarch, 'Lives of the ten Orators', *Moralia*, 845B, and often in Cicero, e.g. *De oratore*, iii. 56.213, *Brutus*, xxxviii. 142. *action*. The five 'parts' of rhetoric were *inventio*, *dispositio*, *elocutio*, *memoria*, and *actio* (or *pronuntiatio*), the last stage teaching the rhetorical gestures to be used in delivering a speech. *that*. That which. In his 'Life' of Demosthenes 11.1–2 Plutarch records that he had by nature a soft voice, a stammer, and short breath, all of which defects he overcame by training. *player*. Actor. *noble*. The intellectually more demanding processes of composition.

362. *taken*. Enthralled. Cf. Aristotle, *Rhetoric*, iii. 1, 1404a1–9. *wonderful like*. Wonderfully similar. *civil business*. Politics, public life. *baseness*. Mediocrity. *weak times*. Moments of weakness. *popular*. Democratic (or demagogic). Reynolds (p. 85) cites Herodotus v. 97 on how Aristagoras 'failed to persuade Cleomenes to attack Persia and give aid to the Ionian revolt; but when he came to Athens he carried the people with him by his boundless promises and assurances of easy success'. *entrance*. Appearance. *mountebanks*. Spurious healers or quacks. *politic body*. 'The nation in its corporate character; the State' (*SOED*). *undertake*. Promise. *grounds of science*. Scientific principles. *hold out*. Carry through to the end. *Mahomet's miracle*. Proverbial in the 16th century already. *never a whit*. Not at all. *slight it over*. Dismiss, make light of. *make a turn*. Change direction. *ado*. Fuss. *a sport*. Source of laughter. *wooden*. Stiff. *in bashfulness . . . spirits . . . stay*. In shy people the vital spirits fluctuate wildly (as in blushing); in bold people they are less volatile or expressive. *stale*. Stalemate. *stir*. Continue. *seeth*. Notices. *counsel*. Deliberation. *use*. Employment.

13. *Of Goodness and Goodness of Nature*

363. *affecting*. Desiring, having a love for. *weal*. Well-being. *that*. That which. *Philanthropia*. Love of mankind. *humanity*. Latin 'humanitas' was sometimes used to translate *philanthropia*, but weakly. *habit... inclination*. Categories used by Aristotle in *Metaphysics*, v. 20; *Eth. Nic.* ii. 1, vi. 13, x. 9; *Eudemian Ethics*, ii. 2; *Politics*, vii. 13. *dignities*. Good qualities. *character*. Expression. *busy*. Meddlesome. *wretched*. Debased, occupied with ignoble matters. *answers*. Corresponds. *no excess*. Bacon places goodness outside the scheme of Aristotle's *Eth. Nic.* in which each virtue occupies a mid-point between excess and deficiency. *but*. Only. *come in danger*. Be endangered. *issue*. Manifest itself. *take unto*. Incline towards. *Turks*. In the Middle Ages and Renaissance they were major opponents of the Church, and thus proverbial for inhumanity. *as*. That. *Busbechius*. Ogier Ghiselain de Busbecq (1522–92), a diplomat at the Turkish court, whose letters, published in 1589, include many anecdotes of the Turks' kindness to animals. *had like*. Was nearly. *waggishness*. Mischievous game. *fowl*. Bird (of any kind). *Errors*. Cf. Aristotle, *Eth. Nic.* vi. 13. *ungracious*. Unkind. *doctors*. Learned men, teachers. *Machiavel*. In *Discourses*, ii. 2. (Bacon distorts Machiavelli's point, which is that Christianity makes men indifferent to worldly affairs by proposing more worthy goals.) *confidence*. Boldness. *spake*. Said. *magnify*. Make great or important. *scandal*. Discredit to religion. *take knowledge of*. Observe, note. *bondage... fancies*. Gratifying their passing whims. *Aesop's cock*. Actually, in a fable by Phaedrus (iii. 12). *He sendeth... unjust*. Matt. 5: 45. *Common*. The property of all.

364. *communicate*. Shared. *peculiar*. Individual. *portraiture*. Portrait. *pattern*. Mould (from which copies are made). '*Sell... me*'. Mark 10: 21; Matt. 19: 21. *sell not... me*. Reynolds (p. 90) cites Aquinas, *Summa Theol.* II. 184.3. *a habit of goodness*. Cf. Aristotle, *Eth. Nic.* vi. 13. *right reason*. In Renaissance ethics, the faculty that should control human behaviour. *disposition*. Inclination. *be*. Are. *affect*. Desire. *crossness*. Contrariness. *frowardness*. Stubbornness. *difficilness*. Awkwardness. *mere*. Real. *in season*. Flourishing. *loading part*. Taking the side which increases other people's burdens. *Lazarus' sores*. Luke 16: 21. *still*. Always. *Timon*. In his extreme misanthropy he invited anyone who so wished to hang themselves from a tree in his garden. See Plutarch's 'Life' of Timon, 70.2, and Shakespeare, *Timon of Athens*, V. ii. 205–12. *errours*. Sports, deformities. Cf. Suetonius, 'Life' of Augustus, 65. *timber*. Material. *politiques*. Politicians. *knee timber*. Growing crooked (like the human knee), and so more resistant. *citizen of the world*. This concept, first formulated by Cicero (*De legibus*, I. xxiii. 61), defines someone receptive to the whole of human life. *no island... lands*. Cf. Donne's *Devotions* (1624): 'No Man is an *Island*, intire of itselfe; every man is a peece of the *Continent*, a part of the *maine*' (Meditation 17). *the noble tree*. The incense tree (see Pliny, *Nat. Hist.* xii. 14) which, like the pelican, was an emblem of self-sacrifice for the good of others. *planted above*. Superior to. *shot*. Hurt. *minds*. Intentions. *trash*. Anything worthless; a cant word for money. *St. Paul's perfection*. Rom. 9: 3, 'For I could wish that

myself were accursed from Christ for my brethren, my kinsmen according to the flesh'. Cf. *Adv. L.*, above, p. 246. *anathema*. A person or thing formally cursed by a church authority.

14. *Of Nobility*

364. *estate*. State. *particular*. Individual.

365. *pure*. Mere. *attempers*. Moderates. *draws . . . aside*. Distracts attention from. *line royal*. Royal family, descendants. *But for*. As for. *stirps of nobles*. Stems of noble families. *business*. Management of public affairs. *fittest*. Best suited to, competent. *flags*. Insignia (as coats of arms). *pedigree*. Genealogy of nobility. *the Switzers*. The Swiss, who by this time had thirteen cantons, both Protestant and Catholic. *last well*. Thrive. *utility*. Mutual benefit. *respects*. Respecting of persons (rank, nobility). *The united provinces*. Holland, Belgium, Flanders. *consultations*. Considerations. *indifferent*. Impartial. *presseth*. Oppresses (since the nobility is a closed class). *too great . . . for justice*. Being above the law. *fast*. Close. *surcharge*. Overcharge. *fall*. Chance, happen. *in time*. Sooner or later. *means*. Riches. *reverend*. Worthy of respect. *fair*. Handsome. *timber tree*. Useful, not ornamental. *weathers*. Tempests. *act of power*. Created by royal decree. *virtuous*. Strong, resourceful (as in the Italian *virtù*). *rising*. Social advancement. *commixture*. Combination. *it is reason*. It is right. *abateth industry*. Blunts initiative. *at a stay*. In the same place. *motions*. Emotions. *passive envy*. Being envied. *of*. Among. *slide*. Access (business will flow more smoothly).

15. *Of Seditions and Troubles*

366. This essay is found in the MS collection preceding the 1612 edn., but was not printed there. (For that version see *Works*, vi. 589.) The detail with which Bacon analyses civil disturbances shows the fear with which they were generally regarded throughout Europe in the 16th and 17th centuries. Many states suffered great and prolonged losses from this cause. *Shepherds*. Rulers, governors. *calendars*. Foreseeable, predictable times. *things . . . equality*. When the various levels of society gain equal power. *Equinoctia*. In the spring and autumn equinoxes, when the days and nights are equal in length, storms are more likely. The word 'equinox' was still rare in English, although occurring in Blundeville's *Exercises*, 1594 (Wright, p. 304). *hollow*. Empty: not yet powerful. *secret*. Inexplicable. *Ille . . . bella*. The sun 'often warns us that dark tumults threaten, and deceits and hidden wars are swelling' (Virgil, *Georgics*, i. 464–5). *libels*. Defamatory writings. *licentious*. Excessively free, provocative. *false news*. Cf. Shakespeare, *2 Henry IV*, the Induction spoken by Rumour. *Illam . . . progenuit*. 'Mother Earth (as they relate), irritated by anger against the gods, brought forth [Rumour] last as sister to Caeus and Enceladus' (*Aeneid*, iv. 178–80): cf. *Adv.L.*, above, pp. 187–8, and the fragmentary Essay 'Of Fame', above, p. 455. *fames*. Rumours. *plausible*. Deserving applause. *traduced*. Misrepresented. *envy*. Discontent. '*conflata . . . premunt*': 'when envy is once roused, good actions are as much assailed as bad' (*Histories*, i. 7, adapted). *going about*. Taking special pains. *make . . . long-lived*. Prolongs curiosity. '*Erant . . . exequi*'.

'They were on duty, but none the less preferred to interpret the orders of their generals rather than to follow them' (*Histories*, ii. 39, adapted). *cavilling*. Making trivial objections. *mandates*. Commands. *directions*. Orders. *assay*. Essay, attempt. *fearfully*. Hesitatingly. *tenderly*. Cautiously. *Machiavel. Discourses*, iii. 27, on the danger of factions. *common parents*. As if the parents of everyone in a state. *party*. Political group, faction.

367. *entered league*. Joined with. The Holy League, formed in 1576 by a group of Catholics to defend their faith and to destroy the Protestants, was expelled by King Henri III (1574–89) from Paris in 1585. *presently after*. Immediately afterwards. *accessary . . . cause*. Participating in some activity, but not as principal. *bands*. Links. *faster*. More closely. *possession*. Office. *carried*. Carried on. *primum mobile*. In Ptolemaic astronomy, the 'first mover' or outermost of the ten planetary spheres, from which the rest derive their motion, was supposedly set in motion by the Creator. For Bacon, as for other Renaissance writers, it was a favourite metaphor for the King's orderly rule of the state. *old opinion*. Ancient theory. Bacon did not fully accept either the Ptolemaic or the Copernican system, holding both unproven. *every*. Each one. *softly*. Gently. *particular*. In the stars, the planet's individual motion; in the state, the (violent) actions of powerful men seeking their own gain. '*liberius . . . meminissent*.' 'More freely than if they had remembered their governors' (*Annals* iii. 4, adapted). *orbs . . . frame*. The system is disordered. *girt*. Fitted out. '*Solvam . . . regum*'. A fusion of two texts, Job 12: 18 and Isa. 45: 1, 'I will loosen the girdles of kings', a threat that presages the break-up of society. *mainly*. Violently. *Counsel*. Advisory and consultative functions, as in the Privy Council. *this part*. Of the discussion. *Materials . . . Motives*. Reynolds and Melchionda both note that here, and later in the essay, Bacon is using the Aristotelian categories of 'material cause' and 'efficient cause' (*Anal. Post*. ii. 1, 94a–b; *Met*. iv. 2, 1013). Material causes, described in the two following paragraphs, are situations generally existing (poverty, discontent); efficient causes, described in the fourth paragraph, are events taking place at a given point in time. *bear*. Permit. *fuel*. Tinder, touchwood. *matter*. Cause. *discontentment*. Public envy. *overthrown estates*. Ruined fortunes; bankrupts. *state*. Condition. *Hinc . . . bellum*. 'Hence devouring usury and interest rapidly compounded, hence shaken credit, and war profitable to many' (*Pharsalia*, i. 181–2; *Promus*, no. 369). *of the belly*. From hunger, in the 'mean people' or labouring classes. *better sort*. Nobility. *politic body*. The state; civil society. *humours*. The four fluids whose proper balance maintained health.

368. *preternatural*. Abnormal. *people*. The people. *griefs*. Grievances. *rise*. Rise up, rebel. '*Dolendi . . . item*'. 'There is an end to suffering, but not to fearing' (Pliny, *Epist*. viii. 17, referring not to political discontent, but to an inundation of the Tiber). *mate*. Beat down, overpower. *fears*. Objects of fear. *secure*. Without care, complacent. *often*. Frequently occurring. *vapour or fume*. Mist or damp cloud arising from the earth or sea. *blow over*. Pass on. *Causes and Motives*. Synonymous: 'efficient causes'. *innovation*. Any alteration to established rituals or beliefs in the period 1520 to 1690 was (rightly) feared as likely to create disturbances. *breaking*. Suspension.

strangers. Grievances arising out of the success of foreigners in trade. *dearths*. Shortages. *knitteth*. Unites. *preservatives*. Safeguards. *just*. Proper, complete. *counsel . . . rule*. Individual discretion, no general rule. *material cause*. The Aristotelian category, discussed above. *estate*. State, country. *cherishing*. Supporting, protecting. *idleness*. The Latin translation reads 'desidiam et otium'. *repressing*. By sanctions. *sumptuary laws*. Laws against extravagance in dress and behaviour. *husbanding*. Careful agricultural treatment. *vendible*. For sale. *tributes*. Payments by subject people. *foreseen*. Provided. *stock*. Available wealth, resources. *wear out*. Weaken, drain. *estate*. State, country. *live lower and gather more*. Satisfied with a lower level of subsistence, such people save more. *multiplying of nobility*. The vast increase in numbers of the aristocracy in the 16th and 17th centuries created many social problems. *quality*. Rank. *bred*. Educated. *preferments*. Appointments, positions. *take off*. Absorb. See Bacon's analysis of this problem in *Works*, xi. 25.

369. *upon the foreigner*. For the benefit of foreign trading-partners. (Reynolds takes 'upon' to mean 'at the expense of'.) *whatsoever . . . lost*. Cf. Aristotle, *Politics*, i. 10, 1258b1–2, and Montaigne, *Essais*, i. 22, 'Le Profit de l'un est dommage de l'aultre'. *commodity*. Raw materials, natural produce. *vecture*. Transportation. *wheels*. Sources of trade. *as in a spring tide*. Abundantly (as tides are fullest at the new moon). '*materiam . . . opus*'. 'The workmanship will surpass the material' (Ovid, *Met*. ii. 5). *notably*. Remarkably. *mines above ground*. The Dutch, despite their lack of raw materials, were celebrated for their industry both in manufacturing and in transportation, which became as profitable as gold mines. *muck*. Manure, fertilizer. *keeping . . . hand*. Controlling. *strait*. Strict. *devouring*. Exploiting. For Bacon's carefully considered views on usury see Essay 41, above, p. 421–4. *ingrossing*. Monopolizing, buying 'in the gross' in order to corner the market. Many statutes were passed in the 16th century forbidding buying in order to resell, but they failed to prevent the practice. *great pasturages*. Enclosures, by which small farmers were denied use of common agricultural land, which was fenced off for sheep pastures to support the increasingly profitable export trade in wool. Bacon introduced a bill in the 1597 Parliament against the depopulation caused by enclosures: *Works*, ix. 82. *nobless*. Nobility. *discontent*. Discontented. *excited*. Provoked, set on. *greater sort*. Upper class. *of themselves*. Spontaneously. *troubling of the waters*. As in the pool of Bethesda, which could cure paralytics: John 5: 2 ff. *declare themselves*. Disclose their grievances. *poets*. Homer, *Iliad*, i. 396–406 (where in fact it is Thetis, not Pallas, who sent for Briareus). *feign*. Tell. *safe*. Salutary, tending to safety. *bravery*. Boastful defiance. *turneth . . . back*. Obstructs, preventing the fluid (e.g. blood) from escaping. *inwards*. Internally. *endangereth*. Risks producing. *pernicious imposthumations*. Malignant tumours. *part*. Role, action. *Epimetheus*. Afterthought. In this version of the fable Jupiter takes revenge for Prometheus' gift of fire to the human race by creating Pandora with her box or jar of calamities, which Epimetheus opens. Cf. *De Sapientia Veterum*, 26 (*Works*, vi. 745 ff.). *become*. Suit. *Prometheus*. Forethought. *provision*. Precaution. *politic*. Artful. *artificial*. Skilful. *entertaining*. Encouraging.

370. *proceeding*. Policy. *peremptory*. Destructive, irremediable. *particular persons*. Individuals. *apt*. Willing. *brave*. Boastfully pretend. *that, which*. The uncorrected copies read only 'that' (Kiernan, p. 49). *fit*. Suitable, qualified. *head*. Leader. *known*. Well-known. *greatness*. High standing in society. *hath confidence with*. Is trusted by. *own particular*. Personal concerns. *won*. Persuaded, won over. *fast*. Firm, binding. *fronted*. Confronted, opposed by. *combinations*. Alliances. *at distance*. Against each other. *hold with*. Agree with, endorse. *sharp*. Sarcastic. *fallen from*. Been uttered by. '*Sulla ... dictare*'. 'Sulla did not know his letters, he could not dictate' (Suetonius, *Caesar*, 77, with a pun on 'act the dictator'. '*legi ... emi*'. 'He selected his soldiers, and did not buy them' (Tacitus, *Histories*, i. 5). *donative*. A gift of money; a bonus. '*si ... militibus*'. 'If I live, the Roman empire shall have no more need of soldiers' (Flavius Vopiscus, *Script. Hist. Aug*. xx. 20). *despair*. Disappointment. *tender*. Delicate, needing tact. *ticklish*. Tricky, unstable. *darts*. Arrows. *large discourses*. Lengthy speeches. *flat*. Dull, uninteresting. *against all events*. In all eventualities. *useth*. Is, in general. *trepidation*. Trembling, alarm. '*Atque ... paterentur*'. 'And such was the condition of their minds, that a few dared the evilest deeds, more desired them, all permitted them' (*Histories*, i. 28).

371. *assured*. Trustworthy. *reputed of*. Of good reputation. *popular*. Relying on popular support (likely to be subversive). *holding ... correspondence*. Bearing a proportion, corresponding.

16. *Of Atheism*

371. *the Legend*. The *Legenda aurea*, a collection of Saints' lives and miracles compiled in Latin by Jacobus de Voragine in the 13th century, which had an enormous influence on Renaissance religion and art. *Talmud*. The collection of Jewish civil and ceremonial law. *Alcoran*. The Koran, the sacred book of Islam. *universal frame*. The universe. *convince*. Refute, overcome in argument. *bringeth ... about*. Converts. *second causes scattered*. God is the first cause: the works of man are second (or efficient) causes, seemingly 'scattered' or unconnected to the divine plan. Cf. *Adv.L.*, above, pp. 123, 125–6. *rest*. Remain contented with. *chain*. Connection. *confederate*. Leagued, united. *fly to*. Turn to, accept. *the school*. The earliest atomists (not properly a 'school', since they lived at different times) believed that matter had been created out of the random combination of small particles. In both classical and Christian times their denial of a divine creation caused all atomists to be judged atheistical. *Leucippus*. fl. *c*.400 BC first of the atomists. *Democritus*. Philosopher, *c*.460–370 BC. *Epicurus*. *c*.342–270 BC, *four mutable elements*. According to Heraclitus (followed by Aristotle), earth, air, fire, water. *immutable fifth essence*. Aristotle added a fifth element, ether, supposedly the substance from which the heavenly bodies are composed. Cf. *De caelo*, i. 4, 270a13 ff. *duly and eternally placed*. Fixed for ever in their correct position. *portions*. Particles. *unplaced*. Not in their proper place. *marshal*. Commander. '*The fool ... God*'. Ps. 14: 1, 53: 1. *by rote*. Mechanically. *would have*. Wants to believe. *throughly*. Wholly. *but*. Except. *for ... maketh*. For whose advantage it is to believe. *appeareth*. It is evident. *in nothing more*.

Above all. *fainted*. Became feeble, doubted. *consent*. Conviction, agreement. *you shall have*. There are. *fareth*. Happens. *most of all*. Most significant. *you shall have of them*. There are some people. *charged*. Accused. Cf. Cicero, *De natura deorum*, I. xliv. 123. *blessed natures*. The gods, whom Epicurus conceived as happy in their own existence, not caring about man (ibid. 123, 139). *having respect to*. Interfering in. *temporize*. Conform to time and circumstances.

372. '*traduced*. Misrepresented, maligned. *Non . . . profanum*'. 'It is not profane to deny the gods of the vulgar, but it is profane to apply the opinions of the vulgar to the gods': according to Diogenes Laertius, x. 123. *he*. Epicurus. *confidence*. Boldness. *administration*. The gods' rule over and care for mankind. *Indians of the west*. The South American Indians, according to Joseph de Acosta, *Naturall and Morall Historie of the East and West Indies* (1590; Engl. tr., 1604), v. 3. *latitude*. Scope. *savages*. Indians. *take part*. Agree. *subtlest*. Most acute. *contemplative*. Speculative. *Diagoras*. Diagoras of Melos (late 5th century BC), and *Bion*, Bio the Borysthenite (*c*.325–*c*.255 BC), both avowed atheists. Cf. Cicero, *De natura deorum*, I. i. 2, xxiii. 63, xlii. 117; III. xxxvii. 89; Diogenes Laertius, iv. 54–7. *Lucian*. Lucian of Samosata (b. *c*. AD 120), rather a mocker than an atheist. *impugn*. Oppose. *received*. Accepted, established. *handling*. Discussing, having to do with. *must needs be*. Will have to be. *cauterized*. Burned, branded. *scandal*. Discredit (to religion). '*Non . . . sacerdos*'. 'One can no longer say the priest is as [bad as] the people, for the people are not [so bad] as the priests': Pseudo-St Bernard, *Sermo ad Pastores*. *scoffing*. Mockery. *deface*. Destroy. *of kin*. Akin. *magnanimity*. Noble, generous feelings. *raising*. Elevating (to a divine level). *mark*. Observe. *maintained*. Supported, looked after. *melior natura*. 'A better nature' (Ovid, *Met*. i. 21). *resteth*. Relies. *assureth*. Trusts. *gathereth*. Derives.

373. '*Quam . . . superavimus*'. 'Rate ourselves as highly as we may, Conscript Fathers, yet we cannot match the Spaniards in numbers, the Gauls in bodily strength, the Carthaginians in craft, the Greeks in art, nor our own Italians and Latins in the home-bred and native patriotism characteristic of this land and nation. But our piety, our religion, and our recognition of the one great truth of the Divine government of all things—these are the points wherein we have surpassed all nations and peoples' (Cicero, *De Haruspicum Responsis*, ix. 19).

17. *Of Superstition*

373. Bacon echoes Plutarch, who in his essay 'On Superstition' (*Moralia*, 164E–171F) also judged superstition to be worse that atheism. Wright (p. 310) cites Bacon's letter to Toby Matthew (who was 'imprisoned for religion' between Aug. 1607 and Feb. 1608, as a convert to Catholicism), for the judgement that 'superstition is far worse than atheism; by how much it is less evil to have no opinion of God at all, than such as is impious towards his divine majesty and goodness' (*Works*, xi. 10). *opinion*. Conception. *contumely*. Insolent abuse. *the reproach of the Deity*. Seneca, *Epist*. cxxiii. 16. *Plutarch*. 'De superstitione', x (*Moralia*, 169F). *poets*. Mythographers. *Saturn*. According to ancient legend Saturn, fearing a prophecy that he would be

deposed by his children, ate them as fast as they were born, only Jupiter escaping. Cf. Ovid, *Fasti*, iv. 197 ff. *leaves*. Directs. *sense*. Experience through the senses. *natural piety*. The links of affection joining mankind. *reputation*. Good name (as a restraint upon behaviour). *were not*. Did not exist. *dismounts*. Dethrones, disqualifies. *wary ... no further*. Concerned wholly with. *Augustus Caesar*. Emperor 31 BC–AD 14, whose reign was unusually 'civil' or peaceful. *confusion*. Cause of chaos. *primum mobile*. The tenth heaven or outer sphere in classical cosmology, which transmitted motion to the other nine. *ravisheth*. Hurries away, disturbs. *spheres*. Domains. *arguments are fitted to practice*. Logical reasoning is (erroneously) adapted to experience, instead of governing it. So the wise men champion the foolish instead of enlightening them. *reversed*. Preposterous (back to front). *gravely*. Authoritatively. (In fact, it was meant as a witticism.) *council of Trent*. Ecumenical Council of the Roman Catholic Church, which met three times between 1545 and 1563 to correct abuses and codify dogma in response to the Protestant Reformation. See Paolo Sarpi, *Istoria del Concilio Tridentino* (London, 1619), ii. 83. *schoolmen*. Medieval scholastic philosophers. *bare great sway*. Had much influence. *eccentrics ... phenomena*. The Ptolemaic philosophy of nine planetary spheres revolving concentrically around the earth did not 'save' (that is account for) all the 'phenomena' of observed (irregular) planetary movement, so it evolved theories of additional types of rotation, imagining the planets moving in circles (epicycles) whose centres themselves moved in circles (eccentrics) at a little distance from the earth. *engines of orbs*. Skilful devices (made with 'ingenium' or intelligence) representing planetary orbits. *axioms and theorems*. Scientific propositions. So, the metaphor argues, the scholastic philosophers had to 'frame' or contrive similarly fictitious principles to account for inconsistencies within the Church. *sensual*. Appealing to the senses. *pharisaical*. Self-righteous, hypocritical.

374. *load*. Burden. *lucre*. Profit. *openeth the gate*. Invites. *conceits*. Fantastic conceptions. *novelties*. Innovations (in belief). *taking an aim ... human*. Conjecturing, trying to understand religion purely in terms of human reason, not divine revelation. *mixture of imaginations*. Confused, fantastic theories. *without a veil*. Undisguised. *similitude*. Comparison. *wholesome*. Healthy, fresh. *corrupteth*. Putriefies. *petty*. Trivial. *go furthest*. Distance themselves as far as possible. *care ... had*. Care should be taken. *fareth*. Happens. *purgings*. Medical treatment, emptying the bowels.

18. *Of Travel*

374. *Travel*. The title of the Latin version is *De Peregratione in Partes Externas*, namely Europe. Cf. Bacon's 'Advice to the Earl of Rutland on his travels', above, pp. 69–80 ff. *entrance into*. Rudimentary knowledge of. *school*. To learn the language. *grave*. Reliable. *allow*. Approve. *so*. Provided. *hath*. Knows. *exercises or discipline*. Forms of instruction and study. *yieldeth*. Offers. *go hooded*. Be closed off from life (metaphor from falconry). *abroad*. Around. *make*. Keep. *sit*. Are in session. *causes*. Cases. *consistories*. Ecclesiastical courts of justice. *disputations*. Formal university debates. *of state and pleasure*. Famous for their magnificence and beauty.

375. *armories.* Arsenals, storage-places for military and naval equipment and weapons. *magazines.* Public storage-houses (e.g. for grain). *exchanges; burses.* Meeting-places where merchants transact business. *cabinets and rarities.* Museums, collections of rare objects. *triumphs.* Triumphant processions. *shows.* Spectacles. *put . . . little room.* Benefit most quickly. *card.* Map, itinerary. *key.* Answer. *adamant.* Magnet. *diet.* Dine. *removes.* Changing places. *recommendation.* Letter of introduction. *of quality.* Of high rank. *abridge.* Shorten, ease. *employed men.* Household staff. *suck.* Absorb. *tell.* Find out. *the life.* Their actual way of living. *agreeth.* Corresponds to. *discretion.* Astuteness. *healths.* Toasts (a common cause of quarrels). *place.* Questions of precedence. *words.* Insults. *engage . . . into.* Involve in. *gesture.* Behaviour. *advised.* Deliberate, judicious. *forward.* Too eager.

376. *country manners.* Those of his own country (like Lat. *patrius*). *prick in.* Transplant. *flowers.* The choicest parts.

19. *Of Empire*

376. *Empire.* Rule, especially by a king or emperor. *of.* With. *want.* Lack. *matter of desire.* Things to desire. *languishing.* Inert, enervated. *representations.* Imaginations, delusions; in the Latin version, 'phantasmata'. In his *History* Bacon noted that Henry VII suffered from such fears: cf. *Works*, vi. 49, 67, 243. *effect.* Fact. '*That . . . inscrutable*'. Prov. 25: 3. *jealousies.* Suspicions. *sound.* Probe, discover the inclinations. *make themselves.* Invent for themselves. *toys.* Trifles. *erecting.* Establishing. *order.* Fraternity, especially religious. The Latin adds: 'or college'. *advancing.* Promotion (as of a court favourite). *feat.* Manual skill. *Nero.* Cf. Suetonius, 'Nero', 20; Dio Cassius, lxiii. 1. *Domitian.* Cf. Suetonius, 'Domitian', 19. *certainty . . . arrow.* Accuracy in archery. *Commodus.* Cf. Aelius Lampidius, *Script. Hist. Aug.* xi. 10 ff.; Herodian, i. 15 ff.; Dio Cassius, lxxii. 10, 22. *Caracalla.* Cf. Dio Cassius, lxxvii. 10, 17. *profiting.* Advancing. *standing at a stay.* Remaining passive, stagnant. *check.* Rebuff. *Alexander.* Cf. Plutarch, 'Of Isis and Osiris', *Moralia*, 466D. *Diocletian.* Roman emperor 284–305, who abdicated because of ill-health rather than melancholy. *Charles the Fifth.* Holy Roman emperor (1519–58) and King of Spain, who abdicated in 1556 in favour of his son, Philip II, and spent the two remaining years of his life in ascetic contemplation. *stop.* Obstacle. *favour.* Self-respect. *temper.* Mixture, temperament: mingling the constituent parts into one balanced whole. *distemper.* Disorder, alteration; interchanging contraries without mingling them. *answer.* In Philostratus, *Life of Apollonius of Tyana*, v. 28. *overthrow.* Ruin. *pins.* Pegs.

377. *unequal.* Unbalanced. *pressed.* Exerted, enforced. *wisdom.* Art (of government). *fine.* Astute. *deliveries.* Means of escape from difficulties. *shiftings.* Subterfuges (to avoid danger). *grounded.* Well-founded. *courses.* Policies of government. *keep them aloof.* Drive them away. *try masteries.* Compete with, as to who shall be master. *suffer.* Allow. *matter.* 'Used like the Latin *materia* in the literal sense of fuel; hence "cause generally"'

(Wright, p. 373). *forbid.* Prevent. *will.* Desire. '*Sunt . . . contrariae*'. In fact Sallust, *Bell. Jug.* cxiii. 1: 'The desires of kings are commonly vehement and contradictory'. *solecism.* Lack of congruence, violating the norms (of syntax; or government). *mean.* Means. *prelates.* Ecclesiastical dignitaries. *second-nobles.* Gentry. *men of war.* Soldiers. *for.* As to. *holdeth.* Is valid. *sentinel.* Lookout. *embracing.* Dominating. *approaches.* Massing forces on the frontier. *standing counsels.* Permanent councils. *During.* Between 1519 and 1556. Bacon reuses here his discussion of the balance of power from 'Considerations Touching a War with Spain' of 1624: cf. *Works*, xiv. 469–505, especially 477, and V. Luciani, 'Bacon and Guicciardini', *PMLA* 62 (1947), 96–113, at 104–7. *palm.* A hand's breadth. *straightways.* Immediately. *if need were.* In case of need. *take up . . . interest.* Accept ('borrow') a present peace at the cost of future loss. *league.* Made in 1480; Guicciardini, *Storia d'Italia*, i. 1. *schoolmen.* Cf. *Works*, xiv. 477–8, quoting Aquinas, *Summa Theol.* II. 40. 1. Reynolds (pp. 139–40) quotes other scholastic views. *received.* Accepted. *precedent.* Previous. *just.* Well-founded. *blow given.* Attack launched. *infamed.* Infamous. Cf. Tacitus, *Annals*, iv. 3. *husband.* Augustus, with poisoned figs: cf. Dio Cassius, lvi. 30.

378. *destruction.* In 1553, to favour her own son. *otherwise.* In other ways. Roxolana supported Bajazet her younger son against his elder brother Selymus. *queen.* Isabella. See e.g. Marlowe's *Edward II.* *hand.* Part. *have.* Lay. *raising.* Advancing. *advoutresses.* Adulteresses. *tragedies.* Unhappy fates. *entering . . . suspicion.* Becoming suspicious. *that.* Which. *line.* Descent. *succession.* Dynasty. *strange.* Impure, 'not of one's kin or family' (*SOED*). *suppositious.* Illegitimate. *Crispus.* Executed in AD 326, at the instigation of his stepmother Fausta, Constantine's second wife. *towardness.* Promise. *house.* Lineage. *Demetrius.* Falsely accused by his brother Perseus, and executed by his father, 179 BC: Livy, xl. 24, insists on his innocence. *Second.* In fact, the Fifth: Livy, xl. 54–6. *had good.* Benefited. *open arms.* Armed rebellion. *Selymus.* Who had his father poisoned by his own physician. *three sons.* Who openly rebelled against their father between 1173 and 1189. *Anselmus.* St Anselm, appointed Archbishop of Canterbury in 1093, and who clashed with two kings. *Becket.* Murdered by four of King Henry II's knights in Canterbury Cathedral in 1170. *crosiers.* Bishop's crook, carried in processions. *try it.* Measure their strength against. *stout.* Bold. *state.* Estate of the realm (here, the clergy). *but.* Except. *dependance . . . authority.* Depends on a foreign power (i.e. the Pope). *come in.* Achieve power. *collation.* The presentation of a clergyman to a benefice. *depress.* Lessen their power; humiliate. *noted.* See *Works*, vi. 242. *came to pass.* Happened. *continued.* Remained. *fain.* Obliged.

379. *second.* Inferior. *dispersed.* Scattered, decentralized. *discourse high.* Talk recklessly. *temper.* Moderate. *commotions.* Disturbances. *vena porta.* The gate-vein, a large vein which was thought to distribute chyle to the liver: similarly, the merchants concentrate the resources of a country in order to redistribute them. *nourish.* Receive nourishment. *imposts.* Import duties. *wins . . . shire.* Gains in a small matter but loses in a larger. *hundred.* A

division of a county in England originally supposed to consist of a hundred families. *rates*. Revenues from tax. *bulk*. Volume. *commons*. Working classes. *heads*. Leaders. *customs*. Taxes, imposts. *state*. Situation. *in a body*. Grouped together. *donatives*. Special additional payments; generous presents. *janizaries*. An élite Turkish corps, formed in 1326. Bacon's source here and elsewhere is Richard Knolles, *The generall historie of the Turks* (1603). *pretorian bands*. Originally an imperial bodyguard, instituted by Augustus. *several*. Separate. *Princes are like . . . but no rest*. Cf. Seneca, *Consolatio ad Polybium*, vii. 3: 'On the day that Caesar dedicated himself to the wide world, he robbed himself of himself; and even as the planets, which, unresting ever pursue their courses, he may never halt or do anything for himself.' '*memento . . . Dei*'. 'Remember that you are a man'; 'Remember that you are a God', or 'God's lieutenant'. *will*. Desire.

20. *Of Counsel*

379. *in other*. Dealing with stewards, tutors, agents, financial advisers. *such as*. Those who. *they*. The counsellors. *obliged*. Bound. *derogation*. Discredit. *sufficiency*. Competence. *without*. Lacking (counsel). *great names*. Major titles. *Counsellor*. Isa. 9: 6. *Salomon*. Prov. 20: 18 (paraphrased). *agitation*. Motion; also debate.

380. *reeling*. Swaying unsteadily from side to side. *Salomon's son*. In 1 Kgs. 12: 1–19 Rehoboam rejects the advice of his fathers' counsellors to be lenient to the Israelites, who rebel at his harshness. *found*. Discovered. *rent*. Torn apart. *ill counsel*. Bad advice. *marks*. Characteristics. *for*. As regards. *matter*. Contents, substance. *in figure*. Emblematically. *intend*. Mean, imply. *Jupiter . . . head*. According to Hesiod (*Theogony*, 886–900) Metis is 'counsel' and Pallas Athena 'wisdom'. Cf. *De Sapientia Veterum*, ch. 30, 'Metis; or Counsel' (*Works*, vi. 761–2). *suffered . . . not*. Did not allow. *stay . . . forth*. Wait until she gave birth. *Pallas armed*. Dressed in armour. *monstrous*. Unnatural. *empire*. Government. *council of state*. The King's main advisory body (the Privy Council in Britain). *elaborate*. Developed. *go through with*. Implement. *direction*. Course of action. *take the matter back*. Recall the issue. *resembled to*. Find their emblem in. *head and device*. Brain and ingenuity. *calling*. Inviting. *less of themselves*. Less capable. *cabinet councils*. Secret councils, held in 'cabinets' (private apartments), often of *ad hoc* advisers, 'recommended chiefly by flattery and affection', as Bacon put it in a MS draft which he did not publish (Kiernan, p. 216). For other contemporary disapproval of this innovation see Melchionda, p. 485. *communicate . . . with*. Impart to, share with.

381. *declare*. Announce. *unsecreting*. Divulging. *plenus . . . sum*. 'I am full of cracks': Terence, *Eunuchus*, i. 2.23–5 (cf. the contemporary term 'to leak' a confidential matter). *futile*. Talkative, indiscreet. *which . . . persons*. Not more than one or two people should know the secret. *constantly*. Steadily. *one spirit of direction*. One consistent path or policy. *distraction*. Disorder. *grind with a hand-mill*. Manage affairs simply, without helpers. *inward*. Intimate. *Henry the Seventh*. King of England 1485–1509. Cf. *Works*, vi. 40.

imparted. Revealed. *Morton and Fox*. Counsellors who became respectively Archbishops of Ely and Exeter. *the fable*. Of Metis and Jupiter, above. Kings should call back the fruits of counsel, which they own as children. *bereaved of his dependances*. Deprived of his dependencies, and so power. *over-greatness*. Excessive power. *over-strict combination*. Rigorous alliance. *holpen*. Remedied. '*non . . . terram*'. 'He will not find faith on the earth': Luke 18: 8. *times*. Epochs. *There be*. There are some people. *involved*. Complicated, devious. *out of faction*. On partisan grounds. *Principis . . . suos*. 'It is a ruler's greatest excellence to know his subjects': Martial, *Epigrams*, viii. 15.8. *side*. Hand. *speculative*. Inquisitive, prying. *skilful*. Expert. *feed his humour*. Indulge his whims. *singular*. Unique. *separately*. In private. *reverent*. Respectful. *in consort*. In company. *obnoxious*. Exposed to, under the influence of; deferential. *both*. (Opinions.)

382. *inferior sort*. Of lower social standing. *matters*. Affairs of state. *life*. Vigour, success. *secundum genera*. Generally, according to types. *idea*. Abstract notion. *most*. Greatest. *optimi . . . mortui*. 'The best counsellors are the dead' (a saying of Alphonso of Aragon, 1416–58). *blanch*. Whiten: either 'blench', or 'blandish, flatter'. *conversant*. Familiar with. *actors upon the stage*. Taken part in public life. *talked on*. Gossiped about. *order or act of counsel*. Decree to be put into action. *matter*. Question. *spoken to*. Debated. *in nocte consilium*. 'Night is the season for counsel': a proverb (found in Erasmus' *Adagia* and elsewhere). *Commission of Union*. In 1604 a committee was formed to discuss the implications of uniting the kingdoms of England and Scotland. Bacon had written a detailed memorandum on the issue in 1603, 'A Brief Discourse touching the happy union of the Kingdoms of England and Scotland' (*Works*, x. 90–9, 241). *grave*. Important. *set*. Fixed, agreed upon beforehand. *estate*. State. *hoc gere*. 'Do this', or 'Mind this', i.e. 'concentrate on the business in hand'. Cf. Plutarch, 'Life' of Coriolanus, 25. *ripening*. Preparing, maturing. *indifferent*. Impartial. *indifferency*. Balance (and possibly stalemate). *standing commissions*. Permanent committees. *suits*. Petitions. *provinces*. Shires. *divers*. Several. *save*. Except. *mintmen*. Makers of coins. *tribunitious*. Violent, turbulent, like the Tribunes in early Rome (cf. Livy, iii. 19, iv. 2; Shakespeare, *Coriolanus*). *clamour*. Disturb. *of form*. Merely formal. *upper end*. With the president at the end of a table, those counsellors near him can 'sway' or control business. *form*. Arrangement. *opens*. Reveals. *take the wind of him*. Gain the advantage. *placebo*. 'I shall please you' (Ps. 116: 4), the Vesper hymn for the dead; here, words of flatterers or timeservers who tell their superior what they know he wants to hear.

21. *Of Delays*

383. On the associative structure of argument in this essay see Vickers, *Francis Bacon and Renaissance Prose* (Cambridge, 1968), 228–31. *stay*. Wait. *Sibylla's offer*. This old woman offered to sell Tarquin at an enormous price nine books containing the oracles of the Gods. When the King refused she burnt three, then another three, but he finally bought the last three at the

original price for all nine. Cf. Aulus Gellius, *Noctes Atticae*, i. 19. *at full*. As a whole. *holdeth up*. Sustains. *the common verse*. Popular saying, found in Erasmus' *Adagia* and elsewhere: *Fronte capillata, post haec occasio calva*, 'Take Time (Occasion) by the forelock, for she is bald behind'. *noddle*. Back of the head. *after*. Afterwards. *beginnings*. Of actions to meet danger. *light*. Trivial. *forced*. Harmed. *nothing*. Not at all. *watch*. Look out. *shone on . . . back*. Illuminated from behind. Cf. Plutarch's 'Life' of Pompeius, 32–6. *buckling*. Putting on armour, preparing to do battle. *ripeness*. Maturity. *Argos*. In classical mythology Argos, who had a hundred eyes, was set to watch over Io. Cf. Aeschylus, *Prometheus Bound*, 567 ff. *Briareus*. A giant with a hundred hands, who helped Jupiter to combat the Titans. Cf. Homer, *Iliad*, i. 403. *Pluto*. Perseus was able to slay the gorgon Medusa by wearing Pluto's helmet, which made him invisible. *politique*. Astute. Cf. Homer, *Iliad*, v. 845.

22. *Of Cunning*

383. *take*. Mean, understand. *sinister*. Dishonest, corrupt. *crooked*. Fraudulent. *pack the cards*. Shuffle the cards deceitfully. *canvasses*. Intrigues. *perfect*. Well versed. *humours*. Whims. *capable*. Having capacity for. *constitution*. Disposition.

384. *practice*. Trickery, scheming. *their own alley*. On their own ground, in familiar surroundings (a metaphor from bowls). '*Mitte . . . videbis*'. 'Send them both naked to those they know not, and you will see [who they are]': a saying of Aristippus recorded by Diogenes Laertius, ii. 73. *scarce hold*. Is hardly valid. *haberdashers*. Dealers in small goods. *set . . . shop*. Display their tricks. *wait upon*. Observe closely. Cf. *Adv.L.*, above, p. 266. *Jesuits*. One of their rules for modesty was not to look directly at the person you talk to, but with a lowered gaze. *would*. Ought to. *demure*. Modest. *abasing*. Lowering. *use*. Practise. *present despatch*. Immediate urgency. *awake*. Alert. *a counsellor*. Probably Sir Francis Walsingham (?1530–90), appointed to a leading post on the Privy Council in 1573, its sole secretary from 1577 on, famous for his cunning. *bills*. Petitions. *put . . . into*. Lead, direct. *estate*. State. *moving*. Proposing, supporting. *that*. That which. *cross*. Thwart. *doubts*. Fears, suspects. *other*. Someone else. *foil*. Bring to naught. *took himself up*. Corrected himself (the rhetorical figure *aposiopesis*, breaking off an utterance uncompleted). *works*. Gives results. *question*. Request of. *yourself*. Spontaneously. *another*. Different. '*And . . . king*'. Neh. 2: 1. *tender*. Delicate, needing tact. *upon*. Following on. *Narcissus*. Tacitus, *Annals*, xi. 29–30.

385. *material*. Important. *bye-matter*. Subordinate point, afterthought. *have speech*. Hold a conversation. *like*. Likely that. *work upon*. Manipulate. *come upon*. Encounter. *to the end*. In order that. *apposed of*. Questioned about. *two . . . secretary's place*. Probably Sir Robert Cecil and Sir Thomas Bodley in 1596; Cecil got the post. *good quarter*. Kept their proper place; were on good terms. *declination*. Decline. *ticklish*. Tricky. *affect*. Like. *straight*. Immediately. '*The turning . . . pan*'. Saying of obscure origin,

meaning 'to turn a situation round to one's own advantage'. *lays.* Imputes, reports. *passed.* Was exchanged. *glance and dart.* Allude, as if reluctantly. *Tigellinus.* A depraved crony of the emperor Nero, commenting on the death in AD 62 of Nero's sometime tutor and adviser, Sextus Aranius Burrus. '*Se... spectare*'. 'He had no divergent aims; but his one object was the safety of the Emperor' (Tacitus, *Annals*, xiv. 57). *keep... in guard.* To better protect themselves. *carry it.* Accept the suggestion. *would have.* Would like to hear. *stick.* Hesitate.

386. *in wait.* In ambush. *far about... fetch.* What a roundabout route they take. *beat over.* Refer to. *lay him open.* Reveal himself. *in Paul's.* St Paul's Cathedral, London, a popular meeting-place. Cf. Jonson, *Every Man out of his Humour* (1599), iii. 1, set in 'The middle aisle of St Paul's'. *straightways.* Immediately. *petty points.* Trivial items. *for that.* Because. *resorts and falls.* Sources (springs) and outcomes; or, beginnings and endings. *sink... main.* Penetrate to the important part. *fair.* Large, spacious. *find out pretty looses.* Make an impressive display (from 'looses', discharging arrows). *wits of direction.* Intellects fitted to manage affairs. *abusing... putting tricks upon.* Deceiving. '*Prudens... dolos*'. 'The wise man giveth heed to his own steps; the fool turneth aside to deceits' (Prov. 14: 8, 15).

23. *Of Wisdom for a Man's Self*

386. *Wisdom... Self.* Selfishness, egoism. *wise.* The epithet traditionally applied to the ant in Prov. 6: 6, 30: 24–5, here used ironically, as Melchionda observes (p. 496). Cf. *Adv.L.*, above, p. 271. *shrewd.* Pernicious, damaging. *waste.* Lay waste, destroy. *divide.* Distinguish. *self-love.* Selfishness. *society.* The claims of other people. The Latin version is 'amorem reipublicae'. *true.* Loyal (holding a straight course). *right earth.* Truly base (earth was the lowest of the four elements in Aristotelian philosophy). *stands fast.* Depends on, has reference to. *his.* Its. *at the peril.* Subject to risks from. *affairs.* In the Latin translation, 'negotia publica'. *crooketh.* Bends. *eccentric.* Removed, distant. *mark.* Characteristic. *mean.* Intend, wish. *the accessary.* Subordinate.

387. *proportion.* Harmony, correspondence (continuing the cosmological analogy). *preferred before.* Promoted above, made superior to. *carry things.* Prevail. *set a bias... bowl.* Cause the bowl to swerve from its straight course: i.e. manipulate affairs to their own advantage. *envies.* Desires. *overthrow.* Ruin. *model.* Pattern (that is, for self-profit). *sell.* Trading his master's disadvantage for his own profit. *as.* That. *and.* If. *hold credit.* Retain esteem. *respect.* Consideration. *rats.* Cf. Pliny, *Nat. Hist.* viii. 28. *digged.* Dug. *crocodiles... tears.* According to the old belief that crocodiles weep in order to disarm their prey. *would.* Wish to. '*sui... rivali*'. 'Lovers of themselves, without rivals' (*Ad Quintum fratrem*, iii. 8). *pinioned.* Clipped.

24. *Of Innovations*

387. *Innovations.* In Renaissance political theory, which was often conservative, changes in the structure or government of a state were usually regarded with fear or suspicion. Bacon adopts a carefully balanced position. *births.* Offspring. *worthy.* Courageous, enterprising. *succeed.* Come after. *attained.* Reached, equalled in excellence. *Ill.* Evil. *as... perverted.* Corrupted as it is by the sin of Adam and Eve. *natural motion.* Like the fall of a heavy body, continually accelerating. *in continuance.* As time passes. *forced motion.* Like the flight of an arrow, continually less rapid. *of course.* In its course, progress.

388. *good.* Valuable. *fit.* Suitable, opportune. *confederate.* Unified. *piece.* Agree, match together. Wright (p. 319) compares Matt. 9: 16, 'No man putteth a piece of new cloth unto an old garment, for... the rent is made worse'. *utility.* Usefulness. *trouble.* Cause trouble. *inconformity.* Want of agreement. *strangers.* Foreigners. *admired.* Wondered at. *round.* Swiftly, uninterruptedly. *froward.* Self-willed, stubborn. *turbulent.* Unruly. *scorn.* Matter for scorn. *otherwise.* Anyway. *ever.* Always. *mends.* Improves the conditions of. *pairs.* Impairs, hurts. *other.* Others. *holpen.* Helped, benefited. *author.* Initiator. *beware.* Take note. *draweth on.* Brings about. *pretendeth.* Puts forward as a pretext. *suspect.* A person or thing suspected (liable to be dangerous). *Scripture.* Jer. 6: 16 (paraphrased).

25. *Of Dispatch*

388. *Dispatch.* Speed in settling business. *Affected dispatch.* Ostentatious haste. *predigestion.* Premature digestion. *crudities.* Undigested matter in the stomach. *times of sitting.* Length or frequency of meetings. *high lift.* The step of a horse. *taking of it.* Dealing with. *speedily.* Successfully. *for the time.* At this point; or, 'in proportion to the time taken'. *false periods.* Divisions which profess to include all the relevant topics, but do not. *because.* In order that. *contracting.* Reducing to essentials. *a wise man.* Sir Amias Paulet, with whom Bacon went to France in 1576: cf. *Works*, vii. 136 for this apophthegm. *by-word.* Proverbial saying. *Stay.* Wait. *that.* So that.

389. *rich.* Precious. *dear hand.* High price. *Spartans... Spaniards.* Both proverbial for prudence and caution. For the Spartans, cf. Thucydides, i. 70, 84. Bacon quotes the proverb elsewhere, more appropriately, in Italian (*Works*, x. 351). *direct.* Instruct. *moderator.* Chairman. *actor.* Orator. *Iterations.* Repetitions. *curious.* Elaborate. *train.* The part of a cloak or dress that trails on the ground. *passages.* Digressions. *excusations.* Rhetorical apologies. *person.* Speaker. *of.* From. *bravery.* Ostentation. *too material.* Coming to the point too abruptly. *pre-occupation of mind.* Being absorbed with problems. *fomentation.* Warm moist cloth used to open skin pores. *unguent.* Oily salve. *distribution.* Classification. *subtle.* Minute. *divide.* Distinguish (in rhetoric the process of *partitio*, dividing up the main points to be considered). *perfection.* Completion. *middle.* The 'debate or examination'. *proceeding.* Basing a discussion. *conceived.* Formulated. *pregnant of direction.* Suggestive, productive. *ashes... dust.* Cf. Pliny, *Nat. Hist.* xvii. 5, and Bacon, *Works*, ii. 525, 546.

26. *Of Seeming Wise*

390. *Having ... thereof.* Paul, in 2 Tim. 3: 5. *power.* Essence. *in point of.* As regards. *sufficiency.* Ability. *solemnly.* Superciliously. *magno ... nugas.* 'Trifles with great effort': Terence, *Heautontimoroumenos* (*The Self-Tormentor*), iv. 1.8 (line 621). *shifts.* Evasive tricks. *formalists.* Impostors, 'solemn pretenders to wisdom' (*SOED*). *prospectives.* Optical glasses which make surfaces appear three-dimensional. *superficies.* Surface. *close and reserved.* Secret. *dark light.* Furtively. *somewhat.* Something. *signs.* Grimaces. '*Respondes ... placere*'. 'You answer, with one eyebrow hoisted to your forehead and the other bent down to your chin, that you do not approve of cruelty' (*In Pisonem*, vi. 14). *bear it.* Succeed, carry it off. *great word.* Portentous utterance. *take by admittance ... good.* Accept as valid something that they cannot prove. *impertinent.* Irrelevant. *curious.* Insignificant. *difference.* Distinction. *subtilty.* Trivial point. *blanch.* Gloss over, evade. '*Hominem ... pondera*'. 'A madman who breaks up weighty matter with verbal niceties'. Bacon misattributes the quotation: it is Quintilian's judgement on Seneca's style (*Inst. Or.* x. i). *Plato. Protagoras*, 337A–C. *in scorn.* Scornfully. *find ease.* Prefer. *to be.* In being. *affect ... object.* Try to gain credit by objecting. *propositions.* Proposals. *allowed.* Approved. *bane.* Ruin, destruction. *inward.* Beneath the surface (his poverty hidden). *make shift.* Manage. *get opinion.* Gain reputation. *were better.* Had better. *absurd.* Deaf to reason, difficult.

27. *Of Friendship*

390. Bacon replaced the 1612 essay on friendship (above, p. 301) with this completely new text, at the urging of his friend Sir Toby Matthew. As Bacon wrote to him in Mar. 1622: 'It is not for nothing that I have deferred my essay *De Amicitia*, whereby it hath expected the proof of your great friendship towards me' (*Works*, xiv. 344); and in June 1623: 'For the essay of friendship, while I took your speech of it for a cursory request I took my promise for a compliment. But since you call for it I shall perform it' (xiv. 429). See also Kiernan, p. lxxix n. 82. Bacon's classical sources probably included Aristotle, *Eth. Nic.* viii and ix, *Eudemian Ethics*, vii, *Magna Moralia*, ii. 11–17, *Rhetoric*, ii. 4, and Cicero, *De amicitia*. *had been.* Would have been.

391. '*Whosoever ... god*'. Aristotle, *Politics*, i. 2, 1253a3 ff., 25 ff. But Bacon's interpretation is unfair. *natural.* Innate. *aversation.* Aversion. *character.* Distinctive trait. *sequester.* Withdraw. *conversation.* Company, way of life. *feignedly.* Pretended. Bacon's strictures only apply to Numa, in fact. *Epimenides.* Teacher and miracle-worker of Crete (*c.*600–500 BC), supposed to have acquired prophetic powers after having fallen asleep in a cave for 57 years. *Numa.* King of the Romans (*c.*715–673 BC), who regularly withdrew from society for mystic meditation with the goddess Egeria. But, according to Plutarch, he feigned this 'conference with the gods' to make his countrymen accept religion (Kiernan, p. 227). *Empedocles.* Philosopher and poet (*c.*493–*c.*433 BC), who claimed divine powers. *Apollonius of Tyana.* A wandering ascetic in the early Christian era, whose miraculous powers were celebrated in Philostratus' *Life* of him. *gallery*. Collection, display. *tinkling cymbal*. 1 Cor. 13: 1. *meeteth with*. Fits, applies. *Magna . . . solitudo*. 'Great city, great solitude': Erasmus' *Adagia*, from Strabo (who applied it to Babylon); *Promus*, no. 268. *less*. Smaller. *mere*. Absolute. *want*. Lack. *frame*. Constitution. *taketh it*. Derives it from. *humanity*. Human nature. *ease*. Relieving. *fullness . . . heart*. Tension, emotional upset (in Renaissance physiology usually located in the heart). *diseases of stoppings*. Physical obstructions. *sarza*. Sarsaparilla, a medicinal plant from tropical America used to regulate the digestion. *open*. Purge, relieve. *steel*. One of the new chemical treatments advocated by Paracelsus. *spleen*. Milt. *flower of sulphur*. 'Amorphous sulphur in a fine powder' (*SOED*). *castoreum*. Castor oil. *receipt*. Medical prescription. *civil*. Non-clerical. *shrift*. Penance for guilt. *high a rate*. Great value. *as*. That. *in regard of*. Owing to. *sorteth to*. Results in. *privadoes*. Spanish word for intimates or favourites. *grace*. Favour. *conversation*. Familiarity.

392. *attaineth*. Arrives at. *participes curarum*. 'Partners in care': Dio Cassius, lviii. 4.3. But, as Reynolds (p. 193) observes, there is 'no authority for Bacon's statement that this is "the Roman name". He seems to have been misled by his double habit of reading Greek authors in a Latin version, and of quoting from memory afterwards.' Here Bacon used Xylander's Latin version of a passage describing the titles which Tiberius conferred on Sejanus, as in fact part of a plan to overthrow a feared rival. This forced application shows how phrases entered in, and quoted from, a commonplace book can become false to their original context. *tieth the knot*. Unites people. *passionate*. Emotional. *received*. Common, current. *that*. Such a. *carried*. Gained. *pursuit*. Competition. *speak great*. Become excited. '*for . . . setting*'. Plutarch's 'Life' of Pompey, 14 f. *Julius Caesar*. See Plutarch's 'Life' of Julius Caesar, 64. *interest*. Ascendancy. *heir . . . nephew*. Heir in succession after Augustus. *power with*. Influence over. *discharged*. Dismissed. *favour*. Standing, popularity. *as*. That. *Cicero*. *Philippics*, xiii. 11. *Maecenas*. See Dio Cassius, liv. 6.5. *friends*. Cf. Tacitus, *Annals*, i. 3; Dio Cassius, lviii. 4 ff., 14; lxxv. 15.2. '*haec . . . occultavi*'. 'These things, for the sake of our friendship, I have not hidden from you': Tacitus, *Annals*, iv. 40, and 74 for the altar to Friendship. *Septimius Severus*. See Dio Cassius, lxxvi. 2.7; Herodian, iii. 12.3 f. *maintain*. Support, uphold. *doing affronts*. Insulting. *words*. Dio Cassius, lxxv. 15.2. *over-live*. Survive. *Trajan*. Marcus Ulpius Trajanus, emperor AD 98–117, a great benefactor of learning. *Marcus Aurelius*. Emperor AD 169–80, whose *Meditations* was one of the most admired works of moral philosophy by a pagan.

393. *wise*. Shrewd. *so*. Such. *an half piece*. Either a work of art left incomplete; or (Reynolds) 'the old practice of cutting silver pennies into halves to make up for the deficiency of smaller coins'. *nephews*. The word also meant 'grandsons'. *Comineus*. Philippe de Commines (*c*.1446–*c*.1511), French diplomat and historian, in his *Mémoires*, v. 3, 5. *communicate*. Share. *closeness*. Secrecy. *perish*. Corrupt, destroy. *parable*. Saying. *Pythagoras*.

According to Diogenes Laertius, viii. 18, and Plutarch, 'On the Education of Children', 12 (*Moralia*, 17E): 'Eat not thy heart; that is to say, offend not thine own soul, nor hurt and consume it with pensive cares'; *Promus*, no. 817. *dark*. Enigmatic. *a hard phrase*. Put it bluntly. *open*. Reveal, disclose. *admirable*. Wonderful. *works*. Produces. *imparteth*. Communicates. *joyeth*. Rejoices. *truth of operation*. Efficacy. *virtue*. Effectiveness. *use*. Are accustomed. *stone*. The alchemists' 'Philosopher's stone', supposedly able to cure all diseases. *praying in aid of*. Calling in, as an advocate. *image*. Figure. *bodies*. Inanimate bodies. *union*. Composition (as in an alloy). *cherisheth*. Favours. *impression*. Impact, force. *sovereign*. Remedial. *affections*. Passions. *fair day*. Good weather. *from*. Out of. *understanding*. Intellect. *fraught*. Occupied, worried. *clarify*. Become clear. *break up*. Open out. *tosseth*. Moves. *marshalleth*. Sorts.

394. *waxeth*. Becomes. *discourse*. Conversation. *meditation*. Continuous thought. For the educative effect of conversation with experts cf. *Advice to Rutland*, above, p. 74. *That speech . . . in packs*. Plutarch's 'Life' of Themistocles, 29; recorded more accurately (as juxtaposing not speech and thought, but the perfect and imperfect expression of thought by language) in Bacon's *Apophthegms* (*Works*, vii. 153). *cloth of Arras*. Embroidered wall-hanging tapestry made in Arras, France. *put abroad*. Spread out. *imagery*. Pattern, picture. *in figure*. In full outline; or, 'according to the design' (Melchionda). *packs*. Heaped up. *restrained*. Limited. *whetteth*. Sharpens (cf. Horace, *Ars poetica*, 304–5). *were better*. Would be better. *relate himself*. Explain, talk to. *in smother*. Concealed, undisclosed. *lieth more open*. Is more manifest. *vulgar*. Common, widespread. *Heraclitus*. Presocratic philosopher (*fl. c.*500 BC); the saying is recorded in Plutarch's 'Life' of Romulus. Cf. Bacon's application of this 'enigma' (obscure saying) in *Adv.L.*, above, p. 125 and note. *infused and drenched*. Soaked, submerged or confused by the passions and habit. *friend . . . flatterer*. Bacon is alluding to the essay in Plutarch's *Moralia* (48E ff.), 'How a man may discern a flatterer from a friend'. *manners*. Morals. *business*. Public, active life. *preservative*. Medicament. *admonition*. Severe counsel. *calling . . . account*. Self-criticism, stringently applied. *flat*. Insipid. *improper*. Inappropriate. *take*. Have a lasting effect. *greater sort*. People of high status. *them*. Their 'errors and absurdities'. *saith*. Jas. 1: 23–4 (paraphrased). *presently*. Immediately. *favour*. Feature. *gamester*. A player at any game. *said over*. Repeated (until his anger calmed down).

395. *the four and twenty letters*. Constituting the alphabet (*j* and *u* were not generally included until after 1630). *rest*. Support for a fire-arm. *fond and high*. Foolish, exaggerated. *all in all*. All things in all respects. *but . . . . by pieces*. From different people. *runneth*. Risks. *bowed and crooked*. Bent, perverted. *unsafe*. Uncertain. *meaning*. Intention. *mischief*. Harmful. *put you in way for*. Give you good chance of. *present*. Immediate. *kind*. Respect. *estate*. Condition. *dasheth upon*. Meets with. *scattered*. Separate, uncoordinated. *kernels*. Seeds. *bearing a part in*. Sharing. *cast*. Reckon, count up. *sparing speech*. Understatement. *say*. As did Pythagoras, according to

Porphyry's *Life*, 33; cf. also Aristotle, *Eth. Nic.* ix. 4, 1166a31, *Magna Moralia*, ii. 15, 1213a12f., and *Eudemian Ethics*, vii. 12, 1245a30; Diogenes Laertius, vii. 23; and Cicero, *De amicitia*, xxi. 80. *their time*. Their appointed time. *in desire*. Still lacking. *take to heart*. Care about. *bestowing*. Settling a patrimony on, placing in life. *in his desires*. So far as his desires are concerned. *offices*. Duties. *exercise*. Carry out. *How many... himself?* Cf. Cicero, *De amicitia*, xvi. 57. *face*. Modesty. *comeliness*. Decorum. *brook*. Bear. *supplicate*. Beg.

396. *and... like*. Et cetera. *are blushing*. Cause embarrassment. *person*. Role (persona) that he has to play in society. *proper*. Belonging to it. *upon terms*. Formally, on agreed conditions. *sorteth with*. Suits. *fitly play*. Perform.

28. *Of Expense*

396. *Expense*. This topic was of acute interest to land-owners in this period, especially the aristocracy, suffering under a huge inflation rate: 'the cost of living rose by about 80 per cent between 1550 and 1600', hence 'competent estate management became increasingly important as the pressure of inflation built up. Careful scrutiny and strict control of leases... could make crucial differences to the fortunes of noble families' (D. M. Loades, *Politics and the Nation 1450–1600* (London, 1974), 344, 340). *for honour*. According to one's social status. *limited*. Measured, appropriate. *voluntary undoing*. Freely giving away one's substance. *kingdom of heaven*. Poverty was valued in the ascetic life, as being more likely to earn a place in heaven. Cf. Essay 13, 'Of Goodness and Goodness of Nature', above, p. 364, which also juxtaposes worldly and other-worldly attitudes to money. *compass*. Limits, control. *abuse*. Larceny. *ordered... shew*. Make the best impression. *that*. So that. *less than... estimation abroad*. Smaller than generally imagined. *of even hand*. Equally balanced. *receipts*. Income. *wax*. Grow. *baseness*. Degradation. *doubting*. Fearing. *in respect*. In case. *broken*. Ruined, bankrupt. *searching*. Examination. *had need*. Had better. *new*. New employees. *subtle*. Deceitful, cunning. *turn... certainties*. Establish a scale of fixed expenses (so denying his employees any chance of cheating). *plentiful*. Lavish. *some*. One. *as... again*. Correspondingly frugal. *diet*. Food and drink. *hall*. Hospitality. *decay*. Undoing. *clearing*. Settling the accounts, outstanding debts. *a man's*. One's. *as well*. Equally. *hurt*. Damage. *sudden*. Precipitous. *interest*. Paying interest on a loan. *clears*. Settles. *at once*. At one go. *straits*. Difficulties. *customs*. Habits. *induceth... frugality*. Gets used to being frugal. *gaineth... upon*. Gets control over. *who*. He that. *state*. Estate.

397. *abridge*. Reduce. *petty*. Trivial. *charges*. Expenses. *return not*. Are 'extraordinary', as discussed at the beginning.

29. *Of the True Greatness of Kingdoms and Estates*

397. Bacon first set down his ideas on this topic in a speech delivered to the House of Commons in Feb. 1607, and further developed in a treatise 'Of the

True Greatness of the Kingdom of Britain' (*Works*, vii. 39–40, 47–64), planned and written in 1608 (*Works*, xi. 73–4) but not published until 1734. An earlier (less warlike) version was included in the 1612 *Essays*, above, pp. 301–3. The revised and expanded English text first appeared in a Latin translation in the *De Augmentis* (viii. 3), with the title 'De proferendis Imperii finibus', 'Of extending the bounds of Empire' (*Works*, i. 793–802; Engl. tr. v. 79–88). The 'true greatness' in the English title, as the second paragraph makes clear, means 'expansive power', or 'the power and forces' of a state to acquire and hold new territory by warfare, a dynamic activity not employed (or needed?) by large but static states. On other occasions (e.g. *Works*, xiii. 20) Bacon recommended the pursuit of peace. For a useful study of his thinking on these issues see Markku Peltonen, 'Politics and Science: Francis Bacon and the True Greatness of States', *Historical Journal*, 35 (1992), 279–305. *Kingdoms and Estates*. As Melchionda shows (p. 518) this paired phrase, which occurs 19 times in the *Essays*, connotes 'Monarchies and Republics', corresponding to the similar terms 'principati e repubbliche', frequently used by Machiavelli. *speech*. Saying, recorded by Plutarch, 'Life' of Themistocles, ii. 4. *had*. Would have. *censure*. Judgement, opinion. *at large*. Generally. *fiddle*. Meaning: (*a*) 'play the instrument', (*b*) cheat. *holpen*. Helped. *metaphor*. By being transferred to politics (the Latin term for metaphor is *translatio*). *business of estate*. Affairs of state. *gift*. Talent. *arts and shifts*. Skills and tricks. *graceful*. Attractive. *weal*. Welfare. *sufficient*. Competent. *negotiis pares*. Equal to the business, just. Cf. Tacitus, *Annals*, vi. 39, xvi. 18. *manage*. Rein in, control (a metaphor from horsemanship). *precipices*. Crises. *inconveniences*. Injuries. *which*. Who. *means*. Resources. *Greatness*. Magnitude; grandeur. *means*. Ways of achieving. *argument*. Topic, subject for consideration. *over-measuring*. Overestimating. *leese*. Lose. *fearful*. Timid. *counsels*. Resolutions. *fall . . . measure*. Can be measured. *appear*. Be displayed, computed. *musters*. Registers. *cards*. Plans, maps.

398. *amongst*. In. *compared*. In Matt. 13: 31; Mark 4: 30–2. *least*. Smallest. *get up*. Sprout, grow. *apt*. Adapted, suitable. *stem*. Stalk, bearing leaves; a race or family. *stored*. Fully stocked. *armories*. Deposits of weapons. *races*. Breeds; stables. *ordnance*. Military stores or supplies. *artillery*. 'Originally any engines of war were called artillery, and the term was retained after the invention of gunpowder' (Wright, p. 356). *sheep . . . skin*. Cf. Matt. 7: 15. *stout*. Bold, courageous. *importeth*. Signifies. *Virgil*. *Eclogues*, vii. 51–2. *wished*. Entreated. *set upon*. Attack. *answered*. Plutarch, 'Life' of Alexander, 31.5. *said*. Plutarch, 'Life' of Lucullus, 27.4. *ambassage*. Embassy. *give . . . chase*. Put him to flight. *odds*. Difference, disproportion. *greatness . . . military men*. Cf. Machiavelli, *Discourses*, ii. 18 and *The Prince*, ch. 10. *trivially said*. A classical commonplace (found in Diogenes Laertius, Cicero, Plutarch, Tacitus), which Bacon probably took from Machiavelli, *Discourses*, ii. 10: 'Money is not the sinews of war, although it is generally so considered.' *Solon*. This saying, found in the same passage in Machiavelli, derives from Lucian's dialogue 'Charon'. *soberly*. Cautiously. *militia*. Originally 'citizen army' (not professional). *otherwise . . .*

themselves. Weak for other reasons. *help.* Remedy. *rest.* Rely. On the dangers of relying on mercenaries see *Discourses*, i. 43, ii. 20, and *The Prince*, chs. 12, 13. *spread.* Display. *mew.* Moult, shed. *blessing.* 'Grace given by God': see Gen. 49: 9, 14.

399. *between burthens.* Loaded down. *overlaid.* Burdened. *estate.* State. *abate.* Beat down, depress. *notably.* Remarkably. *excises.* Taxes on home commodities, especially food, drink, and clothing, which were as high as 50 per cent in the Low Countries at this time. *subsidies.* Contributions to the royal budget voted by Parliament. *tribute.* Levy. *all one.* All the same. *diversely.* Differently. *take heed how.* Beware lest. *peasant.* Serf. *swain.* Peasant. *out of heart.* Disheartened, demoralized. *coppice.* Woods of small growth for periodic cutting. *staddles.* Young trees left standing in a plantation after the removal of the underwood. *base.* Degraded (morally). *hundred.* Hundredth. *poll.* Head. *helmet.* Only one man in a hundred fit to be a soldier. *infantry . . . an army.* Cf. Machiavelli, *Discourses*, ii. 18. *an overmatch.* Stronger. *in regard.* Because. *middle people.* Yeomanry (see below). *good soldiers . . . do not.* V. Luciani ('Bacon and Machiavelli', *Italica*, 24 (1947), 26–40, at 30) cites Machiavelli's *Ritratto di cose di Francia* for this judgement. *device.* Plan. *spoken.* See *Works*, vi. 93–5. *husbandry.* Agriculture. *of a standard.* Up to a reasonable standard, proportional to their undertakings. *convenient.* Suitable. *hirelings.* Hired labourers. Bacon is referring to current economic conditions: 'while the cost of living rose by about 80 per cent between 1550 and 1600, wages had risen by only 50 per cent. At the same time the expanding population, and the necessity for landlords to exploit their estates more efficiently, were multiplying the numbers of the rural poor. . . . Outright depopulation was not common, but the numbers of cottages and wage labourers increased; and they tended to become poorer as those with secure tenures or long leases flourished' (Loades, *Politics and the Nation*, 344–5). *character.* Characterization. *Terra . . . glebae.* 'A land powerful in arms and richness of soil' (*Aeneid*, i. 531). *state.* Social class. *free servants.* As Melchionda makes clear (p. 524), these are paid servants wearing livery supplied by their masters, a practice condemned in *Henry VII* (*Works*, vi. 80, 85–6, 219–20, 224). *no ways.* In no way. *yeomanry.* Farmers owning their land, ranking one grade below gentlemen, and who often served in the army as foot-soldiers. *hospitality.* Liberality. *received . . . custom.* An accepted practice. *conduce.* Contribute.

400. *close and reserved.* Thrifty and private. *penury.* Want, lack. *tree.* Dan. 4: 10–26. *great enough . . . boughs.* Cf. Machiavelli, *Discourses*, ii. 3, and *Works*, x. 96. *natural.* Indigenous, native. *stranger.* Foreign, naturalized. *liberal of.* Easily concede. *fit for empire.* Able to expand. *it.* The dominion. *nice.* Fastidious, particular. Cf. Machiavelli, *Discourses*, i. 6. *in point of.* As regards. *compass.* Limits, bounds. *windfall.* Anything blown down by the wind. *upon the sudden.* Rapidly. *sorted with.* Resulted. *monarchy.* In the sense of 'absolute dominion' (Melchionda). *jus civitatis.* Right of citizenship. *jus commercii . . . honorum.* The all-important legal rights of commerce, marriage, receiving property by will, voting, holding office. *singular.* Single.

colonies. Roman colonies were set up by the state, not by private individuals, and modelled on the government of Rome. *removed*. Transplanted. *constitutions*. Institutions (naturalization and colonization). *marvelled*. Wondered. *contain*. Hold in, control. *natural*. Native. *body*. Trunk. *at the first*. At their beginnings. *militia*. Armed force. *highest*. Many Spanish generals were foreigners. *sensible*. Aware. *Pragmatical Sanction*. A royal proclamation, issued by Philip IV in 1622, offering tax incentives to those who married and bore many children, especially male. *within-door arts*. Crafts practised in a building or factory. *delicate manufactures*. Skilled occupations. *require*. Need the use of. *travail*. Labour, toil. *broken of*. Worn out by. *shall*. Are to.

401. *rid*. Perform; or, 'dispose of'. *manufactures*. Artisanal activities. *abolished*. Slavery was abolished in Europe in the late medieval period, but persisted in the Americas. *received*. Accepted. *contain*. Restrain, direct. *vulgar natives*. Common citizens. *tillers*. Cultivators. *handicraftsmen*. Artisans. *professed*. Having as a profession. *importeth*. Is important. *study*. Desire. *habilitations*. Trainings, means of attaining ability. *intention and act*. Endeavour, putting into practice. *Romulus*. See Plutarch, 'Life' of Romulus, 28; Livy, i. 16.6–8. *sent a present*. Bequeathed (the advice to become soldiers). *intend*. Aim at, dedicate themselves to. *fabric*. Structure. *framed*. Contrived. *scope*. Aim, object. *flash*. Brief moment. *declination*. Decline. *that*. That which. *stood upon*. Insisted on. *point at*. Indicate. *directly*. Straightforwardly, absolutely. *look to*. Expect. *oracle of time*. Prediction proved true by history. *notwithstanding*. Nevertheless. *Incident to this point*. Relevant in this respect. *reach forth*. Provide. *pretended*. Put forward as pretexts. *justice*. Principle of justice. *enter . . . upon*. Engage in. *at the least*. At least. *specious*. Attractive, plausible. *quarrels*. Occasions of complaint. *sect*. Religion (Muhammadanism). *command*. Have at his disposal. *pretend to*. Aspire to. *sensible of*. Sensitive to.

402. *borderers*. Who live in one country and work in another. *politique ministers*. Official representatives of the state. *sit*. Delay, remain passive. *prest*. Prepared. *offered*. Threatened. *severally*. Separately. *party*. Faction. *tacit conformity*. As if the foreign country had the same political system, which would justify interfering in its internal affairs. *Romans*. The 2nd Macedonian war, 200–197 BC. Cf. Livy, xxxiii. 32–3. *Athenians*. The Peloponnesian wars. Cf. Thucydides, i. 19. *awake*. Alert. *slothful peace*. The idea that peace is dangerous to a country, bringing idleness and diseases, goes back to the ancient Romans, especially Cato and Livy. See e.g. Brian Vickers, 'Leisure and Idleness in the Renaissance: The Ambivalence of *otium*', *Renaissance Studies*, 4 (1990), 1–37, 107–54. *courages*. Spirits, resolve. *effeminate*. Grow soft. *it maketh*. It is advantageous. *still for the most part*. Almost always. *veteran*. Skilled, professional. *chargeable*. Expensive. *always on foot*. On permanent service. *the law*. The power of arbitrating, the supremacy. *in*. In the case of. *six score*. 120. *abridgment*. Epitome, a short cut to dominion. *Pompey his*. The older genitive form: Pompey's. '*Consilium . . . potiri*'. 'Pompey's plan is quite Themistoclean; for he thinks that the mastery of the sea means the mastery of the war' (*Epist. ad Att.* x. 8). *had.* Would have. *confidence.* Presumption. *left.* Given up. *way.* Course. *Actium.* Where the fleet of Augustus defeated Antony's, Sept. 31 BC. *Lepanto.* Where the fleet of the Holy League, led by Don John of Austria, crushed the Turkish fleet in 1571. *final.* Decisive. *set... rest.* Risked everything (a metaphor from gambling). *take... little.* Can give or refuse battle when it suits him. *straits.* Difficulties.

403. *with... Europe.* For us Europeans. *vantage.* Advantage. *merely.* Entirely. *girt.* Surrounded. *compass.* Boundaries. *accessary.* A consequence of, dependent. *reflected.* Was reflected. *degrees.* Honorary titles. *promiscuously.* Indifferently. *no soldiers.* Such as knighthood, originally conferred for bravery on the battlefield, but subsequently a civic honour which could also be purchased by a payment into the royal coffers. (When Bacon received his knighthood from James I in 1603, he was one of 300 ennobled that day.) *scutcheon.* Coat of arms or shield, which sometimes recorded military honours. *hospitals.* The earliest hospitals in Europe, founded in the Middle Ages, were for wounded soldiers. *trophies.* Memorials to honour a victory, originally arms or other spoils set up on the battlefield. *laudatives.* Eulogies. *garlands personal.* Crowns granted to individuals: the Romans awarded them to soldiers who had saved the lives of their comrades, or first mounted the wall of a besieged town (cf. Shakespeare, *Coriolanus*, I. ix. 57 ff.). *style of Emperor.* The 'title' of *imperator*, with which Roman soldiers saluted their generals after a victory. *after.* Afterwards. *donatives.* Rewards made by the emperors to victorious soldiers. *largesses.* Generous presents. *disbanding.* Dismissing. *able.* Sufficient. *Triumph.* A triumphal procession awarded for outstanding victories: an institution that attracted much interest in the Renaissance with the rediscovery of Roman customs, as in the *Trionfi* of Petrarch and the 'Triumph of Caesar' by Mantegna, now in Hampton Court, London. *gaudery.* Finery, cheap display. *were.* Would be. *impropriate.* Appropriate. *achieve.* Carry to successful completion. *ensigns.* Badges or flags, insignia. *Scripture.* Matt. 6: 27; Luke 12: 25. *model.* Plan or frame on a small scale: the microcosm. *greatness.* Greater extent, area. *touched.* Briefly discussed. *to.* For the advantage of. *observed.* As they ought to be, according to Bacon's design for the establishment of a science of politics.

30. *Of Regiment of Health*

403. *Regiment.* Control, regimen. *this.* That is, the following observation. *physic.* Medicine. *observation... hurt of.* Cf. Plutarch, 'Advice about keeping well', *Moralia*, 136E f.

404. *owing.* Will be paid for. *still.* Always. *safer... one.* Cf. Machiavelli, *Discourses*, i. 26. *so, as.* In such a way that. *by.* In. *particularly.* Individually. *free-minded.* Open, carefree. *studies.* Concerns. Cf. Bacon's *Historia Vitae et Mortis* (*Works*, ii. 171–2). *fretting inwards.* Suppressed, corrosive. *knotty.* Complicated. *inquisitions.* Investigations. Cf. *Works*, ii. 154. *communicated.* Released. *Entertain.* Cultivate. *fly... altogether.* Completely avoid medicine. *strange.* Unaccustomed. Cf. Plutarch, *Moralia*, 123B. *com-*

mend. Recommend. *be grown . . . custom*. Become a habit. *accident*. Sudden change, symptom. *respect*. Concentrate on recovering. *action*. Activity, physical exercise. *endure*. Bear exertions. *sharp*. Serious. *tendering*. Careful nursing. *Celsus*. Aulus Cornelius Celsus, AD 14–37, a physician and author of *De medicina*. *spoken*. *De medicina*, i. 1; cf. Bacon, *Works*, ii. 153. Reynolds (pp. 230–1) objects that Celsus' advice is misrepresented here. *lasting*. Long life. *a man*. One. *contraries*. Opposed states. *watching*. Staying awake. *sitting*. Repose. *masteries*. Superiority, to overcome difficulties.

405. *pleasing*. Obliging. *conformable to the humour*. Readily adapting treatment to the bodily state, however imbalanced. *press*. Penetrate to, discover. *regular*. Rigid. *art*. Medical precepts. *either*. Each. *faculty*. Professional ability.

31. *Of Suspicion*

405. *guarded*. Kept close. *cloud*. Overshadow. *leese*. Cause the loss of. *check with*. Hinder, interfere with. *whereby*. With the result that. *currently*. In a smooth current. *heart*. Seat of courage. *take place*. Have effect. *stoutest*. Most vigorous. *suspicious*. See *Works*, vi. 243. *stout*. Firm, resolute. *composition*. Combination of character traits (a metaphor from alchemy). *fearful*. Timid. *remedy*. Treat, cure. *in smother*. Covered up, stifled. *moderate*. Restrain. *account upon*. Regard. *bridle*. Restrain. *provide*. Arrange things. *buzzes*. Empty noise. *artificially*. Maliciously. *mean*. Means. *wood*. Tangle. *communicate*. Share. *party*. Person. *truth*. Foundation. *withal*. At the same time. *would*. Should. *base*. despicable.

406. *Sospetto . . . fede*. 'Suspicion licenses faith'. *passport*. Permission to depart. *discharge itself*. Free itself from suspicion by good behaviour.

32. *Of Discourse*

406. *Discourse*. In addition to this essay, see *Adv. L.*, above, pp. 265 ff. and the fragmentary 'Short Notes for Civil Conversation' (*Works*, vii. 109). Reynolds (p. 236) suggests that the essay may be in part a response to Cicero, *De officiis*, i. 37.132: 'There are rules for oratory laid down by rhetoricians; there are none for conversation; and yet I do not know why there should not be.' It was a common topic in Italian courtesy-manuals. The 1597 version (above, pp. 81–2) lacks this element. *wit*. Mental agility. (In Renaissance psychology 'wit' is the associative faculty, 'judgement' the discriminative one.) *hold*. Maintain. *discerning*. Discriminating. *should be thought*. That is, the truth. *common places and themes*. Prepared topics. *want variety*. Cf. Plutarch, 'On the education of children', *Moralia*, 7B–C. *part*. Action. *give the occasion*. Suggest a topic. *moderate*. Sum up, pronounce judgement. *discourse*. Longer utterance. *present occasion*. Of immediate but ephemeral interest. *arguments*. Substantial, more general topics. *reasons*. Statements. *jade*. To tire, weary. (A neologism, first recorded as a verb in 1606: *OED*.) *jest*. Joking. *privileged*. Exempt. *matters of state*. Business of national importance. *great persons*. People in power. *present*. Current. *dart out*. Utter

sharply. *piquant*. Offending. *to the quick*. Penetrating, hurting. *vein*. Inclination. Cf. Cicero, *De officiis*, i. 38.136. *would*. Ought to. *Parce ... loris*. 'Spare, child, the whip and rein the horses hard' (Ovid, *Met*. ii. 27). *saltness*. Satiric wit. *others'*. Other people's. *content*. Please others. *apply*. Adapt. *poser*. An examiner, who 'poses' or puts questions. *reign*. Dominate. *take them off*. Stop them performing. Cf. Cicero, *De officiis*, i. 37.135. *use*. Are accustomed. *galliards*. Lively French court dances, where evidently some dancers 'hogged the stage'. *that*. That which.

407. *self*. Cf. Cicero, *De officiis*, i. 37.134. *with good grace*. Decently. *pretendeth*. Aspires. *touch*. Referring to, affecting others. *field*. An open space, held in common. *coming home*. Affecting, being directed towards. *scoff*. Mockery. *cheer*. Entertainment, feasting. *flout*. Mocking speech. *dry blow*. Scornful jest. *passed*. Happened. *mar*. Spoil. *agreeably*. Fitly, suitably. *continued*. Sustained speech, monologue. *interlocution*. Alternate speaking, conversational exchanges. *the turn*. In turning, manœuvring (cf. *Adv.L.*, above, p. 226). *circumstances*. Introductory comments. *ere*. Before.

33. *Of Plantations*

407. *Plantations*. Colonies. The first English settlements in Roanoke, off the coast of Virginia, were short-lived, from 1585 to 1587 (a search party in 1590 found only four survivors). Bacon had drawn up in 1606 'Certaine Considerations Touching the Plantation in Ireland' (*Works*, xi. 116–26), in which he had expressed scepticism about plans for overseas plantations. But he soon changed his mind, developing an interest in colonizing ventures and subscribing to the refounded Virginia Company of London (1610) and several similar enterprises (for Newfoundland, the North-West Passage, and East India). Kiernan (p. 239) describes this essay as 'essentially a gloss upon the Virginia plantation before 1623', when a bitter internal dispute led to the dissolution of the Company and the imposition of royal control. *heroical*. Performed by heroes. *world was young*. Cf. Lucretius, v. 818 ff. *former*. Already existing. *pure soil*. Uncultivated, unoccupied—i.e. 'not actually possessed of any Christian Prince nor inhabited by Christian people', as laid down in the earliest Letters Patent, to Humphrey Gilbert in 1578, and to Ralegh in 1584 (Reynolds, p. 241). Cf. Bacon, *Works*, xiii. 50 f. *people*. Populations, nations. *displanted*. Displaced. *extirpation*. Weeding out, killing. *make account*. Reckon. *leese*. Lose. *recompense*. Reward. *thing*. Cause. *base*. Sordid. *drawing*. Withdrawing. For complaints about crude profit motives see e.g. Thomas Harriot, *True Report of the New Found Land of Virginia* (1588), in D. B. Quinn (ed.), *The Roanoke Voyages, 1584–1590* (London, 1955), i. 320, 323. *stand with*. Be compatible with. *unblessed*. Accursed. *scum of people*. Several of the early colonies were peopled with criminals: Reynolds, pp. 242–3; Kiernan, p. 240. *condemned men*. Convicts.

408. *plant*. Colonize. *fall to*. Apply themselves to. *spend*. Consume. *victuals*. Food supplies. *certify*. Send reports, as did a group of 30 men (described in the official report as 'that scum of men') in 1610, who, having

stolen a ship, turned pirates but failed, subsequently defaming the Company in England when their food ran out (Kiernan, p. 240). *fowlers*. Bird catchers. *of itself*. Spontaneously. *pine-apples*. Pine cones. *esculent*. Edible. *artichokes of Hierusalem*. A species of sunflower (Ital. *girasole*) with an edible root. *For*. As for. *ask*. Require. *meat*. Food of all kinds (perhaps also 'can supply a main meal'). *increase*. Produce. *store*. Large supply. *biscuit*. Ship's biscuit, hard tack. *meal*. 'The edible grain or pulse, now usu. other than wheat, ground to a powder' (*SOED*). *expended*. Consumed. *certain allowance*. Fixed ration. *to*. For. *common stock*. Public granary. *laid in*. Gathered, harvested. *in proportion*. According to the size of the individual farmer's land. *manure*. Cultivate. *private*. Personal use. *charge*. Initial costs. *so*. Provided that. *prejudice*. Damage. *prejudice ... tobacco*. The financial yields from Virginia tobacco were so great that farmers neglected all other crops (Kiernan, pp. 242–3). *one*. Commodity. *set*. Build, locate. *brave*. Fine. Cf. Harriot, *True Report*, 331–2. *bay-salt*. Coarse-grained salt extracted from sea-water ponds by the sun's heat. *would ... experience*. Ought to be tried. *Growing silk*. A vegetable fibre, resembling silk. Cf. Harriot, ibid. 325–6. *likely*. Sought-after. *sweet woods*. Various South American and West Indian lauraceous trees. *soap-ashes*. Alkaline salts with corrosive qualities (e.g. caustic soda). *moil*. Toil, labour. *mines*. Gold-mines. (The fruitless search for precious metals in the early colonies damaged their proper development: Melchionda, p. 541).

409. *counsel*. Council. *commission*. Authority. *martial laws*. Military laws, with severe penalties. *limitation*. Restriction. *that*. Such. *undertakers*. Shareholders. *temperate*. Moderate, restricted. *present*. Immediate. *custom*. Customs duties. *best*. Greatest profit. *company*. Shipments of people. *hearken*. Ascertain. *waste*. Dwindle. *supplies*. Reinforcements. *surcharge*. Overpopulation. *marish*. Marshy. *carriage*. The cost and bother of transport. *discommodities*. Disadvantages. *still*. Always. *upwards*. Away. *necessary*. (To preserve food.) *entertain*. Deal with; amuse. *gingles*. Cheap jewellery. *guard*. Caution. *defence ... not amiss*. It is right to help them defend themselves. *them*. Some of them. *condition*. Level of existence. *spread into generations*. Grow with the human propagation. *pieced*. Patched up. *destitute*. Abandon (as happened with several early settlements, notably Roanoke). *forwardness*. Prosperity. *guiltiness of blood*. Responsibility for the death of. *commiserable*. Pitiable.

34. *Of Riches*

409. *baggage*. Hindrance. In his *Promus* Bacon had noted 'Divitiae impedimenta virtutis' (no. 67). Cf. Guicciardini, *Ricordi*, no. 65: 'The man who called baggage trains "impediments" could not have expressed it better; the one who invented the saying "more trouble than moving camp" put it very well': ed. cit. 20. *word*. Here, metaphor.

410. *loseth*. Causes the loss of. *distribution*. Spending. *conceit*. Fantasy, imagination. *Where ... eyes?*. Eccles. 5: 11. *fruition*. Pleasure arising from possession. *reach*. Extend. *feel*. Experience. *dole and donative*. Charitable distribution and gift. *feigned.* Fanciful. *because.* In order that. *buy men.* Ransom. *Riches . . . man.* Prov. 18: 11. *sold.* Deceived, ruined. Cf. Juvenal, x. 12–18. *bought out.* Redeemed. *proud.* Ostentatious. *get justly.* Acquire honestly. *abstract.* Ascetic. *friarly.* Like a monk, with his vow of poverty. '*In . . . quaeri*'. 'In striving to increase his wealth, it was apparent that he sought not prey for his avarice, but an instrument for his goodness' (*Pro Rabirio Postumo*, iii. 5). '*Qui . . . insons*'. 'He that maketh haste to be rich shall not be innocent' (Prov. 28: 20). *poets feign.* Lucian, *Timon*, 20 ff. *Plutus.* The Greek god of wealth. *Pluto.* The Greek god of the underworld. *pace.* Proceed. *by the course of.* By way of. *upon speed.* All of a sudden. *enrich.* Get rich. *obtaining.* Gaining. *stoop.* Personally apply themselves. *husbandry.* Managing their own estates (agriculture). *a nobleman.* Probably George Talbot, 6th Earl of Shrewsbury (*c.*1522–90). *greatest.* Biggest. *audits.* Revenues, receipts. *grazier.* Cattle-farmer. *sheep-master.* Owner of sheep. *timber man.* Proprietor of timber. *collier.* Owner of coal-mines. *corn-master.* Owner of corn. *lead-man.* Owner of lead-mines. *points.* Branches. *husbandry.* Here in the general sense of 'productive activity' (*SOED*).

411. *one.* Lampone, a rich Greek merchant, who said, according to Plutarch, 'That aged men ought to govern the commonwealth' (*Moralia*, 787A). *hardly.* With difficulty. *expect the prime.* Wait for the most favourable conditions. *overcome.* Master, make his own. *bargains.* Deals. *greatness.* Size. *money.* Resources. *industries.* Commercial activities. *mainly.* Mightily. *trades.* Handicrafts. *vocations.* Professions. *bargains.* Speculations. *wait upon.* Watch. *broke.* Negotiate, do business. *instruments.* Intermediaries. *put off.* Discourage, deceive. *chapmen.* Buyers. *practices.* Frauds. *naught.* False, despicable. *chopping.* Bartering, speculating. *hold.* Keep, retain. *grindeth double.* Cuts in both directions. *Sharings.* Joint enterprise. *certainest.* Most certain. '*in . . . alieni*'. 'In the sweat of another man's face': a parody of God's sentence on Adam for his disobedience, 'in the sweat of thy face shalt thou eat bread' (Gen. 3: 19). *plough upon Sundays.* Violates religious principles. *flaws.* Weak points. *scriveners and brokers.* Intermediaries between lender and borrower. *value.* Represent as trustworthy; or overvalue (to increase their own commission). *unsound.* Unreliable. *privilege.* Monopoly on foreign trade. *sugar man.* Owner of a sugar plantation. *Canaries.* Canary Islands. *logician.* In Peter Ramus's educational reforms, invention (*inventio*) and judgement (*dispositio*) were transferred from rhetoric to logic. *gains certain.* Safe returns on investments. *adventures.* Risky enterprises. *break.* Go bankrupt. *guard.* Protect. *certainties.* Sure investments. *uphold.* Make up for. *coemption of wares.* Ingrossing, hoarding goods. *restrained.* Restricted, forbidden. *party.* Person concerned. *intelligence.* Inside knowledge. *store.* Make large provisions. *rise.* Origin, source. *feeding humours.* Pandering to whims, appetites. *executorships.* The administrator of a will received a fee. '*testamenta . . . capi*'. 'He seized testaments [of men without heirs] and wards as if with nets' (*Annals*, xiii. 42). *by how much.* Inasmuch as. *that.* Who. *none worse.* No one behaves more badly.

412. *penny-wise*. As in the proverb, 'Penny wise, pound foolish'. *riches have wings . . . themselves*. Prov. 23: 5. *portions*. Amounts. *state*. Estate, fortune. *lure*. Enticement, bait. *better stablished*. Thoroughly settled. *glorious*. Ostentatious. *sacrifices without salt*. See Lev. 2: 13 and Mark 9: 49, texts which Bacon elsewhere interpreted as meaning 'that God is not pleased with the body of a good intention, except it be seasoned with that spiritual wisdom and judgment, as it be not easily subject to be corrupted and perverted: for salt, in the scripture, is a figure both of wisdom and lasting' (*Works*, xi. 249). *painted sepulchres*. Christ described the hypocritical Pharisees as 'whited Sepulchres': Matt. 23: 27. *putrefy*. Rot (salt was used to preserve meat and fish). *advancements*. Gifts. *frame . . . measure*. Give according to need. *liberal . . . of his own*. Distributes generously what belongs rather to others than to himself (perhaps because many wills, including Bacon's own, contained bequests which the estate could not cover).

35. *Of Prophecies*

412. *divine prophecies*. Those involving God's chosen people and the Messiah. *natural*. Made naturally, from known data, as opposed to 'hidden causes'. *Pythonissa*. The name used in the Vulgate (drawing on Greek traditions involving Apollo and the python) for the witch of Endor, whom Saul petitioned to call back the ghost of Samuel. Cf. Chron. 10: 13. *Saul*. 1 Sam. 28: 3–25. *At . . . illis*. 'The house of Aeneas shall reign in all lands, and his children's children, and their generations': Virgil, *Aeneid*, iii. 97–8, itself an adaptation of *Iliad*, xx. 307–8. Bacon uses Eobanus Hessus's Latin translation of the *Iliad* (Basle, 1540), 50 (Wolff, ii. 7). *Venient . . . Thule*. 'In far-off years there shall come the ages when ocean shall loosen the bounds of the world, and the huge earth shall lie revealed, and Tiphys shall disclose new worlds; and Thule shall no longer be the limit of all lands' (*Medea*, ii. 375–9). *Polycrates*. Tyrant of Samos, lured to his death by the Persians. Cf. Herodotus, *Hist*. iii. 124–5. *Philip of Macedon*. Plutarch's 'Life' of Alexander the Great, 2.3. *whereby . . . it*. Which he interpreted to mean. *soothsayer*. Truth-teller, of future events. *phantasm*. Ghost, spirit. '*Philippis . . . videbis*'. 'You shall see me again at Philippi', Plutarch's 'Life' of Brutus, 36.3; Appian, *Bell. Civ*. iv. 134; cf. Shakespeare, *Julius Caesar*, IV. iii. 283 f.

413. '*Tu . . . imperium*'. 'Thou, too, Galba, shalt have thy taste of empire' (Tacitus, *Annals*, vi. 20); Suetonius, 'Life' of Galba, 4, tells it of Augustus, not Tiberius. *Vespasian's time*. Tacitus, *Histories*, v. 13. *went*. Circulated. *of*. As referring to. *Domitian*. Suetonius, 'Life' of Domitian, 23. *succession*. Successors. *Henry the Sixth*. Cf. *Henry VII* (*Works*, vi. 245), and Shakespeare, *3 Henry VI*, IV. vi. 65–76. *in France*. Bacon spent from 1576 to 1579 in the train of Sir Amias Paulet, ambassador to the French court. *Queen Mother*. Wife of King Henri II of France (1547–59). *given*. Prone to credit. *curious*. Occult. *nativity*. Horoscope. *above . . . duels*. A king would not accept such challenges. *a course at tilt*. A joust with a lance; tournament. *beaver*. Front part of a helmet, which had openings for the eyes. *trivial*.

Common. *hempe.* A coarse fibre, sometimes woven (or spun) into a rope. *principial.* Initial, a word coined by Bacon (*OED*). *utter confusion.* Complete chaos. *Britain.* 'Britain, after the OE period, was for long used only as a historical term, but in 1604 James I & VI was proclaimed "King of Great Britain" ' (*SOED*). *eighty-eight.* 1588, when the Spanish Armada was outsailed and outgunned by the English navy, and subsequently destroyed by bad weather. *Baugh . . . May.* The Bass Rock, the isle of May—both islands in the Firth of Forth, Scotland. *Regiomontanus.* Johannes Müller of Königsberg (1436–76), German astronomer, who is said to have written four lines in German foretelling great revolutions in 1588. But these were enlarged and Latinized by Gaspar Bruschius in 1553, and applied to a quite different prediction (Reynolds, p. 262). *octogesimus . . . annus.* 'The wonderful year 88', when a supposedly inauspicious conjunction of the planets was prophesied to bring terrible disasters.

414. *Cleon's dream.* In Aristophanes' *Knights* (197 ff.) Demosthenes expounds the oracle's prediction that Cleon will be overcome by a serpent as referring to a sausage-seller (who indeed triumphs over him). *certain credit.* Definite authenticity. *despised.* Ignored. *winter talk.* Idle chatter. *sort.* Way. *severe laws.* Henry VIII, Edward VI, and Elizabeth all passed laws against fantastical prophecies, especially those involving 'any figure, casting of nativity, or by calculation, prophesying, witchcraft, conjuration, &c'. *grace.* Favour. *mark.* Notice. *coveteth divination.* Earnestly desires predictions. *collect.* Deduce, infer. *Seneca's verse.* From *Medea*, quoted above. *then.* Before the world had been explored. *parts.* Regions. *Atlanticus.* In *Timaeus*, 24E–25D Plato records the tradition of a huge sunken island in the Atlantic. His dialogue *Critias* (sometimes known in Renaissance Latin translations as *Atlanticus*) describes this lost island. *one.* Someone. *last.* (Thing.) *great.* Decisive. *idle and crafty.* Thoughtless and malicious. *merely.* Wholly.

36. *Of Ambition*

414. *Ambition.* Cf. Guicciardini, *Ricordi*, no. 32: 'Ambition is not to be condemned, nor should one revile the ambitious man's desires to attain glory by honourable and worthy means. Such men as these do great and outstanding things, and anyone who lacks this urge is a cold spirit and inclined rather to idleness than to effort. Ambition is pernicious and detestable when its sole end is power. This is usually true of those princes who, when they set it up as an idol to achieve what will lead them to power, set aside conscience, honour, humanity, and all else' (ed. cit. 13). *choler.* Yellow bile, the source of anger, one of the four 'humours' from which the body was thought to be composed. Health depended on their balanced circulation. The analogy with ambition is positive at this point. *stirring.* Energetic. *stopped.* Obstructed. *his.* Its. *adust.* Scorched, parched, causing irritation. *venomous.* Poisonous. *still.* Always. *busy.* Intriguing, prying into people's affairs. *checked.* Blocked. *evil eye.* Enviously. *go backward.* Deteriorate. *property.* Characteristic. *handle it.* Manage things.

415. *progressive*. Going forward. *retrograde*. Moving backwards, as the planets sometimes seem to do when viewed from the earth (thought by astrologers to be an inauspicious omen). *service*. Duties. *take order*. Take measures. *were*. Would be. *be . . . ambitious*. However ambitious they are. *dispenseth with*. Compensates for. *screens*. Protecting walls. *envy*. Discontent. *seeled*. Blindfolded. The eyelids of the bird (usually a falcon) were stitched together by thread, to make it fly higher. *overtops*. Rises beyond their station. *Tiberius . . . Sejanus*. See Dio Cassius, lviii. 9; Tacitus, *Annals*, vi. 48; and Ben Jonson, *Sejanus*, Act v. *there resteth*. It remains. *bridled*. Controlled. *harsh of nature*. Gruff, surly. *gracious*. Endowed with charm. *popular*. Loved by the people (a source of instability). *new*. Recently. *of all . . . the best*. Better than any other. *way . . . favourite*. When the dispensation of favours and disgraces depends on the favourite. *pleasuring . . . displeasuring*. Pleasing and displeasing. *any other*. Anyone else. *middle*. Moderate. *ballast*. Heavy material used to weigh down a ship. *inure*. Habituate, train. *meaner*. Less ambitious. *scourges*. Punishing restraints. *obnoxious*. Exposed to, in danger of. *fearful*. Timid. *do well*. Be successful. *stout*. Bold, courageous. *precipitate*. Bring on prematurely. *in a wood*. Lost, bewildered. *harmful*. Hurtful, pernicious. *confusion*. Chaos. *mars business*. Ruins everything. *great in dependances*. Having many followers.

416. *public*. Community. *figure . . . ciphers*. The only number amid zeros. *Honour*. Social eminence. *vantage*. Superior position. *aspireth*. Wants to make a career. *another*. One. *sensible*. Responsive to. *conscience*. Moral sense. *bravery*. Display, boastfulness. *discern*. Distinguish.

37 *Of Masques and Triumphs*

416. *Masques*. For Bacon's involvement with these hybrid entertainments see the General Introduction, above, pp. xxii ff. *Triumphs*. Chivalric exercises, discussed in the final paragraph. *toys*. Trivialities. *serious observations*. As are the other Essays. *daubed with cost*. Put on at great expense (in the currency of those days, from £700 to over £3,000). *Dancing to song*. That is, in the main masque. *state*. Stateliness, splendour. *in quire*. Choir. *aloft*. Above, in the gallery. *broken music*. A consort or small group containing both wind and string instruments. *ditty . . . device*. Text or words (Lat. *dicta*) of the song matching the masque's plot or theme. *Acting in song*. Sung recitative, as in the operas produced in Florence from the 1590s on. *dancing*. Singing while dancing, as in the popular 'jig'. *would*. Ought to. *treble*. Boy sopranos. *high and tragical*. In the high style, with dignity. *nice or dainty*. Fastidious, elaborate. *Several*. Separate. *against another*. Antiphonally, as in the music written by Gabrieli and Monteverdi for St Mark's Cathedral, Venice. *by catches*. In separate phrases. *figure*. Shape (e.g. a pyramid, or spelling out a name). *curiosity*. Elaborate trick. *take the sense*. Appeal to the senses. *respect*. Are concerned with. *wonderments*. Surprises. *alterations*. Scene-changes, for which the machinery (only possible at this time in some specially adapted private halls, not in the open-air public theatres) was often noisy. *full*. Satiated. *masquers*. The performers, some of whom were professional. *other*. Others. *come . . . scene*. Descend from the stage to the hall, for a dance in which the leading members of the audience took part. *motions*. Movements, gestures which will attract the onlookers' attention. *strangely*. Uncommonly. *that*. That which. *chirpings or pulings*. Singing like birds or in a thin, child-like voice (the 'trebles' that Bacon disliked). *sharp*. Vivacious, cheerful (in a major key). Melchionda (p. 557) quotes from a letter of 1594–5 where Bacon writes that 'in music, I ever loved easy airs, that go full all the parts together, and not these strange points of accord and discord' (*Works*, viii. 356–7). *shew*. Stand out. *candle-light*. As indoor performances, masques depended on artificial lighting.

417. *oes*. The plural of the letter 'o', *oes* or *spangs* were spangles, sequins, circular pieces of glittering metal sewn on to costumes. *glory*. Lustre, splendour. *lost*. Invisible from a distance, lacking adequate lighting. *suits*. Costumes. *become*. Suit. *vizards*. Masks: at the end of the show the masquers often removed their masks and presented themselves to the leading spectators. *attires*. Styles of dress, special clothing. *antimasques*. Grotesque episodes, with music and dancing, as contrast to the stately main masque. *antics*. Grotesque figures, buffoons. *sprites*. Spirits. *Ethiops*. Blacks (as in Jonson's *Masques of Blackness*, 1605). *turquets*. 'Little Turks', a word coined by Bacon (*OED*). *statua's moving*. Statues that move. *not... enough*. Not suitably comic for. *as unfit*. Equally unsuitable. *recreative*. Diverting. *strange changes*. Unexpected transitions, discords suitable to the anti-masque. *odours*. Fine sprays of perfume. *steam*. Humidity, or body-odours. *state*. Splendour. *jousts... tourneys*. Tilts: sporting encounters between armed knights on horseback with blunted lances and swords, respectively. *barriers*. Combats within a fenced-off area, fought on foot with short swords and lances. *devices*. Invented plots. *bravery*. Rich display. *furniture*. Trappings, harness.

38. *Of Nature in Men*

417. *Nature*. Inclination, innate drives. *return*. Rebound (having been suppressed). *doctrine*. Teaching, education. *discourse*. Precepts. *importune*. Importunate, pressing. *custom*. Habit, practice. Melchionda (p. 558) cites similar advice from G. Della Casa, *Galateo* (1558), ch. 25. *practise*. Train. *bladders*. Bladder of an animal used as a float. Cf. *Letter to Savile*, above, p. 119, and *Adv.L.*, above, p. 242. *rushes*. Plants with hollow stems. *disadvantages*. Impediments. *thick*. Heavy. *use*. Normal use. *degrees*. Stages. *stay*. Restrain. *arrest... time*. Slow down nature's workings. *say... letters*. Repeat the alphabet. *forbearing*. Giving up. *healths*. Toasts. *draught*. A glass of wine or beer.

418. *enfranchise*. Set free, liberate. *at once*. At one go. *Optimus... semel*. 'He is the best liberator of his mind who has burst the chains binding his breast, and has done with grieving' (Ovid, *Remedia amoris*, 293–4). *ancient rule*. Cf. Aristotle, *Eth. Nic.* ii. 9, 1109b4 ff., and *Adv.L.*, above, p. 261. *amiss*. Inapplicable. *wand*. Stick. *right*. Straight. *understanding it*. That is Cf. *Eth. Nic.* ii. 6, 1107a9 ff. *abilities*. Capabilities (cf. *Adv.L.*, above, p. 242). *induce... both*. Get used to the same (bad) habits in both. Cf. Cicero,

De oratore, i. 33.149 f. *lay*. Lie. *Aesop's damsel*. In the (post-Aesopic) fable, a cat which had been turned into a girl actually revealed her true nature not at the 'board' (dining-table) but in the bridal chamber. *put*. Apply. *that*. So that. *in passion*. Under strong emotion. *new case*. Unexpected situation. *experiment*. Experience. *sort with*. Agree with. '*multum . . . mea*'. 'My soul has long been a sojourner' (Ps. 120: 6, Vulgate). In the Authorized Version it reads 'My soul hath long dwelt with him that hateth peace' (Geneva Bible, similarly). *converse*. Are engaged. *affect*. Like, enjoy. *studies*. Occupations. *commandeth*. Imposes. *set hours*. Assign fixed times. *so as*. So that. *spaces*. Intervals. *runs . . . to*. Spontaneously produces.

39 *Of Custom and Education*

418. *inclination*. Natural tendencies. *infused*. Inculcated, received. *after as*. According to. *Machiavel*. *Discourses*, iii. 6. *evil-favoured*. Bad, ugly. *bravery*. Courage; also boastfulness. *corroborate*. Strengthened, confirmed.

419. *achieving*. Success. *rest*. Rely, depend. *resolute undertakings*. Brave promises. *formerly*. Already. *Clement . . . Gerard*. Four (Catholic) assassins: Jacques Clement assassinated King Henri III of France in 1589; François Ravillac killed King Henri IV of France in 1610; John Jaureguy wounded, but failed to kill Prince William of Orange in 1582; Baltazar Gerard finally murdered him in 1584. The Latin translation of the *Essays* (1638) adds 'or Guy Fawkes'. *engagement*. Obligation. *forcible*. Effective. *Only*. Only according to. *advanced*. Disseminated. *men of the first blood*. Those committing their first murder. *occupation*. Profession. *votary resolution*. Taking a vow to perform something. *equipollent*. Of equal importance. *dead images*. Lifeless statues. *engines*. Machines. *what it is*. To what it leads. *Indians*. The Gymnosophists: see Cicero, *Tusc. D*. V. xvii. 27–8. *Sparta*. See ibid. II. XIV. 34 and Montaigne, 'Défense de Sénèque et de Plutarque', *Essais*, ii. 32 (Pléiade edn. 701). *queching*. Uttering a sound; or perhaps 'flinching'. Cf. Cicero, *Tusc. D*. II. xiv. 34, V. XXVII. 77. *an Irish rebel*. Brian O'Rourke, in 1597; the incident also appears in the play *Sir John Oldcastle*, pt. 1 (1600). *put up*. Offered as a prayer. *Deputy*. Lord Lieutenant of Ireland. *with*. A band made of twisted twigs. *halter*. Type of rope used to hang criminals, or lead animals. *penance*. Penitence. *engaged*. Held fast, as in a vice. Bacon used Giles Fletcher, *Russe Commonwealth* (1561). *put*. Given. *magistrate*. Schoolmaster, guide. *pliant*. Adaptable. Cf. Montaigne, 'De l'institution des enfans', *Essais* i. 26 (Pléiade edn. 152). *motions*. Movements. *take the ply*. Receive a new shape. *fix*. Grow rigid. *amendment*. Improvement. *simple and separate*. Among individuals. *copulate*. Coupled with. *collegiate*. In company (as in a college). *there*. In such cases. *comforteth*. Strengthens. *quickeneth*. Stimulates. *as*. That. *exaltation*. Highest point, of greatest influence. *multiplication*. Increase.

420. *resteth upon*. Depends on. *societies*. Social institutions. *mend*. Improve. *misery . . . desired*. Some commentators see here a reference to the Jesuits' educational system: Bacon admired its efficiency, but deplored its aims (*Works*, iii. 276–7). To others it is a complaint about the gap between methods and goals in contemporary moral education.

40. *Of Fortune*

420. *but*. That. *outward accidents*. External events. *conduce*. Contribute. *favour*. Kindness, patronage. *occasion fitting virtue*. Circumstances coupled with merit. *mould*. Formation. '*Faber... suae*'. 'Each man is the maker of his own fortune': a saying combined from several classical sources, including Plautus, *Trinummus*, II. ii. 84 (line 363). Cf. *Adv.L.*, above, pp. 217 ff. '*Serpens... draco*'. 'A serpent must eat another serpent before it can become a dragon': Erasmus, *Adagia*; *Promus*, no. 362. *apparent*. Manifest. *deliveries*. Ways of behaving; free use of the limbs. *desemboltura*. Boldness, cheek. *stonds*. Obstructions, hindrances. *restiveness*. Obstinacy. *that*. When. *keep way*. Keep pace. *Cato Major*. Marcus Porcius Cato (234–149 BC), exemplar of Roman austerity and virtue. '*In... videretur*'. 'In this man there was such strength of body and mind that no matter where [in what level of society] he was born he would have made himself a fortune': paraphrase of *Histories*, xxxix. 40. *falleth upon*. Settles on (as an explanation). '*versatile ingenium*'. 'Adaptable nature.' *blind*. (According to the traditional representation.) *not seen asunder*. Invisible separately. *scarce*. Scarcely. *faculties and customs*. Abilities and habits. *little think*. Easily miss. *Poco di matto*. (Ital.) 'A bit of a fool', corresponding to the English proverb, 'Fortune favours fools'. *properties*. Qualities. *extreme*. Unrestrained. *without*. Outside. *hasty fortune*. Sudden riches. *enterpriser*. Adventurer. *remover*. Restless person, always on the move. *exercised*. Acquired with difficulty. *and... but*. If only. *Confidence*. Credit. *felicity*. Prosperity.

421. *decline*. Turn aside. *virtues*. Here, great achievements. *use*. Are accustomed. *assume*. Take possession of, put on. *greatness in*. Motive of greatness for. *care*. Object of attention. '*Caesarem... ejus*'. 'You carry Caesar and his fortune' (Plutarch, 'Life' of Caesar, 38). *Sulla... Magnus*. Spoken at the end of a triumph: Plutarch, 'Life' of Sulla, 6.3 f. *Felix*. Fortunate. *Timotheus*. Athenian general, referred to by Plutarch in the same passage. *interlaced*. Inserted. *there be*. There are those. *slide*. Smoothness of movement. In his 'Life' of Timoleon (36) Plutarch compared the 'singular grace' of Homer's verses, 'easily made', with the wars of Timoleon, in which, 'besides equity and justice, there is also great ease and quietness', thus proving that credit should be given not to Fortune but to his 'most noble and fortunate courage'. *much*. Mostly—that is, *virtù* (human endeavour) is the crucial factor, not fortune.

41. *Of Usury*

421. *Usury*. Money-lending had long been accepted throughout Europe (not just in Jewish hands) as a standard commercial practice in the financing of new businesses, with the shareholders receiving a form of interest. However, traditional ethical attitudes, going back to Aristotle, disapproved of the unnaturalness in making money 'breed' of itself. The Church officially endorsed this disapproval, also invoking the Christian concept of charity against taking advantage of an individual in need. The accepted interest rate in Elizabeth's reign was 10 per cent, but in 1623 Parliament reduced it to 8 per cent. Bacon, like most courtiers and noblemen in this period, lived a life based on credit and died heavily in debt. For Bacon's proposals to control usury, drafted in 1623 and very similar to this essay, see *Works*, xiv. 410, 413–20; Reynolds, pp. 292–5; and Kiernan, pp. 124, 266–9. *pity*. Pitiful. *tithe*. One-tenth of the annual income or produce was supposed to be paid to the Church. The maximum rate of usury under Henry VIII and Elizabeth I was 10 per cent. *sabbath-breaker*. Church-laws forbade working on Sundays, but interest accumulated daily. *Ignavum... arcent*. 'They drive the drones, a lazy herd, from the hives' (*Georgics*, iv. 168). *the first law*. God's commandment to Adam and Eve. *In... alieni*. 'In the sweat of thy face shalt thou eat bread' (Gen. 3: 19); not 'in the sweat of another man's face'. *orange-tawny bonnets*. The Jews in the Venetian Ghetto were compelled to wear a distinguishing dress, red and yellow (dark orange) turbans. *do judaize*. Follow Jewish customs. In fact, despite popular stereotypes and prejudice, Jews were in a minority as money-lenders, and not only in England. *against nature*. This argument goes back to Aristotle, *Politics*, i. 10, 1258b2 ff. See also Shakespeare, *The Merchant of Venice*, I. iii. 94–137, and 'A Devillish Usurer' in Sir Thomas Overbury's *Characters*. '*concessum... cordis*'. 'Because of the hardness of heart' (Matt. 19: 8). *freely*. Gratis; also 'spontaneously'. *suspicious*. Exciting suspicion. *of*. With reference to. *discovery*. Investigation into individuals' wealth. *set before us*. Remind ourselves of. *incommodities and commodities*. Disadvantages and advantages. *weighed out*. Weighed and dispensed in portions accordingly. *culled*. Picked out for rejection. *make forth*. Proceed. *makes fewer*. Reduces the number of.

422. *vena porta*. 'Gate-vein', i.e. the multi-branched vein distributing chyle to the liver. *husband*. Farm, cultivate. *sit at... rent*. Be a tenant paying a high rent. *drive*. Conduct. *sit... usury*. Paying high interest. *incident*. Naturally connected. *decay of customs*. Decline in excise duties. *realm or state*. Monarchy or republic. *being at certainties*. Having guaranteed returns. *box*. The 'bank' or pot in gambling. *spread*. (Like manure.) *beats down*. Collapses. *purchasing*. Permanently acquiring (as of land). *waylays*. Ambushes, obstructs (like a highwayman). *damp*. Suffocates. *slug*. Obstacle. *canker*. Anything that corrupts and destroys. *breeds... poverty*. Creates a general impoverishment. *side*. Hand. *merchandizing*. Trade. *advanceth*. Promotes. *call in*. Demand it back, in repayment. *presently*. Immediately. *stand*. Standstill. *necessities*. Things necessary for daily life. *undoing*. Ruin. *means*. Assets. *under foot*. Below their real value. *gnaw upon*. Consume in part. *bad markets*. Unfavourable transactions. *mortgaging*. Loaning of property or land as security for a debt. *pawning*. Depositing objects as security for payment. *take pawns without use*. Take securities in pledge (and lend money upon them) without exacting interest. *precisely*. Sharply, punctiliously. *forfeiture*. Yielding up a property as a penalty for not paying the interest due. Cf. Shylock in *The Merchant of Venice*, IV. i. 206 ff. *keeps us from*. Denies us. *vanity*. Silly idea. *cramped*. Suppressed. *idle*. Pointless;

a statute passed in 1551 prohibiting usury failed, encouraging the revival of even worse abuses. *rate*. Interest rate. *sent to Utopia*. Dismissed as imaginary, unreal. *reiglement*. Regulation.

423. *grinded*. Blunted; cf. earlier the (traditional) metaphor for usury as 'gnawing'. *invite*. Give incentive to. *quickening of trade*. Giving life to, stimulating commerce. *several*. Distinct. *less*. At a lower rate. *to seek for*. At a loss for, have difficulty finding money. The high profits of commerce can support a higher interest rate. *contracts*. Forms of business. *intentions*. Purposes. *general*. Universally applicable. *five in the hundred*. 5 per cent. *free and current*. Negotiable and official. *shut itself out*. Refrain. *to take*. From taking. *penalty*. Proportion of the interest rate (state tax). *dryness*. Lack of circulation. *purchase*. 'The annual return or rent from land' (*SOED*), so that *sixteen years' purchase* is the purchase price calculated as the revenue for that period. *edge*. Urge, stimulate. *venture... kind*. Invest in this sector. *cautions*. Provisos. *more easy*. Lighter. *ease*. Relief. *common*. Joint. *mislike*. Dislike. Banks were not yet established in England, and were regarded with distrust as centres of financial power. *brooked*. Put up with. *answered... matter*. Guaranteed a small fee (known as 'the King's profit'). *licence*. Official permission. *abatement*. Deduction. *no whit*. Not at all. *give over*. Give up. *gains of hazard*. Risky profits. *colour*. 'Represent as one's own' (*SOED*): lending money they have themselves borrowed, under pretence that it is their own. *send*. The uncorrected copies read 'lend' (Kiernan, p. 128).

424. *in a sort*. In a manner. *declaration*. Official recognition.

42. *Of Youth and Age*

424. *Youth and Age*. This essay owes something to Aristotle's juxtaposition of these two states of life in *Rhetoric*, ii. 12–13, 1389a2–1390a25. *old in hours*. Mature, experienced. *lost no time*. Cf. Guicciardini, *Ricordi*, no. 145: 'Rest assured that, although man's life is short, yet there is plenty of time for those who know how to make capital of it and not to consume it in vain' (ed. cit. 38). *cogitations*. Reflections. *invention*. Inventiveness. *imaginations*. Ideas, projects. *heat*. (An excess of one of the four humours.) *meridian*. Midpoint (of the sun's course; here of life). *Julius Caesar*. Fought the Gallic war in his mid-forties, and became dictator in 46 BC, aged 56. *Septimius Severus*. Proclaimed emperor in AD 193, aged 47. '*Juventutem... plenam*'. 'He spent a youth full of errors, nay more, of madnesses': paraphrase of A. Spartianus, 'Life' of Severus, *Script. Hist. Aug.* ii. *reposed*. Calm, settled. *Augustus Caesar*. Octavius (63 BC–AD 14), first Roman emperor, in power by his early thirties. *Cosmus*. Cosimo I de' Medici (1537–69) became Duke of Florence at the age of 17. *Gaston de Fois*. As Reynolds (p. 301) and Melchionda (p. 572) argue, more likely to refer to Gastone III 'Fébus', Count of Foix and Viscount of Béarn (1331–90), an outstanding military and political leader, than to Gaston de Foix, Duke of Nemours, who died in the battle of Ravenna aged 22. *in age*. Old age. *composition*. Temperament. *business*. Public life. *invent... judge*. Terms from logic: to have ideas, then to evaluate them.

fitter . . . counsel. Cf. Plutarch, 'That aged men ought to govern the commonwealth', *Moralia*, 789E. *settled.* Routine. *of age.* Of old men. *compass.* Range. *directeth.* Keeps on a straight path. *abuseth.* Deceives. *manage.* Management. *stir.* Put in motion. *quiet.* Resolve, settle. *degrees.* Intermediate stages. Cf. Aristotle, op. cit.: 'Young men have strong passions . . . are hot-tempered and quick-tempered. . . . All their mistakes are in the direction of doing things excessively and vehemently . . . they overdo everything' (1389a3 f., 9 f., 1389b3 ff.). *pursue . . . absurdly.* Without good grounds. *care not.* Have no qualms about. *inconveniences.* Damages. *doubleth.* Intensifies. *unready.* Untamed, restive. *object too much . . . adventure too little.* Cf. Aristotle, op. cit.: old men 'are sure about nothing and under-do everything. They "think", but they never "know", and because of their hesitation they always add a "possibly" or a "perhaps", putting everything this way and nothing positively' (1389b17 ff.). *period.* Completion, conclusion. *mediocrity.* A middling amount. *compound.* Combine. *succession.* Continuity, future.

425. *actors.* Protagonists, here, 'directors'. Cf. Plutarch, op. cit. 790D ff. *extern accidents.* Outside events (unpredictable). *moral.* In ethics. Cf. Aristotle, op. cit.: young men 'love honour . . . more than they love money . . . They look at the good side rather than the bad . . . They trust others readily . . . They would always rather do noble deeds than useful ones', showing the dominance of 'character' and 'excellence' over 'reasoning' (1389a12 ff.). *politic.* Politics, but also implying 'prudential self-concern'. Cf. Aristotle, op. cit.: old men are 'distrustful', 'not generous', 'cowardly', and 'are too fond of themselves; this is one form that small-mindedness takes. Because of this, they guide their lives too much by consideration of what is useful and too little by what is noble—for the useful is what is good for oneself, and the noble what is good absolutely' (1389b21 ff.). *rabbin.* The rabbinical scholar Isaac Abravanel (1437–1508). '*Your . . . dreams*'. Joel 2: 28. *intoxicateth.* Poisons, depraves. *profit . . . in.* Derive benefit from. *affections.* Passions. *betimes.* Soon. *brittle.* Weak, thin. *edge.* Acumen, sharpness of mind. *Hermogenes.* A rhetorician of the 2nd century AD, who produced several books between the ages of 18 and 20 but at 24 completely lost his memory. Cf. Philostratus, *de Vitis Sophistarum*, ii. 7. *waxed.* Became. *have better grace.* Are more suitable. *luxuriant.* Exuberant in growth, overabundant (originally: producing foliage without fruit). Cf. Quintilian, XI. i. 31. '*Idem . . . decebat*'. 'He remained the same, when the same was not becoming': paraphrase of *Brutus*, 95. Hortensius was Cicero's early rival. *third.* (Sort.) *as.* That. *take . . . strain.* Make too great an effort. *magnanimous.* Lofty, ambitious. *tract.* Length. '*Ultima . . . cedebant*'. 'His end did not match his beginning': the phrasing is from Ovid, *Heroides*, ix. 23, but Bacon is thinking of Livy, xxxviii. 52–3.

43. *Of Beauty*

425. *Beauty.* In this essay Bacon discusses not so much aesthetics as human beauty and its relationship to other qualities, beginning and ending with virtue. As he wrote in the *Antitheta Rerum*, 'Virtue is nothing but inward

beauty; beauty is nothing but outward virtue' (*Works*, iv. 473). The corollary, for many Renaissance theorists, was that ugly people must be evil. For the background see A. P. McMahon, 'Francis Bacon's Essay "Of Beauty"', *PMLA* 60 (1945), 716–59. *plain set*. Mounted simply. *comely*. Attractive. Reynolds (p. 306) cites Cicero, *De officiis*, I. xxxvi. 130f. *delicate*. Sensual. *presence*. Bearing, deportment. *aspect*. Appearance. *almost*. Generally. *busy*. Concerned. *excellency*. Perfection. *accomplished*. Well endowed. *study rather*. Try to acquire. *behaviour ... virtue*. Manners, outer appearance, not inner worth. *holds*. Is valid. *Augustus Caesar*. Suetonius, 'Life' of Augustus, 79. *Titus Vespasianus*. Emperor AD 69–79, had 'a goodly presence and countenance': Suetonius, 'Life' of Vespasianus, in a passage not found in modern editions, but quoted in Reynolds (p. 306 as 'ch. 3') and Kiernan (p. 272 as 'ch. 2'). *Philip le Bel*. Philip IV, King of France 1285–1314. *Edward the Fourth*. King of England 1461–83. *Alcibiades*. Athenian statesman and general (*c*.450–404 BC). Cf. Plutarch, 'Life' of Alcibiades, I. *Ismael the Sophy*. Shah Ismael (1500–24). *high ... spirits*. Men of great courage. *favour*. Feature, countenance. *colour*. Complexion. Contemporary art theory, influenced by rhetoric, debated whether the highest artistic gift was *disegno* (design, composition) or *colore*. *decent*. Graceful, appropriate (the rhetorical concept of decorum). *gracious*. Graceful. *motion*. Bearing, movement. *nor ... life*. Not even the true likenesses when seen for the first time. *strangeness*. Irregularity. *strangeness ... proportion*. Kiernan (p. 273) cites Cicero, *De inventione*, II. i. 3: nature never makes 'anything perfect and finished in every part ... She bestowes some advantage on one ... but always joins it with some defect'. *Apelles*. The same passage in Cicero tells how Zeuxis (not Apelles) painted a composite portrait of Venus using the best features of five beautiful maidens; cf. also Pliny, *Epist*. xxxv. 36.2. *Durer*. Albrecht Dürer (1471–1528), German painter and engraver, whose *De Symmetria Partium in Rectis Formis Humanorum Corporum* (1532) contains elaborate diagrams of correct bodily proportions.

426. *more trifler*. Greater time-waster. *the one*. Here, the latter. *personage*. Representation of the human face. *Not but*. Not but that. *felicity*. Happy skill. *air*. Melody. *a*. One. *altogether*. All the parts together. *marvel*. Wonder. *though*. If. *many times*. Often. *amiable*. Loveable. '*pulchrorum ... pulcher*'. 'The autumn of beautiful persons is beautiful': Plutarch, 'Life' of Alcibiades, i. 5, and Erasmus, *Adagia*; *Promus*, no. 370. *comely*. Graceful. *but by pardon*. Except by permission (not by right). *make up*. Contribute. *makes a dissolute youth*. Tempts youth to self-indulgence, laxity of morals. *age*. An aged person. *out of countenance*. Upset, discomforted. *light well*. Alight on a worthy owner. *vices blush*. Out of shame at their evil.

44. *Of Deformity*

426. *Deformed persons*. For the tradition that Bacon was alluding to Robert Cecil (1563–1612), see Reynolds, pp. 309–11, and Kiernan, p. 274. This essay was first published in 1612. *even*. Got even with, revenged on. *by*. With respect to. *Scripture*. Rom. 1: 31; 2 Tim. 3: 3 (but not applied to deformed persons). *consent*. Sympathy. *erreth*. Goes astray. *ventureth*. Risks failure. *election*. Liberty of choice. *frame*. Framework, make- up. *discipline*. Education, training. *more deceivable*. Apt to be deceptive. *faileth of*. Misses. *fixed*. Permanent (deformity). *person*. Body. *induce*. Generate. *as*. As if. *repay*. Revenge.

427. *asleep*. Off guard. *in possession*. In office. *upon the matter*. In fact; on the whole. *wit*. Intellect, gifted person. *rising*. Making a career. *present*. Present time. *wont*. Accustomed. *obnoxious*. Submissive, obsequious. *officious*. Dutiful. *spials*. Spies. *whisperers*. Malicious informers, detractors. *reason*. Motive force, rationale. *still*. Always. *of spirit*. Enterprising, talented. *marvelled*. Wondered at. *Agesilaus*. Spartan king (444–360 BC), who had one leg shorter than the other. Cf. Plutarch's 'Life'. *Zanger*. A hunchback, son of the Ottoman ruler, Soliman the Magnificent. *Aesop*. Traditionally ugly and deformed. *Gasca*. Pedro de la Gasca (*c*.1493–1567), ugly and disproportioned. *Socrates*. Who pronounced himself ugly.

45. *Of Building*

427. *Building*. The country house whose construction Bacon describes here is on the grand scale, showing the *Essays*' connection with the 'Advice to Princes' tradition. It is designed on the old-fashioned courtyard plan, suitably fitted out for the 'progress' or country tour of the monarch and his retinue. As Kiernan shows (pp. 276–84), in some respects the building resembles Bacon's family house at Gorhambury. *use*. Usefulness, practicality. *uniformity*. Symmetry. *enchanted palaces*. As in Renaissance epic: Ariosto's *Orlando Furioso*, Spenser's *Faerie Queene*. *ill seat*. A bad site. *air*. Climate. *unequal*. Variable. *seats*. Houses. *knap*. Small hill, knoll. *pent in*. Captured, concentrated. *gathereth . . . troughs*. Collects, as if canalized. *so as*. So that. *several*. Different. *ways*. Roads. *Momus*. Greek god of fault-finding, who blamed Athena's house for not having wheels with which to escape unpleasant neighbours. Wright (p. 337) suggests Aesop, *Fab*. 275 as the source; Reynolds (pp. 319–20) suggests Neveletus' *Mythologia Aesopica* (1610), while Wolff (ii. 173) proposes Natalis Comes, *Mythologiae*, ix. 20. *more*. (Wants.) *want of . . . mixture*. Lacking the proper variety of terrain. *prospect*. View. *lurcheth*. Swallows up (medieval Lat. *lurcare*, to swallow food greedily), merchandise being dearer. *provisions*. Supplies of food and drink. *great living*. Large property. *scanted*. Lacking.

428. *take*. Take into account. *sort*. Arrange, dispose of. *Lucullus*. Plutarch, 'Life' of Lucullus, 39.4. *galleries*. Arcades. *lightsome*. Full of light. *fowl*. Bird (of any kind). *delivers*. Describes. *precepts*. Principles, rules. *perfection*. Completion, perfect realization. *Vatican*. The residence of the Pope, a vast assemblage of private and state apartments, chapels and courtyards. *Escurial*. El Escorial, the huge palace outside Madrid, built by Philip II of Spain in 1563–83. *fair*. Habitable, comfortable. *several sides*. Separate buildings. *banquet*. Dining hall. *Hester*. Esther 7: 8, 'The place of the banquet of wine'. *triumphs*. Magnificent shows. *household*. Domestic offices, residential quarters. *not only returns*. Not just the wings, where the

building turns back from the front, but extended on either side of the central tower at a lower elevation. The tower, extending over two storeys, is about 90 feet high. *severally partitioned*. Separately divided up. *preparing place*. Changing room for the performers in the 'masques and triumphs'. *at the first*. Beginning from the tower. *partition*. One running lengthwise. *state*. Proportion. *bigness*. Size. *parlour*. Smaller room, for conversation and less formal meals. *fair*. Handsome. *privy*. Private. *butteries*. Storerooms for provisions. *pantries*. Rooms for storing and cleaning kitchenware. *a piece*. Each. *goodly leads*. Well covered with lead. *railed*. Adorned. *open newel*. Central staircase built around an open well; a recent innovation in English houses (*c*.1605). *images of wood*. Carved wood figures, painted to look like brass; probably heraldic. *landing-place*. Landing. *point*. Appoint.

429. *tunnel*. Chimney-flue. *room*. Space on the ground floor. *outside*. (The turrets, not the staircases.) Reynolds suggests that Bacon may have had in mind the Great Court of his Cambridge college, Trinity. *striketh up*. Reflects. *But*. 'Let there be' (understood). *side alleys . . . to graze*. The lawn surrounded by stone paths, with two central paths intersecting and dividing it into quarters, the turf being cropped by grazing sheep. *near*. Closely. *row of return*. Side wing and front. *several*. Different. *works*. Designs in the stained-glass windows. *chambers of presence*. Reception rooms, where the king or lord receives his guests. *ordinary entertainments*. The family's normal common rooms. *double house*. Having rooms back and front. *thorough lights*. Through lights, windows placed on both sides of the room. *from*. Away from. *Cast*. Arrange. *become*. Betake oneself. *inbowed*. Bay windows, like balconies. *upright*. Not projecting. *uniformity*. Symmetrical appearance. *pretty*. Useful. *conference*. Conversation. *strike*. Penetrate (the sun's rays). *sides*. Two sides (neither on the back of the main building nor on the façade of the structure that closes the court). *inward*. Interior. *square*. Area. *cloistered*. Surrounded with cloisters. *all sides*. Both (lateral) sides. *decent*. Appropriately sized. *story*. Storey. *grotta*. Grotto, artificial cave. *place of . . . estivation*. Summer-house. *no whit*. Not at all. *work of statua's*. Statuary. *privy lodgings*. Private apartments, bedrooms. *foresee*. Provide. *them*. (Of the lodgings.) *infirmary*. The first appearance of this word in English (*OED*). *antecamera*. Ante-room. *recamera*. Retiring room.

430. *ground story . . . gallery*. At the end of the courtyard facing the garden, a portico. *open gallery*. Loggia, covered terrace. *take*. Partake of, enjoy. *cabinets*. Small rooms for displaying art objects. *daintily*. Elegantly. *richly hanged*. With ornate wall-hangings. *crystalline*. Windows of clear, not coloured glass. *elegancy*. Refinement. *upper*. (On the third floor.) *yield*. Permit. *fine avoidances*. Inconspicuous (or invisible) outlet pipes. *save*. Except. *green court plain*. Wholly turfed courtyard. *of the same*. Similar to the first. *garnished*. Ornamented. *built*. Built up, enclosed. *tarrasses*. Terraces; open spaces (above the cloisters that surround this courtyard). *leaded*. Covered with lead. *offices*. Buildings for the household administration. *galleries*. Arcades.

46. *Of Gardens*

430. *Gardens*. Bacon was closely associated with gardens all his life. Every house that he owned, rented, or built had extensive gardens, and he was responsible for redesigning and developing the gardens of his 'law school', Gray's Inn, where he had rooms for many years. For his alterations to the gardens of the Bacon family house at Gorhambury see the notes in his personal diary of 1608, the *Commentarius Solutus* (*Works*, xi. 76–7). The garden described here, with its 'royal ordering' like the palace in the previous essay, is meant for a prince. But Bacon does not follow the fashions for 'state and magnificence', elaborate structural and visual effects, preferring 'use' (which, he believes, constitutes 'the true pleasure of a garden'), and an environment pleasant to the senses all the year round. 'What is fresh in the essay is the aim of combining landscape architecture and horticulture for as much of the year as the gardener's art will allow' (Kiernan, p. 286). Kiernan's meticulous collation of all extant copies of the 1625 edn. has established that Bacon corrected the proofs of this essay himself (pp. cv cvi). *God... Garden*. Gen. 2: 8. *gross handyworks*. Mere manual contrivances, omitting nature. *civility*. Civilization. *elegancy*. Refinement. *stately*. In a stately manner. *finely*. Skilfully. *hold*. Maintain, believe. *ordering*. Arrangement. *severally*. Separately. *pine-apple trees*. Pine-trees, bearing cones. *germander*. Plants of the genus *Teucrium*, e.g. chickweed. *flags*. The long slender leaves of such plants as lilies or 'sword-flags'. *stooved*. Heated by a stove (in a hothouse). The uncorrected copies read 'stirred' (Kiernan, p. 139). *warm set*. Planted in a sunny spot. *mezereon-tree*. The dwarf bay tree, or German olive. *crocus vernus*. Probably the yellow and white spring saffron, since the yellow crocus was not introduced into England until 1629 (Wright, p. 361). *chamaïris*. The dwarf iris.

431. *cornelian-tree*. The male cornel-tree. *natures*. Kinds. *dammasin*. Damson-tree. *white thorn*. Hawthorn. *pinks*. Carnations. *bugloss*. Name of several boraginaceous plants, e.g. cowslips. *flos Africanus*. Type of marigold. *ribes*. Redcurrants. *rasps*. Raspberries. *satyrian*. Variety of orchis. *herba muscaria*. Type of hyacinth. *lilium convallium*. Lily of the valley. *genitings, quadlins*. Early summer apples. *barberries*. Shrub with sweet-smelling yellow flowers and red berries. *filberds*. Hazelnuts. *musk- melons*. Water-melons. *melocotones*. Melon peaches. *cornelians*. Fruit of the cornel-tree, used in some parts of England for tarts. *wardens*. Winter pears. *come*. Flower. *services*. Small, long brown berries of the service-tree (or sorb), edible when over-ripe. *medlars*. Fruit of the medlar-tree, edible when over-ripe. *bullaces*. Wild plums (black). *ver perpetuum*. 'Perpetual spring' (cf. Virgil, *Georgics*, ii. 149). Bacon altered the text here, his first version having been more expansive: 'Thus, if you will, you may have the *Golden Age* againe, and a Spring all the yeare long' (Kiernan, pp. cv, 140 n.). *place*. Location (in this case, London). *breath*. Scent, odour. *warbling*. Birdlike song. *red*. While in the bud, growing. *fast*. Retentive. *find nothing of*. Not notice. *Bartholomew-tide*. 24 Aug. *which [yield]*. An insertion proposed by Spedding and Wright; the 1625 text reads simply 'which'. The 1629 edn. changed

it to 'with' (Kiernan, p. 140). *cordial*. Stimulating. *dust*. Particles of pollen from the vine's small greenish-yellow flowers. *bent*. A rushlike grass, used for chimney ornaments. *so*. Provided that.

432. *passed by*. When one walks past. *burnet*. Small purplish-brown flower that yields a penetrating scent when crushed. *thirty acres*. As Kiernan (p. 291) notes, these were truly 'royal' dimensions, ten times the size of Henry VIII's Hampton Court garden, four times as big as Lord Burghley's Great Garden at Theobalds. This scale would have made it the largest garden in England, bigger even than Robert Cecil's Hatfield House. *heath or desert*. Part left uncultivated, wild. *going forth*. Outlet, exit. *alleys . . . sides*. Bacon disapproves of paths cutting across a garden. *in front upon*. Towards the garden's entrance. *hedge*. Fence. *buy*. Pay a high price for. *thorough*. Through. *covert alley*. Covered walk (made here from trellises and vines). *carpenter's work*. Wooden construction. *knots . . . earth*. Herbs planted in intricate geometric designs or abstract patterns, sometimes using coloured sands. *tarts*. 'Open pastry cases, with a sweet or savoury filling' (*SOED*). *hedge*. Enclosure. *entire*. Complete. *belly*. A protruding part. *deliver*. Let in, admit.

433. *for letting your prospect*. Which will obstruct the view. *ordering*. Disposition. *device*. Design, contrivance. *busy*. Over-elaborate. *images cut out*. The classical art of topiary (the training and clipping of shrubs into ornamental shapes) was enthusiastically revived in the Renaissance. Bacon was unusual in disliking it. *welts*. Ornamental edgings or borders (of a dress). *pyramides*. Plants cut in this shape. *closer*. Narrower. *mount*. Raised central area, giving a unique perspective on the garden below. *ascents*. Paths leading uphill, circling the mount at various levels, so giving a clear view over the garden. *bulwarks*. Ramparts, projections. *embossments*. Ornaments with raised work. *banqueting-house*. For light meals, or afte-dinner desserts. *chimneys . . . cast*. Ornamental fireplaces. *pools*. Stagnant pools. *mar*. Spoil. *receipt*. Receptacle, basin. *do well*. Are suitable. *stay*. Stand still. *cistern*. Tank. *admit . . . curiosity*. Is capable of elaborate design. *images*. Figures, designs. *of lustre*. Shining. *delivered*. Admitted. *fair spouts*. Well-functioning pipes. *equality of bores*. Of the same size as the spouts filling the basin. *canopies*. Over-arching spouts. *plot*. Project (also 'plot of ground'). *framed*. Arranged, ordered.

434. *amongst*. Intermixed. *set*. Planted. *bear's-foot*. Stinking hellebore. *sweet*. Sweet-smelling. *sightly*. Pleasant to look at. *standards*. Shrubs, trees standing upright. *pricked*. Set, planted. *but*. Only. *kept with cutting*. Kept pruned. *out of course*. Out of bounds, excessively. *alleys, private*. Secluded walks (perhaps covered by pergolas). *frame*. Ordain. *going wet*. To avoid walking on wet grass. *ranges*. Ranks, rows. *would*. Should. *deceive*. Deprive of nutriment. *do not deny but*. Do not exclude that. *thick*. Close together. *rest upon*. Keep to. *make account*. Consider. *overcast*. Cloudy.

435. *turfed.* Covered with grass-roots. *have more scope.* Can fly about more freely. *nestling.* Place to nest. *foulness.* Dirt, ordure. *platform.* Plan, outline. *model.* Strict pattern (graphic design to be followed in construction). *workmen.* Skilled craftsmen. *for state and magnificence.* Ostentatious display. *but nothing.* Adding nothing.

47. *Of Negotiating*

435. *Negotiating.* A modern equivalent might be 'business dealings'. *third.* Third party. *by . . . self.* In person. *draw.* Obtain, elicit. *it.* There. *danger.* Dangerous. *by pieces.* Incompletely. *breedeth regard.* Inspires respect. *tender.* Tricky, delicate. *countenance.* Behaviour, reactions. *reserve.* Keep. *disavow.* Disown words wrongly imputed to you. *expound.* Disclaim wrong interpretations by explaining your real meaning. *instruments.* Go-betweens. *plainer.* Simpler, straightforward. *like.* Likely. *success.* Outcome (good or bad). *somewhat.* Something. *grace.* Ingratiate. *help . . . sake.* Improve upon the facts, so as to please their employer. *affect.* Have a liking for. *quickeneth much.* Is efficacious. *expostulation.* Protest, complaint. *fair-spoken.* Good orators. *crafty.* Astute. *froward.* Perverse, contrary. *absurd.* Unreasonable. *bear out.* Back up, confirm. *prevailed.* Succeeded. *prescription.* Title, reputation. *sound.* Probe. *afar off.* Indirectly. *fall upon.* Come to. *short.* Brusque, unexpected. *in appetite.* Desirous of rising, ambitious.

436. *upon conditions.* Conditionally (expecting a return in service or reward). *performance.* Action performed. *all.* All-important. *demand.* Expect. *which.* That it. *go before.* Be performed first (without guarantees of a return). *party.* Person. *he.* The first agent. *counted.* Reckoned. *honester.* More creditable. *practice.* Plotting. *discover.* Reveal the true nature of someone. *work.* Influence, manipulate. *in trust.* By trusting someone. *passion.* Anger. *at unawares.* Unexpectedly, being surprised. *would.* Want to. *fashions.* Habits. *disadvantages.* Infirmities. *awe.* Influence by awe. *have interest in.* Are involved with. *look for.* Expect. *may not look.* Should not expect.

48. *Of Followers and Friends*

436. *Costly.* Expensive. *train.* Retinue (punning on 'peacock tail'). *which.* Who. *charge.* Burden. *importune.* Importunate. *suits.* Petitions, requests for favour. *challenge.* Claim. *conditions.* Treatment, favours. *countenance.* Approval, moral support. *Factious.* Partisan. *range.* Ally. *ensueth.* Results. *ill intelligence.* Misunderstandings. *glorious.* Boastful, ostentatious. *taint.* Spoil. *export.* Deduct. *honour.* Good reputation. *return.* Import (bringing envy in return). *espials.* Spies. *inquire . . . house.* Cf. Juvenal, iii. 113. *bear tales of them.* Divulge information. *officious.* Forward to do office. *exchange tales.* Trade information about households. *estates.* Classes. *answerable.* Corresponding. *that.* The profession. *him.* The person of high rank, nobleman. *civil.* Decent, orderly. *so.* Provided.

437. *popularity.* Courting popular favour. *apprehendeth.* Studies, strives. *advance.* Promote, reward. *desert.* Merit. *no . . . sufficiency.* When neither is clearly superior. *take with.* Have to do with; or perhaps 'employ'. *passable.*

Generally acceptable, tolerable. *base*. Corrupt. *virtuous*. Able. *in government*. Normal employment, business. *one rank*. The same capacity. *equally*. In the same way. *claim a due*. Demand favours as if by right. *discontent*. Discontented. *contrariwise*. On the contrary. *in favour*. When it comes to distributing favours. *difference ... election*. Discrimination, choice. *officious*. Ready to serve (Lat. *officiosus*). *all ... favour*. All depends on the patron's good grace. *make too much*. Take too much notice of; or 'treat favourably'. *hold out*. Sustain. *proportion*. Level of treatment. *one*. Servant or employee. *softness*. Indulgence. *gives a freedom*. Gives occasion. *disreputation*. Discredit. *immediately*. Directly. *great with*. Acquainted with, friendly. *their*. (That of their master.) *distracted*. Receive conflicting advice. *of the last impression*. Affected by the most recent opinion. Cf. *Adv.L.*, above, p. 258. *gamesters*. Gamblers. *vale ... hill*. The valley gives the best view of the hill: important people are more transparent to those below them. Cf. *Adv.L.*, above, p. 252, and the dedication to Machiavelli, *The Prince*. *wont to be magnified*. Friendship, especially between men, was much celebrated in classical myth and history (Orestes and Pylades, Damon and Pythias, *et al.*) and philosophy. See Diogenes Laertius, viii. 10, and Aristotle, *Eth. Nic.* ix. 8. *comprehend*. Include, reciprocally depend on.

49. *Of Suitors*

437. *Suitors*. In the Jacobean court, as in many governmental hierarchies, a person's main chance of success in a dispute or project (to gain office, say, or a wealthy sinecure) lay in winning the support of some influential figure who could affect the outcome of the affair. As a result, men of influence were pestered with suitors who requested their help and gave them valuable presents or bribes. Bacon had much unhappy experience on both sides of this relationship. Reynolds (pp. 339–40) notes several points of contact with Bacon's letter and his 'Advice' to George Villiers (later Duke of Buckingham): *Works*, xiii. 15, 27–30. *ill matters*. Evil cases. *projects*. Schemes. *undertaken*. Taken on (by a patron). *putrefy*. Corrupt. Private citizens who continually lobby men in public office will damage their moral standards. *performance*. Doing what they promised. *embrace*. The Latin version is more explicit: 'receive, and eagerly promise aid'. *which*. Who. *deal*. Behave, act. *life ... mean*. That the matter will succeed in some other way. *second*. Secondary, less important. *cross*. Thwart, obstruct. *make*. Gain, obtain.

438. *entertainment*. 'Diversion, something which withdraws attention from the main subject' (Wright, p. 364). *to the end*. In order. *sort*. Measure. *equity*. Moral justice. *suit of controversy*. One in a chancery suit or lawcourt, in which a suitor seeks his patron's help to influence the judge's decision (Reynolds). *desert*. Merit. *suit of petition*. One seeking help in obtaining some office for which there were other competitors. *affection*. Liking, partiality. *in justice*. In questions of justice. *countenance*. Influence. *compound*. Settle by agreement, compromise. *carry*. Pursue, prosecute. *depraving*. Misrepresenting, disparaging. *disabling*. Damaging, depreciating. *deal*. Act, take part. *with honour*. Honourably. *referendaries*. Referees, advisers. *led by the nose*. Deceived, manipulated. *distasted*. Disgusted, offended.

abuses. Deceits. *denying*. Refusing. *at first*. From the outset. *reporting... barely*. Giving a frank account of the outcome, or perhaps 'chance of success'. *challenging*. Claiming. *gracious*. Agreeable, deserving thanks. *suits of favour*. Petition (already mentioned). *first coming*. Priority in presenting one's petition. *take... place*. Have little effect, not give precedence. *so far forth*. So far. *his*. The first petitioner. *intelligence*. Knowledge. *note*. Information. *party*. Person (presenting the suit). *discovery*. Disclosure of information. *value*. Relevance, import. *simplicity*. Naïvety, stupidity. *right*. Justice. *mean*. Means, condition. *voicing*. Divulging. *forwardness*. Preparation. *timing*. Choosing the right moment. *which*. Who. *like*. Likely. *mean*. Go-between, instrument. *fittest*. Most suitable. *greatest*. Most powerful. *certain*. Definite, specific. *reparation*. Restoration, reversal. *denial*. Refusal. *first grant*. What was originally requested. '*Iniquum... feras*'. Quintilian, IV. v. 16, 'ask for too much, if you want enough'. *hath... favour*. Enjoys the goodwill of whoever will grant the request. *were*. Had. *rise in his suit*. Gradually increase his requests. *easy*. Sure of succeeding. *in*. For. *general*. All-purpose, indiscriminating. *contrivers*. Devisers of litigations and petitions.

50. *Of Studies*

439. *for ability*. To make men able. *retiring*. Retirement (for individual study). *ornament... discourse*. Effectiveness in conversation and in society. In classical and Renaissance rhetoric 'ornament' could also refer to a soldier's weapons or kit, that is, to effective communication, not mere display. *judgment... business*. Evaluating and organizing the practicalities of life. *expert*. Experienced (in specific fields). *execute*. Do, perform. *counsels*. Arrangements. *plots and marshalling*. Planning and executing. *sloth*. Laziness (that is, avoiding one's duties to society in the *vita activa*). *humour*. Character, style. *scholar*. One living in a university; remote from the active life. *proyning*. Pruning, cultivating. *and*. On the other hand. *directions... large*. Instructions that are too general, vague. *Crafty*. Cunning. *contemn*. Despise (presumably as irrelevant to cunning practices). *simple*. Innocent (but also 'stupid'). *admire*. Wonder at. *wise*. Prudent, judicious. *teach not their own use*. A course of study can convey the content and methods of a discipline, but not its application. *without... above them*. External to, and transcending the scope of any discipline. *Read... to weigh and consider*. Cf. the account of 'vanities in studies' in *Adv.L.*, above, pp. 140–1. *curiously*. With minute attention. *by deputy*. By proxy (someone hired for that purpose). *extracts*. Excerpts. *would*. Should. *arguments*. Matters. *distilled books*. Epitomes, which Bacon, like many Renaissance scholars, regarded as destructive to scholarly enquiry. Cf. his criticism of Ramus, *Adv.L.*, above, p. 236. *distilled waters*. Juices distilled from radishes, sage, and other plants, for medicinal purposes. *flashy*. Insipid, tasteless. *conference*. Consultation. *confer*. Consult. *present wit*. Quick, alert mind. *witty*. Ingenious, full of ideas. *grave*. Serious, weighty. *Abeunt... mores*. 'Studies affect our behaviour': Ovid, *Heroides*, xv. 83; *Promus*, no. 1121. *stond*. Obstruction, obstacle. *wit*. Mind. *wrought out*. Eliminated. *Bowling*. Playing bowls.

stone and reins. Bladderstone, disorders of the kidneys. *shooting*. Archery. *wandering*. Lacking concentration. *demonstrations*. Mathematical proofs. *called away*. Distracted. *never so little*. Even slightly.

440. *differences*. Distinctions (which proliferated in scholastic thought: cf. *Adv.L.*, above, pp. 141, 157). *Schoolmen*. Medieval 'scholastic' philosophers. *cymini sectores*. Quibblers, splitters of cumin seeds: Dio Cassius, lxx. 3.3 and Erasmus' *Adagia*; *Promus*, no. 891. *beat over*. Reflect upon, compare. *prove*. Test (by comparison). *illustrate*. Clarify. *lawyers' cases*. Law reports. *receipt*. Recipe, treatment.

51. *Of Faction*

440. *Faction*. The grouping together of politicians holding a common attitude, often in opposition to the government. It must be remembered that Bacon was writing before the establishment of separate political parties. *not wise*. Bacon agrees with Machiavelli (*Discourses*, iii. 27) on the dangers of a ruler continually trying to exploit rival political groups. *estate*. State. *proceedings*. Actions. *the respect of*. Considering the wishes of. *policy*. Both 'astuteness' and 'politics'. *contrariwise*. On the contrary. *ordering*. Disposing, executing. *general*. Of public advantage. *several*. Distinct, different. *with correspondence to*. According to, suiting the interests of. *Mean*. Of low rank. *rising*. Social or political advancement. *adhere*. Associate, stick together. *great*. Distinguished, powerful. *strength in themselves*. Their own resources. *indifferent*. Impartial. *passable*. Acceptable. *giveth best way*. Opens the best path (to office). *conjunction*. Cohesion. *stiff*. Determined. *moderate*. Pliable. *extinguished*. Eliminated. *Lucullus*. Lucius Licinius Lucullus (*c*.117–56 BC), leader of the opposition to Pompey's bid for power; Pompey responded by making a faction with Julius Caesar. Cf. Plutarch, 'Life' of Lucullus, 38.2, 42.4 ff. *Optimates*. A conservative group from 'the best' families. *brake*. Broke up, quarrelled (Caesar defeated Pompey at Pharsalus). *of wars*. From civil wars. *private*. Not threatening public order. *seconds*. Secondary figures. *cyphers*. Nonentities. *cashiered*. Dismissed in disgrace. *use*. Practice. *placed*. Settled in place, position. *take in*. Join, realign themselves. *enter*. Begin. *belike*. Probably. *sure*. Secured. *purchase*. Acquisition. *lightly . . . it*. Easily comes off the gainer. *balancing*. Equilibrium. *casteth*. Decides (as in a 'casting vote'). *even carriage*. Impartial proceeding.

441. *of*. From. *trueness . . . self*. Loyalty to one's own interests. *end*. Intention. *both*. Both factions. *suspect*. Suspicious. *Padre commune*. 'Father to all men.' *refer*. Direct (as if naturally belonging to). *had need*. Need to. *side*. Take sides, align. *raise*. Give rise to. *obligation*. Binding agreement. *paramount*. Superior. '*tanquam . . . nobis*'. 'Just like one of us' (Gen. 3: 22). *League of France*. The Catholic league, formed in 1576 between Henri III and the Guise. *factions . . . in princes*. Cf. Machiavelli, *Discourses*, iii. 27. *high . . . violently*. Terms also used to describe planetary motion, as in the following sentence, and Essay 11, 'Of Great Place', above, p. 361. *proper*. Own, individual. *quietly*. With tranquillity; in silence. *primum mobile*.

'First mover', the outside sphere in Ptolemaic cosmology, moved by God. See Essay 15, 'Of Seditions and Troubles', above, p. 367.

52. *Of Ceremonies and Respects*

441. *Ceremonies and Respects*. Formal and respectful behaviour, good manners. Bacon provides his own version of themes much treated in the Renaissance conduct-book, such as Della Casa's *Galateo* (1558), or especially Castiglione's *Il Cortegiano* (1528), with its emphasis (e.g. i. 26) on *sprezzatura*, a combination of natural, spontaneous, graceful behaviour, downplaying or dissimulating any conscious exertion. *real*. Sincere, straightforward, lacking social graces. *parts of virtue*. Merits. *foil*. 'A thin leaf of metal placed under the stone to improve its colour and lustre' (Reynolds). *it is in*. The same holds for. *light gains*. Small profits. *light . . . purses*. *Promus*, no. 269. *thick*. Frequently. *small matters*. Minor politenesses. *in note*. Noticed. *on festivals*. Rarely. *Isabella*. Queen of Spain, according to a contemporary book of sayings. *commendatory*. Recommending a person's suitability (e.g. for employment). *forms*. Manners. *attain*. Acquire. *so*. Provided that. *and*. Then. *the rest*. Practising them himself. *For if he labour . . . unaffected*. Advice very close in spirit to Castiglione, as Melchionda indicates (pp. 605–6). *measured*. Counted, calculated. *comprehend*. Deal with. *breaketh*. Trains, accustoms. *again*. In return. *be*. Are. *formal*. Unduly precise. *dwelling upon*. Insisting on.

442. *credit*. Trustworthiness. *conveying*. Insinuating. *imprinting*. Striking, impressive. *passages*. Phrases. *compliments*. Courtly flattery. *a little to keep state*. Be a bit formal, dignified. *too much*. Effusive, exaggerated. *apply*. Adapt, accommodate. *upon regard*. Out of personal respect, affection. *facility*. Pliancy. *seconding*. Agreeing with. *motion*. Proposal. *allow his counsel*. Approve of his advice. *alleging*. Adding, producing. *perfect*. Too meticulous. *sufficient*. Capable. *business*. Affairs, practical life. *full of respects*. Ceremonious. *curious*. Precise. *times and opportunities*. (For making compliments.) '*He . . . reap*'. Eccles. 11: 4. *strait*. Tight, constricting. *point device*. Very precisely fashioned (cf. *Adv.L.*, above, p. 266).

53. *Of Praise*

442. *Praise . . . virtue*. In traditional ethics and rhetoric (the branch known as epideictic), from the Greeks to the Renaissance, praise was the right response to virtue, blame to vice. *it is as*. It varies with, depends on. *naught*. Worthless. *vain*. Conceited, superficial. *work*. Arouse. *sense*. Capacity, mental apprehension. *or*. Correcting Spedding's 'of'. *perceiving*. Perception. *shews*. False appearances. '*species . . . similes*'. 'Outward appearances resembling virtues' (Tacitus, *Annals*, xv. 48). *swoln*. Inflated. A recurrent metaphor in Bacon for the uneven survival of merit: cf. *Adv.L.*, above, p. 145, and *Works*, iii. 503, iv. 72, 77; etc. *serve best*. Impress most. *quality*. Rank, high status. *judgment*. Wisdom, experience. '*Nomen . . . fragrantis*'. 'A good name is like a sweet-smelling ointment' (Eccles. 7: 1, adapted). *not . . . away*. Fade, disappear. *ointments*. Perfumes. *points*. Instances. *hold . . . suspect*. Regard with suspicion. *of*. From.

443. *have*. Hold ready, prepared. *arch-flatterer*. Self-love. *uphold*. Endorse, agree with. *look*. Watch out. *out of countenance*. Discomforted, embarrassed. *entitle*. Impute. *perforce*. Of necessity. *spreta conscientia*. 'Self-knowledge disdained': someone accepts flattery even when conscious of his own weakness. *respects*. Paying compliments. *form*. Mode of behaviour. *laudando praecipere*. 'To praise by teaching' (Pliny, *Epist*. iii. 18). *what they are... should be*. As Melchionda (p. 609) observes, in praising the use of ideal behaviour-types here Bacon reverses his endorsement elsewhere of Machiavelli and all those 'that write what men do, and not what they ought to do' (*Adv.L.*, above, p. 254). The discrepancy shows the extent to which some *Essays* resemble the traditional conduct-book. *stir*. Excite. '*pessimum... laudantium*'. 'The worst kind of enemies, the adulators' (Tacitus, *Agricola*, 41). *Grecians*. Perhaps Theocritus, *Idyll*, ix. 30, xii. 23–4, where the 'push' or pustule is said to be the result of having bestowed praise untruly. *upon... one's tongue... lie. On the liar's tongue*. *with opportunity*. At the right time. *not vulgar*. But appropriate. '*He... curse*'. Prov. 27: 14. *irritate*. Provoke. *a man's*. One's own. *decent*. Proper. *he*. That man (so praising his own profession). *magnanimity*. Generosity, nobility. *which*. Who. *notable*. Remarkable. *civil business*. Civic (as opposed to ecclesiastical) affairs. *embassages*. Diplomatic activities. *sbirrerie*. Bailiffs and constables. *under-sheriffries*. Trivial affairs. *catch-poles*. Petty officers of justice. *interlace*. Insert. '*I... fool*'. 2 Cor. 11: 23. '*magnificabo... meum*'. 'I will magnify my apostolate' (Rom. 11: 13, adapted).

54. *Of Vain-Glory*

443. *Vain-Glory*. Vanity. *prettily devised*. Cleverly imagined. In fact the fabulist was Laurentius Abstemius (L. Bevilacqua), whose fables were included in Renaissance editions of Aesop. *vain*. Empty-headed, silly. *goeth alone*. By its own motion. *greater means*. Superior power. *carry*. Perform. *glorious*. Boastful (cf. the *Miles Gloriosus* or 'boastful soldier' in classical comedy). *factious*. Quarrelsome.

444. *bravery... comparisons*. Boasting depends on a person elevating himself, devaluing others. *vaunts*. Boasts. *not effectual*. Ineffective (the reality beneath the boasting). *bruit*. Cry, fuss. *quality*. Characteristic. '*There... lies*'. Editors have cited various passages in Livy (Wright: xxxvii. 48; Kiernan: xxxv. 12, 17–18), but Reynolds (p. xxxix) and Melchionda (pp. 611–12) show that the correct passage is in fact xxxv. 49.4, where Livy writes: 'if anyone had been ignorant before what cause had brought Antiochus and the Aetolians together, it could now be clear from the speeches of their delegates and of boasts of strength which they did not possess they filled one another's minds, and in turn were filled, with groundless hopes.' Bacon referred to it again in his *Apophthegms* (*Works*, vii. 171–2): 'It was an easy matter to perceive what it was that had joined Antiochus and the Aetolians together; that it appeared to be by reciprocal lying of each, touching the other's forces.' *cross*. Reciprocal, interchanged. *either*. Each. *interest*. Influence over. *somewhat... nothing*. In contradiction to the normal law, that 'nothing will come of nothing'. *opinion*. Reputation. *brings on substance*.

Acquires a (spurious) reality (for 'nothing succeeds like success'). *point.* Characteristic. *as iron sharpens iron.* Prov. 27: 17, and *Promus*, no. 549. *upon charge and adventure.* Subject to expense and risk. *composition.* Combination. *and.* Whereas. *solid.* Steady. *ballast... sail.* Have a steadying, rather than a propulsive effect. *flight.* Ascent, as of an arrow, with its 'feathers'; also referring to Virgil's description of Fame, cited in the fragmentary Essay 'Of Fame', above, pp. 454–5. *'Qui... inscribunt'.* 'Those who write books on scorning glory, sign their names [on the title- page]': Cicero, *Tusc. D.* I. xv. 34. *Socrates, Aristotle, Galen.* None of whom attacked glory. Bacon elsewhere accuses Socrates of false modesty (*Adv.L.*, above, p. 222), Aristotle of confuting every author he named in order to promote his own authority (ibid. 193–4, 204), and Galen of various faults (*Works*, iii. 531). Bacon was presumably following his stated policy of attacking reverence towards antiquity: cf. *Adv.L.*, above, pp. 143–4, *Works*, xi. 64. *beholding.* Indebted. *as.* That. *at the second hand.* This sentence has given much difficulty to editors: Melchionda (p. 613) spends half a page elucidating it. I suggest that Bacon is alluding to the well-known principle from classical philosophy that 'virtue is its own reward', and drawing the consequence that virtue has therefore never been indebted to humanity ('at the second hand') for a reward or recognition, which would in any case be inferior. *had.* Would have. *Plinius Secundus.* Pliny the Younger (*c.* AD 61–115), quoted below. *ceilings.* Panelling, wainscotting. *property.* Characteristic. *Omnium... ostentator.* 'A boaster with a certain art [of setting forth to advantage] all he had said or done' (*Histories*, ii. 80). *magnanimity.* Great-heartedness. *discretion.* Sagacity, judgement. *comely.* Becoming. *gracious.* Graceful. *excusations.* Excuses. *cessions.* Making concessions, being compliant. *governed.* Controlled (with calculation). *liberal.* Generous. *wittily.* Ingeniously. *'In... less'.* *Epist.* vi. 17.

445. *be to be.* Merits being. *scorn.* Objects of scorn... admiration. *idols of parasites.* They are worshipped by those who prey on them: as in Terence's *Eunuchus*, where the foolish Miles Gloriosus Thraso is flattered and exploited by Gnatho (Kiernan, p. 303).

55. *Of Honour and Reputation*

445. *winning.* Obtaining. Wright (p. 344) notes that Bacon's unpublished MS of 1612 reads 'The *true* winning of honour', which would be preferable. *revealing.* Recognition. *disadvantage.* Injustices. *affect.* Pretend to, counterfeit. *inwardly.* In private. *darken.* Obscure, lessen. *given over.* Abandoned. *circumstance.* 'All the surroundings and accompaniments of an action' (Wright, p. 359). *purchase.* Acquire. *virtue.* Ability. *follower.* Imitator. *temper.* Balance, harmonize. *combination.* Group. *husband.* Manager. *entereth into.* Undertakes. *failing... can honour him.* Cf. Suetonius, 'Life' of Augustus, 25. *broken.* Made to break, shine. *upon another.* By comparison with, at the expense of a rival (cf. Shakespeare, *1 Henri IV*, III. ii. 132–52). *quickest.* Sharpest. *facets.* 'Little faces'. *contend.* Strive, exert himself. *bow.* Special accomplishment. *'Omnis... emanat'.* 'All reputation derives from one's servants' (Q. Cicero, *De petitione consultatus*, v. 17).

canker. An eating sore, corruption. *declaring*. Demonstrating. *a man's*. One's. *felicity*. Good fortune. *policy*. Astuteness. *marshalling*. Hierarchy, disposition. *conditores imperiorum*. 'Founders of empires.' *commonwealths*. Republics. *Romulus*. Legendary founder of Rome. *Cyrus*. Founder of the Persian monarchy. *Caesar*. Julius Caesar, who established the basis for the Roman empire. *Ottoman*. Othman or Osman (1259–1326), founder of the Turkish dynasty. *Ismael*. The 'Sophy' of Persia, who became sovereign in 1503, aged 18. *perpetui principes*. 'Perpetual princes.' *ordinances*. Authoritative decrees. *Lycurgus*. Reformer of the Spartan government and mores, 7th century BC. *Solon*. Reformer of the Athenian constitution and laws, 6th century BC. *Justinian*. Roman emperor AD 527–65, codifier of Roman laws. *Eadgar*. King of England, 959–75, codifier of English law: see *Works*, xiv. 361. *Alphonsus of Castile*. King of Castile and Leon (1252–84). *Siete partidas*. The 'Seven parts' of his law code. *liberatores . . . salvatores*. 'Liberators, saviours'. *compound*. Settle, put an end to.

446. *Augustus Caesar*. His defeat of Antony inaugurated a long period of peace for the Roman empire. *Vespasianus*. Delivered the empire from the civil wars following the death of Nero. *Aurelianus*. Emperor in AD 270, victor in many battles. *Theodoricus*. Liberated Italy from foreign dominion, AD 493. *Henry the Seventh*. Put down several rebellions. *Henry the Fourth*. Ended the long wars between Catholics and Protestants, signing the Edict of Nantes in 1598. *propagatores . . . imperii*. 'Enlargers or defenders of empire.' *patres patriae*. 'Fathers of their country', an honorific title given to many Roman emperors, e.g. in Suetonius, 'Life' of Tiberius, 67. *which*. Who. *good*. Happy. *participes curarum*. 'Sharers of care', confidants. But cf. note on Essay 27, above, p. 745. *duces belli*. 'Leaders of war.' *lieutenants*. Second-in-command. *scantling*. Measure, limit. *negotiis pares*. 'Equals in business.' *places*. Duties. *sufficiency*. Ability. *of*. In the case of. *M. Regulus*. Roman general (d. 250 BC) who, although captured, rejected peace terms that would have brought his release. *the two Decii*. Father and son, who, in 340 BC and 295 BC respectively, sacrificed themselves in battle for Rome.

56. *Of Judicature*

446. *Judicature*. Bacon's training, from the age of 15, was in the law, a profession with which he never lost contact. He began studying at Gray's Inn in 1576, was called to the bar in 1582, but continued to serve his Society in several offices: Bencher (on the governing body), Reader (or special lecturer on points of law), and Treasurer (1608–17). In his public career he became Solicitor-General in 1607, Attorney-General in 1613, Lord Keeper in 1617, and Lord Chancellor in 1618, these last two offices making him Chief Judge of the Court of Chancery until his removal in 1621. This Essay, written for the 1612 volume, and only slightly changed in 1625, is concerned more with the duties of judges (the Latin title is 'De officio Judiciis') than the administration of justice, and should be read alongside those professional works deriving from Bacon's own duties as a senior legal official, such as the 'Charge on opening the Court of the Verge', 1611 (*Works*, xi. 265–75), the 'Speech on taking his seat in Chancery' (ibid. xiii. 182–93), the Speeches to

Justice Hutton, Sir John Denham, and Sir William Jones (ibid. 201–7), and the 'Speech to the judges before the Circuit', 1617 (ibid. 211–14). *jus dicere ... not jus dare*. To interpret the words of the law as a whole, not unduly pressing a single point or idiosyncratic version. *stick*. Scruple. *pronounce*. Give authoritative utterance. *novelty*. Innovation. Bacon thought that caution should be observed in amending the laws. *witty*. Ingenious. *reverend*. Worthy of respect. *plausible*. Courting applause. *advised*. Deliberate, acting on reflection. *portion*. Lot, assignment. '*Cursed ... landmark*'. Deut. 27: 17. *mislayer*. One who misplaces. *mere-stone*. Boundary mark (recording land-ownership). *blame*. Blameworthy. *capital*. Chief. *defineth amiss*. Gives an erroneous judgment. *foul*. Wrong, corrupt. The damage so arising is much greater in the English legal system, in which adjudged cases stand as precedents. *sentence*. Verdict. '*Fons ... adversario*'. 'A just man falling down in his cause before his adversary is a troubled fountain and a corrupt stream' (Prov. 25: 26, paraphrased). *may have reference*. May be seen as referring (i.e. does so refer). *parties that sue*. Those who take a case to court. *sovereign or state*. Monarchy or republic. *causes*. Motives, but also 'legal cases'. *There be*. There are those. '*There ... wormwood*'. Amos 5: 7. Wormwood is a bitter plant, prescribed as a drug to expel worms.

447. *force and fraud*. Violence and injurious deceit. According to classical philosophy, these were the two principal destroyers of justice: cf. Cicero, *De officiis*, I. xiii. 41. *close*. Cunningly concealed. *contentious suits*. Legal actions arising out of malice or quarrelsomeness. *spewed out*. Rejected with loathing. *surfeit*. Over-eating, causing disgust. *as God ... hills*. Isa. 40: 3–4; Luke 3: 5. *high hand*. Overbearing use of authority. *violent prosecution*. Rather than the obvious meaning, a 'prosecution using violent methods', Melchionda (p. 619) suggests that Bacon means an unjustified, oppressive prosecution of a private citizen by the state. *combination*. Conspiracy. *great counsel*. Kiernan suggests the sense 'disparity of counsellors' (p. 308), but the phrase could also mean 'prestigious advocates', or 'very able defence lawyers', whose skills can gain an unjust advantage (Melchionda, p. 619). *to make*. In making. '*Qui ... sanguinem*'. 'He who blows his nose vigorously, makes it bleed' (Prov. 30: 33). *where*. When. *wrought*. Worked (too vigorously). *hard constructions*. Extreme interpretations. *torture*. Punning on the original sense, turn or twist. *have care ... rigour*. Lest punishments meant as extreme deterrents become rigidly or mechanically applied in less serious cases. '*Pluet ... laqueos*'. 'He shall rain snares upon them' (Ps. 11: 6). *pressed*. Rigorously applied. *snares*. Traps. *of long*. For a long time. (For this whole sequence cf. Shakespeare, *Measure for Measure*.) *confined*. Restricted. '*Judicis ... rerum*'. 'It is the duty of the judge [to consider] both the deed and the circumstances of the deed' (Ovid, *Tristia*, i. 1.37). *in justice*. In the administration of justice, or 'giving sentence'. *example*. Deed, act. *counsel*. Advocate for the defence. *overspeaking*. One who talks too much and does not listen. *well-tuned cymbal*. Source of harmony (from Ps. 150: 5). *no grace*. Unbecoming. *to find*. To establish by his own efforts. *from the bar*. From the barrister responsible. *conceit*. Understanding. *cutting off ... too short*. Brusquely interrupting. *prevent*. Anticipate. *parts*. Tasks. *direct the*

evidence. Elicit relevant testimony. *moderate*. Control. *impertinency*. Irrelevance. *give the rule*. Pronounce the verdict. *above*. More than. *glory*. Vainglory, vanity. *willingness*. Wishfulness. *to hear*. In hearing. *staid and equal*. Steady, uniform. *they*. The judges. '*represseth . . . modest*'. 'God resisteth the proud, but giveth grace to the humble' (Jas. 4: 6, paraphrased).

448. *noted*. Notorious. *fees*. Bribes. *bye-ways*. Secret ways; devious influences. *gracing*. Compliment. *fair*. Handsomely. *obtaineth*. Attains the victory. *conceit*. High evaluation; but also 'illusion'. *to the public*. For the general good. *civil*. Discreet. *gross*. Grave. *slight*. Insufficient, insubstantial. *pressing*. Insistent, importunate behaviour (as in cross-questioning). *chop*. Bandy words, dispute. *wind . . . into*. Insinuate. *meet . . . half way*. Interrupt a case in progress. *proofs*. Evidence. *ministers*. Officers of the court. *hallowed*. Holy. *foot-pace*. Walking-area near the judge's bench. *precincts and purprise*. The whole area or enclosure of the court. '*Grapes . . . thistles*'. 'Do men gather grapes of thorns, or figs of thistles?': Matt. 7: 16; *Promus*, no. 1450. *catching*. Grasping. *polling*. Stripping, plundering (hence the term 'catch-pole'). *attendance*. Administration of justice. *sowers of suits*. Who bring unjustified legal cases. *swell*. Blow up, overcharge. *pine*. Dwindle, starve. *quarrels of jurisdiction*. Disputes (often lengthy) as to which court should try a case. *amici curiae*. 'Friends of the court'. *parasiti curiae*. 'Parasites of the court'. *puffing*. Extending beyond its proper scope. *scraps*. Petty gains. *sinister*. Left-handed, hence 'inauspicious'. *pervert*. Divert. *poller*. Court-official who extracted fees (often excessive and unnecessary). *resemblance*. Comparison. *bush*. Here, a bramble. *weather*. Bad weather. *ancient*. Experienced. *understanding*. Intelligent. *business*. Proceeding. *finger*. Sign-post. *Twelve Tables*. The earliest code of Roman laws, *c*.451 BC (engraved on tablets in the Forum). *Salus . . . lex*. 'The welfare of the people is the highest law': in fact from Cicero, *De legibus*, III. iii. 8. *in order to that end*. Ordained for that purpose. *captious*. Misleading. *oracles*. Sources of (supposedly) infallible truth.

449. *states*. 'Estates of the realm' (*SOED*), or civic authorities. *intervenient*. Intervening. *business of state*. National affairs. *deduced*. Brought down. *meum and tuum*. 'Mine and yours': the basic principle of ownership in private law. *reason*. Principle. *trench to*. Extend in effect to. *point of*. Matters concerning. *estate*. State; national interests. *parts*. Rights, prerogative. *alteration*. Innovation. *weakly*. Foolishly. *true policy*. Wise government. *antipathy*. Incompatibility. *spirits*. The 'animal spirits' in the blood. *that*. Which. *Salomon's throne . . . lions*. 1 Kgs. 10: 19–20. *circumspect*. Attentive. '*Nos . . . legitime*'. 'We know that the law is good, if only a man use it lawfully' (1 Tim. 1: 8, Vulgate).

57. *Of Anger*

449. *Anger*. For a partial source see Aristotle, *Rhetoric*, ii. 2, on anger. *bravery*. Ostentatious attempt. *Stoics*. See e.g. Seneca's treatise *De ira*, which indeed argues that anger should be wholly suppressed. '*Be angry . . . anger*'. Eph. 4: 26. *race*. Course, scope. *habit*. Mental disposition

or constitution. *attempered*. Regulated, moderated. Cf. Aristotle, *Rhetoric*, ii. 3, on 'growing calm' as 'a settling down or quieting of anger' (1380a5 ff.). *refrained*. Restrained. *throughly*. Completely. *That... falls. De ira* i. 1. *ruin*. Falling brickwork or masonry. *that it*. That which. *To... patience*. Luke 21: 19. *out of patience... soul*. Cf. Cicero, *De officiis*, I. xxxviii. 136 f. *animasque... ponunt*. 'And lay down their lives in the wound' (Virgil, *Georgics*, iv. 238).

450. *baseness*. Degradation. *Anger... sick folks*. Aristotle, op. cit. 1379a16 ff.; Seneca, *De ira*, i. 13, 16; Plutarch, 'On the Control of Anger', *Moralia*, 457B. *Only*. At all events. *beware*. Ensure. *scorn*. Contempt (as being beneath them). *give law... in it*. Set himself rules to be observed in such cases. *causes... chiefly three*. Cf. Aristotle, op. cit.: 'There are three kinds of slighting—contempt, spite, and insolence' (1378b12 ff.). *sensible*. Susceptible. *construction*. Interpretation. *contempt... hurt itself*. Cf. Aristotle, op. cit. 1378a31 f. *putteth an edge upon*. Sharpens, stimulates. *ingenious in*. Have a talent for. *circumstances*. Manifestations. *touch*. Offending. Cf. Aristotle, op. cit. 1378b22 ff. *Consalvo*. Gonzalo Fernandez Consalvo Cordoba (1453–1515), distinguished Spanish general. '*telam... crassiorem*'. 'The honour of a gentleman should be made of a stronger cloth.' Cf. *Promus*, no. 392: 'Tela honoris tenerior', and *Charge touching Duels, above, p.* 311. *refrainings*. Restraining, bridling. *win time*. Cf. Aristotle, op. cit. 1380b6 ff. *make... believe*. Convince. *time*. A propitious moment. *still*. Calm down. *reserve*. Hold back (till later). *contain*. Restrain. *mischief*. Doing damage. *have... caution*. Take care. *aculeate*. Pointed, stinging. *proper*. Specially applicable to the person. *communia maledicta*. 'General curses.' *nothing so much*. Not serious. *business*. Dealings, affairs. *frowardest*. Most recalcitrant. *touched*. Mentioned. *aggravate*. Accentuate. *good times*. Favourable moment. *relate*. Narrate, inform. *angry business*. Matter that will arouse anger. *construction*. Interpretation. *from the point of contempt*. Away from any insinuation of contempt.

58. *Of Vicissitude of Things*

451. *Vicissitude*. Change, mutation. One of Bacon's sources is Machiavelli, *Discourses*, ii. 5, describing how 'the changes of religion and of language, together with the occurrences of deluges and pestilences, destroy the record of things'. '*There... earth*'. Eccles. 1: 9–10, paraphrased. *imagination*. Conjecture. *That... remembrance*. Cf. *Phaedo*, 72E; *Meno*, 81C–D. *sentence*. Judgement. *That... oblivion*. Eccles. 1: 9–11. *Lethe*. The river in Hades: to drink its water produced oblivion. *abstruse*. Recondite, enigmatic. *astrologer*. Perhaps Telesio, whom Bacon quotes to similar effect in *Works*, iii. 98–100. *the matter*. Matter generally (as in 'the air'). *matter... flux... stay*. Cf. *Promus*, no. 992: 'Omnium rerum vicissitudo [est]', from Erasmus' *Adagia*. *flux*. Fluctuation. *stay*. Stand-still. *winding-sheets*. Sheets used to wrap corpses. *merely*. Entirely. *dispeople*. Depopulate. *Phaëton's car*. The chariot which Phaëton stole from his father Helios, the sun-god, but was unable to control: Ovid, *Met*. ii. 35–328. Reynolds (p. 389) and Melchionda

(p. 628) suggest an allusion to Plato, *Timaeus*, 22C–D, which interprets the myth as referring to cosmic conflagrations. *Elias*. 1 Kgs. 17: 1, 18: 1. *particular*. Limited to one region. *West Indies*. In the Renaissance this term referred to the Americas as a whole, the continent as well as islands. *narrow*. Limited. *hap . . . reserved*. Happen to be spared. *mountainous*. Cf. Machiavelli, op. cit.: 'the few that escape [inundations] are chiefly ignorant mountaineers, who, having no knowledge of antiquity themselves, cannot transmit any to posterity'. See also *New Atlantis*, above, p. 468. *all one*. Just the same (i.e. total). *newer . . . old world*. See *New Atlantis*, above, p. 469, where the primitive nature of their civic and political institutions is cited as one proof, another being that they have survived from a later Flood. *heretofore*. Formerly. *earthquakes*. Plato, *Timaeus*, 25C–D. *seldom*. Actually, Bacon's main source of knowledge for the Americas, Acosta's *Naturall Historie of the Indies*, reports frequent earthquakes in the Andes. *pouring*. Huge, torrential. *Andes*. The word means mountains. *whereby*. By the mountains' great height. *generation of men*. The human race. *Machiavel. Discourses*, ii. 5, describing the early Christians' persecution of 'the heathen religion', so extreme that they 'destroyed all its institutions and all its ceremonies, and effaced all record of the ancient theology', and would have extinguished the Latin language if there had been an alternative. *Gregory the Great*. Pope 590–604, who, according to Machiavelli, persecuted with great obstinacy 'all ancient memorials, burning the works of the historians and of the poets, destroying the statues and images and despoiling everything else that gave but an indication of antiquity' (ibid.). *that*. To the effect that. *zeals*. Excesses of zeal; fanaticism. *do*. Work, produce. *Sabinian*. Pope 604–6, who succeeded Gregory and severely criticized him. *Superior Globe*. According to Aristotelian-Ptolemaic cosmology, the concentric spheres surrounding the earth. *Plato's great year*. *Timaeus*, 39D. The 'great year' was the space of time taken by all the stars to return to the places they occupied at the creation of the world; variously estimated as 10,000 to 36,000 years. Reynolds (pp. 392–3) notes similar discussions in Cicero, *De natura deorum*, II. xx. 51–2, and Augustine, *De Civitate Dei*, xii. 13.2.

452. *state*. Condition. *like*. The same. *fume*. Smoke, empty fantasy. *accurate*. Exact, minute. *influences*. Powers by which, astrologers believed, planets affected human life. *in gross*. In general terms. *waited upon*. Watched, observed. *respective*. Particular. *version*. Direction; but perhaps also 'conversion', rotation. *placing*. Position. *region*. Tone. *toy*. Trifle. *given over*. Overlooked. *waited upon*. Observed. *suit*. Sequence. *Prime*. 'The beginning of a period or cycle' (*SOED*); perhaps named by analogy with a lunar cycle of 19 years. *concurrence*. Agreement. *points*. Questions. *orbs*. The spheres, or starry orbs. *'built . . . rock'*. Matt. 16: 18. *To speak*. 'I shall now discuss.' *give stay*. Check, hinder; or perhaps 'estimate, comprehend'. *revolutions*. Changes in affairs. *received*. Accepted. *professors*. Those who profess, practise. *withal*. Besides. *doubt*. Suspect, fear. *extravagant*. Intemperate, excessive. *to make*. That makes. *points held*. Conditions existed. *Mahomet*. Muhammad (AD ?570–632), founder of Islam, whose 'law' is

embodied in the Koran. *properties*. Characteristics. *authority*. Civil power. *popular*. Favoured by the common people (and therefore subversive). *speculative*. Theoretical (matters of thought, not practice). *Arians*. Arius, a 4th-century thinker who taught that Christ was the greatest man ever born, but neither eternal nor equal with God. *Arminians*. Followers of Jacobus Arminius (1560–1609), a Dutch Reformed theologian who emphasized free will, opposing Calvinist notions of predestination and the elect. *civil*. Political, involving many citizens. *manner of plantations*. Modes of disseminating. *signs*. Omens; miracles as evidence of supernatural power or authority.

453. *of*. As regards. *compound*. Compromise over, settle. *take off*. Disarm, placate. *winning*. Persuading. *advancing*. Promoting, giving them office. *stages*. Theatre of operations. *Tartars*. 'The combined forces of central Asian peoples, including Mongols and Turks, who under the leadership of Genghis Khan (1202–27) overran much of Asia and Eastern Europe' (*SOED*.) *people*. Nations. *Gallo-Graecia*. Galatia, in central Asia Minor, conquered by the Gauls *c*.278 BC. *certain*. Stable. *fixed*. The two poles. *southern people... contrariwise*. V. Luciani ('Bacon and Machiavelli', 29) cites Machiavelli's *Istorie fiorentine* (i. 1) for this claim. *northern tract... martial region*. Cf. Lucan, *Pharsalia*, viii. 363–6. *in respect*. Because. *stars*. Astrology divided the world up into regions, each supposedly influenced by the stars overhead. *discipline*. Special training. *hardest*. Best able to endure. *shivering*. Shattering into pieces. *enervate*. Weaken. *resting*. Relying on. *prey*. Victims of other nations. *Almaigne*. Germany. *Charles the Great*. Charlemagne (724–814), King of the Franks and Holy Roman emperor, vastly extended his empire, but it was subsequently divided up among his three grandsons. *were*. Would be. *unlike*. Unlikely. *befall*. Happen. *if ... break*. Should its empire collapse. *accessions*. Acquisitions. *over-power*. Excessive power. *barbarous peoples*. Barbarians. *know means*. Know how to find. *people*. Nations. *shoals*. Masses. *foreseeing*. Providing. *sustenation*. Sustenance.

454. *it*. This topic. *hardly ... observation*. Is difficult to analyse and generalize. *ordnance*. Large guns, artillery. *Oxidrakes*. According to Philostratus, *Life of Apollonius of Tyana*, ii. 14, and Raleigh, *History of the World* (1614), IV. ii. 21. *Macedonians*. In ancient times, inhabitants of Northern Greece, Bulgaria, and present-day Macedonia. *well known*. Cf. Montaigne, 'Des Coches', *Essais*, iii. 6 (Pléiade edn. 886). *hath been*. Has existed. *fetching*. Striking, reaching. *arietations*. Assault using the aries or battering-ram. *commodious*. Versatile. *rested*. Relied. *extremely*. Entirely. *main force*. Superior physical force. *pointing*. Appointing. *pitched fields*. Battles. *trying*. Fighting. *upon... match*. With equal resources. *ignorant*. Cf. *Iliad*, ii. 362–8; iv. 296–305. *ranging and arraying*. Setting in order of battle, according to terrain and tactics. *battles*. Battalions, bodies of troops. *After*. Afterwards. *grew*. Became accustomed. *ordering*. Arrangement. *learning*. Knowledge, the sciences. *his*. Its. *luxuriant*. Uncontrolled in growth. *reduced*. Concentrated, brought within bounds. *exhaust*. Unproductive. *philology*. Perhaps 'stories to which the subject has lent itself'; or

'literary documentation'. *circle of tales.* Perhaps a pun on the Greek phrase for the complete educational system, *enkyklos paideia*, which the Romans rendered as 'encyclopedia'.

59. *A Fragment of an Essay on Fame*

454. *Fame.* This fragment, undoubtedly authentic, was first published in the *Resuscitatio*, p. 281. The word *fame* means 'rumour'. *poets.* As in Virgil, *Aeneid*, iv. 173–90, passages noted in *Promus*, nos. 1080–2. In the translation by C. Day Lewis:

> Rumour, the swiftest traveller of all the ills on earth,
> Thriving on movement, gathering strength as it goes; at the start
> A small and cowardly thing, it soon puffs itself up,
> And walking upon the ground, buries its head in the cloud-base.
> The legend is that, enraged with the gods, Mother Earth produced
> This creature, her last child, as a sister to Enceladus
> And Coeus—a swift-footed creature, a winged angel of ruin,
> A terrible, grotesque monster, each feather upon whose body—
> Incredible though it sounds—has a sleepless eye beneath it,
> And for every eye she has also a tongue, a voice and a pricked ear.
> At night she flits midway between earth and sky, through the gloom
> Screeching, and never closes her eyelids in sweet slumber:
> By day she is perched like a look-out either upon a roof-top
> Or some high turret; so she terrorizes whole cities,
> Loud-speaker of truth, hoarder of mischievous falsehood, equally.
> (*The Eclogues, Georgics and Aeneid of Virgil* (London, 1960), 224–5)

gravely. Weightily. *sententiously.* Full of meaning.

455. *feathers . . . tongues.* The traditional iconograpy, as in *2 Henry IV*, *Induction*: 'Enter *Rumour*, painted full of tongues.' *flourish.* Embellishment. *parables.* Specific attributes, often interpreted allegorically, as by Bacon himself on many occasions. *gathereth . . . going.* The more rumours are disseminated, the stronger they grow. *ground . . . clouds.* Rumours readily circulate, but are intangible. *sitteth . . . watch-tower.* Is constantly alert. *flieth . . . night.* Rumour flourishes in darkness and secrecy. *things not done.* Cf. *2 Henry IV*, *Induction*, lines 6 ff.: 'Upon my tongues continual slanders ride, | . . . Stuffing the ears of men with false reports.' *terror . . . cities.* The disturbances produced by false rumours are greatest where the population is most dense. *passeth.* Surpasses. *they.* The poets. *that.* Who. *anger.* Angry fit. *figured.* Represented. *masculine.* Cf. Essay 15, 'Of Seditions and Troubles', above, p. 366. *fly.* Let fly (transitive) as when starting a falcon after its prey. *ravening fowl.* Birds of prey. *is . . . worth.* Would be valuable. *sad.* Grave. *politics.* Political science. *place.* Topic. *handled.* Discussed. *discerned.* Distinguished (true from false). *fames . . . sown.* Rumours may be planted. *raised.* Encouraged. *Mucianus.* See Tacitus, *Histories*, ii. 80 for his spreading a rumour that troops would be posted to the harsh duties of Germany. *had in purpose.* Purposed. *remove.* Transfer. *unprovided.* By

surprise. *laid asleep*. Neutralized. *cunningly*. See Caesar, *De bello civili*, i. 6 (where, however, the rumour is attributed to Pompey, not Caesar himself). *Livia*. See Tacitus, *Annals*, i. 5. *amendment*. Recuperating from sickness. *Bashaws*. Grandees (the earlier form of *Pasha*). *conceal*. This had recently happened twice, in 1566 and 1574. *Janizaries*. Turkish soldiers who formed the Sultan's guard noted for their devotion. *save*. Prevent. *sacking*. Destruction. *Themistocles*. See Plutarch, 'Life' of Themistocles. *post apace*. Leave hastily. *athwart*. Across.

456. *governors*. Rulers. *the actions*. Political acts.

NEW ATLANTIS

At the end of the volume containing Bacon's *Sylva Sylvarum: or A Natural Historie* (1627; Gibson 170), there appeared, without any prior announcement on the title-page, *New Atlantis*. 'A Worke unfinished'. William Rawley, Bacon's chaplain and literary executor, prefixed to it this note: 'This fable my Lord devised, to the end that he might exhibit therein a model or description of a college instituted for the interpreting of nature and the producing of great and marvellous works for the benefit of men, under the name of Salomon's House, or the College of the Six Days' Works. And even so far his Lordship hath proceeded, as to finish that part. Certainly the model is more vast and high than can possibly be imitated in all things; notwithstanding most things therein are within men's power to effect. His Lordship thought also in this present fable to have composed a frame of Laws, or the best state or mould of a commonwealth; but foreseeing it would be a long work, his desire of collecting the Natural History diverted him, which he preferred many degrees before it. This work of the *New Atlantis* (as much as concerneth the English edition) his Lordship designed for this place; in regard it hath so near affinity (in one part of it) with the preceding Natural History.' Rawley accurately describes the work's hybrid nature: a 'fable', starting like an Elizabethan narrative of a sea voyage, followed by the description of a foreign country, the whole transforming itself into a 'model' of a scientific research institute, the first full outline of such an organization. The Accademia dei Lincei (or 'lynx-eyed') of Rome had been founded in 1603 to pursue research in the mathematical and physical sciences, and the publication of their *Praescriptiones Lynceae Academiae* (1624) may have stimulated Bacon to a timely emulation. More stimulus might have been provided by Campanella's utopian fable, *The City of the Sun*, published at Frankfurt in 1624, with which Bacon's work shares some resemblances, possibly accidental (see E. D. Blodgett, 'Bacon's *New Atlantis* and Campanella's *Civitas Solis*: A Study in Relationships', *PMLA* 46 (1931), 763–80). Bacon's own contextual indications place the *New Atlantis* within his peculiarly eclectic tradition of scientific knowledge: King Salomon, so often cited by him as legitimizing the study of natural history; the 'Six Days' Works' referring to the account in Genesis of God's Creation (around which a vast amount of 'hexaemeral' literature had collected); and the tradition of the lost island of Atlantis described by Plato (*Timaeus*, 25 ff., *Critias*, 113E).

Rawley's report that Bacon had planned to add a 'frame of Laws, or the best . . . mould of a commonwealth', would have made it emulate Plato's late work, the *Laws*. The *New Atlantis* has indeed a distinguished place in the tradition of utopian literature (as shown by J. C. Davis, *Utopia and the Ideal Society: A Study of English Utopian Writing 1516–1700* (Cambridge, 1981), 105–37).

Within this classical and Christian framework Bacon writes, first, a free-flowing narrative of travellers being lost, and found, describing the imaginary society in which they find themselves with that interest in ceremony and display so typical of much of his work. Although much of this part is freely invented, many of the historical and topographical details are the product of careful research, Bacon's major source being the recently published English version of Joseph de Acosta's *Naturall and Morall Historie of the East and West Indies* (1590; Engl. tr. 1604). The transition between the voyage narrative and the description of Salomon's House, which constitutes the work's real substance, owes something to the form of the sacred dialogue, as found for instance in the second book of Esdras (Apocrypha), ch. 3 ff., in which the author describes how a divine spirit reveals to him some portion of hidden wisdom. So the Father of Salomon's House, left alone with the narrator, says: 'God bless thee, my son; I will give thee the greatest jewel I have. For I will impart unto thee, for the love of God and man, a relation of the true state of Salomon's House'. Such dialogues are also a form of apocalypse, an unveiling or disclosure of things ordinarily hidden from human eyes. Once again, Bacon's emphasis is on the open communication of scientific knowledge, as against the secrecy practised by the occult tradition. At one point he reserves to scientists the right to withhold information from the state, probably, as A. B. Gough suggested (below, p. 797), concerning inventions of unusually destructive weapons which he considers would be dangerous if generally known.

As for the 'College of the Six Days' Work', Bacon develops ideas for the organization and division of scientific research that he had first outlined some thirty years earlier in the device *Of Tribute* in 1592 (above, pp. 34–6) and the speeches for the *Gray's Inn Revels* of 1594 (above, pp. 54–5), and mentioned in passing in his *Comentarius Solutus*, a series of memoranda set down in July 1608. Here he recorded his plans to pursue scientific researches in various fields (mineralogy, medicine, physics), and to commission 'an History of marvels, historia naturae errantis or variantis', and 'an History mecanique', to include 'observations of all Mechanicall Arts', studying their materials, instruments, engines. Then he proposed the 'Foundation of a college for Inventors . . . And a Library and an Inginary . . . Vaults, furnaces, terraces for insolation; work houses of all sorts' (*Works*, xi. 63–7). In the *New Atlantis* Bacon retrieves all these ideas from the past, and also draws on new work. He incorporates some of the stages in experimental procedures described as recently as the *Novum Organum* of 1620, and sets down briefly many matters discussed more fully in the *Sylva Sylvarum* or 'Natural History' with which it was published. (As for dating the *New Atlantis*, although some attempts have been made to place it between 1612 and 1615, all the evidence seems to locate it in the early 1620s.) In several ways, then, this work sums up a lifetime's concern with the reformation of natural philosophy. Its form, too, using imaginative means to develop scientific thought, is typical of Bacon's union of literature and science, while its unfinished state, petering out into a list of goals for human knowledge—inchoate in form but expressing great vision—exemplifies a thinker whose reach so often exceeded his grasp.

Much of the fascination of the *New Atlantis* for later periods has been its projection of all manner of inventions into the future, as in the newer genre of science fiction. Bacon seems to have anticipated even relatively recent scientific developments: telephones, gramophones, microphones, amplifiers and synthesizers, lasers, the production of synthetic fibres, and organ transplants. Yet, far-sighted though Bacon was, some of the inventions and processes he described were already known. The telescope was in its early, crude form, while the possibility of constructing a more complex form of magnification than the traditional magnifying glass was already under discussion. Closer to Bacon, as the late Rosalie L. Colie showed ('Cornelis Drebbel and Salomon de Caus: Two Jacobean Models for Salomon's House', *Huntington Library Quarterly*, 18 (1954), 245–69) were two outstanding contemporary scientists and craftsmen, the Dutchman Cornelis Drebbel (1572–1633), and the Frenchman Salomon de Caus (*c.*1576–1626). Drebbel was in England from 1605 to his death, in the service of King James and Prince Henry, while Caus was also in the household of the Prince, later transferring to that of his sister, Princess Elizabeth, the Electress Palatinate. Drebbel had a laboratory at Eltham, where he carried out experiments and gave demonstrations in such areas as meteorological phenomena (according to one visitor, he 'could make it rain, lighten, and thunder . . . as if it had come about naturally from heaven'), and artificial incubation (he could hatch eggs 'without any Ducks or Chickens by . . . even in midwinter'). He demonstrated a camera and a magic lantern, with which he produced optical illusions matching those that Bacon describes; was famous for his accurately ground lenses, which were used to make very efficient telescopes; perfected a scarlet dye, the secret of which he fiercely guarded; developed a remarkably effective drainage pump; and made a 'perpetual motion' machine which demonstrated 'le flux et reflux de la mer' (probably a kind of barometer responding to air pressure).

Two of Drebbel's most spectacular performances were certainly known to Bacon. He gave a demonstration of instruments he had invented to chill the air before King James in the Great Hall of Westminster, making it 'so cold on a summer's day that the King and his nobles and many great lords were forced to flee'. Bacon's remark, in the *De Augmentis*, that 'in the late experiment of artificial freezing, salt is discovered to have great powers of condensing' (*Works*, i. 628; iv. 417), undoubtedly refers to Drebbel's demonstration. The Dutchman achieved an even greater coup in 1620, when he exhibited before the King one of the first successful submarines, disappearing under the Thames for three hours, according to the Dutch scientist Constantin Huygens, present in the crowd, holding 'the King, his court, and several thousand Londoners in excited expectation'. Caus achieved nothing quite so spectacular, but he was famous for experiments with water, involving cataracts, musical fountains, hydraulically activated songbirds, a steam engine, and a player piano. Knowledge of these and other contemporary developments helps to place the *New Atlantis* more firmly in its historical context, showing that Bacon's flights of imagination and fantasy had a solid grounding in fact. Indeed, for its seventeenth-century readers one of its great fascinations must have been the fact that much of what it described had already become true, so that Bacon's often-repeated vision of science's ultimate ability to overcome any problem—memorably expressed here as the goal of Salomon's House, to discover 'the knowledge of causes and secret motions of things; and the enlarging of the bounds of human empire, to the effecting of all things possible'—must have seemed within imminent reach.

For this reason, perhaps, the *New Atlantis* was one of Bacon's most popular and influential works. It was reprinted with the *Sylva Sylvarum* thirteen times between 1627 and 1685; it was translated into French in 1631 (and in 1652, with a continuation by G. B. Rahuet), and into Latin (1643, 1648, 1661). Its celebrity was such that it could be reused for the most disparate purposes, as we can see from R. W. Gibson's pioneering catalogue of Bacon allusions, the 'List of Baconiana' included in his *Bibliography*, items 257–680. Edward Bushell, a mining engineer, reissued the *New Atlantis* in 1659 and 1660 in order to promote his own 'mineral prosecutions' (Gibson 214, 317). One R. H. (most probably Robert Hooke) issued a continuation of it in 1660 in order to complete the gap Bacon had left, 'Wherein is set forth a Platform of Monarchical Government, with a pleasant intermixture of diverse rare Inventions, and wholsom Customs, fit to be introduced into all Kingdoms, States and Commonwealths' (Gibson 417). Two years later, in *The Holy Guide: Leading the Way to the Wonder of the World*, John Heydon plagiarized Bacon's text to describe a 'Journey to the Land of the Rosicrucians', with Salomon's House becoming the 'Temple of the Rosie Crosse' (Gibson 436). Among the *Essays on several important Subjects in Philosophy and Religion* which Joseph Glanvill published in 1676 was one on 'Anti-fanatical Religion, and Free Philosophy. In a Continuation of the *New Atlantis*' (Gibson 404). Other contemporaries took the work entirely seriously as a factual account of 'an Iland of this Southern Continent, discovered by Sir Francis Bacon' (Peter Heylin's *Cosmographie*: Gibson 439b), or referred to it in describing Guinea, or the location of paradise (Gibson 409, 346).

Apart from such bizarre traces, the *New Atlantis* left its greatest mark in the seventeenth century on the many and diverse groups concerned with science. For the Puritan reformers associated with Samuel Hartlib during the Civil War and the interregnum, it was an ever-present source of inspiration. Samuel Platt invoked it in his utopian project *Macaria* (1641), addressed to the Long Parliament, a scheme which Hartlib was still hoping to realize in 1660. Edward Bushell, the indefatigable mining engineer, actually evolved a plan to erect in London an institution to emulate Bacon's College of the Six Days' Work. By May 1654 his project had taken on some kind of form, for he could invite Hartlib and others 'to *Lambeth-Marsh*, to see part of that Foundation or building, which is designed for the execution of my lord

Verulam's *New-Atlantis*'. Bushell was still hoping to complete the project in 1659. It was not only the Puritans, however, who had been inspired by *New Atlantis*: as Joseph Glanvill put it, Bacon's 'name and parts might give credit to any undertaking'. So the College of Physicians, the professional organization promoting university-trained doctors, could also be hailed in 1657 as a centre for scientific experiment where, according to Walter Charleton, 'you may behold Salomon's House in reality'. (See, in *Further Reading* § 6, the books by Jones and Webster.)

The *New Atlantis* took on special significance in the mid-century, when several different attempts to set up collective scientific research groups—Hartlib's 'invisible ... philosophical college' in 1646–7, a London group meeting at Gresham College from 1645 on, some of whom joined the group which, from 1648 onwards, formed around Boyle and Wilkins at Wadham College Oxford—culminated in the founding of the Royal Society in 1662. These collective activities were accompanied by a series of projects, circulated and in some cases published, evidently inspired by Bacon. In 1648 Sir William Petty published *The Advice of W.P. to Mr. Samuel Hartlib, for the Advancement of some particular Parts of Learning*, which advocated the founding of a research and teaching institute in the theory and practice of botany, horticulture, anatomy, chemistry, metallurgy, and other subjects, with an ancillary hospital, zoological garden, 'botanic theatre', museum, collection of models, and other Baconian projects. In 1659 John Evelyn, in a letter to Robert Boyle, proposed the founding of a scientific college, consisting of laboratories for 'the promoting of experimental knowledge', with 'a repository for rarities and things of nature' attached, along with an aviary and gardens. Finally, Abraham Cowley published in 1661 a carefully worked-out *Proposition for the Advancement of Experimental Philosophy* embodying a 'philosophical college' in which the professors would not only teach all the pure and applied sciences but pursue 'all manner of experiments' in chemistry, botany, meteorology, 'and briefly all things contained in the catalogue of Natural Histories annexed to My Lord Bacon's *Organon*'. It was altogether appropriate that Glanvill, in dedicating his *Scepsis Scientifica* (1665) to the Royal Society, should express his confidence that 'The success of those your *great* and *Catholick Endeavours* will promote the *Empire* of *Man* over *Nature*, and bring plentiful accession of Glory to your Nation ... For *You really* are what former Ages could contrive but in *wish* and *Romances*; and *Solomons House* in the NEW ATLANTIS, was a Prophetick Scheam of the ROYAL SOCIETY' (Sig. c). Thomas Sprat, in his officially commissioned *History* of that body (1667), duly praised Bacon as its inspiring figure, acknowledging that 'great Man, who had the true Imagination of the whole extent of this Enterprize, as it is now set on foot'. If he had had his way, Sprat writes, 'there should have been no other Preface to the *History* of the *Royal Society*, but some of his Writings' (pp. 35–6). As it is, he happily records that 'this foundation of the Royal Society ... was a work well becoming the largeness of his Wit to devise' (p. 144).

The *New Atlantis* was printed at least fifteen times in the 17th century attached to *Sylva Sylvarum*, and was issued by Thomas Bushell in his

Abridgment of the Lord Chancellor Bacons philosophical theory in mineral prosecutions (London, 1659). My text is taken from Spedding, and in annotating it I have benefited from several earlier editions: by G. C. Moore Smith (Cambridge, 1900), A. B. Gough (Oxford, 1915, 1924), Arthur Johnston (Oxford, 1974), and the late Enrico de Mas, in his valuable collection, *Scritti politici, giuridici e storici di Francesco Bacone* (Turin, 1971), i. 779–827.

457. *continued.* Stayed. *South Sea.* The 'Mar del Sur', the name given to the Pacific when first seen from the Quareca Mountains, Darien, by Balboa on 25 Sept. 1513. *winds . . . east.* The trade wind blowing from the coast of Peru across the Pacific. *came about.* Shifted. *so as.* So that. *way.* Progress. The 'anti-trade' or contrary winds, which could delay ships for long periods. *a point.* One division on the mariner's compass, *c.*$11\frac{1}{4}^{\circ}$. *good spare.* Thrifty use. *showeth . . . deep.* Paraphrase of Ps. 107: 23–4, 'These see the works of the Lord, and his wonders in the deep'. *discovered.* Revealed; cf. Gen. 1: 9. *within a kenning.* Within 'ken' or sight, which at sea is about 20 miles. *might.* Could. *bent.* Directed. *boscage.* Bushes, thickets. *great.* Large. *gave.* Afforded. *offered.* Showed the intention. *divers.* Several. *bastons.* Truncheons, used as weapons or as symbols of office. *discomforted.* Discouraged. *advising.* Considering. *tipstaff.* Staff with a tip or cap of metal, carried as a badge by officials. *blue.* The most popular colour in the island of Bensalem (see below).

458. *writing tables.* Tablets, thin sheets of ivory or wood used as a notebook. *foremost man.* Leader (as it turns out, the narrator himself). *School.* University (scholastic Latin was an international language then). *provide.* Make preparations. *except.* Unless. *stamp.* Seal, impression. *cherubins' wings.* A biblical symbol for divine blessing. Cf. Exod. 25: 20, 37: 7–9, describing the tabernacle which Moses had made. *denial.* Refusal. *had.* Knew. *humanity.* Kindness. *instrument.* Official document. *for.* As for. *ill case.* Pitiable condition. *particular.* Detail. *deal.* Bargain. *chargeable.* An expense. *pistolets.* Spanish and Italian gold coins. *of place.* Of rank or station. *water chamolet.* A fabric made of goat's hair, with a watered or wavy surface. A costly fabric, much esteemed. *reverend.* Venerable. *gilt.* Painted gold. *flight-shot.* Of arrows: about 280 yards.

459. *presently.* Immediately. *save.* Except. *subscription.* Seal or signature on a document. *no pirates.* Gough (p. 51) compares Homer, *Odyssey*, iii. 71–4, where Nestor asks Telemachus whether he is a trader or a pirate. *shed blood.* Either in a ceremonial sacrifice, or in violence. *licence.* Permission. *act.* Solemn undertaking. *of.* Out of. *greatness.* High rank. *for that.* Because. *Conservator of Health.* Some Renaissance towns, especially in Italy, were designed on hygienic principles, with broader streets, canals, and sewage arrangements; but few (if any) had a Public Health Officer. *accounted.* Considered. *hoped well.* Had good hope. *orange-tawney.* Yellowish-orange. *cast.* Emitted. *after.* Afterwards. *accommodated.* Supplied with. *whole.* Men in good health. *twice paid.* Cf. Plato, *Laws*, 12.948A–B.

460. *prevented*. Come before. *hour*. As appointed. *because*. In order that. *will*. Wish to. *fair*. Broad, handsome. *put . . . abroad*. Stretched their arms out to the sides. *any*. Any one. *cambric oiled*. A fine white linen, made waterproof (and transparent) by the application of oil; cheaper than glass windows. *above stairs*. Upstairs. *chambers*. Rooms. *cast*. Reckoned. *civilly*. Decently, without ostentation. *dorture*. Dormitory. *cells*. Cubicles. *instituted*. Designed. *waxed well*. Recovered. *set forth*. Set apart. *give any charge*. Deliver an order received from a superior.

461. *attend*. Wait on. *abroad*. Out of doors. *affection*. Friendliness. *right*. Very. *viands*. Food. *collegiate*. Institutional. *store*. Supply. *assured*. Certain. *carriage*. Transporting. *quiet*. Allayed. *know ourselves*. Alluding to the Socratic principle 'know thyself'. *Jonas*. The Book of Jonah 1–2. *as*. As it were. *but*. Only. *of*. To. *confusion of face*. Shame; cf. Ezra 9: 7. *in*. Under the. *take some taste*. Observe, sample. *manners*. Moral conduct. *conditions*. Dispositions, temper. *attendance*. Attendants. *withal*. Besides. *weal*. Welfare.

462. *grace*. Favour. *amendment*. Recovery. *divine pool of healing*. The pool of Bethesda, John 5: 2–4. *mended*. Recovered. *kindly*. Naturally, readily. *morrow*. Morning. *tippet*. Cape trimmed with fur. *of*. On. *looking*. Expecting. *avoided*. Quitted. *occasions*. Needs. *ask*. Require. *at this time*. At the moment. *aforehand*. Provided for future needs. *laid up*. Put by, saved. *defray*. Pay your expenses. *used*. Treated. *return*. Equivalent. *all one*. All the same. *make . . . fall*. Disappoint, disconcert you. *admiring*. Wondering at. *usage*. Treatment. *free*. Generous.

463. *before*. In front of. *tongues . . . forget*. Ps. 137: 6. *bounden*. Indebted. *which*. Who. *prevent us*. Anticipate our needs. *familiarly*. In a friendly manner. *meaner sort*. Lower ranks. *set*. Seated. *Bensalem*. The name means in Arabic 'son of peace'. *have this*. Possess this special quality. *by means*. In consequence. *entertainment . . . time*. Employing the time to advantage. *state*. Condition. *met*. Come together. *several*. Different, opposed. *both parts*. On both sides. *in respect*. Considering.

464. *first . . . heaven*. Matt. 6: 33; Luke 12: 31. *within night*. At night. *some*. About (one mile). *into*. Out. *pillar of light*. Cf. Exod. 13: 21–2, God leading the Israelites out of Egypt: 'And the Lord went before them by day in a pillar of a cloud, to lead them the way; and by night in a pillar of fire, to give them light; to go by day and night.' *sharp*. Coming to a point. *Upon*. In consequence of. *apace*. Quickly. *put themselves*. Embarked. *bound*. Spellbound, fixed. *might*. Could. *go about*. Turn round. *theatre*. In a semicircle, like spectators in a theatre (a word which still had the original connotation of 'a beholding-place'). *college*. Institution. *eye*. Intellectual centre. *vouchsafed*. Granted. *our order*. Of Salomon's House (described at pp. 480–8). *discern*. Distinguish. *divine miracles . . . illusions*. This is Bacon's own classification of the forms of creation: divine and miraculous, natural, human (by 'art'), and magical (spurious, deceptive: the Latin translation adds 'illusiones daemonum'). *Finger*. Handiwork. *forasmuch as*. Considering that. *laws of nature... thine own laws*. Cf. *Confession of Faith*, above, pp. 108–9. *exceedest*. Overstep. *upon great cause*. On miracles as the suspension of a law of nature, cf. *Confession of Faith*, above, p. 109. *prosper*. Make prosperous, auspicious. *interpretation*. Application. *in some part*. Partly. *presently*. At once. *softly*. Gently. *cast... abroad*. Dispersed itself. *firmament*. Arch, vault.

465. *swam*. Floated. *fore-end*. Front part. *sindons*. Fine thin linen, used as a wrapper. *canonical books*. Included in the canon or 'rule' of Faith: that is, authentic Scripture. *receive*. Accept. *Apocalypse*. The Revelation of St John (not accepted by all sects). *not at that time written*. At the supposed date of this event, 'about twenty years after the Ascension', i.e. *c*. AD 49, hardly any of the books of the New Testament had been written. 'No one, however, knew this in Bacon's day' (Gough, p. 53). *Bartholomew*. Saint, one of the twelve Apostles, who traditionally travelled to India (i.e. Asia) and spread the Gospel there. According to Eusebius (*Hist. Eccl.* v. 10) Bartholomew left a copy of St Matthew's Gospel with the Indians. *floods*. Currents. *in*. On. *writings*. Written documents. *as well... as*. Both... and. *wrought*. Performed. *conform to*. Similar to. *Gift of Tongues*. The power to understand all languages, granted to Christ's Apostles in the miracle of Pentecost: Acts 2: 1–11. (But that gave the ability only 'to speak with other tongues', not to read them.) *read upon*. Could read. *infidelity*. Heathenism. *remain*. Remnant. *evangelism*. Gospel message. *held*. Considered. *conference*. Conversation. *part*. Side.

466. *hardiness*. Boldness. *observed*. Noted. *formerly*. Previously. *business*. Affairs. *any of*. Gough (p. 53) suggests the right reading to be 'of any'. *for that*. Because. *inter-knowledge*. Mutual knowledge (first recorded use in *OED*). *relation*. Report. *both ways... both parts*. Two nations can gain some knowledge of each other in these ways. *for*. As regards. *rested*. Consisted. *conclave*. Private, lockable chamber (*con clave*, 'with a key'). *condition*. Peculiarity. *propriety*. Property. *open*. Disclosed. *as*. As it were. *imported*. Meant, signified. *humbleness*. Humility. *taking knowledge*. Showing awareness. *merrily*. In jest. *apt*. Ready. *angelical... magical*. The work of angels, not demons. *tender*. Scrupulous, hesitant. *doubtful*. Timid. *conceit*. Thought. *touch*. Hint. *touching*. Concerning.

467. *reserve*. Keep back. *three thousand... more*. There is no evidence for this belief, which expresses the Humanists' over-reverence for antiquity. *navigation*. Navigating. *six-score years*. Counting exactly 120 years from Columbus's discovery of America in 1492 would set these events in 1612. *what it was*. Whatever the cause. *The Phoenicians... fleets*. The Phoenicians were indeed the great traders and navigators of antiquity, dominating sea trade between the 12th and 6th centuries BC, from the Mediterranean to the Indian Ocean and the Scilly Isles. Tyre, their major city, traded with India and Arabia (Gough, p. 54). *their colony*. The Tyrians are said to have founded Carthage in 813 BC. *junks*. Chinese ships, thought by Europeans to be unsuitable for long sea voyages, but in fact large and reliable. *canoes*. The Indians' boats, first observed by English explorers in 1555. *faithful registers*. Reliable records. *content*. Tonnage, capacity. *sparing*. Scanty. *large*. Abundant. *might*. Power. *stirps*. Descendants, offspring. *Pillars of Hercules*. The straits of Gibraltar, for long marking the end of the known world. *Paguin*. Peking (which Bacon imagined to be a sea-port). *with*. As. *Cambaline*. More usually 'Cambalu', the Tartar name for Peking. *Quinzy*. The Quinsai of Marco Polo, later Hangchow. *East Tartary*. The eastern coast of Asia, from Turkey to Mongolia. *age*. Century. *description*. Of the temple, following. *a great man*. Plato, in *Timaeus*, 25 and *Critias*, 113E. *with*. Among. *planted*. Settled as colonists. *environed*. Surrounded. *degrees*. Steps. *scala coeli*. A ladder to heaven. Cf. Gen. 28: 16: 'And he [Jacob] dreamed, and behold a ladder set up on the earth, and the top of it reached to heaven; and behold the angels of God ascending and descending it.' *as well . . . as that*. Including both.

468. *Coya . . . Tyrambel*. These supposed older names for Peru and Mexico were coined by Bacon. Coya is the name for the chief wife of an Inca sovereign in Acosta's *History of the Indies* (Gough, p. 56). Many 16th-century writers identified Atlantis with America. *both*. Each. *through*. Across. Cf. *Timaeus*, 22–8. *Egyptian priest*. From whom Plato (*Timaeus*, 22B) supposedly derived his knowledge. *repulse*. Driving back. *had*. Would have had. *handled*. Managed. *entoiled*. Trapped, ensnared. *render*. Surrender. *without . . . stroke*. Without a blow fought. *Divine Revenge*. The Greek Nemesis. *saith*. Plato, *Timaeus*, 25C–D, *Critias*, 112. *tract*. Region. *inundation*. Rejecting Plato's earthquake, Bacon substitutes a biblical Flood to account for the disappearance of Atlantis. The alteration also brings the Platonic myth into accordance with the geology of South America, which Bacon (wrongly) imagined to be non-volcanic. (His source was Acosta's *History*, iii. 25–6.) Cf. Essay 58, 'Of Vicissitude of Things', above, p. 451. *past*. More than. *vale*. Valley, lower parts. *So as*. Wherefore. *account*. Consider. *your*. The (a grammatical construction known as the 'ethic dative'). *particular*. Partial.

469. *letters*. Literacy. Cf. Plato, *Laws*, 3.677–9. *arts*. Skills, crafts. *civility*. Civilization. *in respect*. On account of. *tigers*. Probably the jaguar, which Acosta calls a tiger. *goats*. Either the alpaca or the vicugna, which Acosta describes as larger than a goat. *heats*. Hot weather. *means*. Way of obtaining. *delight . . . birds*. A common practice in both Americas: the Spaniards admired the Mexicans' use of bird-feathers (Gough, p. 59). *took*. Derived. *invited*. Led, attracted. *infinite*. Very many. *flights*. Flocks. *main*. Chief. *accident of time*. Greatest catastrophe since time began. *traffic*. Trade. *respect*. Consequence. *natural . . . time*. For Bacon's concept of cyclic revolution see Essay 58, above. *the rather*. The more so. *galleys*. Low, flat-built boats, propelled by sail or oars. *brook*. Endure, cope with. *left and omitted*. Left off, abandoned. *yield*. Give. *but*. Otherwise than. *appertain*. Belong. *draw*. Approach. *large heart*. 'And God gave Solomon wisdom and understanding exceeding much, and largeness of heart': 1 Kgs. 4: 29 ff. That Bacon should insert the great lawgiver Solomon (as earlier, the apostle

Bartholomew) into this part-classical part-imaginary narrative, shows his constant wish to legitimize scientific enquiry with biblical authority. *inscrutable for good*. The goodness of which could not be searched out or fathomed. Bacon's direct translation from the Vulgate, Prov. 25: 3, 'cor regum inscrutabile'; in the Authorized Version, 'the heart of kings is unsearchable'. *substantive*. Independent, self-supporting.

470. *set on work*. Employed. *transportations from port to port*. Coastwise traffic. *estate*. Condition. *wanted*. Was lacking. *heroical*. Half-divine. *perpetuity*. Gough (p. 60) notes that the legal code of Bensalem has continued in force for 1,000 years: 'This implies a static concept of society characteristic of antiquity and the Renascence.' *doubting novelties*. Fearing innovations and disturbances. *commixture*. Mingling, contamination. *manners*. Customs. *like*. Same. *against the admission... China*. Such laws were laid down at a very early period by the Chinese government, and only briefly suspended. *yet*. Still. *curious*. Naïve, uninstructed. Bacon's remarks on the bad effects of isolation were just. *fearful*. Unwarlike. *temper*. Constitution, frame. *taking order*. Taking measures. *as reason was*. As was proper. *policy*. Political judgement. *discover*. Disclose. *estate*. State, body politic. *conditions*. Treatment. *several*. Different. *bottoms*. Hulls, ships. *think*. Realize. *restrain*. Prohibit.

471. *open*. Explain. *lanthorn*. Lantern, source of intelligence and discovery. *Creatures*. Both animate and inanimate. *denominate of*. Named after. *with*. Among. *cedar... motion*. 1 Kgs. 4: 33, Bacon's favourite biblical allusion to support his idea for a natural history, with the idiosyncratic reading 'moss' instead of 'hyssop': see the General Introduction, above, p. xxxviii. *Libanus*. The Latin form of 'Lebanon'. *symbolize*. Correspond to, agree with. *true nature... things*. The Latin translation (pub. 1643) adds the characteristic Baconian phrase 'interiora rerum'. *workmanship*. Creation. *fruit... use*. Again Bacon invokes biblical authority to justify his programme for the reform of knowledge, to produce a *scientia operativa*, yielding 'works' from which mankind can benefit. *ordinance*. Decree. *set forth*. Dispatched. *appointed to*. Designated for. *several*. Different, various. *either*. Each. *mission*. Embassy. *only*. Merely. *designed*. Destined, bound. *inventions*. Including discoveries. *patterns*. Directions or models. *fraught*. Equipped.

472. *vulgar*. Ordinary. *contained*. Prevented, restrained. *at land*. On shore. *colour*. Disguise. *practique*. Practice, custom. *commodity of matter*. Material goods or profits. *God's first creature... Light*. *Gen*. 1: 3. Cf. Essay 1, 'Of Truth', above, p. 342. *light... growth*. Intelligence of the growth of knowledge (seen as a substance or commodity, in varying 'growths' or vintages). *all parts... world*. Wherever it is to be found. *so*. Such. *probably*. Plausibly, convincingly. *took us off*. Relieved our embarrassment. *descended*. Condescended. *with*. Among. *scant*. Stint. *suffer*. Allow. *when it came once*. As soon as it was known. *used*. Had the custom of. *conditions*. Stipends, allowances. *presently*. Instantly. *crave*. Request. *took*. Considered. *perdition*. Ruin. *tedder*. Tether, allotted range. *obtaining*. Making. *quality*. Rank. *freedom*. Liberality. *relation*. Retelling, narration. *hold*. Detain. *compounded*. Made up.

473. *to*. With. *assisted*. Attended. *consultation*. Gough (p. 62) notes: 'this resembles the Roman *consilium domesticum* or family council, a legal institution for determining family disputes or other difficulties.' *good estate*. Welfare. *suits*. Quarrels, differences. *compounded*. Settled. *distressed or decayed*. Suffering from hardship, poverty. *order is taken*. Arrangements are made. *take ill courses*. Have bad habits. *direction*. Advice. *advices*. Admonitions. *assisteth*. Is present. *needeth*. Is necessary. *half-pace*. Raised platform, dais. *state*. Canopy. *ivy*. Probably chosen as an evergreen, and so symbolizing 'a green old age' (Gough, p. 62). *asp*. Aspen, poplar-tree. *curiously wrought*. Elaborately embroidered. *broiding*. Plaiting, interweaving. *true*. Natural. *traverse*. Partition, separated by a curtain drawn across the gallery. *privy*. Private. *carved*. Sculptured (ornate, with coloured glass). *leaded*. Framed.

474. *return*. One of the sides of the dais. *Taratan*. An onomatopoeic word, suggesting the sound of a herald's trumpet: but cf. 'Taratantara' in Ennius, *Annales*, ii. 35. *streamed with gold*. Adorned with gold stripes. *curtesies*. Salutations. *inclinations*. Bowing the head. *revenew*. Revenue, profit from leased lands. *privileges*. Advantages (in trade or commerce). *exemptions*. Freedom from service, taxes. *points of honour*. Distinctions. *styled and directed*. Bears this designation. *but*. Except. *set*. Attached. *imbossed*. Set in relief. *expedited*. Issued. *of course*. As a matter of course. *by discretion*. According to circumstances. *supported*. Attended. *thus much*. As much as to say. *daintily*. Delicately. *ensign*. Sign.

475. *except he hap*. Unless he happens. *service of*. Waiting at. *upon the knee*. Kneeling. *bidden*. Invited. *comely order*. Suitable ceremony. *invention*. The rhetorical process *inventio*, finding new material to 'vary' a composition, or present it in new forms. *poesy*. Poetry. *praises*. Bacon partly follows Plato, who restricted poetry in his ideal state to the praise of the gods and of virtuous men. *Father of the Faithful*. Cf. Rom. 4: 11; Gal. 3: 7. *in whose . . . only blessed*. 'In the birth of whom alone the births of all are blessed.' *pilgrimage*. Life seen as a journey towards heaven: Cf. Gen. 47: 9. *every*. Every one. *so*. Provided that. *above*. More than. *either*. Each. *fall to*. Devote themselves to. *By that time*. When. *strait*. Close, intimate. *stirps*. Small tribes.

476. *Surely*. Certainly. *Milken Way*. 'To the Jews the Milky Way was a river flowing through the heavens proceeding from the throne of God' (Johnston, p. 290). *Eliah*. Elijah was expected to herald the Messiah's coming (Mal. 4: 5; Matt. 17: 10). *generations*. Lineage. *Nachoran*. In the Bible Nahor (or Nachor), one of Abraham's brothers, had many sons: see Gen. 11: 22–6, 22: 20–4, Luke 3: 34. *cabala*. Hebrew name for the secret interpretation of the Jewish scriptures in a mystical sense, said to have been handed down from Moses. *dreams*. The narrator dismisses such beliefs as fantasies. *policy*. Political wisdom. *seen*. Versed, informed. *affected*. Moved, impressed. *relation*. Report. *nature*. Natural, blood relationships. *kept*. Supported. *population*. Populating a country. *affected*. Desired, aimed

at. *European books*. De Mas (p. 809) refers to *Sintram* (1820) by La Motte Fouqué (1777–1843), which preserves a later version of this story, but also cites the fable 'Nemesis' in Bacon's *Wisdom of the Ancients* (1609). In it Nemesis, the goddess of retribution, daughter of Night and Ocean, is represented with wings and crown, holding in her left hand a phial containing Ethiops, who represent 'that dark and ominous spectre' which visits mortals 'at the summit of felicity, with images of death, diseases, misfortunes . . . and the like' (*Works*, vi. 737–9). This whole passage, containing the narrator's discussion of marriage, down to the reference to 'masculine love' was omitted by Moore Smith as 'unsuitable for an edition which may be read in schools' (p. v), and by Gough as offending 'modern taste' (p. vi). Both also omitted the reference to More's *Utopia*.

477. *stews*. Hot-air public baths, notorious places for sexual contacts. *dissolute houses*. Brothels. *unlawful concupiscence*. Cf. St Paul, Eph. 2: 3; Rom. 7: 5 ff. *expulsed*. Driven out. *seen*. Observed. *infinite*. Many. *bargain*. Commercial transaction. *alliance*. Union with a prestigious family. *portion*. Dowry. *issue*. Children. *same matter*. Same (corrupt) flesh. *change*. Exchange, infidelities. *meretricious*. Belonging to harlots. *advoutries*. Adulteries. *preposterous*. Back to front. *Lot's offer*. Gen. 19: 4–14. *abusing*. Committing sodomy. *vent*. Outlet. *touch*. Trace.

478. *decent*. Fitting. *pause*. Stopping. *Sarepta*. The Greek form of Zarephath. *Elias*. See 1 Kgs. 17: 8–24; Luke 4: 26. *intermarry*. Marry (reciprocally). *contract*. Become engaged. *without consent of parents*. Formally forbidden in Roman law. *mulct*. Penalize. *admitted*. Permitted. *above*. More than. *a book*. More's *Utopia*, Book 2, from Plato, *Laws*, 6.771E ff. See also Plutarch's 'Life' of Lycurgus, 21–2 for a similar custom in Sparta. *familiar*. Intimate. *severally*. Separately. *conference*. Conversation. *huke*. A cloak or cape with a hood. *in state*. With ceremony. *standing*. Position. *aspect*. Look. *pitied men*. Philanthropy is the guiding force in Bacon's conception of science, from the 1592 Letter to Burghley (above, p. 20), through all his pronouncements on a reformed natural philosophy, down to this final, incompleted work. Cf. M. E. Prior, 'Bacon's Man of Science', in *Essential Articles*, 140–63. *curious*. Of delicate workmanship. *set with stone*. Inset with jewels. *Montera*. A Spanish hunter's helmet-like cap, with ear flaps.

479. *decently*. Becomingly. *chariot*. Carriage. *litter-wise*. As in a sedan-chair. *either*. Each (thus four horses, and four footmen). *trapped*. Caparisoned, fitted out. *crystal*. 'A form of quartz, transparent like glass' (*SOED*). *Peru*. Peruvian emeralds were intensely green. *radiant*. Surrounded by rays. *displayed*. Outspread. *cloth . . . blue*. A blue fabric with gold threads interwoven. *to*. As far as. *Next*. Immediately. *went*. Walked. *girt*. Girded, belted. *crosier*. An archbishop's staff, surmounted by a cross, unlike a bishop's 'pastoral staff', with its curved top, like a shepherd's crook. *balm-wood*. Balsam. *Companies of the City*. The trade guilds, as in Bacon's London: cf. *The History of Henry VII* (*Works*, vi. 32). *as*. As if. *kept*. Arranged. (The Latin translation describes the street as empty, cleared of

crowds.) *placed*. Carefully arranged. *shew*. Spectacle. *attend*. Accompany. *would*. Would like to. *charge*. Duty. *laid*. Imposed. *access*. Interview. *hanged*. Adorned with tapestry. *degrees*. Steps.

480. *state*. Canopied throne. Gough (p. 66) notes the humility shown by the Father of Salomon's House, who sits on the same level as his audience (cf. *Redargutio philosophiarum, Works*, iii. 560). He does, however, allow them to kiss the hem of his tippet, a token of respect refused by the Governor of the Stranger's House. *posture*. Attitude. *warned . . . forth*. Sent out. *relation*. Account. *keep*. Observe. *end*. Aim, goal. *preparations*. Apparatus. *works*. Operations, experiments. *motions*. Motives, causes. *Empire*. Rulership (over nature). *fathom*. 6 feet (approx. 2 metres). *depth . . . same thing*. Provided the vertical height is the same in both cases. *coagulations*. Experiments in thickening liquids. *indurations*. Hardening solids. Cf. *Sylva Sylvarum*, i. 82–90 (*Works*, ii. 374–7). *conservations of bodies*. Preserving substances. *natural mines*. Mineral veins. *compositions*. Compounds of metals and other elements.

481. *accommodated of*. Supplied with. *by*. From. *burials*. Immersing substances (not people). *several*. Various. *earths*. Types of soil. *Chineses . . . porcellain*. Chinese porcelain, known in England since 1506, excited much wonder, its composition were unknown, and fabulous stories were told of its origin (Gough, pp. 67–8). In one of his legal works Bacon said that the Chinese had 'beds of porcelain . . . which porcelain is a kind of plaster buried in the earth and by the length of time congealed and glazed into that fine substance' (*Works*, vii. 528–9). *composts*. Prepared manures. *half a mile*. This would be twice as high as the Eiffel Tower. *vantage*. Height above the plain. *insolation*. Exposure to the sun's rays. *meteors*. Meteorological phenomena. *fiery meteors*. Including lightning and shooting stars. *strain*. Extract. *motions*. Kinds of motion, using the energy obtained by falling water (little exploited as yet). *engines*. Machines. *enforcing*. Increasing the intensity, to make wind-driven engines (e.g. turbines). *set . . . on going*. Set in motion. *as*. For example. *tincted upon*. Tinctured or impregnated with. *vitriol*. The hydrous sulphate of a metal: such springs were said to be good for skin diseases. *steel*. Mineral waters impregnated with iron salts are known as chalybeates. *brass*. As Gough (p. 69) points out, this is an artificial compound, never found naturally. Perhaps Bacon uses the word in the biblical sense of 'copper': cf. Deut. 8: 9. *take the virtue*. Receive the property (potency, medicinal efficacy) of the things infused. *sovereign*. Of great efficacy. *prolongation of life*. A possibility that increasingly interested Bacon: cf. *De Augmentis*, iv. 2, and the *Historia Vitae et Mortis* (*Works*, iv. 390–4, v. 217–335). *demonstrate meteors*. Exhibit meteorological phenomena. *bodies*. Solids. *generations of bodies in air*. A common belief in Bacon's time, to explain the sudden appearance of young frogs after a shower, or swarms of midges. *qualify*. Attemper, adapt.

482. *arefaction*. Drying up, withering (thought to be the cause of many skin diseases). *confirming*. Strengthening. *respect*. Consider. *conclusions*.

Experiments. *inoculating*. Budding of trees. *effects*. Results. *earlier or later*. On the acceleration or retardation of germination see *Sylva Sylvarum*, v. 401–21 (*Works*, ii. 475–80). *greater*. Larger; cf. *Sylva Sylvarum*, v. 422–79 (*Works*, ii. 481–93). *figure*. Shape; cf. *Sylva Sylvarum*, vi. 502–5 (*Works*, ii. 501–2). *order*. Treat. *without seeds*. By spontaneous generation, supposedly: cf. *Sylva Sylvarum*, vi. 563–73 (*Works*, ii. 516–18). *vulgar*. Ordinary. *turn into another*. Cf. *Sylva Sylvarum*, vi. 521–31 (*Works*, ii. 506–9). *for view*. For the pleasure of seeing them. *trials*. Experiments. *take light*. Gain understanding. *wrought*. Performed. *find*. Discover. *taken forth*. Removed. *chirurgery*. Surgery. *physic*. Medicine. *kind*. Nature; cf. *Sylva Sylvarum*, vi. 354 (*Works*, ii. 458–9). *stay*. Stop. *bearing*. Fertile. *commixtures*. Mixtures, breeds. *barren*. Sterile (like the mule, a cross between the horse and the ass). *serpents*. Formerly this term included most vermiform animals, whether vertebrate or not. *putrefaction*. Long thought to cause the generation of new life, such as maggots and worms, which appear to be produced by only heat and moisture. *advanced*. Raised. *perfect*. Complete. *matter*. Substance.

483. *breed*. Breeding. *meats*. Foods. *manna*. 'Sweet hardened flaky exudation from the manna-ash', a white-flowered ash tree, or from the tamarisk (*SOED*). In *Sylva Sylvarum* (vii. 612) Bacon lists manna as one of the natural sources of sweetness and describes the process of gathering it (in a passage taken from George Sandys, *A Relation of a journey begun An: Dom: 1610. Foure bookes. Containing a description of the Turkish empire, of Egypt* . . ., 1615) in Calabria, from 'the leaf of the mulberry-tree' (viii. 781): *Works*, ii. 532, 593. *decocted*. Boiled down, concentrated. *tears*. Sap. *last*. Duration. *fleshes*. Kinds of flesh. *white meats*. Poultry, veal, rabbit, and pork. *meat and drink both*. Chocolate was thought to have this property. Cf. *Sylva Sylvarum*, i. 45–7 (*Works*, ii. 358–60). *divers . . . age*. Some (especially old) people. *with*. On. *parts*. Elementary parts, atoms. *insinuate*. Creep or wind in. *all*. Any. *fretting*. Corrosion, irritation. *with . . . stay*. After a little while. *ripen*. Bring to a state of perfection. *Breads . . . flesh*. Like the pemmican of the North American Indians. *leavenings*. Fermentation. *move*. Arouse. *mortified*. Made tender by hanging or keeping. *chylus*. Chyle, 'the white milky fluid formed by the action of the pancreatic juice and the bile on the chyme' (*OED*). *fast long after*. 'Bacon had probably read in Acosta's *History of the Indies*, pp. 272–3, of the use of coca by the natives of Peru. The leaves of this plant when chewed enable a person to fast three days without feeling hunger, as the drug anaesthetizes the nerves of the stomach. It also temporarily increases the capacity for muscular effort, like the imaginary foods mentioned below' (Gough, p. 72). *sensibly*. Perceptibly. *dispensatories*. Dispensaries. *simples*. Medicinal herbs, natural substances used without elaborate preparation. *long fermentations*. Fermented for a long time. *exquisite*. Carefully prepared, excellent. *separations*. Methods of separating the elements of a substance.

484. *composition*. Union. *incorporate*. Combine into new substances. *as*. As if. *tissues*. Rich cloths, often interwoven with gold or silver. *shops*. Places where any kind of industry is pursued. *yet*. All the same. *principals*.

Originals. (That is, they keep objects that have been invented, even if not subsequently manufactured for general use, in a museum of original designs or inventions.) *diversities*. Varieties. *heats*. Gradations of heat. *blown*. By bellows or blow-pipe (as in blast furnaces). *pass divers ... orbs*. Undergo alterations in their intensity, passing through 'orbs' (orbits) or cycles of variation, like the sun's heat in the course of a year. *progresses*. Onward movement. *returns*. Bending or return movement. *maws*. Inward parts. *laid up*. Stored (moisture helping the chemical changes produced by the activity of living organisms, which yield heat, as with the 'heats of dungs' earlier). *unquenched*. Unslaked (lime only produces heat while it is being 'quenched' or hydrated, to produce calcium hydroxide). *perspective-houses*. Laboratories for the use of optical instruments. *demonstrations*. Experimental tests, proofs. *represent*. Present, produce. *several ... single*. Separate (individual colours, not the spectrum). *multiplications*. Intensifications, concentrations. *sharp*. Penetrating, searching. *points*. Dots. *delusions and deceits*. Illusions. *producing of light originally*. Originating light. *procure*. Produce. *objects afar off*. The telescope, invented in Holland in 1608, was in use in a primitive form in England and elsewhere within a few years. *feigned*. Imaginary. *glasses*. Lenses. *small ... distinctly*. The magnifying glass had been known from antiquity, but the combination of lenses to produce a microscope was beginning to be discussed in the early 17th century. *grains*. Particles, impurities. *urine*. Analysis of urine was a major diagnostic resource in Renaissance medicine, but often according to ancient theories of colour symbolism.

485. *visual beams*. Visible radiations. *crystals*. Kinds of crystal. *vitrificated*. Turned into glass (due to the presence of silicates). *fossils*. Any substances dug out of the earth, not necessarily organic remains. *imperfect*. Minerals in an arrested or incomplete state of development; yet to become metals or precious stones. *virtue*. Efficacy, power. *sound-houses*. Acoustic laboratories. *generation*. Production. *slides*. Quarter-tones and smaller musical intervals. Cf. *Sylva Sylvarum*, ii. 110 (*Works*, ii. 388). *rings*. Chimes or sets of bells. *great*. Loud (by amplification). *extenuate*. Made thin. *tremblings*. Trills, tremolos. *original*. Origin. *entire*. Whole. *set*. Applied. Cf. *Sylva Sylvarum*, iii. 285 (*Works*, ii. 434–5) on the ear-trumpet. *trunks*. Speaking tubes. *strange*. Irregular, not just straight. Cf. *Sylva Sylvarum*, iii. 201–6 (*Works*, ii. 414–15). *practices of taste*. Experiments in tasting. *multiply*. Intensify. *imitate smells*. By synthetic perfumes. *contain*. Include. *confiture-house*. Factory for making sweet-meats. *broths*. Thin soups, concoctions. *sallets*. Salads. *engine-houses*. Engineering-shops. *engines*. Machines. *swifter motions*. Of greater velocity. *small force*. Less expenditure of force. *basilisks*. Heavy artillery. *ordnance*. Engines for discharging missiles.

486. *wildfires*. Inflammable liquids used in antiquity and the Middle Ages for setting enemy ships on fire, their ingredients including naphtha (liquid petroleum). *we ... degrees of*. We have made some progress in; on artificial flight see *Sylva Sylvarum*, ix. 886 (*Works*, ii. 634). *under water*. John Napier, the mathematician, announced in 1596 (when a Spanish invasion was feared)

his plans to build various military inventions, including 'devices of sailing under the water' (Gough, p. 75). The Dutch inventor Cornelis Drebbel demonstrated a submarine in the Thames in 1620. *brooking of seas*. Resisting the force of the waves. *swimming-girdles*. Lifebelts. *supporters*. Supports. *motions of return*. Any machines depending on regularly recurring or cyclical movement. *perpetual motions*. Machines possessing this (miraculous) power. *images*. Automata (already highly developed). *strange ... subtilty*. Remarkable for their evenness of operation, delicate construction, and complexity. *instruments*. Early mathematical instruments were intricate and beautiful. *feats of juggling*. Conjuring tricks, presumably used to expose their deceits. *fallacies*. Refutations, exposures of the illusions involved. *induce*. Produce. *admiration*. Astonishment. *in ... particular*. In a multitude of particular cases. *labour*. Strive. *hate ... impostures*. A constant theme in Bacon's natural philosophy. Cf. *Novum Organum*, i. 87 (*Works*, iv. 85–6); *Sylva Sylvarum*, vi. 501 (*Works*, ii. 501). *insomuch as*. To such a degree that. *work*. Product. *swelling*. Amplified, exaggerated. *pure*. In its true nature. *offices*. Duties. *patterns*. Models. *collect*. Extract. *Depredators*. Summarizers, makers of abstracts. *liberal sciences*. Liberal arts, originally those fit for a free man; not, like mechanical arts, cultivated for the sake of profit. *brought into*. Methodized, systematized as coherent practices. *Mystery-men*. Those who study arts and handicrafts. *Pioners*. Miners (originally soldiers who prepared the way for the main body of troops). *draw ... into*. Classify. *titles*. Heads, related groups. *tables*. In tabular form. *axioms*. Intermediate generalizations of scientific laws. *Compilers*. These workers produce, for instance, 'Tables of Essence and Presence; of Deviation, or of Absence in Proximity; of Degrees, or Comparison', in the specimen analysis of heat. See *Novum Organum*, ii. 11–13 (*Works*, iv. 127–45).

487. *bend*. Apply. *cast about*. Consider. *use and practice*. Capable of being used, put into practice for the benefit of human life: one of Bacon's fundamental criteria for science. *as well for works ... causes*. Both for practical ends and theorizing. *natural divinations*. 'Means of discovering the secrets of nature, forecasting physical changes, &c., without preternatural aid' (Gough, p. 77). *parts of bodies*. Latent or hidden parts; the Latin version reads 'latentas partes'. *Dowry-men*. Endowment men, sources of wealth. *Benefactors*. These perform the 'Commencement of Interpretation, or the First Vintage': *Novum Organum*, ii. 20 (*Works*, iv. 149–55). *consults*. Consulations. *collections*. Compilations. *higher light*. Greater knowledge. *Lamps*. Cf. *Novum Organum*, ii. 38, 'Instances of the Lamp, or of First Information'. *Inoculators*. Those who bud trees. *greater observations*. Higher generalizations of results obtained by observation. *aphorisms*. Scientific laws. The Latin version adds that they only do this after consultation with their colleagues. *Interpreters of Nature*. These have reached the final stage of Bacon's reformed natural philosophy. *experiences*. Experiments. *secret*. It is significant that, although believing that scientific knowledge should be open to all (so rejecting the ideology of the occult), Bacon reserved some forms of knowledge for the scientific community alone, which should not even disclose

it to the state. In *Sylva Sylvarum*, as Gough (p. 77) observes, 'Bacon mentions two alleged inventions which he considers would be dangerous if they were widely known: a white gunpowder that explodes noiselessly (ii. 120), and a method of rendering the flesh of beasts, fowl, or fish poisonous, by feeding them with certain substances (v. 499)': *Works*, ii. 392–3, 498–9. *inventors.* Bacon expressed a lifelong reverence for inventors: cf. *Adv.L.*, above, pp. 153–4, 219–20, and *Novum Organum*, i. 129 (*Works*, iv. 113–15). In the *Comentarius Solutus*, a notebook compiled in 1608, Bacon projected the 'Foundation of a college for Inventors', the building to include 'two galleries with statuas for Inventors past, and spaces or bases for Inventors to come'. He also noted a question to be settled, concerning the 'removes and expulsions' to be made from this college 'in case within a time some invention worthy be not produced. And likewise query of the honors and rewards for inventions' (*Works*, xi. 66–7). *West Indies.* Originally not only the islands but the neighbouring continents of North and South America (thought to be the eastern coast of Asia). *your monk.* Roger Bacon (*c.*1210–92), scholastic philosopher and Franciscan friar, one of those to whom the invention of gunpowder was traditionally ascribed. Cf. also *Adv.L.*, above, p. 220 and note, for Berthold Schwarz's claim to this title. All the other inventors listed here are either unknown, or legendary. In Jewish traditions the inventor of music was Jubal (Gen. 4: 21), that of 'works in metal' was his brother, Tubalcain (Gen. 4: 22), and that of wine Noah (Gen. 9: 20). To the Greeks credit for inventing both letters and glass went to Hermes, wine to Dionysus, and the art of growing wheat to Triptolemus. Bacon's list of inventions owes something to Diodorus Siculus, 1.i. 14, 16. *observations of astronomy.* Perhaps longitude and latitude

488. *tradition.* Historical record. *touch-stone.* Basanite, a black or dark-coloured variety of quartz or jasper, used for testing gold and silver alloys. *gilt.* Gilded. *laud.* Praise. *circuits.* Tours. *publish.* Publicize. *declare.* Announce. *natural divinations.* Warnings, prognostications, not made by supernatural means. *comets.* Traditionally foreboding public disaster. *temperature.* Not just heat but also moisture: any disturbance of the normal balance. *prevention.* Advance provision against anticipated dangers. *value.* Sum. *ducats.* Gold and silver coins of various values, first minted in the duchy (*ducatus*) of Apulia in 1140. *largesses.* Gifts, presents. *The rest . . . perfected.* Rawley's note. The account of Salomon's House seems complete, but the parallel sequence, as described by Rawley ('a frame of Laws, or the best state or mould of a commonwealth') was never written. After that, presumably, Bacon would have gone on to describe the travellers' return to Europe. *Magnalia . . . humanos.* 'The wonderful works of Nature, chiefly such as benefit mankind.' This rather miscellaneous list of desirable discoveries was appended to *New Atlantis* in its first and all subsequent printings. Rossi, p. 869, notes that the term *magnalia naturae* (also used in *Novum Organum*, i. 109; ii. 28) is to be distinguished from the biblical phrase *magnalia Dei* (Ps. 106: 22; Acts 2: 11, in the Authorized Version 'the wonderful works of God'). In the *De Augmentis* Bacon uses the term to

describe his new conception of natural magic, which, purged of occult obscurity and superstition, will discover 'the universal consents of things': a 'science which applies the knowledge of hidden forms to the production of wonderful operations; and by uniting (as they say) actives with passives, displays the wonderful works of nature' (*Works*, iv. 366–7; also 420). In the *Sylva Sylvarum* (vi. 525) he writes that 'this work of the transmutation of plants into one another, is *inter magnalia naturae*: for the transmutation of species is, in the vulgar philosophy, pronounced impossible; and certainly it is a thing of difficulty, and requireth deep search into nature' (*Works*, ii. 507). *purgings*. Evacuations of the bowels; laxatives. *complexions*. Bodily characteristics (e.g. metabolism).

489. *Versions*. Conversions, transformations. *Exhilaration*. Raising, enlivening. *clarifications*. Freeing from impurities. *decoction*. 'Maturation or preparation of a mineral or ore by heat (according to old notions)' (*SOED*). *Impressions*. Pressures; atmospheric influences or phenomena. *emollition*. Making soft. *threads for apparel*. Synthetic fibres.

FURTHER READING

1. *Bibliographies.* For bibliographies of Bacon studies see Brian Vickers, *Francis Bacon and Renaissance Prose* (Cambridge, 1968), 307–12; Paolo Rossi, *Scritti Filosofici di Francesco Bacone* (Turin, 1975, 1986), 49–74 (commendably wide-ranging, but with numerous inaccuracies); J. K. Houck, *Francis Bacon, 1926–1966* (1968: Elizabethan Bibliographies, Supplement); Brian Vickers, *Francis Bacon*, Writers and their Works (London, 1978), 38–46; Wilhelm Totok, *Handbuch der Geschichte der Philosophie*, iii *Renaissance* (Frankfurt, 1980), 473–83; William A. Sessions, 'Recent Studies in Francis Bacon', *English Literary Renaissance*, 17 (1987), 351–71 (limited to works in English); A. Pérez-Ramos, in G. H. R. Parkinson (ed.), *The Renaissance and Seventeenth-Century Rationalism* (London, 1993), 162–7; Brian Vickers, 'Francis Bacon', in Douglas Sedge (ed.), *The Cambridge Bibliography of English Literature*, 3rd edn., vol. ii: *1500–1700* (forthcoming). The listing of primary works in the latest revision of Donald Wing's *Short-Title Catalogue of Books Printed in England ... 1641–1700*, ed. J. J. Morrison *et al.*, vol. i (New York, 1994), is still gravely deficient.

2. *Editions.* The standard edition so far has been that produced by James Spedding, assisted by R. L. Ellis (for Bacon's scientific works) and D. D. Heath (for his legal writings): *The Works of Francis Bacon*, 7 vols. (London, 1857–9). Spedding alone was responsible for editing *The Letters and Life of Francis Bacon, including all his Occasional Works*, 7 vols. (London, 1861–74). Modern photographic reprints include one in reduced format (Stuttgart: Frommann, 1962), and one full-size (New York: Garrett, 1968). The 7-vol. *Works* was also issued in America in a confusing 15-vol. format (Boston: McTaggart and Brown, 1860–4). For a useful concordance enabling the conversion of page-references between the two editions see Julian Martin, *Francis Bacon, the State and the Reform of Natural Philosophy* (Cambridge, 1992), 176–80. A new edition of the works (but excluding, unfortunately, the letters) is under way for Oxford University Press, edited by Graham Rees, Lisa Jardine, Julian Martin, Marta Fattori, Michael Edwards, *et al.*

Although antiquated in some respects, Thomas Fowler's edition of *Bacon's Novum Organum* (Oxford, 1878, rev. edn. 1889) offers an immensely helpful detailed commentary. The Penguin edition of the *Essays*, by John Pitcher (1985), has a bizarre introduction, and is highly unsympathetic to Bacon, making several unfounded allegations. An annotated edition of *The History of Henry VII* by Brian Vickers is forthcoming.

3. *Biography.* The best collection of biographical material is still the documentation provided by Spedding in his commentary to the *Letters and Life*, although superseded by recent scholarship in some areas. There are no

reliable modern scholarly biographies of Bacon. Some help can be gained from F. H. Anderson, *Francis Bacon: His Career and His Thought* (Berkeley and Los Angeles, 1962), and C. D. Bowen, *Francis Bacon: The Temper of a Man* (London, 1963). J. L. Marwil, *The Trials of Counsel: Francis Bacon in 1621* (Detroit, 1976) purports to be a biographical study of Bacon after his fall, but is a biased and unfair indictment of his whole life and work.

4. *General Studies.* Anthony Quinton's short guide in the Past Masters series, *Francis Bacon* (Oxford, 1980), while unsympathetic to his life and personality, has some valuable comments on his work. Useful treatments can be found in Hiram Haydn, *The Counter-Renaissance* (New York, 1950); Hardin Craig, *The Enchanted Glass: The Elizabethan Mind in Literature* (Oxford, 1935, 1960); and R. S. Woolhouse, *The Empiricists* (Oxford, 1988).

5. *Reference Works.* Peter Beal, *Index of English Literary Manuscripts*, i/1, *1450–1625: Andrewes–Donne* (London and New York, 1980); R. W. Gibson (ed.), *Francis Bacon: A Bibliography of his Works and of Baconiana to the year 1750* (Oxford, 1950); *Supplement* (Oxford, 1959); David W. Davies and Elizabeth S. Wrigley, *A Concordance to the Essays of Francis Bacon* (Detroit, 1973); Marta Fattori, *Lessico del 'Novum Organum' di Francesco Bacone*, 2 vols. (Rome, 1980).

6. *Reception and Reputation.* Few authors have experienced such dramatic extremes of glorification and denigration as Bacon. For his early impact as a philosopher see R. F. Jones, *Ancients and Moderns: A Study of the Rise of the Scientific Movement in Seventeenth-Century England* (2nd edn., repr. Berkeley, 1962), a pioneering study still unsurpassed, and Charles Webster, *The Great Instauration: Science, Medicine and Reform 1626–1660* (London, 1975)—although Webster tends to claim as specifically 'Puritan', developments which rather represent Bacon's wider influence. T. M. Brown, 'The Rise of Baconianism in Seventeenth-Century England', in *Science and History: Studies in Honor of Edward Rosen* (Warsaw, 1978), 501–22, offers a critical reassessment of recent discussions of this issue. For later periods see Michel Malherbe, 'Bacon, l'Encyclopédie et la Révolution', *Les Études philosophiques*, 3 (1985), 387–404; H. Dieckmann, 'The Influence of Francis Bacon on Diderot's *Interprétation de la Nature*', *Romanic Review*, 24 (1943), 303–30; P. G. Gates, 'Bacon, Keats and Hazlitt', *Fifty Years of the 'South Atlantic Quarterly'*, ed. W. B. Hamilton (Durham, NC, 1952), 331–43; W. O. Scott, 'Shelley's Admiration for Bacon', *PMLA* 73 (1958), 228–36; D. E. Habben, 'The Reputation of Sir Francis Bacon among the English Romantics', Ph.D. diss., New York University, 1976 (*Dissertation Abstracts International*, 38: 804A); E. Douka-Kabitoglou, 'Bacon, the British Plato: A Romantic Reappraisal', *Working Papers in Linguistics and Literature*, ed. A. Kakouriotis and R. Parkin-Gounelas (Thessaloniki, 1989), 137–69; Jonathan Smith, *Fact and Feeling: Baconian Science and the Nineteenth-Century Literary Imagination*

(London, 1994); Richard Yeo, 'An Idol of the Market-place: Baconianism in Nineteenth-Century Britain', *History of Science*, 23 (1985), 251–98; A. Pérez-Ramos, 'Bacon's Legacy', in M. Peltonen (ed.), *The Cambridge Companion to Bacon* (Cambridge, 1996). Two wide-ranging surveys of Bacon's reputation are Enrico de Mas, *Francis Bacon* (Florence, 1978), and Paolo Rossi, 'Baconianism', in P. P. Wiener (ed.), *Dictionary of the History of Ideas*, 2 vols. (New York, 1968, 1973), i. 172–9.

7. *Philosophical Works.* For many years the standard book was Fulton H. Anderson, *The Philosophy of Francis Bacon* (Chicago, 1948), which consists of detailed summaries of Bacon's philosophical scheme, rather than analytical evaluations, and lacks any historical context. (Its usefulness was also limited by its quoting from the non-standard 15-vol. edition.) Benjamin Farrington, *The Philosophy of Francis Bacon: An Essay on its Development from 1603 to 1609 with New Translations of Fundamental Texts* (Liverpool, 1964), includes helpful, if rather unstylish versions of three texts not translated by Spedding (*Temporis Partus Masculus*, *Cogitata et Visa*, and *Redargutio Philosophiarum*). A recently discovered fragmentary manuscript of an uncompleted work by Bacon, 'De viis mortis', was edited by Graham Rees, assisted by Christopher Upton, as *Francis Bacon's Natural Philosophy: A New Source* (Chalfont St Giles, 1984).

The first modern scholarly study to place Bacon in the historical context of Renaissance science was Paolo Rossi, *Francesco Bacone: dalla magia alla scienza* (Bari, 1958; rev. edn. Turin, 1974), tr. S. Rabinovitch as *Francis Bacon: From Magic to Science* (London, 1968). Two volumes of Rossi's essays include discussions of Bacon: *Philosophy, Technology and the Arts in the Early Modern Era*, tr. S. Attanasia, ed. B. Nelson (New York, 1970); and *La Scienza e la Filosofia dei Moderni. Aspetti della rivoluzione scientifica* (1971; rev. edn. Turin, 1989). Two recent books have greatly advanced the historical and analytical study of Bacon's natural philosophy: Peter Urbach, *Francis Bacon's Philosophy of Science* (La Salle, Ill., 1987), and—the outstanding modern treatment—Antonio Pérez-Ramos, *Francis Bacon's Idea of Science and the Maker's Knowledge Tradition* (Oxford, 1988). For an evaluation of these and other recent studies see Brian Vickers, 'Francis Bacon and the Progress of Knowledge', *Journal of the History of Ideas*, 53 (1992), 495–518, and 'Bacon among the Literati: Science and Language', *Comparative Criticism*, 13 (1991), 249–71. Two important accounts of the history of science which place Bacon in a wider context are Norma E. Emerton, *The Scientific Reinterpretation of Form* (Ithaca, NY, 1984), and A. C. Crombie, *Styles of Thinking in the European Tradition*, 3 vols. (London, 1994). Relevant shorter studies include Mary Hesse, 'Francis Bacon's Philosophy of Science', and Moody E. Prior, 'Bacon's Man of Science', both repr. in Brian Vickers (ed.), *Essential Articles for the Study of Francis Bacon* (Hamden, Conn., 1968; London, 1972), 114–39, 140–63; Mary Horton, 'In Defence of Francis Bacon: A Criticism of the Critics of the Inductive Method', *Studies in the History and Philosophy of*

Science, 4 (1973), 241–78; T. S. Kuhn, 'Mathematical versus experimental traditions in the development of physical science', in Kuhn, *The Essential Tension: Selected studies in scientific tradition and change* (Chicago, 1977), 143–81; L. Jonathan Cohen, 'Some Historical Remarks on the Baconian Conception of Probability', *Journal of the History of Ideas*, 41 (1980), 219–31; and Brian Vickers, 'Bacon's so-called "Utilitarianism": Sources and Influence', in the Fattori volume (below), 281–313. A pioneering account of Bacon's attitude towards medicine is J. Boss, 'The Medical Philosophy of Francis Bacon (1561–1626)', *Medical Hypotheses*, 4 (1978), 208–20.

Useful studies of Bacon's natural philosophy are included in some recent collections of essays: M. Malherbe and J. M. Pousseur (eds.), *Francis Bacon: Science et méthode* (Paris, 1985); the special Bacon issues of *Les Études philosophiques* (no. 3, 1983) and *Revue internationale de philosophie* (vol. 40, 1986); Marta Fattori (ed.), *Francis Bacon: Terminologia e Fortuna nel XVII secolo* (Rome, 1984); William A. Sessions (ed.), *Francis Bacon's Legacy of Texts* (New York, 1990); Markku Peltonen (ed.), *The Cambridge Companion to Bacon* (Cambridge, 1996).

8. *Political and Historical Works*. Here, also, there is no reliable modern study, based on first-hand knowledge of the historical materials. Despite its promising title, B. H. G. Wormald, *Francis Bacon: History, Politics and Science, 1561–1626* (Cambridge, 1994) consists of lengthy summaries of Bacon's views, neither subjected to detailed analysis, nor placed within relevant contexts. J. J. Epstein, *Francis Bacon: A Political Biography* (Athens, Oh., 1977) relates the main events in Bacon's career rather briefly, drawing on modern parliamentary histories, but does not adequately register the fluctuating political alliances and pressure-groups around Bacon, and shows little insight into his character. A better picture can be obtained from the standard histories of Jacobean politics: Menna Prestwich, *Cranfield: Politics and Profits under the English Stuarts* (Oxford, 1966); R. Zaller, *The Parliament of 1621: A Study in Constitutional Conflict* (Berkeley, 1971); Kevin Sharpe (ed.), *Faction and Parliament: Essays on Early Stuart History* (Oxford, 1978); Conrad Russell, *Parliaments and English Politics 1621–1629* (Oxford, 1979); and M. A. Phillips, 'Francis Bacon', in P. W. Hasler (ed.), *The House of Commons 1558–1603*, 3 vols. (London, 1981), i. 374–9. Julian Martin, *Francis Bacon, the State, and the Reform of Natural Philosophy* (Cambridge, 1992), has some helpful material on Bacon's early legal and parliamentary career, but cynically reduces all Bacon's efforts to reform natural philosophy to an attempt to ingratiate himself with King James. Useful shorter studies include T. K. Rabb, 'Francis Bacon and the Reform of Society', in Rabb and J. Seigel (eds.), *Action and Conviction in Modern Europe* (Princeton, 1969), 161–93; Karl R. Wallace, 'Discussion in Parliament and Francis Bacon', repr. in Vickers (ed.), *Essential Articles*, 195–210; Robert C. Johnson, 'Francis Bacon and Lionel Cranfield', *Huntington Library Quarterly*, 23 (1960), 301–20—which quotes (from the Saville MSS) some important Bacon letters not

included in Spedding; and Markku Peltonen, 'Bacon's Political Philosophy', in *The Cambridge Companion to Bacon*, ed. Peltonen (Cambridge, 1996).

Bacon's *History of Henry VII*, and his theories of historiography, are treated in several general books: F. S. Fussner, *The Historical Revolution: English Historical Writing and Thought, 1580–1640* (London, 1962); W. H. Greenleaf, *Order, Empiricism and Politics: Two Traditions of English Political Thought 1500–1700* (London, 1964); F. J. Levy, *Tudor Historical Thought* (San Marino, Calif., 1967); Judith H. Anderson, *Biographical Truth: The Representation of Historical Persons in Tudor–Stuart Writing* (New Haven, 1984). Useful individual studies include L. F. Dean, 'Sir Francis Bacon's Theory of Civil History-Writing', and George H. Nadel, 'History as Psychology in Francis Bacon's Theory of History', both repr. in Vickers (ed.), *Essential Articles*, 211–35, 236–50; Stuart Clark, 'Bacon's *Henry VII*: A Case-Study in the Science of Man', *History and Theory*, 13 (1974), 97–118; D. R. Woolf, 'John Selden, John Borough and Francis Bacon's *History of Henry VII*', *Huntington Library Quarterly*, 47 (1984), 47–53.

9. *Legal Works.* The best overall study is Daniel R. Coquillette, *Francis Bacon*, Jurists: Profiles in Legal Theory (Edinburgh, 1992), which discusses Bacon's legal theories and practice in a largely non-technical manner, copiously documented. J. H. Baker, *The Legal Profession and the Common Law* (London, 1986), includes a translation of Bacon's argument in Slade's Case (pp. 396–409). Mark Neustadt, 'The Making of the Instauration: Science, Politics, and Law in the Career of Francis Bacon', Ph.D. diss., Johns Hopkins University, 1987 (*Dissertation Abstracts International*, A48/05 p. 1302) includes the text and translation of the *Aphorismi de Jure gentium maiore, sive de fontibus Justitiae et Juris*, found by Peter Beal in 1980. Another useful English version is by John C. Hogan and Mortimer D. Schwartz, 'A Translation of Bacon's Maxims of the Common Law', *Law Library Journal*, 77 (1984–5), 707–18, a text which they discussed ibid. 76 (1983), 48–77. See also Peter Stein, *Regulae Juris: From Juristic Rules to Legal Maxims* (Edinburgh, 1966); Louis A. Knafla, *Law and Politics in Jacobean England* (Cambridge, 1977); B. McCabe, 'Francis Bacon and the Natural Law Tradition', *Natural Law Forum*, 9 (1964), 111–21; Paul Kocher, 'Francis Bacon on the Science of Jurisprudence', in Vickers (ed.), *Essential Articles*, 167–94; B. J. Shapiro, 'Sir Francis Bacon and the Mid-Seventeenth Century Movement for Law Reform', *American Journal of Legal History*, 24 (1980), 331–62.

10. *Literary Works.* On Bacon's knowledge of rhetoric Karl R. Wallace, *Francis Bacon on Communication and Rhetoric* (Chapel Hill, NC, 1943) was a pioneering work, but did little more than summarize Bacon's views, and failed to set them in their historical context in classical and Renaissance rhetoric. Other studies include M. B. McNamee, 'Literary Decorum in Francis Bacon', *Saint Louis University Studies*, 1 (1950), 1–52; Marc Cogan,

'Rhetoric and Action in Francis Bacon', *Philosophy and Rhetoric*, 14 (1981), 212–33; and Brian Vickers, 'Bacon and Rhetoric', in M. Peltonen (ed.), *The Cambridge Companion to Bacon* (Cambridge, 1996). Neal W. Gilbert, *Renaissance Concepts of Method* (New York, 1960) is a wide-ranging study which gives due credit to Bacon's originality in formulating the principle of experimentation. Lisa Jardine, *Francis Bacon: Discovery and the Art of Discourse* (Cambridge, 1974), has a useful account of 16th-century dialectic teaching, but unfortunately overvalues its place in intellectual culture and downgrades rhetoric to a purely ornamental role. Her generally negative evaluations of Bacon ('intellectually a provincial', having no conception of 'systematic experimenting', who 'dismisses as secondary and provisional much of what we now regard as science') have not been endorsed by modern research.

On Bacon's style see Brian Vickers, *Francis Bacon and Renaissance Prose* (Cambridge, 1968), which discusses his use of the aphorism, imagery in argument, philosophical image-patterns, syntactical symmetry, and literary revisions, and, like R. M. Adolph, *The Rise of Modern Prose Style* (Cambridge, Mass., 1968), rejects the theories of Morris W. Croll and George Williamson that Bacon had a 'Senecan' or 'anti-Ciceronian' style. On attempts by R. F. Jones and others to identify Bacon with an anti-rhetorical style, supposedly infecting the new science with a fear of metaphor and a distrust of language in general, see Brian Vickers, 'The Royal Society and English Prose Style: A Reassessment', in *Rhetoric and the Pursuit of Truth. Language Change in the Seventeenth and Eighteenth Centuries* (Los Angeles, 1985), 3–76. Other studies include three essays in Vickers (ed.), *Essential Articles*: R. S. Crane, 'The Relation of Bacon *Essays* to his Program for the Advancement of Learning' (272–92), John L. Harrison, 'Bacon's View of Rhetoric, Poetry, and the Imagination' (253–71), and Anne Righter's incisive and sensitive study, 'Francis Bacon' (300–21). For specific studies of Bacon's metaphors see Walter R. Davis, 'The Imagery of Bacon's late work', *Modern Language Quarterly*, 27 (1966), 162–77; Brian Vickers, 'Bacon's Use of Theatrical Imagery', in W. A. Sessions (ed.), *Francis Bacon's Legacy of Texts* (New York, 1990), 171–213.

On Bacon's *De Sapientia Veterum* see Charles W. Lemmi, *The Classic Deities in Bacon* (Baltimore, 1953), and Barbara C. Garner, 'Francis Bacon, Natalis Comes, and the Mythological Tradition', *Journal of the Warburg and Courtauld Institutes*, 33 (1970), 264–91.

INDEX OF NAMES

GENERAL INDEX